Springer Proceedings in Materials

Volume 27

Springer Proceedings in Materials publishes the latest research in Materials Science and Engineering presented at high standard academic conferences and scientific meetings. It provides a platform for researchers, professionals and students to present their scientific findings and stay up-to-date with the development in Materials Science and Engineering. The scope is multidisciplinary and ranges from fundamental to applied research, including, but not limited to:

- Structural Materials
- Metallic Materials
- Magnetic, Optical and Electronic Materials
- Ceramics, Glass, Composites, Natural Materials
- Biomaterials
- Nanotechnology
- Characterization and Evaluation of Materials
- Energy Materials
- Materials Processing

To submit a proposal or request further information, please contact one of our Springer Publishing Editors according to your affiliation:

European countries: **Mayra Castro** (mayra.castro@springer.com)
India, South Asia and Middle East: **Priya Vyas** (priya.vyas@springer.com)
South Korea: **Smith Chae** (smith.chae@springer.com)
Southeast Asia, Australia and New Zealand: **Ramesh Nath Premnath** (ramesh.premnath@springer.com)
The Americas: **Michael Luby** (michael.luby@springer.com)
China and all the other countries or regions: **Mengchu Huang** (mengchu.huang@springer.com)

This book series is indexed in **SCOPUS** database.

Zishan Husain Khan · Mark Jackson ·
Numan A. Salah
Editors

Recent Advances in Nanomaterials

Select Proceedings of ICNOC 2022

Springer

Editors
Zishan Husain Khan
Department of Applied Sciences
and Humanities
Jamia Millia Islamia
New Delhi, Delhi, India

Mark Jackson
Department of Engineering Technology
Kansas State Polytechnic
Salina, KS, USA

Numan A. Salah
Centre of Nanotechnology
King Abdulaziz University
Jeddah, Saudi Arabia

ISSN 2662-3161 ISSN 2662-317X (electronic)
Springer Proceedings in Materials
ISBN 978-981-99-4877-2 ISBN 978-981-99-4878-9 (eBook)
https://doi.org/10.1007/978-981-99-4878-9

This Springer imprint is published by the registered company Springer Nature Singapore Pte Ltd.
The registered company address is: 152 Beach Road, #21-01/04 Gateway East, Singapore 189721, Singapore

Paper in this product is recyclable.

Preface

International Conference on Nanotechnology: Opportunities and Challenges (ICNOC-2022) was organized by the Department of Applied Sciences and Humanities, Jamia Millia Islamia, New Delhi, during November 28–30, 2022, in virtual mode. The main aim of the conference was to provide platform to the scientists, researchers, academicians, and students to discuss their ideas on the opportunities and challenges in the domain of nanotechnology and to share and disseminate their research with the delegates joining the conference from all over the world. The underlying motive behind organizing the conference was to share knowledge of the subject among researchers, who are looking for collaborative ventures, empirical learning, and innovative technological breakthroughs. More than 45 internationally acclaimed experts from India as well as abroad delivered their talks and 600 participants presented their research outputs in the conference.

The conference was divided into three broad tracks, i.e., nanoscience and nanotechnology, nanotechnology for energy and environment, and nanotechnology for medical sciences. The presentation under the track 'nanoscience and nanotechnology' was focused on the research works carried out to understand the science behind the novel nanoscale materials and systems. The understanding of the behavior of nature at the nanoscale is evolving day by day with the minute observations of the properties of the materials at this scale. The development of knowledge at nanoscale science has opened up new possibilities for the development of more efficient, precise, and effective technologies in various fields. Research outcomes on various topics especially nanocomposites, nanofabrication and nanoengineering, sensors and actuators, nanoelectronics, nanophotonics, nanochemistry, nanoscale modeling and simulation, and societal concerns and ethical issues were disseminated in the different technical sessions under this track.

Energy and environmental issues are closely linked as the production and use of energy often have significant impacts on the environment. Nanotechnology has the potential to revolutionize the field of energy and environment. Through the manipulation of materials at the nanoscale, new solutions for global challenges such as energy production, storage, and conservation can be created. In the area of energy production, nanomaterials can be used to improve the efficiency of solar cells and fuel cells. Nanoscale thermoelectric materials can also be utilized for waste heat recovery. Energy storage can also benefit from nanotechnology, as nanomaterials can be used to develop more efficient and high-capacity batteries, supercapacitors, and hydrogen storage. Nanomaterials can also contribute to energy conservation, as they can be used to improve the efficiency of lighting, insulation, and windows. In the area of pollution control and remediation, nanomaterials can be used for air and water purification, as well as soil remediation. Keeping in mind the importance of this area of research, we dedicated one of the three tracks of ICNOC-2022 to the applications for nanotechnology and environment. Several topics in this domain were covered in the conference especially renewable energy, hydrogen production, storage and transportations, energy generation and storage, solar cells, fuel cells, batteries and supercapacitor and water treatment and filtration.

The nanotechnology has significant potential in medical science and can contribute to transforming healthcare by creating innovative solutions for diagnosis, treatment, and prevention. By designing and manipulating materials at the nanoscale, researchers can develop new tools and devices that can interact with biological systems in novel ways. Nanotechnology can be utilized in many areas of medical science, including medical imaging, drug delivery, tissue engineering, and regenerative medicine. Nanoscale imaging agents can provide high-resolution images of internal organs and tissues for early detection and diagnosis of diseases. Nanoparticles can be used to deliver drugs precisely to targeted cells or tissues, which can improve the efficacy of treatments while reducing side effects. In tissue engineering and regenerative medicine, nanomaterials can be used to create scaffolds that mimic the structure and function of natural tissues, promoting the regeneration of damaged or diseased tissues. Nanoparticles can be utilized to deliver growth factors and other signaling molecules that stimulate tissue regeneration. Nanotechnology can also be used to develop advanced prosthetics and medical implants. Nanotechnology can also contribute to disease prevention through the development of vaccines and nanoscale sensors that can detect and respond to pathogens. In cancer treatment, nanotechnology-based therapies can deliver therapeutic agents directly to cancer cells while minimizing damage to healthy cells. In ICNOC-2022, a track on nanotechnology for medical science was created to bring together researchers, scientists, and professionals from various fields to discuss the latest advances, opportunities, and challenges in this field. The major topics included in this track were nanomaterial for denture applications, nanotechnology in medicine, medicinal nanochemistry, nanobiosensors, targeted cellular therapies, nanomedicine-fusing therapy and diagnostics, regenerative medicine and tissue engineering and nanoscale formulation and targeted drug delivery systems.

The conference was organized with 1 inaugural session, 3 planetary sessions, and 71 technical sessions (4 parallel sessions at a time). We received more than 1000 abstract altogether, out of which 600 were selected for the presentation in the conference after a peer review. Full papers were invited from participants who presented their work in the conference, and out of 500 manuscripts, we accepted 181 papers for the publication in these proceedings after a strict peer-reviewed process based on the originality of work, novelty of research, and scientific relevance of the research outcomes. This proceeding is the balanced mix of the topics from all the three tracks of the conference, and readers will be benefited from the research outcomes on the recent applications of nanotechnology in different areas of these domains.

We hope this proceedings will serve as a source of inspiration and knowledge for all those who read it, and that it will encourage them to explore the fascinating domain of Nanotechnology further.

New Delhi, India — Prof. Zishan Husain Khan
Jeddah, Saudi Arabia — Prof. Numan A. Salah
Salina, USA — Prof. Mark Jackson

Contents

About the Editors

Dr. Zishan Husain Khan is currently Professor and Head at the Department of Applied Sciences and Humanities, Faculty of Engineering and Technology, Jamia Millia Islamia, New Delhi. He has almost 25 years of research experience in semiconductor physics and nanotechnology. He has published more than 140 research papers in various international reputed journals and guided several Ph.D. Students. He has completed several research projects on various topics in nanotechnology. He has worked at several positions in the universities abroad. He is actively involved in designing various courses in nanotechnology and energy sciences for graduate and research students He is also the regular reviewer for many international journals of high repute In addition, he has edited several special issues for reputed international journals Dr. Khan has edited many books for reputed publishers including Springer Nature and published many book chapters with reputed publishers His present research interests include hybrid solar cells, OLEDs, 2-D transition metals chalcogenides, carbonaceous nanomaterials, nano sensors and nano biosensors.

Prof. Mark Jackson is a Professor of Mechanical Engineering Technology at Kansas State University, Aerospace and Technology Campus, USA. He received his Ph.D. in Mechanical Engineering at Liverpool. He has been the active researcher at Oxford and Cambridge University and at the University of Liverpool. His areas of expertise in mechanical engineering include space systems and operations, advanced manufacturing, nanotechnology. Prof. Jackson co-edited the book *Advances in Medical and Surgical Engineering*, which demonstrates the connection between engineering and medical science. He has 4887 citations with h-index of 33.

Prof. Numan A. Salah is a Professor at the Center of Nanotechnology, King Abdul Aziz University, Jeddah Saudi Arabia and heads a research group on Carbon Nanostructures at the Centre. His research interest includes nanomaterials (synthesis, characterization, application) such as carbon nanostructures, optical and thermoelectric nanomaterials. He authored/co-authored more 170 research papers and filed more than 17 US patents. He teaches several courses in the department of nuclear engineering at King Abdul Aziz University. He has guided many Ph.D. students and has been the mentor of many Post Doctoral Fellows.

Investigation of Microstructural, Optical, and Electronic Properties of Hydrothermally Synthesized MoS_2 Decorated SnO_2

Priya Pradeep Kumar and Vinod Singh

Abstract In this work, MoS_2/SnO_2 heterostructures have been synthesized using the hydrothermal method, and structural, optical, and compositional properties were studied by various characterizations revealing the confirmation of the co-existence of MoS_2 and SnO_2 phase in heterostructures with changed properties than the pristine ones, which makes them beneficial for gas sensing applications with a unique combination of structural and optical properties.

Keywords Two-dimensional (2D) materials · Hydrothermal synthesis · Lattice strain · Gas sensing

1 Introduction

In recent times, extensive interest has been gained in 2D-layered materials due to the versatility of their structure and properties as well as in different applications such as catalysts, sensing, water splitting, and solar cells [1]. The distinctive characteristics of MoS_2 direct it to be promising sensing material with high performance. A significant band gap (1.2–1.8 eV) and high surface-to-volume ratio provide rich adsorption sites for gas analytes. Multilayered MoS_2 nanomaterials have been chosen for the discerning of harmful gas analytes, for example, NO_2, NH_3, ethanol, and triethylamine. Despite many advantages, MoS_2-based gas sensors still face many major issues, such as inadequate sensitivity and poor recovery. As per reports, the synthesis of MoS_2-based heterostructures can be employed to improve the characteristics of pristine MoS_2 to amplify gas sensing properties using the synergistic effects [2]. Here, the properties of MoS_2/SnO_2 heterostructure have been proposed in terms of achieving better gas sensing performance at room temperature. SnO_2, consisting of a wide bandgap (3.6 eV), is an n-type semiconductor. It is quite known that SnO_2 characteristics can be altered by microstructural features such as particle size and shape to

P. P. Kumar · V. Singh (✉)
Department of Applied Physics, Delhi Technological University, New Delhi, India
e-mail: vinodsingh@dtu.ac.in

Z. H. Khan et al. (eds.), *Recent Advances in Nanomaterials*, Springer Proceedings in Materials 27, https://doi.org/10.1007/978-981-99-4878-9_1

make it more promising for gas sensing [3]. This paper is an effort to synthesize MoS_2 nanosheets and SnO_2 nanoparticles to produce a MoS_2/SnO_2 nanocomposite using a hydrothermal method. As an out-turn, it is presumed that SnO_2 addition to MoS_2 nanocomposite could enhance the microstructural and optoelectronic properties.

2 Synthesis and Characterizations

For MoS_2 synthesis, in 70 ml of DI water, ammonium molybdate, and thioacetamide were mixed and kept for 1 h stirring to obtain a homogeneous mixture. Subsequently, the collected mixture was placed in the hot air oven for 24 h at 200 °C in a Teflon-lined autoclave. The sedimented black precipitates were collected and cleaned consecutively by centrifugation at 7600 rpm with DI water and ethanol. To obtain the resultant powder, then the washed slurry was kept for drying at 80 °C for 12 h in the hot air oven. Further for the composite synthesis, $SnCl_2.5H_2O$ was dissolved into 40 ml of DI water. The solution was adjusted to pH 12 by adding a sufficient amount of NaOH, which resulted in a thick whitish slurry. Simultaneously, 0.1 g of MoS_2 was kept for ultrasonication for 45 min to have better dispersion in 40 ml of DI water. Then both solutions were mixed and kept for another 15 min stirring. The resultant solution was heated for 24 h at 200 °C. The obtained precipitates were collected and kept in the oven at 80 °C for 12 h for drying. For studying the basic properties of the synthesized materials, various characterizations such as X-ray diffraction (XRD), Raman spectroscopy, X-ray photoelectron spectroscopy (XPS, Al K_α, 1486.6 eV), and UV-vis absorption spectra were performed.

3 Results and Discussion

The XRD patterns of all the synthesized materials such as bare MoS_2, bare SnO_2, and nanocomposite of MoS_2-SnO_2 have been shown in Fig. 1a. The diffraction peaks available at 14.03°, 33.60°, and 59.57°, belonging to (002), (100), and (110) planes of MoS_2 respectively correspond to the hexagonal phase. A hump around 23° indicates the presence of residual oxides in MoS_2. The diffraction peaks of SnO_2 nanocrystal are noticed at 26.77°, 34.01°, and 52.07° at different planes corresponding to (110), (101), and (211) respectively confirm the tetragonal rutile phase [4].

Crystallite size is calculated using the intercept, and using the slope from W-H plots, strain is estimated [5] (Table 1).

The main distinctive peaks of MoS_2 at 376 cm^{-1} and 402 cm^{-1}, i.e. $E^1{}_{2g}$ and A_{1g} modes (in and out-plane vibrations), are shown in Fig. 1b. In SnO_2, the peak at 631 cm^{-1} shows the A_{1g} vibrations correspond to Sn-O planes. The peak at 581 cm^{-1} belongs to O-O vibrations in SnO_2. For MoS_2/SnO_2, the positions of $E^1{}_{2g}$ and A_{1g} have some blue shifts compared to that of MoS_2, where A_{1g} corresponding to Sn-O

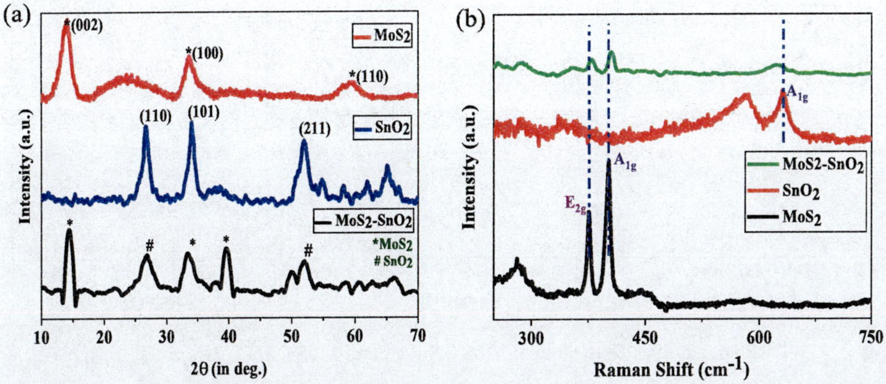

Fig. 1 **a** XRD and **b** Raman spectra of MoS_2, SnO_2, and MoS_2-SnO_2

Table 1 Tabulated chart of size, microstrain, and dislocation density values of MoS_2, SnO_2, and MoS_2-SnO_2

Sample	Debye Scherrer's crystallite size (nm)	W-H crystallite size (nm)	Microstrain	Dislocation density (nm^{-2})
MoS_2	4.2	4.3	4.82×10^{-3}	0.055
SnO_2	7.2	8.2	4.9×10^{-3}	0.014
MoS_2-SnO_2	5.3	5.0	–ve	0.039

vibration shows the redshift in the nanocomposite. The UV-DRS is used to investigate the optical properties and bandgap of MoS_2 nanosheets, SnO2 nanoparticles, and MoS_2/SnO_2 nanocomposite and the bandgap is calculated by extrapolationg the straight line portion of curves obtained using the graph plotted between energy (E=1240/λ) in eV and (αhv)^n; where n=1/2 for indirect bandgap and n=2 for direct bandgap. The corresponding urbach energies have been calculated using the slope between ln(α) and energy; here α is the absorption coefficient which is calculated using 2.303* absorption from the UV-DRS spectra (Fig. 2 and Table 2).

The high-resolution Mo 3d spectrum from MoS_2 is shown in Fig. 3a, which clearly shows two distinct peaks of Mo^{4+} available at binding energies 228.7 eV and 231.9 eV signifying 3d 5/2 and 3d 3/2 states, respectively. In addition, in Fig. 3c two isolated peaks ascribed to S 2p 3/2 and 2p ½ are reflected at 161.6 eV and 162.8 eV, respectively, supporting MoS_2 synthesis. Two distinct spin-orbit peaks belonging to Sn 3d 5/2 and 3/2, located at 486.6 eV and 494.9 eV attributed to Sn^{4+} states.

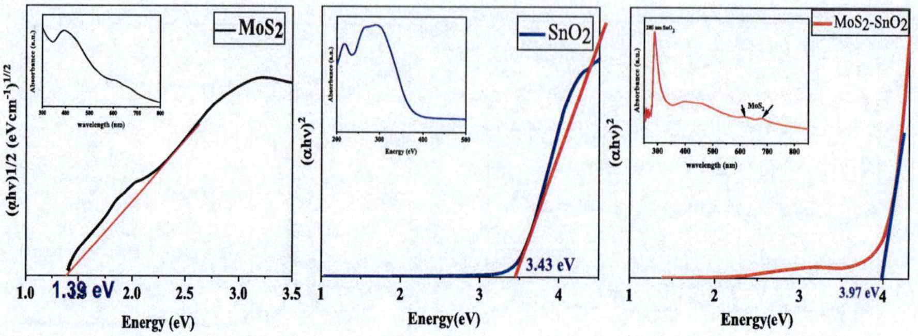

Fig. 2 UV-DRS spectra and Tauc plots of MoS_2, SnO_2, and MoS_2-SnO_2 (from left to right)

Table 2 Band gap and Urbach energy values of MoS_2, SnO_2, and MoS_2-SnO_2

Sample	Type of transition	Bandgap (eV)	Urbach energy (meV)
MoS_2	Indirect	1.39	749
SnO_2	Direct	3.43	331
MoS_2/SnO_2	Direct	3.97	386

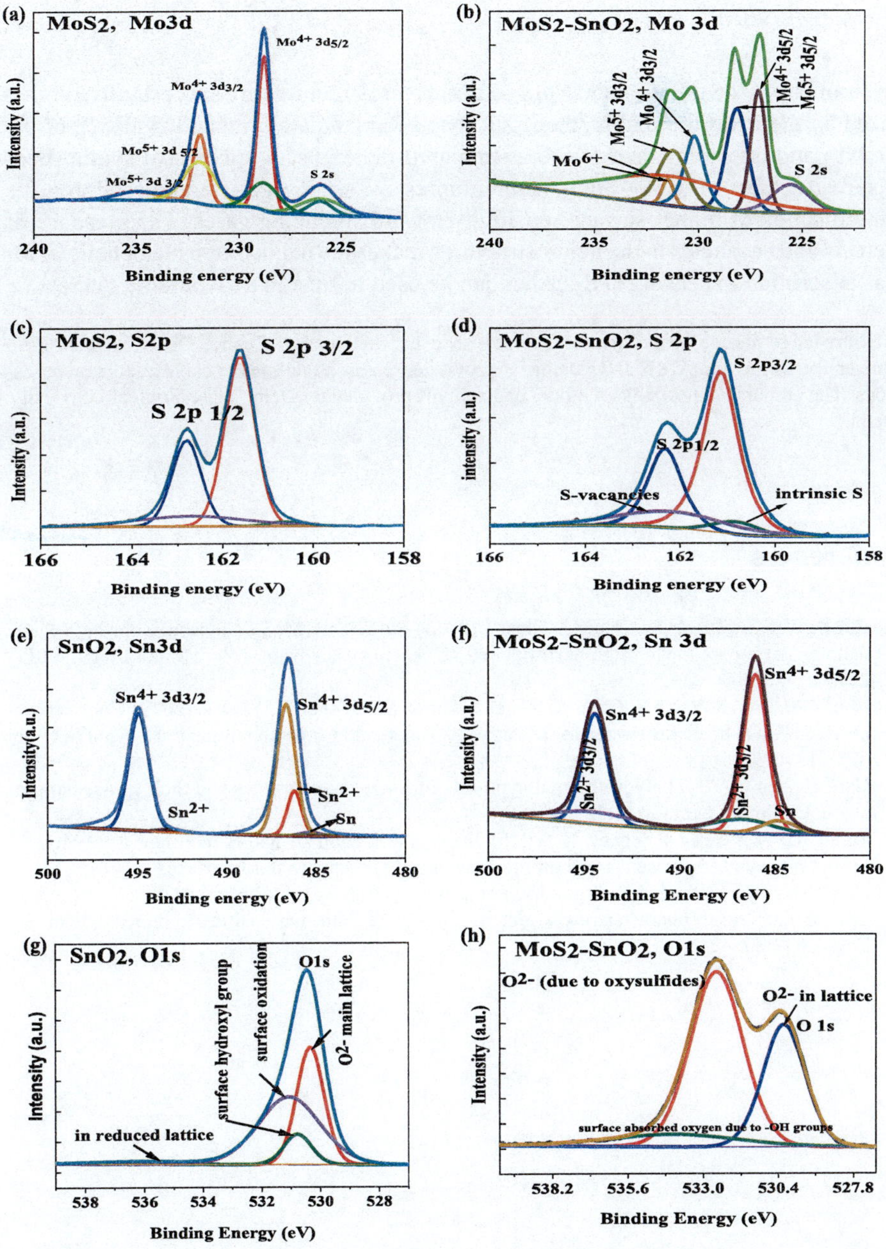

Fig. 3 XPS core level spectra of **a** survey spectra; **b** Mo 3d in MoS_2; **c** Mo 3d in MoS_2/SnO_2; **d** S 2p in MoS_2; **e** S 2p in MoS_2/SnO_2; **f** Sn 3d in SnO_2; **g** Sn 3d in MoS_2/SnO_2; **h** O 1s in SnO_2; **i** O 1s in MoS_2/SnO_2

4 Conclusion

In summary, MoS_2, SnO_2, and MoS_2-SnO_2 heterostructures are successfully synthesized by the hydrothermal method. XRD and Raman data provided evidence of the growth and crystallite size of the as-prepared materials. A successful synthesis of heterostructures of MoS_2-SnO_2 with compressive strain in a lateral direction gives an indication of higher surface sensitivity towards reducing gases. The presence of defects and vacancies in the heterostructures makes the nanocomposite beneficial for gas adsorption in sensing and further can be used to tune in a favourable manner.

Acknowledgements We are thankful to Advance Instrumentation Centre, Delhi Technological University, Delhi, and AIRF, JNU Delhi for providing research facilities to carry out characterizations. The authors acknowledge the financial support provided by Delhi Technological University, Delhi.

References

1. Cui S, Wen Z, Huang X, Chang J, Chen J (2015) Stabilizing MoS_2 nanosheets through SnO_2 nanocrystal decoration for high-performance gas sensing in air. Nano-Micro Small 11(19):2305–2313
2. Han Y, Ma Y, Liu Y, Xu S, Chen X, Zeng M, Hu N, Su Y, Zhou Z, Yang Z (2019) Construction of MoS_2/SnO_2 heterostructures for sensitive NO_2 detection at room temperature. Appl Surf Sci 493:613–619
3. Chiu H, Yeh C (2007) Hydrothermal synthesis of SnO_2 nanoparticles and their gas-sensing of alcohol. J Phys Chem C 111:7256–7259
4. Zhang D, Sun Y, Li P, Zhang Y (2016) Facile fabrication of MoS_2-modified SnO_2 hybrid nanocomposite for ultrasensitive humidity sensing. ACS Appl Mater Interfaces 8:14142–14149
5. HajyAkbary F, Sietsma J, Böttger JA, Santofimia JM (2015) An improved X-ray diffraction analysis method to characterize dis-location density in lath martensitic structures. Mater Sci Eng A 639:208–218

Synthesis of Titania Nanoparticles and Assessment of Antioxidant Activity

Yepuri Venkatesh, Patchamatla Satyanarayana Raju, and Putchakayala Yanna Reddy

Abstract Nanotechnology has inspired a plethora of research subjects and applications. Nanoparticles, in particular, have been manufactured utilizing a number of physical, chemical, and biological techniques, allowing them to be applied in various engineering and scientific applications. The sol-gel method, a wet chemical approach, was used to synthesize titania nanoparticles in this study; these particles were further characterized using an X-ray diffractogram (XRD), which revealed an anatase phase with a strong peak identified at Bragg angle 25 ^{O}. Fourier transform infrared spectroscopy (FTIR) analysis revealed TiO_2 vibrational modes as well as their hydroxyl group linkages at 932 and 3247 cm^{-1}, respectively. Analysis of the produced particles using FESEM characterization revealed the aggregation of particles with irregular morphologies. The diffuse reflectance examination of the particles supported light absorption in the ultraviolet range and 90% reflection in the infrared region. Finally, antioxidant studies on titania nanoparticles using the DPPH assay revealed improved scavenging activity.

Keywords Sol-gel synthesis · Nanoparticles · XRD · FTIR · FESEM · Antioxidant

1 Introduction

Nanotechnology is a breakthrough field of research that is quickly expanding, with applications in all fields of science and engineering [1, 2]. Nanoparticles of diverse shapes and sizes improve the electrical, electronic, optical, and medicinal properties of many materials. Metal oxide particles, in particular, have piqued the interest

Y. Venkatesh (✉) · P. Y. Reddy
Department of Electrical and Electronics Engineering, Swarnandhra College of Engineering and Technology, Seetharampuram, Narsapur, Andhra Pradesh 534280, India
e-mail: venkatesh.yepuri555@gmail.com

P. S. Raju
Department of Mechanical Engineering, Swarnandhra College of Engineering and Technology, Seetharampuram, Narsapur, Andhra Pradesh 534280, India

Z. H. Khan et al. (eds.), *Recent Advances in Nanomaterials*, Springer Proceedings in Materials 27, https://doi.org/10.1007/978-981-99-4878-9_2

of scientific community due to their extraordinary optical, electrical, and medicinal properties [3]. Titania nanoparticles are notable among metal oxides because of their non-toxicity and chemical stability, which are important in medicinal and electrical applications. Titania nanoparticles are also eco-friendly and have a significant antibacterial property. The large number of papers that have been published on the synthesis of titania nanoparticles showed that there has been extensive safety research, much of which is particularly useful in the medical disciplines. Titania's refractive value of $n = 2.4$ and insoluble nature in water make it suitable to be employed as a white pigment. Anatase (A), rutile (R), and brookite (B) are the three types of crystalline polymorphs of titania. With the exception of rutile's reduced weight and corrosion resistance, the three polymorphic forms of rutile and anatase generally share similar characteristics. Excellent research papers on the advantages of nanoparticles and their characteristics in the nanometre domain have been published. Quantum confinement governs the drift of electrons and holes in semiconductor nanostructures more than any other property. The major advantage of size shrinking is the increased specific surface area with lower size. This high surface area is the primary benefit for synthesizing titania-based devices since it facilitates the physical contact between the interacting media and devices, it happens on the surface and is heavily dependent on the material's surface area. Thus, the size of titania particles at the nanometre scale has a considerable effect on the performance of titania-based devices. Titania, being the most promising biocompatible material, is important in a variety of medical difficulties. It also broadens its use in addressing energy crisis challenges by enhancing solar energy usage. The increased microbial resistance to antibiotics, metal ions, multi-resistance drugs, and the emergence of resistant strains has attracted the current research, and titania nanoparticles have exhibited strong antibacterial activity. Titania, due to its strong photocatalytic activity, can be utilized in sunscreen lotions to protect exposed skin from UV-A and UV-B irradiation [4]. Titania, due to its antioxidant property can serve as an important role in preventing food's oxidative rancidity by scavenging free radicals produced during the oxidation process. Oxidative stress is caused by the release of free radicals or reactive oxygen molecules during their metabolic activities that exceed a biological system's antioxidant capability. It is linked to heart disease, cancer, neurological illnesses, and their ageing process. Oxidative stress is a new, universal mechanism underpinning nanoparticle toxicity.

These nanoparticles can be synthesized by various chemical and physical approaches like microwave [5], hydrothermal [6], solution route [7], sol-gel [8], and green synthesis [9]. The rutile titania nanoparticles were tested for morphological alterations, compromised antioxidant system, their reduction in cell viability, significant DNA damage, intracellular production of reactive oxygen species (ROS), and potential of these nanoparticles to induce geno and cytotoxicity in cultures human amnion epithelial cells [10]. Balachandran et al. synthesized Titania (TiO_2) nanoparticles using wet chemical sol-gel synthesis and tested their performance in treating lung cancer [11]. TEM investigation on the particles demonstrated that the particles in the nanometre range and the blue shift in the absorbance and increase in the energy band gap suggested that the particles are well suitable for photocatalytic applications and inhibit 85% of A549 cancer cells in a four-hour reaction time. Chaurasiya et al. used a wet chemical process to create titania nanorods and investigated their performance for photovoltaic and humidity sensing applications [12]. The nanorods application investigation confirmed a 0.3909% efficiency. Furthermore, the sensitivity of a titania nanorods (TiO_2 NRs)-based sensor was determined to be 2.93 pF/%RH for capacitance and 0.42 M/%RH for impedance at ambient conditions. Kignelman et al. used wet chemical sol-gel synthesis to synthesize nanoparticles and investigated the effects of CH_3COOH and HNO_3 on the nanoparticles [13]. The study proved the synergistic action of both acids and advised that they be used for controlled hydrolysis. Under moderate circumstances, a rise in acid content implied an increase in anatase content. Lal et al. synthesized titania nanoparticles and studied the effect of annealing temperature on morphology and particle size [14]. The increase in the temperature evidenced the enhancement of crystallite size and also suggested the decrement in the energy band gap with a temperature increment.

In this chapter, we discuss the wet chemical sol-gel method used to create titania nanoparticles and their antioxidant capabilities. The materials and procedure involved in the synthesis process are discussed in Sect. 2 of the manuscript. The results from field emission scanning emission microscopy (FESEM), diffused reflectance spectroscopy (DRS), Fourier transform infrared spectroscopy (FTIR), X-ray diffraction (XRD), and their antioxidant property were analysed are reported in Sect. 3 of the manuscript.

2 Synthesis and Characterization

2.1 Synthesis

In order to synthesize titania gel, 1.2 ml of solvent ethanol was put into a properly cleaned beaker, followed by 0.03 ml of catalyst HCL and 0.3 ml of CH_3COOH. Finally, 0.3 ml of TTIP was stirred for 15 s to finish the synthesis process. The prepared solution was kept for ageing at ambient temperature for 24 h to allow gelation to take place. The gel was then dried and sintered for 5 h in a muffle furnace at 500 degrees Celsius to eliminate the excess chemical residues and enhance the crystallinity of the particles synthesized. The sol-gel method to synthesize the titania nanoparticles was presented in Fig. 1.

2.2 Characterization

In the following stage, the above-mentioned titania particles were characterized using X-ray diffractograms (XRD-Rigaku, smart lab, Japan) for phase identification, FTIR (Bruker vertex 70, Germany) for functional bond analysis, FESEM (MIRA3

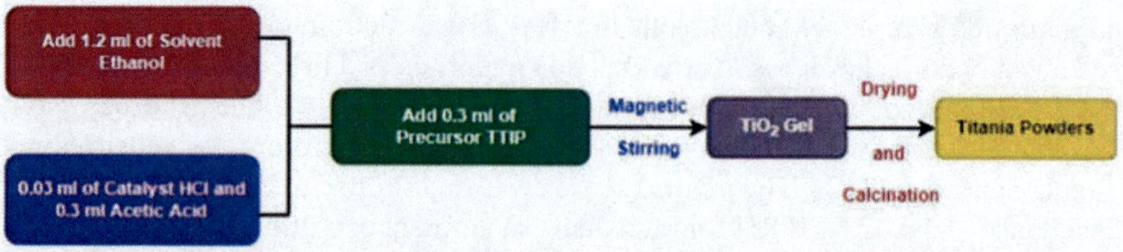

Fig. 1 Flow chart of synthesis process of titania nanoparticles

TESCAN) for morphology study, diffused reflectance analysis for band gap analysis (UV1800 Shimadzu, Japan), and tested with DPPH assay method for antioxidant property and presented in the results and discussion section.

3 Results and Discussion

Figure 2a demonstrates the XRD spectra of the titania particles synthesized by sol-gel method. The patterns were analysed by the powder diffractometer using Cu-Kα radiation source of wavelength $\lambda = 0.1541$ nm. The patterns recorded endorse the synthesized titania nanoparticles, calcined at 500° C belong to the anatase (A) phase, and the crystallite size was calculated by using Debye-Scherrer formula [15]:

$$D = \frac{K\lambda}{\beta}\text{Cos}\theta$$

where D represents grain size, K constant value is 0.94, λ is the wavelength of X-rays, β represents the full-width half maximum, and θ is the angle of diffraction. The sharp diffraction patterns and the values of full-width half maxima represent that the

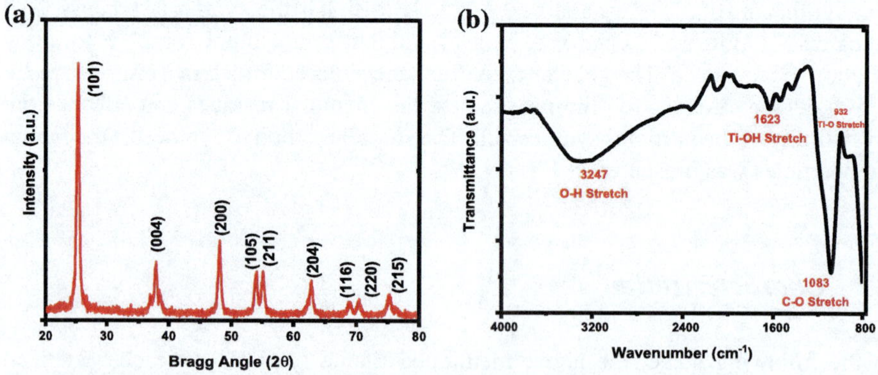

Fig. 2 XRD pattern of titania nanoparticles (a) and FTIR spectrum of titania nanoparticles (b)

synthesized titania nanoparticles are pure and crystalline with a crystallite size about 18 nm.

The peaks represented at Bragg angles 25 ^{O}, 37 ^{O}, 48 ^{O}, 54 ^{O}, 55 ^{O}, 62 ^{O}, 69 ^{O}, 70 ^{O}, and 75 ^{O} were indexed to the planes (101), (004), (200), (105), (211), (204), (116), (220), and (215), respectively. The results were correlated with the other published reports of anatase (A) titania particles [16, 17] and in good accord with the JCPDS card no. 1272. No characteristic peaks were further identified in the XRD pattern and demonstrate the pure anatase (A) phase of titania nanoparticles.

FTIR analysis was used to further examine the titania nanoparticles produced by wet chemical sol-gel synthesis and calcined at 500 ^{O}C in order to validate the presence of functional linkages as shown in Fig. 2b. A broad spectral vibration is observed at 3247 cm^{-1} representing the characteristic O-H stretching that is present on the surface of titania. The transmission peak at 1623 and 1083 cm^{-1} demonstrates the Ti-OH and C-O stretching vibrations. The sharp absorption peak demonstrated at 932 cm^{-1} endorses the Ti-O stretching vibration.

Titania nanoparticles synthesized using a wet chemical process and subsequently calcined at 500 °C were analyse its morphology using FESEM. The findings are shown in Fig. 3a and b. The micrographs examination at 2 µm and 500 nm revealed that the particles are aggregated; sphere-like formations with a diameter of around 60 to 70 nm were seen, and these formations were in excellent accord with the published study [18].

Figure 4a and 4b represent the UV–Vis-NIR diffused reflectance spectrum (DRS) and antioxidant property of synthesized titania nanoparticles. Band gap was calculated for the estimation of energy band gap of the titania nanoparticles. A strong absorption peak was observed in the DRS at 420 nm. The optical band gap was

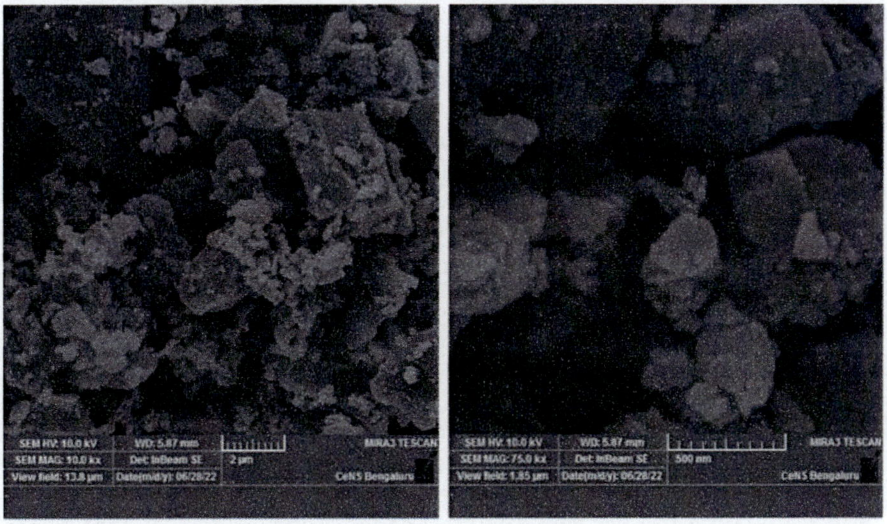

Fig. 3 FESEM image of the titania nanoparticles

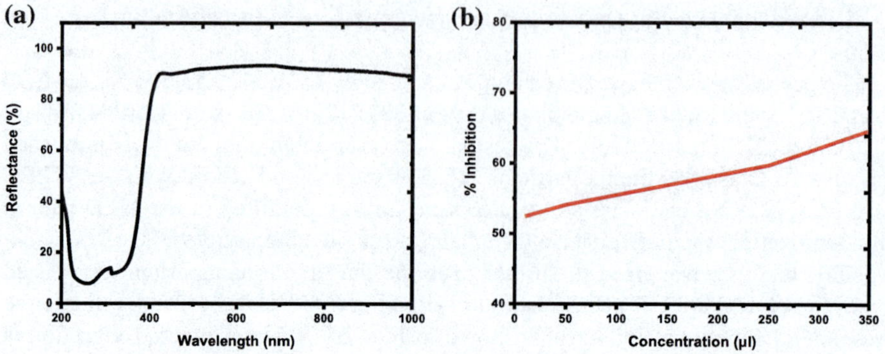

Fig. 4 Reflectance spectra (**a**) and antioxidant property evaluation on titania nanoparticles (**b**)

calculated using the Kubelka-Munk function and found to be 3.6 eV and agrees with the published reports [19].

The antioxidant analysis on the wet chemical sol-gel synthesized titania nanoparticles calcined at 500 °C is demonstrated in Fig. 4b. To stop the oxidation process, the antioxidant titania nanoparticles combat free radicals. Titania nanoparticles that were produced via a wet chemical process were tested for antioxidant activity against DPPH, and the results showed strong efficacy.

4 Conclusions

In this paper, titania nanoparticles were synthesized using a simple wet chemical sol-gel method and calcined at 500 °C. The synthesized titania nanoparticles were characterized by XRD and found anatase phase with a crystallite size about 18 nm. FTIR analysis on the particles demonstrated the vibration modes of hydroxyl and titania functional bonds. FESEM investigations on the titania nanoparticles endorsed that the particles were agglomerated with a spherical shape about 60 to 70 nm. UV-Vis spectroscopy study using the diffused reflectance setup demonstrated a bandgap about 3.6 eV. Further, antioxidant property on the titania nanoparticles evidenced that the synthesized particles have antioxidant activity against DPPH, and the results showed strong efficacy.

5 Competing Interests

The authors declare no competing interests.

Acknowledgements We acknowledge centre for nano and soft matter sciences (CeNS), and Avinashlingam University for their continuous support in providing the measurement facilities.

References

1. Zhu S, Meng H, Gu Z, Zhao Y (2021) Research trend of nanoscience and nanotechnology—a bibliometric analysis of Nano Today. Nano Today 39:101233
2. Gottardo S, Mech A, Drbohlavová J, Małyska A, Bowadt S, Riego J, Rauscher H (2021) Towards safe and sustainable innovation in nanotechnology: state-of-play for smart nanomaterials. Nano Impact 21:100297
3. Tortella GR, Pieretti JC, Rubilar O, Fernandez-Baldo M, Benavides-Mendoza A, Diez MC, Seabra AB (2022) Silver, copper and copper oxide nanoparticles in the fight against human viruses: progress and perspectives. Crit Rev Biotechnol 42(3):431–449
4. Fourtanier AF, Bernerd C, Bouillon L, Marrot D, Moyal S, Seite (2006) Protection of skin biological targets by different types of sunscreens. Photodermatol Photoimmunol Photomed 22(1):22–32
5. Ragupathi C, Vijaya JJ, Narayanan S, Kennedy LJ, Ramakrishna S (2013) Catalytic properties of nanosized zinc aluminates prepared by green process using Opuntiadilenii haw plant extract. China J Catal 34(10):1951–1958
6. Hayash H, Hakuta Y (2010) Hydrothermal synthesis of metal oxide nanoparticles in supercritical water. Materials 3(7):3794–3817
7. Chava RK, Kang M (2017) Improving the photovoltaic conversion efficiency of ZnO based dye sensitized solar cells by indium doping. J Alloy Compd 692:67–76
8. Liu H, Xu J, Liu G, Wang M, Li J, Liu Y, Cui H (2018) Building an interpenetrating network of Ni (OH)2/reduced graphene oxide composite by a sol-gel method. J Mater Sci 53(21):15118–15129
9. Devi Priya D, Roopan SM (2017) Cissusquadrangularis mediated ecofriendly synthesis of copper oxide nanoparticles and its antifungal studies against Aspergillus Niger, Aspergillus-favus. Mater Sci Eng C 80:38–44
10. Hu CW, Li M, Cui YB, Li DS, Chen J, Yang LY (2010) Toxicological effects of TiO_2 and ZnO nanoparticles in soil on earthworm Eisenia fetida. Soil Biol Biochem 42(4):586–591
11. Balachandran K, Mageswari S, Preethi A (2020) Photocatalytic decomposition of A549-lung cancer cancer cells by TiO_2 nanoparticles. Materials Today 37(2):1071–1074
12. Chaurasiya N, Kumar U, Sikarwar S, Yadav BC, Yadawa PK (2021) Synthesis of TiO_2 nanorods using wet chemical method and their photovoltaic and humidity sensing applications. Sens Int 2:100095
13. Kignelman G, Thielemans W (2021) Synergistic effects of acetic acid and nitric acid in water-based sol-gel synthesis of crystalline TiO_2 nanoparticles at 25 °C. J Mater Sci 56(30):16877–16886
14. Lal M, Sharma P, Ram C (2021) Calcination temperature effect on titanium oxide (TiO_2) nanoparticles synthesis. Optik 241:166934
15. Holzwarth U, Gibson N (2011) The Scherrer equation versus the Debye-Scherrer equation. Nat Nanotechnol 6:534
16. Wei X, Zhu G, Fang J, Chen J (2013) Synthesis, characterization, and photocatalysis of well-dispersible phase-pure anatase TiO_2 nanoparticles. Int J Photoenergy, 1–6
17. Chu L, Qin Z, Yang J, Li X (2015) Anatase TiO_2 nanoparticles with exposed {001} facets for efficient dye-sensitized solar cells. Sci Rep 5(1):12143
18. Arafat M, Haseeb A, Akbar S (2014) A selective ultrahigh responding high temperature ethanol sensor using TiO_2 nanoparticles. Sensors 14(8):13613–13627

19. Balakrishnan A, Gopalram K, Appunni S (2021) Photocatalytic degradation of 2, 4-dicholorophenoxyacetic acid by TiO_2 modified catalyst: kinetics and operating cost analysis. Environ Sci Pollut Res 28(25):33331–33343

Structural and Magnetic Studies of Nanocrystalline $La_{0.8-x}Ag_{0.2}Bi_xMnO_3(x = 0, 0.05)$

Priyanka Bisht and Rabindra Nath Mahato

Abstract We report the exploration of structural and magnetic properties of nanocrystalline $La_{0.8}Ag_{0.2}MnO_3$ and $La_{0.75}Ag_{0.2}Bi_{0.05}MnO_3$ compound, synthesized using the citrate sol-gel method. By analyzing X-ray diffraction data, it has been shown that both the nanocrystalline samples crystallized into a rhombohedral structure ($R\overline{3}c$ space group). The average crystallite size was evaluated to be ~41 and 35 nm by employing Williamson–Hall (W–H) method. SEM images reveal that the particles have an average particle size of 137 and 162 nm which are almost uniform in size. The SEM micrographs reveal the homogeneity of the nanocrystalline compound. At 299 K, paramagnetic to ferromagnetic transition is observed in nanocrystalline $La_{0.8}Ag_{0.2}MnO_3$ based on the temperature dependence of magnetization measurements. $La_{0.75}Ag_{0.2}Bi_{0.05}MnO_3$ sample shows transition temperature above room temperature which will be beneficial for magnetic refrigeration application.

Keywords Nanocrystalline · Sol-gel · Magnetic refrigeration

1 Introduction

The manganite exhibits notable properties, including phase transitions between ferromagnetic to paramagnetic, charge ordering, orbital ordering, and metal to insulator transitions depending on atomic structure and charge density, among other features [1]. Double-exchange mechanisms and Jahn–Teller effect can explain the unique features of these systems that rely on their structural, magnetic, and electrical properties. A significant amount of research has been conducted in the field of perovskite manganites and their consequent technical applications such as in the fields of colossal magnetoresistance (CMR), storage devices, magnetic field

P. Bisht · R. N. Mahato (✉)
School of Physical Sciences, Jawaharlal Nehru University, New Delhi 110067, India
e-mail: rnmahato@jnu.ac.in

Z. H. Khan et al. (eds.), *Recent Advances in Nanomaterials*, Springer Proceedings in Materials 27, https://doi.org/10.1007/978-981-99-4878-9_3

sensors, and spintronics [2]. When an increasing concentration of silver (Ag, monovalent ion) is doped into lanthanum manganite system, a large percentage of magnetoresistance (MR%) and temperature coefficients of resistance were observed [3], which makes these materials suitable for a wide range of applications. The substitution of Ag at A-site (AMnO3) in $La_{0.7}Sr_{0.2}Ag_{0.1}MnO_3$, $La_{0.5}Ca_{0.5-x}Ag_xMnO_3$, and $La_{0.7}Ag_{0.2}Bi_{0.1}MnO_3$ results into an enhanced temperature coefficient of resistance, magnetoresistance, as well as magnetocaloric properties [4–6]. As a result of monovalent ions being substituted at R-site, Mn ions can change their valency, despite the fact that these ion exhibits the same properties as a half-doped divalent ion. As a consequence of the lone pair effect of Bi^{3+} ions, substitution of bismuth will have different electrical and magnetic properties in lanthanum-based manganites. The work in this paper focuses on the synthesis and crystal structure, as well as the magnetic properties of $La_{0.8}Ag_{0.2}MnO_3$ and $La_{0.75}Ag_{0.2}Bi_{0.05}MnO_3$.

2 Experimental Details

In order to synthesized nanocrystalline samples $La_{0.8}Ag_{0.2}MnO_3$ (LAM) and $La_{0.75}Ag_{0.2}Bi_{0.05}MnO_3$ (LABM), citrate sol-gel technique [4] was used in which stoichiometric quantities of $La(NO_3)_3.6H_2O$ (99.999%, alfa aesar), $Bi(NO_3)_3.xH_2O$ (99.999%, alfa aesar), $C_4H_6MnO_4.4H_2O$ (99.999%, alfa aesar), $AgNO_3$($\geq$ 99.7%, Fisher Scientific) were taken and schematically presented in Fig. 1. Using X-ray diffractometer (Rigaku-made Miniflex 600 powder diffractometer), a crystal structure of the synthesized samples was analyzed at 300 K. Using FULLPROF, the refined lattice parameters were obtained by Rietveld refinement method. Scanning electron microscope (SEM, Carl Zeiss Evo-40) and energy dispersive spectroscopic (EDS, Bruker X flash detector 4010) analysis were conducted on the nanocrystalline samples to determine the surface morphology, average particle size, and elemental mapping, respectively. Magnetic measurements were conducted in vibrating sample magnetometer (PPMS, Cryogenic Ltd.). A magnetic field maximum up to 0.05 T at a rate of 2 K min^{-1} was applied within a temperature range between 2 K to 300 K for magnetization versus temperature measurements.

3 Results and Discussions

3.1 Structural Properties

In Fig. 2, we show the refined X-ray diffraction (XRD) patterns for the nanocrystalline LAM and LABM samples. Sample crystalizes in rhombohedral perovskite crystal structure (space group $R\overline{3}c$). Both samples consist of very small amount of metallic silver phase, as reported earlier in silver-doped manganite samples [6]. With

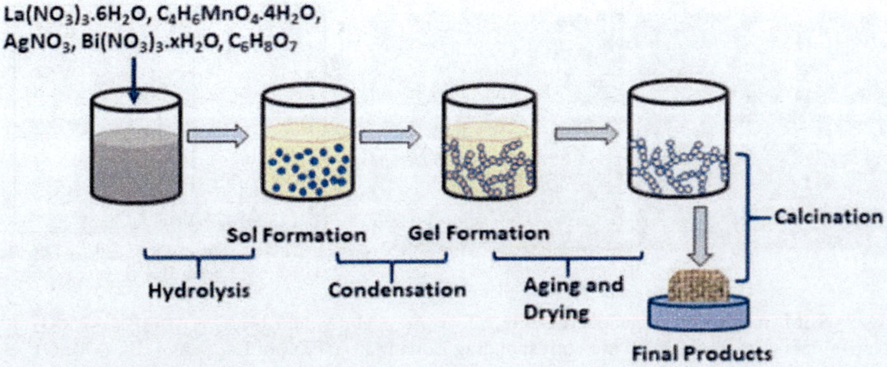

Fig. 1 Schematic diagram of sol-gel synthesis method

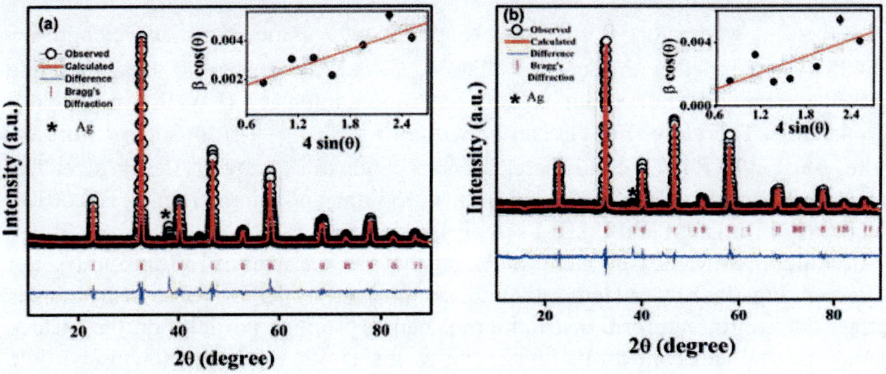

Fig. 2 Refined XRD pattern of **a** $La_{0.8}Ag_{0.2}MnO_3$ and **b** $La_{0.75}Ag_{0.2}Bi_{0.05}MnO_3$

Table 1 Nanocrystalline sample's (LAM and LABM) structural parameters, lattice cell volume, and convergence factor are presented in table

Sample	Crystal structure	Space group	a(Å)	b(Å)	c(Å)	V(Å^3)	χ^2
LAM	Rhombohedral	$R\bar{3}c$	5.5139 (3)	5.5137 (9)	13.3654 (5)	351.405(7)	2.03 (5)
LABM	Rhombohedral	$R\bar{3}c$	5.5166 (9)	5.5167 (3)	13.3917 (3)	352.686(2)	2.36 (4)

FULLPROF software, the pseudo-Voigt function was used to refine XRD data using the Rietveld method.

Table 1 lists the results of refinement parameters. Furthermore, average crystallite size was evaluated using W–H formula.

The line broadening in XRD is due to both the factors, i.e., crystallite size (β_s) and lattice strain (β_1):

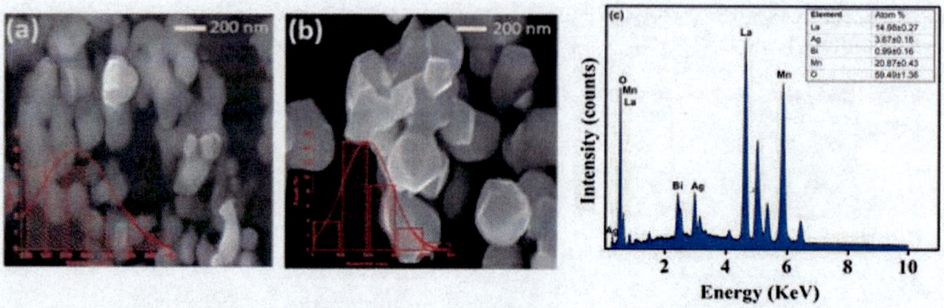

Fig. 3 SEM image for the nanocrystalline **a** $La_{0.8}Ag_{0.2}MnO_3$ and **b** $La_{0.75}Ag_{0.2}Bi_{0.05}MnO_3$ at room temperature. Insets showing corresponding Lorentz fit. **c** EDS of $La_{0.75}Ag_{0.2}Bi_{0.05}MnO_3$

$$\beta = \beta_s + \beta_1 \tag{1}$$

$\beta_s = \frac{K\lambda}{D\cos\theta}$, where $K \sim 0.9$ denotes shape factor, λ is the X-ray wavelength (= 1.5406 Å) equipped with Cu-Kα radiation, θ (degree) represents Bragg's angle and β corresponds full width at half maximum intensity (FWHM, in radian). $\beta_1 = 4\varepsilon\tan\theta$. Therefore, Eq. (1) can be written as, $\beta = \left(\frac{K\lambda}{D\cos\theta}\right) + 4\varepsilon\tan\theta$; furthermore, $\beta = \varepsilon\tan\theta + K\lambda / D\cos\theta$; here, D corresponds to average crystallite size. The calculated average crystallite size is 41 nm and 35 nm, and microstrain (ε) is 0.00138 and 0.00114 for LAM and LABM samples, respectively.

In addition, SEM measurements indicated that particle sizes of 130 nm and 162 nm (shown in Fig. 3a, b) were larger than those observed by XRD. These SEM images further confirm the uniform distribution of nanocrystalline particles on the surface of the material and confirm its homogeneity. Insets of Fig. 3a, b showing particle size histogram and corresponding Lorentz fit. Atomic percentage and elemental composition were calculated by EDS spectra (Fig. 3c).

3.2 Magnetic Properties

The temperature-dependent DC magnetization (M) in zero field cooled (ZFC) and field cooled (FC) manner under an applied magnetic field of (H) 0.05 T for the nanocrystalline LAM and LABM samples presented in Fig. 4a and b. The similar trend of M-T data after Bi doping has been shown elsewhere [6]. As shown in Fig. 3c, the differential magnetization plot has minima around 299 K, which confirms that the LAM sample has ferromagnetic (FM) to paramagnetic (PM) transition as temperature increases. From Fig. 4, we can see that from x = 0 to 0.05, the bifurcation increases as Bi-doping increases and such significant divergence between FC and ZFC has been ascribed to the magnetic frustration due to the coexistence of insulating antiferromagnetic (AFM) and metallic ferromagnetic (FM) phases or from both the

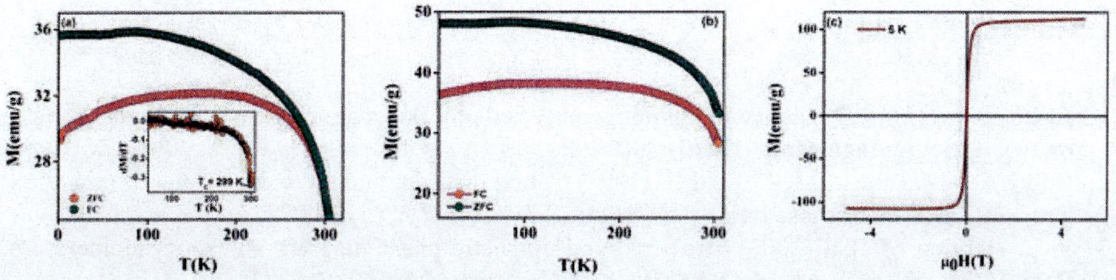

Fig. 4 **a** M-T (main panel) for the nanocrystalline LAM under the field strength of 0.05 T. Inset presenting curve of dM-$(dT)^{-1}$ vs. T. **b** Magnetization FC and ZFC curves in temperature range 5 K-300 K for LABM. **c** M-H loop at low temperature for LABM sample

Table 2 Comparison of different compounds to the nanocrystalline sample's LAM and LABM

Composition	Reference	T_C (K)
$La_{0.75}Bi_{0.05}Ag_{0.2}MnO_3$	This work	< 300
$La_{0.8}Ag_{0.2}MnO_3$	This work	299
$La_{0.8}Ag_{0.2}MnO_3$	[7]	306
$La_{0.62}Bi_{0.05}Ca_{0.33}MnO_3$	[8]	242
$La_{0.6}Bi_{0.1}Sr_{0.15}Ca_{0.15}MnO_3$	[9]	291

competing phases. Table 2 presents the comparison of the LAM and LABM to the other known samples.

Figure 4b shows that there is a finite magnetization in LABM sample even at 300 K and T_C is above room temperature. T_C around room temperature makes it suitable candidate for room temperature magnetic refrigerants. Figure 4c presents M-H (isothermal magnetization) loop at low temperature over a field range of $\pm$ 5 T which confirms the FM nature and found saturation magnetization is determined to be around 107.8 emu-g^{-1} and coercivity is 0.00558 T near the origin for nanocrystalline LABM sample. Observed low hysteresis loss confirmed the second order of magnetic phase transition [10].

4 Conclusions

In this study, structural and magnetic studies of LAM and LABM have been conducted. Both the samples successfully synthesized by citrate sol-gel method. On application of a magnetic field, nanocrystalline samples undergo phase transitions from PM to FM in rhombohedral crystal structure with $R\overline{3}c$ space group. M-H loop confirms the FM nature at low temperature. Magnetic transition around room temperature makes these materials suitable for magnetic refrigeration technology.

Acknowledgements Authors would like to express our gratitude to AIRF, JNU for providing magnetic, SEM, and EDS measurements. PB acknowledges UGC for the scholarship.

References

1. Ju HL, Sohn H (1997) Magnetic inhomogeneity and colossal magnetoresistance in manganese oxides. J Magn Magn Mater 167(3):200–208
2. Rodriguez-Martinez LM, Attfield JP (1998) Cation disorder and the metal-insulator transition temperature in manganese oxide perovskites. Phys Rev B 58(5):2426
3. Pi L, Hervieu M et.al (2003) Structural and magnetic phase diagram and room temperature CMR effect of $La_{1-x}Ag_xMnO_3$. Solid State Commun 126(4):229–234
4. Ayadi F et al (2014) Effect of synthesis method on structural, magnetic and magnetocaloric properties of $La_{0.7}Sr_{0.2}Ag_{0.1}MnO_3$ manganite. Mater Chem Phys 145(1–2):56–59
5. Smari M, Hamouda R, Walha I, Dhahri FJ et al (2015) Magnetic and magnetoresistance in half-doped manganite $La_{0.5}Ca_{0.5}MnO_3$ and $La_{0.5}Ca_{0.4}Ag_{0.1}MnO_3$. J Alloy Compd 644:632–637
6. Bisht P, Nagpal V, Singh G, Mahato RN (2022) Observation of Griffith like phase and large magnetocaloric effect in nanocrystalline $La_{0.7}Ag_{0.2}Bi_{0.1}MnO_3$. J Appl Phys 132(2):023901
7. Hien NT, Nguyen PT (2002) Preparation and magneto-caloric effect of $La_{1-x}Ag_xMnO_3$ (x = 0.10–0.30) perovskite compounds. Phys B Condens Matter 319(1–4):168–173
8. Gencer HÜSEYİN et al (2005) Magnetocaloric effect in the $La_{0.62}Bi_{0.05}Ca_{0.33}MnO_3$ compound. Phys B Condens Matter 357(3–4):326–333
9. Liu L, Zou Z, He B, Mao B, Xie Z (2022) Effect of Bi doping on the crystal structure, magnetic and magnetocaloric properties of $La_{0.7-x}Bi_xSr_{0.15}Ca_{0.15}MnO_3$ (x = 0, 0.05, 0.10, 0.15) manganites. J Magn Magn Mater 549:169006
10. Selmi A, M'nassri R, Cheikhrouhou-Koubaa W, Chniba Boudjada N, Cheikhrouhou A (2015) Ceram Int 41:10177–10184

Synthesis of Polymeric Nanoparticles Encapsulating Extract of *Datura Stramonium* and Study of Its Various in Vitro Activities

Rani Usha, Rani Asha, and Thakur Rajesh

Abstract In the present work, the extract from *Datura stramonium* is commonly used for its various pharmacological properties and was used for nanoparticles synthesis by ionic gelation method using complex of tragacanth and chitosan as polymers. The synthesized nanoparticles were characterized using techniques like PSA, FTIR, SEM, and TEM. The particle size of nanoparticles thus obtained ranged from 100 to 300 nm. SEM micrograph of nanoparticles showed the roughly spherical structures which appear to be encapsulated in the polymer. TEM images showed somewhat sphere-shaped nanoparticles that might be due to the encapsulation in tragacanth-chitosan complex. Synthesized nanoparticles showed considerable antibacterial action against various bacteria such *Pseudomonas aeruginosa* and *Escherichia coli*. Both Datura and chitosan-tragacanth NPs showed clear zone of inhibition showing that both possess antibacterial activity but the zone of inhibition formed by extract-loaded chitosan-tragacanth NPs is better than extract alone which shows they are more promising agents for antimicrobial activity. Before going to further studies the hemocompatibility of the synthesized nanoparticles was also investigated, and after it, the DPPH radical scavenging assay was used to measure the antioxidant activity and HRBC method was used to test the in vitro anti-inflammatory activity. It was found that the synthetic nanoparticles exhibit adequate amount of antioxidant activity, and it was observed by the results that the synthesized nanoparticles provided more protection than the individual plant extract and blank nanoparticles.

Keywords *Datura stramonium* · Nanoparticles · DPPH assay · Hemocompatibility

R. Usha · R. Asha · T. Rajesh (✉)
Department of Bio and Nano Technology, Guru Jambheshwar University of Science & Technology, Hisar, Haryana 125001, India
e-mail: rthakur99@gmail.com

Z. H. Khan et al. (eds.), *Recent Advances in Nanomaterials*, Springer Proceedings in Materials 27, https://doi.org/10.1007/978-981-99-4878-9_4

1 Introduction

Global demand for natural products is rising, which encourages the investigation of more and more alternatives and in these investigations' plants from Solanaceae family play a major role. The cultivation of this plant is practiced in many places around the world [1]. It is having large number of medicinal benefits like antiasthmatic, antifungal, antibacterial, sedative, expectorant, and demulcent [2]. Datura combats irrationality and reduces allergies [3]. Datura is also used in several scientific uses for example paste from the leaves of Datura can be used to treat injuries sustained outside, pain, and baldness [4]. Physical aches and pains can be relieved by the fruit's oil [5]. Seeds and leaves both have anticholinergic properties [6]. Gout and rheumatism can both be relieved by the vapors of a leaf infusion [7]. Additionally, it works well for the treatment of Parkinson's disease and for relaxing bronchial smooth muscle [8]. The plant's aqueous extract is a naturally occurring source of antioxidants, having a considerable amount of antioxidant activity [9]. The polyunsaturated carboxylic acid, oleic acid, palmitic acid, and lipoic acid made up most of the 55 g/kg methyl seeds [10].

The objective of the present work was to study the various in vitro activities of the synthesized nanoparticles and to explore how these polymeric nanoparticles can be used for pharmacological perspectives.

2 Materials and Methods

Tragacanth and chitosan were purchased from Sigma Aldrich. Plant leaves were collected from various locations in Hisar and Fatehabad (India). All the chemicals used for experimental purpose were of analytical grade, and these were autoclaved before use.

2.1 *Preparation of* D. Stramonium *Plant Extract-Loaded Tragacanth-Chitosan Nanoparticles*

Leaves of the *Datura stramonium* were collected from the various locations, and these were washed and shade dried. Once the leaves were completely dried, these were crushed and a powder was formed. Crushed leaves powder was taken in Duran bottle along with water in pre heated hot water bath. After 2 h, the sample was taken out and extract was separated and this plant extract is used for the preparation of nanoparticles.

2.2 *Characterization of* D. Stramonium *Extract-Loaded Tragacanth-Chitosan Nanoparticles*

2.2.1 Particle Size and Zeta Potential

Plant extract-loaded chitosan-tragacanth nanoparticles were analyzed by dynamic light scattering technique for particle size and zeta potential of the optimized nanoparticles. Zetasizer Nano ZS90, Malvern, UK, was used to analyze the optimized samples.

2.2.2 SEM and TEM

To analyze the morphology of the optimized plant extract-loaded chitosan-tragacanth nanoparticles scanning electron microscopy (SEM) as well as transmission electron microscopy (TEM) were used.

2.2.3 Fourier Transform Infrared Spectrophotometer (FTIR)

Plant extract-loaded chitosan-tragacanth nanoparticles, blank nanoparticles, and plant extract alone were lyophilized, and after lyophilization, the powdered form of nanoparticles was analyzed in the range of 4500–500 cm^{-1} using Fourier transform infrared spectrophotometer(FTIR) (IR Affinity-1, Shimadzu, Japan).

2.3 *Hemocompatibility*

The plant extract-loaded chitosan-tragacanth nanoparticles were also checked for their hemocompatibility by using volunteer's blood before going for the next activities.

2.4 *In Vitro Anti-Inflammatory Activity*

By using the HRBC membrane stabilization method, the anti-inflammatory activity of the plant extract-loaded tragacanth-chitosan nanoparticles was determined by the given formula [11–15].

$$\%\ \text{Hemolysis Protection} = \frac{\text{optical density control} - \text{optical density sample}}{\text{optical density control}} \times 100$$

2.5 Antioxidant Activity

1,1-Diphenyl-2-picrylhydrazyl hydrate (DPPH) radical scavenging activity was used to gauge the antioxidant potency of plant extract-loaded chitosan-tragacanth nanoparticles. Each test tube's assay mixture contained DPPH ethanol solution (0.1 M) and separate sample solution was incubated at a temperature of 37 °C for 30 min. Absorbance was determined by spectrophotometer at 517 nm. The following equation was used to determine the sample's % free radical scavenging.

$$\text{Free radical scavenging (\%)} = \frac{\text{Absorbance control} - \text{Absorbance sample}}{\text{Absorbance control}} \times 100$$

2.6 Antimicrobial Activity

The in vitro antibacterial activity was performed using the agar well diffusion method.

3 Results and Discussion

3.1 Particle Size and Zeta Potential

The concentration of polymers had impact on the particle size (nm) and encapsulation efficiency of *D. stramonium* plant extract in the polymers and may be due to this the size and the encapsulation efficiency of the nanoparticles changes between 100 to 400 and 30% to 60%, respectively (Fig. 1).

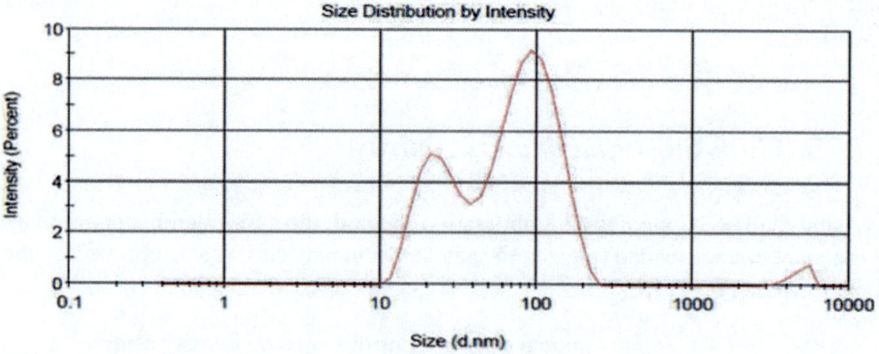

Fig. 1 PSA image of the plant extract-loaded chitosan-tragacanth nanoparticles

3.2 SEM and TEM

The SEM images of the synthesized polymeric nanoparticles of plant extract show the spherical structures, and the similar results were observed by the Ilium [13] and Kolahi et al. [16] (Figs. 2 and 3).

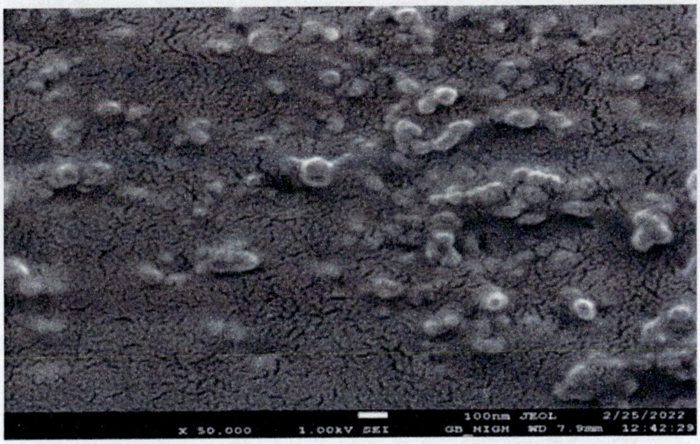

Fig. 2 SEM image of the plant extract-loaded chitosan-tragacanth nanoparticles

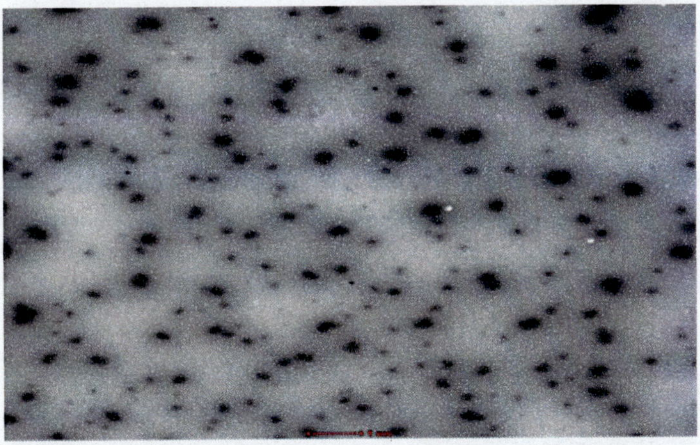

Fig. 3 TEM image of the plant extract-loaded chitosan-tragacanth nanoparticles. In the transmission electron microscopy, the prepared nanoparticles appeared spherical in shape

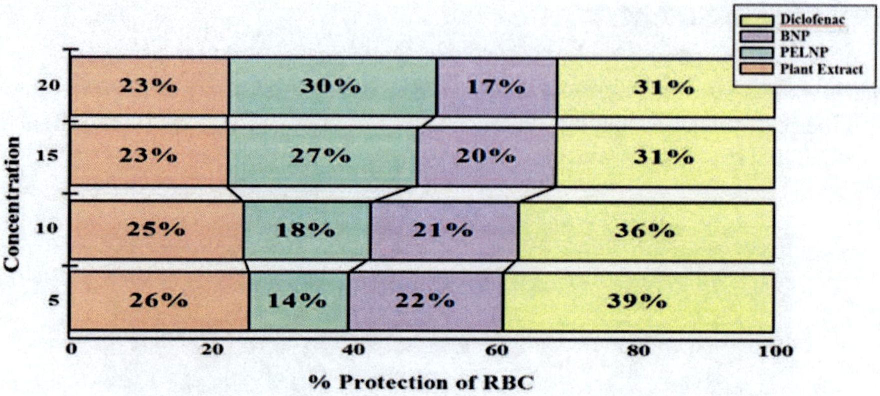

Fig. 4 In vitro % RBC protection by plant extract-loaded nanoparticles, plant extract, blank nanoparticles, and diclofenac

3.3 *FTIR*

Plant extract-loaded chitosan-tragacanth nanoparticles observed infrared spectra peaks may be identified as the straightforward superposition of their individual components.

3.4 *Hemocompatibility*

When the resulting samples were seen, it was found that the red blood cells pelleted and that the supernatant had not been lysed. The incubated sample's lack of lysis indicated that the formulation was best for blood circulation (Fig. 4).

The results were observed, and it has been demonstrated that nanoparticles loaded with plant extract were hemocompatible and might be used for various applications.

3.5 *In Vitro Anti-Inflammatory Activity of the Plant Extract Incorporated Tragacanth-Chitosan Nanoparticles*

The results compiled in Table 1 indicate that the percentage protection by the tragacanth-chitosan nanoparticles carrying the plant extract increased with their concentration, and that the concentration 20 μg/mL imparted protection higher than that of other concentrations. Earlier reports also suggest that the percentage protection of aqueous extract is concentration dependent, and it increases at higher concentrations [17, 18].

Table 1 Percentage of membrane stabilization by plant extract-loaded nanoparticles, plant extract, blank nanoparticles (BNP), and sodium diclofenac

S. no	Conc. (μg/ml)	% protection by HRBC method			
		Plant extract	Plant extract-loaded nanoparticles	BNP	Diclofenac
1	5	47.92 ± 0.02	26.13 ± 0.03	40.53 ± 0.094	71.85 ± 0.16
2	10	53.03 ± 0.03	37.83 ± 0.011	44.71 ± 0.122	76.47 ± 0.42
3	15	62.04 ± 0.07	73.66 ± 0.016	54.00 ± 0.03	85.12 ± 0.32
4	20	65.42 ± 0.01	85.48 ± 0.015	48.53 ± 0.06	89.10 ± 1.21

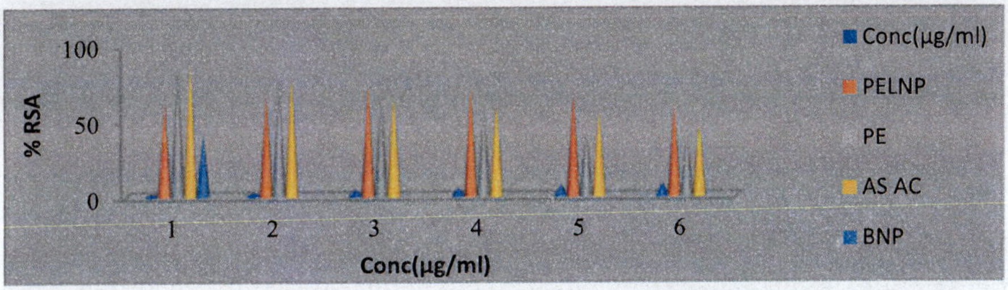

Fig. 5 % Radical scavenging activity shown by plant extract-loaded nanoformulations (PELNP), plant extract (PE), ascorbic acid (AS AC), and blank nanoparticles (BNP)

3.6 *Antioxidant Activity*

The maximum percentage radical scavenging capacity of the plant extract-loaded tragacanth-chitosan nanoparticles was 73.87 ± 0.042% as compared to the blank nanoparticles 42.41 ± 0.015% (Fig. 5).

3.7 *Antimicrobial Activity*

To check the antibacterial activity various gram-positive and gram-negative bacteria were used, and it was observed that the leaf extract-loaded tragacanth-chitosan nanoparticles show considerable antimicrobial potency using the agar well diffusion method [19].

4 Discussion

Size of the synthesized nanoparticles lies between 100 to 400 nm, and the encapsulation efficiency was found to be between 30 and 60% which is considered good for the pharmacological perspective. The nanoparticles appeared spherical in SEM and TEM images. FTIR peaks show a clear superposition of their individual components which shows that its constituents did not show any chemical interaction. The hemocompatibility results show that the synthesized nanoparticles are safe for use, and it was observed that at the concentration of 20 μg/mL aqueous extract shows higher anti-inflammatory activity than that of blank nanoparticles. The radical scavenging capacity of the plant extract-loaded tragacanth-chitosan nanoparticles was high with comparison to the blank nanoparticles, and the antimicrobial potency was of considerable level.

5 Conclusion

Datura stramonium leaf extract-loaded chitosan-tragacanth nanoparticles were synthesized by using ionic gelation method but in this process no stabilizers were used. It was observed that the optimized nanoparticles were of spherical shape, and the particle size and encapsulation efficiency varied with change in concentration of polymers. Anti-inflammatory, radical scavenging activity, and antimicrobial activity of leaf extract-loaded chitosan-tragacanth nanoparticles were considerable. Hence, it can be concluded that this bipolymeric *Datura stramonium* leaf extract-loaded chitosan-tragacanth nanoparticles can be further used for in vivo studies to explore more options because it can be used for various pharmaceutical applications due to its important properties.

6 Declaration of Interest Statement

The authors declare that they have no conflict of interests.

Acknowledgements The authors are thankful to the Department of Bio and Nano Technology, Guru Jambheshwar University of Science and Technology, Hisar, for providing the research facilities. Ms. Usha Rani acknowledges the University Grant Commission Junior Research Fellowship from UGC, Ministry of Education, India.

References

1. Moore RL (1972) Spiritualism and science: reflections on the first decade of the spirit rappings. Am Q 24(4):474–500
2. National Pharmacopoeia Committee, Pharmacopoeia of People's Republic of China (2015) Part 1, Chemical Industry Press, Beijing, p 267
3. Savithramma N, Sulochana C, Rao KN (2007) Ethnobotanical survey of plants used to treat asthma in Andhra Pradesh, India. J Ethnopharmacol 113(1):54–61
4. Njoroge G (2012) Traditional medicinal plants in two urban areas in Kenya (Thika and Nairobi): diversity of traded species and conservation concerns. Ethnobot Res Appl 10:329–338
5. Vijendra N, Kumar KP (2010) Traditional knowledge on ethnomedicinal uses prevailing in tribal pockets of Chhindwara and Betul Districts, Madhya Pradesh, India. Afr J Pharm Pharmacol 4(9):662–670
6. Wazir SM, Dasti AA, Shah J (2004) Common medicinal plants of Chapursan valley, Gojal II, Gilgit, Pakistan. J Res (Sci) 15(1):41–43
7. Abul HM, Tabassum R, Razzak A, Rai P, Biswas B, Mesbaus S, Robiul IM (2020) Allelopathic effects of aqueous leaf extracts of datura on growth and yield of lentil. J Biol Res 4(1):48–57
8. Ivancheva S, Nikolova M., Tsvetkova R (2006) Pharmacological activities and biologically active compounds of Bulgarian medicinal plants. Phytochem Adv Res 37661:87–103
9. Akhtar N, Mirza B (2018) Phytochemical analysis and comprehensive evaluation of antimicrobial and antioxidant properties of 61 medicinal plant species. Arab J Chem 11(8):1223–1235
10. Ramadan MF (2011) Bioactive phytochemicals, nutritional value, and functional properties of cape gooseberry (Physalis peruviana): an overview. Food Res Int 44(7):1830–1836
11. Dai T, Tanaka M, Huang YY, Hamblin MR (2011) Chitosan preparations for wounds and burns: antimicrobial and wound-healing effects. Expert Rev Anti Infect Ther 9(7):857–879
12. Islam MA, Firdous J, Choi YJ, Yun CH, Cho CS (2012) Design and application of Chitosan microspheres as oral and nasal vaccine carriers: an updated review. Int J Nanomed 7:6077
13. Ilium L (1998) Chitosan and its use as a pharmaceutical excipient. Pharm Res 15(9):1326–1331
14. Pinheiro AC, Bourbon AI, Medeiros BGDS, da Silva LH, da Silva MC, Carneiro-da-Cunha MG et al (2012) Interactions between κ-carrageenan and chitosan in nanolayered coatings—structural and transport properties. Carbohydr Polym 87(2):1081–1090
15. Shah NC (1982) Herbal folk medicines in Northern India. J Ethnopharmacol 6:293–301
16. Kolahi P, Shekarchizadeh H, Nasirpour A (2022) Stabilization of Pickering emulsion using tragacanth nanoparticles produced by a combination of ultrasonic and anti-solvent methods. J Sci Food Agric 102(4):1353–1362
17. Nagaharika Y, Rasheed S (2013) Anti-inflammatory activity of leaves of Jatropha gossypifolia L. by HRBC membrane stabilization method. J Acute Dis 2(2):156–158
18. Kiani K, Wang Q (2012) On the interaction of a single-walled carbon nanotube with a moving nanoparticle using nonlocal Rayleigh, Timoshenko, and higher-order beam theories. Eur J Mech-A/Solids 31(1):179–202
19. Hossain MA, Shah MD, Charles G, Iqbal M (2011) In vitro total phenolics, flavonoids contents and antioxidant activity of essential oil, various organic extracts from the leaves of tropical medicinal plant Tetrastigma from Sabah. Asian Pac J Trop Med 4:717–721

Synthesis, Characterization and Various in Vitro Activities of Essential Oil-Loaded Polymeric Nanoformulations

Choudhary Asha, Rani Usha, Salar Raj Kumar, and Thakur Rajesh

Abstract The main objectives of the present research were to synthesize cumin essential oil-loaded polymeric nanoformulations using the ionic gelation method and use it in a variety of in vitro activities. *Cumminium cyminum* is well known in world among the spices and if we discussed its seeds that also contain good amount of oil content having various bioactive compounds like cumin aldehyde, *P*-cymene, D-limonene, terpinene and eugenol with considerable potential in various in vitro activities. Various techniques were used to characterize the synthesized nanoformulations. Using a particle size analyzer, the nanoformulations obtained had diameters ranging from 20 to 400 nm. The nanoformulations could be aggregated with a nearly spherical shape, as seen by SEM analysis. The nanoformulations, cumin oil and ascorbic acid exhibit strong antioxidant activity, ranging from 27 to 80% as assessed by the DPPH assay.

Keywords *Cumminium cyminum* essential oil · Polymeric nanoformulations · Methyl bromide · Cuminal · Phosphate-buffered saline (PBS)

1 Introduction

Essential oils are aromatic components that have a distinct odour and fragrance and are rich in biologically active compounds such as terpenoids, terpenes and phenolic components [1, 2]. As a result, essential oil extracted from cumin seeds has various functional properties like antimicrobial and antioxidant properties and all these are due to the various bioactive compound present in it for example terpinene, p-cymene, pinene, cumin aldehyde, safranal, eugenol and cuminal [3, 4].

C. Asha (✉) · R. Usha · S. R. Kumar · T. Rajesh
Department of Biotechnology, Chaudhary Devi Lal University, Sirsa, Haryana 125055, India
e-mail: choudharyasha260@gmail.com

Department of Bio and Nano Technology, Guru Jambheshwar University of Science & Technology, Hisar, Haryana 125001, India

Z. H. Khan et al. (eds.), *Recent Advances in Nanomaterials*, Springer Proceedings in Materials 27, https://doi.org/10.1007/978-981-99-4878-9_5

Due to the various limitations such as hydrophobic in nature, volatility and lack of uniform dispersion, these essential oils are encapsulated in the various suitable biodegradable polymer so that these can be easily accessible to check the various in vitro activities [5]. Encapsulating essential oils in polymer-based nanoformulations can be an effective way to improve the life and potential so that we can use it for long term. As a naturally biocompatible and biodegradable polymer, chitosan nanoparticles are now widely used in drug delivery studies [6, 22]. Natural polysaccharides such as pectin (Pec) and chitosan (CS), which are both biodegradable and biocompatible, non-toxic to mammals, hydrophilic and able to synthesize emulsions, are widely utilized in pharmaceuticals. Due to its outstanding gelling property, pectin, among the primary structural water-soluble polysaccharides obtained from plants, used in a variety of applications. It is composed of many poly-D-galacturonic acid residues linked by -1,4-glycosidic linkages. Chitosan, on the other hand, is derived from the deacetylation of Chitin and is made of linear chains of glucosamine and N-acetyl-D-glucosamine. It is an excellent encapsulating agent. It was found that the solubility of Pec and CS in aqueous solutions decreased when they were combined, preventing the encapsulated drug from being released very early [7–10].

There is currently a new trend to use nanotechnology to solve the problem of volatility and stability by developing numerous nanoformulations. These nanotechnologies or nanoformulations are major concerns regarding the use of essential oils in modern agriculture and pest management [11].

Stored grain pests are a major issue, accounting for 10 to 40% of global grain loss, according to the Food and Agriculture Organization (FAO). Insect and other bioagent infestation of stored products have many disadvantages, including loss of product weight, nutritive content, commercial and aesthetic value, and it may pose a health risk [12, 13]. There are many fumigants like methyl bromide, phosphine and sulfuryl fluoride which have been used commonly from decades for stored grain pests management but it was observed that these are having disastrous effects on environment and also lead to ozone depletion [14].

To resolve these issues, natural compounds derived from plants, such as essential oil, have been proposed as novel and environmentally friendly substitutes for synthetic pesticides, recently. Although many active compounds are hydrophobic by nature, they must be formed to enhance their physicochemical features, for example using an oil-in-water nanoformulations method. Recently, nanotechnologies have been proposed to address these issues by encapsulating essential oils in nanoformulations and improving the durability of essential oil-based insecticides [15, 16].

2 Materials and Methods

2.1 Chemicals

All the chemicals, including chitosan (CS), acetic acid, dichloromethane, calcium chloride (CaCl2), pectin and Tween 80, were of analytical grade and were purchased from HiMedia Laboratories Pvt. Ltd. (India).

2.2 Preparation of Cumin Essential Oil-Loaded Polymeric Nanoformulations

Cumin essential oil-loaded polymeric nanoformulations (CEO/P/NPs) were synthesized by ionic gelation process [17]. In our experiment, we used 1.5% solution of pectin and chitosan, and in this solution, we add Tween 80 (5): Span80 (2) and then it was placed on magnetic stirrer 900 rpm for 3 h at room temperature. *Cuminium cyminum* essential oil was added to the mixture while it was still being mixed, and the mixture was then stirred for additional 10 min.

2.3 Characterization of CEO/P&CSNPs

Characterization was done by using various techniques as mentioned below.

2.3.1 Particle Size Analyzer (PSA)

Size of particle was determined after an appropriate dilution by (Malvern Instruments, UK).

2.3.2 Scanning Electron Microscope (SEM)

Lyophilized samples of CEO/P&CSNPs were used to check the morphology by using the SEM (JEOL Model JSM—6390LV).

2.3.3 Transmission Electron Microscopy (TEM)

TEM was used to check the size of the CEO/P&CSNPs, and the images were taken using a TECNAI (HR-TEM).

2.4 *In Vitro Release Study*

In our experiment to study the in vitro release of the prepared nanoformulations containing essential oil, we made several modifications to the dialysis method and used by Natrajan et al. [18]. The membrane which was used for the release activity was soaked for 24 h in distilled water so that the preservative if used removed prior to the experimental set-up. After 24 h, 3 ml of the nanoformulation solution was taken out from and the same amount of PBS buffer was added to maintain the volume. The use of ethanol aids in the reduction of coalescence and the uniform release of oil. The time-dependent release study was carried out from 0 to 14 h. All the sets were incubated with gentle shaking. After each 1 h, 3 ml of the medium was taken out and replaced with the same amount of fresh media and spectrophotometrically evaluated. The percentage release was calculated by using the below-given equation:

$$\text{Release } (\%) = \frac{\text{Released oil}}{\text{Total oil}} \times 100$$

2.5 *Antioxidant Activity*

2.5.1 DPPH Assay

Using the DPPH Assay, the antioxidant potency of Cumin CEO/P/NPs and ascorbic acid was determined [5, 19]. The concentrations required were added to a 4 mL DPPH solution. The sample's absorbance was tested after 30 min. The formula below was used to calculate the per cent radical scavenging activity.

$$\%\text{RSA} = \frac{\text{Absorbance of blank} - \text{Absorbance of sample}}{\text{Absorbance of sample}} \times 100$$

2.6 *Haemocompatibility*

The haemolysis test can be used to determine the haemocompatibility of essential oil-loaded nanoformulations. To check the haemocompatibility of the prepared samples, the blood was withdrawn from the healthy person. The CEO/P&CSNPs introduced to diluted blood, and they were then incubated at 37 °C for the next hour. Saline was used as the negative control while distilled water was placed as the positive control. Later that, samples were taken and centrifuged at 4000 rpm for 5 min Natrajan et al. [18].

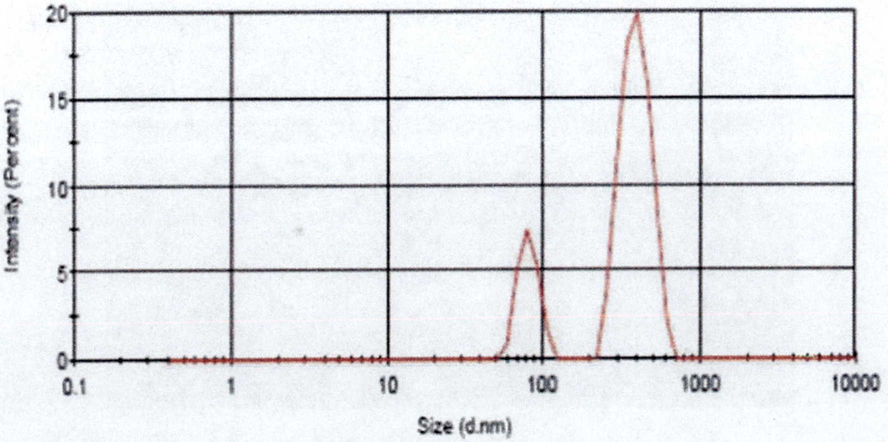

Fig. 1 Graphical representation of CEO/P&CSNPs

3 Results

3.1 Preparation of CEO/P&CSNPs

CEO/P&CSNPs were synthesized using the ionic gelation method.

3.2 Characterization of CEO/P&CSNPs

Synthesized nanoformulations were characterized using various techniques.

3.2.1 Particle Size Analyzer (PSA)

The size of the nanoformulations is varied with concentration, reaching 220 nm at 1.5% biopolymer (Fig. 1).

3.2.2 Scanning Electron Microscope (SEM) and

SEM was used to examine the morphology of CEO/P/NPs, and according to Fig. 2, SEM images revealed that the prepared nanoformulations seem to be mostly spherical in shape, but the precise shape could not be determined.

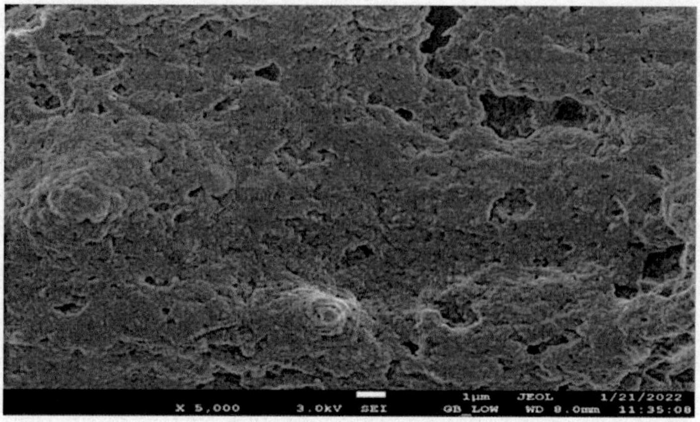

Fig. 2 Scanning electron microscopic images of CEO/P&CSNPs

3.2.3 Transmission Electron Microscopy (TEM)

According to Fig. 3, TEM images show that the synthesized nanoformulations were core-shaped with cross-linking.

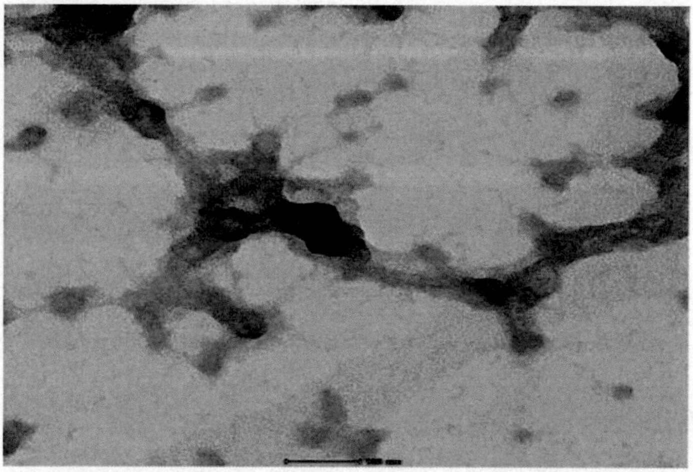

Fig. 3 TEM micrographs of CEO/P&CSNPs

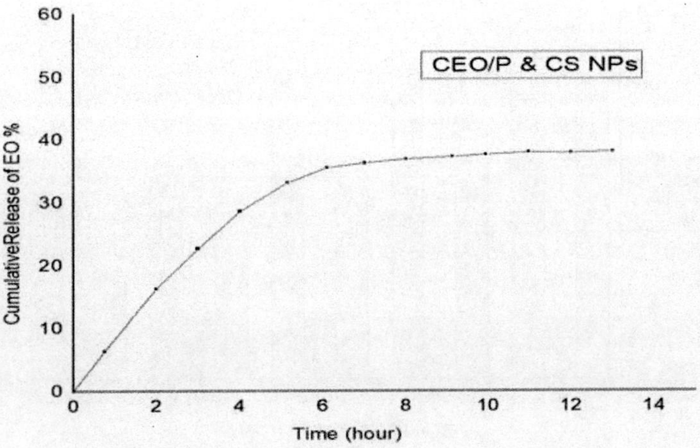

Fig. 4 Essential oil release kinetics of CEO/P&CSNPs

3.3 *In Vitro Release Study*

As shown in Fig. 4, the percentage of CEO release from CEO/P&CSNPs was found to be 5–38% after a 14-h release study.

3.4 *Antioxidant Activity by DPPH Assay*

In our experiment, we found that the antioxidant properties of CEO/P&CSNPs (29–81%), and for blank nanoformulations (16%), ascorbic acid was significantly different from the range of 48–87% as shown in Fig. 5.

3.5 *Haemocompatibility*

In the present experiment, distilled water was used as a positive control to induce red blood cell lysis and saline was used as negative control, which had zero lysis percentages. There was no haemolysis in the blood samples after incubation with essential oil loaded (Table 1).

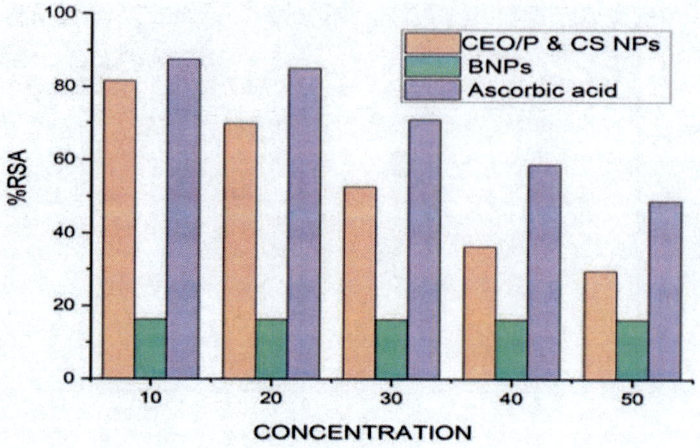

Fig. 5 %Radical scavenging activities of (CEO/P&CSNPs), blank nanoformulations (BNPs) and ascorbic acid

Table 1 Percentage of membrane stabilization by CEO/P&CSNPs, CEO, BNPs and sodium diclofenac

Sr. no	Conc. (μg/ml)	% protection by HRBC method			
		EO	CEO/P&CSNPs	Sodium diclofenac	BNPs
1	10	48.07 ± 0.18	52.50 ± 0.02	68.766 ± 0.19	46.16 ± 0.17
2	20	58.57 ± 0.11	61.47 ± 0.03	77.73 ± 0.03	56.16 ± 0.25
3	30	70.90 ± 0.07	84.07 ± 0.28	84.38 ± 0.01	67.54 ± 0.153
4	40	77.96 ± 0.06	75.33 ± 0.05	86.86 ± 0.03	44.74 ± 0.21

4 Discussion

Essential oil-loaded polymeric nanoformulations were synthesized, characterized and evaluated successfully. The particle size range for all experiments, according to Attallah et al. [7], was 468.5–698.3 nm. According to the previously reported work on the oil-loaded nanocapsules by Natrajan et al. [18], it was observed that the particle size of the various nanocapsules was in the range of 256 and 226 nm.

According to Froiio et al. [20], the prepared dispersions were mainly comprised of spherical particles, but due to the small size and low density of the polymer employed in the study exact shape was not determined.

Furthermore, Jiang et al. [21] observed that TEM images revealed a structure that was almost core-shaped, which does not exactly match what happened in the sample before it dried on the sample holder or could be delamination caused by cumin essential oil.

According to Natrajan et al. [18], their research demonstrates that at pH 7.4, encapsulated turmeric oil released over 90% of its contents, compared to encapsulated lemongrass oil, which released 42% over a 48-h period. The release rates of encapsulated turmeric oil and lemongrass oil were 70% and 38%, respectively, at pH 1.5. The findings suggested that this might be because of a different pH level. Moreover, the release study demonstrates the first quick release of *Petroselinum crispum* essential oil and chitosan nanoemulsion loaded with *Petroselinum crispum* essential oil, which proceeded by sustained and prolonged release [22].

Additionally, the antioxidant activity of Cur-NEs was not impacted by the amount of water (20.3–69%) but increased dramatically with concentration of surfactant. Increasing in the amount of surfactant may help curcumin dissolve in the oil phase, increasing antioxidant activity proposed by Joung et al. [23].

According to Natrajan et al. [18], both the turmeric oil- and lemongrass oil-loaded nanocapsules were incubated with blood samples, and it was observed that there is the absence of lysis where haemocompatible further uses.

5 Conclusion

CEO/P&CSNPs were synthesized using an ionic gelation method, with particle sizes ranging from 20 to 400 nm and good stability. According to SEM results, the morphology of essential oil-loaded nanoformulations was spherical-like. TEM images revealed a structure that was nearly core-shaped. The absence of cell lysis in the haemocompatibility assay also confirmed that the prepared oil-loaded nanoformulations were haemocompatible, implying that they can be used as a carrier for future biomedical applications. The free radical scavenging activity of CEO/PNs was (29–81%), while blank nanoformulations (16%) and ascorbic acid (48–87%) were significantly different. In the 14-h drug release experiment, the prepared nanoparticles showed the steady and sustained release of the encapsulated EO from the prepared nanoparticles (5–38%).

The encapsulation of cumin essential oil into polymer was thus found to be a stable and controlled release of essential oil, and it was subsequently used in various in vivo studies to analyse more alternative choices because of its significant properties that can be used for various pharmaceutical and food industry applications. The usefulness of such pectin nanomaterials for several uses has to be investigated in further explanation [24].

6 Declaration of Interest Statement

The authors declare that they have no conflict of interests.

Acknowledgements The authors are thankful to Chairperson, Department of Biotechnology, Chaudhary Devi Lal University, Sirsa, and Department of Bio and Nanotechnology, Guru Jambheshwar University of Science and Technology, Hisar, for providing the research facilities.

References

1. Esmaeili A, Asgari A (2015) In vitro release and biological activities of Carum copticum essential oil (CEO) loaded chitosan nanoparticles. Int J Biol Macromol 81:283–290
2. Rizwan-ul-Haq M, Aljabr AM (2015) Rhynchophorus ferrugineus midgut cell line to evaluate insecticidal potency of different plant essential oils. In Vitr Cell Dev Biol-Anim 51(3):281–286
3. Amiri A, Mousakhani-Ganjeh A, Amiri Z, Guo YG, Singh AP, Kenari RE (2020) Fabrication of cumin loaded-chitosan particles: characterized by molecular, morphological, thermal, antioxidant and anticancer properties as well as its utilization in food system. Food Chem 310:125821
4. Karimirad R, Behnamian M, Dezhsetan S (2019) Application of chitosan nanoparticles containing Cuminum cyminum oil as a delivery system for shelf-life extension of Agaricus bisporus. LWT 106:218–228
5. Das S, Singh VK, Dwivedy AK, Chaudhari AK, Upadhyay N, Singh P et al (2019) Encapsulation in chitosan-based nanomatrix as an efficient green technology to boost the antimicrobial, antioxidant and in situ efficacy of Coriandrum sativum essential oil. Int J Biol Macromol 133:294–305
6. Alipanah H, Rasti F, Zarenezhad E, Dehghan A, Sahebnazar B, Osanloo M (2022) Comparison of anticancer effects of carvone, carvone-rich essential oils, and chitosan nanoparticles containing each of them. Biointerface Res Appl Chem 12(4):5716–5726
7. Attallah OA, Shetta A, Elshishiny F, Mamdouh W (2020) Essential oil loaded pectin/chitosan nanoparticles preparation and optimization via Box-Behnken design against MCF-7 breast cancer cell lines. RSC Adv 10(15):8703–8708
8. Chaiwarit T, Ruksiriwanich W, Jantanasakulwong K, Jantrawut P (2018) Use of orange oil loaded pectin films as antibacterial material for food packaging. Polymers 10(10):1144
9. Zhang H, Li X, Kang H (2019) Chitosan coatings incorporated with free or nano-encapsulated Paulownia Tomentosa essential oil to improve shelf-life of ready-to-cook pork chops. LWT 116:108580
10. Tiwari S, Upadhyay N, Singh BK, Singh VK, Dubey NK (2022) Chemically characterized nanoencapsulated Homalomena aromatica Schott. essential oil as green preservative against fungal and aflatoxin B1 contamination of stored spices based on in vitro and in situ efficacy and favorable safety profile on mice. Environ Sci Pollut Res 29(2):3091–3106
11. Mossa AT, Afia SI, Mohafrash SM, Abou-Awad BA (2019) Rosemary essential oil nanoemulsion, formulation, characterization and acaricidal activity against the two-spotted spider mite Tetranychus urticae Koch (Acari: Tetranychidae). J Plant Prot Res 59(1)
12. Nattudurai G, Baskar K, Paulraj MG, Islam VIH, Ignacimuthu S, Duraipandiyan V (2017) Toxic effect of Atalantia monophylla essential oil on Callosobruchus maculatus and Sitophilus oryzae. Environ Sci Pollut Res 24(2):1619–1629
13. Ikawati S, Himawan T, Abadi AL, Tarno H (2021) Toxicity nanoinsecticide based on clove essential oil against Tribolium castaneum (Herbst). J Pestic Sci 46(2):222–228
14. Kanda D, Kaur S, Koul OA (2017) Comparative study of monoterpenoids and phenylpropanoids from essential oils against stored grain insects: acute toxins or feeding deterrents. J Pest Sci 90(2):531–545
15. Yang E, Lee JW, Chang PS, Park IK (2021) Development of chitosan-coated nanoemulsions of two sulfides present in onion (Allium cepa) essential oil and their nematicidal activities against the pine wood nematode, Bursaphelenchus xylophilus. Environ Sci Pollut Res 28(48):69200–69209

16. Ziaee M, Moharramipour S, Mohsenifar A (2014) MA-chitosan nanogel loaded with Cuminum cyminum essential oil for efficient management of two stored product beetle pests. J Pest Sci 87(4):691–699
17. Kala S, Sogan N, Naik SN, Agarwal A, Kumar J (2020) Impregnation of pectin-cedarwood essential oil nanocapsules onto mini cotton bag improves larvicidal performances. Sci Rep 10(1):1–12
18. Natrajan D, Srinivasan S, Sundar K, Ravindran A (2015) Formulation of essential oil-loaded chitosan–alginate nanocapsules. J Food Drug Anal 23(3):560–568
19. Alsaraf S, Hadi Z, Akhtar MJ, Khan SA (2021) Chemical profiling, cytotoxic and antioxidant activity of volatile oil isolated from the mint (Mentha spicata L.,) grown in Oman. Biocatal Agric Biotechnol 34:102034
20. Froiio F, Ginot L, Paolino D, Lebaz N, Bentaher A, Fessi H, Elaissari A (2019) Essential oils-loaded polymer particles: preparation, characterization, and antimicrobial property. Polymers, 11(6):1017
21. Jiang Y, Wang D, Li F, Li D, Huang Q (2020) Cinnamon essential oil Pickering emulsion stabilized by zein-pectin composite nanoparticles: characterization, antimicrobial effect and advantages in storage application. Int J Biol Macromol 148:1280–1289
22. Chaudhari AK, Singh A, Das S, Dubey NK (2021) Nanoencapsulated Petroselinum crispum essential oil: characterization and practical efficacy against fungal and aflatoxin contamination of stored chia seeds. Food Biosci 42:101117
23. Joung HJ, Choi MJ, Kim JT, Park SH, Park HJ, Shin GH (2016) Development of food-grade curcumin nanoemulsion and its potential application to food beverage system: antioxidant property and in vitro digestion. J Food Sci 81(3):N745–N753
24. Jonassen H, Treves A, Kjøniksen AL, Smistad G, Hiorth M (2013) Preparation of ionically cross-linked pectin nanoparticles in the presence of chlorides of divalent and monovalent cations. Biomacromol 14(10):3523–3531

Energy Gap Dependence on the Hydrostatic Pressure and Temperature of GaAs Quantum Wire

Priyanka and Rinku Sharma

Abstract In this study, we examine the energy gap dependence on the hydrostatic pressure and temperature with the existence of an intense magnetic field and Rashba spin–orbit interaction. First, we employ diagonalizing method to work out the Schrödinger equation. After that, we calculate the energy gap for a GaAs quantum wire. Our numerical outcomes illustrate that the energy gap dependence on the hydrostatic pressure and temperature of a GaAs quantum wire. We also focus on the effect of magnetic field and Rashba spin–orbit interaction on the curve of energy gap vs pressure/temperature.

Keywords Quantum wire · Energy gap · Hydrostatic pressure · Temperature

1 Introduction

A short time ago, there is considerable evolution in the quantum physics of low-dimensional semiconductor devices [1]. Many theories and experimental works have been admiring the empathetic of the energy states in quantum dots and quantum wires. The wire-like semiconductor devices with a size of nanometer have considerable interest among researchers [2]. On the other hand, the spin-based phenomena studied in nanostructures have importance by a reason of their uses in spintronic devices just like spin filters, spin waveguides, spin transistors, and spin valves [3]. Spin–orbit interaction (SOI) is an important example of spin-related phenomena. The SOI pairs the degree of freedom of spin of electron with its orbiting motion, which can precisely operate and regulate the electron's spin along the voltage gate or electric field. The foremost SOIs are Rashba and Dresselhaus SOIs, which rise because of the inversion symmetry of the structure and bulk, respectively. The physical properties of nanostructures have been affected by these SOIs [4].

Priyanka · R. Sharma (✉)
Department of Applied Physics, Delhi Technological University, NewDelhi 110042, India
e-mail: rinkusharma@dtu.ac.in

Z. H. Khan et al. (eds.), *Recent Advances in Nanomaterials*, Springer Proceedings in Materials 27, https://doi.org/10.1007/978-981-99-4878-9_6

In this inquiry, we will analyze the impact of hydrostatic pressure and temperatures upon the energy gap of GaAs parabolic QW with various value of magnetic field, and Rashba SOIs. This paper contains the following sections: In Sect. 2, the numerical expressions for the energy are obtained using the diagonalizing method. In Sect. 3, we will focus on the energy gap dependence on various physical parameters in the GaAs parabolic QW. In last, a summary of our result is presented in Sect. 4.

2 Theory

In the context of effective-mass approximation for a quantum wire in which an electron is allowed to move only in the x-direction and the external magnetic field (**B**), and electric field (**E**) is applied along the z-direction and x-direction, respectively, the Hamiltonian with parabolic confining potential and Rashba SOI is [5].

$$\hat{H} = \frac{(p_x^2 + (p_y + eBx)^2)}{2m_e^*(P,T)} + \frac{1}{2}m_e^*(P,T)\omega_o^2(P)x^2 + eEx + \frac{1}{2}gU_B\vec{\sigma}\cdot\vec{B} + \frac{\alpha_R}{\hbar}\sigma_x(p_y + eBx) - \sigma_y p_x) \quad (1)$$

where $\omega_o(P)$ is known as oscillator strength, σ and U_B are the Pauli matrix and Bohr magnetron, respectively. $m_e^*(P,T)$ is the, $m_e^*(P,T)$ known as effective mass for an electron which dependences on hydrostatic pressure and temperature, for GaAs quantum wire $m_e^*(P,T)$ is given by

$$m_e^*(P,T) = \left[E_P^\Gamma\left(\frac{2}{E_g^\Gamma(P,T)} + \frac{1}{E_g^\Gamma(P,T) + \Delta_0}\right) + 1\right]^{-1} m_o \quad (2)$$

where m_o represents the free electron mass, $\Delta_0 = 0.341\text{eV}$ & $E_P^\Gamma = 7.51eV$ $E_g^\Gamma(P,T)$ hydrostatic pressure and temperature-dependent energy band gap at Γ-point for GaAs QW, given by

$$E_g^\Gamma(P,T) = E_g^\Gamma(0,0) - \frac{\alpha_i T^2}{T + 204} + \beta_i P + \gamma_i P^2 \quad (3)$$

Here, $E_g^\Gamma(0,0) = 1.519eV$, $\alpha_i = 5.405\times10^{-4}\text{eV}/K$, $\beta_i = 1.26\times10^{-2}\text{eV}/k\text{bar}$ and $\gamma_i = 3.771.26 \times 10^{-2}\text{eV}/k\text{bar}^2$ From Eq. (1), energy eigenvalues and eigenvectors of $\hat{H}$ are specified by

$$\hat{H}\Psi_{n\sigma}(x) = E_{n\sigma}\Psi_{n\sigma}(x) \quad (4)$$

and

$$\Psi_{n\sigma}(x) = \frac{1}{(\sqrt{\pi} a_i 2^n n!)^{1/2}} H_n\left(\frac{x - x_i}{c_i}\right) \exp\left(-\frac{1}{2}\left(\frac{x - x_i}{a_i}\right)^2\right) \chi_\sigma \quad (5)$$

Here, $a_i = \sqrt{\frac{\hbar}{m_e^*(P,T)\omega(P)}}$ is the characteristic length of a harmonic oscillator with the n = 0, 1, 2… $H_n(x)$ is known as Hermite polynomial.

After diagonalizing Eq. (1), we get energy eigenvalues and eigenvectors for our system.

3 Result

This work is related to the energy gap between the two energy levels' dependence on the hydrostatic pressure and temperature for a quantum wire. In Fig. 1, we focus on the energy gap dependence on the hydrostatic pressure. The energy gap increases with the increase in the hydrostatic pressure as shown in Fig. 1(a). However, the preliminary cell which present in the lattice of quantum wire can be squashed with the hydrostatic pressure, which increases the distortion energy of quantum wire. The binding energy per bond increases, and the bond length decreases when the hydrostatic pressure is applied. Moreover, high-pressure initiatives to more tight packed structures and higher electron confinement than primary states (without any pressure), as well as its impacts on the semiconductor band gap energy of a GaAs quantum wire. When we increase the magnetic field strength then the energy gap between the two consequent energy levels also increases. Figure 1(b) shows the same behavior of energy gap variation with esteem to pressure for different magnetic field values but in this case when Rashba SOI differs then the energy gap increases with an increase in pressure but a small amount [6].

In Fig. 2, we study the energy gap variation concerning the temperature. Figure 2 a represents that when we increase the temperature the energy gap decreases. The band gap energy of GaAs quantum wire drops as the hydrostatic temperature rises, as can be seen. This is happened as a consequence of rise in the thermal energy which increases the amplitude of atomic vibrations, due to which the interatomic distance increases, and the band gap energy deceases. Moreover, when there is an increment in the magnetic field value the energy gap between the levels increases. Figure 2 b shows the variation in the energy gap due to Rashba SOI as the α increases energy gap increases but in a small amount compared to Fig. 2 a.

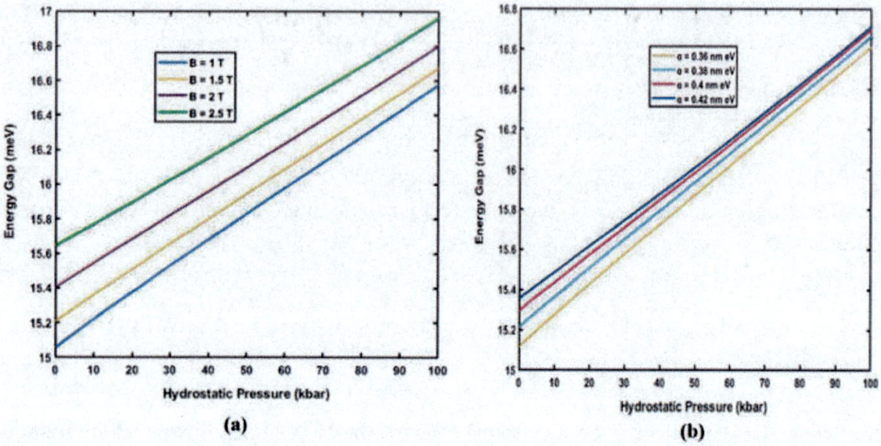

Fig. 1 Energy gap variation concerning hydrostatic pressure for various values of **a** magnetic field, and **b** Rashba SOI

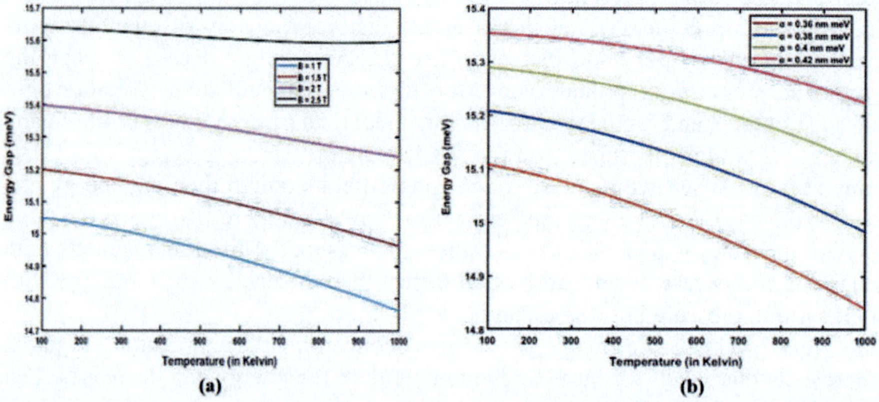

Fig. 2 Energy gap variation concerning temperature for various values of **a** magnetic field, and **b** Rashba SOI

4 Conclusion

In this work, we have offered a detailed study of the energy gap dependence on hydrostatic pressure and temperature of a GaAs quantum wire in the existence of a magnetic field and Rashba SOI is verified. It was initiated that the hydrostatic pressure and temperature changes the energy gap. It can be clinched from this comprehensive hypothetical study that the energy gap between the two energy levels is strongly affected by the pressure and temperature and thus can be controlled by these parameters. This study helps in the study of optical and electronic properties of nanostructures' devices.

Acknowledgements Priyanka acknowledges the financial support from University Grants Commission and R. Sharma is thankful to the Delhi Technological University for the research facilities.

References

1. Khordad R (2016) Thermodynamical properties of triangular quantum wires: entropy, specific heat, and internal energy. Continuum Mech Thermodyn 28(4):947–956. https://doi.org/10.1007/s00161-015-0429-2
2. Khordad R, Firoozi A, Sedehi HRR (2021) Simultaneous effects of temperature and pressure on the entropy and the specific heat of a three-dimensional quantum wire: tsallis formalism. J Low Temp Phys 202(1–2):185–195. https://doi.org/10.1007/s10909-020-02536-w
3. Pramanik S, Bandyopadhyay S, Cahay M (2003) Spin dephasing in quantum wires. Phys Rev B Condens Matter Mater Phys 68(7). https://doi.org/10.1103/PhysRevB.68.075313
4. Ashrafi-Dalkhani V, Ghajarpour-Nobandegani S, Karimi MJ (2019) Effects of spin–orbit interactions, external fields and eccentricity on the optical absorption of an elliptical quantum ring. Eur Phys J B 92(1). https://doi.org/10.1140/epjb/e2018-90691-5.M
5. Kumar M, Lahon S, Jha PK, Gumber S, Mohan M (2014) Spin-orbit interaction effect on nonlinear optical rectification of quantum wire in the presence of electric and magnetic fields. Physica B 438:29–33. https://doi.org/10.1016/j.physb.2013.12.044
6. Owji E, Keshavarz A, Mokhtari H (2016) The effects of temperature, hydrostatic pressure and size on optical gain for GaAs spherical quantum dot laser with hydrogen impurity. Superlattices Microstruct 98:276–282. https://doi.org/10.1016/j.spmi.2016.08.037

Effect of Annealing Temperature on Microstructural, Optical and Magnetic Properties of spinel-$ZnFe_2O_4$ Nano Particles

Mohd Rehan Ansari and Koteswara Rao Peta

Abstract In this work, zinc ferrite nano-particles (ZF NPs) were synthesized using co-precipitation method and annealed at different temperatures 400, 600 and 800 °C in air ambient. It is observed that temperature significantly effects the structural, optical and magnetic properties of ZF NPs. The high-resolution X-ray diffraction (HR-XRD) spectra of ZF NPs is matched with JCPDS card No. 00–022-1012, which shows a single-phase cubic spinel structure with Fd-3 m space group. The estimated average crystallite size is in the range of 11.8–18.4 nm. The scanning electron microscope images shows that the particle size of ZF NPs is increased due to increase in the nucleation rate of particles during the annealing which is good agreements with the XRD results. EDX elemental analysis showed that the ZF NPs consists of Zn, Fe and O atoms and no other impurities are present. The Fourier transform infrared spectroscopy (FTIR) spectrum having absorption peaks in fingerprint region at 541 cm^{-1} and 607 cm^{-1} showing the stretching vibrations of Fe–O and Zn–O bond at tetrahedral and octahedral lattice sites respectively. The optical energy bandgap is decreased from 2.1 eV to 1.9 eV with increase in annealing temperature calculated by diffuse reflectance spectroscopy using Kubelka–Munk function due to increase in the crystallite size. In addition, vibrating sample magnetometer (VSM) showed the superparamagnetic nature of the ZF NPs as M-H curve passing through origin with zero coercivity and remanence.

Keywords Zinc ferrite · Nano-particles · Kubelka–Munk · Crystallite size · Tetrahedral · Octahedral

---

M. R. Ansari · K. R. Peta (✉)
Department of Electronic Science, University of Delhi South Campus, Benito Juarez Road, New Delhi 110021, India
e-mail: krao@south.du.ac.in

Z. H. Khan et al. (eds.), *Recent Advances in Nanomaterials*, Springer Proceedings in Materials 27, https://doi.org/10.1007/978-981-99-4878-9_7

1 Introduction

In recent years, spinel ferrites nano structures have been intensively investigated because of their superior, physical, magnetic, optical and electrical properties [1, 2]. The spinel ferrites having the general molecular formula MFe_2O_4 (M = Zn, Co, Cu, Ni, Mn, ….) where M is divalent cation [3–5]. Particularly, zinc ferrite oxide ($ZnFe_2O_4$) attracted more attention due to their application in the field of water treatment by photocatalysis [4, 6], microwave absorber [7], supercapacitors [8], magnetic recording [8], ferrofluids [9], magnetic resonance imaging (MRI) [10], drug delivery [9], toxic gas-sensor [11], and antibacterial agent [4, 6]. According to this formula divalent cation (Zn^{2+}) can occupy either tetrahedral (A) lattice sites and trivalent cation (Fe^{3+}) occupies the octahedral (B) sites in a unit cell with 32 oxygen atoms of cubic close pack (ccp) structure [4, 6]. Zinc ferrite (ZF) shows semi-conductor nature having the narrow bandgap 1.9 eV for bulk [4, 11].

The nano-size particles are synthesized broadly two methods (1) top to down method and (2) bottom-up method uses following techniques such as co-precipitation [12], solution-combustion [6], sol–gel [13], hydrothermal [7], surfactant-assisted [14], solid-state method [11], microwave-assisted [4], using CTAB (Cetyl Trimethylammonium Bromide) [15] and ball milling [16]. Among all these methods, co-precipitation method is easy, economical, gives ultrafine and high purity nanomaterial with crystalline structures in shortest duration [12]. In this work, we synthesized ZF NPs using co-precipitation method and studied the influence of temperature on the structural, optical and magnetic properties of $ZnFe_2O_4$ NPs.

2 Experimental Details

The precursor materials zinc nitrate hexahydrate (99% purity) [$Zn(NO_3)_2.6H_2O$] and iron nitrate nano hydrate (99% purity) [$Fe(NO_3)_3.9H_2O$] were purchased from Sigma Aldrich, India and were used as received without any further purifications. The aqueous solution of zinc nitrate and iron nitrate was prepared in 1:2 molar ratio. Under the constant magnetic stirring, sodium hydroxide (NaOH) solution (2 M) was added in the mixture till the pH of the solution become 12. After 1 h of constant stirring at 80 °C , the solution was cool down at room temperature without disturbing it. The obtained precipitate was filtered out through Whatman no. 1 filter paper and dry at room temperature for 24 h. The dried powder was annealed at different temperatures and milled using agate mortar before characterizations.

The material was characterized by using standard characterization techniques such as HR-XRD by using Bruker, D8 Discover diffractometer, SEM images were taken by JEOL Japan Model: JSM 6610LV, FTIR by using Thermo scientific Model: Nicolet iS50 FT-IR, diffuse reflectance recorded by UV/Vis/NIR spectrophotometer PerkinElmer Model: LAMDA 1050 and the magnetic properties were characterized by vibrating sample magnetometer Model: ADE-EV9.

3 Results and Discussions

3.1 Morphological Studies

The surface morphology of the synthesized ZF NPs at different temperature was observed using scanning electron microscope (SEM) images Fig. 1(a–d). The shape of the particles is non-spherical and non-homogeneously distributed. SEM micrographs showed that the most of the particles were agglomerated due to the magnetic natures of the particles, absorbed water molecules and hydroxides of zinc and iron present in the sample before annealing at high temperatures [14, 17, 18]. It has been seen that the increase of in the grain size with respect to the annealing temperature due to increase of nucleation rate. The SEM image of sample 1 (**a**) and (**b**) shows irregular microstructures with fairly small grains highly agglomerated which indicates the temperature insufficient for the complete formation of the structure. As the annealing temperature increases to 800 °C, octahedral like shaped appeared with presence of intra-granular pores (grain boundary pores) resulting from discontinuous grain growth (Fig. 1(d)).

The energy dispersive-ray analysis (EDX) showed the composition of the synthesized ZF insets of Fig. 1(a–d). The synthesized samples mainly consist of the Zn, Fe and O atoms, no other impurity was found in the spectrum. It is interestingly noted that the preparation conditions were favours the formation of ZF. These results are very good agreement with the corresponding XRD results.

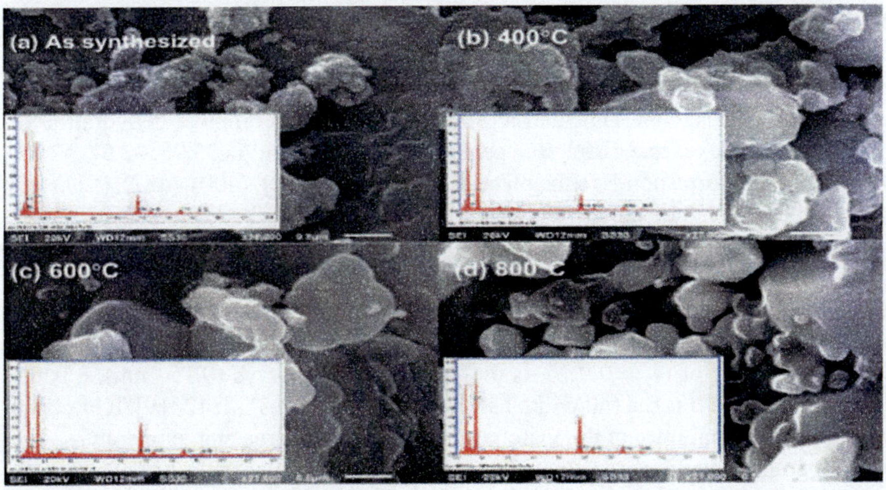

Fig. 1 SEM morphology and EDX spectrum of ZF NPs, **a** as synthesized, **b** 400 °C sample, **c** 600 °C sample, **d** 800 °C sample

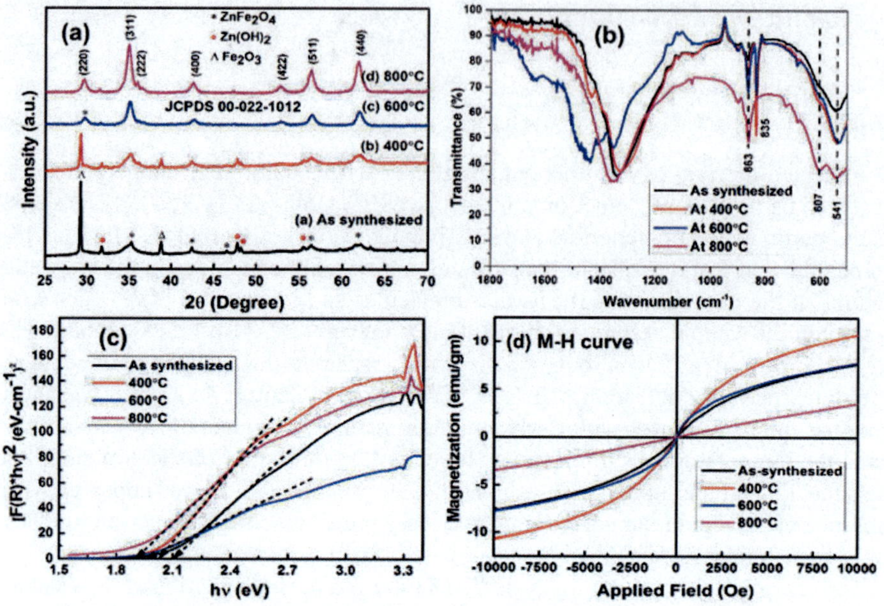

Fig. 2 **a** HR-XRD. **b** FTIR. **c** UV-DRS. **d** M-H curve of ZF NPs annealed at different temperature.

3.2 *Structural and Optical Studies*

Figure 2(a) showed the HR-XRD pattern of the samples annealed at different temperatures. The peaks of the as synthesized sample showed that the sample contain hydroxides of zinc and iron which dehydrated and decomposed with increasing temperature. The sample annealed at 800 °C having the complete dehydration of hydroxide complexes and diffraction peaks (2θ) at 29.68, 34.92, 35.96, 42.52, 52.82, 56.3 and 61.88° corresponds to the planes (220), (311), (222), (400), (422), (511) and (440) respectively. The pattern is matching with the JCPDS card no 00–022-1012 confirms the formation of single-phase cubic spinel structure of ZF NPs with Fd-3 m (227) space group [5, 6, 11]. No secondary phase is observed in the sample annealed at 800 °C further confirms the successful formation of ZF NPs. The crystallite size was calculated for the plane (311) using Scherrer equation $D = k\lambda/\beta cos\theta$ where k is the shape factor ($k = 0.9$), λ is the wavelength of x-rays (0.154 nm), θ is the Bragg's angle and β is the full width half maxima (in radians) [4, 12, 19]. It has been seen that the crystallinity of the material increases with increasing temperature. The crystallographic parameters were calculated and tabulated in Table 1.

Figure 2(b) shows the Fourier transform infrared spectroscopy (FTIR) of the synthesized samples at different annealing temperatures in the range 500–1200 cm^{-1}. The ZF having the two fundamental absorption peaks ν_1 and ν_2 as per previous studies which attributes the metal–oxygen band at octahedral and tetrahedral lattice sites respectively [4, 6, 14]. In the spectrum the only sample which annealed at

Table 1 Crystallographic parameters and optical bandgap of 400, 600 and 800 °C annealed ZF NPs

Temperature (Celsius)	Crystallite size (nm)	Lattice constant (Å) $a = d\sqrt{h^2+k^2+l^2}$	Volume(Å^3) $V = a^3$	Dislocation density (m^{-2}) $\delta = 1/D^2$	Bandgap (eV)
As-synthesized	–	–	–	–	2.1
400 °C	11.8	8.4739	608.48	7.18×10^{15}	2.05
600 °C	14.5	8.4840	610.66	4.75×10^{15}	1.98
800 °C	18.4	8.5104	616.38	2.95×10^{15}	1.9

Table 2 Elemental composition of the synthesized $ZnFe_2O_4$ by co-precipitation method

	As synthesized		400 °C		600 °C		800 °C	
Elements	wt %	at %	wt %	at %	wt %	at %	wt %	at %
O K	49.8	78.43	55.12	81.79	38.63	69.69	32.09	63.44
Fe K	33.8	15.25	30.89	13.13	42.69	22.06	44.8	25.38
Zn K	16.4	6.32	13.99	5.08	18.68	8.25	23.11	11.18

800 °C having the two absorption peaks at 607 and 541 cm^{-1}ensures the Zn–O and Fe–O bond at tetrahedral and octahedral sites [4, 6, 12, 14]. The other peaks attribute the presence of carboxyl groups and water molecules absorbed at the surface of the synthesized ZF [6, 14, 17].

Figure 2(c) shows the ultra-violet diffuse reflectance spectroscopy of the ZF NPs annealed at different temperatures. The spectrum was recorded in the range of 400–800 nm. The spinel ZF absorbs the visible light radiation and excites their electron from valence band to conduction band (O-2p level to Fe-3d level) [6]. The optical energy bandgap of the material was calculated using the Kubelka–Munk function, F(R) = $(1\text{-}R)^2/2R$, where R is reflectance [4, 6]. The estimated bandgap of ZF NPs decreases with increasing annealing temperature and matching with the already reported values [4, 6, 11]. The blue shift in the bandgap was observed as compared to their bulk having the bandgap of 1.9 eV [4, 6]. The shifting of bandgap energy of the ZF NPs with increase in the crystallite size due to the quantum confinement effect which was arising due to small size regime.

3.3 Magnetic Properties

Figure 2(d) shows the magnetization curve of ZF NPs annealed at different temperature recorded by VSM at room temperature. The M-H curve crossing revealing that thermal energy is higher than the anisotropic energy of ZF NPs [12]. The zero remanence and coercivity field reveals the superparamagnetic nature of the ZF NPs [14,

15, 20]. The superparamagnetic nature appears when the particle size less than the critical value [14]. The saturation magnetization values decreased from 10.75 to 3.52 emu/gram with respect to increasing temperature due to the reduction in surface to volume ratio, cation disordering and increase in the crystallite size of ZF [14, 21]. In the spinel structures, the main cause of magnetization is due to the different magnetic moment of magnetic Fe^{3+} and the non-magnetic Zn^{2+} ions at tetrahedral and octahedral lattice sites. The magnetic properties of the ferrite materials depend upon the method of preparation, crystallite size, annealing temperature and the oxygen vacancies present in the lattice [22].

4 Conclusion

The ZF NPs has been successfully synthesized by co-precipitation method. The annealing temperature effects the structural, morphological, optical and magnetic properties of the ZF NPs. The HR-XRD study we found that the clear formation of ZF NPs spinel with cubic structure. The presence of tetrahedral and octahedral lattice sites in the synthesized ZF were revealed by FTIR analysis. The SEM micrographs shows the agglomerated and crystalline surface morphology of the annealed samples. The elemental compositional and the purity of the samples were characterized by EDX analysis which is good agreement with the XRD analysis. The superparamagnetic nature of ZF NPs was confirmed by VSM studies and it has been seen from the M-H loop that the magnetic saturation is decreases with respect to the annealing temperature.

Acknowledgements This research work was funded by Institute of Eminence (IoE), University of Delhi under faculty research programme (IoE/2021/12/FRP). The author would like to express his gratitude to University Grant Commission for providing Junior Research Fellowship.

Declaration of Interest Statement
The authors have declared that there is no conflict of interest.

References

1. Zhao Q, Yan Z, Chen C, Chen J (2017) Spinels: controlled preparation, oxygen reduction/evolution reaction application, and beyond. Chem Rev 117(15):10121–10211
2. Kefeni KK, Mamba BB, Msagati TA (2017) Application of spinel ferrite nanoparticles in water and wastewater treatment: a review. Sep Purif Technol 188:399–422
3. Haghniaz R, Rabbani A, Vajhadin F, Khan T, Kousar R, Khan AR, Wahid F (2021) Anti-bacterial and wound healing-promoting effects of zinc ferrite nanoparticles. J Nanobiotechnol 19(1):1–15
4. Naik MM, Naik HB, Nagaraju G, Vinuth M, Naika HR, Vinu K (2019) Green synthesis of zinc ferrite nanoparticles in *Limonia acidissima* juice: characterization and their application as photocatalytic and antibacterial activities. Microchem J 146:1227–1235

5. Paz-Díaz B, Vázquez-Olmos AR, Almaguer-Flores A, García-Pérez VI, Sato-Berrú RY, Almanza-Arjona YC, Garibay-Febles V (2021) $ZnFe_2O_4$ and $CuFe_2O_4$ Nanocrystals: synthesis, characterization, and bactericidal application. J Cluster Sci. 1–9
6. Patil SB, Naik HB, Nagaraju G, Viswanath R, Rashmi SK (2018) Sugarcane juice mediated eco-friendly synthesis of visible light active zinc ferrite nanoparticles: application to degradation of mixed dyes and antibacterial activities. Mater Chem Phys 212:351–362
7. Qiao Y, Xiao J, Jia Q, Lu L, Fan H (2019) Preparation and microwave absorption properties of $ZnFe_2O_4$/polyaniline/graphene oxide composite. Results Phys 13:102221
8. Bohra M, Alman V, Arras R (2021) Nanostructured $ZnFe_2O_4$: an exotic energy material. Nanomaterials 11(5):1286
9. Sharifi I, Shokrollahi H, Amiri S (2012) Ferrite-based magnetic nanofluids used in hyperthermia applications. J Magn Magn Mater 324(6):903–915
10. Chaudhary R, Roy K, Kanwar RK, Walder K, Kanwar JR (2016) Engineered atherosclerosis-specific zinc ferrite nanocomplex-based MRI contrast agents. J nanobiotechnol 14(1):1–17
11. Wu J, Gao D, Sun T, Bi J, Zhao Y, Ning Z, Xie Z (2016) Highly selective gas sensing properties of partially inversed spinel zinc ferrite towards H2S. Sens Actuators B Chem 235:258–262
12. Sathiyamurthy K, Rajeevgandhi C, Bharanidharan S, Sugumar P, Subashchandrabose S (2020) Electrochemical and magnetic properties of zinc ferrite nanoparticles through chemical co-precipitation method. Chem Data Collections 28:100477
13. Zhang X, Chen Z, Liu J, Cui S (2021) Synthesis and characterization of $ZnFe_2O_4$ nanoparticles on infrared radiation by xerogel with sol-gel method. Chem Phys Lett 764:138265
14. Rameshbabu R, Ramesh R, Kanagesan S, Karthigeyan A, Ponnusamy S (2014) Synthesis and study of structural, morphological and magnetic properties of $ZnFe_2O_4$ nanoparticles. J Supercond Novel Magn 27(6):1499–1502
15. Sinthiya MMA, Ramamurthi K, Mathuri S, Manimozhi T, Kumaresan N, Margoni MM, Karthika PC (2015) Synthesis of zinc ferrite ($ZnFe_2O_4$) nanoparticles with different capping agents. Int J Chem Tech Res 7:2144–2149
16. Rajini R, Ferdinand AC (2022) Structural, morphological and magnetic properties of (c-$ZnFe_2O_4$ and t-$CuFe_2O_4$) ferrite nanoparticle synthesized by reactive ball milling. Chem Data Collections 38:100825
17. Naik MM, Naik HS, Nagaraju G, Vinuth M, Vinu K, Rashmi SK (2018) Effect of aluminium doping on structural, optical, photocatalytic and antibacterial activity on nickel ferrite nanoparticles by sol–gel auto-combustion method. J Mater Sci Mater Electron 29(23):20395–20414
18. Din MI, Jabbar S, Najeeb J, Khalid R, Ghaffar T, Arshad M, Ali S (2020) Green synthesis of zinc ferrite nanoparticles for photocatalysis of methylene blue. Int J Phytorem 22(13):1440–1447
19. Kem A, Ansari MR, Prathap P, Jayasimhadri M, Peta KR (2022) Eco-friendly green synthesis of stable ZnO nanoparticles using citrus limon: x-ray diffraction analysis and optical properties. Physica Scripta
20. Rameshbabu R, Neppolian B (2016) Surfactant assisted hydrothermal synthesis of superparamagnetic $ZnFe_2O_4$ nanoparticles as an efficient visible-light photocatalyst for the degradation of organic pollutant. J Cluster Sci 27(6):1977–1987
21. Tomar D, Jeevanandam P (2022) Synthesis of $ZnFe_2O_4$ nanoparticles with different morphologies via thermal decomposition approach and studies on their magnetic properties. J Magn Magn Mater 564:170033
22. Mouhib Y, Belaiche M, Elansary M, Ferdi CA (2022) Effect of heating temperature on structural and magnetic properties of zinc ferrite nanoparticles synthesized for the first time in presence of Moroccan reagents. J Alloy Compd 895:162634

Studies on Zinc Oxide Thin Film and Nanoparticles Synthesized by Chemical Bath Deposition

S. Pandya, V. K. Pathak, P. D. Lad, and M. P. Deshpande

Abstract Zinc Oxide (ZnO) nanoparticles (NPS) and thin film are synthesized by chemical bath deposition technique. Synthesized nanoparticles and thin film are characterized by X-ray diffraction (XRD), Transmission electron microscope (TEM), UV–Visible spectroscopy, Photoluminescence spectroscopy (PL) and Raman Spectroscopy. XRD analysis shows that ZnO NPS and thin film are formed in hexagonal wurtzite structure. A particle size of NPS varies from 40–60 nm, which is confirmed with TEM. UV–Vis absorbance spectrum shows absorption peak at ~ 369.2 nm (3.36 eV) for thin film and at ~ 378.8 nm (3.27 eV) for nanoparticles. Determined band gap from Uv–Vis absorption are 3.11 and 2.81 eV for ZnO thin film and NPS respectively. PL spectra of thin film and NPS show multiple peaks for violet, blue and green emissions which may be attributed to various defect levels present in material. Raman spectrum of ZnO nanoparticles shows presence of different Raman – active modes. Presence of sharp peak for non-polar optical phonons E_2 (high) mode confirms good quality hexagonal wurtzite crystal phase of ZnO nanoparticles.

Keywords ZnO thin film · ZnO nanoparticles · Bath deposition · Characterizations

1 Introduction

Transparent conducting oxide (TCO) is a material that exhibits high conductivity and transparency simultaneously. This unique combination of properties makes TCOs useful in the optoelectronic industry [1]. Zinc oxide (ZnO) is a well-known TCO material that has diverse applications in various industries. Many research studies are focused on synthesizing ZnO-based thin films and nanoparticles (NPs) using cost-effective techniques. Researchers continue to fine-tune the properties of ZnO for desired applications using a variety of synthesis methods with different doping

S. Pandya (✉) · V. K. Pathak · P. D. Lad · M. P. Deshpande
Department of Physics, Sardar Patel University, Vallabh Vidyanagar, Anand, Gujarat 388120, India
e-mail: Swati.j.pandya@gmail.com

Z. H. Khan et al. (eds.), *Recent Advances in Nanomaterials*, Springer Proceedings in Materials 27, https://doi.org/10.1007/978-981-99-4878-9_8

in nanostructured ZnO. In this work, ZnO thin films and NPs were synthesized using the chemical bath deposition method, also known as the chemical solution method. The detailed synthesis process is discussed in this study, along with the structural and optical studies of the prepared ZnO thin films and NPs.

2 Experimental

2.1 *Materials and Methods*

The chemicals used to synthesize ZnO thin films and NPs are zinc acetate [Zn $(CH_3COO)_2{\cdot}2H_2O$], triethylamine (TEA), sodium hydroxide (NaOH), acetone, and methanol. Distilled water was used as a solvent throughout the synthesis process.

To prepare the ZnO thin film, an aqueous solution of 0.25 M zinc acetate [Zn $(CH_3COO)_2{\cdot}2H_2O$] in 80 mL double distilled water with two drops of triethylamine (TEA) was made, and the solution was stirred for two hours using a magnetic stirrer at room temperature. The glass substrate was cleaned using distilled water and acetone for 10 min in an ultrasonic cleaner and then placed inside the solution with the help of a suitable holder. The solution was continuously stirred, and the temperature was maintained at 75–80 °C. The pH of the solution was maintained at ~ 11 by adding NaOH to the solution. The solution was stirred for three hours in the presence of the glass substrate. After three hours of deposition, the glass substrate was removed from the bath and dried at 60 °C for 15min on a hot plate. The substrate was coated with a milky white deposition. Finally, the film was annealed at a temperature of 300 °C for 1.5 h in a box furnace.

The by-product of the solution was used for the NPs synthesis. The white powder obtained from the solution was filtered and washed three times with distilled water and then twice with methanol. It was then dried in an oven at 80 °C for two hours. The final calcination of the ZnO NPs was done at a temperature of 600 °C for three hours.

2.2 *Characterization*

X-ray diffraction was used for the structural analysis of the synthesized ZnO thin film and ZnO NPS using a Philips X-ray diffractometer (model: Xpert MPD). The size of the synthesized nanoparticles was determined using a Philips Transmission electron microscope (model: Technai 20) operated at a 200 kV accelerating voltage. Absorption studies were conducted using a JASCO V-730 UV–VIS spectrophotometer. Photoluminescence studies were carried out using a Perkin Elmer Spectro-fluorophotometer with a 325 nm excitation wavelength. Raman spectroscopy was employed to study active Raman modes using a Horiba Xplora Micro Raman Spectrometer with an excitation wavelength of 532 nm using a diode laser.

3 Results and Discussion

Figure 1 shows the XRD pattern of ZnO thin film and NPS. All diffraction peaks were indexed to ZnO with hexagonal wurtzite crystal structure. XRD of ZnO NPS was analyzed using Rietveld refinement through fullprof software with goodness of fit 1.17. The fitted XRD pattern is shown in Fig. 2. Refined lattice parameters of ZnO nanoparticles are: $a = b = 3.2500$ Å and $c = 5.2071$ Å with space group $P6_3mc$, which are in well agreement with reported results [2].

Figure 3 shows TEM image of ZnO nanoparticles. Small amount of ZnO NPS were dispersed in methanol and ultra-sonicated for 10 min to avoid agglomeration of particles and a drop of the solution was placed on copper grid and used for TEM. TEM images show that the particles size varies from 40 to 60 nm.

Figure 4 shows UV–Vis spectrum for ZnO thin film and NPS in the range of 300 to 600 nm. UV–Vis absorbance spectrum shows absorption peak at ~ 369.2 nm (3.36 eV) for thin film and at ~ 378.8 nm (3.27 eV) for NPS. Direct band gap (E_g) is determined using Tauc relation for direct allowed transition, $\alpha h\nu = A (h\nu - E_g)^{1/2}$,where α is the optical absorption coefficient, $h\nu$ is photon energy, E_g is band gap and A is a constant

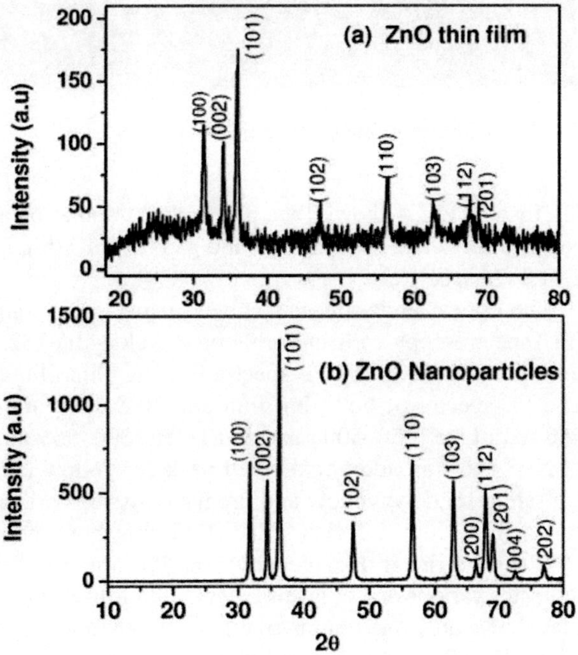

Fig. 1 X-ray diffraction of ZnO. **a** Thin film. **b** Nanoparticles

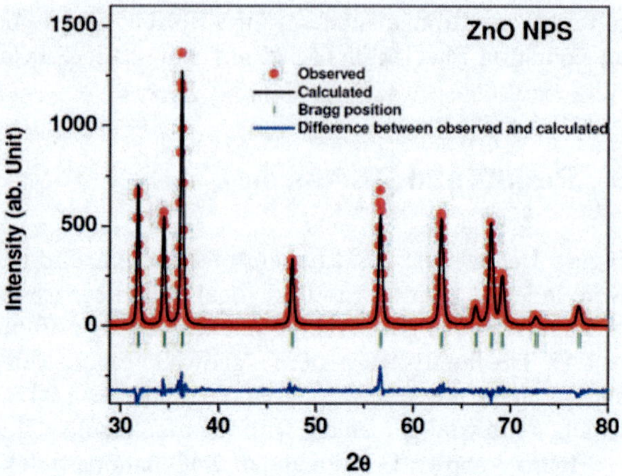

Fig. 2 Rietveld refined XRD of ZnO nanoparticles

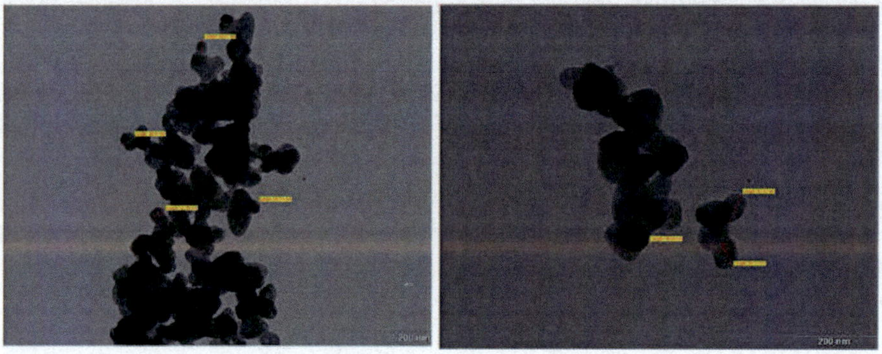

Fig. 3 TEM image of ZnO nanoparticles

[3]. Inset of Fig. 4 shows Tauc plot with extrapolation of linear portion to absorption equal to zero. The determined band gap for ZnO thin film and NPS are 3.11 eV and 2.81 eV respectively.

The optical emission was studied using room temperature photoluminescence (PL) spectroscopy with an excitation wavelength of 325 nm for the synthesized ZnO thin film and NPS. The PL spectra for ZnO thin film and NPS are shown in Fig. 5. The PL spectra of both thin film and NPS show multiple peaks for violet (400–450 nm), blue (450–500 nm), and green (500–565 nm) emissions. The PL spectra of NPS show an additional small peak for yellow (572 nm) and orange (592 nm) emission. Usually, yellow and orange emissions are caused by oxygen interstitials and Zinc vacancy defects [4]. The thin film shows no peak in the UV region, while NPS show a rise in PL intensity near 350 nm. The presence of multiple PL peaks indicates various defect levels in ZnO thin film and NPS. The PL spectra of the thin film shows only one dominating peak at ~ 433 nm, and other peaks are very small, whereas ZnO NPS show multiple strong peaks in the UV, violet, blue, and green

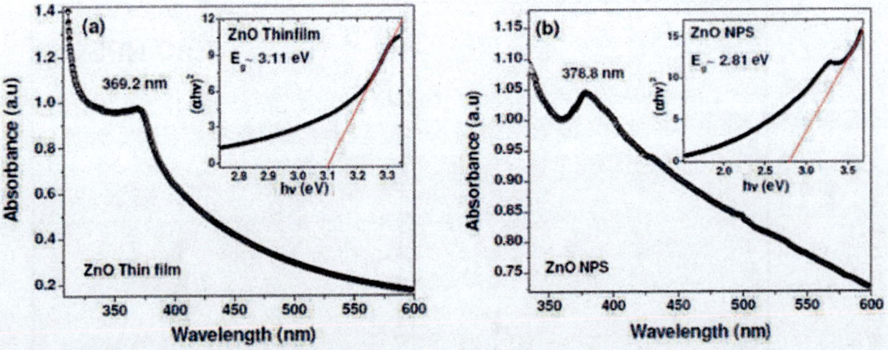

Fig. 4 UV–VIS absorption spectra of ZnO. **a** Thin film. **b** Nanoparticles. Inset shows Tauc plot for thin film and NPS

regions. This observation indicates that the ZnO thin film has comparatively fewer defect levels than the ZnO NPS. A detailed analysis of the position of peaks may provide information about the types of defects [4].

Figure 6 shows the Raman spectra of ZnO NPS. Group theory predicts that the active Raman modes for wurtzite ZnO are $A_1 + E_1 + 2E_2$. The phonons of A_1 and E_1 symmetry are polar phonons, and they exhibit different frequencies for the transverse-optical (TO) and longitudinal-optical (LO) phonons. The E_2 mode is a non-polar mode composed of two modes from low and high frequency [5]. The Raman spectrum of ZnO NPS shows the presence of different Raman-active modes at 99, 199, 330, 382, 437, and 581 cm^{-1}, which are indicated with E_2 (low), 2TA; $2E_2$ (low), E_2(high)-E_2(low), A_1(TO), E_2(high), and E_1(LO), respectively. These prominent peaks are assigned with possible Raman modes with the help of reported ZnO Raman spectra [5]. The presence of a sharp peak for non-polar optical phonons E_2 (high) mode verifies the good quality hexagonal wurtzite crystal phase of ZnO NPS. The E_1 (LO) mode is associated with the presence of oxygen vacancies and interstitial Zn [5].

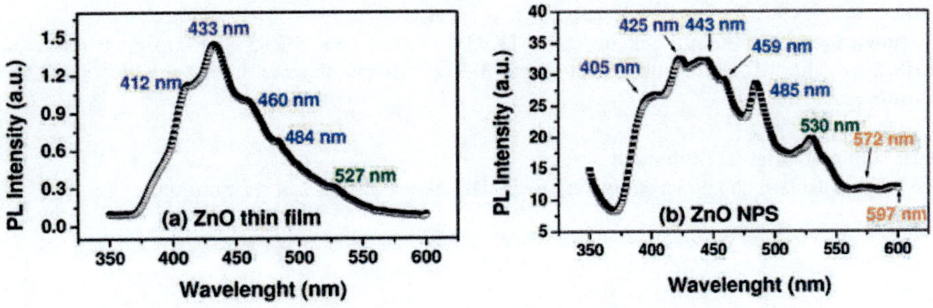

Fig. 5 Photoluminescence spectra of ZnO. **a** thin film. **b** and ZnO NPs

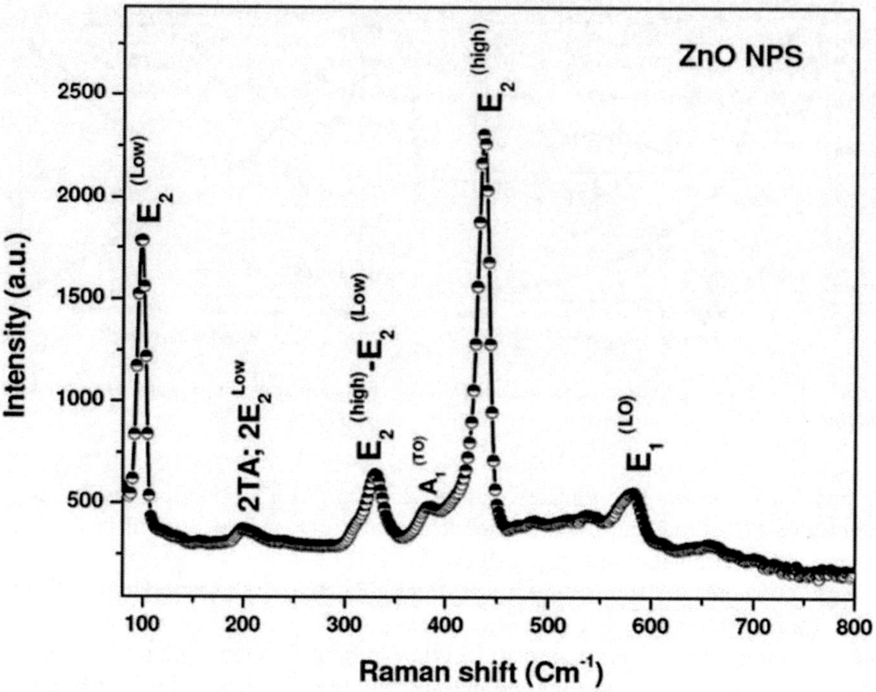

Fig. 6 Raman spectra of ZnO NPS

4 Conclusion

ZnO thin film and NPS were successfully synthesized using a simple and cost-effective chemical bath deposition technique. XRD studies confirmed a wurtzite hexagonal crystal structure. The size of NPS was found to vary from 40 to 60 nm, as observed from TEM images. The optical band gap was determined to be 3.11 eV and 2.81 eV for ZnO thin film and NPS, respectively, from UV–Vis absorption studies. The presence of a sharp peak for the non-polar optical phonons E_2 (high) mode confirmed the good quality hexagonal wurtzite crystal phase of ZnO nanoparticles.

Acknowledgements Author acknowledges UGC, India for UGC FRPS Start-Up grant for year (2018-2020) and SICART, Vallabh Vidyanagar-388120, Anand, Gujarat, India for XRD and TEM measurements.

Declaration of Interest Statement
The authors declare no known competing financial interest or personal relationship.

References

1. G Haacke 1976 J Appl Phys 47 4086
2. KG Chandrappa TV Venkatesha 2012 Nano-Micro Lett 4 1 14 24
3. Chithra MJ, Sathya M, Pushpanathan K (2015) Acta metallurgica sinica (English Letters) 28(3):394404
4. V Koutu L Shastri MM Malik 2016 Mater Sci-Pol 34 4 819 827
5. R Raji KG Gopchandran 2017 J Sci Adv mat Devices 2 51 58

Down-Conversion Fluorescence Study of Non-metal Co-doped Carbon Dots

Rajnee Yadav, Sanjay, and Vikas Lahariya

Abstract In this work, a down-conversion fluorescence study of carbon dots is presented. A facile microwave irradiation route is used to prepare the carbon dots by carbonization of citrus limetta juice in an aqueous medium. Phosphonic acid and ethylenediamine are taken as initial phosphorus (P) and nitrogen (N) sources, respectively. The optical and photoluminescence properties are investigated with different ratios of P to N. UV–Visible absorption spectroscopy and Raman spectroscopy are employed for the optical study. Fluorescence spectra are noted by spectrofluorometer in various excitation wavelengths. The observed absorbance peaks in the ultraviolet region represent n and π molecular transitions. The molar ratio of P to N modifies the electronic transitions. From PL spectra, the excitation-dependent emission spectra are observed. The enhanced down-conversion fluorescence in the green to the red region is blue-shifted with changing initial P/N ratio in carbon dots. The change in emission color with P/N is observed due to changes in emitting states and electronic transitions.

Keywords Fluorescent nanomaterials · Photoluminescence · Carbon dots

1 Introduction

Carbon dots are leading-edge carbon nanomaterials having remarkable fluorescence properties and versatile applications. Carbon dots are biocompatible, non-toxic, hydrophilic, and zero-dimension nanomaterial. In which a strong quantum confinement effect significantly offers size-dependent optoelectronic properties. Hence, these are widely used in optoelectronics applications such as photovoltaic, photocatalytic, optical sensor, QD-LED, and photodetectors [1]. Moreover, ease of synthesis, highly fluorescent, size-tunable properties, and non-hazardous nature make it a promising material for bioimaging, fluorometric sensing, anti-microbial, and drug-delivery agent [2]. The study of fluorescent material started with semiconductor

R. Yadav · Sanjay · V. Lahariya (✉)
Amity University Haryana, Gurugram 122413, India
e-mail: vlahariya@ggn.amity.edu

Z. H. Khan et al. (eds.), *Recent Advances in Nanomaterials*, Springer Proceedings in Materials 27, https://doi.org/10.1007/978-981-99-4878-9_9

quantum dots and then later in 2004 carbon dots was discovered. During the purification of single-walled carbon nanotubes by Xu et al. some explicitly fluorescent material was identified [3]. Since then, continuous efforts are being made to modify and enhance their fluorescence. Among such efforts, surface passivation and doping/co-doping of heteroatoms are widely used. In the recent past, extensive research work has been done on tailoring the optoelectronic properties and fluorescence characteristics by surface functionalization or doping of heteroatom [4]. In the literature, nitrogen (N), phosphorous (P), boron (B), and sulfur (S) are the most common dopant atoms for doped carbon dots, by which the photoluminescence emission has been modified [5]. Since the PL process depends on electron–hole recombination, dopant atom and energy states play a major role in photoexcited electrons and recombination. Carbon dots have been prepared by some bottom-up routes such as hydrothermal synthesis, thermal decomposition, template route, and microwave irradiation [6–8]. However, microwave irradiation is the most reliable, easy, quick, and monitorable technique to synthesize carbon dots with desired properties [9]. In addition, the use of organic natural materials including plant extract, plant seeds, fruit, bio-mass, and organism waste for the initial carbon source makes it environment benign synthesis route. The work aims to study the fluorescence properties of P, N co-doped carbon dots prepared by microwave irradiation using a natural carbon source. The effect of the dopant ratio P to N on the optical and fluorescence of carbon dots is presented and discussed.

2 Materials and Methods

For preparation, AR grade phosphonic acid (PA) and ethylenediamine (EDA) purchased by Merck India were used for P and N sources, respectively. Citrus limetta juice is used as a carbon source. MilliQ water was used as the solvent. The synthesis of carbon dots is reported in our previous published work [6]. For P and N doping, an initial 50 ml of citrus limetta juice was mixed with three different ratios of PA/EDA (50:50, 60:40, and 80:20) and named C, B, and D, respectively. The mixture was heated in the microwave, and the final solution was filtered several times to separate impurities.

The prepared samples were characterized by Raman, UV–Visible absorption, and photoluminescence (PL) spectroscopies. Raman spectra were recorded in Horiba JY- LABRAM HR Raman Spectrophotometer using 785 nm lasers over the range of 1800–1100 cm^{-1}. The UV–Visible absorption spectra were taken on an Agilent spectrophotometer from 250 to 800 nm. The PL emission response was recorded on an Agilent spectrofluorometer.

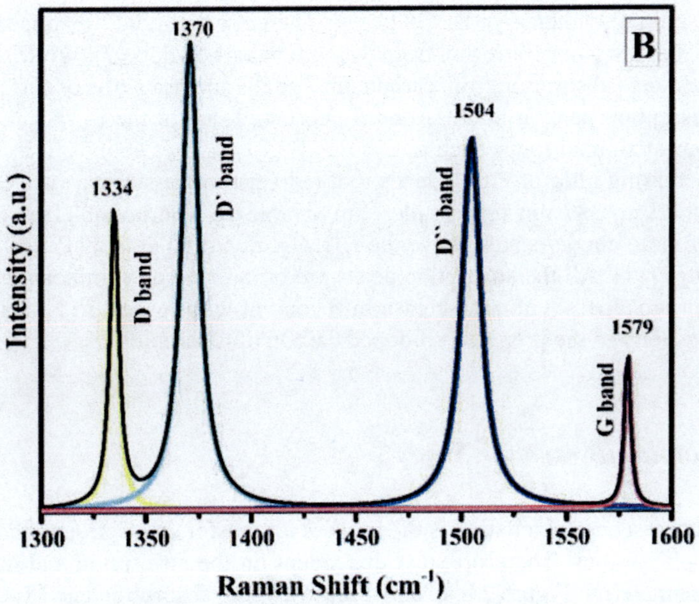

Fig. 1 Deconvoluted Raman spectra of sample B

3 Results and Discussions

3.1 Raman Spectroscopy

Raman spectroscopy is used to study whether the prepared samples are crystalline or amorphous in nature. Disordered graphite exhibits two modes in its Raman spectra, the graphitic G peak at around 1580 cm^{-1}, and the disorder D peak at about 1350 cm^{-1}, which are attributed to the zone center phonons of E2g vibrations and the K-point phonons of A_{1g} vibrations, respectively. The characteristic G and D bands for prepared sample B are observed at 1579 and 1334 cm^{-1}. The obvious high D band peak from Fig. 1 indicates the amorphous nature of the carbon dots sample. Also, the presence of D‘ and D“ bands at 1370 and 1504 cm^{-1} specify the high disorder in the sample. Previously, such disorders have been reported for graphene quantum dots and carbon nanomaterials [10].

3.2 UV–Visible Absorption Spectroscopy

The UV–Visible absorption spectra of diluted P, N co-doped carbon dots are presented in Fig. 2(a). All samples show a broad absorption band from 280 to 360 nm. However, samples C and D display two intense peaks around 290 and 320 nm. Whereas sample

B exhibits only one intense peak at 322 nm. The peak around 290 nm is attributed to the C=C bond's π-π* transition from the core of carbon dots [7]. While the peaks 300 nm are due to absorption from carbon dots' shells and transition of surface states [7], the absorption peak around 320 nm is observed due to the n-π* transition of the surface functional groups such as C=O/C=N/N = P/P= O bonds. It is observed that the increasing ratio of P to N leads to a red-shift in the surface state transition peak from 319 to 337 nm for sample C to sample D. The possible reason is the creation of more energy states above the HOMO of carbon dots by P-related functional groups [11]. All the absorption peaks are broad, and no significant change is observed in broadness with an increase in P content with respect to N. It indicates the heterogeneity in the prepared co-doped carbon dots samples.

3.3 Photoluminescence Study

In photoluminescence, excitation is the process to transfer an electron to the excited state through photons. Therefore, it is dependent on the absorption and molecular states of the material. Figure 2 b, c, and d represent the absorption and PL emission spectra of prepared P, N co-doped carbon dots. Table 1 gives information about the sample's λ_{max} and corresponding excitation and emission wavelengths. From Table 1, it is clear that the prepared doped carbon dots samples are exhibiting down-conversion luminescence. Down-conversion luminescence is the process of emission, in which a lower energy photon is emitted by the excitation of higher energy radiation. During the radiative transition, electrons from singlet excited states recombine at the singlet ground state. PL emission is red-shifted with higher excitation wavelength indicating wider size distribution. As electrons in different-sized atoms have different excitation energy, and dependency on excitation wavelength is seen [2].

From the table, this is assigned to the better passivation of samples B and D by the higher content of P against N. This is further confirmed by the blue shift in the PL emission for samples B and D. For 290 nm excitation, sample C has an emission at 579 nm (yellow), whereas sample D exhibits an emission at 489 nm (blue). The passivation and doping of carbon dots can modify the surface state structure and electronic energy level which further lead to a change in PL emission intensity and a shift of PL emission peak. Moreover, enhancement in the PL emission intensity is also observed. It is attributed to the passivation of dangling bonds on carbon dots' surfaces and the change in electronic orbitals [8].

4 Conclusion

A simple, quick, and controllable microwave irradiation method was used to synthesize P, N co-doped carbon dots. As prepared samples are found to be amorphous in nature. The P, N co-doped carbon dots exhibited down-conversion fluorescence. The

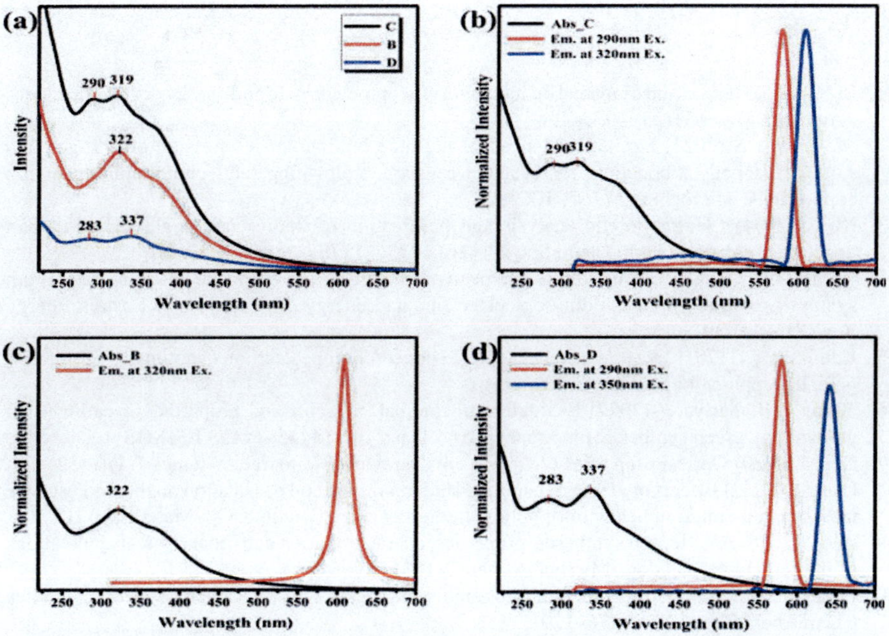

Fig. 2 **a** UV–Visible absorption spectra of P/N co-doped carbon dots in respect of change in P to N ratio and **b**, **c**, and **d** combined UV–Visible absorption (Abs) and emission (Em.) spectra at respective excitation (Ex.) of as-prepared samples C, B, and D of carbon dots respectively.

Table 1 Absorption, excitation, and emission parameters of prepared carbon dots

Sample	Abs. λ_{max} (nm)	λ_{ex} (nm)	λ_{em} (nm)	Region
C	290 319	290 320	579 608	Orange, Yellow Red
B	322	320	548	Green
D	283 337	290 350	489 550	Blue Green

excitation-dependent PL emission was observed in all samples. The PL emission from green to red and PL intensity is tuned by changing the ratio of initial P and N content. It is attributed to surface state transitions. The PL peak was blue-shifted for higher P content, indicating better surface passivation of carbon dots by PA and EDA. Therefore, the study on P to N ratio-dependent fluorescence properties of carbon dots suggests, the optimum ratio of dopants has a considerable effect on determining the enhanced and narrow PL emission.

References

1. Li X (2015) Carbon and graphene quantum dots for optoelectronic and energy devices: a review. Adv Funct Mater 31(25):4929–4947
2. Mohandoss S (2021) Excitation-dependent multiple luminescence emission of nitrogen and sulfur co-doped carbon dots for cysteine sensing, bioimaging, and photoluminescent ink applications. Microchem J (167):106280
3. Xu X (2004) Electrophoretic analysis and purification of fluorescent single-walled carbon nanotube fragments. J Am Chem Soc 40(126):12736–12737
4. Ghosh D (2021) Current and future perspectives of carbon and graphene quantum dots: From synthesis to strategy for building optoelectronic and energy devices. Renew Sustain Energy Rev (135):110391
5. Kandasamy G (2019) Recent advancements in doped/co-doped carbon quantum dots for multi-potential applications. J Carbon Res 2(5):24
6. Yadav R, Lahariya V (2022) Evaluation of thermal behavior and properties of carbon dots prepared by green synthesis. Electrochem Soc Trans 107:14445–14453 ICTSGS
7. Liu M (2020) Optical properties of carbon dots: a review. Nanoarchitectonics 1(1):1–12
8. Liang S (2022) Effects of elemental doping, acid treatment, and passivation on the fluorescence intensity and emission behavior of yellow fluorescence carbon dots. Opt Mat (128):112471
9. Zuo P (2016) A review on syntheses, properties, characterization and bioanalytical applications of fluorescent carbon dots. Microchim Acta 2(183):519–542
10. Pimenta M (2007) Studying disorder in graphite-based systems by Raman spectroscopy. Phys Chem Chem Phys 11(9):1276–1291
11. Ding H (2020) Surface states of carbon dots and their influences on luminescence. J Appl Phys 23:(127)

Green Synthesis of Carbon Dot (CD_s) and Sensing of Metal Ion

Momina and Ahmad Kafeel

Abstract Carbon dots (CDs) have demonstrated significant promise in the application of ion sensing, water/waste treatment, photo-catalysis, biological imaging, supercapacitor, heavy metal detection, and membrane filtration. In this paper, fluorescent carbon dots (CDs) have been synthesized by solvothermal treatment using fruit waste (apple peel) as a raw material. Ecological solvents such as distilled water and ethanol have been used as solvent to produce CDs. The prepared fluorescent CDs have been characterized by PL spectra, UV–Vis, zeta potential, FTIR, and TEM. Some metal ions like Cr (VI) have human health mutagenic and carcinogenic consequences which can penetrate cells membrane and also lead to DNA mutation. Thus, the green synthesized CDs have been studied for the detection of Cr (VI) ions with detection limit of 3.82 μM.

Keywords Carbon dots (CDs) · Green synthesis · Fruit waste · Metal ion sensing

1 Introduction

Existence of heavy metal ions in water and wastewater is becoming a serious issue nowadays. These pollutants arrive from various industrial effluents. Among them, chromium is considered as one of the toxic pollutants which is non-biodegradable and highly soluble in nature. Hexavalent Cr (VI) is generally used in textile industry as a catalyst in the dyeing processes [1]. Its higher mobility, oxidizing potential, and higher permeability through biological membrane are more toxic and lead to carcinogenic and mutagenic effects to human beings as well as animals. As per US Environmental Protection Agency (EPA) guidelines, the minimal concentration of Cr (VI) is 0.1 mg/L [2]. Hence, it is necessary to detect and know the concentration of Cr (VI) ions in water samples to reduce their repercussive effect due to migration and subsequent properties.

Momina (✉) · A. Kafeel
Department of Civil Engineering, Jamia Millia Islamia, New Delhi 110025, India
e-mail: momina1910086@st.jmi.ac.in

Z. H. Khan et al. (eds.), *Recent Advances in Nanomaterials*, Springer Proceedings in Materials 27, https://doi.org/10.1007/978-981-99-4878-9_10

Fluorescence sensing is a simple with high sensitivity, real-time monitoring on site, and fast analysis for the detection of metal ions in more reliable and consistent form [3–5]. Very few studies have been reported on fluorescence sensing of Cr (VI) ion.

Recently carbon dots (CDs) have come to be recognized as a possible fluorescence senor for the detection of cationic and anionic ions. CDs have been generally synthesized by chemical precursors. While green synthesis of CDs requires natural sources which contain carbohydrates, lipids, proteins, etc., [6] which reduces the use of toxic or expensive chemicals and complicated post-treatment methods. The strong fluorescent, high stability, ease of synthesis and functionalization, biocompatibility makes green synthesized CDs novel sensing materials in the application of ion sensing [7]. Moreover, hydrothermal/solvothermal process for the synthesis of green CDs provides great advancement over other physical methods because of simplicity and good quantum yield [8].

As a result, we reported the synthesis of luminous CDs using a solvothermal approach using fruit waste in this work (apple peel). The resulting CDs were examined using UV–Vis, FTIR, zeta potential, PL spectra, and TEM. Further, the CDs were also applied for the detection of Cr (VI).

2 Materials and Methodology

Different metal ions from nickel nitrate hexahydrate (Ni $(NO_3)_2{\cdot}6H_2O$), calcium chloride ($CaCl_2$), iron sulphate (Fe_2SO_4), lead nitrate (Pb $(NO_3)_2$), magnesium chloride ($MgCl_2$), sodium chloride (NaCl), cadmium nitrate (Cd $(NO_3)_2{\cdot}4H_2O$), cobalt chloride hexahydrate ($CoCl_2.6H_2O$), potassium chromate (K_2CrO_4), manganese sulphate ($MgSO_4$), strontium nitrate (Sr $(NO_3)_2$) were prepared in double distilled water. Ethanol (C_2H_5OH) was purchased from generic.

2.1 *Preparation of CDs*

CDs were prepared from facile green solvothermal method using apple peel as a carbon source. Briefly, apple peel was chopped into small pieces and dried in oven at 80 °C. Dried peels were grinded, mixed with 75:25 v/v ratio of DDW:C_2H_5OH, and transferred to autoclave reactor heated at 180°C for 5 h and then gradually brought to room temperature. Purification of dots was performed by 0.45 μm filter paper.

2.2 Instruments

Fourier transform infrared spectroscopy (FTIR) was performed by PerkinElmer version 10.4 for the determination of functional groups present on CDs. A UV–Visible spectrophotometer (U3900, Hitachi) was used to investigate the optical characteristics of dots and time-resolved fluorescence lifetime measurement spectrometer (Delta Flex 01-DD, Horiba). Size of dots was investigated by transmission electron microscopy (TEM) at 200 kV and zeta potential by zeta potential analyzer (Zeta sizer Nano ZS, Malvern).

2.3 Sensing of Cr (VI)

The fluorescence approach was used to detect Cr (VI) in aqueous solution at room temperature. The concentration of Cr (VI) was prepared from stock solution of 1000 ppm and varied from 500 to 0.01 ppm. The Cr (VI) solutions (50 mL) were mixed with 2 mL of CDs, and fluorescence spectra were measured after 30 min. The selectivity for Cr (VI) was also analysed by using different types of metal ion (Ca^{2+}, Cd^{2+}, Cr^{6+}, Fe^{2+}, Mg^{2+}, Mn^{2+}, Na^{+}, Ni^{2+}, Pb^{2+}, Sr^{2+}) with 100 ppm of concentrations. The PL intensity was measured at a wavelength of excitation of 360 nm. For all experiments, the excitation emission slit was kept at 10 nm.

3 Result and Discussion

3.1 Characterization of CDs

The broad band was detected at 3447 cm^{-1} owing to amine (-NH) and hydroxyl groups (-OH) stretching vibrations, as shown in Fig. 1(a). The C-H stretching, C= C stretching (alkene), stretching vibration of C–O–C, and $_{\gamma}$C–O epoxy were observed at 2864, 1641, 1177, 1029, and 904 cm^{-1}. All of these peaks are thought to represent the typical peaks of dots. These functional groups of synthesized CDs facilitates the water solubility and can be functionalized for different applications.

Figure 1(b) and (c) depict the UV–Vis absorbance, excitation, and emissions data. The absorption peak at 280 nm corroborated the π–π^* transition of C= C, but the absorption peak of 310 nm was ascribed to n–π^* transition of C = O. At an excitation wavelength of 360 nm and with quinine sulphate as a reference, the quantum yield of dots was determined to be 23%. The dots show emission at 470 nm when it is excited at excitation wavelength of 360 nm which is higher than PL intensity of emission at excitation wavelength of 280 nm as depicted in Fig. 1(d). The fluorescence spectra of dots were also investigated at different excitation wavelengths, indicating that when the excitation wavelength increases from 270 to 380 nm, the emission wavelength

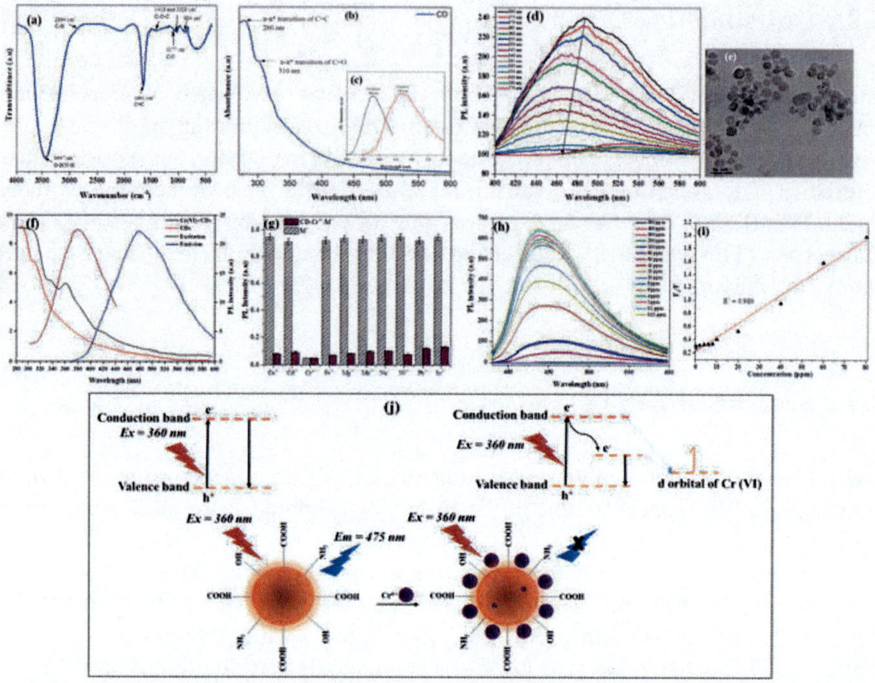

Fig. 1 **(a)** FTIR analysis. **(b)** UV–Vis absorption. **(c)** Excitation and emission intensity. **(d)** Fluorescence spectra of CDs. **(e)** TEM analysis with particle size distribution of CDs. **(f)** UV–Vis absorption of CDs with/without Cr (VI), emission and excitation spectra of CDs with Cr (VI). **(g)** Sensitivity of CDs and selectivity of Cr (VI) with different types of metal ions selectivity. **(h)** PL spectra of CDs with multiple concentration of Cr (VI) ions **(i)** SV plot for concentration ranging from 0.2 –1500 **(j)** Mechanism of sensing of Cr (VI) by CDs

increases and shifts (460 to 485 nm). The existence of photoinduced electrons and holes on CDs at various energy surfaces might cause this shifting [9]. Larger particle size can result in a greater excitation wavelength for fluorescence spectra.

The nanostructure and size of synthesized CDs were found to be spherical, uniform in size with diameter distribution ranging from 8 to 10 nm as depicted in Fig. 1(e). This shows that synthesized CDs are within range which possess excellent luminescence properties. Meanwhile, the zeta potential of CDs was negative (-18.1 mV), possibly due to the occurrence of carboxyl and amine groups on dot's surface.

3.2 Sensing of Cr^{6+}

The CDs were used as a fluorescence probe for the several metal ions including Ca^{2+}, Cd^{2+}, Cr^{6+}, Fe^{2+}, Mg^{2+}, Mn^{2+}, Na^{+}, Ni^{2+}, Pb^{2+}, Sr^{2+} of 100 ppm solution each. The fluorescence of CDs was reduced after the addition of Cr (VI), whereas rest of the metal ion shows slight variation in their fluorescence behaviour as shown in Fig. 1(g). This result indicates that synthesized CDs are more selective towards Cr (VI) ion and therefore could be used as fluorescent probe for Cr (VI). To evaluate the sensitivity of Cr (VI), fluorescence analysis was performed on the availability of other metal ions (M^+), as shown in Fig. 1(g). There was no interference in the detection of Cr (VI) in the presence of other metal ions, as can be shown. The PL spectra of CDs was gradually quenched by varied doses of Cr (VI) ranging from 500 to 0.01 ppm. As the Cr (VI) ion concentration increases, the emission intensities of CDs reduce continuously (depicted in Fig. 1(h)) which shows the turn-off [10]. Additionally, the shifting of emission peaks from 430 to 440 nm was observed in the presence of Cr (VI). This behaviour can occur due to surface defect states, functional groups, various surface energy traps and/or due to particle size [11]. The fluorescence quenching mechanism was additionally explored by Stern–Volmer (SV) plot representing the dynamic and static quenching and can be described by Eq. 1.

$$\frac{F_o}{F} = 1 + K_{SV}[Q] \tag{1}$$

F_o and F represent fluorescence of intensities of CDs in the availability and non-availability of Cr(VI), K_{SV} is SV constant, and Q is the concentration of Cr(VI). Figure 1(i) represents the SV plot linear plot with regression coefficient of $R^2 =$ 0.989 for a concentration ranging from 0.01 to 80 ppm. The limit of detection (LOD) was measured by Eq. 2

$$\text{LOD} = \frac{n\sigma}{S} \tag{2}$$

where n is total number of test ($n = 11$), S is the slope of curve, and σ is the standard deviation of the signals. The LOD was found out to be 0.162 ppm (3.82 μM).

3.3 *Mechanism for Sensing of Cr^{6+}*

The UV–Vis absorbance of dots in the presence of Cr (VI) aids in explaining the sensing of Cr (VI) ions by CDs, as depicted in Fig. 1(f). Following the inclusion of dots, the absorption peaks of CDs at 280 nm changed somewhat, but the peak of oxygen-related n-π* transition of CDs relocated to 353 nm after sensing. It is noticeable that excitation and emission wavelength of CDs after addition of Cr (VI) were 360 nm and 430 nm, respectively. It can happen due to inner filter effect (IFE) which occur due to some extent overlapping of quencher absorption with excitation [7, 12]. During the treatment of Cr (VI), the formation of hydroxide metal complex or any other type of complex can lead to accumulation of metal ion on surface of CDs. The functional groups of CDs (act as an electron donor) can also form coordination bond with Cr (VI) ion. This coordination excited the electron from CDs surface to half-filled 3d orbital Cr (VI), since the outer shell orbital of Cr (VI) is 3d5 which is half-full [10]. Therefore, electron from valence band in CDs is excited and reaches to conduction band. It releases some energy and undergoes transition to the d-orbital of Cr (VI) ions. This process results in fluorescence quenching. The illustrative fluorescence quenching mechanism is shown in Fig. 1(j).

4 Conclusion

In conclusion, apple peel-based CDs were synthesized via one step solvothermal process using ecological solvents which is facile, green, and economically viable. The synthesized CDs showed excellent fluorescent nature, consisting of carboxyl, hydroxyl, and amine groups, and size of CDs ranging from 8 to 10 nm. The synthesized showed good sensing of Cr (VI) using fluorescence sensing method. This method was found to highly sensitive with detection limit of 3.82 μM. Furthermore, the intensity of CD emission was discovered to be quenched with Cr(VI) ions on the basis of overlapping of quencher absorption with excitation and CDs emission.

Declaration of Interest Statement The authors declare that they have no conflict of interests.

References

1. Yaseen DA, Scholz M (2019) Textile dye wastewater characteristics and constituents of synthetic effluents: a critical review. Int J Environ Sci Technol 16(2):1193–1226
2. Ming F. et al. (2019) One-step synthesized fluorescent nitrogen doped carbon dots from thymidine for Cr (VI) detection in water. Spectrochim Acta Part A Mol Biomol Spectrosc 222:117165
3. Feng S. et al. (2019) Feasibility of detection valence speciation of Cr(III) and Cr(VI) in environmental samples by spectrofluorimetric method with fluorescent carbon quantum dots. Spectrochim Acta Part A Mol Biomol Spectrosc 212:286–292
4. Liu X et al (2017) Carbon nanodots as a fluorescence sensor for rapid and sensitive detection of Cr(VI) and their multifunctional applications. Talanta 165:216–222
5. Ravindran A et al (2012) Selective colorimetric detection of nanomolar Cr (VI) in aqueous solutions using unmodified silver nanoparticles. Sens Actuators B Chem 166–167:365–371
6. Pooja D et al (2019) Green synthesis of glowing carbon dots from *Carica papaya* waste pulp and their application as a label-freechemo probe for chromium detection in water. Sens Actuators B Chem 283:363–372
7. Qianqian H et al (2022) Carbon dots derived from *Poria cocos* polysaccharide as an effective "on-off" fluorescence sensor for chromium (VI) detection. J Pharm Anal 12(1):104–112
8. Mehta VN et al (2015) One-step hydrothermal approach to fabricate carbon dots from apple juice for imaging of mycobacterium and fungal cells. Sens Actuators B Chem 213:434–443
9. Ramar V et al (2018) Metal free, sunlight and white light based photocatalysis using carbon quantum dots from *Citrus grandis*: a green way to remove pollution. Sol Energy 169:120–127
10. Sekar A et al (2021) Fluorescence quenching mechanism and the application of green carbon nanodots in the detection of heavy metal ions: a review. New J Chem 45(5):2326–2360

11. Nagaraj M et al. (2022) Detection of Fe3+ ions in aqueous environment using fluorescent carbon quantum dots synthesized from endosperm of *Borassus flabellifer*. Environ Res 212:113273, PB
12. Kamali SR et al (2021) Sulfur-doped carbon dots synthesis under microwave irradiation as turn-off fluorescent sensor for Cr(III). J Anal Sci Technol 12(1):1–11

Synthesis of Monosized Silica Microparticles and Fabrication of Size-Controlled Silicon Microwires

Anjali Saini, Premshila Kumari, Sanjay K. Srivastava, and Mrinal Dutta

Abstract The present study has been carried out to fabricate size-controlled monodisperse silica microparticles and SiMW arrays using a silica template via the nanosphere lithography (NSL) technique in accumulation to the metal-catalysed electroless etching (MCEE) technique. Considering the need, importance, and benefits of monosized SiO_2 microparticles as proclaimed in the scientific area with diverse industrial applications, monodispersed silica particles were synthesized by using modified Stober process. Synthesis of monosized microparticles has been achieved by using a silica precursor tetraethyl orthosilicate (TEOS) in an ethanol solution underneath basic conditions and assimilating a surfactant. In this process, a single monolayer closed-pack configuration of SiO_2 microparticles via self-assembled mechanism formed on the 2-inch silicon wafer by using a simple spin coating technique, and further, non-closed-packed structure was obtained from the closed-packed structure using the reactive ion etching (RIE) process. Using the MCEE technique, the size-controlled SiMW arrays were synthesized. The size and morphology of the microparticles and SiMW arrays were examined by the field-emission scanning electron microscope (FESEM). The SiMW arrays showed less than 15% reflection in the wide range of spectrum (300–1000 nm). The surface composition of the microparticles and SiMWs was studied using FT-IR. Such SiMW arrays may have several potential applications including solar cells.

Keywords Silica microparticles · Metal catalysed electroless etching · Nanosphere lithography

---

A. Saini · P. Kumari · S. K. Srivastava (✉)
CSIR-National Physical Laboratory (NPL), New Delhi 110012, India
e-mail: srivassk@nplindia.org

Academy of Scientific and Innovative Research (AcSIR), Ghaziabad 201002, India

M. Dutta (✉)
National Institute of Solar Energy, Gurgaon 122003, India
e-mail: duttamrinal@nise.res.in

Z. H. Khan et al. (eds.), *Recent Advances in Nanomaterials*, Springer Proceedings in Materials 27, https://doi.org/10.1007/978-981-99-4878-9_11

1 Introduction

Silicon (Si) nanostructure manufacturing is crucial because it might be used in cutting-edge optical and electrical device applications. Si nanostructures may find use in the domains of chemical sensors [1], photovoltaics [2], fuel cells, nanoelectronics, and optoelectronics [3], according to several research efforts. Since the amounts of c-Si (reduced thickness) that can be used in photovoltaic applications is limited, a lot of research activity has been carried out on Si nanowires (NWs) arrays for the futuristic c-Si solar cells, which are predicted to have good efficiency. The SiNW arrays that are vertically aligned have a black hue and do not reflect incoming light, which leads to better light absorption. The surface of the Si NW arrays scatters the light that is injected in the Si [4]. Even in the NWs that are comparatively small in length as comparison to the thick c-Si substrates, the scattered light is effectively absorbed. The SiNWs offer good solar cell efficiency in conjunction to superior light absorption due to their one-dimensional geometries. The greatest path that photo-generated carriers must travel to reach the radial junction when it forms on the surface of the SiNW is equal to half of the diameter of the NW arrays. In comparison to traditional solar cells, an exceptionally small carrier transit distance is permitted, allowing for efficient carrier collection. However, the SiNW array-based solar cells have been observed to have a much lower efficiency than traditional c-Si solar cells. This is attributed to the totally depleted p-n junctions. In addition, the relatively high surface area leads to significant surface recombination, which lowers the solar cell's efficiency despite its outstanding radiation absorption capacity. As a potential replacement structure for the wire-based silicon solar cells, Si-microwires (SiMWs) have been studied [5, 6]. The SiMWs may be implemented to the radial p-n junctions with ease by employing customary doping techniques since they share the same one-dimensional structure as the SiNW arrays. A SiMW array-based solar cell [7] recently reported by the Seo group has the efficiency of 19%, which was due to both high (99%) sunlight absorption rate and the efficient photocarrier accumulation at the radial junction. The SiMWs are also more flexible mechanically than the bulk c-Si. The MWs have remarkable flexibility because of their 1D construction with a huge aspect ratio, which allows the mechanical tension brought on by bending the SiMWs to be discharged efficiently through the void between the MWs array. The SiMW array-based flexible solar cells have drawn a lot of interest for uses like wearable technology. Due to the stability of material (e.g. c-Si) and its high performance, the SiMW array-based flexible solar cells have demonstrated strong potential as upcoming flexible solar cells, in contrast to traditional flexible solar cells based on organic materials, which have less stability and efficiency. Consequently, several techniques have been investigated and developed. A reliable and reproducible approach for fabricating periodic Si nanostructures is currently highly difficult. This necessitates a highly time-consuming and expensive procedures (such as electron-beam lithography), which poses a hurdle to their widespread application. Therefore, efforts are being made to develop a large-area nanofabrication technique that is less expensive, takes less time, and has adequate structural precision and repeatability. Various innovative nanosphere lithography (NSL) techniques and production processes have been created and researched as part of this endeavour. As a result, the NSL has received a lot of attention since it makes it simple to change the lithographical scale by simply placing various-sized nanospheres directly onto the specimen surface to act as etch shields. A Si nanosphere or microsphere (SNS or SMS) monolayer may be uniformly deposited on large Si substrates, using spin coating technique. This technique results in coverage of more than 90% [5]. Reactive ion etching (RIE) is a method that may be used to reduce the diameter of the SMS and achieve different nanostructure/microstructure dimensions over time. For efficient light diffraction or scattering, dimension management of the silicon nanostructures over a certain time period is a crucial factor in disciplines such as optoelectronics [8, 9]. The development of a useful nanofabrication technique with improved lithographical precision of both period and size has been prompted by the rising interest in solar cells. This is due to the fact that regulated periodic nanostructures on the subwavelength scale may significantly increase light absorption in Si absorbers, the theoretical absorption barrier of Si [10]. The architecture of the textured Si surface produced by electrochemical etching is evident from earlier investigations [11, 12].

In this work, a method for the synthesis of monosized SiO_2 microparticles is reported with the formation of closed-pack monolayer of SiO_2 particles in wafer scale by using the simple spin coating technique. Later, this monolayer of the SiO_2 particles was used as a mask for the fabrication of the periodic SiMW arrays in the wafer scale. This SiMW array reduces the reflection to less than 15% in the 300–800 nm wavelength range in comparison to the more than 35% reflection of the planar wafer in the same range. This study has been done by considering the fact that the SIMW arrays could play a crucial role in forthcoming applications in the PV devices to achieve higher power conversion efficiency at lower-cost. Both the PV industry and academic researchers will be benefited by the work presented in this paper.

2 Materials and Methods

2.1 Materials

KCl (99.9%) was bought from Merck Life Science Private Limited (India). Hydrofluoric acid (48%), hydrogen peroxide (30%, emplura), nitric acid (69%, emparta), N, N-dimethylformamide (DMF), ethanol (EtOH), sulfuric acid (H_2SO_4) were purchased from Merck Life Science Private Limited (India). Tetraethyl orthosilicate (TEOS, 99%) was purchased from Sigma-Aldrich.

2.2 Methods

SiO_2 microparticles synthesis via modified Stober method: Ammonia solution was used as a catalyst to synthesize monodisperse SiO_2 microspheres in ethanol. For this, two types of solutions were prepared: (i) (solution I) KCl (7.5 mg) with ethanol, D. I. water, and ammonia were mixed together in 20:2:1 and 20:3.4:1 volume ratios, respectively. These mixtures were stirred for 15 min at room temperature. (ii) Solution II containing ethanol and TEOS mixed together (11:1 volume ratio) by stirring at room temperature for a certain time. After this, these solutions were mixed together by using a syringe pump (see Fig. 1a). After further reaction, the final product was sterilized by centrifugation and followed by three times ethanol washing. The final product was then vacum-dried at room temperature. In this work, amount of all other parameters were kept constant except H_2O.

Fabrication process of Si microwire by nanosphere lithography techniques: p-type (100)-oriented 2"Si wafers (1–5 Ω-cm) were cleaned ultrasonically in acetone and isopropyl alcohol (IPA) for 10 min and then immersed in a piranha solution (3:1 ratio of H_2SO_4 and H_2O_2). The wafers were treated for 10 min at a temperature of 110 °C. After that, the wafers were washed in deionized (D.I.) water. This was followed by a 5 min immersion in a 2% HF solution and two rounds of D. I. water rinsing. After these, the surface of the Si wafers was made hydrophilic by using a mixture of H_2O_2, D.I. water, and NH_4OH at fixed temperature for 1 h. The SiO_2 microparticles were dispersed in N, N-dimethylformamide (DMF). This solution was then slowly and carefully dropped onto the wafer's surface while rotating at a constant speed. After injecting enough amount of the solvent, a closed-packed monolayer was developed. A flowchart of the experimentation method is shown in Fig. 1b. Reactive ion etching with power of 200 W and CF_4 flow of 20 sccm was used for the reduction of the size of SiO_2 microparticles and the formation of a non-closed-packed arrangement. The next step was utilizing an e-beam evaporator to deposit silver (Ag) on the top of non-closed-pack structure. Then the Si wafers were ultrasonically cleaned to eliminate the microparticles from the wafer surface in order to create an Ag mask layer. The as-prepared wafers were then immersed in an etching solution containing hydrogen peroxide (H_2O_2), ethanol (EtOH) and HF in a 1:1:10 ratio to form the SiMWs array by metal catalysed electroless etching (MCEE). The etched wafers were then cleaned with D. I. water and immersed for 5 min in both concentrated and diluted HNO_3 to remove the Ag particles from the surface of the Si. These wafers were then dried at room temperature after being cleaned with D. I. water.

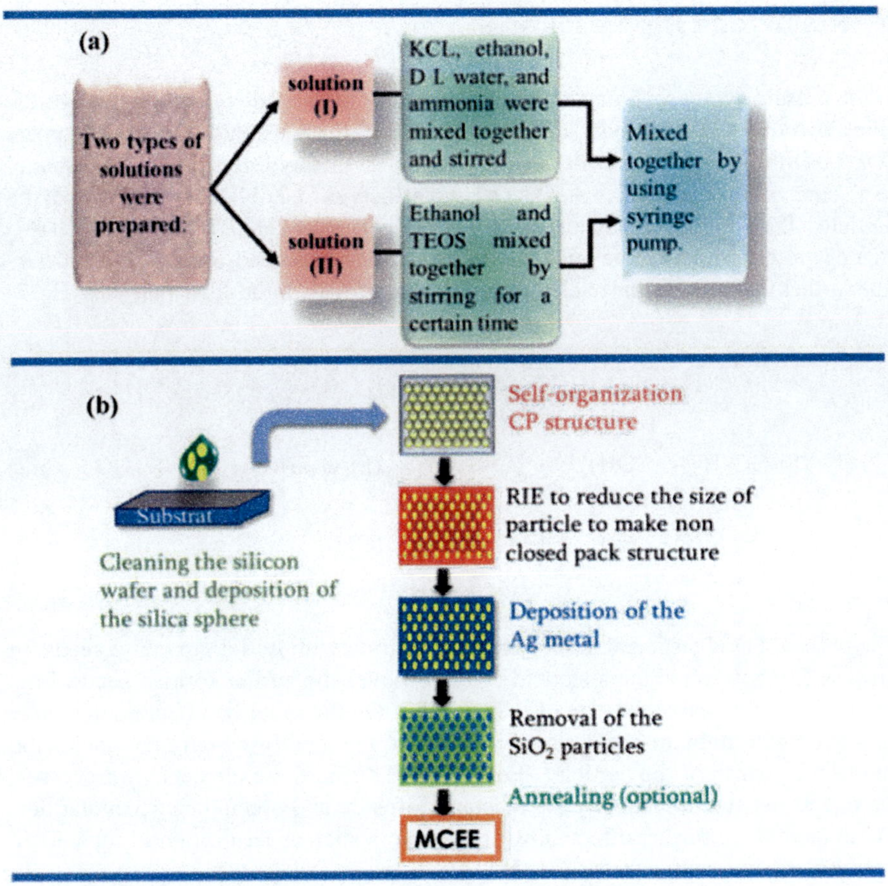

Fig. 1 **a** Block diagram for the synthesis process of SiO_2 particles. **b** Schematic of nanosphere lithography technique for fabrication of SiMW arrays

2.3 Characterization

The high resolution field-emission scanning electron microscope (FESEM, Hitachi S-4800) was used to analyse the morphology of the SiMW arrays and SiO_2 microparticles. The reflectance spectra of the samples were captured using a UV–Vis–NIR spectrophotometer (PerkinElmer, Lambda 1050) in the range of 300–1400 nm. Fourier transform infrared spectroscopy (FT-IR) was recorded by PerkinElmer Optica (spectral range from 525 to 4000 cm^{-1}) to identify different functional groups.

3 Results and Discussion

In the present synthesis process, two consecutive reactions dictate the growth of the silica particles (1) hydrolysis of TEOS (i.e., $Si(OC_2H_5)_4$) into silanol monomers (Eq. 1). During the synthesis of the silica particles, this hydrolysis occurs in alcohol (e.g., ethanol) in the presence of NH_4OH, which acts as a catalyst for the growth of the particles. (2) Simultaneously, the silanol monomers start condensation between two silanol groups which create siloxane clusters in the branched form. These clusters link further that triggers the nucleation and hence growth of the silica particles (Eq. 2) [13].

$$Si(OC_2H_5)_4 + {}_xH_2O \leftrightarrow Si(O\,C_2H_5)_{4-x}(OH)_x + xC_2H_5OH \quad (1)$$

$$2Si(OC_2H_5)_{4-x}(OH)_x \leftrightarrow (C_2H_5O)_{8-2x}(Si\text{-}O\text{-}Si)(OH)_{2x-2} + H_2O \quad (2)$$

$$\begin{aligned} &Si(O\,C_2H_5)_4 + Si(O\,C_2H_5)_{4-x}(OH)_x \\ &\leftrightarrow (C_2H_5O)_{7-x}(Si\text{-}O\text{-}Si)_1(OH)_{x-1} + C_2H_5OH \end{aligned} \quad (3)$$

The process (1) is responsible for earlier nucleation with initial growth of seeds. In process (2), new monomers start to condense over the earlier formed seeds. This condensation increases the size of the particles. On the other hand, monomers may also react with unhydrolyzed ethoxyl groups of the TEOS to start new nucleation and hence growth of the particles (Eq. 3) [13]. During these elementary processes, the rate of reaction affected by the reactant diffusion and precipitated coagulations. These factors are further affected by other factors such as adsorption of ions, ionic strength, solvent compositions, or pH [14].

Monosized silica microspheres synthesized by the modified Stober method [15] with average diameters from 1.1 μm to 2.2 μm must form a monolayer closed-packed (MCP) configuration in order to fabricate the periodic SiMW arrays at the wafer scale using the NSL technique. These silica microparticles closed-pack structures were created using a straightforward spin coating method. On a 2-inch silicon wafer, the fabrication of such a MCP structure was accomplished. The FESEM images of closed-pack monolayer structures of the silica microparticles are shown in Fig. 2a. The SiO_2 microspheres, which define the diameter of the MWs and the pitch, underwent consistent size reduction with a circular shape through the RIE process. By eliminating the non-closed-pack (NCP) monolayer of the SiO_2 microparticles, a uniform metal mask was formed prior to the creation of the SiMW array using the MCEE approach. The cross-sectional FESEM image of the representative SiMW arrays is shown in the inset of Fig. 3.

The FESEM image (shown in Fig. 2 c and d) of the SiMW arrays shows that the tips are of spongy nature produced by the NSL technique followed by the MCEE technique. The average diameters of these microwires were estimated to be 0.95 and 2 μm which is similar to the average diameter of the microparticles after the RIE

Fig. 2 FESEM images of **a** and **b** closed-pack monolayer structure of SiO_2 microparticles, **c** and **d** Top views of FESEM images of the SiMW arrays corresponding to (**a**) and (**b**), respectively

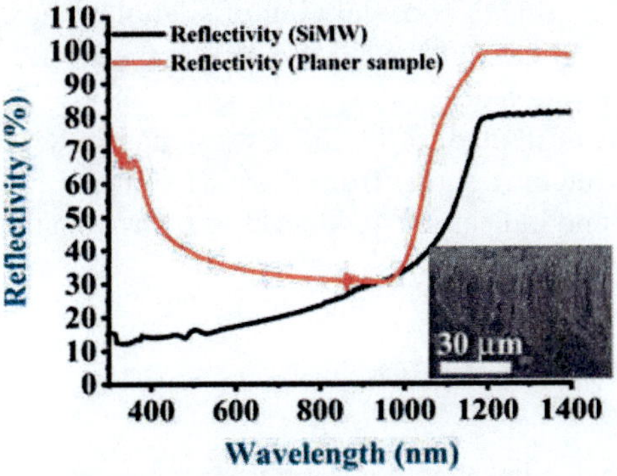

Fig. 3 UV-Vis-NIR reflection spectra of planar polished Si wafer and the SiMW array. Inset shows the cross-sectional FESEM image of SiMW arrays

process. Thus, by controlling the reduction of average size of the microparticles by RIE process, the diameter and pitch of the microwires could be controlled. The UV–Vis-NIR reflectance spectra of the planer Si wafer and the SiMW arrays are shown in Fig. 3 in a wide spectral region of 300–1400 nm. The band edge of Si may be identified by the sudden transition in wavelength between 1000 and 1200 nm [16]. While the reflectivity of the glossy planar Si wafer ranges between 75 and 30% in the range of 300–1000 nm, it is considerably subdued by the SiMW arrays over a wide spectral range from 300 to 1000 nm. The median reflectivity of the SiMW arrays reduced to less than 15% in the 300–800 nm wavelength range due to multiple light scattering and the subwavelength radiation trapping within the SiMW arrays, as opposed to an average reflectance of 30% of the planer-polished surface of the Si wafer. The

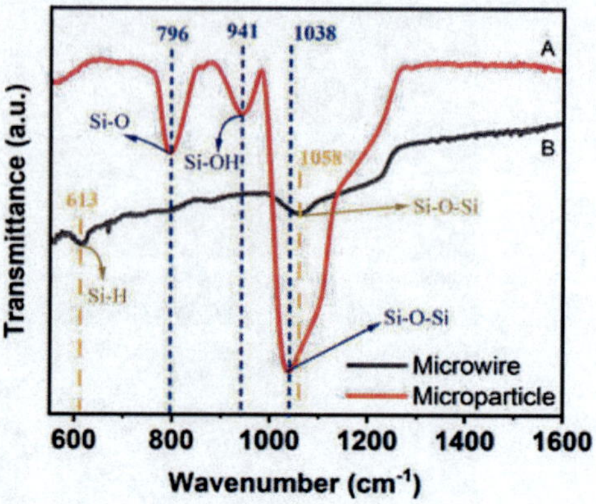

Fig. 4 FT-IR spectra of prepared SiO_2 microparticles and the SiMW arrays

investigation of various functional groups can be obtained by FT-IR spectroscopy. Figure 4 shows the FT-IR spectra of the prepared SiO_2 microparticles (curve *A*) and the SiMW arrays (curve *B*). By using FT-IR, vibration peaks of Si–O, Si–H, Si–OH and Si–O–Si were confirmed. In case of SiO_2 microparticles, the strong peak at 1038 cm^{-1} is attributed to siloxane Si–O–Si group (*A*). The Si–O band was observed at 796 cm^{-1}. Band at 941 cm^{-1} is the characteristic for the Si–OH stretching vibration due to high concentration of silanol groups, which is clearly observed in the curve A. In case of the SiMW arrays, the FT-IR spectrum shows the Si–H and Si–O-Si vibration peaks at 613 cm^{-1} and 1058 cm^{-1}, respectively. The peak at 613 cm^{-1} could be attributed to the Si–H wagging mode vibration on surface and the asymmetric stretching signal from Si–O-Si shows a peak at 1058 cm^{-1}. The presence of Si–H bond indicates a good level of hydrogenation.

4 Conclusion

Size-controlled SiO_2 microparticles were synthesized by using modified Stober method. By using this method, monosized SiO_2 microparticles of average diameter 1.1–2.2 μm were synthesized. Using the SiO_2 microparticles, a MCP layer was formed using the spin coating method. Further, in combination to ICP-RIE, the NSL approach has been utilized to produce non-closed-pack structure of the SiO_2 microspheres on a wafer scale. Ag coating on the top of the SiO_2 microsphere patterning was done by an e-beam evaporator. A chemical etching process was used to form SiMW arrays. FESEM observation revealed that the average size of the microwires is 0.95 and 2.0 μm which is almost same to the average diameter of the microparticles after RIE process. The reflection characteristics of the SiMW arrays were investigated by UV–Vis-NIR spectroscopy tool. For the SiMW arrays, reflection was reduced to as little as 15% over a broad spectrum region of 300–800 nm in comparison to the 30% reflectance of the planar Si wafer in the same range. The FT-IR spectrum confirmed the presence of Si–O, Si–H, Si–OH, and Si–O-Si vibrational modes in SiO_2 microparticles and presence of Si–H and Si–O–Si vibrational modes of the SiMWs. Thus, pitch and size-controlled SiMW arrays are fabricated using the NSL technique followed by the MCEE technique. This work is valuable for the industrial Si solar cell manufacturing process since SiMW arrays might be employed as a light-trapping surface rather than adding additional antireflection coating layers.

Acknowledgements The PhD research scholarships is made possible by the Council of Scientific and Industrial Research (CSIR), and MNRE-NREF, Govt. of India which are acknowledged by the authors, AS and PK respectively. DST-SERB Ramanujan Fellowship (File No. SB/S2/RJN-077/2017) provided funding for this work.

Declaration of Interest Statement
The authors declare that they have no conflict of interests.

References

1. Patolsky F, Zheng G, Lieber CM (2006) Fabrication of silicon nanowire devices for ultra-sensitive, label-free, real-time detection of biological and chemical species. Nat Protoc 1(4):1711−1724
2. Peng K, Xu Q, Wu Y, Yan Y, Lee ST, Zhu J (2005) Aligned single-crystalline Si nanowire arrays for photovoltaic applications. Small 1(11):1062−1067
3. Tian B, Zheng X, Kempa TJ, Fang Y, Yu N, Yu G, Huang J, Lieber CM (2007) Coaxial silicon nanowires as solar cells and nanoelectronic power sources. Nature 449(7164):885−889
4. Muskens OL, Rivas JG, Algra RE, Bakkers EPAM, Lagendijk A (2008) Design of light scattering in nanowire materials for photovoltaic applications. Nano Letters 8(9):2638–2642
5. Jung JY, Guo Z, Jee SW, Um HD, Park KT, Hyun MS, Yang JM, Lee JH (2010) Awaferscale Si wire solar cell using radial and bulk p-n junction. Nanotechnol 21(44):445303
6. Hwang I, Um HD, Kim BS, Wober M, Seo K (2018) Flexible crystalline silicon radial junction photovoltaics with vertically aligned tapered microwires. Energy Environ Sci 11(3):641–647
7. Choi JY, Alford TL, Honsberg CB (2014) Solvent-controlled spin-coating method for large-scale area deposition of two-dimensional silica nanosphere assembled layers. Langmuir 30(20): 5732−5738
8. Spinelli P, Verschuuren MA, Polman A (2012) Broadband omnidirectional antireflection coating based on subwavelength surface Mie resonators. Nat Commun 3:692
9. Kim I, Jeong DS, Lee WS, Kim W M, Lee T-S, Lee D-K, Song J-H, Kim J-K, Lee K-S (2014) Silicon nanodisk array design for effective light trapping in ultrathin c-Si. Opt Express 22(S6):A1431−A1439
10. Yu Z, Raman A, Fan S (2010) Fundamental limit of nanophotonic light trapping in solar cells. In:proceedings of the national academy of sciences of the United States of America 107(41):17491−17496
11. Li S, Ma W, Zhou Y, Chen X, Ma M, Xiao Y, Xu Y (2014) Influence of fabrication parameter on the nanostructure and photoluminescence of highly doped P-porous silicon. J Lumin 146:76–82
12. Saini A, Abdelhameed M, Rani D, Jevasuwan W, Fukata N, Kumari P, Srivastava SK, Pathi P, Samanta A, Dutta M (2022) Fabrication of periodic, flexible, and porous silicon microwire arrays with controlled diameter and spacing: effect on optical properties. Opt Mat 134:113181
13. Han Y, Lu Z, Teng Z, Liang J, Guo Z, Wang D, Han M-Y, Yang W (2017) Unravelling the growth mechanism of silica particles in Stöber method: In-situ seeded growth model. Langmuir 33(23):5879–5890

14. Nagao D, Nakabayashi H, Ishii H, Konno M (2013) A unified mechanism to quantitatively understand silica particle formation from tetraethyl orthosilicate in batch and semi-batch processes. J Colloid Interface Sci 394:63–68
15. Lei X, Yu B, Cong HL, Tian C, Wang Y-Z, Wang Q-B, Liu C-K (2014) Synthesis of monodisperse silica microspheres by a modified Stober method. Integrated Ferroelectrics 154(1):142–146
16. Dutta M, Fukata N (2015) Low-temperature UV ozone-treated high efficiency radial p-n junction solar cells: N-Si NW arrays embedded in a p-Si matrix. Nano Energy 11:219–225

Thermoelectric Properties of LiYSi Half-Heusler Alloy

Grewal Savita and Kumar Ranjan

Abstract Half-Heusler compounds are a broad class of materials that have attracted attention as possible high-temperature thermoelectric materials due to their favourable electrical transport behaviour and other features useful for device manufacturing. Although half-Heusler compounds exhibit an incredibly wide range of thermal conductivities, these are nonetheless often higher than other cutting-edge thermoelectric materials, which present a unique difficulty. It was observed that the chosen half-Heuslers were direct bandgap semiconductor with high power factors equivalent to the art of thermoelectric materials. Since the computed phonon is positive, the substance is dynamically stable. The high power factor and high Seebeck coefficient of the half-Heusler alloy are also revealed by the density of states. The stability of the material is also revealed by the formation energy. The optimized structure of LiYSi has band gap of 0.70 eV. The calculated power factor of n-type LiYSi is $7.31 \times 10^{11} Wm^{-1} K^{-2} s^{-1}$ per relaxation time. The calculated power factor per relaxation time of p-type LiYSi is $1.44 \times 10^{12} Wm^{-1} K^{-2} s^{-1}$.The power factor increases with increase in temperature. High n-type thermoelectric performance is produced by the interaction of low lattice thermal conductivity and excellent electrical transport characteristics. This suggests that at high temperatures, LiYSi is a potential n-type half-Heusler thermoelectric material. It is clear that the electronic structure has high valence band degeneracy, which contributes for good transport properties and results good thermoelectric material.

Keywords Transport properties · DOS · Partial density of states

G. Savita (✉) · K. Ranjan
Department of Physics, Panjab University, Chandigarh 160014, India

K. Ranjan
Department of Physics, Faculty of Science, King Abdulaziz University, P.O. Box 80203, Jaddah 21589, Saudi Arabia

Z. H. Khan et al. (eds.), *Recent Advances in Nanomaterials*, Springer Proceedings in Materials 27, https://doi.org/10.1007/978-981-99-4878-9_12

1 Introduction

In the twenty-first century, major challenges include the limited supply of petrochemical fuel and the ensuing environmental degradation. In this context, thermoelectric (TE) materials are helpful since they can transform heat directly into power without the usage of moving parts or emissions [1]. However, improving conversion efficiency is necessary before TE materials are widely used $ZT = S^2 \sigma T/k$ This is determined by electrical conductivity, absolute temperature, and thermal conductivity, where S, σ, T, and k are the Seebeck coefficients correspondingly. Lattice thermal conductivity (k_l) and electronic thermal conductivity (k_e) both have an impact on thermal conductivity [2, 3]. It is challenging to get both high electrical conductivity and the Seebeck coefficient because they are frequently inversely correlated and have very low thermal conductivity. Thermoelectric performance can be improved using a variety of techniques. One strategy is to improve electrical transport performance by band engineering to decouple S and σ, for example, by large band degeneracy (Nv) and/or anisotropic carrier pockets [4, 5].

Half-Heusler (HH) compounds are a prospective future genus of thermoelectric materials with generally good thermodynamic and mechanical characteristics [6]. Importantly, the HH compound class of materials is quite diverse, providing plenty of flexibility for searching for desirable properties and for optimization. They usually have significant band degeneracies because of their excellent cubic crystal symmetry [7–9]. This may lead to beneficial electrical properties due to the consequent decoupling of S and σ, especially if the degenerate carrier pockets are also anisotropic, as is typically the case. Furthermore, n-type has advantageous thermoelectric characteristics.

2 Materials and Methods

The QE Software program is used based on DFT to optimize the LiYSi crystal structure. The trade correlation potential is approximated by Perdew-Generalized Burke's Ernzerhof Gradient Approximation (GGA) [10, 11]. For the electron–ion interactions, we employ the projector augmented wave technique. [12]. The kinetic energy cutoff number is fixed to 80 Ry for the formalization of the expansion of the electronic wave functions in a plane wave basis set. The Monkhorst–Pack (MP) technique carries out the Brillouin zone integration on a mesh of 12 × 12 × 12 k-points. The DOS and TE properties of LiYSi are studied using a tightly packed MP type 40 × 40 × 40 k-point mesh. The transport properties are produced using first-principle eigen values mesh and the BoltzTraP code's implementation of the semiclassical Boltzmann theory [13].

3 Results and Discussion

A substance termed h-Heusler LiYSi has a quasi Mg-Ag-As type cubic crystal structure, which stands out due to its eight valence electrons in group number 216 and space group F-43 m. Four interpenetrating fcc sublattices make up this type of crystal structure, with three of them filled and the fourth empty. In this case, the lithium (Li), yttrium (Y), and silicon elements occupy the Wyckoff positions 4c (0.25, 0.25, 0.25), 4a (0, 0, 0), and 4b (0.5, 0.5, 05.), respectively, while the 4d (0.75, 0.75, 0.75) sites are left empty [14]. At absolute zero pressure, LiYSi has an optimized lattice constant of $a_0 = 6.57$ Å.

LiYSi has a direct band gap in the electronic band structure as shown in Fig. 1a. It is clear that the valence band maximum (VBM) of LiYSi is located at k-points 'X' of the Brillouin zone. Additionally, CBM is located at the point X. The phonon dispersion graph is shown in Fig. 1b Phonon with positive frequency shows that the compound is dynamically stable. Calculated formation energy also declares the stability of compound. Figure 2a shows that the power factor will increase as the density of states with large peaks increases. The contribution of Y-4d orbital is more in CBM, and Si-3p orbital is more in VBM for LiYSi alloy.

Plots of the p-type and n-type LiYSi half-Heusler alloy's electronic thermal characteristics S, $S^2\,\sigma$, σ, and K_e computed are shown in Fig. 3a–d using the BoltzTraP code, respectively. The BoltzTraP code determines the electrical conductivity and electronic thermal conductivity, which are dependent on and are represented as K_e/τ and σ/τ, respectively. Figure 3a shows that variation in Seebeck coefficient rises with increase in temperature conventionally. According to Mott equation which is given as.

$$S = \frac{\Pi^2 k^2 T}{3e}\left\{\frac{\mathrm{d}\ln\sigma(E)}{\mathrm{d}E}\right\}|E = E_F$$

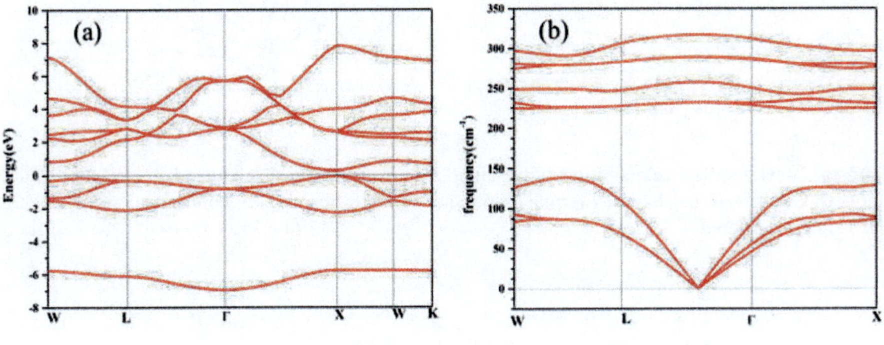

Fig.1 **a** Electronic band structure of LiYSi. **b** Phonon dispersion curve of LiYSi

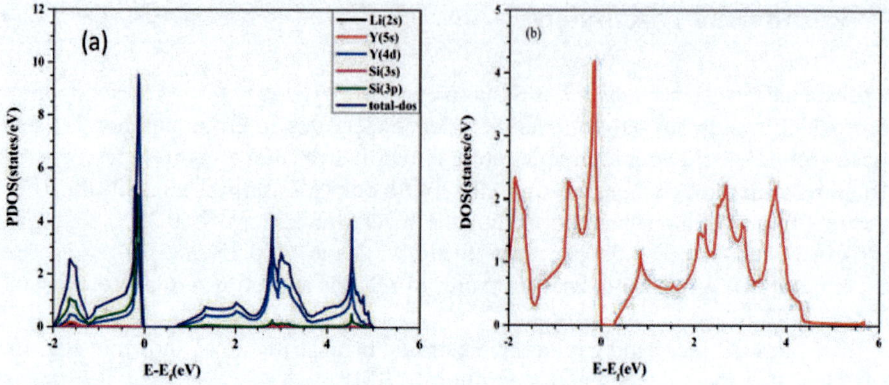

Fig.2 **a** Total density of states of LiYSi. **b** Partial density of states of LiYSi

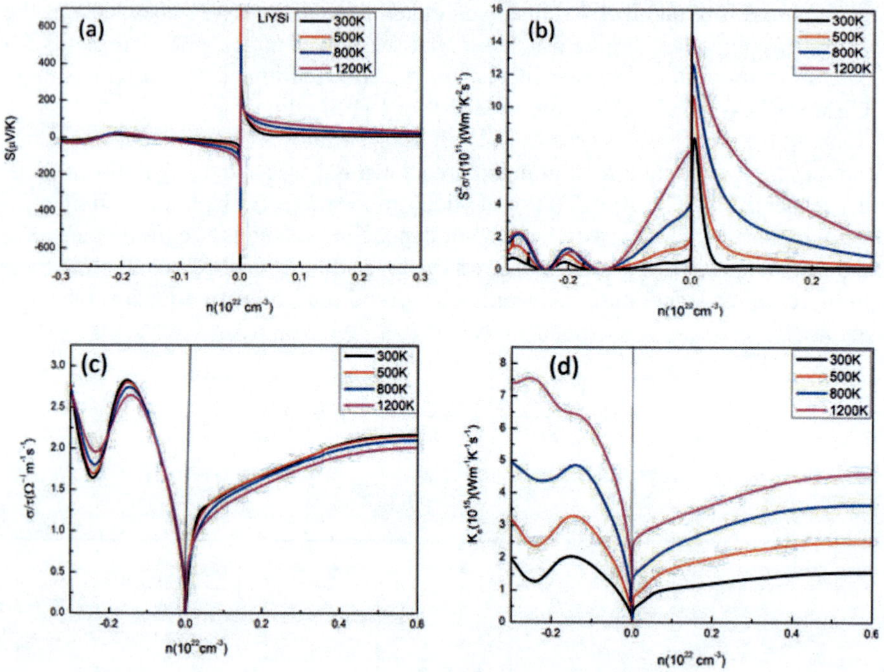

Fig.3 **a** Calculated Seebeck coefficient. **b** Calculated power factor. **c** Calculated electrical conductivity. **d** Calculated electronic thermal conductivity with respect to concentration of LiYSi half-Heusler compound

E_f is the Fermi energy, e is the electronic charge, k is the Boltzmann constant, E is energy, σ is the electrical conductivity, and T is the absolute temperature. It is predicted that high peak in DOS slop around the Fermi level will result in higher n-type Seebeck coefficient values. Figure 3b depicts the behaviour of power factor ($S^2\sigma$), which indicates that power factor rises as temperature rises. This is a result of the S and σ enhancement. Figure 3c depicts the behaviour of electrical conductivity (σ). Figure 3d provides electronic thermal conductivity, which is the shift in electronic input to the thermal conductivity. From the Wiedemann–Franz rule, we know this.

$$K_e = L\sigma T$$

Electrical conductivity changes inversely correlated with k_e. The Lorentz number, L, has a value of 2.44×10^{-8} V^2K^{-2} for semiconductors that are degenerate or highly doped. As a result, an increase in σ causes k_e to decrease or increase. Temperature increases result in a decrease in the σ, measured in units of τ. The reduction in electrical conductivity that occurs as temperature rises is what causes an increase in charge carrier dispersion. As n, e and μ here represent the concentration of charge carriers, the electrical charge, and the mobility of the charge carriers, respectively, according to the formula $n = ne\ \mu$. This results increase charge carrier scattering at high temperatures, μ diminishes. The electrical thermal conductivity (k_e) v is shown in Fig. 3d.

4 Conclusion

In the current research, the transport coefficient of the hH compound LiYSi was calculated using DFT calculations and BoltzTraP transport theory. LiYSi has been found to have a 0.70 eV direct semiconducting band gap. LiYSi Half-Heusler material are dynamically stable. It can be seen by phonon dispersion curves. For n-type LiYSi, the transport coefficient S, $S^2\ \sigma$ higher than p-type. Seebeck is negative for n-type and positive for p-type which increases with rise in temperature. Electronic thermal conductivity decreases with increase in temperature. The calculated power factor of n-type LiYSi is 7.31×10^{11} $Wm^{-1}\ K^{-2}\ s^{-1}$ per relaxation time. The calculated power factor per relaxation time of p-type LiYSi is 1.44×10^{12} $Wm^{-1}\ K^{-2}\ s^{-}$ which will be responsible good thermoelectric material.

Acknowledgements The authors thankful to Department of Physics, Panjab University and University Grants Commission (UGC) for financial support.

Declaration of Interest Statement
There is no conflict of interests.

References

1. AF Ioffe LS Stil'Bans EK Iordanishvili TS Stavitskaya A Gelbtuch G Vineyard 1959 Semiconductor thermoelements and thermoelectric cooling Phys Today 12 5 42
2. C Wood 1988 Materials for thermoelectric energy conversion Rep Prog Phys 51 4 459
3. A Roy 2016 Estimates of the thermal conductivity and the thermoelectric properties of PbTiO 3 from first principles Phys Rev B 93 10 100101
4. Q Zhang Q Song X Wang J Sun Q Zhu K Dahal Z Ren 2018 Deep defect level engineering: a strategy of optimizing the carrier concentration for high thermoelectric performance Energy Environ Sci 11 4 933 940
5. Y Pei X Shi A LaLonde H Wang L Chen GJ Snyder 2011 Convergence of electronic bands for high performance bulk thermoelectrics Nature 473 7345 66 69
6. H Zhu R He J Mao Q Zhu C Li J Sun Z Ren 2018 Discovery of ZrCoBi based half Heuslers with high thermoelectric conversion efficiency Nat Commun 9 1 1 9
7. G Rogl A Grytsiv M Gürth A Tavassoli C Ebner A Wünschek P Rogl 2016 Mechanical properties of half-Heusler alloys Acta Mater 107 178 195
8. H Shi W Ming DS Parker MH Du DJ Singh 2017 Prospective high thermoelectric performance of the heavily p-doped half-Heusler compound CoVSn Phys Rev B 95 19 195207
9. S Li H Zhu J Mao Z Feng X Li C Chen Q Zhang 2019 N-Type TaCoSn-based half-Heuslers as promising thermoelectric materials ACS Appl Mater Interfaces 11 44 41321 41329
10. P Giannozzi S Baroni N Bonini M Calandra R Car C Cavazzoni RM Wentzcovitch 2009 QUANTUM ESPRESSO: a modular and open-source software project for quantum simulations of materials J Phys: Condens Matter 21 39 395502
11. JP Perdew K Burke M Ernzerhof 1996 Generalized gradient approximation made simple Phys Rev Lett 77 18 3865
12. JJ Mortensen LB Hansen KW Jacobsen 2005 Real-space grid implementation of the projector augmented wave method Phys Rev B 71 3 035109
13. Madsen GK, Singh DJ (2006) Program title: BoltzTrap Catalogue identifier: ADXU_v1_0 distribution format: tar. gz. J Reference: Comput Phys Commun 175:67
14. T Gruhn 2010 Comparative ab initio study of half-Heusler compounds for optoelectronic applications Phys Rev B 82 12 125210
15. Mott NF, Jones H, Jones H, Jones H (1958) The theory of the properties of metals and alloys. Courier Dover Publications

Encapsulation of Polyphenols from *Murraya Koenigii* by Using Two Different Polymer Matrices

A. Noor, S. P. Khillar, S. Dasgupta, and R. Basu

Abstract Plant polyphenols are secondary metabolites that have recently gained attention for their several therapeutic properties. Nevertheless, they are unstable and hence prone to oxidation under light, heat, pH and low bioavailability once ingested which limits their applications. Effective encapsulation techniques have been implemented to overcome these drawbacks and mask the unpleasant flavor and odor also. Delivery systems with natural biopolymers have attracted attention for applications in food and pharmaceuticals. In this study, microspheres of *Murraya koenigii* polyphenols were synthesized through the extrusion method using two different polymer matrices, alginate and caseinate. The effect of encapsulation on their antioxidant activity and stability was studied for a six-month period. Further, they were evaluated for their size, stability, yield and surface morphology by using SEM and FTIR. In addition, the effects of different temperatures at different time intervals on the release of polyphenols were also studied. The results showed that the caseinate matrix was more effective in the encapsulation process of polyphenols at 80 °C at 7th h than the alginate matrix, and additionally, the polyphenols were maintained from degradation over a six-month period. In addition, the antioxidant property was retained better after encapsulation. Overall, the findings showed that the caseinate matrix was more effective than the alginate in improving the stability of polyphenols encapsulated from *Murraya koenigii* leaves and could be used as a potential technology for drug delivery techniques in the food industry/ nutraceuticals.

Keywords Microencapsulation · Bioactive constituents · *Murraya koenigii* · Caseinate

---

Supplementary Information The online version contains supplementary material available at https://doi.org/10.1007/978-981-99-4878-9_13.

A. Noor (✉) · S. P. Khillar · S. Dasgupta · R. Basu
Centre for Bio-Separation Technology (CBST), Vellore Institute of Technology (VIT), Vellore, Tamil Nadu 632014, India
e-mail: ayeshanoor17@yahoo.co.in; ayeshanoor@vit.ac.in

Z. H. Khan et al. (eds.), *Recent Advances in Nanomaterials*, Springer Proceedings in Materials 27, https://doi.org/10.1007/978-981-99-4878-9_13

1 Introduction

Murraya koenigii (M koenigii), belonging to the family Rutaceae, is popularly referred as curry leaf [1]. It is a desirable plant for its medical benefits and unique aroma. It contains flavonoids, polyphenols, coumarins, carbazole alkaloids, thiamine, calcium, iron, phosphorous, vitamin C and essential oil [2]. These bioactive constituents are accountable for the numerous therapeutic activities, which include antioxidant, antidiabetic, antidysentric, anti-anemic, anti-inflammatory properties, etc. [3]. Additionally, the leaves of *M koenigii* are rich in phenolic compounds [1]. Polyphenols are extremely potent antioxidants that act as protective barrier against oxidative stress caused by an abundance of reactive oxygen species [4]. These polyphenols may have bitter and caustic taste and are susceptible to degradation once extracted [4]. However, there is a rising interest in protein/polysaccharide biopolymers that facilitate the formation of a broad range of microstructure based on the composition of the solution, nature of the biopolymers, as well as the environmental circumstances [5]. One such biopolymer, milk protein such as casein that is biodegradable, edible, as well as free of adverse effects [6]. Additionally, it has been reported that the interaction of polyphenols with caseins might lessen their astringency [6]. Microencapsulation using sodium alginate has acquired prominence in the pharmaceutical as well as food industries due to its high biocompatibility, low toxicity and inexpensive cost [7]. The extrusion approach is used in the present study to synthesize microspheres from alginate and caseinate that is environmentally beneficial, efficient and inexpensive. The stability, efficacy studies, total polyphenol and antioxidant content, and scavenging properties of the polyphenols encapsulated within both the matrices were assessed for six months.

2 Methods

2.1 Collection, Preparation and Estimation of Total Polyphenolic, Flavonoid and Antioxidant Content of Murraya Koenigii *Extract*

Murraya koenigii leaves (Voucher specimen–VIT/CBST/09/2019/05) were authenticated by Dr. T. Shekar, Pachaiyappa's College, Chennai, Tamil Nadu, India. The extract was prepared. The total polyphenols, flavonoids and antioxidant content of *M koenigii* were estimated [8].

2.2 *Identification of Phenolic Compounds by Using Reverse Phase High-Performance Liquid Chromatography (RP-HPLC)*

HPLC water pumps (binary pump system-1525) and C18 column (waters, 150 3.9 mm, d., 5 m) with gradient system ($\lambda = 290$ nm) were utilized for the analysis [9].

2.3 *Preparation of Caseinate and Alginate Microspheres with* Murraya Koenigii *Leaf Extract, and Scanning Electron Microscopy (SEM) and FTIR Analysis*

The preparation of caseinate and alginate microspheres of *M koenigii* was carried out [10, 11]. In 5% calcium chloride solution, sodium alginate mixture was gradually added on a magnetic stirrer [10]. Sodium caseinate, xanthane, distilled water and glycerol were mixed with extract and without extract as a control to prepare caseinate beads [11]. At magnifications of 86X and 90X, the SEM (Carl Zeiss, EVO/18 Research) images were obtained. FTIR analysis of *M koenigii* was carried with a 0.5 cm^{-1} resolution and a detection range of 400–4000 cm^{-1} [2].

2.4 *Destabilization of Casein and Alginate Encapsulated Microspheres*

The destabilization was performed for both the microspheres of *M koenigii* at various temperatures (40, 50, 60, 70 and 80 °C) and at 1, 4 and 7 h [8].

2.5 *Stability Study and Antioxidant Activities of* Murraya Koenigii *Extract During a Six-Month Period*

Estimation of total polyphenol was performed using FC reagent [10] and antioxidant activity was carried out by phosphomolybdenum method after destabilization [8]. The DPPH and ABTS radical scavenging activity, H_2O_2 scavenging activity, metal chelating activity as well as reducing power activity of both the matrices of *M. koenigii* were studied for over six-month period [9].

2.6 *Statistical Analysis*

Two-way ANOVA (Triplicates; mean ± SD; at $p < 0.05$) was used for the statistical analysis through Graph Pad Prism 6 (Bonferroni method).

3 Results and Discussion

3.1 *Identification of Polyphenols Through RP-HPLC*

The total polyphenol, flavonoid and antioxidant content was 48 ± 0.6 mg/g GAE, 28 ± 0.6 mg/g QE and 33 ± 0.6 mg/g AAE, respectively. Ascorbic acid, pelargonidin chloride, gallic acid, catechin, luteolin, naringin, taxifolin, myricetin, quercetin, naringenin and kaempferol were the polyphenols identified in the *M koenigii* extract based upon the retention time of the standards (Fig. S1 and Table S1).

3.2 *Encapsulation of* M. Koenigii *Extract, Encapsulation Efficiency, SEM and FTIR Analysis*

Higher encapsulation efficiency of caseinate encapsulated microspheres (CEM) (70.41%) than alginate encapsulated microspheres (AEM) (56.38%) may be due to the complex bond formation between the polyphenols and proteins which may lead to the improved bioavailability. The SEM images of both the matrices are shown in Fig. 1. In the CEM, the surface was smoother than the AEM. The FTIR analysis of both the matrices showed alterations in the bond formation (Fig. S2). The alginate and casein polymers adhere to the functional groups and form a bond with the matrix by entrapping the polyphenols within the encapsulated microspheres [9, 11].

3.3 *Effect of Temperature and Time on the Polyphenol Release from AEM and CEM Containing* Murraya Koenigii *Extract*

The highest polyphenol release was observed at 7th h of 50 °C in AEM [see Fig. 2 (i)] compared to CEM of *M. koenigii* [see Fig. 2 (ii)]. The antioxidant and polyphenol content of the unencapsulated extract decreased as the time progressed, but in the encapsulated microspheres, phenolic content was maintained by preventing the oxidation reactions as they are encased in the matrix [see Fig. 3(i) and (ii)].

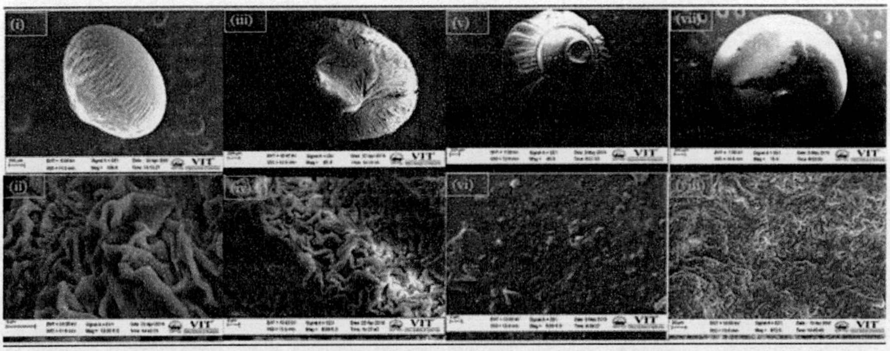

Fig. 1 SEM image of (i), (ii) alginate control microspheres (iii), (iv) AEM (v), (vi) casein control microspheres (vii), (viii) CEM of *M. koenigii* extract

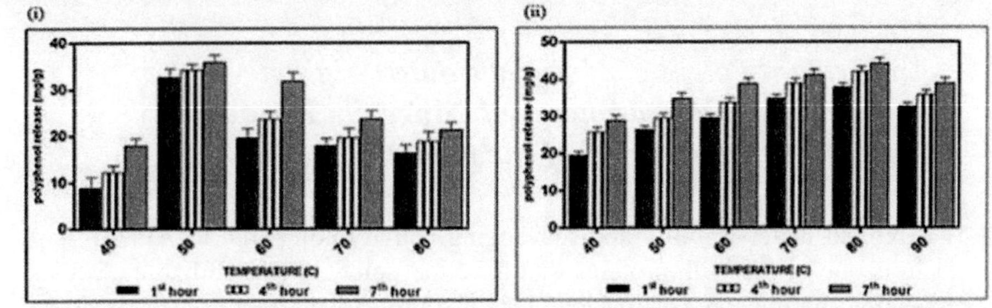

Fig. 2 Effect of temperature and time on the polyphenol release from (i) AEM (ii) CEM

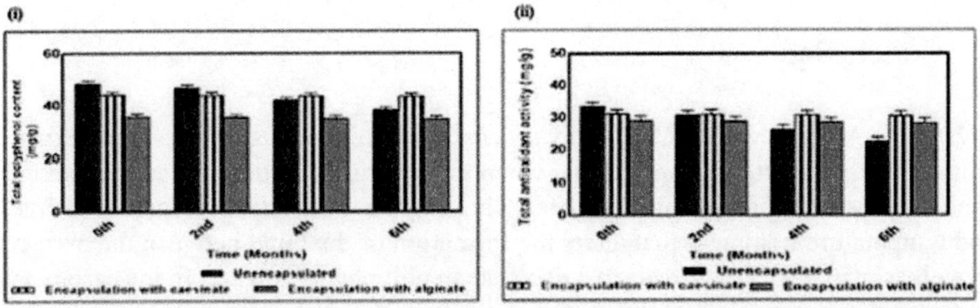

Fig. 3 Comparative stability study of (i) total polyphenol content, (ii) total antioxidant content of AEM, CEM, and unencapsulated microspheres of *M. koenigii* extract for over six-month period. Values are expressed as mean ± SD, $n = 3$ ($p < 0.001$)

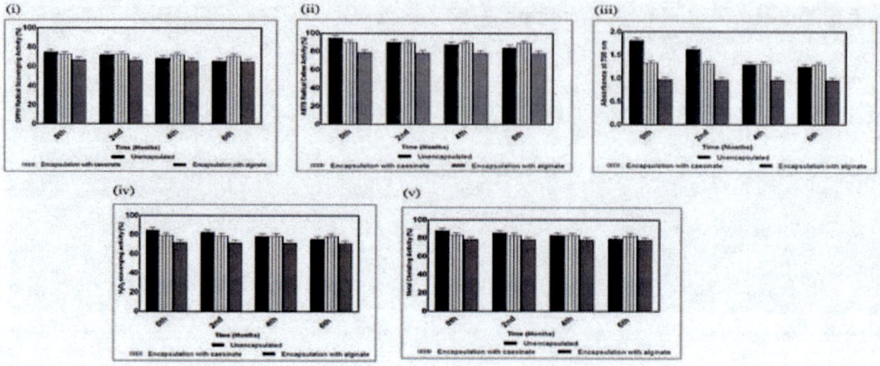

Fig. 4 Antioxidant stability in unencapsulated extract and AEM, CEM for a period of six months by (i) DPPH, (ii) ABTS, (iii) reducing power, (iv) H_2O_2 and (v) metal chelating activity. Values are expressed as mean ± SD of Ascorbic acid equivalents, $n = 3$ ($p < 0.001$)

3.4 Antioxidant Potential of Encapsulated Alginate and Caseinate Microspheres of Murraya Koenigii *Extract During a Period of Six months*

CEM exhibited better stability and scavenging activity compared to AEM Fig. 4 as casein is one of the milk proteins that have the potential to stabilize molecules by preventing the transport of oxygen [11]. The protective effect of AEM may be attributed to the phenolic compounds' amphiphilic characteristics [8, 9].

4 Conclusion

M koenigii AEM as well as CEM were successfully prepared using the simple extrusion method. This study revealed that when temperature and time exceed threshold limit, the release of polyphenol decreases. This might be as result of the increased time and temperature treatment that alters the formation of the bond between the extract as well as matrix and thus leads to a decrease in polyphenol release. In the caseinate encapsulated microspheres, the surface was smoother than the alginate encapsulated microsphere which may be due to high proline concentration of casein, which prevents the synthesis of organized structures like α-helixes or ß-sheets. This makes the structure relatively open as well as rheomorphic, which refers to the tendency of structures to change shape in response to various environmental changes. Additionally, the caseinate-containing hydrogel microspheres were bigger than those without caseinate, indicating that the protein also enhanced the size of the present in the solution binding the polymer matrices together with hydrophobic bonds, imparting the microspheres. Further research is required to determine the polyphenols' and antioxidants' long-term stability. Future research should focus on the bioavailability in animal studies, since temperature and time have a crucial impact on the release of polyphenols.

Acknowledgements The authors acknowledge the management of VIT for providing the facilities and funding "VIT SEED GRANT" and DST-FIST cum VIT funded SEM and FT-IR facility.

References

1. Noolu B, Ajumeera R, Chauhan A, Nagalla B, Manchala R, Ismail A (2013) *Murraya koenigii* leaf extract inhibits proteasome activity and induces cell death in breast cancer cells. BMC Complement Altern Med 13(7):1–17
2. Saini SCGBSR (2015) A review on curry leaves (*Murraya koenigii*): versatile multi-potential medicinal plant. Oral Biol Med 3:363–368
3. Desai SN, Patel DK, Devkar RV, Patel PV, Ramachandran AV (2012) Hepatoprotective potential of polyphenol rich extract of *Murraya koenigii L.*: An in vivo study. Food Chem Toxicol 50(2):310–314
4. Sandoval-Acuña C, Ferreira J, Speisky H (2014) Polyphenols and mitochondria: an update on their increasingly emerging ROS-scavenging independent actions. Arch Biochem Biophys 559:75–90
5. Rehman A, Ahmad T, Aadil RM, Spotti MJ, Bakry AM, Khan IM, Zhao L, Riaz T, Tong Q (2019) Pectin polymers as wall materials for the nano-encapsulation of bioactive compounds. Trends Food Sci Technol 90(March):35–46
6. Sadiq U, Gill H, Chandrapala J (2021) Casein micelles as an emerging delivery system for bioactive food components. Foods 10(8):1965
7. Jana S, Kumar Sen K, Gandhi A (2016) Alginate based nanocarriers for drug delivery applications. Curr Pharm Des 22(22):3399–3410
8. Dasgupta S, Lal M, Govindarajan S, Babu SN, Noor A (2022) Effect of microencapsulation by calcium alginate on the anti-oxidant properties of *Swietenia Macrophylla* polyphenols. Curr Trends Biotechnol Pharm 16(1):123–132
9. Noor A, Al Murad M, Jaya Chitra A, Babu SN, Govindarajan S (2022) Alginate based encapsulation of polyphenols of Piper betelleaves: development, stability, bio-accessibility and biological activities. Food Biosci 47(3):101715
10. Basu R, Lal M, Govindarajan S, Babu SN, Noor A (2022) Preservation of antioxidant activity and polyphenols in *Mentha spicata L.* with the Use of Microencapsulation by Calcium Alginate. Curr Trends Biotechnol Pharmacy 16(1):28–37
11. Chandrasekhar Reddy B, Noor A, Sarada NC, Vijayalakshmi MA (2011) Antioxidant properties of *Cordyline terminalis (L.) Kunth* and *Myristica fragrans Houtt.* encapsulated separately into casein beads. Curr Sci 101(3):416–420

Study of Lattice Dynamics of the Graphene Along Highly Symmetry Directions

Mohammad Imran Aziz and Quddus Khan

Abstract Graphene is a two-dimensional material that has successfully been separated into single- or few-layer sheets from bulk graphite. The graphene structure has interesting features which is the good reason for studying its lattice vibrational properties. Graphene is one of the important areas of research due to its potential for integration into future-generation electronic devices. Graphene exhibits soft nature and hence could be easily integrated with current technology in electronic devices on substrates in comparison to stanene, silicene and germanene. We focus on the lattice dynamical of graphene and try to understand them from their honeycomb and buckled lattice structures. Lattice dynamical properties of single-layer two-dimensional honeycomb lattices exhibit interesting features. We, at present, find the phonon frequencies at Γ points along symmetry directions with the help of Python program. The acoustical and optical contributions to the phonon frequencies are also discussed. We hope that phonon frequencies along Γ–M of graphene, 2D materials will have reasonably similar result obtained by other researchers.

Keywords Hamiltonian mechanics · Dynamical matrix · Lattice dynamics of graphene

1 Introduction

Graphene is a single layer of carbon atoms arranged in the manner of hexagonal which has become practically available today [1–3] (Fig. 1).

The atoms in the 2D mono-layer graphene are capable of executing oscillations about their equilibrium position (n, l). In oscillating states, the instantaneous position of atoms (n, l)) is denoted by: $r(n, l) = x(n, l) + u(n, l)$.

M. I. Aziz (✉)
Physics Department, Shibli National College, Azamgarh, India
e-mail: azizimran33@gmail.com

Q. Khan
Department of Applied Sciences and Humanities, Faculty of Engineering and Technology, JMI, New Delhi, India

Z. H. Khan et al. (eds.), *Recent Advances in Nanomaterials*, Springer Proceedings in Materials 27, https://doi.org/10.1007/978-981-99-4878-9_14

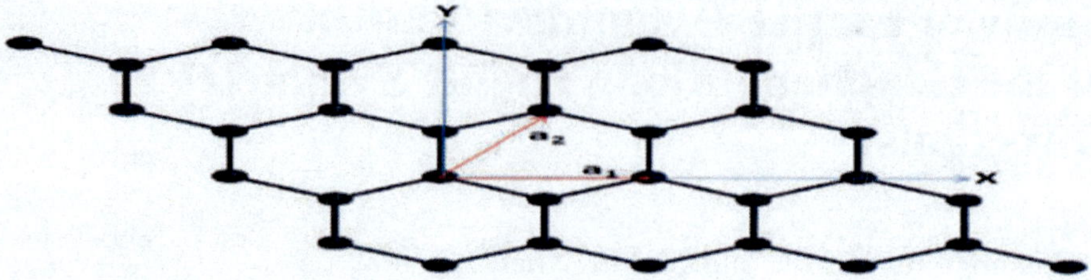

Fig. 1 Structure of graphene sheet

Thus, the Hamiltonian of the graphene is

$$H = \sum_{nli} \frac{M_n}{2} \dot{u}_i^2(n, l) + \frac{1}{2} \sum_{nli} \Phi_{ij}\begin{pmatrix} mn \\ l, l' \end{pmatrix} \sum_{ml'j} u_i(n, l) u_j(m, l') \tag{1}$$

Where

$$\Phi_{ij}\begin{pmatrix} mn \\ l, l' \end{pmatrix} = \left[\frac{\partial^2 U}{\partial u_i(n, l) \partial u_j(m, l')} \right]_0 \tag{2}$$

The force constants are defined as:

$\Phi_{ij}\begin{pmatrix} mn \\ l, l' \end{pmatrix} = -\gamma e_i e_j$, where e_i & e_j are unit vectors; it is the resultant force acting on the nth atom due to a unit displacement of the mth atom.

The equation of motion as the result of above force is given by:

$$M_n \ddot{u}_i(n, l) = -\sum_{ml'j} \Phi_{ij}\begin{pmatrix} mn \\ l, l' \end{pmatrix} u_j(m, l') \tag{3}$$

The solution of above equation is modified by the periodicity of lattice thus a wave-like solution of type

$$u_i(n, l) = M_n^{-\frac{1}{2}} u_{in} \exp[i\{q.r(n, l) - \omega(q)t\}] \tag{4}$$

where u_{in} is the amplitude of oscillation under harmonic approximation along ith direction of the nth atom, ω is the angular frequency and q is wave vector [4].

The equation of motion in matrix notation is

$$\omega^2(q) MU(q) = D(q)U(q) \tag{5}$$

The condition for non-trivial solution:

$$\left| D(q) - \omega^2(q) MI \right| = 0 \tag{6}$$

The elements of dynamical matrix are defined as:

$$D_{ij} = \sum_{l'} \Phi_{ij}\begin{pmatrix} mn \\ l, l' \end{pmatrix} \exp(iq.r(mn, ll') \tag{7}$$

The above equation in matrix form is solved by MATLAB program. And the result is investigated along hexagonal Brillouin zone with symmetry points Γ (0,0), M ($\frac{2\pi}{a\sqrt{3}}$, 0) [5, 6]. The vibrational frequencies along symmetry line with coupling constant $\gamma = \gamma_j$ for graphene are deduced as:

Γ−M

$$\omega_1^2 = \gamma_j\left[1 - \cos\left(\frac{\sqrt{3}}{2} q_y a\right)\right] \tag{8}$$

$$\omega_2^2 = 3\gamma_j\left[1 - \cos\left(\frac{\sqrt{3}}{2} q_y a\right)\right] \tag{9}$$

$$\omega_3^2 = \left[\frac{3}{2}\gamma_j - \left(\frac{5}{4}\right)\gamma_j - \gamma_j \cos\left(\frac{\sqrt{3}}{2} q_x a\right)\right] \tag{10}$$

$$\omega_4^2 = \left[\frac{3}{2}\gamma_j + \left(\frac{5}{4}\right)\gamma_j + \gamma_j \cos\left(\frac{\sqrt{3}}{2} q_x a\right)\right] \tag{11}$$

$$\omega_5^2 = 0 \tag{12}$$

$$\omega_6^2 = 3\gamma_j \tag{13}$$

2 Phonon Dispersion Curve

Using the phonon dispersion at $a = 2.47$ Å, we computed the different mode as shown in Table 1 with help of Python Program.

The lattice dynamical relations have been evaluated by solving the secular equation for the six vibrational frequencies corresponding to the phonon wave vectors q along the high symmetry direction **Γ−M**. The phonon dispersion curve has been attained by plotting the vibrational frequencies (ω) against the wave vector (q) with the help of Python program, and the following points are investigated from the thorough analysis of phonon dispersion curves of single-layer graphene along high symmetry direction **Γ−M**. The dispersion of the longitudinal phonon shows oscillatory behavior in higher q region. In contrast, the $\omega - q$ curves for transverse phonons

Table 1 Calculated phonon frequencies (THz) for grapheme

Wave vector (q)	Transverse-acoustic mode TA(THz)	Longitudinal -acoustic mode LA(THz)	Z-direction-acoustic mode ZA(THz)	Longitudinal-optical mode LO(THz)	Z-direction-optical mode ZO(THz)	Transverse -optical mode TO(THz)
0.0	0.0	0.0	3.9642	5.8030	0	5.1903
0.1	0.4022	0.6967	3.9845	5.7890	0	5.1903
0.2	0.8009	1.3872	4.0443	5.7474	0	5.1903
0.3	1.1923	2.0652	4.1396	5.6791	0	5.1903
0.4	1.5730	2.7245	4.2649	5.5857	0	5.1903
0.5	1.9394	3.3592	4.4132	5.4693	0	5.1903
0.6	2.2884	3.9636	4.5773	5.3327	0	5.1903
0.7	2.6166	4.5322	4.7499	5.1795	0	5.1903
0.8	2.9213	5.0598	4.9243	5.0140	0	5.1903
0.9	3.1995	5.5418	5.0943	4.8412	0	5.1903

the oscillatory behavior seems less dominating for higher q- value. This behavior suggests that the TA and TO modes phonons undergo large-scale thermal motion than do for LA and LO modes phonons. The $\omega - q$ curves for longitudinal phonons attain maxima at a higher q-value. Figure 2 shows the calculated phonon dispersion curve of graphene which is in agreement with previous work [7, 8]. Similar to other two-dimensional materials like stanene, silicene and germanene, the longitudinal-acoustic and transverse-acoustic branches of group-IV materials are straight line near at the Γ point. For wave vectors along the $\Gamma(0, 0)$, $M\,(\frac{2\pi}{a\sqrt{3}}, 0)$ symmetry directions, the LA and TA modes are both degenerate. There is apparent near-crossing in mid-points, called anti-crossing of the LO and TO modes along the $M\,[\frac{2\pi}{a\sqrt{3}}, 0]$. This phenomenon is dominant in graphene. In high symmetry situations, it is possible to separate LO modes from TO modes. The main characteristic of the dispersion curve is to notice that there is a segregation of optic and acoustic mode frequencies across the wide range of phonon wave vectors. This characteristic is arising because of association of optical vibrations with electric moments. The transverse modes indicate a separation of the optic and acoustic modes but there is a crossing of LA and TA modes at 4.5322 THz and 5.0598 THz. Optical vibrations dominate in the single -layer graphene because of the strong electric moments. Phonon dispersion curve with wave vector $M\,[\frac{2\pi}{a\sqrt{3}}, 0]$ is showing longitudinal-optical modes moving in opposite directions parallel to $M\,[\frac{2\pi}{a\sqrt{3}}, 0]$, and the transverse-optical modes moving in opposite directions perpendicular to $M\,[\frac{2\pi}{a\sqrt{3}}, 0]$. At $\Gamma\,[00]$, both types of modes become exactly same; in this case, longitudinal-optical and transverse-optical modes frequencies would be equal at 5.1795 THz. But as moving from [00] to $[\frac{2\pi}{a\sqrt{3}}, 0]$, the long-wavelength LO and TO modes generate electric fields which have prominent effect on the frequency of the mode. The phonon dispersion curve for single-layer graphene has been plotted and utilized to study the phonon group velocities in these structures [8]. We derived the solution for the Γ and M points and numerically investigated the buckling effect on the material for phonon properties.

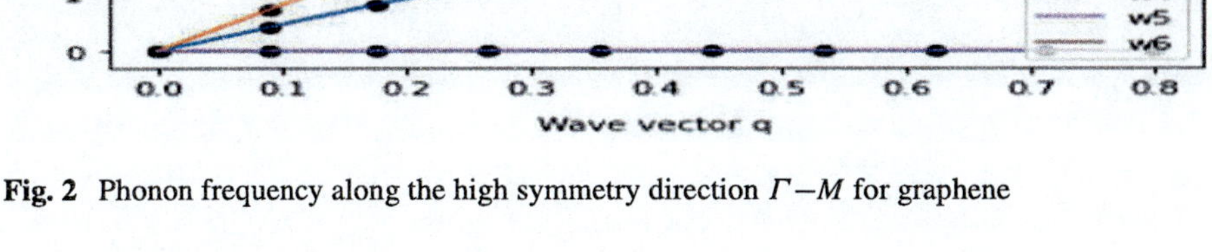

Fig. 2 Phonon frequency along the high symmetry direction Γ–M for graphene

3 Results and Discussions

We have studied the relation between lattice structure and lattice dynamical properties. Using the quasi-harmonic approximation (QHA), we obtained the vibrational frequencies of single-layer graphene, 2D materials. For the lattice dynamical properties, we need the long-range interactions, which are important for the long-wavelength, low-frequency phonons near Γ. The vibrational frequencies are computed using Python–QHA script. The computed phonon dispersion curves along high symmetry directions within the first Brillouin zone are shown in Fig. 2. The dispersion lines are similar due to the honeycombed lattice structures. The ZA and ZO phonon modes do not couple with other phonon modes, resulting in inter-system crossings of dispersion lines in single-layer graphene along high symmetry direction. This is happening because of larger buckling. This results in decrement of phonon group velocity. Further, they reduced the phonon thermal conductivity. Thus, the dispersion of phonons is helpful in understanding the heat transfer in 2D materials. Interestingly, the large buckling in single-layer graphene results in a larger Γ point ZO frequency, and the ZA mode is very low near Γ points. This denotes that the applied strain should not be very large; otherwise, harmonic approximation will not be valid. In this paper, we have systematically reported phonon dispersion curves of graphene. The agreement between theory and experimental data at Γ point is very good. Another salient feature of the present study is noticeable that best result of almost all phonon frequency branches. Here in this paper, lattice vibrational properties of graphene are compared with other researchers, Das et al. [9]. The theoretical predictions achieved for the vibrational frequencies of graphene are in reasonably good agreement with other researchers.

APPENDIX A

PYTHON Program along the high symmetry direction Γ-M for Graphene

```
import matplotlib.pyplot as plt
import numpy as np
import math

points  =  np.array([0, 0.1, 0.2, 0.3, 0.4, 0.5, 0.6, 0.7, 0.8])
points=  np.linspace(0.0,0.8,10)

#vj  =  gamma
a = 2.47
vj  =8.98

def solve(fun):
  op = []
  for i in points:
    op.append(fun(i))

  return op

def ws1f(qy):
  return (math.sqrt(vj*(1-math.cos((math.sqrt(3)*qy*a)/2))))

def ws2f(qy):
 return (math.sqrt(3*vj*(1-math.cos((math.sqrt(3)*qy*a)/2))))

def ws3f(qx):
  return (math.sqrt(((3/2)*vj)  +  ((5/4)*vj) - (vj*math.cos((math.sqrt(3)*qx*a/2)))))

def ws4f(qx):
  return (math.sqrt(((3*vj)/2) + ((5/4)*vj)  +  vj*math.cos((math.sqrt(3)*qx*a/2))))

def ws5f(qx):
  return 0

def ws6f(qx):
  return (math.sqrt(3*vj))

ws1  =  solve(ws1f)
ws2  =  solve(ws2f)
ws3  =  solve(ws3f)
ws4  =  solve(ws4f)
ws5  =  solve(ws5f)
ws6  =  solve(ws6f)
plt.plot(points,  ws1,  color='k',  linewidth1,=  marker  ±o')
plt.plot(points,  ws2,  color='k',  linewidth1,=  marker  ±o')
plt.plot(points,  ws3,  color='k',  linewidth1,=  marker  ±o')
plt.plot(points,  ws4,  color='k',  linewidth1,=  marker  ±o')
plt.plot(points,  ws5,  color='k',  linewidth1,=  marker  ±o')
plt.plot(points,  ws6,  color='k',  linewidth1,=  marker  ±o')

plt.plot(points,  ws1,  label='w1')
plt.xlabel("Wave  vector  q")
plt.ylabel("Frequency  (THz)")
plt.title("Phonon  Dispersion  Curve")
print  (ws1)
#  Show  the  plot

#plt.show()
```

```
plt.plot(points, ws2, label='w2')
plt.xlabel("Wave vector q")
plt.ylabel("Frequency (THz)")
plt.title("Phonon Dispersion Curve")
print (ws2)
# Show the plot

#plt.show()

plt.plot(points, ws3, label='w3')
plt.xlabel("Wave vector q")
plt.ylabel("Frequency (THz)")
plt.title("Phonon Dispersion Curve")
print (ws3)
# Show the plot

#plt.show()

plt.plot(points, ws4, label='w4')
plt.xlabel("Wave vector q")
plt.ylabel("Frequency (THz)")
plt.title("Phonon Dispersion Curve")
print (ws4)
# Show the plot

#plt.show()

plt.plot(points, ws5, label='w5')
plt.xlabel("Wave vector q")
plt.ylabel("Frequency (THz)")
plt.title("Phonon Dispersion Curve")
print (ws5)
# Show the plot

#plt.show()

plt.plot(points, ws6, label='w6')
plt.xlabel("Wave vector q")
plt.ylabel("Frequency (THz)")
plt.title("Phonon Dispersion Curve")
print (ws6)
# Show the plot
plt.legend()
plt.show()
```

References

1. Evans JW, Thiel PA, Bartelt MC (2006) Morphological evolution during epitaxial thin film growth: formation of 2d islands and 3d mounds. Surf Sci Rep 61(1):1–128
2. Novoselov KS, Geim AK, Morozov SV, Jiang D, Zhang Y, Dubonos SV, Grigorieva IV, Firsov AA (2004) Electric field effect in atomically thin carbon films. Science 306(5696):666–669
3. Geim AK, Novoselov KS (2007) The rise of graphene. Nat Mater 6(5):183–191
4. Woods LM, Mahan GD (2000) Phys Rev B 61:10651
5. Ge X-J, Yao K-L, Lü J-T (2016) Phys Rev B 94:165433

6. Kumar K, Aziz MI, Anas KA (2022) Am J Nanosci 8(2):13–18
7. Mann S, Kumar R, Jindal VK (2017) RSC Adv 7:22378
8. Aziz MI, Ahmad N (2023) Am J Nanosci 8(4):43–47
9. Dasa GP, Raghuvanshi PR, Bhattacharya A (2019) In: 9th International conference on materials structure and micromechanics of fracture phonons and lattice thermal conductivities of graphene family, vol 23, pp 334–341

The Effect of Sintering Temperature on the Photocatalytic Activity of Nickel Ferrite on Methylene Blue

Anju Ganesh, Richu Rajan, and Smitha Thankachan

Abstract Nickel ferrite is synthesized by sol–gel autocombustion method with ethylene glycol as fuel. The thermal stability of the sample is studied using thermogravimetric analysis. The prepared ferrite sample is sintered at 500, 600 and 700 °C. The structure of the ferrite powder hence obtained is characterized by XRD and TEM. The grain size is found to increase with sintering temperature as expected. The photocatalytic degradation of methylene blue is studied using all the samples under sunlight and the degradation was studied using UV–Visible spectrophotometry.

Keywords Nickel ferrite · Methylene blue · Photocatalytic · Sintering temperature

1 Introduction

One of the major issues encountered by mankind in this new era is pollution of water resources by effluents from industries such as drug, textile, fertilizers, cosmetics petroleum products. Water contamination has resulted in imbalance of ecological equilibrium. Many organic pollutants in water resources have adversely affected reproductive, developmental, behavioural, neurologic, endocrinal and immunologic human health systems. One of the major textile dye pollutants is methylene blue.

Many research groups have reported biological, thermal and chemical degradation methods to treat these pollutants. Development of a novel and efficient waste water management system is very significant in this present scenario. Organic pollutants like dyes get detoxified using advanced oxidation processes which is simple and effective method due to absence of secondary pollution and reusability of catalyst [1, 2]. Ferrite nanoparticles with chemical formula MFe2O4 may be used as a catalyst for the removal of pollutant via adsorption or photodegradation. They possess good adsorption capacity for both anions and cations which make them to behave as a perfect photocatalytic material.

A. Ganesh · R. Rajan · S. Thankachan (✉)
Research Department of Physics, Mar Athanasius College (Autonomous), Kothamangalam, India
e-mail: smithathankachan@macollege.in

Z. H. Khan et al. (eds.), *Recent Advances in Nanomaterials*, Springer Proceedings in Materials 27, https://doi.org/10.1007/978-981-99-4878-9_15

Nickel ferrite, with its chemical formula NiFe2O4, is a mixed metal oxide. Its traits such as typical ferromagnetic properties, high electrochemical stability, catalytic behaviour, abundance in nature, low conductivity and thus lower eddy current losses make nickel ferrite one of the versatile and technologically important soft ferrite materials. Various wet chemical methods to fabricate nickel ferrite nanoparticles reported till now are co-precipitation, sol–gel autocombustion, microwave synthesis, solvothermal method, thermal decomposition, etc. Nickel ferrite prepared by sol–gel autocombustion method has already been reported as a photocatalyst [3]. Here, the effect of sintering temperature on the photocatalytic activity of nickel ferrite nanoparticles synthesized using sol–gel autocombustion method on degrading methylene blue under sunlight is being studied.

2 Materials and Methods

98% pure ferric nitrate nonahydrate, nickel nitrate hexahydrate and ethylene glycol were used for synthesizing nickel ferrite nanoparticles, and methylene blue alkaline solution was used for photodegradation studies.

2.1 Preparation of Nickel Ferrite Nanoparticles

Nickel ferrite nanoparticles were synthesized by sol–gel autocombustion method as seen elsewhere with ethylene glycol as fuel [4, 5]. For this, ferric nitrate nonahydrate and nickel nitrate hexahydrate in the ratio 2:1 were dissolved in minimum amount of ethylene glycol by stirring magnetically at room temperature. The solution obtained so was heated to a temperature of 60 °C to obtain a gel. The obtained gel was heated to 120 °C which self-ignites to form a fluffy product, which was then crushed to powder [6, 7]. The as-prepared sample was then sintered at three different temperatures—500 °C, 600 °C and 700 °C for three hours to obtain the samples S1, S2 and S3, respectively.

3 PhotocatalyticStudies

Methylene blue solution was made by dissolving 50 μL of alkaline methylene blue in 300 mL of distilled water, and sample was drawn and labelled as a reference. The photodegradation of methylene blue was studied by keeping the methylene blue solution containing each of the three samples S1, S2 and S3 under sunlight. Solution was drawn at two hours, four hours and six hours, and those drawn from solution containing S1 were labelled S1-1, S1-2 and S1-3, respectively, those drawn from solution containing S2 were labelled S2-1, S2-2 and S2-3, respectively, and those drawn from solution containing S3 were labelled S3-1, S3-2 and S3-3, respectively. All the samples including reference were characterised by UV–Visible spectrophotometry.

4 Results and Discussion

Thermal properties of as-prepared sample were studied by TGA along with DSC. The structural characterization of the three samples—S1(500 °C), S2(600 °C) and S3 (700 °C)—was done by XRD, and the morphology was studied by taking TEM of S1 (500 °C). The photocatalytic activity of nickel ferrite on MB was studied using UV–Visible spectrophotometer.

4.1 Thermal Studies

The thermogravimetric analysis (TGA) measures change in weight in relation to change in temperature, whereas differential scanning calorimetry measures the heat flow in relation to temperature. Simultaneous measurement of these two material properties improves productivity but also simplifies interpretation of result. The TGA–DSC of as-prepared nickel ferrite shown in Fig. 1a indicates formation of thermally stable nickel ferrite at about 400 °C. The sudden decrease in TGA curve indicates the evaporation of by products such as oxides of nitrogen, water or carbon dioxide. A small bump starting from 400 °C extending to 800 °C may be attributed to the formation of iron oxide which is supported by XRD data.

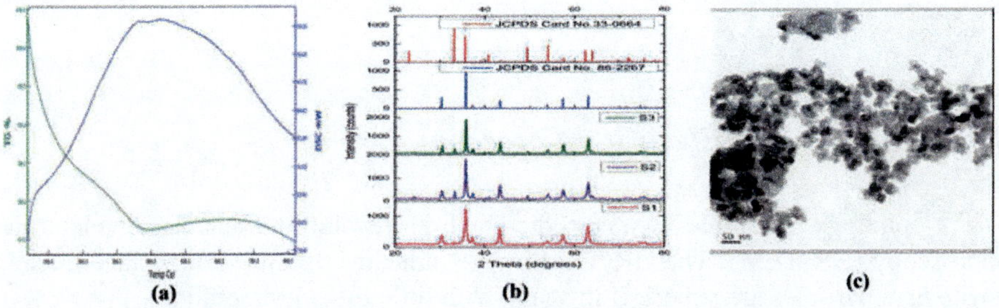

Fig. 1 **a** DTA- DSC of as-prepared $NiFe_2O_4$; **b** XRD of samples S1, S2 and S3; **c** TEM image of S1

4.2 X-Ray Diffraction Studies

Powder X-ray diffraction was carried out for all the three sintered samples, and the obtained data was compared with JCPDS Cards for identification as well as to know the phase composition. The XRD data of each samples along with JCPDS data is shown in figure.

The XRD of the sample sintered at 500 °C, (S1) exhibits peaks at 2θ values—18.570, 30.360, 35.740, 37.354, 43.439, 53.919, 57.460, 63.083, 71.5418, 74.820 representing the planes (111), (220), (311), (222), (400), (422), (511), (440), (620) and (533), respectively. This is in agreement with JCPDS card no 86-2267 confirming the presence of nickel ferrite. But a single peak at 2θ of 33.249 corresponding to the plane (104) showing the presence of alpha Fe_2O_3 in agreement with JCPDS card no 33-0664. The XRD data of samples sintered at 600 °C (S2) with reference to the same JCPDS card of nickel ferrite reveals the existence of planes (111), (220), (311), (222), (400), (422), (511), (440), (620), (533) and (622) at 2 theta values 18.487, 30.361, 35.738, 37.377, 43.440, 73.864, 57.496, 63.066, 71.523, 74.646 and 75.572, respectively. But additional 7 planes corresponding to that of α-Fe_2O_3 viz (012), (104), (113), (024), (116), (300) and (1 0 10) at 2 theta values 24.214, 33.220, 40.921, 49.528, 49.528, 54.117, 64.052 and 72.006, respectively is also present in sample S2. For the sample S3 also peaks indicate the presence of the nickel ferrite and α-Fe_2O_3.

The number of peaks corresponding to alpha increases with sintering temperature indicating the conversion of ferrite to ferric oxide. The crystalline size was calculated for the peak of maximum intensity in each of the three cases using the Debye Scherrer equation. The D value was obtained as 18.3017 nm, 24.773 nm and 31.6273 nm for the samples S1(500 °C), S2(600 °C), S3(700 °C), respectively. The crystallite size increases with increase in sintering temperature in agreement with the reported literatures [8–10].

4.3 Transmission Electron Microscopy

The morphology of sample S1 was studied by high-resolution transmission electron microscopy (HR-TEM). The HR-TEM image indicates that the synthesized nickel ferrite nanoparticles are spherical in shape with little bit agglomeration.

4.4 UV–Visible Spectrophotometry

The UV–Visible spectra of Reference indicate presence of two absorbance peaks around 286 and 640 nm. The intensity of peak decreases with passage of time for all of the three samples—S1, S2 and S3. This indicates that as irradiation time increases,

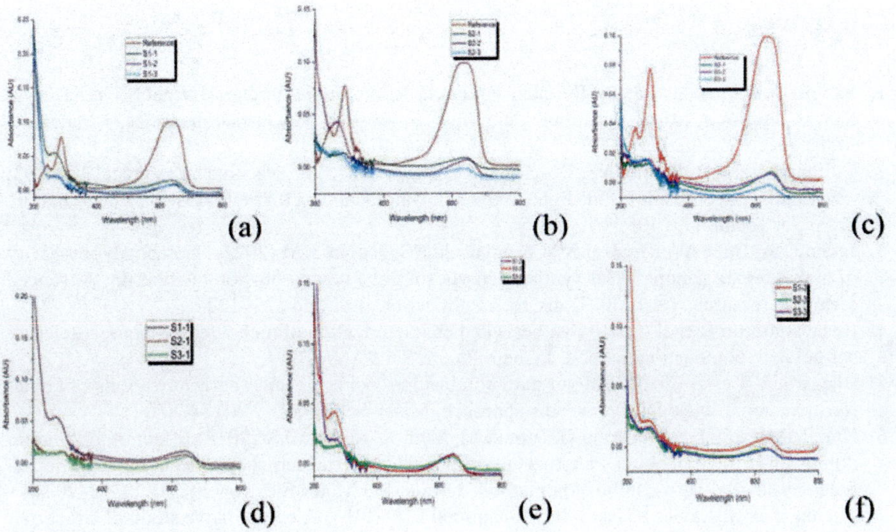

Fig. 2 UV–Visible spectra of samples. **a–c** shows the effect of time on S1, S2 and S3, respectively; and **d–f** shows the effect of temperature with samples drawn at 2 h, 4 h and 6 h, respectively

the concentration of MB decreases indicating that nickel ferrite nanoparticles are good photocatalyst as reported by literature [6, 11, 12]. This is clear from Fig. 2a–c. Also, the concentration of methylene blue on the solution drawn at same time of irradiation decreases from S1 to S3, i.e. the concentration of methylene blue decreases with increase in sintering temperature for the same irradiation time which is proved by Fig. 2d–f. The effect of temperature is less remarkable which may be attributed to the fact that the difference between the temperatures is small.

5 Conclusions

The TGA–DSC of as-prepared nickel ferrite indicates formation of thermally stable nickel ferrite at about 400 °C. A small bump starting from 400 °C extending to 800 °C may be due to the formation of alpha ferric oxide which is supported by XRD data. The formation of nickel ferrite in all samples is confirmed by XRD pattern with reference to the JCPDS Card No. 86-2267. As sintering temperature increases, crystallite size increases, and also, more amount of impurity, α-Fe_2O_3 is formed (JCPDS Card No. 33-0664). HR-TEM image indicates the formation of spherical nickel ferrite nanoparticles. From UV–Visible spectra as irradiation time increases, the concentration of MB decreases indicating that nickel ferrite nanoparticles are good photocatalyst. The concentration of MB for same irradiation time decreases with increase in sintering temperature. The effect is less significant which may be due to the fact that the temperature difference is small.

References

1. Santos-Beltran M, Paraguay-Delgado F, Garcia R, Antunez-Flores W, Ornelas-Gutierrez C, Santos-Beltran A (2017) Fast methylene blue removal by MoO3 nanoparticles. J Mater Sci Mater Electron 28:2935–2948
2. Chithambararaj A, Sanjini NS, Bose AC, Velmathi S (2013) Flower-like hierarchical h-MoO3: new findings of efficient visible light driven nano photocatalyst for methylene blue degradation. Catal Sci Technol 3:1405
3. Jadhav SA, Raut AV, Khedkar MV, Somvanshi SB, Jadhav KM (2022) Photocatalytic activity of nickel ferrite nanoparticles synthesized via sol-gel auto combustion method. In: Advanced materials research, vol 1169. Trans Tech Publications Ltd., pp 123–127
4. Kesavamoorthi R et al (2016) Synthesis and characterization of nickel ferrite nanoparticles by sol-gel auto combustion method. J Chem Pharm Sci 9:160–162
5. Bhagwat VR et al (2019) Sol-gel auto combustion synthesis and characterizations of cobalt ferrite nanoparticles: Different fuels approach. Mater Sci Eng: B 248:114388
6. Naik MM, Naik HS, Nagaraju G, Vinuth M, Vinu K, Rashmi SK (2018) Effect of aluminium doping on structural, optical, photocatalytic and antibacterial activity on nickel ferrite nanoparticles by sol–gel auto-combustion method. J Mater Sci Mater Electron 29(23):20395–20414
7. Thankachan S, Xavier S, Jacob B, Mohammed EM (2013) A comparative study of structural, electrical and magnetic properties of magnesium ferrite nanoparticles synthesised by sol-gel and co-precipitation techniques. J Exp Nanosci 8(3):347–357
8. Joshi S, Kumar et al, Structural, magnetic, dielectric and optical properties of nickel ferrite nanoparticles synthesized by co-precipitation method. J Mol Struct 1076:55–62
9. Vepulanont K et al (2021) Nickel ferrite ceramics: combustion synthesis, sintering, characterization, and magnetic and electrical properties. J Asian Ceram Soc 9(2):639–651
10. Shirsath SE, Kadam RH, Gaikwad AS, Ghasemi A, Morisako A (2011) Effect of sintering temperature and the particle size on the structural and magnetic properties of nanocrystalline Li0. 5Fe2. 5O4. J Magn Magn Mater 323(23):3104–3108
11. Casbeer E, Sharma VK, Li XZ (2012) Synthesis and photocatalytic activity of ferrites under visible light: a review. Sep Purif Technol 87:1–14
12. Lassoued A, Lassoued MS, Dkhil B, Ammar S, Gadri A (2019) Substituted effect of Al3+ on structural, optical, magnetic and photocatalytic activity of Ni ferrites. J Magn Magn Mater 476:124–133

Manipulating Superconductivity in Superconductor/Ferromagnet Hybrid Nanostructures

Asif Majeed, Junaid Ul Ahsan, and Harkirat Singh

Abstract In this paper, we have simulated a hybrid structure consisting of a conventional BCS superconductor placed in proximity with a ferromagnet possessing intrinsic spin–orbit coupling (due to crystal mismatch at the interface) to study the effect of spin–orbit interaction on density of states near Fermi energy and the critical temperature. We report that the relationship between the in-plane and out-of-plane magnetic exchange fields (inside the ferromagnet) and the superconducting critical temperature reveals that a single homogenous ferromagnet possessing intrinsic spin–orbit coupling regulates the existence of triplets. We also report that inclusion of spin–orbit coupling in the hybrid structure manifests itself in the form of an energy gap in the density of states in the vicinity of Fermi energy which is attributed to the spin 1 triplets. Our research shows that spin–orbit coupling actively controls the triplets, which is a crucial step in new superconducting spintronic devices being made.

Keywords Usadel equation · Diffusive limit · Critical temperature · Rashba–Dresselhaus spin–orbit coupling · Short-range triplet (SRT) · Long-range triplet (LRT)

1 Introduction

The homogenous ferromagnetic materials associated with property of intrinsic spin–orbit coupling can substitute as a source for LRT generation [1]. Such a scenario where in SF structures the F-layer exhibiting intrinsic SO coupling serves as the LRT source has been theoretically assumed to exist in the presence of a heavy metal being added to the system. However, for this to be possible, it is necessary for the ferromagnetic exchange field to possess one component in the plane of hybrid nanostructure and one component out of the plane [2]. Accessing such a situation

A. Majeed · J. U. Ahsan · H. Singh (✉)
Department of Physics, National Institute of Technology, Srinagar, Jammu and Kashmir 190006, India
e-mail: harkirat@nitsri.net

Z. H. Khan et al. (eds.), *Recent Advances in Nanomaterials*, Springer Proceedings in Materials 27, https://doi.org/10.1007/978-981-99-4878-9_16

practically is feasible [3–5], but this makes it cumbersome to explicitly differentiate spin-polarized cooper pairs and drastically limits the range of materials displaying a customized out-of-plane anisotropy. Magnetoresistance measurements [6] and spin pumping studies [7–9] have recently discovered further experimental support for the existence of LRTs in superconductor/ferromagnet hybrid devices owing to the existence of spin–orbit coupling at the interface. The critical temperature of such hybrid structures is much below than its bulk value or even can be zero depending on the precise specifications of the hybrid system [10]. In reality, the critical temperature of an SF hybrid structure is always smaller as compared to the bulk value. Also, the density of states near Fermi energy gets modified inside the F-layer which can either appear as zero energy peak or an energy gap at zero energy in the vicinity of Fermi energy as reported earlier by Asif et al. [11].

In this article, we have studied the density of states (DOS) and critical temperature of superconductor/ferromagnet hybrid nanostructures in presence of spin–orbit coupling present at the interface. We report enhancement in the physical quantities like density of states and critical temperature with only a single layer of homogenous ferromagnet, exhibiting intrinsic spin–orbit coupling, in proximity with a conventional superconductor. We consider Rashba and Dresselhaus type of spin–orbit coupling that can be attributed to the lack of center of inversion inside the crystal lattice or due to inversion asymmetry that arises as a result of mismatch in the crystal structures meeting at the interface between different materials. We also report that the presence of intrinsic spin–orbit coupling in SF hybrid structures has an influence on the critical temperature of the setup. We compare the behavior of various $T_c(d_f)$ curves, with and without spin–orbit coupling, and validate the notion that critical temperature is regulated by manipulating ferromagnetic magnetization orientation. We also report enhancement of DOS near Fermi energy in the F-layer which is credited to the presence of short-ranged triplets (SRTs) or long-ranged triplets accordingly if either an energy gap or an energy peak appears corresponding to zero energy, respectively.

2 Computational Methodology

The investigation of critical temperature of superconductor/ferromagnet nanostructures exhibiting intrinsic spin–orbit coupling was conducted employing simulations via the Genius software package, which has been developed by Quassou [12]. It is a set of numerical techniques for the equilibrium and non-equilibrium solution of the Usadel diffusion equation in one-dimensional superconducting hybrid devices. In addition to magnetic components, spin–orbit coupling, spin-dependent scattering, highly polarized magnetic surfaces, voltage gradients, and temperature gradients, the suite is capable of handling a broad range of hybrid configurations. In this paper, our main aim is to illustrate and describe how the presence of intrinsic SO coupling at the superconductor/ferromagnet interface modifies the critical temperature of the bilayer setup (shown in Fig. 1) in the vicinity of the SF interface.

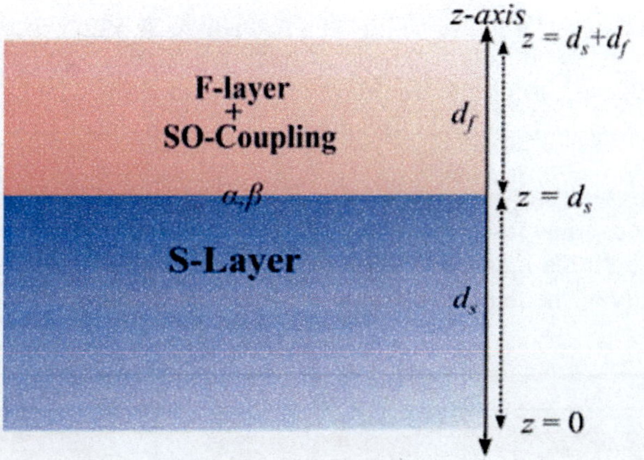

Fig. 1 SF hybrid structure. The region $0 \leq z \leq d_s$ represents the superconductor, and the region $d_s \leq z \leq d_s + d_f$ represents the ferromagnet, and the F-layer layer is assumed to have an in-plane (here *xy*-plane) type of Rashba(α)–Dresselhaus(β) SO coupling

Table 1 Material parameters we have used in the simulations involving DOS. Here, 'a' is the SOC strength, and 'χ' is the SOC angle

Superconductor	
Length (d_s)	1.5
Ferromagnet	
Length (d_f)	0.5
Magnetization	(0,0,3) or (0,3,0) or (3,0,0)
α	$-a\sin(\chi)\xi_s$
β	$a\cos(\chi)\xi_s$

3 Results and Discussion

3.1 *Effect on DOS*

To study the effect of SO coupling on the DOS, we employ the material parameters as shown in Table 1. Here, the lengths of the different layers have been normalized with the superconducting coherence length of S-layer (ξ_s). Also, we have taken both the cases of the magnetization, i.e., purely out-of-plane and purely in-plane normalized to superconducting gap 'Δ'. The variable 'a' is the coupling strength which can take different values. The SOC is purely in-plane with alignment depending upon the angle 'χ'. The cases $\chi = 0$ and $\chi = \pm\frac{\pi}{2}$ corresponds to purely Dresselhaus and purely Rashba type, respectively, whereas the cases $\chi = \pm\frac{\pi}{4}$ give equal magnitude for α and β. Any deviations from these values will give unequal values for α and

β. The effects of these type of variations on the density of states are summarized in Fig. 2.

From Fig. 2, we can clearly see that without spin–orbit coupling, there is no generation of either SRTs or LRTs because the DOS vs ε is featureless in the vicinity of Fermi energy ($\varepsilon = 0$). For the case when α and β have the same magnitude, there is formation of an energy gap inside F-layer near $\varepsilon = 0$ for only purely out-of-plane magnetization case, whereas when they differ in magnitude, zero-energy gap is seen in both the magnetization profiles, i.e., in-plane and out-of-plane. The detailed analysis of the density of states with different values of α and β can be

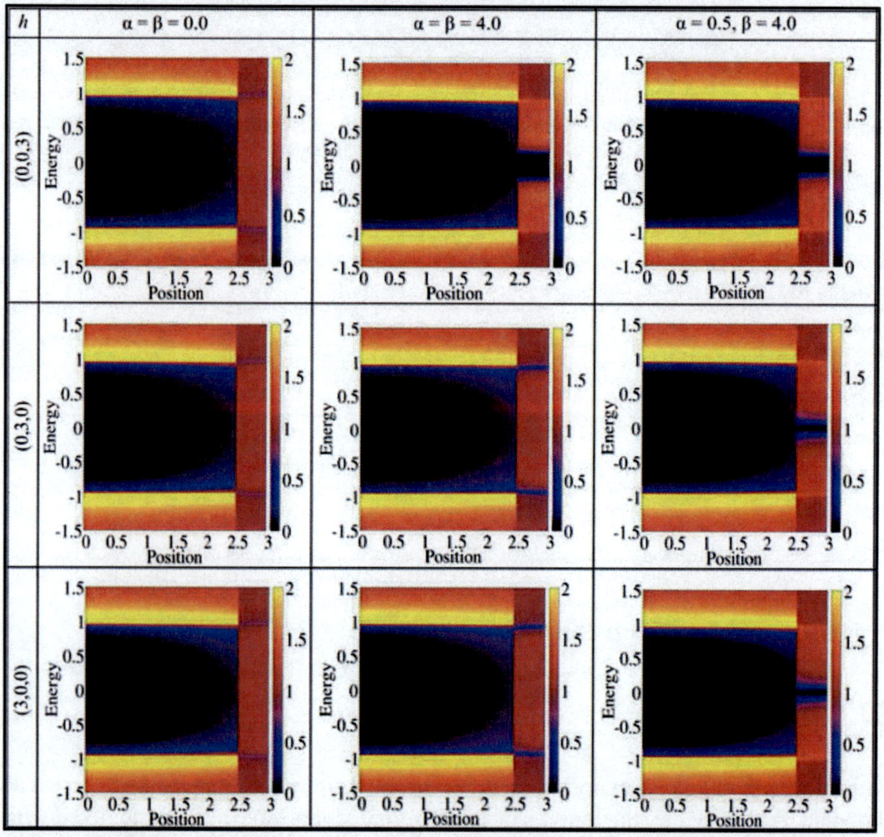

Fig. 2 Effect of varying Rashba and Dresselhaus coefficients for both in-plane and out-of-plane magnetization orientations. Here, the magnitude of exchange field is 3Δ and the length of the ferromagnet is $0.5\xi_f$, while the S-layer thickness is $2.5\ \xi_s$. Also, the energies and the coefficients α, β are expressed in unit less form. It is clearly reflected from the figure that in absence of SOC, there is no effect on DOS for either configuration; as SOC is introduced within the system, the DOS of states are modified near the Fermi energy and an energy gap appears at zero energy for out-of-plane magnetization orientation irrespective of equal or unequal magnitudes of SOC coefficients, whereas for the in-plane magnetization, it appears only when $\alpha \neq \beta$

found in Asif et al. [11]. The occurrence of the zero-energy gap is attributed to the generation of SRTs ($\uparrow\downarrow + \downarrow\uparrow$) within the system that is resistant against the pair-breaking mechanism of the exchange field of the F-layer. Thus, we conclude that in presence of spin–orbit coupling at the SF interface, there is modification of the density of states near Fermi energy.

3.2 *Effect on Critical Temperature*

We have studied the effect of varying S-layer thickness and spin–orbit on the critical temperature of the setup involving the same material parameters as given in Table 1. The simulations involving different parameters were carried, and the results are summarized in Fig. 3a and b respectively. We have taken $d_s = 2.5\xi_s$ and $h = (0, 0, 3\Delta)$. We have normalized the critical temperature T_c with bulk value T_{cs} and plotted it with the ferromagnetic thickness d_f normalized to the superconducting coherence length ξ_s. Increasing S-layer thickness shows that superconductivity is established for higher lengths. The effect of SOC on the critical is seen from different $T_c(d_f)$ curves, as shown in Fig. 3b which correspond to different values of SOC magnitude. For the case $\alpha = \beta = 0.0$, there is a non-monotonic decrease of the T_c with the increase in ferromagnetic thickness which has been reported in many experiments before. But we can see that as spin–orbit coupling is introduced within the system, there is increase in the critical temperature of the setup which is attributed to the SRTs present in the system. Also, for some values of α and β, the effect is pronounced and then it saturates as the values increase further ($\alpha = \beta = 5.0$). Thus, we can say that due to proximity effect between S-layer and F-layer, when placed in contact to each other, enhance the critical temperature of the system close to the bulk value.

4 Conclusion

We studied the proximity effect in the superconductor/ferromagnet hybrid structures in the presence of spin–orbit coupling intrinsic to the F-layer. We carried simulations for two physically observable quantities, i.e., the density of states and the critical temperature. In the absence of SOC, we saw there is no modification of DOS inside the ferromagnet, whereas as soon as SOC is introduced within the system, there is occurrence of an energy gap in the neighborhood of Fermi energy implying superconducting nature and the same is attributed to the generations of SRTs within the system. Also, with the introduction of spin–orbit coupling in the system, the critical temperature of the setup is enhanced which is once again credited to the exchange field resistant short-ranged triplets. The conventional SF proximity theory cannot explain these findings without taking into consideration the spin–orbit interaction present at the interface where the proximity effect happens.

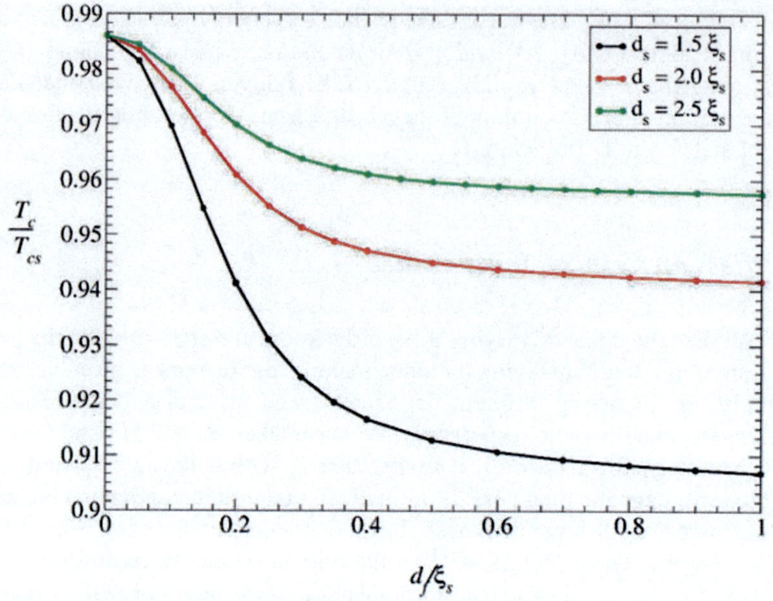

(a)

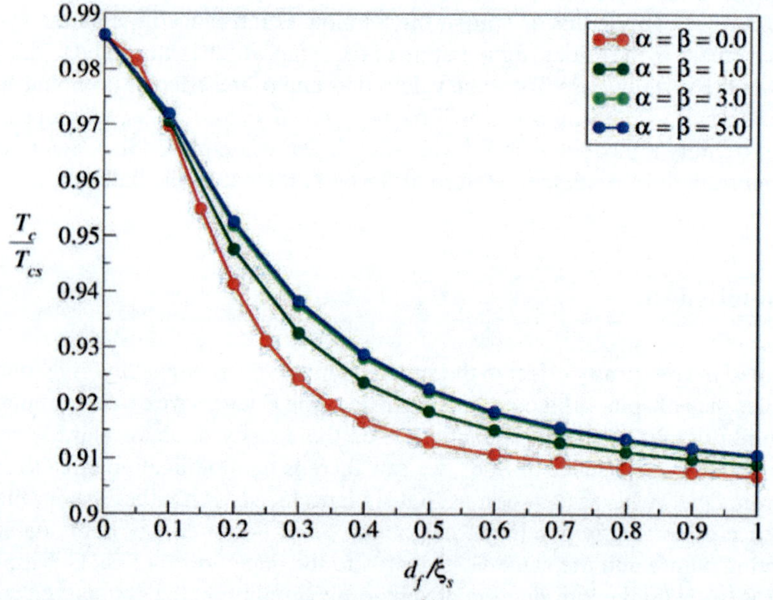

(b)

Fig. 3 Variation of critical temperature as a function of ferromagnetic thickness with **a** varying S-layer thickness and **b** spin–orbit coupling magnitude α and β. Here, we have taken $d_s = 2.5\xi_s$ and exchange field as (0, 0, 3Δ). The $T_c(d_f)$ curves show a non-monotonic behavior with F-layer thickness. Also, with the inclusion of SO coupling, we can see enhancement in the critical temperature of the hybrid structure

Acknowledgements The authors would like to thank DST/SERB, Govt. of India, for financial support (Grant No. MTR/2021/000524).

Declaration of Interest Statement The authors declare that they have no conflict of interests.

References

1. Bergeret FS, Tokatly IV (2014) Spin-orbit coupling as a source of long-range triplet proximity effect in superconductor-ferromagnet hybrid structures. Phys Rev B 89(13):134517
2. Jacobsen SH, Ouassou JA, Linder J (2015) Critical temperature and tunneling spectroscopy of superconductor-ferromagnet hybrids with intrinsic Rashba-Dresselhaus spin-orbit coupling. Phys Rev B 92(2):024510
3. Satchell N, Birge NO (2018) Supercurrent in ferromagnetic Josephson junctions with heavy metal interlayers. Phys Rev B 97(21):214509
4. Satchell N, Loloee R, Birge NO (2019) Supercurrent in ferromagnetic Josephson junctions with heavy-metal interlayers II. Canted magnetization. Phys Rev B 99(17):174519
5. Banerjee N, Ouassou JA, Zhu Y, Stelmashenko NA, Linder J, Blamire MG (2018) Controlling the superconducting transition by spin-orbit coupling. Phys Rev B 97(18):184521
6. Mart´ınez I, Ho¨gl P, Gonza´lez-Ruano C, Cascales JP, Tiusan C, Lu Y, Hehn M, Matos-Abiague A, Fabian J, Zˇ utic´ I, Aliev FG (2020) Interfacial Spin-orbit coupling: a platform for superconducting spintronics. Phys Rev Appl 13(1):014030
7. Jeon KR, Ciccarelli C, Ferguson AJ, Kurebayashi H, Cohen LF, Montiel X, Eschrig M, Robinson JWA, Blamire MG (2018) Enhanced spin pumping into superconductors provides evidence for superconducting pure spin currents. Nat Mater 17(6):499–503
8. Jeon KR, Ciccarelli C, Kurebayashi H, Cohen LF, Komori S, Robinson JWA, Blamire MG (2019) Abrikosov vortex nucleation and its detrimental effect on superconducting spin pumping in Pt/Nb/Ni80Fe20Nb/Pt proximity structures. Phys Rev B 99(14):144503
9. Jeon KR, Montiel X, Komori S, Ciccarelli C, Haigh J, Kurebayashi H, Cohen LF, Lee C-M, Blamire MG, Robinson JWA (2019) Tunable pure spin supercurrents and the demonstration of a superconducting spin-wave device
10. Majeed A, Singh H (2022) Effect of the interface transparency and ferromagnetic thickness on the critical temperature of NbN/Gd/NbN hybrid structure. Physica C Supercond Appl 1354127
11. Majeed A, Singh H (2023) Triplet proximity effects in superconductor/ferromagnet hybrid structures exhibiting intrinsic spin–orbit coupling. Physica B Condens Matter 414749
12. Ouassou JA "GENEUS", Ouassou, Jabir Ali https://github.com/jabirali/geneus

Studies on Solution-Processed Cu_2ZnSnS_4 Nanoparticles

K. G. Deepa and Praveen C. Ramamurthy

Abstract Cu_2ZnSnS_4 (CZTS) is identified to be a suitable candidate as an absorber layer for solar cells. In this work, CZTS nanoparticles are synthesized by wet-chemical method with non-toxic solvents and characterized for application in solar cells. The precursor concentrations are varied to achieve phase pure CZTS having a Cu-poor and Zn-rich composition, which is ideal for high-efficiency solar cells. Formation of tetragonal kesterite structured CZTS phase is confirmed from the X-ray diffraction analysis. Study also revealed the presence of shoulder peaks corresponding to Cu_2S phase in Cu-rich samples. Sample with Cu/Zn+Sn ratio of 0.85 having a band gap of 1.48 eV and particle size ~ 20 nm is opted for further studies. Minority carrier lifetime of this sample is estimated to be 6 ns from time-resolved photoluminescence measurement. The work function and valence band maximum of CZTS nanoparticle system are also estimated using UPS spectrum. The study demonstrates the suitability of CZTS nanoparticle in solar cells.

Keywords Nanoparticle · Carrier lifetime · Hole transport material · Work function · Solar cell

1 Introduction

Cu_2ZnSnS_4 (CZTS) is an attractive material having properties applicable in versatile fields such as solar cells, photocatalysis, non-volatile memory, thermoelectrics and photodetector. The non-toxicity and abundancy of the constituent elements make CZTS an ecofriendly and economic material. CZTS has a direct band gap of ~ 1.5 eV and absorption coefficient > 10^4 cm^{-1} which are advantageous in photovoltaic as well as photocatalytic applications. Both physical and chemical methods have

K. G. Deepa (✉) · P. C. Ramamurthy
Interdisciplinary Centre for Energy Research, Indian Institute of Science, Bangalore 560012, India
e-mail: deepachaithanya@gmail.com

P. C. Ramamurthy
Department of Materials Engineering, Indian Institute of Science, Bangalore 560012, India

Z. H. Khan et al. (eds.), *Recent Advances in Nanomaterials*, Springer Proceedings in Materials 27, https://doi.org/10.1007/978-981-99-4878-9_17

been used in the fabrication of CZTS-based solar cell, and conversion efficiency up to 12.6% was obtained from the solution-processed CZTSSe solar cell [1, 2]. The synthesis procedure of this champion solar cell involves the reaction with hydrazine which is highly toxic and not economical. Hydrazine-free synthesis method yielded an efficiency of 10% in which an amine–thiol solvent was used [3]. Conversion efficiencies of 7% and 4.92% were achieved with CZTS nanoparticles synthesized using ball-milling method and microwave method, respectively [4]. Lowering the band gap of CZTS with partial replacement of S with Se was found to enhance the photogenerated current which in turn improves the conversion efficiency significantly [5].

It is important to synthesize CZTS nanoparticles using hydrazine-free non-toxic method which could be used in various applications. With this aim, we have synthesized CZTS nanoparticles using cost-effective and non-toxic solvents such as oleic acid and 1-octadecane and investigated the properties systematically using different techniques to see their suitability as absorber layer in solar cell.

2 Materials and Methods

Solution processing method using oleic acid and1-octadecane was employed to prepare Cu_2ZnSnS_4 nanoparticles as given elsewhere [6]. 10 mM $CuCl_2$, 10 mM $ZnCl_2$, 10 mM $SnCl_2$ and 30 mM sulfur were used as the initial precursors (sample 1), and the precursors were dissolved in equal volumes of oleic acid and1-octadecane (30 ml each). The molarity of S was reduced to 20 mM and 10 mM, respectively (Sample 2 and Sample 3), keeping the other parameters same. Samples were also prepared by reducing the molarity of $CuCl_2$ to 7.5 and 6.5 mM (samples CZTS4 and CZTS5). The nanoparticles were characterized using XRD (Bruker D8) operated at 40 kV and 40 mA. Cu Kα line of wavelength 1.5405 Å was used as the radiation source. The absorption spectra of samples were recorded using spectrophotometer (PerkinElmer λ-750). Energy-dispersive X-ray analysis (EDX) and scanning electron microscopy (SEM) images were obtained using FEI ESEM quanta-200. TEM measurements were carried out using TITAN Themis transmission electron microscope. Time-resolved photoluminescence measurement is taken for the optimized sample using Horiba Jobin Yuvon Fluorocube -01-NL Fluorescence Lifetime System. Picosecond laser diode was used as the excitation source. Ultraviolet Photoelectron Spectroscopy (UPS) measurement is carried out using Axis DLD photoelectron spectrometer at a pressure of 2.6×10^{-7} Pa. He I of energy 21.22 eV was used as the excitation line.

3 Results and Discussion

Composition of the CZTS nanoparticles is analyzed using energy-dispersive X-ray spectroscopy (Table 1).

Samples CZTS1 to CZTS3 are rich in Cu with atomic percentage above 30. Despite the low molarity of the initial precursor $CuCl_2$, the composition is appeared to be Cu-rich. There was only a nominal increase in the sulfur concentration with increase in sulfur molarity from 20 to 40 mM. In the meantime, atomic percentage of Cu is reduced from 38 to 30% with increase in the sulfur concentration. This effect could be attributed to the charge compensation in CZTS. Cu-rich film may result in defects such as CuZn site which acts as acceptor with a net charge of − 2. In order to compensate this extra charge, the donor defect V^{S2+} forms in the sample. Hence, when S concentration increases, Cu concentration decreases accordingly as observed in the EDX. Cu percentage is reduced to 23.98 with reduction in the Cu precursor molarity from 10 to 6.5 mM (CZTS 5). The resulting film was rich in Zn and slightly Cu-poor with Cu/Zn+Sn ratio of 0.85 which falls in the range for high-efficiency CZTS solar cells (between 0.75 and 1) [7].

Figure 1 shows the XRD patterns of CZTS nanoparticle with variation in composition. The high-intensity peak appears at 28.5° in these samples which corresponds to (112) plane of tetragonal kesterite CZTS (JCPDS No. 34-1246). Peaks are also observed at planes (020), (220), (312) and (332) of kesterite CZTS.

Shoulder peaks are observed at 46.17° and 27.06° in samples CZTS1 to CZTS3 which could be attributed to Cu_2S (JCPDS No. 12-0227). The probability of formation of Cu_2S phase in these samples is high due to the non-stochiometric and Cu-rich composition as revealed from EDX analysis. Broad XRD peaks are observed in these samples probably due to the smaller size of CZTS particles. Based on the composition and XRD analysis, CZTS5 with Cu/Zn+Sn ratio of 0.85 is optimized for further studies. Band gap of the optimized sample CZTS5 is estimated to be 1.48 eV from the Tauc plot Fig. 2a. The band gap is calculated by extrapolating the linear portion of the Tauc plot to *x*-axis. The band gap of 1.48 eV enables CZTS to absorb in the visible region which is desired for the photovoltaic application. TEM image of the CZTS nanoparticle is shown in Fig. 2b. Particle size ranges from ~ 15 to 21 nm. Ellipsoidal-shaped particles with various aspect ratio ranging from 15:12 to 21 to

Table 1 EDX results of CZTS nanoparticles

Sample name	Atomic percentage of elements			
	Cu	Zn	Sn	S
CZTS1	38.92	6.23	9.29	45.56
CZTS2	30.81	9.86	12.07	47.26
CZTS3	31.92	7.30	11.79	48.99
CZTS4	26.26	14.05	12.38	47.31
CZTS5	23.98	16.31	11.79	47.91

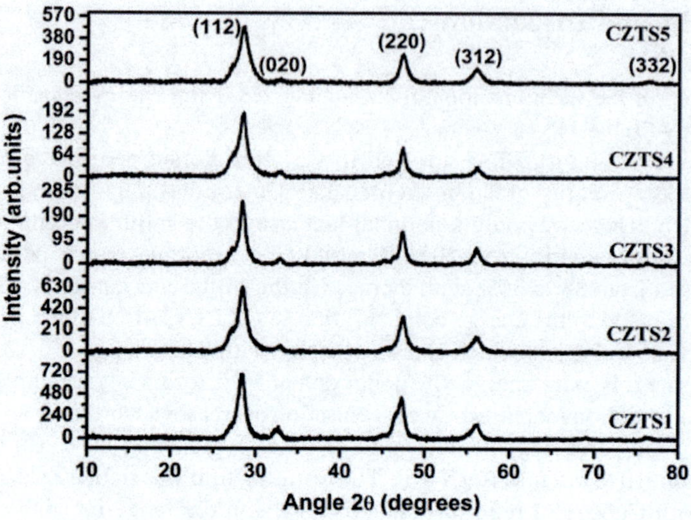

Fig. 1 XRD pattern of CZTS nanoparticles

17 nm are observed in TEM. The minority carrier lifetime in CZTS is evaluated using time-resolved fluorescence measurement. Minority carrier lifetime is an important parameter in determining the quality of the absorber layer in a solar cell. It is an indication of the recombination rate of minority carrier in a material. Here, picoseconds laser diode is used as the excitation source. Figure 3 shows log (intensity) vs decay time for the CZTS5. The graph is fitted with single exponential decay given by equation:

$$y(t) = A * \exp(-(t - t_0)/\tau, \qquad (1)$$

where τ is the minority carrier lifetime which is calculated to be 6.2 ns from the Gaussian fit. The carrier lifetime value is comparable with that obtained from CZTS samples prepared using different methods [8]. Reducing the Shockley–Read–Hall recombination at the bulk and surface may help to increase the carrier lifetime which demands further study [9]. The HOMO level obtained from electrochemical measurement of CZTS nanoparticles is compared with the valence band maximum (VBM) determined from the UPS measurement. VBM line which cuts the *X*-axis at 4.95 eV is shown in Fig. 3b. Secondary electron cut-off in UPS measurement helps to determine the work function of the material. Here, the Fermi energy is set as zero binding energy. The work function of the material is obtained by subtracting the secondary electron offset (16.88 eV) from the excitation energy value, which is 21.22 eV in this case. The work function of CZTS is evaluated as 4.34 eV using the UPS spectrum. The work function of indium-doped tin oxide (ITO) is reported to be ~ 4.5 eV which is higher than that of CZTS which promotes the ohmic contact between CZTS and ITO [10].

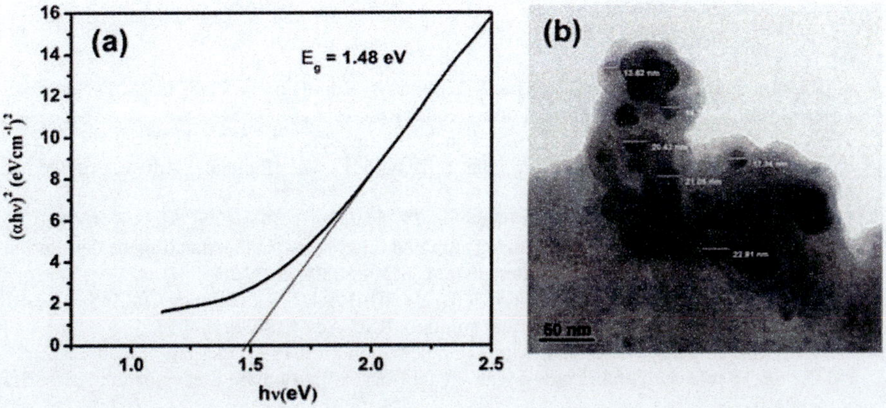

Fig. 2 **a** Tauc plot and **b** TEM image of CZTS5

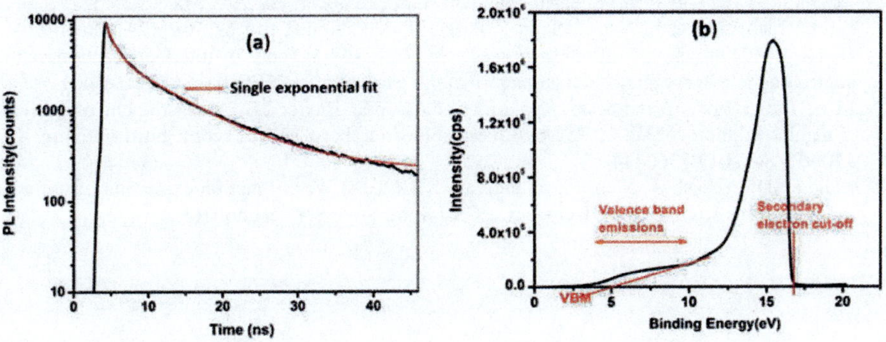

Fig. 3 **a** Time-resolved photoluminescence spectrum and **b** UPS spectrum of CZTS5

4 Conclusion

Phase pure Cu_2ZnSnS_4 nanoparticles with a Cu/Zn+Sn ratio of 0.85 are prepared using solvothermal method with ecofriendly solvents. The minority carrier lifetime is estimated to be 6.2 ns in the sample. The work function and valence band maximum are found to lie at 4.34 eV and 4.95 eV, respectively.

Acknowledgements This work was supported by the Science and Engineering Research Board (SERB), DST, Government of India, through Young Scientist Start-up grant (YSS/2014/000805).

References

1. Green MA, Dunlop ED, Hohl-Ebinger J, Yoshita M, Kopidakis N, Hao X (2021) Solar cell efficiency tables (version 58). Prog Photovoltaics Res Appl 29(7):657–667
2. Wang W, Winkler MT, Gunawan O, Gokmen T, Todorov TK, Zhu Y, Mitzi DB (2013) Device characteristics of CZTSSe thin-film solar cells with 12.6% efficiency. Adv Energy Mater 4(7):1301465
3. Lowe JC, Wright LD, Eremin DB, Burykina JV, Martens J, Plasser F, Malkov AV (2020) Solution processed CZTS solar cells using amine–thiol systems: Understanding the dissolution process and device fabrication. J Mater Chem C 8(30):10309–10318
4. Wang W, Shen H, Wong L.H., Su Z, Yao H, Li Y (2016) A 4.92% efficiency Cu2ZnSnS4 solar cell from nanoparticle ink and molecular solution. RSC Adv 6:54049–54053
5. Woo K, Kim Y, Yang W, Kim K, Kim I, Oh Y, Moon J (2013) Band-gap-graded Cu_2ZnSn (S1-X, Sex)4 solar cells fabricated by an ethanol-based, particulate precursor ink route. Sci Rep 3(1):1–7
6. Deepa KG, Ramamurthy PC, Singha MK (2019) Mesoporous Cu_2ZnSnS_4 nanoparticle film as a flexible and reusable visible light photocatalyst. Opt Mater 98:109492
7. Chen S, Walsh A, Gong X, We S (2013) Classification of lattice defects in the kesterite Cu_2ZnSnS_4 and $Cu_2ZnSnSe_4$ earth-abundant solar cell absorbers. Adv Mater 25:1522
8. Liu F, Huang J, Sun K, Yan C, Shen Y, Park J, Pu A, Zhou F, Liu X, Stride JA, Green MA, Hao X (2017) Beyond 8% ultrathin kesterite Cu2ZnSnS4 solar cells by interface reaction route controlling and self-organized nanopattern at the back contact. NPG Asia Mater 9:e401
9. Li S, Lloyd MA, Golembeski AA, McCandless BE, Baxter JB, Measurement of carrier dynamics in photovoltaic CZTSe by time-resolved terahertz spectroscopy. https://doi.org/10.1109/PVSC.2017.8366444
10. Nehate SD, Prakash A, Mani PD, Sundaram KB (2018) Work function extraction of indium tin oxide films from MOSFET devices. ECS J Solid State Sci Technol 7(3)

Studies of $Se_{85}Te_{12}Bi_3$ and $Se_{85}Te_9Bi_6$ Nanochalcogenide Thin Films at Different Working Pressures

Aditya Srivastava, Zubair M. S. H. Khan, Zishan H. Khan, and Shamshad A. Khan

Abstract In this research work, we have synthesized amorphous $Se_{85}Te_{12}Bi_3$ and $Se_{85}Te_9Bi_6$ Chalcogenide Glasses (ChGs) using melt quenching technique. The physical vapour condensation technique has been adopted for the synthesis of nanothin films of thickness 30 nm of synthesized samples at two different working pressures (1 torr and 3 torr) of ambient argon gas on the ultrasonically cleaned glass substrates. The temperature of the substrates was kept at 77 K (constant) using liquid nitrogen. The morphological analysis of the prepared nanochalcogenide thin films using FESEM confirmed that the particle size on the films was from 20 to 80 nm. We also observed more aggregation and reduction in particle size on the substrates with an increase in working pressure for both $Se_{85}Te_{12}Bi_3$ and $Se_{85}Te_9Bi_6$ nanothin films. HRXRD pattern confirmed the non-crystalline texture of synthesized nanothin films. Based on UV–Visible spectroscopy, the parameters related to optical characteristics like absorption coefficients and optical direct band gaps were measured for synthesized $Se_{85}Te_{12}Bi_3$ and $Se_{85}Te_9Bi_6$ nanochalcogenide thin films. A slight increment in the absorption coefficient and band gap with an increase in working pressure is observed for both $Se_{85}Te_{12}Bi_3$ and $Se_{85}Te_9Bi_6$ nanostructured thin films.

Keywords ChGs · Nanochalcogenide · Thin films · FESEM · HRXRD · UV–visible spectroscopy

A. Srivastava (✉) · S. A. Khan
Materials Science Research Lab, Department of Physics, St. Andrew's College, Gorakhpur, Uttar Pradesh 273001, India
e-mail: adishakti2709@gmail.com

Z. M. S. H. Khan · Z. H. Khan
Department of Biosciences, Jamia Millia Islamia, New Delhi 110025, India

Z. H. Khan
Department of Applied Science and Humanities, F/O Engineering and Technology, Jamia Millia Islamia, New Delhi 110025, India

Z. H. Khan et al. (eds.), *Recent Advances in Nanomaterials*, Springer Proceedings in Materials 27, https://doi.org/10.1007/978-981-99-4878-9_18

1 Introduction

In this era of nanorevolution, materials based on chalcogenides (Sulphides, Selenides and Tellurides) turned out to be potential and promising contenders for many applications in the emerging field of nanotechnology [1]. Chalcogenides composite materials or more specifically ChGs are acknowledged for their large nonlinear refractive index and low maximum phonon energy [2]. Chalcogenide-based materials have high thermal and chemical stability which facilitates their applications in optical devices. The possibility for modification in the refractive index of ChGs and their high photosensitivity makes them a suitable candidate for their application in writing channel waveguides. The availability of ChGs as thin films is another advantage of them in the field of integrated optics [3, 4]. Nowadays, many research groups are interested in the study of chalcogenides at the nanoscale. The materials exhibit phenomenal optical, electrical, thermal, physical, and chemical properties at the nanoscale. The applications of nanochalcogenides in memory and switching devices, optoelectronic devices, and UV lasing diodes are very fascinating [5]. Such nanochalcogenides can be used in photonic circuits, optical memory devices, optical imaging, optical recording, optical communications, and microelectronics [6]. In recent years, considering global issues research by renowned scientists have been focused their attention on the synthesis of nanochalcogenides at low temperatures and exploring its applications in superconductors, energy storage, and many other fields [7, 8]. Oke et al. [9] reviewed the preparation of thin films of chalcogenide-based materials through the atomic layer deposition technique and explored their properties and applications. Khan et al. [10] have synthesized the CdS nanothin films doped with manganese on glass substrates using the spray pyrolysis technique with enhanced photovoltaic performance for visible-light sensing applications. Ghobadi et al. [11] have synthesized CdSe nanostructured thin films through the chemical solution deposition method and calculated their optical band gap and Urbach energy for their application in photocatalysis. Algarasan et al. [12] studied the photo-sensing applications of nanostructured CdSe thin films annealed at different temperatures and synthesized by the thermal evaporation method. The effect of the addition of an electron reflector layer in nanostructured ultra-thin CdTe solar cells has been done by Rashwan et al. [13]. Synthesis and thickness-dependent studies of ultra-thin bismuth and antimony chalcogenide thin films using physical vapour condensation technique with their applications in thermoelectric generators were done by Andzane et al. [14], Elyamny and Kashyout [15] have studied the Bi-Te thin films synthesized by thermal evaporation technique. Many research groups across the globe have studied chalcogenide-based materials at the nanoscale in the form of thin films and explored their applications [16, 17].

In this research work, our obvious choice for parent material is selenium (Se) due to its high photoconductivity and photovoltaic effect. Selenium-based materials have a broad range of utility in rectifiers, xerography, photographic exposure metres, and anticancer agents [18, 19]. The ageing effect, short lifetime, and low sensitivity of Se required some additives in pure Se. Several efforts have been done to remove or minimize these shortcomings by adding less-toxic appropriate materials such as tellurium (Te) and bismuth (Bi) [20]. Te shows various applications as machining additives, thermoelectric devices, catalysts, photoreceptors, and many chemical uses due to their physical properties. Se–Te-composed alloys are frequently used in phase change memory devices as Te improves the crystallinity and corrosion resistance of Se. The addition of Te also improves the efficiency of solar cells for electricity generation as Te is a heat-resistant and highly corrosive metalloid [21]. Bismuth is chosen as the dopant in Se–Te network; due to its low-melting point, Bi is considered in general for making of low-melting alloys [22]. Taking into account the prominence of such materials, we have decided to analyse the structural pattern and optical characteristics of $Se_{85}Te_{12}Bi_3$ and $Se_{85}Te_9Bi_6$ ChGs as nanostructured thin films synthesized at two disparate working pressures 1 torr and 3 torr with a constant substrate temperature of 77 K.

2 Materials and Methods

To meet the main objective of our study, melt quenching technique has been adopted with the requirement of high-purity (5N) elements (Se, Te and Bi) for synthesizing bulk chalcogenide $Se_{85}Te_{12}Bi_3$ and $Se_{85}Te_9Bi_6$ glasses. The exact proportions of high-purity elements were measured according to their atomic percentages. These materials containing ampoules were further sealed using a molecular turbo pump under a high vacuum (10^{-5} torr). In a programmable furnace, we put these materials containing sealed ampoules where an increment in temperature in three steps will take place, firstly at 723 K for 180 min then 923 K for 180 min, and finally 1073 K for next 240 min. During the heating process, the ingots are continuously shaken at regular intervals as possible for the consistency of the homogenous mixture. After the completion of the heating process, the amorphous specimens were obtained through quenching of sample ingots in ice-cooled water. The fine powder of these sample ingots was made using a pestle and mortar set. $Se_{85}Te_{12}Bi_3$ and $Se_{85}Te_9Bi_6$ nanostructured thin films were made using the physical vapour condensation technique (PVCT). In this technique initially, the specimen-containing chamber was evacuated to 10^{-5} torr followed by the supply of non-reactive ambient argon gas in the chamber at two different working pressures 1 torr and 3 torr. The evaporation of $Se_{85}Te_{12}Bi_3$ and $Se_{85}Te_9Bi_6$ alloys kept in a boat made up of a material having a high-melting point (like Molybdenum) was done for deposition of the nanostructured thin films on the ultrasonically cleaned glass substrates. For cooling the glass substrate, liquid nitrogen (77 K) was used. The Edward model FTM 7 Quartz crystal monitor was used for the determination of the thickness of the deposited films on ultrasonically cleaned glass substrates. The structural studies of the synthesized nanochalcogenide thin films have been done using high-resolution X-ray diffractometer (HRXRD) (Regaku X-ray diffractometer Ultima IV). For morphological studies, QUANT FEG 450, Amsterdam, Netherlands Field Emission Scanning Electron Microscopy (FESEM) was performed for the analysis of nanochalcogenide particle size. The optical studies of nanostructured $Se_{85}Te_{12}Bi_3$ and $Se_{85}Te_9Bi_6$ thin

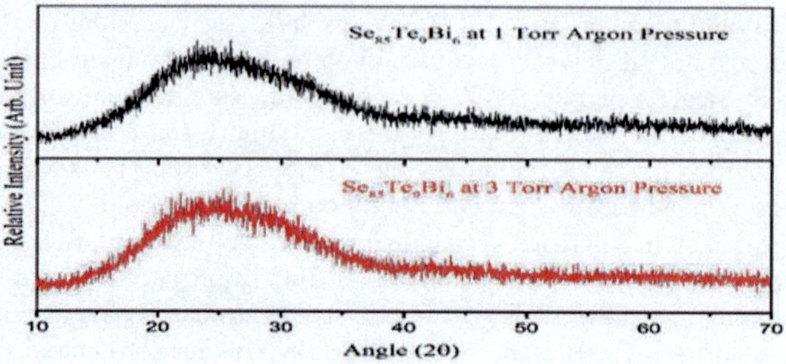

Fig. 1 HRXRD profile of $Se_{85}Te_9Bi_6$ nanostructured thin films synthesized at two different working pressures

films were done for the wavelength region of 400–1000 nm through the LAMBDA 365 UV/VIS spectrophotometer.

3 Results and Discussions

3.1 *Structural Analysis*

The nature of nanochalcogenide $Se_{85}Te_{12}Bi_3$ and $Se_{85}Te_9Bi_6$ thin films was studied by HRXRD measurements by high-resolution X-ray diffractometer (using X-ray source of Cu target with $\lambda = 1.54$ Å($CuK\alpha_1$)) for the glancing region of 10–70° with scanning and chart speed of 2°/min and 1 cm/min, respectively. The non-appeared prominent peaks in the HRXRD pattern at disparate working pressures suggest the non-crystalline texture of the synthesized nanothin films (Fig. 1). The FESEM images of $Se_{85}Te_{12}Bi_3$ and $Se_{85}Te_9Bi_6$ nanothin films at working pressure 1 Torr and 3 Torr (Fig. 2a, b respectively) indicate that the thin films consist high yield of nanochalocogenide particles of dimension 20–80 nm. The increment in the chamber pressure leads to a reduction in the size of nanochalcogenide particles and more accumulation observed on the glass substrates.

3.2 *Optical Studies*

The parameters related to optical studies of nanochalcogenide $Se_{85}Te_{12}Bi_3$ and $Se_{85}Te_9Bi_6$ thin films at disparate chamber pressures have been obtained by the UV/VIS spectrophotometer. The value of the absorption coefficient (α) and the optical direct band gap (E_g) from the absorption spectra were determined for the wavelength region of 400–1000 nm. The value of α can be evaluated by the given relation [23].

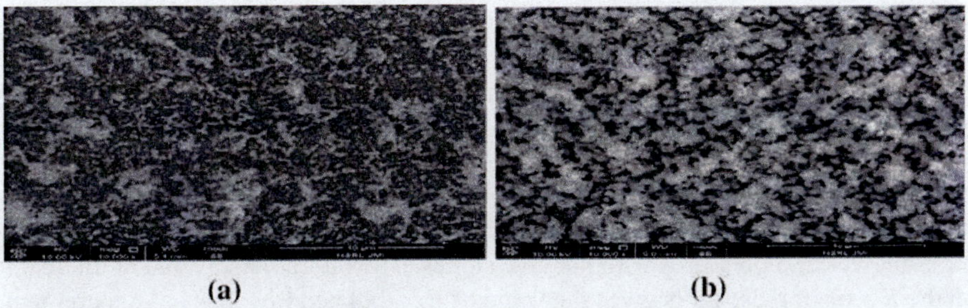

Fig. 2 **a** FESEM image of $Se_{85}Te_9Bi_6$ at 1 torr working pressure, **b** FESEM image of $Se_{85}Te_9Bi_6$ at 3 torr working pressure

$$\text{Absorption coefficient } (\alpha) = \text{Optical Density/thickness of film.} \quad (1)$$

The plot of α and photon energy $(h\nu)$ is shown in Fig. 4. The figure shows that for both working pressures, α increases with $h\nu$ for both samples. The nanothin films prepared at higher working pressure (3 Torr) show higher absorption due to more accumulation of nanochalcogenide particles on the glass substrates. The value of E_g for synthesized nanothin films $Se_{85}Te_{12}Bi_3$ and $Se_{85}Te_9Bi_6$ at two different working pressures can be estimated by an equation given by J. Tauc:

$$(\alpha\, h\nu)^p = A(h\nu - E_g). \quad (2)$$

Here. the value of A is constant and depends on the transition probability. and p is an exponent whose value relies on the different electronic transitions while the

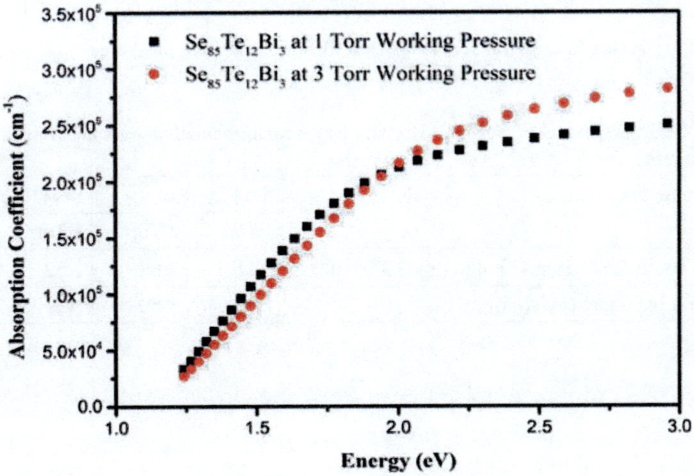

Fig. 4 $A\alpha$ versus $h\nu$ of $Se_{85}Te_{12}Bi_3$ at different working pressures

absorption process. For $p = 2$, the direct transition rule has been followed by the present synthesized system of nanochalcogenide thin films. Putting the value of p in Eq. 2, we can get a direct dependent relation between $(\alpha\, h\nu)^2$ and E_g as:

$$(\alpha\, h\nu)^2 \propto (h\nu - E_g). \tag{3}$$

Using Eq. 3, we get the straight-line plots (Fig. 5) for both the films synthesized at different working pressures with positive slopes. The interpolated value of intercept on the X-axis of these plots gives the value of E_g tabulated in Table 1. An increment in the value of E_g is observed with an increase in chamber pressure as the size of the nanochalcogenide particles is reduced at higher working pressure which was further confirmed by FESEM images.

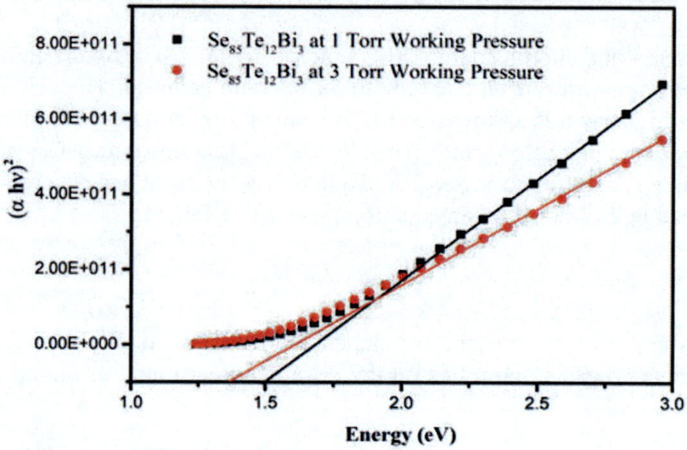

Fig. 5 $A(\alpha h\nu)^2$ versus Energy for $Se_{85}Te_{12}Bi_3$ at different working pressures

Table 1 Optical parameters of $Se_{85}Te_{12}Bi_3$ and $Se_{85}Te_9Bi_6$ nanochalcogenide films at different working pressures

Optical parameters	$Se_{85}Te_{12}Bi_3$		$Se_{85}Te_9Bi_6$	
	1 Torr	3 Torr	1 Torr	3 Torr
Absorption coefficient (α) at $\lambda = 480$ nm (10^5) cm^{-1}	2.41	2.68	2.52	2.77
Optical direct band gap (E_g) (eV)	1.36	1.50	1.39	1.52

4 Conclusion

The presented research work is mainly focused on the analysis of structural patterns and calculation of optical parameters of $Se_{85}Te_{12}Bi_3$ and $Se_{85}Te_9Bi_6$ nanothin films synthesized at two different chamber pressures of argon gas by PVCT. The synthesis of bulk $Se_{85}Te_{12}Bi_3$ and $Se_{85}Te_9Bi_6$ ChGs was done using the melt quenching technique. The prepared samples are amorphous which was revealed by the HRXRD pattern. A high concentration of nanochalcogenide particles accumulated on the glass substrates with dimensions ranging from 20–80 nm was found as analysed by FESEM. UV–Visible spectroscopy measurements show a slight increment in the estimated value of E_g with an increase in the chamber pressure of ambient argon gas due to the change in the size of nanochalcogenide particles. The measured value of E_g proposes that our synthesized nanochalcogenide thin films are promising candidates in solar cell applications and many other photonic devices.

Conflict of Interest There are no such conflicts for the declaration in this manuscript.

References

1. Seddon AB (1995) Chalcogenide glasses: a review of their preparation, properties, and applications. J Non-Cryst Solids 184:44–50
2. Savage JA (1982) Optical properties of chalcogenide glasses. J Non-Cryst Solids 47(1):101–115
3. Zakery A, Elliott SR (2003) Optical properties and applications of chalcogenide glasses: a review. J Non-Cryst Solids 330(1–3):1–12
4. Ahluwalia GK (2017) Fundamentals of chalcogenides in crystalline, amorphous, and nanocrystalline forms BT. In: Ahluwalia GK (ed) Applications of chalcogenides: S, Se, and Te. Springer International Publishing, Cham, pp 3–60
5. Khan SA, Tiwari G, Tripathi RP, Alvi MA, Khan ZH, Al-Agel FA (2014) Structural, optical, and structural characterization of polycrystalline $Ga_{15}Te_{85\text{-}X}Zn_X$ nano-structured thin films. Adv Sci Lett 20:1715–1718
6. Khan SA, Al-Agel FA, Al-Ghamdi AA (2010) Optical characterization of nanocrystalline $Se_{85}Te_{10}Pb_5$ and $Se_{80}Te_{10}Pb_{10}$ chalcogenides. Superlattices Microstruct 47:695–704
7. Bhatt VS, Yadav AK, Dixit D, Tomy CV (2022) High-temperature solution growth of large size chalcogenide FeT_XSe (T:Fe, Co) superconducting single crystals. Superconductivity 3:100016
8. Ni J, Bi X, Jiang Y, Li L, Lu J (2017) Bismuth chalcogenide compounds Bi_2X_3 (X=O, S, Se): applications in electrochemical energy storage. Nano Energy 34:356–366
9. Oke JA, Olotu OO, Jen T-C (2022) Atomic layer deposition of chalcogenide thin films: processes, film properties, applications, and bibliometric prospect. J Mater Res Technol 20:991–1019
10. Khan ZR, Revathy MS, Shkir M, Khan A, Sayed MA, Umar A, Alshammari AS, Vinoth S, Marnadu R, Yousef ES, Algarni H, AlFaify S (2022) Noticeably enhanced opto-electronic and photodetection performance of spray pyrolysis grown Mn:CdS nanostructured thin films for visible-light sensor applications. Surf Interfaces 28:101586
11. Ghobadi N, Sohrabi P, Hatami HR (2020) Correlation between the photocatalytic activity of CdSe nanostructured thin films with optical band gap and Urbach energy. Chem Phys 538:110911

12. Alagarasan D, Varadharajaperumal S, Kumar KDA, Naik R, Arunkumar A, Ganesan R, Hegde G, Massoud EES (2021) Optimization of different temperature annealed nanostructured CdSe thin film for photodetector applications. Opt Mater 122 (Part B):111706
13. Rashwan O, Sutton G, Ji L (2021) Optical modeling of periodic nanostructures in ultra-thin CdTe solar cells with an electron reflector layer. Superlattices Microstruct 149:106757
14. Andzane J, Felsharuk A, Sarakovskis A, Malinovskis U, Kauranens E, Bechelany M, Niherysh KA, Komissarov IV, Erts D (2021) Thickness-dependent properties of ultrathin bismuth and antimony chalcogenide films formed by physical vapor deposition and their application in thermoelectric generators. Mater Today Energy 19:100587
15. Elyamny S, Kashyout AB (2019) Preparation and characterization of the nanostructured bismuth telluride thin films deposited by thermal evaporation technique. Mater Today Proc 8(2):680–689
16. Mochalov LA, Kuznetsov YM, Dorokhin MV, Fukina DG, Knyazev AV, Kudryashov MA, Kudryashov YP, Logunov AA, Mukhina OV, Zdoroveyshchev AV, Zdoroveyshchev DA (2022) Thermoelectrical properties of ternary lead chalcogenide plumbum-selenium-tellurium thin films with excess of tellurium prepared by plasma-chemical vapor deposition. Thin Solid Films 752:139244
17. Ali SA, Ahmad T (2022) Chemical strategies in molybdenum based chalcogenide nanostructures for photocatalysis. Int J Hydrogen Energy 47(68):29255–29283
18. Tripathi RP, Akhtar SM, Khan SA (2018) Thermally deposited $Se_{85}In_{15-x}Sb_x$ chalcogenide thin films: structural electrical and optical properties. Materials Focus 7(2):251–258
19. Tripathi RP, Alvi MA, Khan SA (2020) Investigations of thermal, optical and electrical properties of $Se_{85}In_{15-x}Bi_x$ glasses and thin films. J Therm Anal Calorim 146:2261–2272
20. Srivastava A, Tiwari SN, Lal JK, Khan SA (2019) Phase transformation in $Se_{75}Te_{13}In_{12}$ chalcogenide thin films. Glass Phys Chem 45:111–118
21. Srivastava A, Khan SA, Sahani RM, Tripathi RP, Akhtar MS (2021) Influence of gamma-irradiation on the optical and structural properties of $Se_{85}Te_{15\text{-}x}Bi_x$ nano-thin chalcogenide films. Radiat Phys Chem 188:109659
22. Tripathi RP, Zulfequar M, Khan SA (2016) Structural, optical and electrical properties of $Se_{85}In_9Bi_6$ nanochalcogenide thin film. Curr Nanomaterials 1(3):176–182
23. Al-Agel FA, Suleiman J, Khan SA (2017) Studies of silicon quantum dots prepared at different working pressure. Result Phys 7:1128–1134

Influence of Gamma Irradiation on Structural and Optical Parameters of $Se_{85}Te_9Ag_6$ Nanochalcogenide Thin Films

Archana Srivastava, Zishan H. Khan, and Shamshad A. Khan

Abstract In this research work, we have done a systematic study of the influence of γ-irradiations on the parameters related to the optical characteristics of $Se_{85}Te_9Ag_6$ nanochalcogenide thin films. The conventional melt quenching technique has been adopted for the synthesis of the bulk $Se_{85}Te_9Ag_6$ alloy. The non-isothermal DSC measurements of $Se_{85}Te_9Ag_6$ chalcogenide glasses were done for the confirmation of their glassy and amorphous nature. Thin films (40 nm) were deposited by the physical vapor condensation method on glass and Si substrates. γ-radiation of doses 5, 10, and 15 kGy was exposed on nano-thin films at a rate of 2 kGy hr^{-1}. The JASCO UV/VIS/NIR spectrophotometer was used for optical studies of $Se_{85}Te_9Ag_6$ nano-thin films. The optical measurement data confirms the indirect transition rule in the $Se_{85}Te_9Ag_6$ glass. The observed value of the extinction coefficient and absorption coefficient has shown an increment with the dose of gamma radiation, whereas the increment in the crystal defects may lead to a decrement in the value of the estimated optical band gap.

Keywords Chalcogenide glasses · DSC · Gamma irradiation · Absorption coefficient · Optical band gap

1 Introduction

Many comprehensive experimental and theoretical works have shown net worthy nature of chalcogenides and found its usage in the advancement of exciting techniques for the usefulness of mankind in last few decades. The multicomponent chalcogenide

A. Srivastava (✉) · S. A. Khan
Materials Science Research Lab, Department of Physics, St. Andrew's College, Gorakhpur, Uttar Pradesh 273001, India
e-mail: 31archana84@gmail.com

Z. H. Khan
Department of Applied Science and Humanities, F/O Engineering and Technology, Jamia Millia Islamia, New Delhi 110025, India

Z. H. Khan et al. (eds.), *Recent Advances in Nanomaterials*, Springer Proceedings in Materials 27, https://doi.org/10.1007/978-981-99-4878-9_19

glasses have unique features of composition-dependent variability of their thermal, optical, and structural properties. Due to this, chalcogenide glasses exhibit a range of values for their parameters and they are established as multitasking materials used to design significant technological devices, e.g., diffraction grating, optical fiber, optical switching and lasers, etc. [1–5]. A renewed attention has been developed in these vitreous semiconductors because of their distinguishing optical nonlinear and mid-infrared properties [6–8]. The higher transmittance in mid-IR range and sensitivity to visible light of chalcogenides make them suitable for IR sensors [9]. Current progress in photonics based on chalcogenide has been motivated by industrial and technological encounters tackled in many fields [10]. Amorphous chalcogenides based on Selenium (Se) have found many applications in storage media. Chalcogenide glasses are reactive to external influences, such as annealing, electric field, and irradiation due to their supple configuration. Optical parameters of chalcogenide glasses changed due to gamma irradiation and its dosage [11]. Accordingly, this adaptableness of chalcogenides to γ radiation that resulted as variation in their optical parameters makes them interesting for numerous usages. Ahmad et al. [12] have studied the change in optical parameters of amorphous Se–Hg thin films due to gamma radiation. Halyan et al. [13] have done EPR of defects induced by gamma radiation and their influence on photoluminescent properties of Ag–Ga–Ge–S–Er–S glassy alloys. Abdul-Kader et al. [14] have studied the effect of γ-radiation on parameters related to the optical characteristics of the deposited Ag–Sb–Se films. The systematic study of gamma irradiation effect on structural and optical parameters of Se–Te–Bi nano-thin films has been done by Khan et al. [15]. Various researchers have studied the gamma radiation impact on the optical characteristics of nano-thin films based on chalcogenides [16–18].

Keeping in view above mentioned remarkable works, our aim in this piece of research work is to study the effect of gamma radiation on the optical parameters of $Se_{85}Te_9Ag_6$ nanochalcogenide thin films. We have chosen selenium as a parent material due to its reversible phase change properties, good photoconductivity, and photovoltaic effect [19]. Unfortunately, selenium has little inferiority, e.g., lower sensitiveness and small life span that may be improvised by certain dopants in Se. We have added Te as Se–Te network-based alloys have generally used as memory device active layer due to their lesser melting temperature which makes them utmost promising material for phase change memory devices [20]. We have doped Ag in Se–Te alloy as it has been described by many researchers that Ag-doped chalcogenide alloys can be used as recording layer due to its excellent performance [21].

2 Materials and Methods

We have used melt quenching technique for the synthetization of $Se_{85}Te_9Ag_6$ chalcogenide glasses. DSC was done for $Se_{85}Te_9Ag_6$ chalcogenide alloys at 25 K min^{-1} for the confirmation of their glassy and amorphous nature. Thin films (thickness 40 nm) of $Se_{85}Te_9Ag_6$ glasses were deposited by physical vapor condensation technique (PVCT) on glass and Si wafer (10^{-5} Torr pressure). Further, we have radiated deposited thin films by γ radiation of doses of 5, 10 and 15 kGy (dose rate 2 kGy hr^{-1}) using GC-5000 (gamma chamber) for phase transition studies in $Se_{85}Te_9Ag_6$ chalcogenide glasses. We have used JASCO/UV/VIS/NIR spectrophotometer for optical absorbance measurements of as-prepared and irradiated $Se_{85}Te_9Ag_6$ nanochalcogenide thin films in 400–1100 nm wavelength span.

3 Results and Discussion

The DSC measurement in the non-isothermal mode of the synthesized $Se_{85}Te_9Ag_6$ alloy was done at the heating rate of 25 K min^{-1}. The endothermic peak shows T_g (glass transition) of alloy that is due to the reason that during glass transition heat is required due to relaxation. The exothermic peak is crystallization temperature (T_C). The sharp exothermic and endothermic peaks in DSC thermogram (Fig. 1) of $Se_{85}Te_9Ag_6$ chalcogenide glass confirm its glassy and amorphous nature.

We have used a JASCO spectrophotometer (photometric accuracy 0.004) for determining the optical absorption data of $Se_{85}Te_9Ag_6$ nanochalcogenide thin films. The optical analysis of both deposited as well as radiated $Se_{85}Te_9Ag_6$ chalcogenide thin films has been performed using optical absorbance data between the wavelength span of 400 nm–1100 nm. The absorption coefficient (α) can be calculated by using the given Equation [22]:

$$\alpha = \text{Optical absorbance/Thickness of the film} \quad (1)$$

Fig. 1 DSC thermogram of amorphous $Se_{85}Te_9Ag_6$

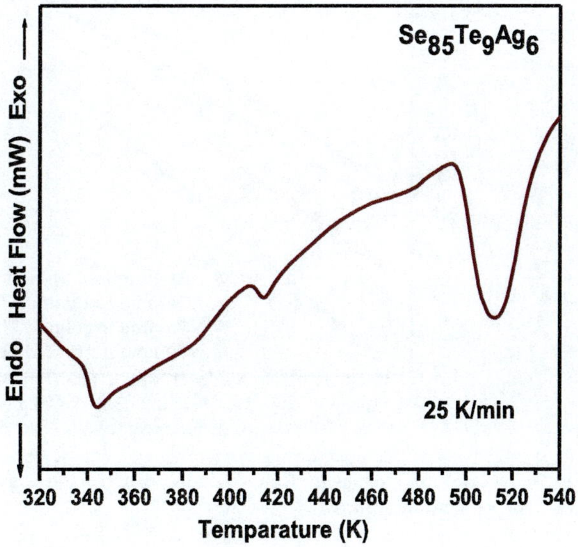

Figure 2 shows the graph between the α and the energy of incident radiation ($h\nu$) for as-prepared and radiated (5, 10 and 15 kGy) $Se_{85}Te_9Ag_6$ thin films. Figure 3 suggests that α increases with the dose of radiation which is also shown by prior works [23]. The values of α for as-prepared and irradiated nano-thin films (at $\lambda =$ 690 nm) with various doses of radiation are given in Table 1.

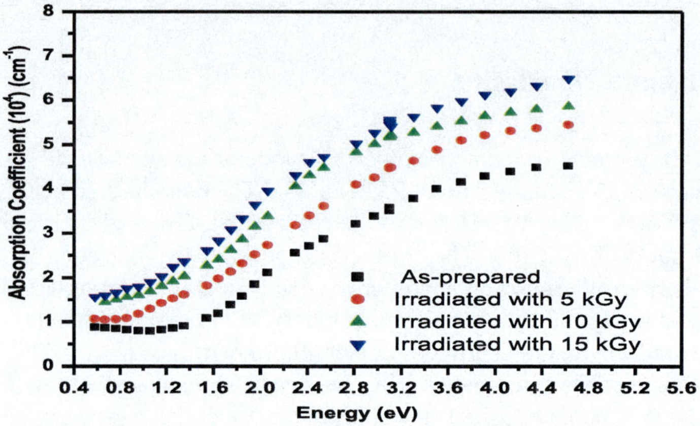

Fig. 2 α versus $h\nu$ of as-deposited and γ-irradiated $Se_{85}Te_9Ag_6$ nanochalcogenide thin films

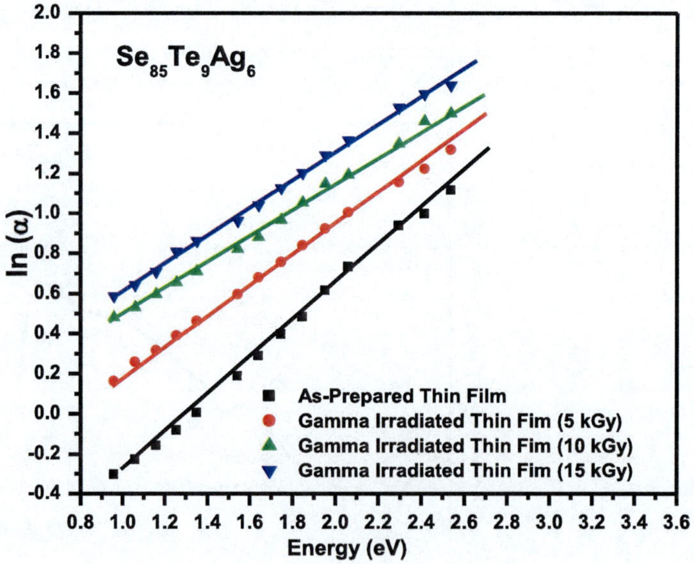

Fig. 3 Plot of ln (α) versus $h\nu$ for calculation of Urbach energy of as-deposited and γ-irradiated $Se_{85}Te_9Ag_6$ nano-thin films

Table 1 Parameters related to optical characteristics of $Se_{85}Te_9Ag_6$ as-prepared and gamma irradiated nanochalcogenide thin films

Sample	Radiation dose (kGy)	α (10^4) (cm^{-1}) at $\lambda = 690$ nm	E_g (eV)	E_u (eV)
$Se_{85}Te_9Ag_6$	As-prepared	2.27	1.50	1.07
	5	2.77	1.43	1.36
	10	3.36	1.32	1.49
	15	3.83	1.61	1.44

Many crystalline and amorphous materials show that absorption coefficient depends exponentially on the incident radiation energy given as [23]:

$$\alpha(\nu) = \alpha' \exp(h\nu/E_t) \tag{2}$$

or

$$\ln(\alpha) = \ln(\alpha') + (h\nu/E_t). \tag{3}$$

Here, α' is a constant, h is the Planck's constant, and E_t is known as Urbach energy or sometimes called as band tail. The Urbach energy arises due to the localized states in the forbidden gap which corresponds to the non-crystalline texture of the sample [24]. The presence of localized states can be verified through the exponential absorption edges corresponding to the fluctuation of internal field and structure disorderness [25]. We have plotted graphs between ln (α) and $h\nu$ (shown in Fig. 3). The plots of Fig. 3 are linear which confirms that Urbach relation holds well in present case. Further, we have determined the Urbach energy for as-prepared and irradiated nano-thin films by taking the reciprocal of these linear graphs (shown in Table 1). An increment in the value of E_t has been observed with the dose of radiation which may be due to structural fluctuations.

For nanochalcogenide thin films, the transition possibilities and the optical band gap can be given by Tauc's relation [26]:

$$(\alpha h\nu)^k = A'(h\nu - E_g). \tag{4}$$

Here, E_g is the band gap and the value of A' and k depends upon the probabilities of transitions during the absorption process. The value of k may be equal to 2/3, 2, 1/2, or 1/3, depending on the type of transition [27]. We have found that the best fit of Eq. 4 is for $k = ½$ for $Se_{85}Te_9Ag_6$ chalcogenide thin films which indicates the indirect transition; Fig. 4 represents the plot of $(\alpha h\nu)^{1/2}$ versus photon energy ($h\nu$) for $Se_{85}Te_9Ag_6$ nanochalcogenide thin films (as-deposited and γ-irradiated). We have extrapolated the plots of $(\alpha h\nu)^{1/2}$ and $h\nu$ of incident radiation of Fig. 4, for the determination of E_g of as-prepared and γ-irradiated $Se_{85}Te_9Ag_6$ nano-thin films. It is noticeable that an increment in the dose of incident γ-radiation leads to a decrement in the value of E_g (Table 1). Many researchers reported a similar result

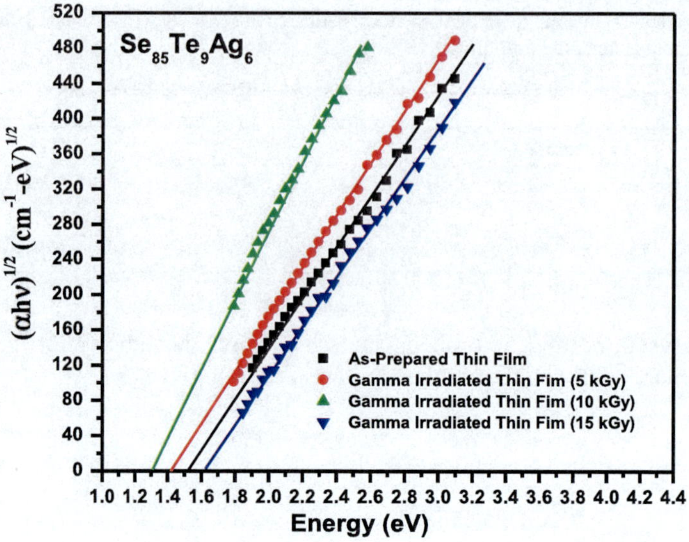

Fig. 4 Plot of $(\alpha h\nu)^{1/2}$ versus $h\nu$ for calculation of indirect optical band gap of as-deposited and γ-irradiated $Se_{85}Te_9Ag_6$ nano-thin films

[28]. The reason for the change in E_g might be based on the possibilities of various types of defects and disorderliness in the crystal structure and lattice strain.

4 Conclusion

In present research work, the synthesized $Se_{85}Te_9Ag_6$ alloy was found glassy and homogeneous in nature by DSC measurements. Optical measurement data of chalcogenide nano-thin films confirm the indirect transition rule in this $Se_{85}Te_9Ag_6$ glass. Obtained results are indicating that gamma irradiation affects the optical parameters of $Se_{85}Te_9Ag_6$ glassy alloys. The value of E_g for synthesized $Se_{85}Te_9Ag_6$ nanochalcogenide thin films recorded with a decrement with the dose of incident γ radiation which may be because of an increase in crystallite size as well as changes in the microstructure. The outcomes of these studies are indicating that gamma irradiation on nanochalcogenide thin films can be utilized for the tuning purpose of their optical band gap which can be used for various optoelectronic devices.

Acknowledgements We are thankful to Dr. R. M. Sahani, Nuclear Radiation Management and Application Division, Defence Laboratory (DRDO), Ratanada, Jodhpur, Rajasthan, India, for providing us gamma irradiation facilities in their research lab.

Declaration of Interest Statement The authors declare that they have no conflict of interests.

References

1. Nyakotyo H, Sathiaraj TS, Muchuweni E (2017) Effect of annealing on the optical properties of amorphous $Se_{79}Te_{10}Sb_4Bi_7$ thin films. Opt Laser Technol 92:182–188
2. Hassanien AS, Aly KA, Akl AA (2016) Study of optical properties of thermally evaporated ZnSe thin films annealed at different pulsed laser powers. J All Comp 685:733–743
3. El-Nahass MM, Ali MH, Zedan IT (2014) Photoinduced changes in the linear and non-linear optical properties of $Ge_{10}In_{10}Se_{80}$ thin films. J Non-Cryst Solids 404:78–83
4. Shockley W (1950) Electrons and holes in semiconductors: with applications to transistor electronics. R. E. Krieger Pub. Co.
5. Lojek B (2007) History of semiconductor engineering. Springer Science & Business Media, pp 321–333
6. Harbold JM, Omerllday F, Wise FW, Aitken BG (2002) Highly nonlinear Ge–As–Se and Ge–As–S–Se glasses for all-optical switching. IEEE Photonics Technol Lett 14(6):822–824
7. Sojka L, Tang Z, Zhu H, Beres-Pawlik E, Furniss D, Seddon AB, Benson TM, Sujecki S (2012) Study of mid-infrared laser action in chalcogenide rare earth doped glass with Dy^{3+}, Pr^{3+} and Tb^{3+}. Opt Mater Express 2(11):1632–1640
8. Popescu M (2006) Chalcogenides past, present future. J Non-Cryst Solids 352(9–20):887–891
9. Tejaswini ML, Kumar A (2017) Study of chalcogenides for sensor applications. In: Third international conference on current trends in engineering science and technology, pp 1014–1016
10. Fatome J, Fortier C, Nguyen TN, Chartier T, Smektala F, Messaad K, Traynor N (2009) Linear and nonlinear characterizations of chalcogenide photonic crystal fibers. J Lightwave Technol 27(11):1707–1715
11. Sahani RM, Kumari C, Pandya A, Dixit A (2019) Efficient alpha radiation detector using low temperature hydrothermally grown ZnO: Ga nanorod scintillator. Sci Rep 9(1):1–9
12. Ahmad S, Islam S, Nasir M, Asokan K, Zulfequar M (2018) Effects of gamma-ray irradiation on the optical properties of amorphous $Se_{100\text{-}x}Hg_x$ thin films. J Phy Chem Solids 117:122–130
13. Halyan VV, Konchits AA, Shanina BD, Krasnovyd SV, Yukhymchuk VO (2015) EPR of γ-induced defects and their effects on the photoluminescence in the glasses of the $Ag_{0.05}Ga_{0.05}Ge_{0.95}S_2$–$Er_2S_3$ system. Radiat Phys Chem 115:189–195
14. Abdul-Kader AM, El-Gendy YA (2013) Influence of γ-irradiation on the optical properties of AgSbSe2 thin films. Nucl Instrum Methods Phys Res Sect B 305:22–28
15. Khan SA, Sahani R, Tripathi RP, Akhtar MS, Srivastava A (2021) Influence of gamma-irradiation on the optical and structural properties of $Se_{85}Te_{15\text{-}x}Bi_x$ nano-thin chalcogenide films. Radiat Phys Chem 188:109659
16. Ahmad S, Khan MS, Asokan K, Zulfequar M (2015) Effect of gamma irradiation on the structural and optical properties of thin films of a-CdSe. Optik Int J Light Electron Opt 126(23):3501–3505
17. Balboul MR (2008) Optical effects induced by gamma and UV irradiation in chalcogenic glass. Radiat Meas 43(8):1360–1364
18. Shpotyuk M, Kovalskiy A, Shpotyuk O (2018) Phenomenology of γ-irradiation-induced changes in optical properties of chalcogenide semiconductor glasses: a case study of binary arsenic sulfides. J Non-Cryst Solids 498:315–322
19. Dey T (2021) Role of earth-abundant selenium in different types of solar cells. J Electr Eng 72:132–139
20. Srivastava A, Tiwari SN, Alvi MA, Khan SA (2018) Studies on $Se_{75}Te_{25\text{–}x}In_x$ chalcogenide glasses; a material for phase change memory. Mater Res Express 5(1):015206
21. Wfigner T, Frumar M, Kasap SO, Vlcek M (2001) New Ag-containing amorphous chalcogenide thin films—prospective materials for rewriteable optical memories. J Optoelectron Adv Mater 3(2):227–232
22. Tripathi RP, Akhtar MS, Alvi MA, Khan SA (2016) Influence of annealing treatment on phase transformation of Ga15Se77Tl8 thin films. J Mater Sci Mater Electron 27(8):8227–8233
23. Cheng C, Wang X, Xu T, Sun L, Chen W (2016) Optical properties of Ag- and AgI-doped Ge–Ga–Te far-infrared chalcogenide glasses. Infrared Phys Technol 76:698–703

24. Ghayebloo M, Tavoosi M, Rezvani M (2017) Compositional modification of Se-Ge-Sb chalcogenide glasses by addition of arsenic element. Infrared Phys Technol 83:62–67
25. Dongol M, Elhady AF, Ebied MS, Abuelwafa AA (2018) Impact of sulfur content on structural and optical properties of $Ge_{20}Se_{80-x}S_x$ glasses thin films. Opt Mater 78:266–272
26. Mir FA (2014) Transparent wide band gap crystals follow indirect allowed transition and bipolaron hopping mechanism. Results Phys 4:103–104
27. Srivastava A, Tiwari SN, Alvi MA, Khan SA (2018) Phase change studies in $Se_{85}In_{15-x}Zn_x$ chalcogenide thin films. J Appl Phys 123:125105
28. Mansour BA, Gad SA, Eissa HM (2015) Effect of γ-irradiation on the optical and electrical properties of $Pb_xGe_{42-x}Se_{48}Te_{10}$. J Non-Cryst Solids 412:53–57

Structural and Optical Properties of Ba and Co-Doped Lanthanum Ferrite at Room Temperature

Shraddha Agrawal and Azra Parveen

Abstract Ferrites are ferri-magnetic materials having opposite spins of unequal magnetization (non-zero magnetization). Lanthanum ferrites (perovskite oxides) are candidate materials for application as oxygen carriage membranes, cathode materials for chemical sensor applications, the chemical and food industries, and for environmental monitoring. The citrate autocombustion method has been used to synthesize 10% (Ba and Co-doped) $LaFeO_3$ ferrites. Microstructural analyses were carried out by XRD and SEM. The chemical composition of prepared nanoparticles was also analyzed with the help of energy dispersive X-ray spectrum (EDS). The FTIR spectra exhibit an absorption band corresponding to Fe–O stretching vibration. Optical measurements were studied by UV–visible technique.

Keywords XRD · SEM · FTIR

1 Introduction

Ferrites can be categorized into four groups based on the crystalline structure and the magnetic arrangement; spinel, garnets, magneto-plumbite, and ortho-ferrites. Among these, ortho-ferrites perovskites ferrites general formula $MFeO_3$ where $M =$ La are the ones used for energy storage applications [1]. The mixed oxides of rare earth and transition metal oxides establish a great family of ortho-ferrites perovskites motivated from the fundamental as well as application features, e.g., in the field of the colossal magneto-resistance. Recently, a lot of interest was focused on the synthesis and characterization of nanosized, rare earth orthorhombic perovskites type transition Metal oxides, e.g., $LaFeO_3$ having an absorption of visible light, in the range of 2.1–2.0 eV [2, 3]. Properties of nanomaterials can be altered by transition metal and

S. Agrawal
Department of Physics, Gurugram University, Gurugram, Haryana 122018, India

A. Parveen (✉)
Department of Applied Physics Z.H. College of Engineering and Technology, Aligarh Muslim University, Aligarh 202002, India
e-mail: azrap2001@gmail.com

Z. H. Khan et al. (eds.), *Recent Advances in Nanomaterials*, Springer Proceedings in Materials 27, https://doi.org/10.1007/978-981-99-4878-9_20

rare earth doping [4]. Triyono et al. and Hao et al. reported synthesis of Mg and Ba-doped $LaFeO_3$ nanoparticles prepared by a sol–gel method and hydrothermal method, respectively [5, 6]. The present research work reconnoiters the room temperature structural and optical studies of Ba and Co-doped lanthanum ferrite NPs.

2 Materials and Methods

Nano-crystalline lanthanum ferrite powder has been synthesized through the auto-combustion route. The chemicals lanthanum nitrate hexahydrate, iron nitrate hexahydrate, and citric acid monohydrate along with doping nitrates as rare earth elements barium nitrate and transition metal Cobalt nitrate were of analytical grade and used as obtained. The mixture was added to a citric acid and ammonia solution. The obtained transparent solution was put on magnetic stirring at 100 °C for 2 h. The 'gel' obtained with the help of a magnetic stirrer is kept in heat bath at a temperature of 140–150 °C and dried at 200 °C with the help of self-combustion reaction and finally calcinated at ~ 800 °C for 6 h.

3 Results and Discussion

3.1 Structural Analysis

The XRD spectra of 10% (Ba and Co-doped) $LaFeO_3$ are shown in Fig. 1a. XRD patterns show that the product is pure perovskite oxide with an orthorhombic structure. The peaks were indexed using Powder X software, and all peaks were well coordinated with an orthorhombic structure of $LaFeO_3$ using the standard data (JCPDF Card No 37-1493).The orthorhombic lattice parameter obtained for the as-synthesized powder is shown in Table 1. The crystallite size (D) was calculated with the help of Debye Scherrer's size rule [7]. The surface studies of Ba and Co-doped $LaFeO_3$ nanoparticles were analyzed by SEM (Fig. 1b). Spherical surface morphology with particle sizes in the range of 20–28 nm was confirmed from SEM. The EDS analysis shown in Fig. 1c demonstrated that La, Ba, Fe, and Co metals elements are present in the sample of $LaFeO_3$. Weight % measurements also confirmed the doping of 10% Ba and Co-doped $LaFeO_3$ (Fig. 1d).

3.2 Fourier Transform Infrared (FTIR)

The FTIR spectra of 10% (Ba and Co) doped La FeO_3 (Fig. 2). The peaks at 1620–1640 cm^{-1} and the broad peaks at 3100–3600 cm^{-1}, respectively, due to the adsorbed

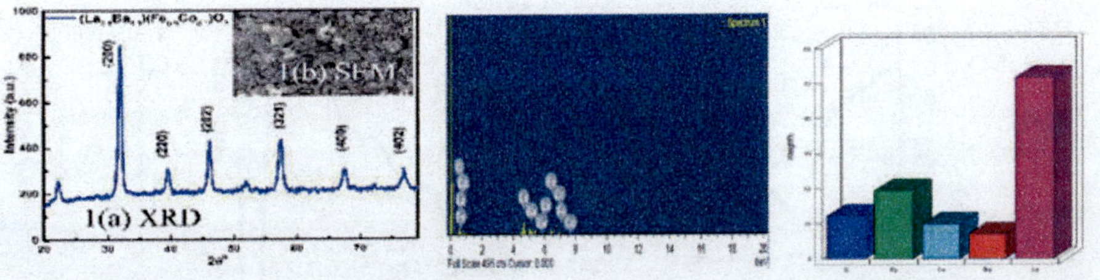

Fig. 1 **a** XRD pattern, **b** SEM, **c** EDAX, and **d** weight % image of 10% (Ba and Co) doped $LaFeO_3$

Table 1 Variation of lattice parameter and crystalline size with dopant concentration

Dopant concentration	Lattice parameter (a) A ^{0}	Lattice parameter (b) A ^{0}	Lattice parameter (c) A ^{0}	Crystalline size (nm)
(La 0.9 Ba 0.1)(Fe 0.9 Co 0.1) O 3	5.5275	7.7991	5.5137	13.62

water molecules and hydroxyl ions. Some small peaks at 1370 cm^{-1} (C–O) arise due to the adsorption of CO_2 presence in the air [8]. There is a sharp characteristic absorption band between 400 to 800 cm^{-1} which corresponds to stretching and bending vibrations of metal oxide ions. The small absorption band at 1370 cm^{-1} is ascribed to the stretching vibrations of C–C. The strong band at 600 cm^{-1} shows the stretching of metal oxide. A broad peak in the range of 3451 cm^{-1} is due to the H_2O, which specifies the presence of water adsorbed on the surface of NPs [9].

3.3 Optical Properties

The UV–Visible absorption studies along with band gap calculations with the help of Tauc relationship for 10% (Ba and Co) doped $LaFeO_3$ nanoparticles were noted (Fig. 3a and b). The absorbance spectra with absorption peak at around 550 nm generally depend on some factors such as lattice strain, band gap, grain size, and impurity. An absorption peak arises due to the photo-excitation of e-from the valence band (VB) to conduction band (CB).

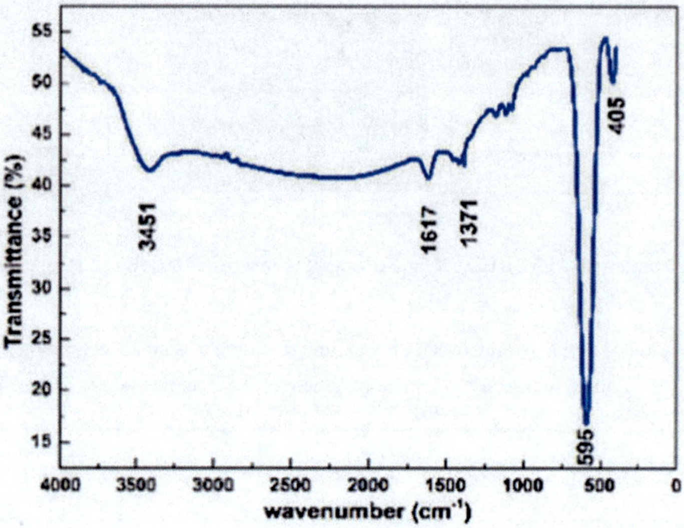

Fig. 2 Transmittance spectra

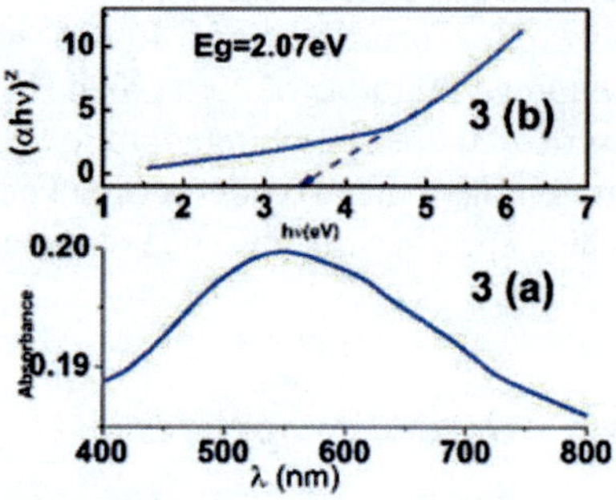

Fig. 3 **a** Absorption spectra and **b** Optical band gap plot of 10% (Ba and Co) doped $LaFeO_3$

4 Conclusion

We reported the structural and optical properties of 10% (Ba and Co) doped $LaFeO_3$ synthesized by self-combustion method. The absence of extra peaks in the XRD spectra shows the purity of the prepared nanoparticles. Spherical surface morphology was confirmed from SEM, and EDS shows the successful doping of Ba and Co without any impurity. The energy band gap estimated from the Tauc relationship comes out to be 2.07 eV. The FTIR images show the sharp peak at 600 cm^{-1}which shows the stretching of metal oxide.

Acknowledgements There has been no significant support for this work.

References

1. Natali Sora I, Fontana F, Passalacqua R, Ampelli C, Perathoner S, Centi G, Parrino F, Palmisano L (2013) Photoelectrochemical properties of doped lanthanum orthoferrites. Electrochim Acta 109:710–715
2. Apostolova IN, Apostolov AT, Trimper S, Wesselinowa JM (2021) Multiferroic properties of pure and transition metal doped $LaFeO_3$ nanoparticles. Phys Status Solidi 258:2000482
3. Matin MA, Hossain MN, Rhaman MM, Mozahid FA, Ali MA, Hakim MA, Islam MF (2019) Dielectric and optical properties of Ni-doped $LaFeO_3$ nanoparticles. SN Appl Sci 1:1479
4. Parveen, A., Agrawal, S., Naqvi, A.H.: Structural, optical and transport properties of transition metals doped (A: Co, Ni and Cu) BiFe0.9A0.1O3. AIP Conference Proceedings., 1665, p. 050016 (2015)
5. Triyono D, Hanifah U, Laysandra H (2020) Structural and optical properties of Mg-substituted $LaFeO_3$ nanoparticles prepared by a sol-gel method. Results Phys 16:102995
6. Hao P, Qu GM, Song P et al (2021) Synthesis of Ba-doped porous $LaFeO_3$ microspheres with perovskite structure for rapid detection of ethanol gas. Rare Met 40:1651–1661
7. Agrawal S, Parveen A, Naqvi AH (2015) Auto-combustion synthesis and characterization of Mg doped $CuAlO_2$ nanoparticles. AIP Conf Proc 1665:050080
8. Kaewpanha M, Suriwong T, Wamae W, Nunocha P (2019) Synthesis and characterization of Sr-doped $LaFeO_3$ perovskite by sol-gel auto-combustion method. J Phys Conf Ser 1259:012017
9. John Berchmans L, Sindhu R, Angappan S, Augustin CO (2008) Effect of antimony substitution on structural and electrical properties of $LaFeO_3$. J Mater Process Technol 207:301–306

X-Ray Absorption Fine Structure Spectroscopic Investigation at Ge K-Edge of AuGe/Ni/AuGe Ohmic Contact to GaAs/AlGaAs

Preeti and Md. Ahamad Mohiddon

Abstract X-ray absorption fine structure spectroscopy (XAFS) has been employed to investigate the crystallographic structural changes that develop at the AuGe-based metal interface to the GaAs/AlGaAs heterostructures. A stack of AuGe/Ni/AuGe is sequentially deposited over the GaAs/AlGaAs substrate by thermal and electron beam evaporation techniques without breaking the vacuum in between the deposition. These stacks were annealed at 100°C and 300°C in a nitrogen-rich atmosphere. XAFS measurements were carried out at Ge K-edge to investigate the crystallographic structural changes developed at the interface as a function of annealing temperature. Further, the effect of Ni layer thickness on the different crystal structures developed at the interface is also reported.

Keywords X-ray absorption fine structure spectroscopy (XAFS) · GaAs/AlGaAs heterostructure · Ohmic contact

1 Introduction

The demand for high-speed electronic circuits in modern technologies has led to the development of GaAs-based transistor devices [1]. Low contact resistance is an important desirable criterion for achieving high-speed electronic devices. The renowned tactic to achieve the Ohmic contact with the GaAs/AlGaAs heterostructure is the deposition of AuGe/Ni/AuGe multilayers followed by rapid thermal annealing for alloying [2–4]. AuGe ratio of 88:12 which is the eutectic point of AuGe alloy is reported to show low contact resistance with appropriate Ni thickness [3, 5]. However, the reason behind such low resistance Ohmic contact is not yet explored. So, in the present investigation, an attempt was made to understand the change in Ge chemistry which is responsible for low resistance Ohmic contact. X-ray absorption

Preeti · Md. Ahamad Mohiddon (✉)
Center for Nanoscience and Nanotechnology, Department of Physics, Faculty of Science and Humanities, SRM University Delhi-NCR, Sonepat, Haryana 131029, India
e-mail: ahamed.vza@gmail.com; mahamad.ahamad@srmuniversity.ac.in

Z. H. Khan et al. (eds.), *Recent Advances in Nanomaterials*, Springer Proceedings in Materials 27, https://doi.org/10.1007/978-981-99-4878-9_21

fine structure spectroscopy (XAFS) is one of the prestigious tool to investigate the local chemistry around the central absorbing atom [6–8]. Here, XAFS measurements were carried out at Ge K-edge to investigate the structural changes developed at the interface.

2 Experimental

The metal contact to GaAs/AlGaAs heterostructure was grown through thermal evaporation and electron beam evaporation techniques. A 50 nm AuGe thin layer was deposited over the GaAs/AlGaAs heterostructure by thermal evaporation followed by Ni thin film (25 nm/10 nm) by electron beam evaporation. Then the blanket layer of the 50 nm AuGe was deposited over Ni resulting in AuGe/Ni/AuGe stack (will be called ANA). The details of the deposition were reported elsewhere [4]. In the present study, the eutectic composition of AuGe (88:12) was used as it provides low contact resistance [5]. The deposited AuGe/Ni/AuGe stack was then subjected to rapid thermal annealing at different temperatures of 100°C (ANA100) and 300°C (ANA300) in the nitrogen-rich ambience for 2 hrs. X-ray absorption fine structure spectroscopy (XAFS) measurements of the deposited stack were carried out at Ge K-edge at 11.1 keV at the Elettra Synchrotron Light Source, Trieste, Italy. XAFS measurements were carried out at a 2° incident angle to see the crystallographic changes developed in the sample as a function of annealing temperature.

3 Results and Discussion

3.1 EXAFS Investigation of Au88Ge12(50 nm)/Ni(25 nm)/ Au88Ge12(50 nm)

The magnitude of the Fourier transform of Ge K-edge EXAFS signal of Au88Ge12(50 nm)/Ni(25 nm)/Au88Ge12(50 nm) multilayers annealed at different temperatures along with reference spectra of Ge and Ni_3Ge is shown in Fig. 1a. From Fig. 1a, it is evident that the EXAFS features of the ANA100 sample have predominant Ge features; however, at higher *R* values, the features of Ni_3Ge are well pronounced. On increasing the annealing temperatures to 300°C, i.e., ANA300 sample, the Ni_3Ge features are found much closer than the Ge features.

A fraction of GeO_2 is also expected in both ANA100 and ANA300 EXAFS spectra. The EXAFS spectra of Au88Ge12(50 nm)/Ni(25 nm)/Au88Ge12(50 nm) multilayers annealed at 100°C and 300°C temperatures and their best fit (dotted lines) along with the theoretical scattering paths of different reference samples are presented in Fig. 1b and c, respectively. In both the cases, the extracted EXAFS spectra perfectly overlap with the best-fit data, which confirms the good choice of

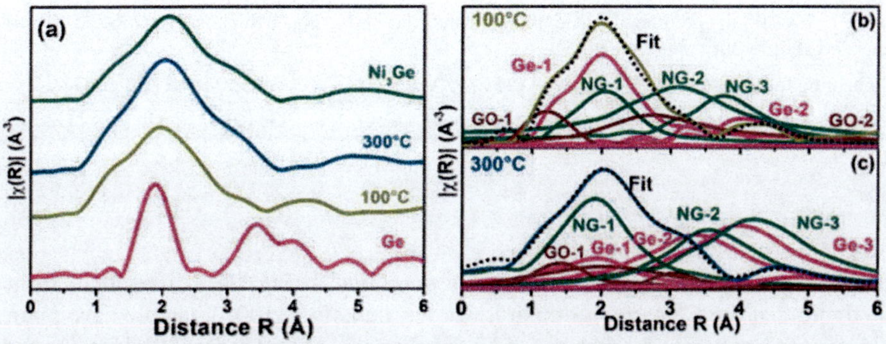

Fig. 1 **a** Magnitude of EXAFS signal of AuGe/Ni(25 nm)/AuGe annealed at 100°C and 300°C along with reference spectra. EXAFS experimental data (solid line), best-fit data (black dotted line) and theoretical paths of above stack annealed at **b** 100°C **c** 300°C

theoretical paths from different reference samples. Firstly, from the qualitative study, it is anticipated that the ANA100 will have dominated Ge and Ni_3Ge reference paths. Seven theoretical FEFF paths were used to fit the ANA100 sample. (i) Ge-1, $R =$ 2.45 Å, $N = 4$; (ii) Ge-2, $R = 3.8$ Å, $N = 12$; (iii) N_3G-1, $R = 2.4$ Å, $N = 6$; (iv) N_3G-2, $R = 2.53$ Å, $N = 12$; (v) N_3G-3, $R = 4.3$ Å, $N = 24$; (vi) GO_2-1, $R = 1.90$ Å, $N = 4$; (vii) GO_2-2, $R = 1.94$ Å, $N = 2$. where R is the distance from the central absorbing atom and N is the coordination number. The paths used here are the highest amplitude FEFF paths out of all calculated theoretical paths. Initially, the best fit was attempted to achieve using a substantially minimum number of theoretical paths, and based on the best-fit outcome, these paths were identified. In the case of ANA300, three Ge–Ge and three Ge-Ni paths and one Ge–O paths were used for the fitting. It is evident from Table 1 that the 100°C sample has a concentration of Ge content (44%) with the remaining concentration restructured into Ni_3Ge (54%) and GeO_2 (2%). On increasing the annealing temperature to 300°C, a systematic decrease in Ge is observed with the rise of Ni_3Ge concentration. From this investigation, it is clear that for 300°C sample, more Ge is converted into Ni_3Ge concentration. The concentration of GeO_2 is same as ANA100 sample within the difference of error limit. From this work, we propose that the Ni_3Ge may be responsible for Ohmic contact. Further, GeO_2 is structured along with the Ni_3Ge, and it may result to increase in the contact resistance. The similar investigation is repeated with 10 nm Ni intermediate layer to observe the effect Ni layer thickness on the GeO_2 and Ni_3Ge phase formation.

3.2 EXAFS Investigation of Au88Ge12(50 nm)/Ni(10 nm)/ Au88Ge12(50 nm)

The Ge K-edge EXAFS signal Fourier transform magnitude of AuGe/Ni(10 nm) / AuGe annealed at different temperatures 100°C and 300°C along with the reference

Table 1 EXAFS fit outcome of AuGe/Ni/AuGe multilayers having Ni thickness (25 nm and 10 nm) annealed at different temperatures

Annealing temperature (°C)	AuGe/Ni (25 nm)/AuGe			AuGe/Ni (10 nm)/AuGe		
	Ge (%)	Ni_3Ge (%)	GeO_2 (%)	Ge (%)	Ni_3Ge (%)	GeO_2
100	44	54	2	45.2	54.8	–
300	14.5	84	1.5	27.8	72.2	–

sample Ge and Ni_3Ge EXAFS spectra is presented in Fig. 2a. It is evident from the figure that the EXAFS spectra at lower R value of the 100°C sample have main features of Ge whereas the features of Ni_3Ge are noticeable at higher R values. On the other hand for the sample annealed at 300°C, Ni_3Ge features are more dominant than Ge features. EXAFS spectra of Au88Ge12(50 nm)/Ni(10 nm)/Au88Ge12(50 nm) multilayers annealed at 100°C and 300°C and the best fit (dotted lines) along with the theoretical scattering paths of different reference samples are presented in Fig. 2b and c, respectively. From both the figures, it is shown that the experimental data (solid lines) exactly overlaps the fitted data (black dotted lines) which confirms the better fitting quality. From the EXAFS qualitative study, it is observed that the 100°C multilayer will have majority Ge and Ni_3Ge reference paths. Five theoretical FEFF paths were used to fit the 100°C annealed sample. (i) Ge-1, $R = 2.45$ Å, $N = 4$; (ii) Ge-2, $R = 3.8$ Å, $N = 12$; (iii) N_3G-1, $R = 2.4$ Å, $N = 6$; (iv) N_3G-2, $R = 2.53$ Å, $N = 12$; (v) N_3G-3, $R = 4.3$ Å, $N = 24$. In the case of 300°C, one Ge–Ge and three Ge-Ni paths were used for the EXAFS fitting, and the fit outcome is incorporated in Table 1.

From Table 1, it is evident that the ANA100 sample has a large concentration of Ni_3Ge content (~55%) with the remaining concentration restructured into Ge (~55%). A high reduction in Ge percentage is observed with the sharp rise of Ni_3Ge concentration is observed for the sample annealed at 300°C. From both these observations,

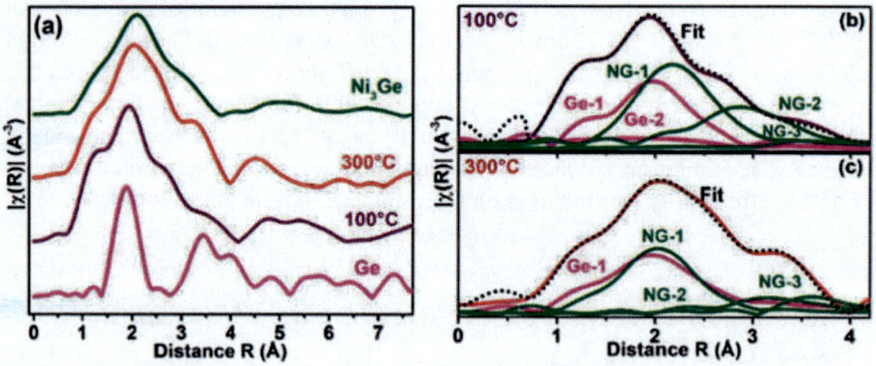

Fig. 2 **a** Fourier transfer magnitude of EXAFS signal of AuGe/Ni(10 nm)/AuGe annealed at 100°C and 300°C along with reference spectra. EXAFS experimental data (solid line), best-fit data (black dotted line) and theoretical paths of above stack annealed at **b** 100°C **c** 300°C

we strongly confirmed that Ni_3Ge is responsible for low resistance Ohmic contact. No oxide formation is recorded in Au88Ge12(50 nm)/Ni(10 nm)/Au88Ge12(50 nm) multilayers. So, we can conclude that, by reducing the thickness for Ni layer to 10 nm one can reduce the formation of excess oxide at the interface, which thereby results to low resistance Ohmic contact.

4 Conclusions

A stack of AuGe/Ni/AuGe is deposited on GaAs/AlGaAs by thermal evaporation and electron beam evaporation techniques. EXAFS investigation was carried out at Ge K-edge to investigate the crystallographic structural changes at the interface. The stack was heat treated at 100°C and 300°C to facilitate the interdiffusion of the metal constituents across the interface. This diffusion results in the formation of new phases Ni_3Ge and GeO_2. An optimum choice of Ni thickness will result in the development of a Ni_3Ge sandwiched layer without GeO_2 residues. From this work, we propose that Ni_3Ge with no oxide residue will be responsible for the low resistance Ohmic contact of AuGe/Ni/AuGe metallization to GaAs/AlGaAs heterostructure.

Acknowledgements The authors acknowledge DST (20160323) and Elettra Synchrotron Centre (20160323) for providing travel facilities and Elettra XRF beamline access. The authors acknowledge Dr. Ch. Ravi Kumar, Prof. Rajaram and Prof. Ghanashyam Krishna, School of Physics, University of Hyderabad for providing the samples required and discussion during this work. Authors also acknowledge Prof. Francesco Rocca, University of Trento, Italy, for his kind support during the experimentation.

References

1. Braslau N (1981) Alloyed ohmic contacts to GaAs. J Vac Sci Technol 19:803–807
2. Abhilash TS, Kumar CR, Rajaram G (2009) Influence of nickel layer thickness on the magnetic properties and contact resistance of AuGe/Ni/Au ohmic contacts to GaAs/AlGaAs heterostructures. J Phys D Appl Phys 42(12):125104
3. Abhilash TS, Ravi Kumar C, Rajaram G (2011) Magnetic, electrical and surface morphological characterization of AuGe/Ni/Au ohmic contact metallization on GaAs/AlGaAs multilayer structures. J Nano Electron Phys 3:396–403
4. Abhilash TS, Ravi Kumar CH, Rajaram G (2010) Nickel dissolution into AuGe in alloyed AuGe/Ni/Au ohmic contacts on GaAs/AlGaAs multilayer structures. Thin Solid Films 518(19):5576–5578
5. Abhilash TS, Ravi Kumar C, Rajaram G (2012) Magnetic evidence for solid-state solubility-limited dissolution of Ni into AuGe in alloyed AuGe/Ni/Au ohmic contacts to GaAs. J Appl Phys 112
6. Rothe J, Léon A (2008) X-ray absorption fine structure (XAFS) spectroscopy. Hydrogen Technol 603–622

7. Ravel B (2000) Introduction to EXAFS experiments and theory, Matthew Newville, consortium for advanced radiation sources, University of Chicago. 1–41
8. Newville M (2014) Fundamentals of XAFS. Rev Miner Geochem 78(1):33–74

A Study of NanoTitanium Dioxide-Silver Ferrite Composite: Synthesis, Characterization, and Band Gap Evaluation

T. Kondala Rao and Y. L. N. Murthy

Abstract Multiferroic materials are the materials that exhibit both ferromagnetic and ferroelectric phases. This type of phenomenon is rare in materials and so the coupling of these properties is useful to change the magnetic ordering of materials by applying electric field and to change the electric polarization of materials by applying magnetic field. These are novel materials of having a variety of applications such as microwave components, energy harvesters, photovoltaics, data storage devices. In this paper, we made a study on two different methods of synthesis of titanium dioxide-silver ferrite nanocomposite. The characterization reveals that the hybrid nanocomposite particles of sizes less than 100 nm with homogeneous silver ferrite distribution on nanoTiO_2 and also coexistence of orthorhombic silver ferrite on tetragonal rutileTiO_2 in both synthetic approaches. The band gaps of this nanocomposites are evaluated by UV–visible spectroscopy while the magnetic properties are from VSM. Results shows there is decrease in the band gap, ferroelectric, and ferromagnetic behaviors. Hence, the synthesized TiO_2-$AgFeO_2$ nanocomposite is multiferroic magnetoelectric material. Therefore, it may be a promising candidate for photo catalysis, photovoltaics, optoelectronics, and nanoscale magnetic storage devices.

Keywords Multiferroic · Titanium dioxide-Silver ferrite nanocomposite · Band gap · Ferroelectric · Ferromagnetic

T. Kondala Rao (✉)
Sri Sivani College of Engineering, Srikakulam, India
e-mail: kondalaraotata@gmail.com

Y. L. N. Murthy
Andhra University, Visakhapatnam, India

Z. H. Khan et al. (eds.), *Recent Advances in Nanomaterials*, Springer Proceedings in Materials 27, https://doi.org/10.1007/978-981-99-4878-9_22

1 Introduction

Titanium dioxide-based composites have a variety of applications in several areas such as paints, polymer pigments, foods, medicines, ultraviolet light protectors, photocatalysis, and storage of energy or its conversion due to their special electronic arrangement, good inertness, and high refractive index [1]. Investigation of finding TiO_2 hybrid materials with enhanced properties is a subject of extensive research in literature which is due to exciting practical uses of these materials in the evolution of hydrogen and in the removal of pollutants from the environment [2, 3]. One of the best methods is to band gap decrease TiO_2 for improving its photoactive nature by the inclusion of oxides of metals or pure metals. Iron (III) oxide (Fe_2O_3)-TiO_2 composites have more optical and morphological properties due to greatly diminished band gap. Some studies also reveal that the unique potential of Fe_2O_3 doped TiO_2 materials has their application as catalysts [4]. Nanoferrite composites consist of multiphase materials with required proportions in each. Composition of the materials should be optimized in ordered to achieve desirable properties for their application as chlorine gas sensors, high audio frequency transformers, microwave absorbers, magneto-optical displays, and high storage of information [5, 6] etc. Ceramic nanocomposites based on ferrites are recently studied widely because of their importance both in research and engineering fields, in particular for various electronic materials due to their multiferroic nature [7]. Several researchers have reported studies on transition metal-doped Bismuth Ferrite [8, 9]. Ferrites are mainly prepared by several techniques including solid-state chemical reaction [10, 11], sol–gel process [12, 13], hydrothermal synthesis [14], combustion [15] etc. Optical properties and band gap of perovskite and spinel ferrites are widely reported, for example, nanoNickel Ferrite powders and its films. Till date, very few studies have appeared on the synthesis and band gap evaluation of perovskite–spinel ferrite nanocomposite powders. The band gap of the nanocomposite material decreases with the reduction in its size [16]. Silver ferrite and other ferrites are found to exhibit good ferromagnetic properties [17, 18]. Rutile titanium dioxide was synthesized and characterized [19]. Rutile form of TiO_2 is an incipient ferroelectric material for which several theoretical studies shown that a ferroelectric transition can be enforced [20]. However, the experimental studies of ferroelectric TiO_2 (110) surfaces would have great impact while tuning the electric permittivity would also be of interest [21]. We made an attempt to synthesize nanocomposite of titanium dioxide-silver ferrite and studied its characterization, magnetic nature, and band gap determination.

2 Materials and Methods

2.1 *Materials*

Main chemicals used in this method are Titanium tetra chloride, Silver nitrate, Ferric nitrate, Ammonium hydroxide glycerol. They are of BDH company (India) grade, used as such for the synthesis of titanium dioxide-silver ferrite nanocomposite.

2.2 *Synthesis of Titanium Dioxide-Silver Ferrite Nanocomposite Involves Three Stages*

(i) **Preparation of nanocrystalline Rutile TiO_2:** It was synthesized by hydrolysis of $TiCl_4$ by ammonium hydroxide. In this typical synthesis, $TiCl_4$ was diluted with ice-cold distilled water and mixed with NH_4OH in a 1:1 molar ratio with continuous magnetic stirring. Since there was no precipitation during mixing, the pH of the solution was not varied. The solution was stirred mechanically for 2 h in a water bath at 60 °C, a white gel was produced after the evaporation of water. Subsequently, the gel was further heated at 600 °C for 2 h to form nanoparticles of rutile TiO_2.

(ii) **Preparation of Silver Iron hydroxide slurry:** Nitrate salts of silver and iron were dissolved in stoichiometric ratio double distilled water separately and mixed (stirring at 60–70 °C); sodium hydroxide solution was added to obtain precipitate of silver and iron hydroxides; the precipitate obtained was filtered and washed with double distilled water thoroughly, and dried in air for 70 °C to form amorphous silver iron hydroxide slurry.

(iii) **Synthesis of TiO_2@ Silver ferrite nanocomposite:** The nanocomposite of titanium dioxide-silver ferrite was synthesized by using the chemical route. Mixing of nanotitanium dioxide and silver iron hydroxide slurry was done by two methods. In the first method, functionalization of nanoTiO_2 was done by glycerol which was then coated with amorphous silver iron hydroxide and left the mixture about 24 h for aging. In the other method, nanoTiO_2 and amorphous silver iron hydroxide were thoroughly mixed for 2 h and then left for 24 h for aging. The obtained mixtures of both methods were thermally treated at 600 °C for 3 h separately to form powders of titanium dioxide-silver ferrite nanocomposites.

(iv) **Characterization**: X-ray diffraction technique was used to identify the crystalline phases and estimate the crystallite size of the titanium dioxide-silver ferrite nanocomposite derived from the chemical method. The XRD patterns were recorded with 2θ in the range of 20°–80°, EDX data. The scanning electron micrograph (SEM) and Transmission electron micrographs (TEM) images were used to know the morphology of the TiO_2-$AgFeO_2$ nanocomposite. UV–VIS spectroscopy in the range of 200–1000 nm in transmittance mode is used to determine the band gap. The magnetic nature of the annealed samples obtained by different methods was measured at 27 °C by using a vibrating sample magnetometer (VSM).

3 Results and Discussion

XRD patterns of synthesized TiO_2-$AgFeO_2$ nanocomposite shows that of Both the peaks of nanorutileTiO_2 [TiO_2 (110), TiO_2 (111) at 27.5°, 41.3° as in JCPDS card 21–1276] [19] and XRD peaks of nano$AgFeO_2$ [$AgFeO_2$ (110) and $AgFeO_2$ (111)] [17] appear at the same 2θ region in XRD patterns (Fig. 1a and b) and the diffraction patterns confirm crystallinity as well as phase purity. The calculated results of particle sizes of TiO_2-$AgFeO_2$ nanocomposite from Scherer's formula showed as 53 nm to 60 nm for functionalized by glycerol and no functinalization methods with the perovskite-rambhohydral phase respectively. EDX data confirms formation of the nanocomposite (Fig. 2). Scanning electron micrographs indicate that the nanocomposites have polygonal agglomeration of the two oxides synthesized. Transition electron microographs suggest interconnected agglomeration of polygonal silver ferrite on the tetragonal nanotitanium dioxide as shown Figs. 3, 4 and 5.

UV–VIS spectroscopic studies are performed for synthesized nanoTiO_2-$AgFeO_2$ sample in the range of 200–1000 nm. The test sample of nanoTiO_2-$AgFeO_2$ is ultrasonicated in de-ionized water for 50 min. The band gaps of prepared samples by both methods are deduced from Tauc relation $(\boldsymbol{\alpha}\mathbf{h}\,\vartheta) = \boldsymbol{B}\,(\mathbf{h}\,\vartheta - \boldsymbol{Eg})^{\mathbf{n}}$ where $n = 0.5$ for the allowed direct transition. $(\boldsymbol{\alpha}\mathbf{h}\,\vartheta)^2$ versus Energy $\mathbf{h}\,\vartheta$ plots (Figs. 6 and 7) help to calculate the direct band gap. Linear region is extrapolated to the zero absorption on the X-axis provides the band gap (allowed band gap). The band gaps obtained for the nanocomposite prepared by different methods are, 3.0 eV and 3.1 eV, respectively. The band gap obtained for the nanocomposite is in the range of semiconductors. It is found that present synthesis routes and the optimized parameters for synthesis of TiO_2-$AgFeO_2$, effect on the particle size of the composite as well as on the band gap.

It also indicates that the decrease in the band gap to 3.0 ev, 3.1 ev is observed depending on the nature of the functionalizing material used in the synthesis of TiO_2-$AgFeO_2$ nanocomposite which suggesting that micro strain, presence of oxygen defects, and exchange interactions, show considerable impact on the band gap.

The band gap values are greater than 2.7 ev, more splitting XRD spectral lines in synthesized nanocomposite and rutile TiO_2 component has ferroelectric phase [20, 21] indicate TiO_2-$AgFeO_2$ nanocomposite may have ferroelectric nature and the M-H analysis of the TiO_2-$AgFeO_2$ nanocomposite heated at 600 °C for 3 h gave hysteresis loops indicating soft ferromagnetic behavior (Fig. 8 and Table 1).

Overall study of titanium dioxide-silver ferrite nanocomposite reveals that it has both ferromagnetic and ferroelectric natures, so it is multiferroic material, hence, it

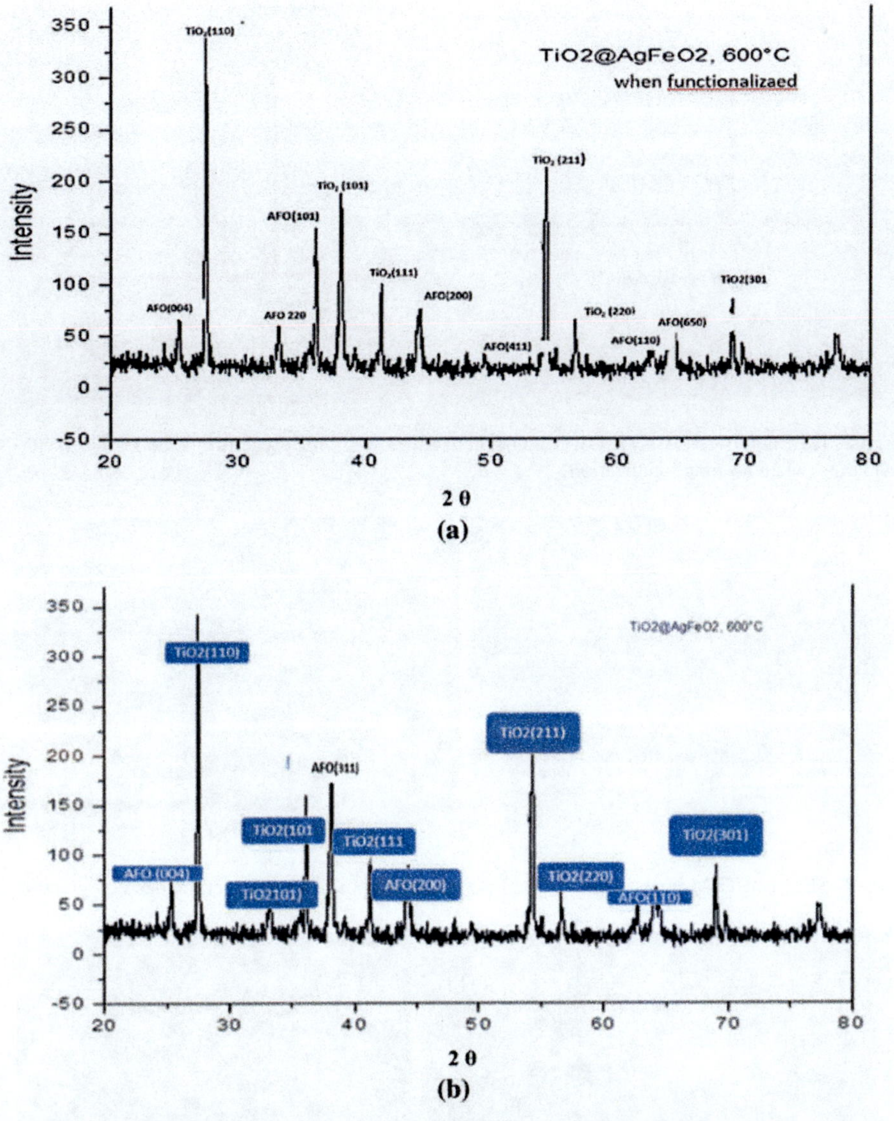

Fig. 1 **(a)** XRD of TiO2-AFO (titanium dioxide-silver ferrite) nanocomposite heat treated at 600 °C (when functionalized by glycerol). **(b)** XRD of TiO2-AFO (nanotitanium dioxide-silver ferrite) composite heat treated at 600 °C (not functionalized)

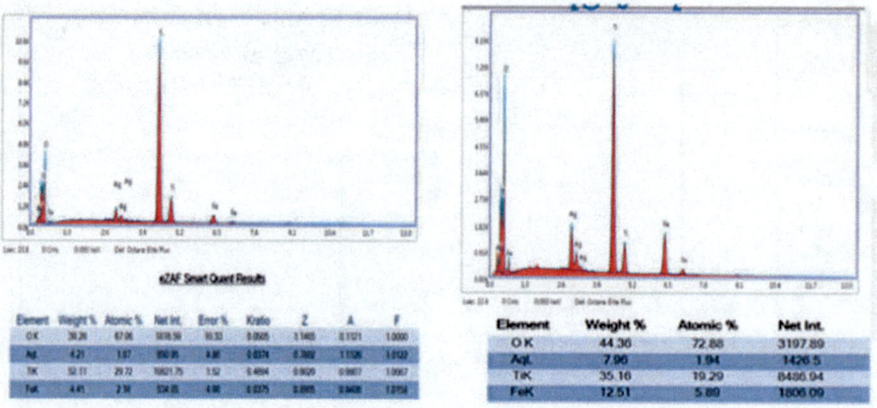

Fig. 2 (**a**) EDX of TiO_2-$AgFeO_2$ (when functionalization with glycerol). (**b**) EDX of TiO_2-$AgFeO_2$ (when no functionalization)

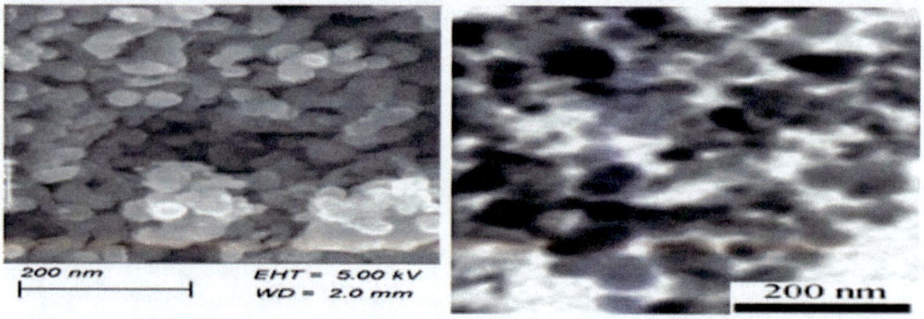

SEM TEM

Fig. 3 SEM and TEM images of TiO_2-$AgFeO_2$ nanocomposite when functionalized by glycerol

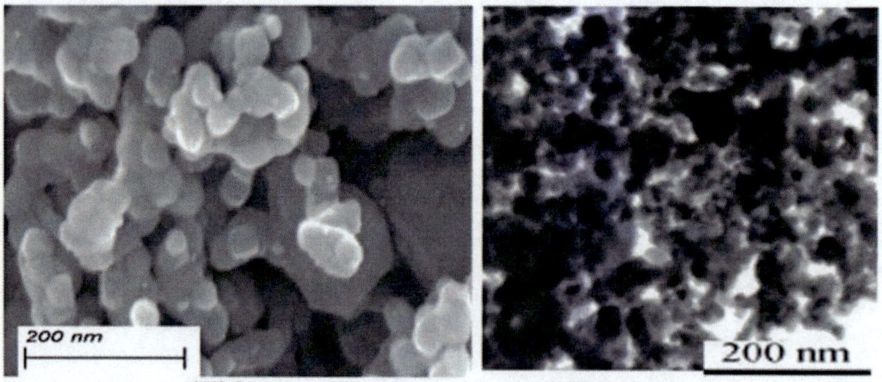

SEM TEM

Fig. 4 SEM and TEM images of TiO_2-$AgFeO_2$ nanocomposite when no functionalization

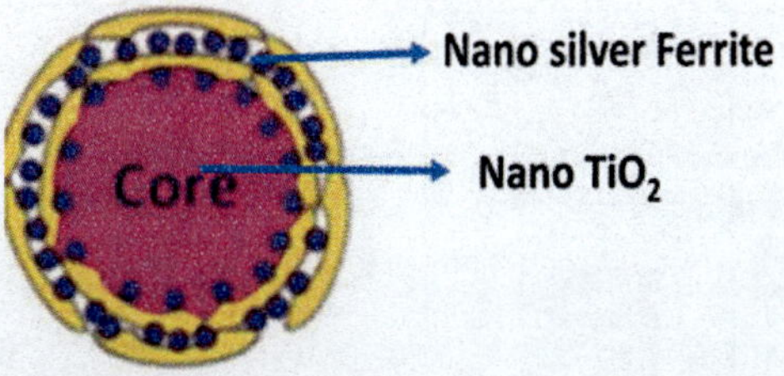

Fig. 5 Pictorial representation of nanocomposite

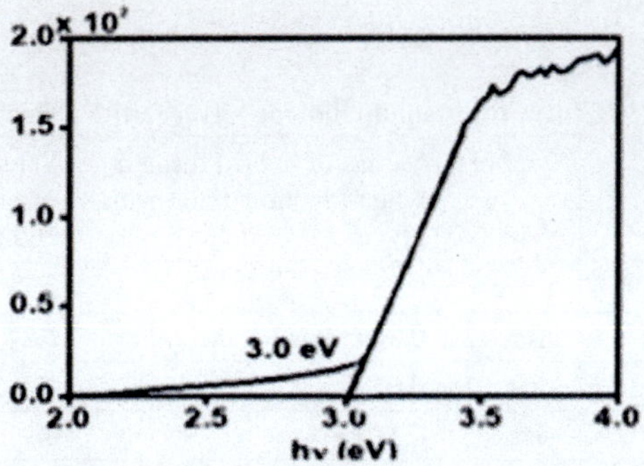

Fig. 6 Band gap of the nanocomposite prepared when functionlized by glycerol

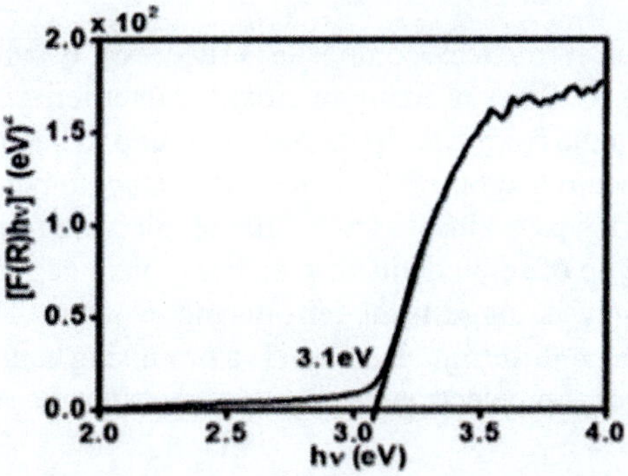

Fig. 7 Band gap of the composite prepared without functionalization

may be promising material for photocatalysis, photovoltaic, optoelectronics, sensors, and some important multiferroic nanoscale magnetic storage devices.

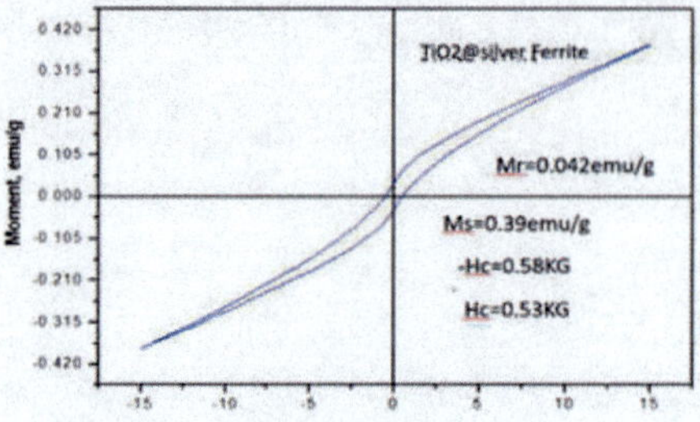

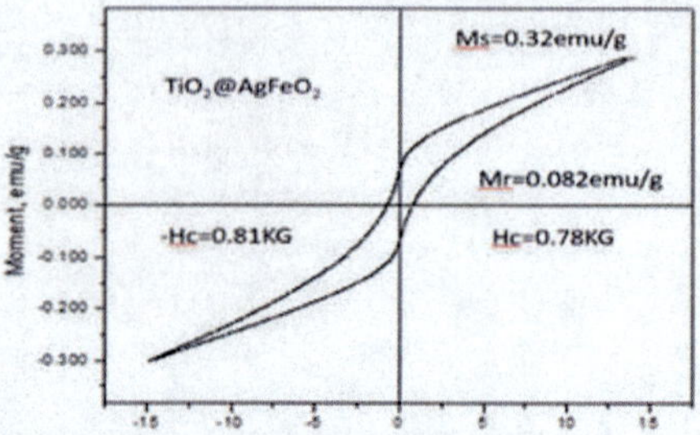

Fig. 8 *M-H* curve of titanium dioxide-$AgFeO_2$ composite functionalized with glycerol (53 nm) and no functionalization (60 nm)

Table 1 Parameters of *M-H* curve for titanium dioxide-silver ferrite nanocomposite

Magnetic parameter	Symbol	Values of hybrid formed when functionalized with glycerol	Values of hybrid formed when functionalized without glycerol
Coercivity	+ Hc	0.53 kg	0.78 kg
Saturated magnetization	Ms	0.39 emu/g	0.32 emu/g
Remanence magnetization	Mr	0.042 emu/g	0.082 emu/g
Coercivity	−Hc	0.58 kg	0.81 kg

4 Conclusion

Titanium dioxide-silver ferrite nanocomposite is successfully synthesized by a facile chemical route at 600 °C. Sizes of titanium dioxide-silver ferrite particles are calculated by Scherer's formula from XRD patterns, are found to be in the range of 53 nm to 60 nm after 3 h annealing at 600 °C. Interconnected agglomeration is observed in the samples. The band gaps evaluated from Tauc relation are found to be 3.0 ev, 3.1 ev, close to the band gap of semiconducting material band gap indicating ferroelectric behavior. *M-H* analysis suggests the ferromagnetic nature. The afore mentioned nanocomposite may be multiferroic. Hence, it is a promising nanomaterial for photocatalysis, photovoltaics, optoelectronics, sensors, multiferroic nanoscale magnetic storage devices, etc.

Acknowledgements The author is grateful to the Management, Principal, Head (Humanities and Basic Sciences) of Sri Sivani College of Engineering for the support and encouragement.

Declaration of Interest Statement The author declare that they have no conflict of interests such as financial and competing interests.

References

1. Ma Y, Wang X, Jia Y, Chen X, Han H, Li C (2014) Titanium dioxide-based nanomaterials for photocatalytic fuel generations. Chem Rev 114:9987–10043
2. Daghrir R, Drogui P, Robert D (2013) Modified TiO_2 for environmental photocatalytic applications: a review. Ind Eng Chem Res 52(10):3581–3599
3. Kim JW, Shim JW, Bae JH, Han SH, Kim HK, Chang IS, Kang HH, Suh KD (2002) Titanium dioxide/poly (methyl methacrylate) composite microspheres prepared by in situ suspension polymerization and their ability to protect against UV rays. Colloid Polym Sci 280(6):584–588
4. Kundu TK, Mukherjee M, Chakravorty D, Sinha TP (1998) Growth of nano-a-Fe_2O_3 in a titania matrix by the sol–gel route. J Mater Sci 33(7):1759–1763
5. Pankhurst QA, Connolly J, Jones SK, Dobson J (2003) Applications of magnetic nanoparticles in biomedicine. Phys D Appl Phys 36(13):R167–R181
6. Tartaj P, del-Puerto-Morales M, Veintemillas-Verdaguer S, Gonzalez-Carreno T, Serna CJ (2003) The preparation of magnetic nanoparticles for applications in biomedicine. J Phys D Appl Phys 36:R182–R197
7. Pullar RC (2012) Hexagonal ferrites: a review of the synthesis, properties and applications of hexa ferrite ceramics. Prog Mater Sci 57:1191–1334
8. Luo B-C, Chen C-L, Xu Z, Xie Q (2010) Effect of Cr substitution on the multiferroic properties of $BiFeO_3$ compounds. Phys Lett A 374(41):4265–4268
9. Kumar A, Yadav KL (2011) Magnetic, magnetocapacittance and dielectric properties Cr doped Bismuth ferrite nanoceramics. Mater Sci Eng B 176:227–230
10. Li J, Duan Y, He H, Song D (2001) Crystal structure, electronic structure, and magnetic properties of bismuth-strontium ferrites. J Alloy Compd 315:259–264
11. Valant M, Axelsson KA, Alford N (2007) Peculiarities of a solid-state synthesis of multiferroic polycrystalline $BiFeO_3$. Chem Mater 19:5431–5436
12. Sen K, Singh K, Gautam A, Singh M (2012) Dispersion studies of La substitution on dielectric and ferroelectric properties of multiferroic $BiFeO_3$ ceramic. Ceram Int 38:243–249
13. Kumar M, Yadav KL (2007) Magnetoelectric characterization of $x\mathrm{Ni_{0.75}Co_{0.25}Fe_2O_4}-(1-x)\mathrm{BiFeO_3}$ nanocomposites. J Phys Chem Solids 68(9):1791–1795
14. Du Y, Cheng YZ, Shahbazi M, Collings WE, Miotell, Dou XS, Wang LX (2010) Enhancement of ferromagnetic and dielectric properties in lanthanum doped $BiFeO_3$ by hydrothermal synthesis. J Alloy Compd 490:637–641
15. Dixit G, Singh PJ, Srivastava CR, Agarwal MH, Chaudhary JR (2012) Structural, magnetic and optical studies of nickel ferrite thin films. Adv Mater Lett 3(1):21–28
16. Chaturvedi S, Das R, Poddar P, Kulkarni S (2015) Tunable band gap and coercivity of bismuth ferrite–polyaniline core–shell nanoparticles: the role of shell thickness. RSC Adv 5:23563–23568
17. Murthy YLN, Kondala Rao T, Kasi Viswanath IV, Rajendra Singh (2010) Synthesis and characterization of nano silver ferrite composite. J Magn Magn Mater 322:2071–2074
18. Kondala Rao T, Jagadeeswara Rao Ch, Magdelono T, Kasi Viswanath IV, Murthy YLN (2012) Synthesis and characterization of nanocrystalline cobalt ferrite spheres with silver metal core. Int J Nanosci Nanotechnol 3(3):139–147
19. Haider A, Jameel ZN, Taha Y (2015) Synthesis and characterization of TiO_2 nanoparticles via sol gel method by pulse laser ablation. Eng Tech J 33(5) Part (B)
20. Lee C, Ghosez P, Gonze X (1994) Lattice dynamics and dielectric properties of incipient ferroelectric TiO_2 rutile. Phys Rev B 50:13379
21. Montanari B, Harrison NM (2004) Pressure-induced instabilities in bulk TiO_2 rutile. J Phys Condens Matter 16(3):273

Understanding the Redox Mechanism of Layered Transition Metal Oxide During Electrochemical Cycling in Sodium-Ion Batteries

Nikita Bhardwaj, Mohammed Saquib Khan, Deependra Jhankal, Deepika Choudhary, Preeti, Himmat Singh Kushwaha, and Kanupriya Sachdev

Abstract From the perspective of the lithium resources situation on earth, it is very vital to develop sodium-ion batteries (SIB) as lithium-ion technologies alternatives. Delivering high capacity along with adequate capacity retaining is difficult in designing favorable cathode material for SIB. Among all of the positive electrode (cathode) materials for SIB, layered transition metal oxides (LTMO) with remarkable properties have attracted intensive interest. In this work, we intend to enhance the electrochemical behavior of O3 $Na[Fe_{0.27}Cu_{0.27}Mn_{0.46}]O_2$ (NFCM) in SIB with graphene oxide in reduced form (rGO), as the negative electrode (anode) material. X-ray diffraction spectrum confirms the presence of expected phase of LTMO. Electrochemical impedance spectroscopy (EIS), Galvano static charge–discharge (GCD), and Cyclic voltammetry (CV) are performed, and material delivers a excellent specific capacity value of 116 $mAhg^{-1}$ with efficient capacity retention at the current rate of 1C.

Keywords Rechargeable sodium-ion batteries · Cyclic voltammetry · X-ray diffraction · Layered transition metal oxides · Raman spectroscopy

1 Introduction

Due to the decline of traditional energy sources, lithium-ion technology has become essential in the domain of energy storage, much like solar and wind energy being alternatives to conventional energy sources [1]. Despite the vital role that lithium

N. Bhardwaj · D. Jhankal · D. Choudhary · K. Sachdev (✉)
Department of Physics, Malaviya National Institute of Technology Jaipur, JLN Marg, Jaipur 302017, India
e-mail: Ksachdev.phy@mnit.ac.in

M. S. Khan · Preeti · H. S. Kushwaha · K. Sachdev
Material Research Center Malaviya National Institute of Technology Jaipur, JLN Marg, Jaipur 302017, India

Z. H. Khan et al. (eds.), *Recent Advances in Nanomaterials*, Springer Proceedings in Materials 27, https://doi.org/10.1007/978-981-99-4878-9_23

technology plays in the areas of transportation and electronic devices, there are significant problems with its scarcity and geographic distribution on earth, which has prompted researchers to look for alternatives [2]. SIBs are an appealing alternate option to lithium technology owing to their minimal cost, abundant supply of sodium resources, and similarities in electrochemical properties between sodium and lithium ions [3]. Almost all the sodium layered oxides, Na_xMO_2, are unstable in the presence of moisture (either they can be oxidized when they are in contact with water, or carbon dioxide or water molecules can enter the layer of alkali metal) [4]. This results in increasing the substantial cost of storing materials, transportation, and production of batteries with instability in the performance of batteries. Accordingly, it is essential in the extreme to increase air or moisture stability of layered oxides in the point of view for practical applications [5]. Although cobalt and nickel incorporation gives better electrochemical performance but they are toxic and the fact that they have widespread used in lithium- and sodium-ion batteries would drive up their prices, it is crucial to develop alternatives that do not rely on cobalt or nickel in their battery materials [6]. So it is necessary to develop cobalt and nickel free layered oxide material as cathode with increased structure stability and electrochemical properties. Among all layered oxide materials, Mn-based oxides hold an important role due to their good electrochemical behavior. Incorporation of copper and iron provide them structure stability and high redox potential resulting in enhanced electrochemical performance [7]. Based on literature reports and the current situation, we have examined the electrochemical behavior of NFCM in full cell configuration using rGO as negative electrode material. NFCM was prepared using the coprecipitation method, and XRD spectra validated the occurrence of required phase in the sample. The I_D/I_G ratio of rGO was calculated using the D and G bands intensities. NFCM delivered 93 $mAhg^{-1}$ specific capacity after 25 cycles.

2 Materials and Methods

2.1 Chemicals

$FeCl_3$, $MnCl_2$, $CuCl_2$, NaOH, H_3PO_4, H_2SO_4, H_2O_2, and HCl were purchased from Merck. $NaClO_4$, propylene carbonate, sodium metal foil, ethylene carbonate, NH_4OH, $KMnO_4$, glycerol, and ethanol from Loba chemicals and graphite flakes (mesh size ~ 325) were purchased from Alfa Aesar.

2.2 Synthesis of O3 Type $Na[Fe_{0.27}Cu_{0.27}Mn_{0.46}]O_2$ and Reduced Graphene Oxide

Using the coprecipitation method followed by two-step calcination NFCM was synthesized. First, 1 M NaOH and 3 M NH_4OH were mixed in a beaker with stirring. Dropwise, 0.27 M $CuCl_2$ and $FeCl_3$ precursor solutions were added to the earlier solution at 700 rpm. After 10 min, 50 mL glycerol was added to prevent agglomeration and stirred for 30 min. The solution was cleaned with centrifugation and washing two–three times with ethanol, deionized water (DI), dried overnight at 80 °C and after drying, calcinated at 500 °C for 4 h to remove chloride and nitrogen impurities, and at 900 °C for 10 h to get the O3 phase. The final product cooled to room temperature.

GO was synthesized using Toor's method. In a mortar pestle, graphite flakes and $KMnO_4$ powder (1:9) were mixed and ground. Then, a mixture of 360 ml H_2SO_4 and 40 ml H_3PO_4 was slowly poured with stirring. The solution was stirred at 50 °C for 12 h at 300 rpm after mixing. Ice cubes of 400 ml of deionized water was poured in the solution cooled to room temperature. After dissolving the ice cubes, 3 ml of 35% H_2O_2 was added while the beaker was covered with aluminum foil. The solution was cleansed three times with hydrochloric acid, deionized water (DI), and ethanol (each 200 ml) and dried out overnight at 80 °C. To reduce graphene oxide, it was ground and heated at 100 °C in deionized water with 10 g of citric acid.

3 Results and Discussion

3.1 Structure Analysis Results

The XRD pattern for NFCM, calcinated at 700 °C in Fig. 1a demonstrate that it is highly crystalline in nature, as indicated by sharp and strong peaks. This highly crystalline nature is responsible for easy de-intercalation and intercalation of Na^+ ions, which leads toward improved electrochemical performance [8]. It was possible to index the pattern using the $NaMnO_2$ model structure, which possessed monoclinic symmetry and the *C2/m* space group (PDF No: 00-025-0845). The XRD plot demonstrates that the majority of significant peaks can be attributed to planes with indexes including (001), (20-1), (002), (200), (−111), (−202), (111), (−113), and (310) that are in agreement with what was previously reported [9]. In Fig. 1b Raman spectrum of NFCM, the bands detected on 290 and 600 cm^{-1} can be accredited to Na–O and TM–O vibrations (Ag + Bg), while a peak at 290 cm^{-1} in the low-frequency approach is to be expected caused by Na–O bond vibrations [10]. The presence of the TM–O bond stretching mode in the cubic lattice structure of oxygen is the source of the peak that can be found at a frequency of 600 cm^{-1} [11]. In Fig. 1c the Raman spectrum of GO covers peaks for *D* and *G* bands at 1353 cm^{-1} and 1586 cm^{-1}, and a 2D band peak at 2701 cm^{-1}, whereas rGO covers peaks for *D*, *G*, and 2D at

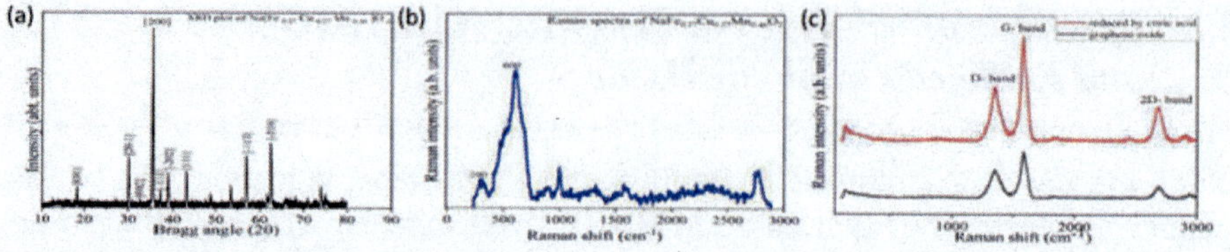

Fig. 1 **a** XRD spectra of synthesized O3 type $Na[Fe_{0.27}Cu_{0.27}Mn_{0.46}]O_2$. **b** Raman spectra of NFCM. **c** Raman spectra of GO and rGO

1341 cm^{-1}, 1552 cm^{-1}, and 2689 cm^{-1} [12]. GO and rGO have excellent exfoliation, as evidenced by the 2D peak [13]. The I_D/I_G ratios for GO and rGO were calculated 0.63 and 0.52, respectively, indicating GO's greater defect density [14]. I_{2D}/I_G ratios for GO and rGO were 0.22 and 0.33, indicating the samples had few layers [15].

3.2 *Electrochemical Results*

Electrochemical measurements as well as CV and GCD tests were conducted. The CV curves are shown in Fig. 2b scanned at 0.1 mVs^{-1} rate between a range of voltages 2.5 and 3.8 V. The peaks at 2.9 and 3.05 V potentials correspond to the oxidation reactions of Fe^{3+}/Fe^{4+} pair are responsible for the oxidation activity [16]. The NFCM electrode distributes a eminent initial specific capacity of about 116 mA h g^{-1}, along with a suitable capacity retaining of 90% after 25 cycles. From this it can be assumed, Cu^{2+} can act as a stabilizing agent by obstructing the reshuffling of Na^+ ion vacancy, easing the Jahn–Teller effect of Mn^{3+} ions center on further sodiation/desodiation (reducing the Mn^{3+} concentration), and overwhelming the irreversible relocation of Fe atoms on the Na sites when the battery is charged to a high voltage [7, 17].

The EIS plot in Fig. 1c entails a semicircle at high frequency value going along with a straight line curve at values of low frequencies [18]. In the region of high frequency, the semicircle represents the sodium-ion movement charge transfer resistance in the electrodes and electrolyte interface [19]. The Warburg zone is due for the diffusion of ions of sodium atoms in the electrodes, indicating the straight gradient line situated in the low-frequency area [20]. Charge transfer resistance (R_{ct}) gives the all over resistance of full cell configuration consisting of resistance because of the electrodes, electrolyte, and the separator [21]. The semicircles are not perfect, and it could be an indication of roughness of the electrode surface. Series resistance (R_s) and charge transfer resistance (R_{ct}) for full cell configuration before and after the cycling calculated from EIS plots which are 16.24 Ω, 737.4 Ω and 0.9547 Ω, 22.17 Ω, respectively.

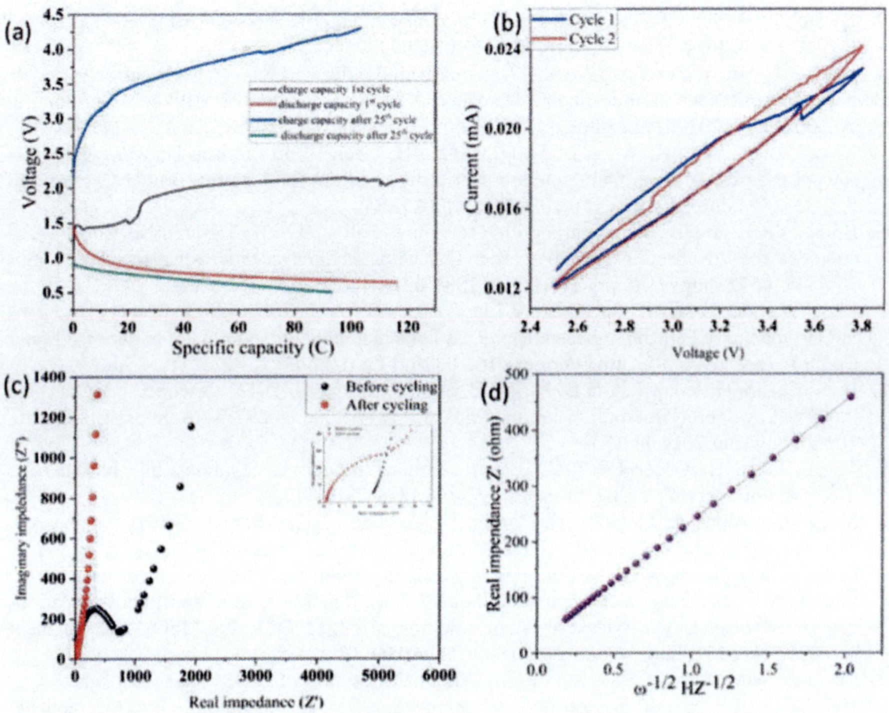

Fig. 2 **a** Specific capacity versus voltage plots. **b** CV profiles of full cell configuration $Na[Fe_{0.27}Cu_{0.27}Mn_{0.46}]O_2$||rGO at 0.1 mV scan rate. **c** EIS plot. **d** Plot Z' (real impedance) versus $\omega^{-1/2}$(sqr root of 2*pi*freq.) in the low-frequency region

4 Conclusion

In conclusion, using the coprecipitation method, we were able to successfully synthesize O3 type $Na[Fe_{0.27}Cu_{0.27}Mn_{0.46}]O_2$ and investigate its potential for Na-ion batteries. XRD confirmed the presence of phase in the material. After nearly 25 cycles, full cell was able to achieve a value of specific capacity of 93 $mAhg^{-1}$ when operated at a good current rate of 1C.

References

1. Nayak PK, Yang L, Brehm W, Adelhelm P (2018) From lithium-ion to sodium-ion batteries: advantages, challenges, and surprises. Angew Chem Int Ed 57:102–120. https://doi.org/10.1002/anie.201703772
2. Yabuuchi N, Kubota K, Dahbi M, Komaba S (2014) Research development on sodium-ion batteries. Chem Rev 114:11636–11682. https://doi.org/10.1021/cr500192f

3. Hwang J-Y, Myung S-T, Sun Y-K (2017) Sodium-ion batteries: present and future. Chem Soc Rev 46:3529–3614. https://doi.org/10.1039/C6CS00776G
4. Wang P-F, You Y, Yin Y-X, Guo Y-G (2018) Layered oxide cathodes for sodium-ion batteries: phase transition, air stability, and performance. Adv Energy Mater 8:1701912. https://doi.org/10.1002/aenm.201701912
5. Xiao J, Li X, Tang K, Wang D, Long M, Gao H, Chen W, Liu C, Liu H, Wang G (2021) Recent progress of emerging cathode materials for sodium ion batteries. Mater Chem Front 5:3735–3764. https://doi.org/10.1039/D1QM00179E
6. Boddu VRR, Puthusseri D, Shirage PM, Mathur P, Pol VG (2021) Layered Na_xCoO_2-based cathodes for advanced Na-ion batteries: review on challenges and advancements. Ionics 27:4549–4572. https://doi.org/10.1007/s11581-021-04265-w
7. Liu Y, Wang D, Li H, Li P, Sun Y, Liu Y, Liu Y, Zhong B, Wu Z, Guo X (2022) Research progress in O3-type phase Fe/Mn/Cu-based layered cathode materials for sodium ion batteries. J Mater Chem A 10:3869–3888. https://doi.org/10.1039/D1TA10329F
8. Han SC, Lim H, Jeong J, Ahn D, Park WB, Sohn K-S, Pyo M (2015) Ca-doped Na_xCoO_2 for improved cyclability in sodium ion batteries. J Power Sources 277:9–16. https://doi.org/10.1016/j.jpowsour.2014.11.150
9. Ma X, Chen H, Ceder G (2011) Electrochemical properties of monoclinic $NaMnO_2$. J Electrochem Soc 158:A1307. https://doi.org/10.1149/2.035112jes
10. Manzi J, Paolone A, Palumbo O, Corona D, Massaro A, Cavaliere R, Muñoz-García AB, Trequattrini F, Pavone M, Brutti S (2021) Monoclinic and orthorhombic $NaMnO_2$ for secondary batteries: a comparative study. Energies 14:1230. https://doi.org/10.3390/en14051230
11. Sato T, Sato K, Zhao W, Kajiya Y, Yabuuchi N (2018) Metastable and nanosize cation-disordered rocksalt-type oxides: revisit of stoichiometric $LiMnO_2$ and $NaMnO_2$. J Mater Chem A 6:13943–13951. https://doi.org/10.1039/C8TA03667E
12. Lucchese MM, Stavale F, Ferreira EHM, Vilani C, Moutinho MVO, Capaz RB, Achete CA, Jorio A (2010) Quantifying ion-induced defects and Raman relaxation length in graphene. Carbon 48:1592–1597. https://doi.org/10.1016/j.carbon.2009.12.057
13. López-Díaz D, López Holgado M, García-Fierro JL, Velázquez MM (2017) Evolution of the Raman spectrum with the chemical composition of graphene oxide. J Phys Chem C 121:20489–20497. https://doi.org/10.1021/acs.jpcc.7b06236
14. Pimenta MA, Dresselhaus G, Dresselhaus MS, Cançado LG, Jorio A, Saito R (2007) Studying disorder in graphite-based systems by Raman spectroscopy. Phys Chem Chem Phys 9:1276–1290. https://doi.org/10.1039/B613962K
15. Childres I, Jauregui LA, Park W, Cao H, Chen YP. Raman spectroscopy of graphene and related materials 20
16. Shen Q, Zhao X, Liu Y, Li Y, Zhang J, Zhang N, Yang C, Chen J (2020) Dual-strategy of cation-doping and nanoengineering enables fast and stable sodium-ion storage in a novel Fe/Mn-based layered oxide cathode. Adv Sci 7:2002199. https://doi.org/10.1002/advs.202002199
17. Zhou D, Zeng C, Xiang J, Wang T, Gao Z, An C, Huang W (2022) Review on Mn-based and Fe-based layered cathode materials for sodium-ion batteries. Ionics 28:2029–2040. https://doi.org/10.1007/s11581-022-04519-1
18. Li J-Y, Lü H-Y, Zhang X-H, Xing Y-M, Wang G, Guan H-Y, Wu X-L (2017) P2-type $Na_{0.53}MnO_2$ nanorods with superior rate capabilities as advanced cathode material for sodium ion batteries. Chem Eng J 316:499–505. https://doi.org/10.1016/j.cej.2017.01.109
19. Ledwoch D, Komsiyska L, Hammer E-M, Smith K, Shearing PR, Brett DJL, Kendrick E (2022) Determining the electrochemical transport parameters of sodium-ions in hard carbon composite electrodes. Electrochim Acta 401:139481. https://doi.org/10.1016/j.electacta.2021.139481
20. Dong X, Chen L, Liu J, Haller S, Wang Y, Xia Y (2016) Environmentally-friendly aqueous Li (or Na)-ion battery with fast electrode kinetics and super-long life. Sci Adv 2:e1501038. https://doi.org/10.1126/sciadv.1501038
21. Chandra M, Khan TS, Shukla R, Ahamad S, Gupta A, Basu S, Haider MA, Dhaka RS (2020) Diffusion coefficient and electrochemical performance of $NaVO_3$ anode in Li/Na batteries. Electrochim Acta 331:135293. https://doi.org/10.1016/j.electacta.2019.135293

Microwave Assisted Synthesis of N,S-Doped Carbon Quantum Dots as a Fluorescent Sensor for Silver(I) Ions

Bony K. John, Neenamol John, and Beena Mathew

Abstract Using–cyclodextrin, melamine, and thiourea, a simple and affordable method based on microwave treatment was established to generate nitrogen and sulphur co-doped carbon quantum dots (N,S-CQDs). Upon UV exposure, they displayed intense blue fluorescence with a 24% quantum yield (QY). Several characterization approaches were used to examine the optical, structural, and morphological characteristics of the N,S-CQDs. The average diameter of the spherical particles was 4 nm. A fluorescence sensor for Ag(I) ions with a limit of detection (LOD) of 45 nM and a linear detection range of 0–20 M was successfully built using N,S-CQDs. The quenching procedure follows a static quenching mechanism.

Keywords Carbon quantum dots · Fluorescence · Silver(I) ions

1 Introduction

Silver-based materials are widely used in several industrial fields and products, and they are well known for their applications as catalysts, and antimicrobial agents [1]. The frequent use of these materials can result in their accumulation in the air, soil, and water pollution. The excess concentration of Ag(I) ions can affect the animal and plant cells and the surroundings. Several advanced methods such as inductively coupled plasma-atomic emission spectrometry, electrochemical methods, and atomic emission spectroscopy are available to detect Ag(I) ions [2, 3]. However, these methods are complicated, expensive, and time-consuming which limits their everyday applications. Thus, it is critically necessary to devise an easy, quick, and affordable approach for identifying Ag(I) ions.

Generally, zero-dimensional carbon quantum dots (CQDs), can be synthesized using hydrothermal treatment, microwave heating, electrochemical methods, acidic oxidation, and arc discharge methods [4, 5]. They have found several applications

B. K. John · N. John · B. Mathew (✉)
School of Chemical Sciences, Mahatma Gandhi University, Kottayam, Kerala 686560, India
e-mail: beenamathew@mgu.ac.in

Z. H. Khan et al. (eds.), *Recent Advances in Nanomaterials*, Springer Proceedings in Materials 27, https://doi.org/10.1007/978-981-99-4878-9_24

and are extensively used as a photoluminescent (PL) sensors for the identification of different metal ions due to the quenching or enhancement in PL intensity [6, 7].

In the present work, we developed quantum dots using a simple microwave approach from β cyclodextrin, melamine, and thiourea. The Ag(I) ions and N,S-CQDs interacted to produce a PL quenching effect, which was further developed to design a selective and sensitive PL-based sensor for Ag(I) ions in the aqueous medium.

2 Experimental

N,S-CQDs were fabricated by employing the microwave method using β cyclodextrin, melamine, and thiourea as the carbon precursor. 0.5 g of β-cyclodextrin, 0.3 g melamine, 0.2 g thiourea, and 0.2 M sulphuric acid are taken and dispersed in 20 mL of deionized water. The solution is then poured to a conical flask, then placed in a microwave oven for 10 min in a cyclic mode. After the completion of the reaction, a brown solid which is the N,S-CQDs formed and then it is dissolved in 50 mL water. The suspension was subjected to lyophilization to obtain solid particles. The final 0.15 mg/mL solution was kept in a refrigerator for future studies.

3 Result and Discussion

3.1 Optical Behaviour

The UV–Vis. and fluorescence methods was employed to evaluate the optical properties of the quantum dots. An absorption band at 335 nm has been assigned to the n–π^* electronic transitions of C = O bonds [8]. The brown coloured carbon dot solution emits blue light when exposed to UV light as seen in the inset of Fig. 1a. Figure 1b shows the fluorescence spectra of N,S-CQDs excited at wavelengths spanning from 310 to 380 nm. This verifies the excitation-dependent emission behaviour, where the peak with the highest intensity occurred at 340 nm. The QY of the reaction was evaluated using quinine sulphate as a reference, and the yield was 24%.

3.2 Morphological and Surface Analysis

The surface and morphological behaviour of N,S-CQDs were analysed using TEM, XRD, FT-IR, and zeta potential measurements. The size, nature, and morphology were examined through TEM analysis. The N,S-CQDs are well separated and exhibited spherical shape as shown in Fig. 2a. The SAED pattern in Fig. 2b clearly indicates

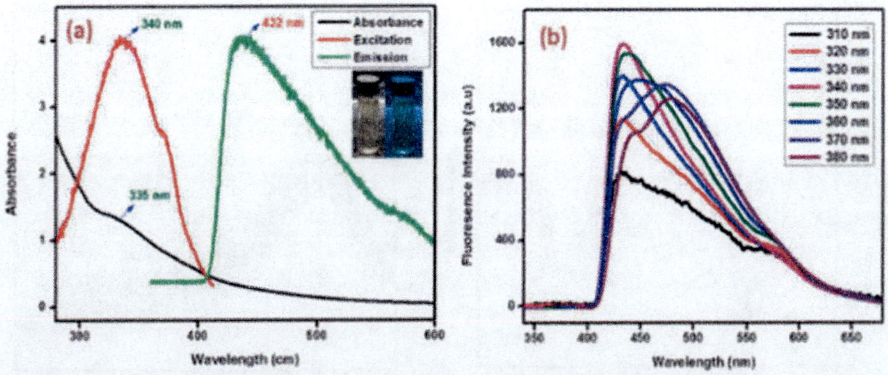

Fig. 1 **a** UV–vis. and PL analysis, inset: pictures of quantum dots with normal light, and Ultraviolet light, and **b** PL emission diagram upon various wavelength of excitation

the amorphous behaviour of the developed sensor. The mean particle size was 4 nm, calculated from the size distribution diagram (Fig. 2c).

FT-IR studies was used to examine the functional moieties existing on the N,S-CQD surface (Fig. 3a). Two intense broad vibrational bands at 3395 and 3115 cm^{-1} can be ascribed to the -NH and -OH bonds, respectively. Another band at 2426 cm^{-1} confirms the existence of S–H bond. The vibrations of C = O and C = C/C = N bonds generates bands at 1724 and 1652 cm^{-1}, respectively. The C = N and C–N bonds peaks at around 1510 cm^{-1} and 1408 cm^{-1} are in conformity with the vibrations of C = N and C–N bonds. A band centred at 1358 cm^{-1} is assigned to the C–H bending vibrations, and the C–S bond produces a band at 1135 cm^{-1} [9]. The band stretching of C–O was obtained at 1060 cm^{-1}. Overall, FT-IR data suggest that the N,S-CQDs contains surface groups -SH, -NH, -OH, and -COOH, which allows excellent solubility in water without any further treatment [9]. The X-ray diffraction (XRD) studies again confirmed the amorphous nature of the as-prepared CQDs. The XRD pattern exhibited a single broad peak at $2\theta = 26.04$ which is ascribed to the (002) Bragg Reflection as shown in Fig. 3b.

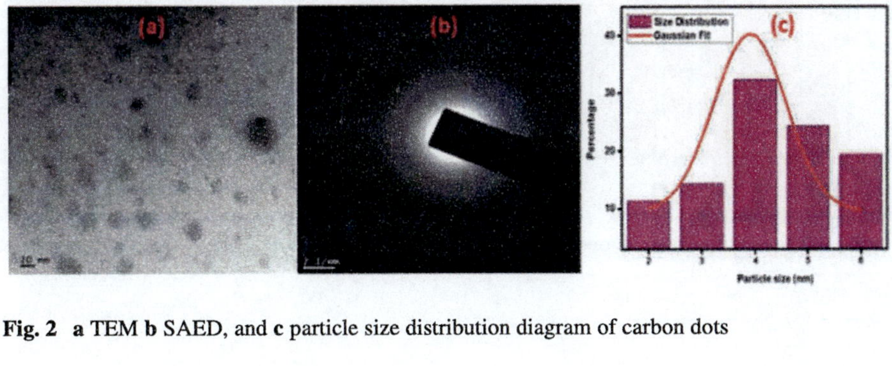

Fig. 2 **a** TEM **b** SAED, and **c** particle size distribution diagram of carbon dots

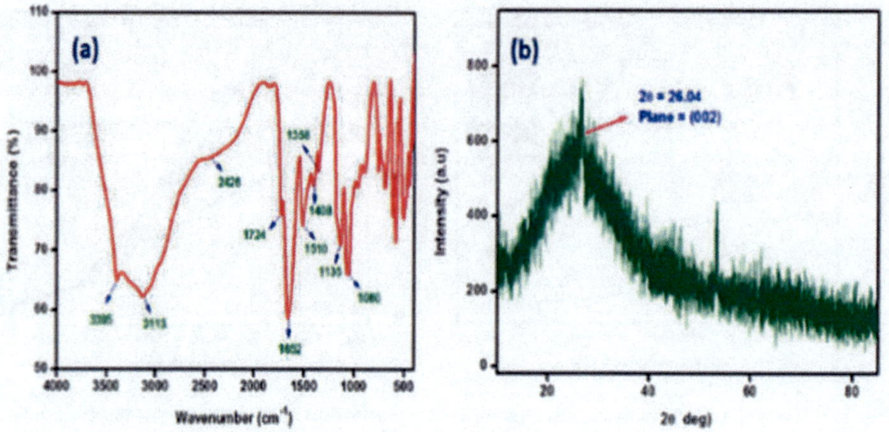

Fig. 3 **a** FT-IR, and **b** XRD spectra of N,S-CQDs

3.3 *Optimization of Sensing Conditions*

To improve sensing ability of the detector, the reaction parameters such as pH and contact time were optimized. A plot of Fo/F against Ag(I) ions concentration (30 μM) at various pH shows that supreme quenching was obtained at a pH of 7 (Fig. 4a). Since functional groups on N,S-CQDs protonate under acidic conditions and Ag(I) ions partially hydrolyse under basic conditions, the interaction between CQDs and Ag(I) ions is reduced in the acidic and basic conditions. After mixing Ag(I) ions with the N,S-CQDs, a contact period of 2 min was found to be sufficient for complete interaction Fig. 4b.

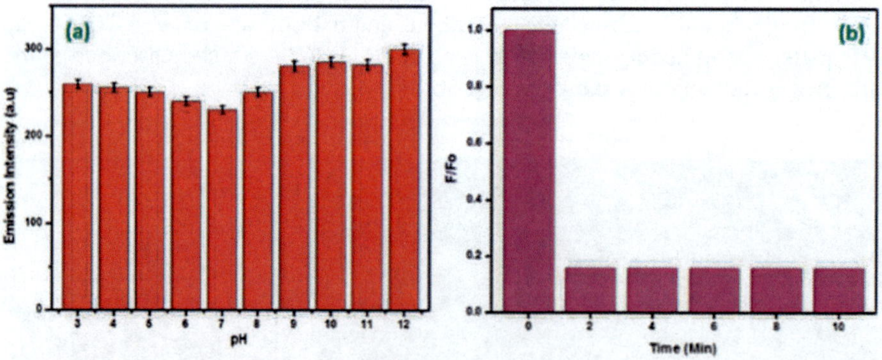

Fig.4 **a** pH, and **b** contact time optimization

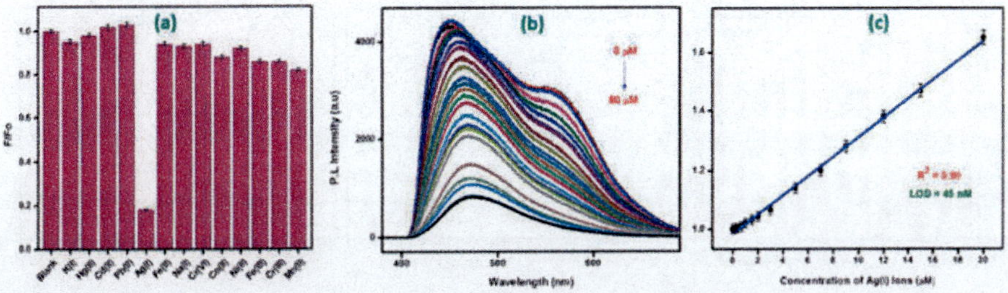

Fig. 5 **a** Sensitivity study, **b** concentration study, and **c** Stern–Volmer plot

3.4 Fluorescent Sensing of Ag(I) Ions

By combining N,S-CQDs and 30 μM various metal ions K(I), Hg(II), Cd(II), Ni(II), Pb(II), Zn(II), Co(II), Fe(III), Fe(II), Cr(VI), Ag(I), Na(I), Mn(II), and Cr(III) the change in the fluorescence intensity was measured. A considerable fluorescence quenching was observed in the case of Ag(I) only, as shown in Fig. 5a. This makes it possible to use the N,S-CQDs as a very accurate PL-based sensor for identifying Ag(I) ions. The detection capability of the N,S-CQDs for Ag(I) ions was investigated by adding different concentrations of Ag(I) ions (0–80 μM) into the N,S-CQD solution at a contact period of 2 min and pH of 7. As the Ag(I) ion concentration increased, as seen in Fig. 5a, b quenching in the fluorescence intensity was recorded. Using the Stern–Volmer equation $F0/F = 1 + \text{Ksv}\ [Q]$, the effect of concentration on the PL intensity of N,S-CQDs was assessed. A plot of *Fo*/*F* against the concentration of Ag(I) ions reveals a linear graph, where Fo and F indicates PL intensities of CQDs without and with Ag(I) ions, [Q] the Ag(I) concentration, and Ksv indicates the Stern–Volmer constant (Fig. 5c). The limit of detection (LOD) was obtained from the equation 3*s*/*m*, where *s* is the standard deviation and *m* is the slope. The LOD of the sensing was 45 nM.

3.5 Interference Studies

The efficiency of the N,S-CQD-Ag(I) sensor was examined by mixing 60 μM interfering metal ions with 30 μM under optimal conditions. As depicted in Fig. 6a, the developed sensor demonstrated exceptional selectivity for Ag(I) ions even in the vicinity of other ions.

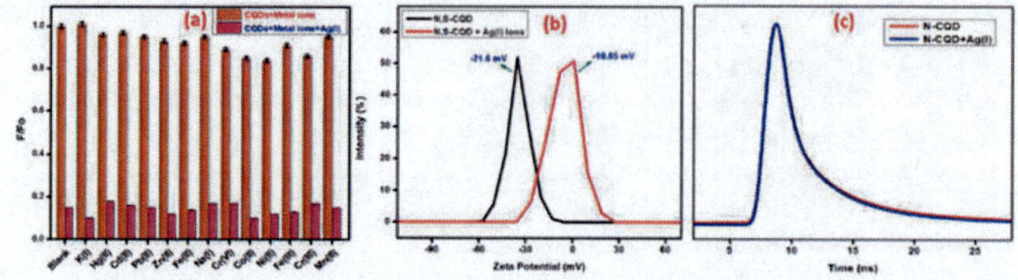

Fig. 6 **a** Interference study, **b** zeta potential analysis of N,S-CQDs and N,S-CQD-Ag(I) mixture, and **c** fluorescence life time analysis of N,S-CQDs, with and without Ag(I) ions

3.6 Mechanism

Due to the presence of functional groups such as -NH, -SH, -OH, and -C = O, the surface of N,S-CQDs is negatively charged (−21.6 mV) as determined by zeta potential (Fig. 6b). The successful interaction of positively charge Ag(I) ions with these negatively charged functional groups causes a lowering in the surface charge (−10.65 mV). There are two phenomena which are essentially responsible for the PL quenching phenomenon; static quenching and dynamic quenching. The fluorescence life time of N,S-CQDs and the N,S-CQD-Ag(I) mixture were 5.62 ns and 5.46 ns, respectively (Fig. 6c). This observation supports the generation of a non-fluorescent ground-state complex that does not alter the fluorescence lifetime. The involvement of a static quenching mechanism was therefore verified [10, 11].

4 Conclusion

In this work, N,S-CQD-based highly efficient and selective fluorescence sensor was designed for the identification of Ag(I) ions. The optical, structural, and morphological were performed using various characterizations methods. N,S-CQDs displayed an excitation and pH depending on fluorescence emission with a quick and excellent sensing ability towards silver ions. The LOD of the detection method was 45 nM. The sensor showed outstanding selectivity without any interferences.

Acknowledgements Bony K. John (the first author) gratefully appreciates the financial assistance provided as a Senior Research Fellowship (SRF) by the University Grants Commission, Government of India.

References

1. Chernousova S, Epple M (2013) Silver as antibacterial agent: ion, nanoparticle, and metal. Angew Chem Int Ed 52(6):1636–1653
2. Mcneil FE, O'meara JM (1999) The in viva measurement of trace heavy metals by K X-ray fluorescence. Adv X-Ray Anal 41:910–921
3. Mimendia A, Legin A, Merkoçi A, del Valle M (2010) Use of sequential injection analysis to construct a potentiometric electronic tongue: application to the multidetermination of heavy metals. Sens Actuators B Chem 146(2):420–426
4. Sun S, Zhang L, Jiang K, Wu A, Lin H (2016) Toward high-efficient red emissive carbon dots: facile preparation, unique properties, and applications as multifunctional theranostic agents. Chem Mater 28(23):8659–8668
5. Bottini M, Balasubramanian C, Dawson MI, Bergamaschi A, Bellucci S, Mustelin T (2006) Isolation and characterization of fluorescent nanoparticles from pristine and oxidized electric arc-produced single-walled carbon nanotubes. J Phys Chem B 110(2):831–836
6. Qi YX, Zhang M, Fu QQ, Liu R, Shi GY (2013) Highly sensitive and selective fluorescent detection of cerebral lead(II) based on graphene quantum dot conjugates. Chem Commun 49(90):10599–10601
7. Ran X, Sun H, Pu F, Ren J, Qu X (2013) Ag nanoparticle-decorated graphene quantum dots for label-free, rapid and sensitive detection of Ag^+ and biothiols. Chem Commun 49(11):1079–1081
8. Ghereghlou M, Esmaeili AA, Darroudi M (2021) Green synthesis of fluorescent carbon dots from elaeagnus angustifolia and its application as tartrazine sensor. J Fluoresc 31(1):185–193
9. John BK, Mathew S, John N, Mathew J, Mathew B (2023) Hydrothermal synthesis of N,S-doped carbon quantum dots as a dual mode sensor for azo dye tartrazine and fluorescent ink applications. J Photochem Photobiol A 436:114386
10. Mathew S, John BK, Thara CR, Korah BK, Mathew B (2023) One-pot synthesis of sustainable carbon dots for analytical and cytotoxicity studies. Biomass Convers Biorefin 1–14
11. Korah BK, John N, John BK, Mathew S, Bijimol D, Mathew B (2022) Carbon dots as a fluorescent ink and dual-mode probe for the efficient detection of doxycycline and Hg(II) ions. J Mater Res 37(18):3060–3070

Evolution of Structural and Electronic Properties in AlN: A DFT Study

Nitika and D. S. Ahlawat

Abstract In this paper, a comparative study has been carried out for structural and electronic properties of Aluminium nitride in three-dimension (wurtzite and hexagonal) and two-dimension honeycomb monolayer structure using density functional theory. The calculation of lattice constants was done using PBEsol exchange correlation functional within GGA approximation. The result for electronic band structures have been obtained using PBEsol, mBJ and local modified Becke-Johnson approximations. The result shows that there exists a changeover from direct energy band gap to indirect energy band gap, as dimensionality of crystal reduces from bulk to two dimensions.

Keywords Monolayer · Density functional theory · Structure optimization · Electronic properties

1 Introduction

As technological applications such as high-temperature diodes, blue LEDs and transistors are of high importance, therefore Group-III nitride semiconductor have been widely studied over the past decade [1–4]. Novel low-dimensional forms of III–V binary compounds, such as 2D hexagonal BN (h-BN) [5], 2D GaN [6], had been emerged with these technological advances. Among all nitrides, in recent years an emanating interest has been seen towards AlN-based nanostructures such as thin film [7], nano-tubes [8], nanoribbons [9] and nanowires [10]. Tsipas et al. had successfully achieved experimental realization of graphite–like hexagonal AlN nanosheets [11]. The stability of perfect h-AlN sheet is just short by 6% as in comparison to bulk wurtzite-AlN [12]. In addition, Onen et al. [13], studied the band line-up, electron confinement and quantum structures of lateral and vertical heterostructures using two-dimensional (2D) single-layer h-GaN and h-AlN. They observed that depending upon stacking sequence one can obtain diverse properties. The magnetic properties

Nitika (✉) · D. S. Ahlawat
Department of Physics, Chaudhary Devi Lal University, Sirsa 125055, Haryana, India
e-mail: nitikaphd295@cdlu.ac.in

Z. H. Khan et al. (eds.), *Recent Advances in Nanomaterials*, Springer Proceedings in Materials 27, https://doi.org/10.1007/978-981-99-4878-9_25

of undoped and TM doped AlN nanosheets had been calculated by Shi et al. [14] using ab-initio calculations. Their finding revealed that doping even with a single 3d TM atom can bring large local magnetic moments in nonmagnetic AlN nanosheets. In present study, we report the ab-initio calculations for structural properties and electronic band structure for 3D AlN and 2D AlN.

2 Materials and Methods

All theoretical calculations have been executed in WIEN2k code [17] using FP-LAPW method. To include the electron exchange and correlations part PBEsol [18] approximation has been used for optimizing the lattice constants [19, 20] as suggested by Perdew et al. for solids. For wave function expansion, plane wave cut off selected to be $R_{mt}K_{max} = 7$. A $2 \times 2 \times 1$ supercell of AlN nanosheets containing 8 atoms was employed. Along the vertical direction, a vacuum space of 10 Å has been created to avoid the interaction of the periodic cells. Geometrical optimizations are performed until the force convergence become lower than 0.01 mRyd/a.u. and maximum energy change between two consecutive steps is smaller than 0.0001 eV. The Brillouin zone is sampled using $10 \times 10 \times 10$ and $5 \times 5 \times 1$ K-point meshes for 3D and 2D AlN nanosheet, respectively. Along with PBEsol approximation, mBJ potential has been applied also for electronic properties only [21] and local mBJ potential (lmBJ) [22] scheme for 3D and 2D AlN crystal respectively to overcome the band gap problem of GGA potential. Figures 1 and 2 shows the crystal structure of wz- and h-AlN, respectively, whereas Fig. 3 shows the monolayer structure of AlN.

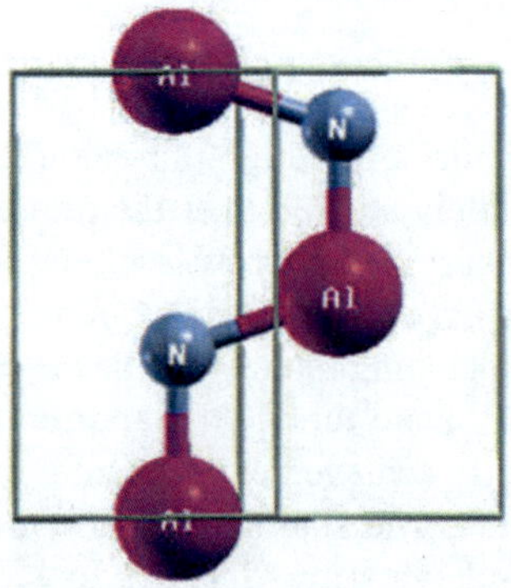

Fig. 1 Optimized structure of AlN in wurtzite phase

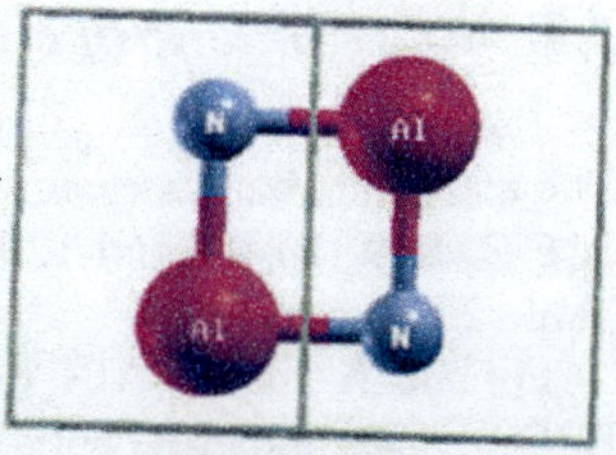

Fig. 2 Optimized structure of AlN in hexagonal phase

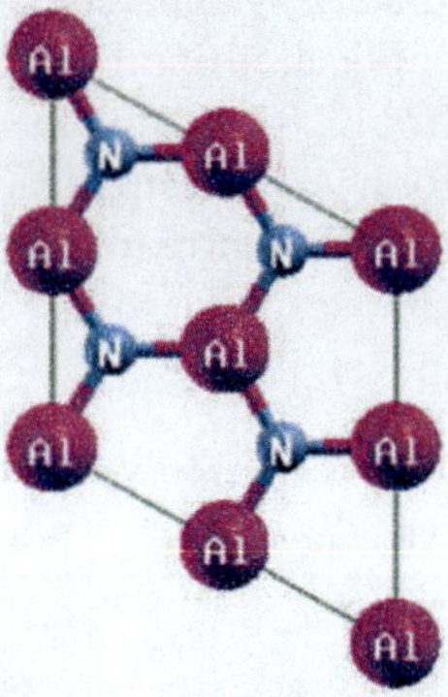

Fig. 3 Optimized structure of AlN monolayer

3 Results and Discussion

3.1 Structural Parameters

The calculated lattice parameters of three-dimensional AlN in wurtzite (wz) and hexagonal planar along with two-dimensional layered structure are presented in Table 1. The results suggest that there exist decrease in lattice constant when we move from hexagonal planar to two-dimensional monolayer structure. Further the comparison of calculated lattice parameters shows a good agreement with the experimental determined values thereby proving the PBEsol a good approximation for optimizing the geometry of two-dimensional materials as well.

Table 1 Optimized lattice constants (in Å) for AlN in 3D and 2D structures

Structure	This work	Other works	Experiment
Wurtzite	*a* 3.116	3.087 [24], 3.110 [13]	3.125 [23]
	c 4.98	4.962 [24], 4.978 [13]	4.99 [23]
Hexagonal	*a* 3.290	$a = 3.30$ [27]	
	c 4.113	$c = 4.15$ [27]	
Monolayer	*a* 3.12	3.13 [25]	3.13 [11]

3.2 Electronic Properties

The electronic band structure plots for wz, h- and 2D AlN are displayed in Figs. 4, 5, 6, 7, 8, 9, 10, 11 and 12 while the calculated band gap values are presented in Table 2.

For both h- and wz-AlN, the band structure (Figs. 4, 5, 6, 7, 8 and 9) obtained using PBEsol, mBJ and lmBJ approximations shows direct band gap with conduction band minima (CBM) and valence band maxima (VBM) both lying at Γ point of Brillouin zone. It is apparent from the band gap results that the band gap increases with the application of mBJ approximation thereby decreasing the deviation from experiments from 2.07 eV obtained with PBEsol approximation to 0.57 eV in case of wz while for h-AlN the application of mBJ increase the band gap by 1.51 eV as obtained with PBEsol approximation. The present mBJ calculations at optimized lattice constant is almost similar to that calculated (5.6 eV) at experimental lattice constant as reported in Ref. [25].

Table 2 Energy band gaps for the AlN in 3D (both wz- and h-) and 2D monolayer structures

Structure	PBEsol	mBJ	lmBJ	Other works
Wurtzite	4.20	5.71	5.57	5.6 [13], 6.28 [28]
Hexagonal	3.44	4.95	4.78	3.4 [27]
Monolayer	2.94	4.04	4.03	2.91 [25], 4.04 [26]

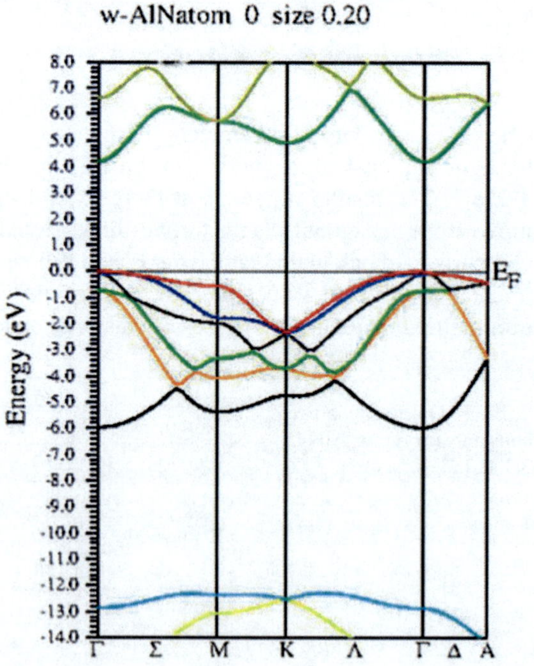

Fig. 4 Band diagram for 3D wz-AlN using PBEsol

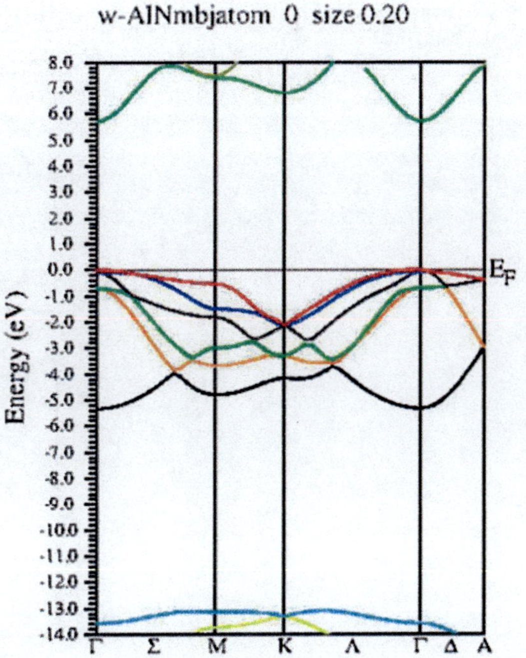

Fig. 5 Band diagram for 3D wz-AlN using mBJ

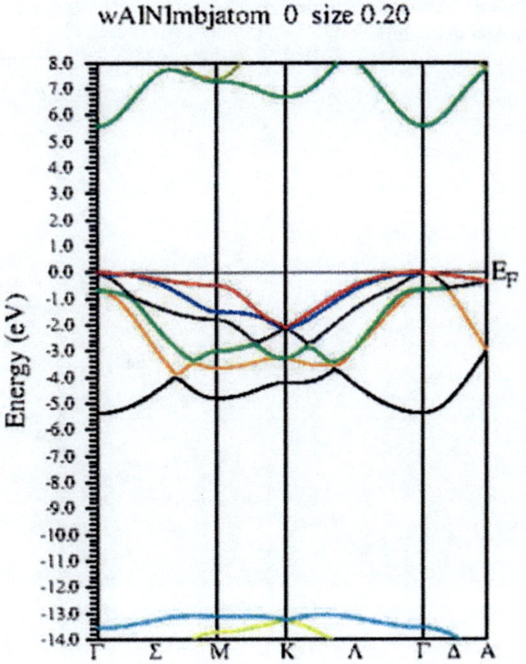

Fig. 6 Band diagram for 3D wz-AlN using lmBJ

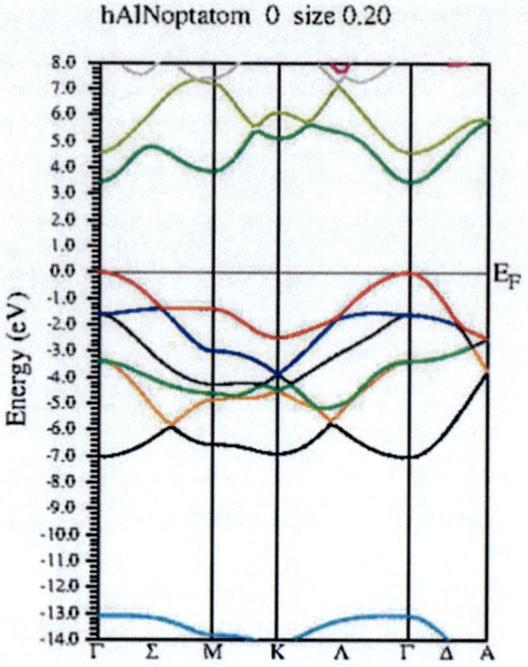

Fig. 7 Band diagram for 3D h-AlN using PBEsol

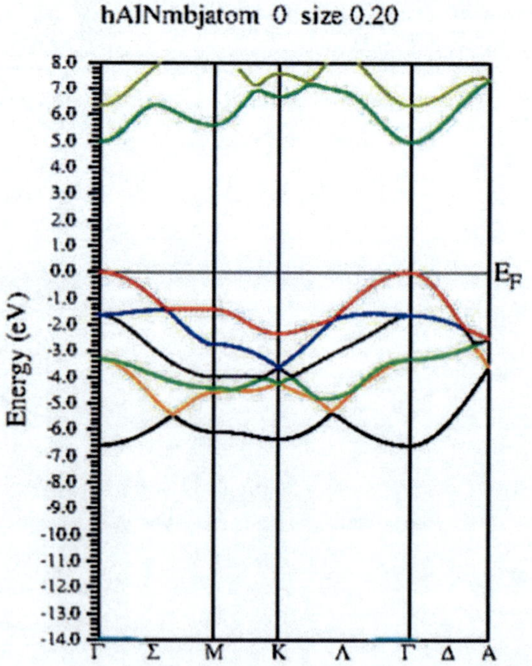

Fig. 8 Band diagram for 3D h-AlN using mBJ

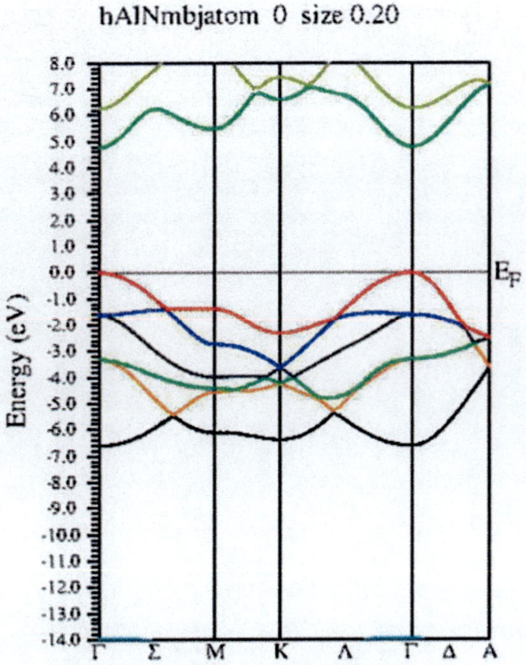

Fig. 9 Band diagram for 3D h-AlN using lmBJ

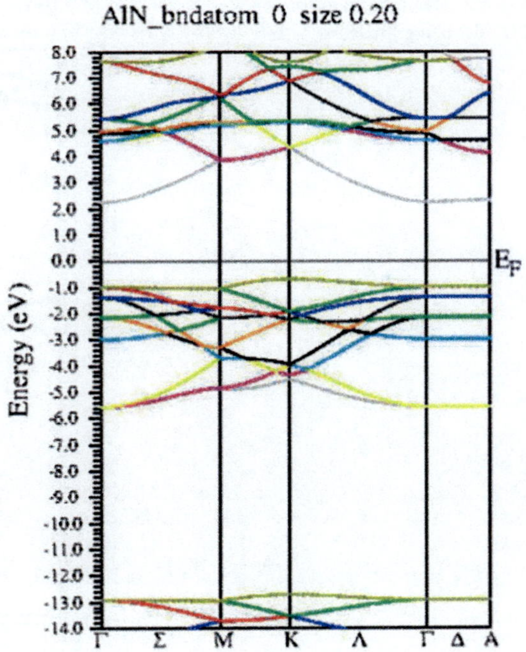

Fig. 10 Band diagram for 2D AlN using PBEsol

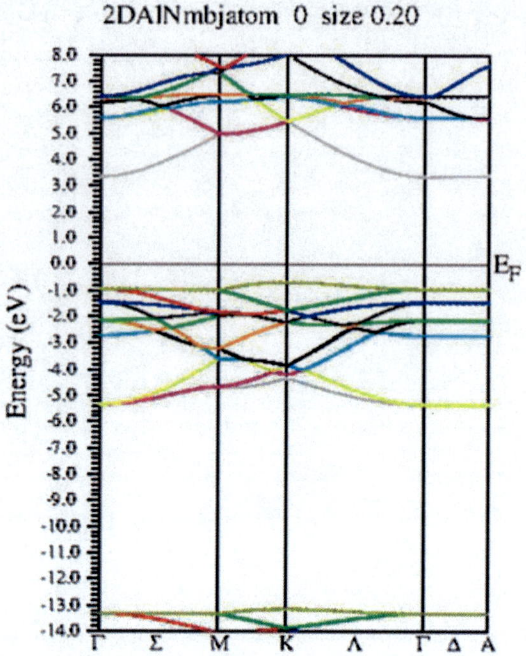

Fig. 11 Band diagram for 2D AlN using mBJ

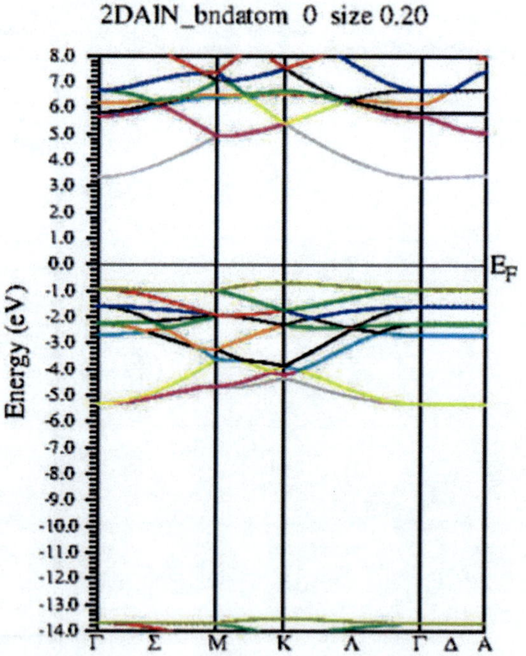

Fig. 12 Band diagram for 2D AlN using lmBJ

In Figs. 10, 11 and 12, the electronic band structures of 2D AlN is shown. From figures it can be observed that the valence band maxima (VBM) occur at the K-point, whereas the conduction band minima (CBM) appear at the Γ-point. Accordingly, the energy bands calculated by PBEsol predict an indirect band gap $E_{\Gamma\text{-}K} = 2.94$ eV. With corrections by using lmBJ method, indirect band gap increases by 1.091 eV. All calculated results are in aligned with the previous theoretical results [25, 26], indicating the authencity of our computational approaches. Further, mBJ approximation provides a better agreement with experimental band gap values in case of 3D structure, however lmBJ seems to be better in case of 2D structure as suggested in previous studies. [28]. The calculated band gap (4.03 eV) for 2D AlN using lmBJ is of same accuracy as with earlier reported result (4.03 eV) using more expansive HSE06 functional [29].

From findings it can be observed that going from 3D to 2D the nature of band gap changes and converted from direct to indirect band gap in UV region of electromagnetic spectrum which make these materials suitable for optoelectronic device applications.

4 Conclusion

In our work, we have performed structural and electronic properties calculations for 2D AlN crystal along with 3D AlN crystal based on DFT. A comparative study is also being carried out to unveil how the physical properties changes with dimensionality. From our findings, it may be observed that on going from 3D to 2D AlN, there exist a transition in the nature of band structure from direct to indirect band gap.

References

1. Morkoc BH (1994) Large-band-gap SiC, III-V nitride, and II-VI ZnSe-based semiconductor device technologies. J Appl Phys 76(3):1363–1398
2. Vurgaftman I (2001) Band parameters for III–V compound semiconductors and their alloys. J Appl Phys 89(11):5815–5875
3. Miwa K (1993) First-principles calculation of the structural, electronic, and vibrational properties of gallium nitride and aluminum nitride. Phys Rev B 48(11):7897
4. Christensen NE (1994) Optical and structural properties of III-V nitrides under pressure. Phys Rev B 50(7):4397
5. Golberg D (2010) Boron nitride nanotubes and nanosheets. ACS Nano 4(6):2979–2993
6. Onen A (2016) GaN: from three-to two-dimensional single-layer crystal and its multilayer van der Waals solids. Phys Rev B 93(8):085431
7. Jones AC (1994) The deposition of aluminum nitride thin films by metal-organic CVD—an alternative precursor system. Adv Mater 6(3):229–231
8. Balasubramanian C (2004) Scanning tunneling microscopy observation of coiled aluminum nitride nanotubes. Chem Phys Lett 383(1–2):188–191
9. Yu L (2011) Vapor–liquid–solid growth route to AlN nanowires on Au-coated Si substrate by direct nitridation of Al powder. J Cryst Growth 334(1):57–61

10. Mitran TL (2011) Ab initio vibrational and thermal properties of AlN nanowires under axial stress. Comput Mater Sci 50(10):2955–2959
11. Tsipas P (2013) Evidence for graphite-like hexagonal AlN nanosheets epitaxially grown on single crystal Ag (111). Appl Phys Lett 103(25):251605
12. de Almeida EF (2012) Defects in hexagonal-AlN sheets by first-principles calculations. Eur Phys J B 85(1):1–9
13. Onen A (2017) Lateral and vertical heterostructures of h-GaN/h-AlN: electron confinement, band lineup, and quantum structures. J Phys Chem C 121(48):27098–27110
14. Shi C (2014) Magnetic properties of transition metal doped AlN nanosheet: first-principle studies. J Appl Phys 115(5):053907
15. Hohenberg P (1964) Inhomogeneous electron gas. Phys Rev 136(3B):B864
16. Blaha P (2001) Wien2k.An augmented plane wave plus local orbital program for calculating the crystal properties 60:1–302 (2001)
17. Perdew JP (2008) Restoring the density-gradient expansion for exchange in solids and surfaces. Phys Rev Lett 100(13):136406
18. Haas P (2009) Erratum: calculation of the lattice constant of solids with semilocal functionals. Phys Rev B 79(20):209902
19. Arora S (2018) Estimation of lattice constants and band gaps of group-III nitrides using local and semi local functionals. Orient J Chem 34(4):2137
20. Tran F (2009) Accurate band gaps of semiconductors and insulators with a semilocal exchange-correlation potential. Phys Rev Lett 102(22):226401
21. Rauch T (2020) Local modified Becke-Johnson exchange-correlation potential for interfaces, surfaces, and two-dimensional materials. J Chem Theory Comput 16(4):2654–2660
22. Yim WM (1974) Thermal expansion of AlN, sapphire, and silicon. J Appl Phys 45(3):1456–1457
23. Zhang CW (2012) First-principles study on electronic structures and magnetic properties of AlN nanosheets and nanoribbons. J Appl Phys 111(4):043702
24. Ahmed B (2021) Structural and electronic properties of AlN in rocksalt, zinc blende and wurtzite phase: a DFT study. Digest J Nanomater Biostruct (DJNB) 16(1)
25. Liu XF (2019) Structural, mechanical, and electronic properties of 25 kinds of III–V binary monolayers: a computational study with first-principles calculation. Chin Phys B 28(8):086105
26. Bacaksiz C (2015) Hexagonal AlN: dimensional-crossover-driven band-gap transition. Phys Rev B 91(8):085430
27. Orton JW (1998) Group III nitride semiconductors for short wavelength light-emitting devices. Rep Prog Phys 61(1):1
28. Tran F (2021) Bandgap of two-dimensional materials: thorough assessment of modern exchange-correlation functional. J Chem Phys 155:104103
29. Abdullah NR (2022) Electronic and optical properties of metallic nitride: a comparative study between the MN (M = Al, Ga, In, Tl) monolayers. Solid State Commun 346:114705

Highly Ordered 1D $NiCo_2O_4$ Nanorods: An Efficient Hybrid Material for Electrochemical Energy Storage Application

Mahvesh Yousuf, Reyaz Ahmad, Asif Majeed, Malik Aalim, Arshid Mir, Aamir Sohail, Ab Mateen, and M. A. Shah

Abstract Here, we presented the regularly arrayed nickel–cobalt oxide nanorods for use as a supercapacitor electrode. According to the findings of our research, the morphology of the nickel–cobalt oxides is highly dependent on the reaction time as well as the ratio of Ni/Co. When $NiCo_2O_4$ deposited on nickel foam, the fabricated electrodes show a specific capacitance of 950.6 Fg^{-1}, when scanned at 2 mVs^{-1} in an aqueous electrolyte containing 4 M KOH and also reveals good charge/discharge stability as its initial capacitance dropped by just 15% even after 4000 cycles. In addition to this the energy and power density were found to be 16.7 $Whkg^{-1}$, 233.2 Wkg^{-1} respectively at 7 $mAcm^{-2}$ current density and demonstrating low impedance confirmed by electrochemical impedance spectroscopy (EIS). The electrode based on $NiCo_2O_4$ demonstrates excellent electrochemical characteristics all of which point to its promising application in supercapacitor devices.

Keywords $NiCo_2O_4$ · Nanorods · Energy density · Supercapacitor

1 Introduction

Supercapacitors (SCs) have a greater cycle performance and outstanding power output despite their small size. The SCs may be readily incorporated into the electrical circuit of a wide variety of consumer electronic products and produce almost little heat during operation [1, 2]. Energy may be stored in SCs in two different ways: through Faradic redox reaction or by producing an electric double layer (EDLC). The materials' redox behavior results in a capacitance value that is 100 times larger than that of EDLCs, which drives the research and development of pseudocapacitive materials [3–5].

M. Yousuf · R. Ahmad (✉) · A. Majeed · M. Aalim · A. Mir · A. Sohail · A. Mateen · M. A. Shah
Department of Physics, National Institute of Technology, Srinagar, Jammu and Kashmir 190006, India
e-mail: reyazah566@gmail.com

Z. H. Khan et al. (eds.), *Recent Advances in Nanomaterials*, Springer Proceedings in Materials 27, https://doi.org/10.1007/978-981-99-4878-9_26

Transition metal-based oxides were employed as the pseudocapacitive electrode material due to the significant Faradic redox processes [6, 7]. RuO_2 has excellent capacitive behavior, but its high cost and poisonous nature prevent it from being used in commercialized device construction. Thus, many researchers continually push themselves to find low-cost materials with high specific capacitance and extended cycle stability. Maximizing the effective surface area of the electrode materials employed is a key strategy for boosting capacitance. As a result, researchers have looked into effective electrode materials with unique morphologies to boost capacitive performance in terms of cycle stability, charge/discharge rate capability, and user safety [8]. Materials such as $Ni(OH)_2$ hollow spheres, three-dimensional bubble-like graphene frameworks, porous graphene frameworks, and composites of MoO_2, reduced graphene oxide, and NiO have all been employed as electrodes in supercapacitors. Nanostructured materials like carbon-based materials and conducting polymers are two examples of those studied for this purpose. While carbon-based materials have low capacitance due to their high resistivity, conducting polymers have low capacitance because they are easily mechanically degraded.

NiO, MnO_2, $FeCo_2O_4$, Co_3O_4, and CuO, on the other hand, have been regarded as viable replacements because to their high energy density, outstanding electrochemical performance, thermal and chemical stability, and other features [9–12]. Because of their low cost, ease of manufacture, high theoretical specific capacitance (3750 Fg^{-1}), and outstanding electrochemical reversibility, materials based on nickel oxide are among the most promising choices currently accessible. To boost electrochemical performance, researchers have tried nickel oxide-based nanomaterials with varying morphologies such microparticles, nanoparticle-wires, micro sheets, and nanorods [13, 14]. The study addresses the field of energy storage using $NiCo_2O_4$ nanorods that were manufactured hydrothermally. $NiCo_2O_4$ nanorods display outstanding electrochemical characteristics which demonstrate the prospective practical functioning of the material.

2 Synthesis

In order to prepare the precursor solution, 0.34 M nickel nitrate hexahydrate, 0.10 M cobalt nitrate hexahydrate, and 0.7 g urea were dissolved in 40 mL of deionized water. The solution stirred for 30 min in order to ensure that it was completely mixed and then transferred into a 100-mL autoclave and heated at 150 °C for 12 h. After waiting for the temperature inside the autoclave to return at room temperature, the powder was removed from it using a centrifuge and washed with ethanol to remove any unreacted components. Finally, they were dried in the open air at a temperature of approximately 50 °C for 10 h and calcinated for a further 2 h at a temperature of 500 °C in the air in order to get the desired end result.

3 Results and Discussion

The as-prepared samples were analyzed by XRD and EDS to determine their crystal structure and phase purity. The XRD pattern of $NiCo_2O_4$ displayed in Fig. 1c, in which seven diffraction peaks can be seen at 2θ values of 30.6°, 37°, 38.3°, 43.7°, 56.2°, 58.8°, and 64.7°. All of the peaks in this case are neatly catalogued with Miller indices of (220), (311), (222), (400), (422), (511), and (440) for the spinel crystalline of $NiCo_2O_4$ [15, 16]. The as-synthesized materials' excellent purity is shown by the lack of secondary peaks in their spectra. The matching EDX patterns of sample, which are given in Fig. 1d, further show that the samples are formed of Ni, Co, and O, which also demonstrates that the nanostructures are constituted of NiO and Co. Further evidence that the $NiCo_2O_4$ composite was successfully prepared is provided by the EDS elemental mapping analysis shown in the inset of Fig. 1d.

Figure 1a, b displays a nanorods of $NiCo_2O_4$ nanomaterial with an average diameter of around 20 nm which are densely organized and confined nanopores of 6 nm. Adjusting the processing time and temperature allows one to control the aspect ratio (the ratio of length to diameter). The addition of Co causes the smooth nanostructures to aggregate their material into nanoparticles while still maintaining the linear shape of nanorods clearly visible from Fig. 1a. Nanorods, which can be seen to be made up of very small nanoparticles, are bundled together to form a cluster of nanorods. As the Co concentration was raised, the nanosized particles fused and produced new

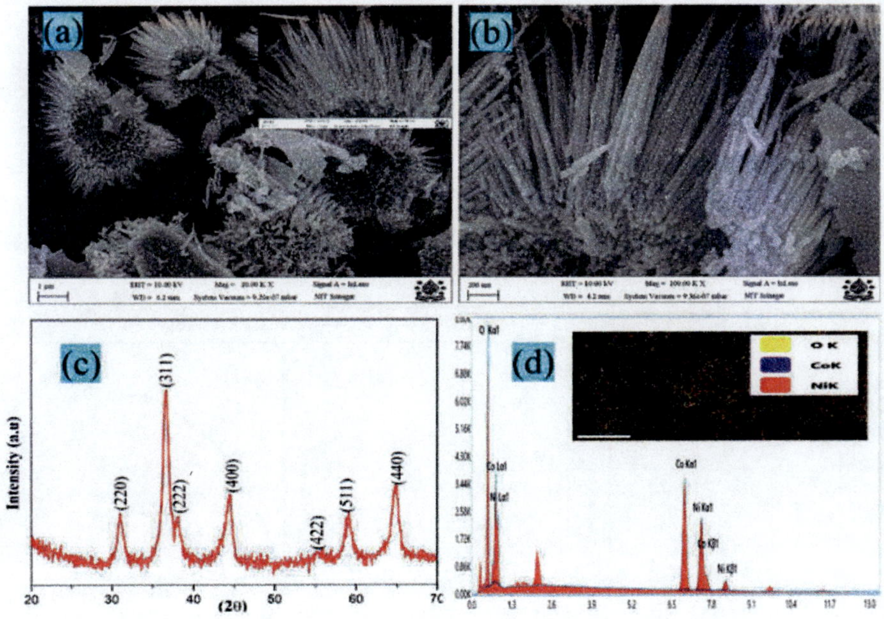

Fig. 1 **a**, **b** FESEM images of $NiCo_2O_4$ nanorods at high and low magnification, **c** XRD pattern of nanorod, **d** EDS and inset is elemental mapping if $NiCo_2O_4$ nanorods

morphologies because the Brownian motion of the particle nuclei was perturbed by the as-formed Co nuclei in the precursor solution. Due to the hydrothermal treatment, the nanoparticles' active collision degrees were significantly increased.

4 Electrochemical Measurements

Cyclic voltammetry, galvanostatic charge–discharge, and electrochemical impedance spectroscopy are all used to analyze the $NiCo_2O_4$ electrode's electrochemical properties. (CV) curves of electrodes are depicted in Fig. 2a. The anodic peak and the cathodic peak are the two redox peaks that are found in CV curves. These peaks shown up because of oxidation and reduction processes. Scanning faster moves the anodic peak toward higher potential, whereas the cathodic peak moves toward lower potential. As scan rates rise, electric polarization also rises, leading to irreversible reactions. Using Eq. (1),

$$C_s = \frac{1}{\Delta V m} \int_{v_i}^{v_f} I(V)\mathrm{d}V \tag{1}$$

where V is the scan rate (mVs^{-1}) and m is the mass of active material, we can determine the C_s of electrodes from the CV plots. $NiCo_2O_4$ electrode exhibit a specific capacitance of 950.6, 604.1, 560, 433.4 Fg^{-1} at a scan rate of 2, 5, 7, and

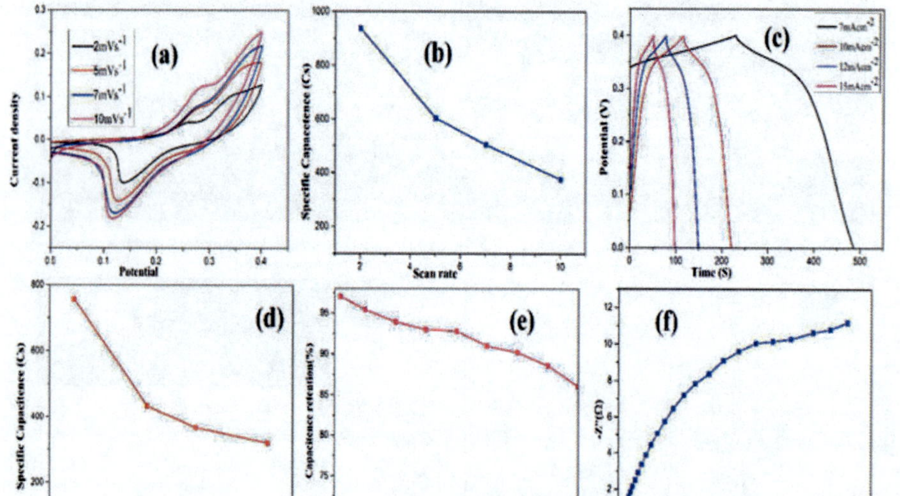

Fig. 2 **a** CV curves of $NiCo_2O_4$. **b** Plot of the scan rate versus specific capacitance, **c** GCD curves of $NiCo_2O_4$, **e** cyclic stability, **f** Nyquist plots in 4 M KOH electrolyte

10 mVs^{-1}, respectively. Figure 2b shows graph of specific capacitance versus scan rate. At low scan rate dispersed ions from the electrolyte may more easily access the electrode surface, more surface intercalation–deintercalation of ions occurs, leading to a high specific capacitance. However, with a fast scan rate, the likelihood of electrolyte ions being adsorbed to the inner surface decreases.

Furthermore, the GCD measurements were performed as depicted in Fig. 2c and Eq. (2) used to calculate the Cs of electrode from the GCD data.

$$C_s = \frac{I \times t_d}{m \times \Delta v} \quad (2)$$

where I is the current density (mA), and t_d is the discharge duration. The C_s were found to be 755.4 Fg^{-1}at a current density of 7 $mAcm^{-2}$. However, with increasing current density a decrease in C_s is seen. Reduced C_s with increasing current density may be explained by the fact that at high current densities, the inner surface area available for charge storage is not as active due to the limited diffusion and mobility of the electrolytic ions. Figure 2d displays the relationship between current density and C_s.

The stability of electrode is evaluated at a current density of 10 $mAcm^{-2}$ in a 4 M KOH electrolyte as shown in Fig. 2e. Almost 85% of the initial C_s was still present in the $NiCo_2O_4$ electrodes after 4000 cycles. The electrodes' EIS graph is displayed in Fig. 2f. Analysis of the EIS is performed from 10 Hz to 100 kHz. The diameter of the semicircle is a representation of the charge transfer resistance (R_{ct}) that exists at the interfaces between the electrodes and the electrolytes in the high frequency zone. It is clear from EIS graph that the electrode did not form a semicircle, which indicates that the charge transfer resistance is extremely low. While the bulk solution resistance (R_s) is represented by the intercept at the real axis and the corresponding values of (R_s) is also very small.

5 Conclusion

The hydrothermal synthesis of $NiCo_2O_4$ nanorods for supercapacitors is shown here. Through the use of XRD and EDS, the crystal structures and elemental compositions were validated. The sample had a larger surface area and smaller pore size, both of which herald the presence of many active sites. With a high concentration of active sites, the electrons may move through the material more quickly and efficiently, improving the material's electrochemical characteristics. Ni-foam loaded with $NiCo_2O_4$ has an extremely high capacitance (950.6 Fg^{-1}) and retains 85% of the initial value after 4000 cycles. Furthermore, at 7 $mAcm^{-2}$ current density, the manufactured electrode had energy and power densities of 16.7 $Whkg^{-1}$ and 233.2 Wkg^{-1}, respectively. Excellent specific capacitance, high stability, high power density, and low impedance are exhibited by the $NiCo_2O_4$-based new supercapacitor electrode produced in this work, indicating promising practical usefulness.

References

1. Ahmad R, Shah MA (2022) Hydrothermally synthesised nickel oxide nanostructures on nickel foam and nickel foil for supercapacitor application. Ceram Int. https://doi.org/10.1016/j.ceramint.2022.10.179
2. Mishra D et al (2022) Bitumen and asphaltene derived nanoporous carbon and nickel oxide/carbon composites for supercapacitor electrodes. Sci Rep 1–12. https://doi.org/10.1038/s41598-022-08159-3
3. Kumar R, Youssry SM, Joanni E, Sahoo S, Kawamura G, Matsuda A (2022) Microwave-assisted synthesis of iron oxide homogeneously dispersed on reduced graphene oxide for high-performance supercapacitor electrodes. J Energy Storage 56(PA):105896. https://doi.org/10.1016/j.est.2022.105896
4. Adalati R et al (2022) Metal nitrides as efficient electrode material for supercapacitors: a review. J Energy Storage 56(PB):105912. https://doi.org/10.1016/j.est.2022.105912
5. Nagaraju YS et al (2022) Single-step hydrothermal synthesis of ZnO/NiO hexagonal nanorods for high-performance supercapacitor application. Mater Sci Semicond Process 142:106429. https://doi.org/10.1016/j.mssp.2021.106429
6. Li J et al (2021) Polypyrrole-wrapped $NiCo_2S_4$ nanoneedles as an electrode material for supercapacitor applications. Ceram Int 47(12):16562–16569. https://doi.org/10.1016/j.ceramint.2021.02.227
7. Wang X, Yan C, Sumboja A, See P (2014) High performance porous nickel cobalt oxide nanowires for asymmetric supercapacitor. Nano Energy 3:119–126. https://doi.org/10.1016/j.nanoen.2013.11.001
8. Zhao Y, Zhang X, He J, Zhang L, Xia M, Gao F (2015) Morphology controlled synthesis of nickel cobalt oxide for supercapacitor application with enhanced cycling stability. Electrochim Acta 174:51–56. https://doi.org/10.1016/j.electacta.2015.05.162
9. Ahmad R, Shah MA (2022) Nickel oxide (NiO) nanoflakes prepared through hydrothermal method and integration into acetone gas sensing application. Chem Pap. https://doi.org/10.1007/s11696-022-02448-x
10. Li K et al (2022) Engineering active sites on nitrogen-doped carbon nanotubes/cobaltosic oxide heterostructure embedded in biotemplate for high-performance supercapacitors. J Energy Storage 53:105094. https://doi.org/10.1016/j.est.2022.105094
11. Li K et al (2021) A multidimensional rational design of nickel–iron sulfide and carbon nanotubes on diatomite via synergistic modulation strategy for supercapacitors. J Colloid Interface Sci 603:799–809. https://doi.org/10.1016/j.jcis.2021.06.131
12. Kumar S, Saeed G, Zhu L, Hui KN, Kim NH, Lee JH (2021) 0D to 3D carbon-based networks combined with pseudo capacitive electrode material for high energy density supercapacitor: a review. Chem Eng J 403:126352. https://doi.org/10.1016/j.cej.2020.126352
13. Devan RS, Patil RA, Lin JH, Ma YR (2012) One-dimensional metal-oxide nanostructures: recent developments in synthesis, characterization, and applications. Adv Funct Mater 22(16):3326–3370. https://doi.org/10.1002/adfm.201201008
14. Hong W, Lin L (2019) Studying the substrate effects on energy storage abilities of flexible battery supercapacitor hybrids based on nickel cobalt oxide and nickel cobalt oxide @ nickel molybdenum oxide. Electrochim Acta 308:83–90. https://doi.org/10.1016/j.electacta.2019.04.023
15. Zou R et al (2013) Chain-like $NiCo_2O_4$ nanowires with different exposed reactive planes for high-performance supercapacitors. J Mater Chem A 1(30):8560–8566. https://doi.org/10.1039/c3ta11361b
16. Samantara AK, Kamila S, Ghosh A, Jena BK (2018) Highly ordered 1D $NiCo_2O_4$ nanorods on graphene: an efficient dual-functional hybrid materials for electrochemical energy conversion and storage applications. Electrochim Acta 263:147–157. https://doi.org/10.1016/j.electacta.2018.01.025

Influence of SiO_2 in PANI Matrix as an Electron Transport Layer for OLEDs

Gobind Mandal, Ram Bilash Choudhary, Debashish Nayak, Sanjeev Kumar, Jayanta Bauri, and Sarfaraz Ansari

Abstract Polyaniline-Silicon dioxide (PANI-SiO_2) nanocomposites were synthesized via chemically oxidative polymerization process. The formation of PANI, and PANI-SiO_2 nanocomposites with their corresponding changes in nanostructured properties were examined by Field Emission Scanning Electron Microscope (FESEM), X-ray Diffraction Spectroscopy (XRD) and Fourier Transform Infrared Spectroscopy (FTIR). From the FESEM analysis it was observed that the PANI have nanofiber like structure whereas SiO_2 has granular structure. The agglomeration of PANI nanofibers over the surface of SiO_2 revealed the formation of PANI-SiO_2 nanocomposites. The UV–Vis. analysis exhibited PANI-SiO_2 nanocomposite possess lower band gap and higher absorption intensity than pristine PANI. The PL spectra of PANI and PANI-SiO_2 nanocomposite reveals four bands as green bands due to radiative recombination of electron–hole pair. The electrical properties for PANI-SiO_2 nanocomposite showed 37% enhancement in current density in comparison to pure PANI. The PANI-SiO_2 nanocomposite owed enhanced carrier density and excellent rate of electron–hole recombination that proposed to be suitable candidate for electron transport layer in OLED devices.

Keywords OLEDs · Electron transport layer · Photoluminescence · Organic–inorganic nanocomposite

1 Introduction

The introduction of organic light emitting diodes (OLEDs) to the lighting and display technologies such as televisions and smartphones have attracted tremendous number of interests. Being thin, flexible, low cost and solution-based manufacturing techniques still makes them highly approached topic in academic as well as industrial research [1–4]. In present lighting and display technology era, OLEDs are still struggling in terms of quantum efficiency, color purity, luminescence, long term stability

G. Mandal · R. B. Choudhary (✉) · D. Nayak · S. Kumar · J. Bauri · S. Ansari
Department of Physics, IIT (ISM) Dhanbad, Dhanbad, India

Z. H. Khan et al. (eds.), *Recent Advances in Nanomaterials*, Springer Proceedings in Materials 27, https://doi.org/10.1007/978-981-99-4878-9_27

and high fabricating costs of the device. For the enhancement in efficiency, brightness and lifetimes, organo-inorganic hybrid materials haven been widely used as one of the layers in OLEDs. OLEDs are multi-layered construction that has different layers of materials that perform following features: anode and cathode, the electron transport layer (ETL) and hole transport layer (HTL), emissive layer (EML) where the radiative recombination of holes and electrons takes place [5, 6]. In this work, SiO_2 has been selected to increase the electrical and thermal stability of polyaniline. Here, we have synthesized a binary nanocomposite of PANI-SiO_2 via in situ oxidative polymerization method. Their chemical interaction and structural morphology were elucidated by XRD analysis and FESEM images. For optical and electrical analysis, UV–Vis. and photoluminescence spectroscopy were employed which showed the absorption of light in visible range and great electron–hole recombination rate in green band region. PGS nanocomposites possess high electrical conductivity and mobility of electrons as calculated for J–V characteristic curve which firmly confirms it as a potential candidate for electron transport layer in OLED applications.

2 Materials and Methods

Aniline (MW, 93.13 g/mol), hydrochloric acid (MW, 36.46 g/mol) (38%), (99%) (Merck), ammonium persulfate (APS) (MW, 228.2 g/mol) (98%) (Merck), silicon dioxide (SiO_2) (MW, 60.08 g/mol) (98%) (SRL), acetone (Merck), ethanol (Merck) and other chemical reagents were procured for the synthesis of PANI, SiO_2 and PANI–SiO_2 nanocomposites with varying concentration of SiO_2. Reagents were used without any further purification for the chemical synthesis.

2.1 Synthesis of PANI-SiO_2 Nanocomposite

In the process of obtaining PANI-SiO_2 nanocomposite, 2 ml Aniline was dissolved into solution of 8 ml HCl and 70 ml DI water with stirring of 30 min. After then another solution made by dissolving 0.25 gm SiO_2 into 30 ml DI water. It was mixed with the above solution. Further separately 4 gm APS was dissolved into 30 ml DI water. This solution was slowly added into final solution. Final solution kept in the ice bath chamber at 0 °C for 5–6 h. Obtained solution was washed by distilled water and ethanol and dried at 60 °C in the oven for 24 h. The schematic method for the synthesis of PANI–SiO_2 nanocomposite using in-situ chemically oxidative method is Fig. 1.

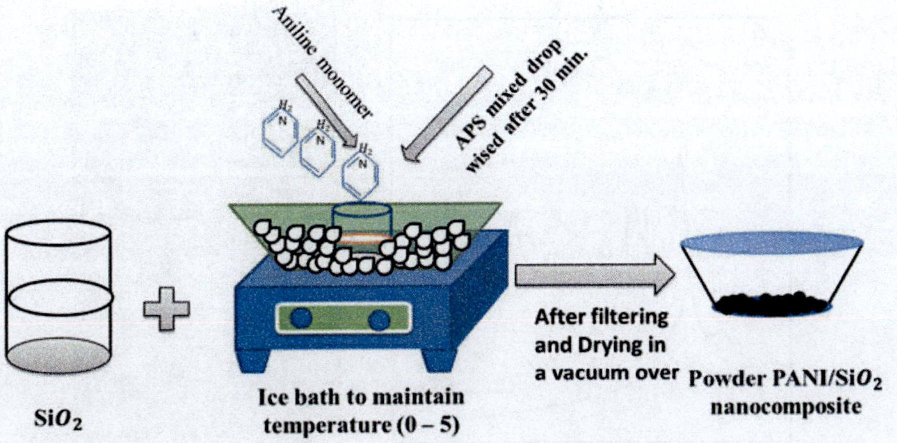

Fig. 1 Schematic route for the synthesis of PANI-SiO_2 nanocomposite

3 Results and Discussion

3.1 Structural Study (XRD and FESEM Analysis)

The XRD spectra of the synthesized polyaniline, SiO_2 and PANI- SiO_2 nanocomposite is shown in Fig. 2a. The XRD pattern for PANI showed three distinct peaks at 15°, 18°, 25°, ascribed to the (001), (020) and (200) planes, respectively. This showed the semi crystalline nature of polyaniline. However, the XRD pattern of SiO_2 showed peaks at 26°, 33°, 51° and 65° having planes (050), (220), (228) and (3210), respectively [7]. In case of PANI-SiO_2 nanocomposite all the prominent peaks are present which confirms the formation of the nanocomposite. Field Emission Scanning Electron Microscopy (FESEM) was employed to find out the surface morphological study of the as-synthesized nanocomposites. The corresponding FESEM images of PANI, SiO_2, and PANI-SiO_2 were shown in Fig. 2b–d, respectively. These images revealed that the PANI has agglomerated nanofiber like structure and SiO_2 nanoparticles have granular structure. In case of PANI-SiO_2 nanocomposite it was observed that the SiO_2 nanoparticles get coated to the tip of the nanofibers of PANI which justifies the complete formation of PANI-SiO_2 nanocomposite.

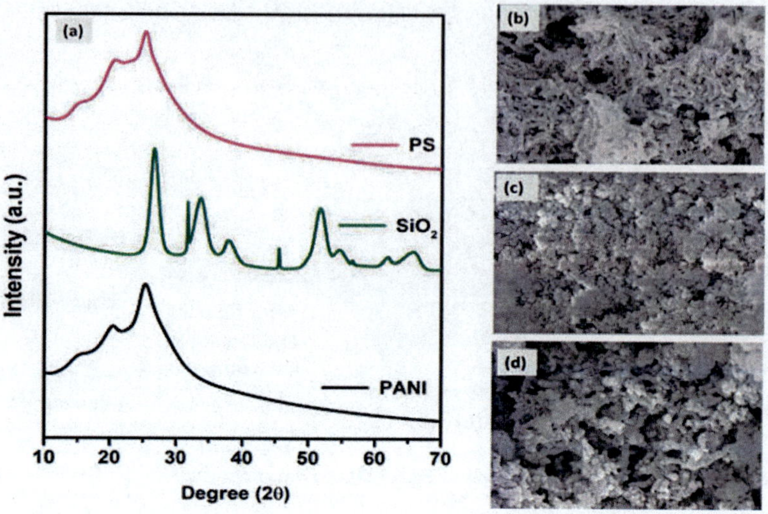

Fig. 2 **a** XRD spectra and **b–d** FESEM images of PANI, SiO_2 and PANI-SiO_2 nanocomposites

3.2 *Optical Study (UV–Vis. Spectroscopic and Photoluminescence Analysis)*

The UV–Visible spectra and corresponding graphs for optical band gap have been shown in Fig. 3a–d. In order to evaluate the optical band gap (E_g) for the as-synthesized PANI, SiO_2 and PANI-SiO_2 nanocomposites. The absorption coefficients α were extracted through their respective absorption spectra by the following equation: $= \frac{2.3.3A}{d}$, where A–the absorbance and d–thickness of the specimens (~ 0.175 mm). Then curves between $(h)^{\ 2}$ and h have plotted using following Tauc's relation [8]. $h = A\left(h - E_g\right)^n$, where h–photon energy and n represent the density of states which may take values 1/2, 3/2, 2, 3 corresponding to direct allowed, direct forbidden as well as indirect allowed, indirect forbidden optical band gap, respectively. The values of direct optical band gap for PANI, SiO_2 and PANI-SiO_2 nanocomposites, were evaluated by Tauc's plot between $(h)^{\ 2}$ and h. The value of pure PANI was calculated as 2.79 eV, however optical band gap for PANI-SIO_2 nanocomposite get increased to 2.94 eV due to addition of SiO_2 nanoparticles.

Photoluminescence spectroscopy is employed for of PANI, SiO_2 and PANI-SiO_2 nanocomposites at have been recorded in the wavelength range 340–420 nm as shown in Fig. 4a. Photoluminescence spectra can be used to study the quantum yield and transition of photogenerated electron–hole pairs. Therefore, a low photoluminescent emission intensity indicates a low photogenerated electron–hole pair recombination. The incorporation of Sio2 increases the photoluminescent intensity of pure PANI as shown in Fig. 4a. This refers that PANI-SiO_2 provides higher electron–hole recombination rate. The PANI-SiO_2 nanocomposite photoluminescence spectra depict a prominent peak in the visible region at 352 nm in the blue color region due to

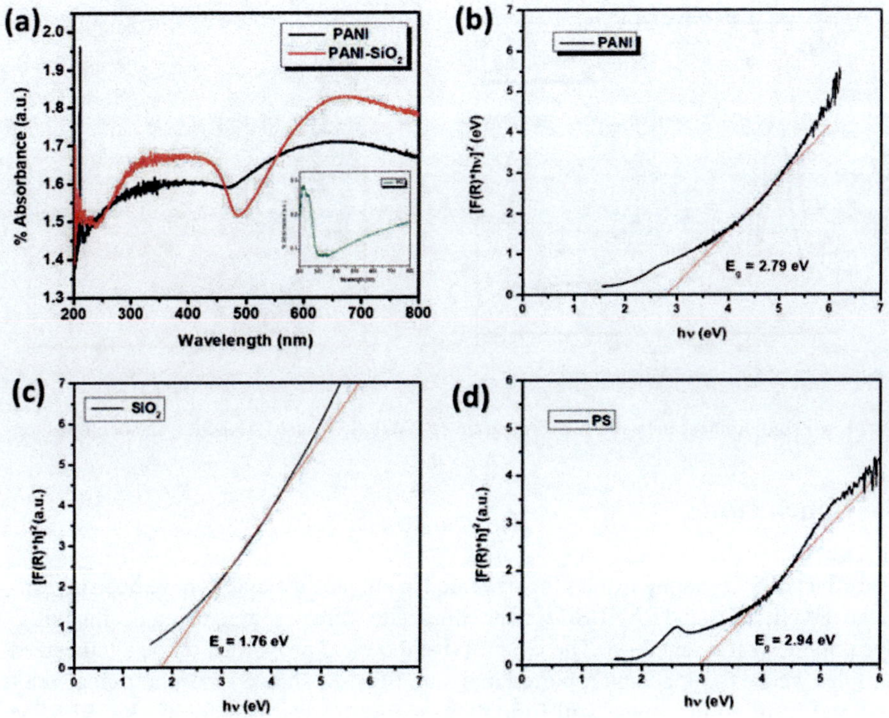

Fig. 3 **a–d** UV–Vis. spectra and photoluminescence of PANI, SiO_2, and PANI-SiO_2 nanocomposites

charge transfer between SiO_2 nanoparticles and PANI matrix. The color coordinates of PANI, SiO_2 and PANI-SiO_2 nanocomposites was shown in Fig. 4b. The approximated coordinates of color, was obtained for PANI, SiO_2 and PANI-SiO_2 nanocomposites at (0.21, 0.26), (0.18, 0.20) and (0.15, 0.32) respectively when deep blue light exposed. Color purity calculated for PANI-SiO_2 nanocomposites obtained 56.1% it displays a good enough color purity.

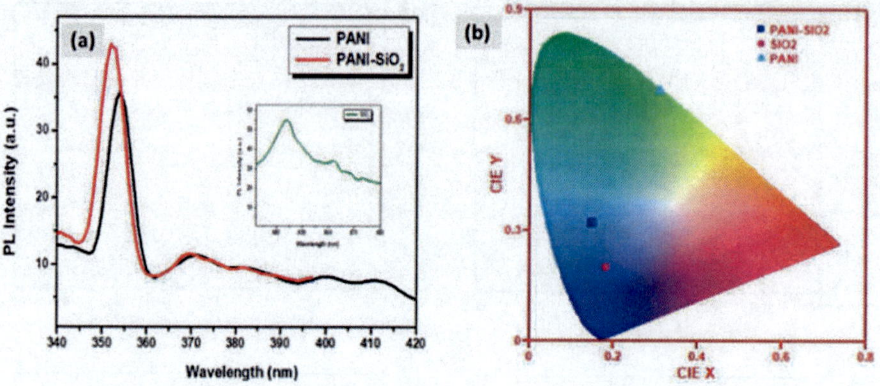

Fig. 4 **a** Photoluminescence and **b** CIE diagram of PANI, SiO_2 and PANI-SiO_2 nanocomposites

4 Conclusion

The effect of SiO_2 nanoparticles incorporated in polyaniline matrix have been investigated. PANI, SiO_2 and PANI-SiO_2 nanocomposite were prepared using in situ oxidative polymerization method. The structural and optical properties of the synthesized samples were examined and reported in detail. FESEM showed morphological analysis and justified the successful formation of PANI-SiO_2 nanocomposite. FESEM showed that the SiO_2 nanoparticles seem to be embedded in polyaniline matrix. The incorporation of SiO_2 into polyaniline matrix enhanced the optical stability of PANI-SiO_2 nanocomposite as revealed by UV–Vis. and PL analysis. The absorbance peaks were observed in UV–Vis. region ascribed the red shift in contrast to pure PANI. The optical band gap of the PANI-SiO_2 nanocomposite was estimated to be 2.94 eV by extrapolation of Tauc's plot using Kubelka Monk's function from UV–Vis. reflectance data. The photoluminescence of the synthesized PANI-SiO_2 nanocomposites exhibited the blue light at ($\lambda = 481$ nm) emission at an excitation wavelength of λexc. = 352 nm. Therefore, hereby the optimized PANI-SiO_2 nanocomposite has good optical properties that favors the material as a potential candidate to be an electron transport layer (ETL) for OLEDs.

Acknowledgements The authors express sincere thanks to the Indian Institute of Technology (Indian School of Mines), Dhanbad, India.

Declaration of Interest Statement The authors declare that they have no conflict of interests.

References

1. Mobin R, Rangreez TA, Chisti HTN, Inamuddin, Rezakazemi M (2019) Organic inorganic hybrid materials and their applications, pp 1135–1156
2. Jeon SO, Jang SE, Son HS, Lee JY (2011) External quantum efficiency above 20% in deep blue phosphorescent organic light-emitting diodes. Adv Mater 23:1436–1441
3. Ha H, Shim YJ, Lee DH, Park EY, Lee I-H, Yoon S-K, Suh MC (2021) Highly efficient solution-processed organic light-emitting diodes containing a new cross-linkable hole transport material blended with commercial hole transport materials. ACS Appl Mater Interfaces 13:21954–21963
4. Mandal G, Choudhary RB (2022) MnO_2 integrated emeraldine polyaniline (PANI-MnO_2) nanocomposites with inflated opto-electricaltraitsas ETLs for OLED applications. Mater Sci Semicond Process 151:107000
5. Kulkarni AP, Tonzola CJ, Babel A, Jenekhe SA (2004) Electron transport materials for organic light-emitting diodes. Chem Mater 16:4556–4573
6. Singh S, Khan ZH, Khan MB, Kumar P, Kumar P (2022) Quantum dots-sensitized solar cells: a review on strategic developments. Bull Mater Sci 45(2):1–13
7. Bauri J, Choudhary RB, Mandal G (2021) Recent advances in efficient emissive materials-based OLED applications: a review. J Mater Sci 56:18837–18866
8. Choudhary RB, Kumar S (2022) Optimum chemical states and localized electronic states of SnO_2 integrated PTh–SnO_2 nanocomposites as excelling emissive layer (EML). Opt Mater (Amst) 131:112736

Strategy to Synthesize Tunable Emissive ZnS QDs Enabled by Cobalt Doping

Urosa Latief and Mohd. Shahid Khan

Abstract In recent years, tunable fluorescent emissive zinc sulfide quantum dots (ZnS QDs) have increased the usefulness and significance of the white light-emitting diode. Herein, we reported the synthesis of cobalt-doped ZnS QDs by co-precipitation method. Examinations using X-ray diffraction (XRD) technology supported the cubic crystalline structure. We provide a novel approach with improved photoluminescence resulting in a strong yellowish-orange emission due to cobalt (Co) doping. The synthesis method used in the current work gives the possibility of producing emission-tuned, low-cost, rare-earth free nanophosphors for creating high-performance light-emitting devices.

Keywords Quantum dots · Phosphor · LEDs

1 Introduction

Semiconductor nanostructures have piqued the interest of researchers over the last decade due to their exceptional electrical, optical and optoelectronic capabilities. Despite the fact that most of these features of semiconductor nanostructures can be controlled by adjusting their size and shape, their structures of electronic bands can be changed by adding an appropriate dopant to fulfill the needs of particular opto-electronic applications. The optical characteristics of semiconductor nanostructures from the II–VI groups, like nanoparticles of ZnS, CdSe, ZnTe, CdS and quantum dots have undergone extensive research for use in optoelectronic applications [1, 2]. Additionally, they have employed consistently in the production of light-emitting diodes, which primarily emit light in the spectral range (400–640 nm) with the potential for additional tuning through the management of their size and impurity [3, 4]. Metal ion doping of these nanostructures could control their optically induced radiative emissions, but controlling their size, or taking use of the quantum confinement effect, can easily change their bandgap energy. Whether generated naturally or as

U. Latief · Mohd. S. Khan (✉)
Department of Physics, Jamia Millia Islamia, New Delhi 110025, India
e-mail: mskhan@jmi.ac.in

Z. H. Khan et al. (eds.), *Recent Advances in Nanomaterials*, Springer Proceedings in Materials 27, https://doi.org/10.1007/978-981-99-4878-9_28

a result of metal ion doping, the sub-bandgap electronic states of semiconductor nanostructures determine their radiation emissions. However, interband transitions of the inserted ions or atoms of rare earth and transition metals can also result in emissions [5]. Despite the fact that nanophosphor emission activators known as rare-earth ions can induce powerful emissions in specific spectrum areas when included in small quantities, they are pricey due of their limited availability on the earth [6]. Therefore, it is critical to produce rare-earth free materials.

Recently, there has been a lot of interest in zinc-based nanostructured systems, especially ZnS. The broad blue emission of ZnS may be tuned to other wavelengths of the visible spectral region because it possesses intrinsic intra-bandgap electronic states [7]. Incorporating dopants or other impurities into the crystalline lattice of ZnS nanostructures is one of the most popular strategies to alter their optical characteristics [2]. Environmentally friendly transition metal ions doped QDs, such as Co, are an excellent choice for use in white LEDs due to their advantages like cheap raw materials, ease of acquisition and superior photoluminescence abilities [8]. In this paper, we describe the production of highly-dispersed Co doped ZnS nanoparticles of size in the range of 1.5–2.5 nm as well as their PL emission characteristics at room temperature. Co-precipitation was used to produce the well-crystalline semiconductor nanoparticles, which were then examined using X-ray diffraction (XRD), room temperature PL spectroscopy and UV–Visible spectroscopy. The visible spectral range could be targeted for the PL emissions of the nanoparticles by the addition of Co^{2+} ions, which is essential for their application in the technology of white light-emitting displays.

2 Materials and Methods

2.1 *Materials*

Zinc acetate dihydrate ($Zn(CH_3COO)_2 \cdot 2H_2O$, 98.0%), cobalt (II) nitrate hexahydrate ($Co(NO_3)_2 \cdot 6H_2O$, 97.7%), sodium sulfide (Na_2S, > 99.5%) and all other solvents were supplied by Alfa Aesar. Unspecified analytically pure reagents were utilized in the exact same manner as they were acquired after being bought from commercial sources. Ultrapure water was used in every synthesis and study.

2.2 *Synthesis*

The co-precipitation approach was employed for the synthesis of ZnS QDs. 0.2 M of $Zn(CH_3COO)_2 \cdot 2H_2O$ and 0.2 M of Na_2S were dissolved in 20 ml double-distilled water separately, and to make sure that all of the ingredients were dissolved, solutions were continually stirred with a magnetic stirrer. In a common flask, each solution was then added one at a time and dropwise while being continuously stirred for 8 h at a speed of 1000 rpm. To maintain a pH of 7, the NaOH solution was dripped into the mixture. White precipitates that were separated by filtering were found at the bottom of the flask after 8 h. To remove any unwanted impurities created during the production process, the precipitates were then washed with deionized water and methanol. Finally, the moist sample was dried in a furnace at 60 °C for 7 h.

By preparing separate solutions of $Zn(CH_3COO)_2{\cdot}2H_2O$, $Co(NO_3)_2{\cdot}6H_2O$, and Na_2S in 20 ml double-distilled water, ZnS QDs doped with Co were prepared using a similar synthesis process. Co was retained at a 6% doping concentration.

2.3 *Characterizations*

After purification, the as-prepared QDs were characterized using a variety of characterization techniques. X-ray diffraction (XRD) examinations was produced utilizing the CuKα radiation source (= 0.15418 nm) and the Rigaku Smart Lab. Agilent Cary Eclipse Spectrophotometer was used to measure the fluorescence spectra. A Lcd LMSP-UV1900S twin beam was used to obtain UV–Vis absorption spectra.

3 Results and Discussion

3.1 *Structural Characterization*

XRD spectra of pure ZnS and Co/ZnS QDs are shown in Fig. 1a. There are three diffraction peak that may be seen to correlate the (111), (220) and (311) planes of miller indices. The JCPDS card number 05-0566 matches all of the diffraction peaks, which are all strongly consistent with cubic crystalline structure [9]. Peak positions for doped ZnS QDs moved toward higher angles, indicating that Co^{2+} are successfully substituted into the host lattice of the parent ZnS without altering its cubic crystalline structure. Also, the crystalline size are calculated by Debey Schrer formula [10]. The average crystallite size of the samples are estimated to be 1.92 nm and 2.06 nm for pure and doped ZnS QDs, respectively.

3.2 *Optical Properties*

The UV–Vis absorption spectra of ZnS QDs with and without Co^{2+} ion incorporation are shown in Fig. 1b. The insertion of Co^{2+} into the ZnS host lattice obviously increases the absorption peak value. Furthermore, for doped samples, position of the absorption edge is pushed toward longer wavelength which means red shift is

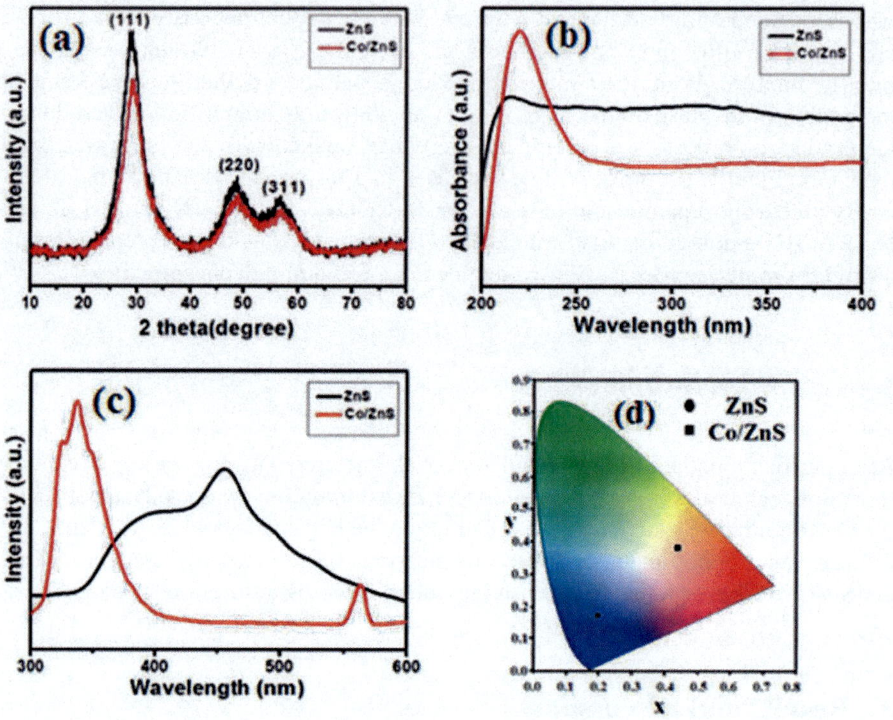

Fig. 1 **a** XRD patterns, **b** UV–Vis absorption spectra, **c** PL spectra at an excitation wavelength of 300 nm, **d** CIE chromaticity diagram of ZnS and Co/ZnS QDs

observed. Integrated Co^{2+} ion manipulation in the ZnS host lattice and the quantum size effect are both responsible for this red shift. [11]. Figure 1c shows photoluminescence spectra of pure ZnS and Co/ZnS QDs at an excitation wavelength of 300 nm. In the case of ZnS QDs, the emission peak is found at 455 nm and this wide emission has commonly been linked to shallow and deep defect states associated with sulfur (V_S) and zinc (V_{Zn}) vacancies in ZnS [12]. It should also be mentioned that the intensity of this emission is improved for doped sample, most likely as a result of the introduction of electronic states (shallow) in the ZnS band gap. Higher sulfur vacancies and zinc interstitials are formed as a result of the insertion of the Co^{2+} ion into the ZnS host lattice. These zinc interstitials act as traps for electrons activated by the valence band edge. It's possible that the trapped electrons may travel to lower S-vacancy states, where they will recombine with the holes found at Co^{2+} centers, resulting in yellow-orange emission.

3.3 Photometric Properties

Co dopant can be used to modify the PL emission of ZnS QDs. To define tunability, several colorimetric characteristics are determined, including the Comission Internationale de L'Eclairage (CIE) color coordinates, color correlated temperature (CCT) and color purity (CP). Several photometric properties of as-synthesized ZnS QDs and doped ZnS QDs are calculated. Figure 1d shows the CIE chromatic diagram, which signifies the specific color coordinate loci of as-synthesized QDs. Color coordinates are shown to be placed in the blue for pure ZnS (0.20, 0.17) and yellow-orange zones for Co/ZnS (0.44, 0.38), showing an evolution of color coordinates because of dopant introduction and thus by combining these QDs with blue chips, we can demonstrate an alternative possibility for white light emission. Additionally, the relation (1) is used to determine the color purity of the samples [10]

$$\text{CP} = \frac{\sqrt{(x_s - x_i)^2 + (y_s - y_i)^2}}{\sqrt{(x_d - x_i)^2 + (y_d - y_i)^2}} \times 100\% \tag{1}$$

where the (x, y) represents the color coordinates and the indices i, d and s indicate illuminated point, dominant wavelength and the sample, respectively. From the data, it can be shown that the color purity is improved by the addition of dopant to ZnS QDs (46.1–61%). Additionally, the McCamy empirical formula is used to determine the CCTs [10]. The estimated CCT values (2651 K for Co/ZnS) are found to be below 3000 K, indicating their suitability for warm white LEDs [13]. According to these findings, by the introduction of dopants, the photoluminescent colors of ZnS quantum dots can be tuned, making them appropriate for solid-state lighting applications as color-tuning components in white LEDs.

4 Conclusion

The non-toxic direct aqueous approach was successfully used to produce pure and cobalt-doped ZnS QDs. X-ray diffraction (XRD) investigation verified the cubic crystalline structure. We describe an innovative method that uses Co/ZnS QDs to enhance photoluminescence, producing a strong yellow-orange emission. The CIE plot demonstrates that the work opens up a new avenue for the prospective employment of doped QDs in solid-state lighting.

5 Declaration of Interest Statement

The authors declare no conflict of interest.

References

1. Li H, Shih WY, Shih W-H (2007) Non-heavy-metal ZnS quantum dots with bright blue photoluminescence by a one-step aqueous synthesis. Nanotechnology 18:205604
2. Latief U, Islam SU, Khan Z, Khan MS (2022) Luminescent manganese/europium doped ZnS quantum dots: Tunable emission and their application as fluorescent sensor. J Alloys Compounds 910:164889
3. Zhang Q, Nie C, Chang C, Guo C, Jin X, Qin Y, Li F, Li Q (2017) Highly luminescent red emitting CdZnSe/ZnSe quantum dots synthesis and application for quantum dot light emitting diodes. Opt Mater Express 7:3875–3884
4. Shen H, Bai X, Wang A, Wang H, Qian L, Yang Y, Titov A, Hyvonen J, Zheng Y, Li LS (2014) High-efficient deep-blue light-emitting diodes by using high quality ZnxCd1-XS/ZnS core/shell quantum dots. Adv Funct Mater 24:2367–2373
5. Tripathi LN, Mishra CP, Chaubey BR (1982) Luminescence in ZnS: La and ZnS: (Mn, La) phosphors. Pramana 19:385–398
6. Jadwisienczak WM, Lozykowski HJ, Xu A, Patel B (2002) Visible emission from ZnO doped with rare-Earth ions. J Electron Mater 31:776−784
7. Fang X, Zhai T, Gautam UK, Li L, Wu L, Bando Y, Golberg D (2011) ZnS nanostructures: from synthesis to applications. Prog Mater Sci 56:175–287
8. Rodriguez S-G, Pl U, Sánchez-Zeferino R, Álvarez-Ramos ME (2020) Tunable white-light emission of Co^{2+} and Mn^{2+} Co-doped ZnS nanoparticles by energy transfer between dopant ions. J Phys Chem C 124(6):3857–3866
9. Zaeimian MS, Gallian B, Harrison C, Wang Y, Zhao J, Zhuac X (2018) Mn doped AZIS/ZnS nanocrystals (NCs): effects of Ag and Mn levels on NC optical properties. J Alloys Compd 765:236–244
10. Islam SU, Latief U, Ahmad I, Khan SM (2022) Novel NiO/ZnO/Fe_2O_3 white light-emitting phosphor: facile synthesis, color-tunable photoluminescence and robust photocatalytic activity. J Mater Sci: Mater Electron 33:23137–23152
11. Su YH, Teoh LG, Tu SL, Lee JH, Hon MH (2009) Photoelectric characteristics of natural pigments self-assembly fabricated on TiO_2/FTO substrate. J Nanosci Nanotechnol 9:960–964
12. Rodriguez S-G, Carrillo Torres RC, Sánchez Zeferino R, Álvarez Ramos ME (2019) Stabilized blue emitting ZnS@SiO quantum dots. Opt Mater 89:396−401
13. Zhou Y, Ge X, Zhang Z, Luo W, Xu H, Li W, Zhu J (2019) Design and realization on orange-red emitting of samarium activated sodium lanthanum metaphosphate with low CCT and high CP. J Alloy Compd 811:0925–8388

Physical Properties of Pure MoS_2 Thin Films Grown on a Large Area Using a CVD Process in a Single-Zone Furnace

Sharmistha Dey, Santanu Ghosh, and Pankaj Srivastava

Abstract We report the large-area growth of pure-phase, high-quality MoS_2 thin films by the Chemical Vapor Deposition (CVD) process in a single-zone furnace. All the parameters, like temperature, gas flow rate, precursor quantity, substrate type, and the distance between precursor and substrate, have been optimized. The XRD confirms the formation of pure 2-H phase MoS_2, and Raman spectroscopy suggests the deposition of a relatively thick film. FESEM images clearly show the formation of the vertical nano-wall structure. The analysis of core level spectra of Mo 3d and S 2p shows that Mo is predominantly in the 4+ valence state and indicates the formation of a pure MoS_2 phase. The film shows diamagnetic behavior and very low reflectance ($\leq 15\%$) at room temperature.

Keywords Pure phase · MoS_2 thin films · Single zone furnace · CVD · Low reflectance

1 Introduction

MoS_2 is a fascinating material for electrical, optical, and optoelectronic applications as it has some special characteristics like intra-layer covalent bonding and inter-layer van der Waals interaction [1]. It is a semiconductor at room temperature with a thickness-dependent band gap (1.2 eV indirect band gap for bulk and 1.9 eV direct band gap for a monolayer) [2]. To grow pure-phase large-area MoS_2 thin films, CVD is a very good technique. However, the optimization of all CVD parameters is a challenging task. Here, we report the optimization of critical parameters of CVD for the growth of pure-phase MoS_2 thin film. Generally, a double-zone or multi-zone furnace has been used to grow MoS_2 but in the present work, a single-zone programmable furnace has been used. Investigations on morphology, crystal and electronic structure, and magnetic and optical properties have been reported in the present study.

S. Dey (✉) · S. Ghosh · P. Srivastava
Nanostech Laboratory, Physics Department, IIT Delhi, New Delhi 110016, India
e-mail: phz198691@physics.iitd.ac.in

Z. H. Khan et al. (eds.), *Recent Advances in Nanomaterials*, Springer Proceedings in Materials 27, https://doi.org/10.1007/978-981-99-4878-9_29

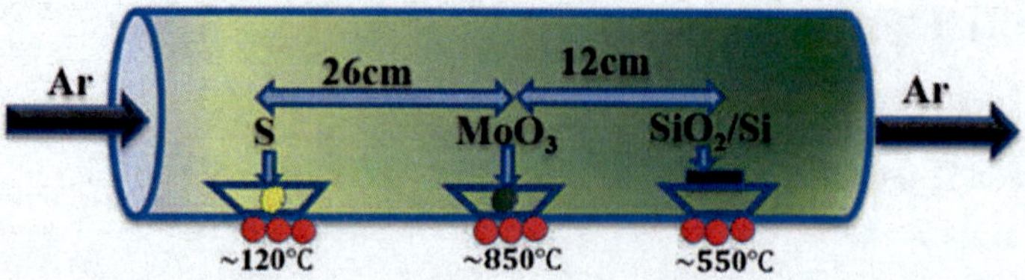

Fig. 1 Schematic diagram of CVD set up for MoS_2 deposition

2 Materials and Methods

Figure 1 shows the schematic diagram for growing large areas of high-quality, pure-phase MoS_2 thin films. Generally, a double-zone or multi-zone furnace is used to grow MoS_2 by CVD; however, growth was achieved in a single-zone furnace in the present work. All the parameters, such as precursor quantity, gas flow rate, distances between precursors and substrate, temperature, and substrate type, were optimized according to a single-zone furnace. We use Sulfur (150 mg) and Molybdenum trioxide (15 mg) powder as precursors. MoO_3 powder was kept in the furnace's central zone, where 850 °C temperature was maintained and S powder was put at 120 °C so that both evaporated at the same time. The temperature of the furnace was increased at a rate of 12 °C/minute to reach 850 °C and the deposition time was 20 min. Then the deposited film was allowed to cool naturally. When the substrate was kept on the MoO_3 powder, we did not obtain a pure phase of MoS_2. We placed the SiO_2/Si substrate face up, 12 cm away from the MoO_3 powder (~ 550 °C). We tried with Si substrates also, but there was no deposition of MoS_2. The substrate should be kept in the dense gas flow region, so the substrate height is also important here. We placed the substrate 1.2 cm from the bottom. As the distance between S and MoO_3 is 26 cm, we need a little higher gas flow rate. We use 50 sccm of Ar gas flow in this case. We flow Ar gas for 30 min before starting the deposition to create an inert atmosphere.

3 Results and Discussion

Figure 2 exhibits FESEM images of the surface morphology of MoS_2 films. From this, we can clearly see that the vertical and horizontal nano-wall structure formation. This surface morphology differs from that of monolayer MoS_2, which grows in a triangular shape, since our MoS_2 is in bulk form.

Figure 3a shows the XRD pattern of MoS_2 thin film. The peak situated at 14.29° indicates the (002) plane. As we get only one peak, it can be inferred that the film is oriented along the (002) direction. Also, the XRD pattern confirms the formation of pure 2-H phase MoS_2. Figure 3b depicts the Raman spectrum of MoS_2 thin films. The peaks situated at 383.2 and 408.8 cm^{-1} correspond to E_{2g} and A_{1g} modes. The peak distance between these two peaks is 25.6 cm^{-1} which confirms bulk deposition.

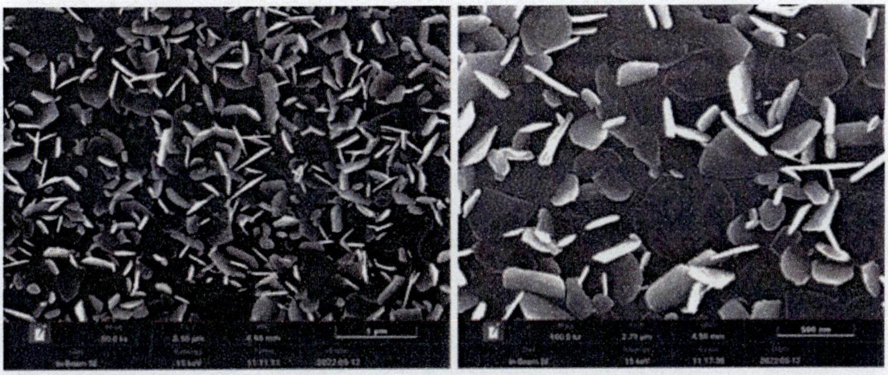

Fig. 2 FESEM images of surface morphology of MoS_2 at 50 and 100Kx

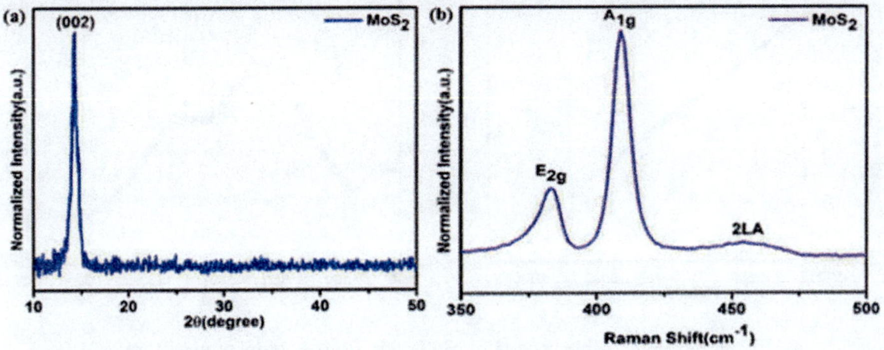

Fig. 3 **a** XRD pattern and **b** Raman spectrum of MoS_2 thin film

Figure 4 shows the XPS spectra of Mo3d and S2p core levels. The peaks at 229.85 and 233 eV correspond to Mo^{4+} $3d_{5/2}$ and $3d_{3/2}$, respectively. The peak situated at 236.32 eV corresponds to the Mo^{6+} state. The peak at 227 eV represents the contribution of S 2 s. It is clear from the deconvolution of the Mo3d core level that Mo is predominantly present in the 4+ valence state in the grown film. The peak positions of the S 2p spectrum correspond to S^{2-} states. So we can clearly conclude from XPS data that there is only pure-phase MoS_2 formation as Mo is mostly in Mo^{4+} valence state.

From Superconducting Quantum Interference device (SQUID) data, it is clear that the MoS_2 film shows diamagnetic behavior at room temperature (Fig. 5a). Magnetism is frequently attributed to S vacancies present in MoS_2 [3]. The XPS analysis suggests that Mo is present in the 4+ valence state and S vacancies are not present, the MoS_2 film exhibits diamagnetic behavior. Generation of defects in these films and study their magnetic properties will be an interesting topic of research. Figure 5b depicts the reflectance of the MoS_2 film in the 200–800 nm wavelength range. It shows very

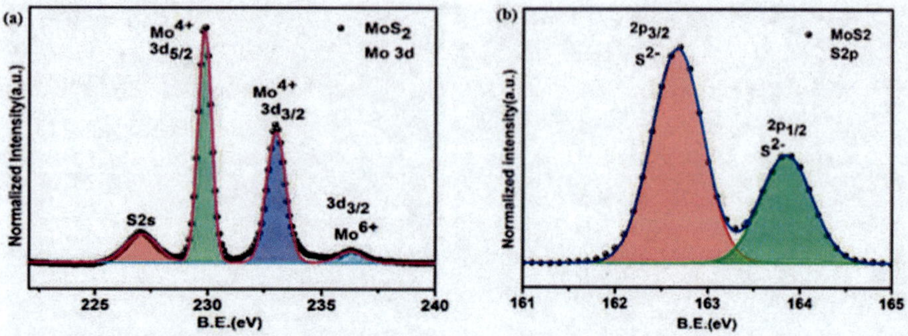

Fig. 4 XPS spectra of **a** Mo3d, **b** S2p core shells of MoS_2 thin films

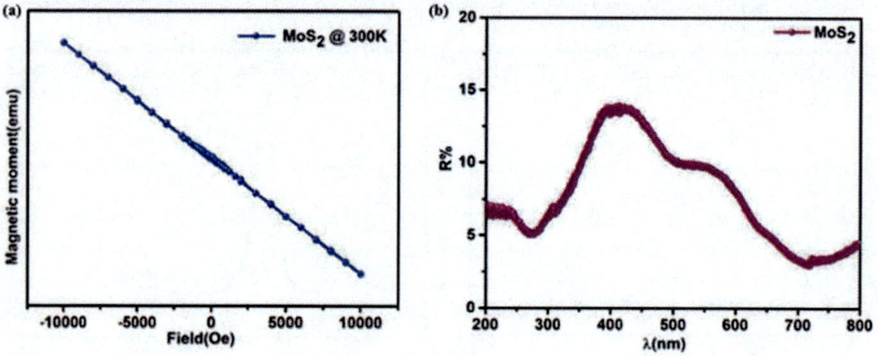

Fig. 5 **a** SQUID data **b** UV–Visible spectra of MoS_2 thin films at room temperature

low reflectance (within 15%) at room temperature suggest its possible application in solar cell. The most efficient TMD monolayer-based solar cells have been discovered is n-type MoS_2 on p-Si [4].

4 Conclusion

To conclude, we optimized all the parameters to get pure-phase, large-area, high-quality MoS_2 thin films. Based on XRD and Raman results, we can conclude that the formation of the 2-H phase and oriented growth of MoS_2 film. From FESEM images, we can clearly see the formation of the vertical nano-wall structure. The SQUID data shows the film is diamagnetic at room temperature. Because MoS_2 is a semiconductor at room temperature, it will be very useful for spintronics applications if, after the formation of some defects, it behaves like a ferromagnet at room temperature. Also, as it shows very low reflectance at room temperature, it may be tried as a potential material for solar cell applications.

5 Declaration of Interest Statement

The authors declare that they have no conflict of interest.

Acknowledgements One of us (SD) would like to acknowledge Dr. Harsh Gupta and the lab members for useful discussions. The authors would like to thank the technical personnel associated with the departmental (XRD, Raman) and the CRF (FESEM, XPS, and SQUID) of IIT Delhi.

References

1. Singh A, Moun M, Singh R (2019) Effect of different precursors on CVD growth of molybdenum disulfide. J Alloy Compd 782:772–779
2. Kaushik V, Varandani D, Mehta BR (2015) Nanoscale mapping of layer-dependent surface potential and junction properties of CVD-grown MoS_2 domains. J Phys Chem C 119:20136–20142
3. Mathew S, Gopinadhan K, Chan TK, Yu XJ, Zhan D, Cao L, Rusydi A, Breese MBH, Dhar S, Shen ZX, Venkatesan T, John TL Thong (2012) Magnetism in MoS_2 induced by proton irradiation. Appl Phys Lett 101:102103
4. Tsai M-L, Su S-H, Chang J-K, Tsai D-S, Chen C-H, Wu C-I, Li L-J, Chen L-J, He J-H (2014) Monolayer MoS_2 heterojunction solar cells. ACS Nano 8:8317–8322

Co-surfactant Modulates Nanoparticle Dimensions Synthesized in Normal Microemulsion

Sundar Singh, Nancy Jaswal, Purnima Justa, Hemant Kumar, Sujeet Kumar Chaurasia, Balaram Pani, and Pramod Kumar

Abstract In this study, we have examined micellar systems with varying amounts of co-surfactant. CMC was found to be more or less independent of the amount of co-surfactant. This indicates that the variation in the amount of the co-surfactant was found not to be playing any significant role in defining the dimension of the micellar aggregates. The same argument was extended when nanoparticles were synthesized within the micellar systems with varying co-surfactant amounts. It was seen from SEM/DLS studies that with an increasing amount of co-surfactant, the size of the NPs increased, suggesting that the dimensions of NPs synthesized in this medium is dependent on the amount of co-surfactant. Growth kinetic studies indicated faster nanoparticle growth in the microemulsion carrying higher co-surfactant, which can be directly correlated with increased fluidity of the micelles.

Keywords Microemulsion · Co-surfactant · CMC · Fluidity · Nanoparticle size

---

S. Singh
Department of Physics, Bareilly College, Bareilly 243005, India

N. Jaswal · P. Justa · P. Kumar (✉)
Department of Chemistry and Chemical Science, School of Physical and Material Sciences, Central University of Himachal Pradesh, Dharamshala 176215, India

S. K. Chaurasia
Center for Nanoscience and Technology, Prof. Rajendra Singh Institute of Physical Sciences for Study and Research, V.B.S. Purvanchal University, Jaunpur, U.P. 222003, India

H. Kumar
Department of Chemistry, Ramjas College, University of Delhi, New Delhi, Delhi 110007, India

Department of Chemistry, University of Delhi, New Delhi, Delhi 110007, India

H. Kumar · B. Pani
Department of Chemistry, Bhaskaracharya College of Applied Sciences, University of Delhi, New Delhi, Delhi 110007, India

Z. H. Khan et al. (eds.), *Recent Advances in Nanomaterials*, Springer Proceedings in Materials 27, https://doi.org/10.1007/978-981-99-4878-9_30

1 Introduction

Microemulsions are stable and isotropic dispersions of oil-in-water (o/w, normal micelles), or *vice-versa* (w/o, reverse micelles), stabilized by an amphiphile monolayer of surfactant [5, 7]. They form surfactant-bound nanometric droplets, which have been extensively researched as templates for the dissolution of insoluble molecules, as well as the synthesis of diverse organic and inorganic nanostructured materials [3, 4, 8–10]. Sodium bis (2-ethylhexyl) sulfosuccinate, usually called aerosol OT or AOT, is one of the most popular surfactants used in microemulsions. However, it has limited solubility in water, which necessitates the addition of co-surfactants for the formation of oil-in-water (normal) microemulsion.

Here, we have investigated the role of the co-surfactant n-butanol (n-BuOH) in the formation of pseudo-ternary (AOT + n-BuOH)/n-hexane/water normal microemulsion system, and the modulation of dimensions of nanoparticles synthesized therein. Three different concentrations of n-BuOH were chosen, each resulting in the formation of stable microemulsions. The experiments involving these microemulsions include study of the critical micelle concentrations (CMCs). After that, ormosil nanoparticles, encapsulating water-insoluble fluorophore Nile Red (NR), has been synthesized within the microemulsion nanoreactors. Dynamic light scattering (DLS) was performed to measure the growth kinetics of these nanoparticles, as a function of reaction time. Scanning electron microscopy (SEM) study confirmed the final size of these nanoparticles.

2 Materials and Methods

Surfactant aerosol OT (AOT) was purchased from Merck. Co-surfactant n-BuOH and liq.NH_3 were obtained from Merck. Hexane (AR Grade) and benzil were purchased from Rankem. Ormosil precursors, vinyltriethoxysilane (VTES), and (3-Aminopropyl) triethoxysilane (APTES) were procured from Sigma Aldrich. Pyrene was purchased from Sigma Aldrich (USA). Doubly distilled water (DDW) was used throughout the experiments.

3 Methods

3.1 Determination of Critical Micellar Concentration (CMC)

Herein, the critical micelle concentrations (CMCs) were determined using fluorimetric analysis of the microemulsions containing pyrene as the fluorescence probe. Here, a stock solution of 1.0 mM pyrene in methanol (solution A) was mixed with the EM/n-hexane/water microemulsions containing various concentrations of the EM.

Pyrene's first and third vibrational peaks, which corresponds to its excitation wavelength of 334 nm and emission wavelengths of 373 and 384 nm, respectively, were observed. This experiment was carried out for three batches of AOT + n-BuOH with ρ values (mass ratios, or m_{11}/m_{12} of AOT/n-BuOH), namely 0.679, 0.452, and 0.339.

3.2 Synthesis of Ormosil Nanoparticles

The dye-encapsulated ormosil nanoparticles have been prepared using similar procedures as described previously [2, 6], Gubala et al., 2020). The synthesized nanoparticles were purified using 12 kDa dialysis membrane and centrifuged afterward, followed by washing with water.

3.3 Study of Size of Nanoparticles

In order to investigate the possible role of the amount of co-surfactant on the kinetics of nanoparticle formation, the time-dependent nanoparticle growth for the varying amounts of n-BuOH (ρ values of 0.679, 0.452, and 0.339) was observed using the technique of dynamic light scattering (DLS). The final size of the nanoparticles was confirmed by scanning electron microscope (SEM) analysis. A film of the nanoparticles was deposited over the glass slide and was allowed to dry over the night. The dried samples were coated with gold and were then observed under a MERA3 TESCAN Scanning electron microscope operating at 5.0 kV.

4 Results and Discussion

The CMC for all the three microemulsion systems (ρ values of 0.679, 0.452, and 0.339) have been studied and calculated using fluorescence emission from the probe molecule pyrene. Figure 1 shows the fluorescence intensity ratio (I_I/I_{III}) of pyrene emission spectra *versus* AOT concentrations (0.001–0.025 M). Above a particular surfactant concentration, the intensity ratio starts to abruptly fall, signaling the start of micelle formation. As a result, the CMC is identified as inflection point seen in the low concentration range. From Fig. 1, it can be seen that the CMCs for all the three systems lie in the range of 3–5 mM. Thus, the CMC is more or less independent of the amount of co-surfactant added ($\rho = 0.679$, 0.452, and 0.339).

The above inference is supported by the structural arrangement of the co-surfactant molecules in the micelles (Fig. 2). As, more and more amount of n-BuOH is added, the distance between the head groups increases. Thus, this solubilization of n-BuOH may result in increased fluidity of micellar interface.

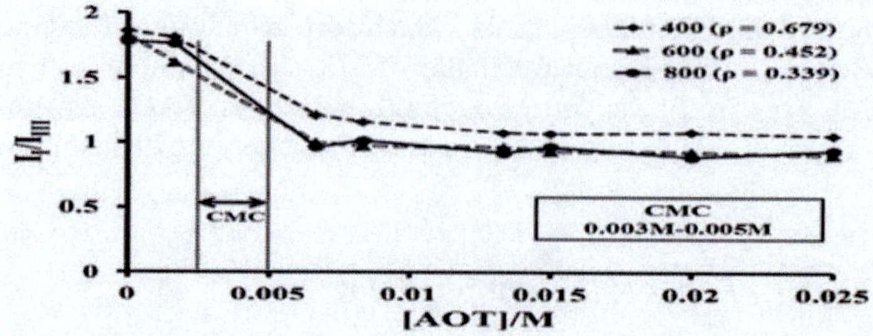

Fig. 1 Plots of I_I/I_{III} for AOT + n-BuOH/n-hexane/water pseudo ternary system as a function of [AOT]

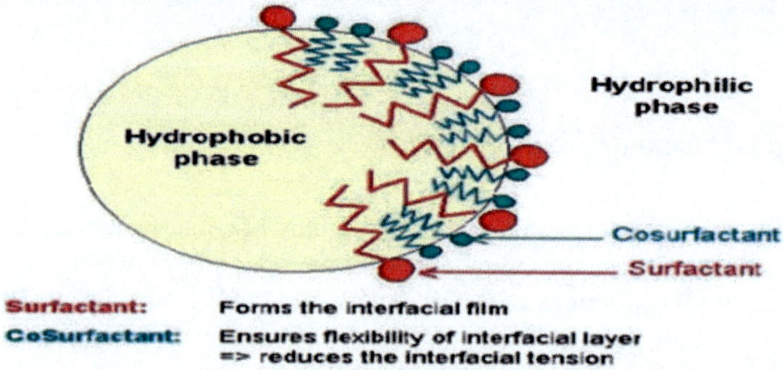

Fig. 2 Structural arrangement of the co-surfactant in o/w microemulsions

The micellar systems, studied above, have been used as template for synthesizing ormosil nanoparticles. Figure 3 (A1, B1, C1) depicts SEM images for ormosil nanoparticles synthesized from micelles with ρ values of 0.679, 0.452, and 0.339, with sizes of 30 nm ± 3.1, 50 nm ± 4.3and, 80 nm ± 6.1, respectively. The three batches of nanoparticles have been designated as ORM-4, ORM-6, and ORM-8, according to the nanoparticle sizes. SEM images show spherical shape of nanoparticles formed.

The final size of the nanoparticles was confirmed by DLS measurement for ORM-4, ORM-6, and ORM-8, as shown in Fig. 3 (A2, B2, C2). The data shows that the average size for ORM-4, ORM-6, and ORM-8 have been found to be 60 nm, 100 nm, and 120 nm, respectively. It can be noticed that the size determined by DLS is a little larger than that measured by SEM. This is because while DLS portrays the hydrodynamic size, with a hydration layer surrounding the particle core, thus accounting for rise in the overall size, SEM depicts the nanoparticles real size.

DLS study was also used to study the growth kinetics of the ormosil nanoparticles in real time (Table 1). It can be observed from the data that soon after the addition of all the reactants, large colloidal aggregates with micrometer dimensions and high

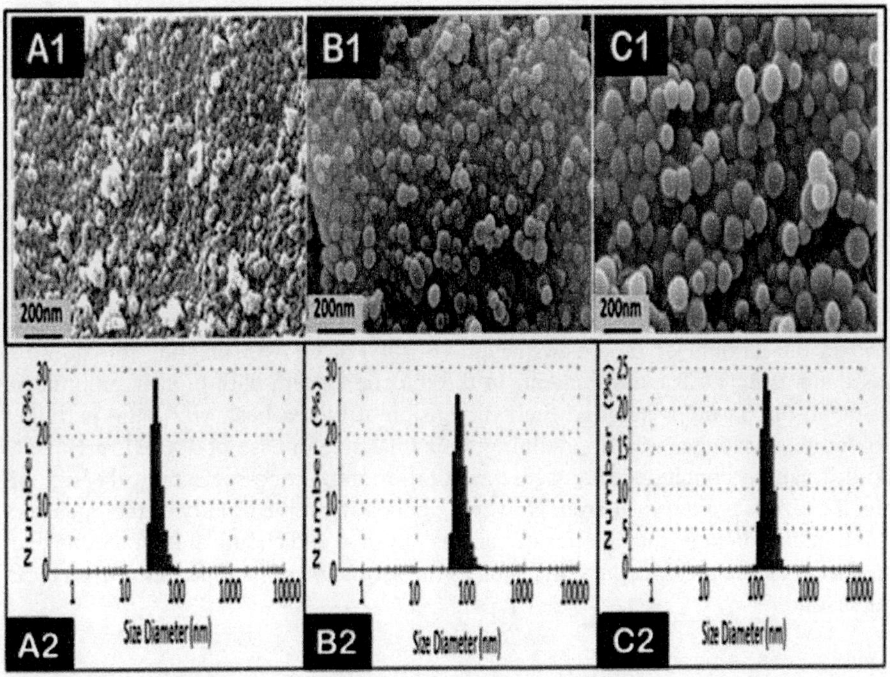

Fig. 3 SEM micrographs of ormosil for (A1) 400µL, (B1) 600 µL, and (C1) 800 µL n-BuOH. DLS graphs for (A2) 400µL, (B2) 600 µL, and (C2) 800 µL n-BuOH

polydispersity index (PDI) are detected. This suggests that faster nanoparticle formation leads to synthesis of bigger nanoparticles, which in turn is driven by the amount of co-surfactant present in the micellar system.

Table 1 Growth kinetic data obtained using dynamic light scattering measurements

$BuOH(\mu L) \longrightarrow$	400		600		800	
Time (min)	Size (nm)	PDI	Size (nm)	PDI	Size (nm)	PDI
5	5312	0.811	5184	1.000	3855	1.000
10	1578	0.437	4009	0.559	3397	0.560
15	208.6	0.695	4643	0.543	382	0.417
30	127.7	0.746	69	0.668	211	0.031
60	32.6	0.797	91	0.057	220	0.023
120	64	1.000	97	0.057	220	0.029
240	58	–	99	0.021	220	–
1 day	54	0.036	99	0.116	235	0.034

5 Conclusion

Pseudo-ternary microemulsion system of AOT + n-BuOH/hexane/water has been studied, and the nature of the excess co-surfactant interaction with the micelles has been reviewed. CMCs were found out to be more or less independent of the amount of co-surfactant. It inferred that the amount of the co-surfactant is not playing a significant role in defining the dimension of the micellar aggregates at the microemulsion level. The three different micellar systems were used to synthesize ormosil NPs. It was seen from SEM/DLS studies that with increasing the amount of co-surfactant, the size of the NPs increased, suggesting the dimension of NPs synthesized is dependent on the amount of the co-surfactant (n-BuOH). Growth kinetics for the NPs was done using DLS measurement. In the micelle system with highest amount of n-BuOH, the particles attained their dimension at the earliest, while the particles took relatively longer time in attaining their final size in case of 400 μL n-BuOH. In all it can be concluded that with variation of amount co-surfactant, the size of micellar size is not being altered. However, it does affect the fluidity of the micelles, which in turn dictates the kinetics of various physical and chemical interactions that may occur inside the micellar core, such as the formation of nanoparticles from their precursors.

6 Declaration of Interest Statement

No conflicts of interest are disclosed by the authors.

Acknowledgements Sincere thanks of authors go out to the SERB (SRG/2020/000381), India, for providing start-up funding.

References

1. Gubala V, Giovannini G, Kunc F, Monopoli MP, Moore CJ (2020) Dye-doped silica nanoparticles: synthesis, surface chemistry and bioapplications. Cancer Nanotechnol 11(1):1–43
2. Kumar H, Agnihotri S, Roy I, Pani B, Kumar P (2020) Microemulsion mediated multifunction of doxorubicin encapsulated core-shell iron oxide/ormosil nanoparticles as efficient magnetically-guided delivery, bioimaging and in-vitro studies. Adv Sci, Eng Med 12:1–8
3. Pileni MP (1997) Nanosized particles made in colloidal assemblies. Langmuir 13(13):3266–3276
4. Ranjan R, Vaidya S, Thaplyal P, Qamar M, Ahmed J, Ganguli AK (2009) Controlling the size, morphology, andaspect ratio of nanostructures using reverse micelles: a case study of copper oxalate monohydrate. Langmuir 25(11):6469–6475
5. Rees BGD, Robinson BH (1993) Microemulsions and organogels: properties and novel applications 9
6. Roy I, Kumar P, Kumar R, Ohulchanskyy TY, Yong KT, Prasad PN (2014) Ormosil nanoparticles as a sustained-release drug delivery vehicle. RSC Adv 4(96):53498–53504

7. Ryan LD, Kaler EW (1999) Microstructure properties of alkyl polyglucoside microemulsions. Langmuir 15(1):92–101
8. Tabor RF, Eastoe J, Dowding PJ, Grillo I, Heenan RK, Hollamby M (2008) Formation of surfactant-stabilized silica organosols 6:12793–12797
9. Tojo C, De Dios M, López-Quintela MA (2009) On the structure of bimetallic nanoparticles synthesized in microemulsions. J Phys Chem C 113(44):19145–19154
10. Vaucher S, Li M, Mann S (2000) Synthesis of Prussian blue nanoparticles and nanocrystal superlattices in reverse microemulsions. Angew Chem Int Ed 39(10):1793–1796

Obtaining Dye-Modified Silica Nanoparticles and Their Characterization as Immunoanalytical Markers

A. A. Bulanaya, N. A. Taranova, A. V. Zherdev, and B. B. Dzantiev

Abstract Colored silica particles with encapsulated bromophenol blue dye were synthesized and characterized as potential markers for lateral flow immunoanalytical systems. Initial concentrations of monomers (aminopropyltriethoxysilane and tetraetoxysilan) and time of their polymerization were varied, and the influence of these parameters on size and optical properties of the obtained particles was studied. It was found that increase of the aminopropyltriethoxysilane concentration in the range from 6.2 to 62 μM accorded to increase of the average particles' diameter from 2 to 5 μm. On another hand, prolonged incubation with tetraetoxysilan during 2–24 h caused lower average diameters of the obtained product—near 230 nm—on contrast with short incubation times (< 1 h). The competitive advantages of the given particles as immunoanalytical markers consist in intense coloration in blue part of spectrum (absorption peak near 600 nm) and large size, which provide higher optical signal from single labeled immune complex formed during the analysis in comparison with traditional high-dispersed nanosized labels. The obtained dye-modified particles were conjugated with antibodies for further bioanalytical application. Aggregation stability and stored antigen-binding properties for the synthesized conjugates were shown.

Keywords Silica nanoparticles · Encapsulation of dyes · Bromophenol blue · Nanolabels

A. A. Bulanaya · N. A. Taranova · A. V. Zherdev · B. B. Dzantiev (✉)
A.N. Bach Institute of Biochemistry, Research Center of Biotechnology of the Russian Academy of Sciences, Leninsky Prospect 33, 119071 Moscow, Russia
e-mail: dzantiev@inbi.ras.ru

Z. H. Khan et al. (eds.), *Recent Advances in Nanomaterials*, Springer Proceedings in Materials 27, https://doi.org/10.1007/978-981-99-4878-9_31

1 Introduction

1.1 A Subsection Sample

Immunochromatographic (lateral flow) test systems, in which the analyte-antibody complexes are formed during the movement of reagents along the membranes of the test strip and are detected by coloration of certain zones of the strip with marker particles, are efficient and widely used tools for rapid out-of-laboratory detection of various compounds [1]. However, analytical parameters of immunochromatographic analysis, first of all—its detection limit, largely depend on the properties of the markers used. This fact causes interest in evaluating capabilities of new particles as potential immunoanalytical markers [2]. Silicate particles are promising candidate preparations due to their large surface area (per one particle) and easy modification with various compounds. Examples of successful applications of silica colored particles (SCPs) in immunochromatography are described [3], but the question of choosing the best preparation remains open. This paper presents the synthesis and characterization of silica particles with bromophenol blue that is used as an encapsulated dye.

2 Materials and Methods

2.1 Chemicals and Materials

Sigma provided tetraetoxysilan (TEOS), aminopropyltriethoxysilane (APTES), and bromophenol blue (BPB) (St. Louis, MO, USA). Chimmed provided other compounds (such as solvents and analytically pure salts) with chemical pure grade (Moscow, Russia). Water cleaned by the Arium® pro system (Sartorius, Göttingen, Germany) and having a minimum resistivity of 18.2 M cm was used to prepare all solutions for syntheses.

3 Syntheses of Dye-Loaded Silica Particle

By polymerizing APTES with TEOS in the presence of BPB, SCPs were created [4]. 0.7 mg of BPB were mixed with 6.3–63 mM (0.003–0.03 mL) of APTES in 0.2 mL of anhydrous ethanol. For 17 h, this mixture was mixed at room temperature in the dark. The prepared substance and TEOS (0.03 mL) were then combined with 1.7 mL of ethanol and 0.09 mL of a 29% solution of ammonium hydroxide in water. The mixture's total volume was changed to 5 mL. During 5–1500 min, this reaction mixture was shaken at room temperature. By sonicating, rinsing the solution three times with alcohol, and centrifuging it at 3000 g for three minutes, unbound dye was eliminated. The finished product was kept at + 4 °C after being redissolved in phosphate-buffered saline (pH 7.4, KH_2PO_4, K_2HPO_4 50 mM, and NaCl 60 mM).

4 Characterization of SCPs

Particle solutions are applied on 300 mesh grids (Pelco International, California, USA) covered with a polyvinyl formal film for transmission electron microscopy (TEM) [5]. Using the Image Tool application, images captured with a JEM CX-100 microscope (Jeol, Japan) at 80 kV were processed (University of Texas Center for Higher Education, Texas, USA). Nanoparticle aggregation was assessed using Zetasizer Nano (Malvern Pananaltical, UK). On a Libra S80 spectrophotometer, the absorption spectra of particles with a wavelength between 300 and 700 nm were captured (Biochrom, UK).

5 Results and Discussion

SCPs were obtained by polymerization of APTES and TEOS in the presence of the BPB in an alkaline environment. At the first stage, APTES micelles were formed with the BPB dye, which is embedded inside the micelles. When TEOS was added to the mixture, a rigid structure was formed on the surface of the micelles (Fig. 1a). Ammonia acted as an initiator of the polymerization reaction. Ten preparations were totally obtained, when conditions of their syntheses were differed in amounts of APTES and/or BPB. The final particle size varies from 100 nm to 2 m depending on the polymerization settings; for an example, see TEM image at Fig. 1b.

Spectrophotometry and dynamic light scattering (DLS) methods were used to describe SCPs. It was demonstrated that the APTES polymerization time had an impact on the quantity of micelles produced but not their size. The initial number of micelles was considered to be equal to the initial total number of particles and was

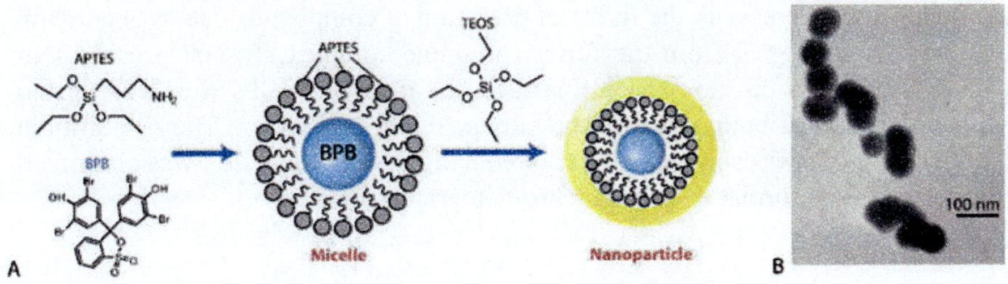

Fig. 1 Scheme of SCPs synthesis (**a**) and TEM image of SCPs (**b**)

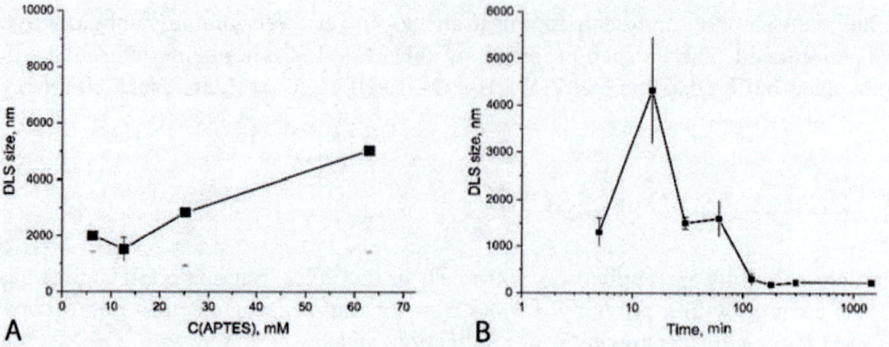

Fig. 2 Influence of the synthesis conditions on the average size (DLS data) of SCPs. **a** Varying APTES concentration at the first step of the synthesis. **b** Varying time of the second step (incubation with TEOS)

calculated using their weight after drying. It was found that changing the time of micelles formation from 10 min to 24 h was led to varying the mass concentration of SCPs in the range 0.5–3 mg/mL. The average hydrodynamic particle diameter increased from 2 to 5 μm in the accordance with the growth of APTES concentration (Fig. 2a). Thus, final particle size depends on the initial micelle size. At the second stage of synthesis, with an increase in the polymerization time with TEOS, in the first 20 min, a jump in the growth of the particle diameter to 5 μm occurred, followed by a decrease to 230 nm (Fig. 2b). The given effects are similar to earlier published studies of SCPs [6].

SCPs are characterized by a higher contrast at membrane-based tests compared to gold nanoparticles, the most widely used immunoanalytical markers, due to large amount of encapsulated dye. We have found that under the synthesis conditions, SCPs encapsulated up to 50% of the added BPB (Fig. 3). When comparing the absorption spectra of the synthesized SCPs and BPB, it was shown that the absorption maximum of the nanoparticles coincides with the absorption peak of the pure dye. The broad profile of the spectrum of nanoparticles indicates SCPs scattering. It was also shown that, after encapsulation in SCPs, the dye retains indicator properties: the absorption spectra have two peaks at 370 and 595 nm.

The large size of silicate particles, and, consequently, the high concentration of the dye leads to decrease in the limits of detection of compounds due to generation of higher optical signal from the formed immune complex. To test processes of physical adsorption on the particles surface, the obtained SCPs were conjugated with monoclonal antibodies against the fatty acid binding protein. Their stability in colloidal solutions was shown and the stored antigen-binding ability was confirmed when tested in the format of immunochromatographic analysis.

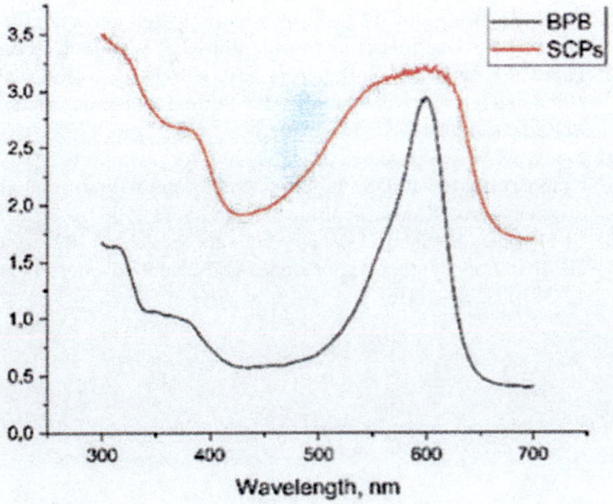

Fig. 3 Absorption spectra of free BPB (black curve) and BPB encapsulated in SCPs (red curve)

6 Conclusion

SCPs as potential markers for immunochromatographic systems have been synthesized and characterized. The dimensions and optical properties of the preparations obtained by varying the concentration of the initial monomers and the polymerization time have been measured. The competitive advantages of these particles are intense coloration and large sizes, which increase the optical signal from single immune complex formed during the analysis. The conjugates of SCPs with antibodies for further bioanalytical application were synthesized; their colloidal stability and stored antigen-binding properties in immunochromatography were shown.

7 Declaration of Interest Statement

The authors declare that they have no conflict of interests.

Acknowledgements This research was financially supported by the Russian Science Foundation, grant 19-14-00370.

References

1. Manmana Y, Kubo T, Otsuka K (2021) Recent developments of point-of-care (POC) testing platform for biomolecules. TrAC, Trends Anal Chem 135:116160
2. Wang Z, Zhao J, Xu X, Guo L, Xu L, Sun M, Li A (2022) An overview for the nanoparticles-based quantitative lateral flow assay. Small Methods 6(1):2101143

3. Su Z, Zhao G, Dou W (2020) Preparing high chroma colored silica nanoparticles based on layer-by-layer self-assembled technique. J Sol-Gel Sci Technol 101:562
4. Pack C-G, Paulson B, Shin Y, Jung MK, Kim JS, Kim JK (2020) Variably sized and multi-colored silica-nanoparticles characterized by fluorescence correlation methods for cellular dynamics. Materials 14(1):19
5. Panferov VG, Safenkova IV, Zherdev AV, Dzantiev BB (2018) Post-assay growth of gold nanoparticles as a tool for highly sensitive lateral flow immunoassay. Application to the detection of potato virus X. Microchimica Acta 185(11):506
6. Li H, Chen X, Shen D, Wu F, Pleixats R, Pan J (2021) Functionalized silica nanoparticles: classification, synthetic approaches and recent advances in adsorption applications. Nanoscale 13(38):15998–16016

Effect of Methyl Substitutions on the Ionization Energy of $OH_{3-n}(CH_3)_n^+$

Harshita Srivastava, Jitendra Kumar Tripathi, and Ambrish Kumar Srivastava

Abstract The superalkali clusters are renowned for having lower ionization energies than alkali metal atoms (i.e., < 5.39 eV). The four decades of superalkalis included the exploration of numerous types of these uncommon species together with their applications. In this work, we designed new series of superalkali clusters as $OH_{3-n}(CH_3)_n^+$ with $n = 0$–3 and scrutinized the role of successive addition of the CH_3 group. We observed that successive addition of the CH_3 group brings a decrease in the VEA of the, respectively, designed cations such that their VEA values are 4.46 eV and 3.77 eV for $OH_2CH_3^+$ and $OH(CH_3)_2^+$, respectively. It further decreases to 2.87 eV for $O(CH_3)_3^+$. These molecules consequently belong to the family of superalkalis. Moreover, we demonstrate that designing novel superatomic systems and exploring their physicochemical features might be used to create desirable functional materials.

Keywords Methyl substitutions · Superalkalis · Physicochemical

1 Introduction

Superalkali cations were predicted to exist by Gutsev and Boldyrev [1] who also proposed their empirical relation, XM_{k+1}^+, here X is an electronegative atom with valence k, likewise F, O, or N, and M is an alkali atom with valence k, such as Li, Na, or K. FLi_2^+, OLi_3^+, NLi_4^+, and other well-known instances of such superalkali cations are available. The superalkali, or their cations, are distinguished from alkali atoms by having lower ionization energies (IE) or vertical electron affinities (VEA) (5.39–3.89 eV) [2] compared to those of first group elements (alkali metal atoms). The exploration of the several superalkalis has advanced significantly. Tong et al. [3] have described novel binuclear superalkali cations with the formula $M_2Li_{2k+1}^+$.

Furthermore, alkali-monocyclic(pseudo)-oxocarbon moieties [4] and other functional groups ($Y = CO_3$, SO_3, SO_4, etc.) have been used to create and study polynuclear superalkali cations. Sun and Wu have recently evaluated the advancements

H. Srivastava (✉) · J. K. Tripathi · A. K. Srivastava
Department of Physics, Deen Dayal Upadhyaya Gorakhpur University, Uttar Pradesh, Gorakhpur 273009, India

Z. H. Khan et al. (eds.), *Recent Advances in Nanomaterials*, Springer Proceedings in Materials 27, https://doi.org/10.1007/978-981-99-4878-9_32

achieved in this sector [5]. Superalkalis have also been designed to behave as charge transfer clusters like supersalts [6, 7], superbase [8, 9], alkalides [10–12], and so on. Further, in contrast to Zintl constituent-based superalkali cations, the significance of alkyl ligand swapping in causing the superalkali character was evident [13]. Upon appropriate alkyl group substitutions, organic-heterocyclic compounds entrenched on pyrrole (C_4NH_5) and imidazolium ($C_3N_2H_5$) [14, 15] have been demonstrated to behave as superalkalis. Recently, we have reported the novel series of superalkalis clusters as $XH_{4-x}(CH_3)_x$ where X = N, P and As [16]. The reduction of CO_2 has been done by the application of superalkali clusters [17]. Despite all of this advancement, the underlying problem with the engineering of superalkalis remained unsolved. Whether a non-metal element or group may be used in place of the metal (M) to create a superalkali cation with the general formula $XM_{k+1}{}^+$. In this regard, a re-examination of the well-known $OH_3{}^+$ cations [18] was attempted, highlighting their low VEA characteristic. By using methyl (CH_3) ligands rather than M, we introduce the $OH_{3-n}(CH_3)_n{}^+$ family of superalkali cations, in this letter. We will carefully investigate how the replacement of CH_3 for H results in a reduction of affinities of the cations as well as also the resultant superalkali nature.

2 Computational Detail

The second-order Møller-Plesset (MP2) perturbation method [19] along with 6-311++ G(d, p) basis set [20] in a Gaussian 09 program [21] was used to generate the geometries of $OH_{3-n}(CH_3)_n{}^+$ cations and similar species without the use of symmetry constraints. A valid minimum on the potential surface must be found for the optimized structures in order for them to incorporate all real frequency values, vibrational frequency calculations were performed after the structural optimization of the original clusters. The charge on atoms was determined using the natural population analysis method [22]. When comparing the total energies of an optimized cation with its corresponding neutral complements at its cationic configuration, the vertical electron affinities for cations have in fact been found.

In earlier research, the current approach had already been used. For VEA of $OH_3{}^+$ cations is 4.39 eV and is consistent with published values of 4.36–4.50 eV from prior research [18]. This may imply the trustworthiness of the results presented in this work as well as the reliability of our scheme.

3 Results and Discussion

We begin by discussing the geometry of $OH_{3-n}(CH_3)_n{}^+$ cations, considering the $OH_3{}^+$ cations depicted in Fig. 1. The geometry of OH^+ comprises quasi-planar C_{3v} (where O–H = 0.978 Å). Then, in the $OH_3{}^+$ cation, we sequentially swap CH_3 for H. Figure 1 also depicts the optimum structure of $OH_2CH_3{}^+$. Also O–H bond length is not affected by the substitution of a single H ligand, whereas the distance of an O–C bond was estimated as 1.510 eV. Further, CH_3 group substitutions result in a decrease in O–C bond distance as 1.487 Å in $OH_{3-n}(CH_3)_n^+$. The optimized configurations of $O(CH_3)_3^+$ cations are shown in Fig. 2. In these $OH_{3-n}(CH_3)_n^+$ cations, the bond lengths O–C reach 1.476 Å. As a result, the consecutive substitution of a CH_3 ligands shortens an O–C bond distances. Figure 2 depicts a usual superalkali complexes, such as OLi_3^+. In contrast to the trigonal planar D_{3h} configurations of OLi_3^+, the geometry of an $O(CH_3)_3^+$ has C_{3v} symmetry. Our findings regarding these cations are consistent with our earlier investigation. The computed VEA of each of the cations is mentioned in (Table 1) because we are curious about the superalkali behavior of clusters created by a serial substitution of CH_3 groups. In line with expectations, the VEA of the popular superalkali cation, OLi_3^+, is lower compared to the C_s atom (3.89 eV). Moreover, OH_3^+ has a lower VEA than Li (5.39 eV). A VEA of $OH_2CH_3^+$ as well as of $OH(CH_3)_2^+$ is estimated as 4.46 eV and 3.77 eV, respectively, with $O(CH_3)_2^+$ having a VEA as lower as 2.87 eV after the CH_3 groups are substituted.

Hence, most cations having an O core and having one or more than one CH_3 ligand show the superalkali nature because of their lower value of VEAs as compared to the alkali atoms. The VEA of $O(CH_3)_3$ complex is lower than that of the superalkali cations $C_xH4_{x+1}^+$ (4.35–2.96 eV for x = 1–5) [23] and 1-methyl-3-methylimidazolium [15] previously reported.

By reducing the VEA of $OH_{3-n}(CH_3)_n^+$, ligands play a key function. When CH_3 is substituted, the VEA of $OH_{3-n}(CH_3)_n^+$ drops in comparison with that of alkali atoms, resulting in superalkali cations. Contrary to popular belief, the vertical electron

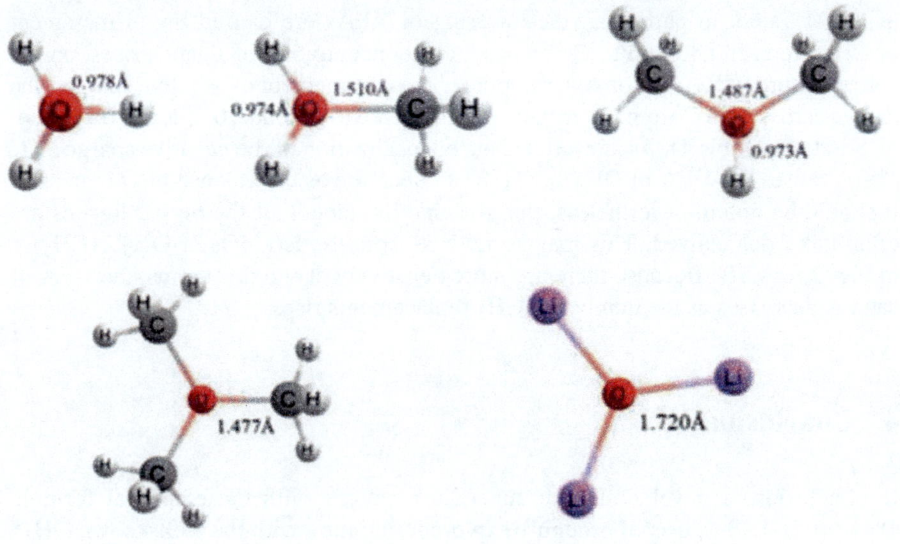

Fig. 1 At the MP2/6-311++ G(d, p) level, equilibrium structures of the $OH_{3-n}(CH_3)_n$ + and OLi_3 + cations. Further, displays include chosen bond lengths

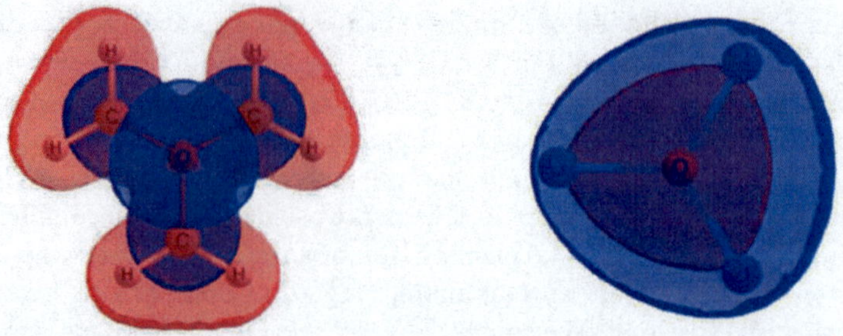

Fig. 2 Highest occupied molecular orbitals (HOMOs) of $O(CH_3)_3^+$ and OLi_3^+cations

Table 1 Vertical electron affinity (eV) as well as the NPA(*e*) charge on O in the optimized structures of $OH_{3-n}(CH_3)_n^+$ and its respective cation

S. No.	Systems	VEAs (eV)	NPA (*e*)
1	OH_3^+	5.16	– 0.75
2	$OH_2(CH_3)^+$	4.46	– 0.66
3	$OH(CH_3)_2^+$	3.77	– 0.56
4	$O(CH_3)_3^+$	2.87	– 0.49
5	OLi_3^+	3.46	– 1.84

affinities of an $O(CH_3)_3^+$ are even less than those of OLi_3^+. Thus, in these species, CH_3 substitutions might result in VEAs that are analogous to cations having Y = Na or K (not reported here).

We calculated and examined the NPA charge on O to understand the pattern of VEAs value, in particular, the lowering of VEAs are caused by an increment of CH_3 group in addition to O valence. As a concentration of ligands rises, on the core structure (CH_3), C atoms share some of the confinement of negative charge. The charge across each C atom, for instance, is estimated to be 0.06*e*, 0.09*e*, and 0.26*e* (not mentioned in Table 1). As a result, enhanced localization of the negative charge on C atoms causes the VEA of $OH_{3-n}(CH_3)_n^+$ to decrease as the valence of O increases. It should be noted, nevertheless, that the electron clouds of the bigger ligands are often more delocalized. This may be noticed from the HOMOs of $OH_{3-n}(CH_3)_n^+$ in Fig. 2 for CH_3. Because there are more negatively charged C atoms, the VEA of cations decreases as the number of CH_3 replacements rises.

4 Conclusion

In conclusion, a novel family of superalkali cations with the empirical formula $OH_{3-n}(CH_3)_n^+$ has been thoroughly explored. Initiating with the well-known OH_3^+ cations as well as gradually replacing the H atoms with CH_3 groups, we looked at the structure as well as the superalkali behavior of the resultant cations. We discovered that the majority of these cations have VEA values that are lower compared to those of alkali atoms, suggesting that they may act like superalkalis. NPA charges and also HOMO surfaces were used to explain how a VEA of cations had decreased. However, the VEA $O(CH_3)_3^+$ cation is surprisingly found lower than that of the normal OLi_3^+ superalkali cation. This work, in our opinion, should start a new era in the study of superalkalis.

References

1. Gutsev GL, Boldyrev AI (1982) DVM Xα calculations on the electronic structure of "superalkali" cations. Chem Phys Lett 92:262–266
2. Lias SG, Bartmess JE, Liebman JF, Holmes JL, Levin RD, Mallard WG (1988) Gas-phase ion and neutralthermochemistry. J Phys Chem Ref Data (Suppl. 170):1
3. Tong Y, Li D, Wu ZR, Li XR (2011) Huang: record low ionization potentials of alkali metal complexes with crown ethers and cryptands. J Phys Chem A 115:2041–2046
4. Tong J, Wu Z, Li Y, Wu D (2013) Prediction and characterization of novel polynuclear superalkali cations. DaltonTrans. 42:577
5. Sun WM, Wu D (2019) Recent progress on the design, characterization, and application of superalkalis. Chem A Eur J 25:9568–9579
6. Srivastava AK, Misra N (2014) Novel $(Li2X)^+(LiX2)^-$ supersalts (X = F, Cl) with aromaticity: a journey towards thedesign of a new class of salts. Mol Phys 112(19):2621–2626
7. Giri S, Bahera S, Jena P (2014) Superalkalis and superhalogens as building blocks of supersalts. J Phys Chem A 118(3):638–645
8. Srivastava AK, Misra N (2015) Superalkalis-hydroxides as strong bases and superbases. New J Chem 39(9):6787–6790
9. Srivastava AK, Misra N (2016) Oli_3O^- anion: designing the strongest base to date using OLi_3 superalkali. Chem Phys Lett 648:152–155
10. Chen W, Li Z-R, Wu D, Li Y, Sun C-C, Gu G-L (2005) The structure and the large nonlinear optical properties of Li@Calix[4]pyrrole. J Am Chem Soc 31(127):10977–10981
11. Sun WM, Fan LT, Li Y, Liu JY, Wu D, Li ZR (2014) On the potential application of superalkali clusters in designing novel alkalides with large nonlinear optical properties. Inorg Chem 53(12):6170–6178
12. Srivastava AK, Misra N (2015) Ab initio prediction of novel alkalides FLi2 -M- Li2F (M = Li, Na, and K). Chem Phys Lett 639:307–309
13. Giri S, Reddy GN, Jena P (2016) Organo-zintl clusters [P7R4]: a new class of superalkalis. J Phys Chem Lett 7(5):800–805
14. Reddy GN, Giri S (2016) Organic heterocyclic molecules become super alkalies. Phys Chem Chem Phys 18(35):24356–24360
15. Srivastava AK (2021) 1-Alkyl-3-methylimidazolium belongs to superalkalis. Chem Phys Lett 778:138770
16. Srivastava H, Srivastava AK (2022) Role of central core and methyl substitutions in XH4-x(CH3)x (X = N, P, As; x = 0–4) superalkalis: an ab initio study. Struct Chem. https://doi.org/10.1007/s11224-022-02003-0
17. Srivastava H, Srivastava AK (2022) Superalkalis for the activation of carbon dioxide: a review. Front Phys 10:870205
18. Srivastava AK, Misra N, Tiwari SN (2020) Superalkali behavior of ammonium ($NH4^+$) and hydronium ($OH3^+$) cations: a computational analysis. SN Appl Sci 2:307
19. Møller C, Plesset MS (1934) Note on an approximation treatment for many-electron systems. Phys Rev 46:618
20. Ditchfield R, Hehre WJ, Pople JA (1971) Self-consistent molecular-orbital methods. IX. An extended Gaussian-type basis for molecular-orbital studies of organic molecules. J Chem Phys 54:724

21. Frisch MJ, Trucks GW, Schlegel HB, Scuseria GE, Robb MA, Cheeseman JR, Scalmani G et al (2009) Gaussian 09, Revision C02. Gaussian Inc., Wallingford, CT
22. Reed AE, Weinstock RB, Weinhold F (1985) Natural population analysis. J Chem Phys 83:735
23. Srivastava AK (2020) $CXH4X+1^{+}$ (x=1-5): a unique series of organic superalkali cations. Mol Phys 118(4):1615648

Eu^{3+}-Doped Ca_2GdSbO_6 Double Perovskite Phosphor for Solid-State Lighting Applications

Chandni Kumari, Jairam Manam, and S. K. Sharma

Abstract A set of Eu^{3+} activated double perovskite Ca_2GdSbO_6 (CGSO) phosphors has been prepared using the facile high-temperature solid-state method. The phase and crystal structure of the sample was studied via XRD. Surface morphology and micrographs were obtained via FESEM images. Chemical composition and oxidation state were attributed by XPS analyses. Optical properties were investigated via UV–Visible-NIR diffuse reflectance spectroscopy (DRS). Luminescence properties were investigated by photoluminescence (PL) excitation and emission spectra. The optical band gap of the sample was calculated using the Kubelka–Munk function along with the Tauc equation. Obtained bandgap energy was 4.72 and 4.82 eV for pure CGSO and CGSO: $0.02Eu^{3+}$. Further, the PL excitation spectrum was recorded for the emission wavelength of 612 nm, while PL emission spectra were monitored for the excitation wavelength of 395 nm. Emission spectra show the characteristic peaks of the f-f transitions of the Eu^{3+} ion. Moreover, the photometric studies revealed high color purity (CP) of 100% and lesser required correlated color temperature (CCT) values. These observed results illustrate that the proposed phosphor could be a convenient material for solid-state lighting applications.

Keywords Double perovskite · Optical bandgap · Photoluminescence · Photometric studies

1 Introduction

At present unique properties of white light-emitting diodes (wLED), such as low power consumption, higher efficiency, environmental friendliness, compactness in size, and excellent operational lifetime, are getting much attention. But currently, available LEDs suffer from a lack of red color purity and have high CCT values, which reduce their applications in medical fields and indoor lighting [1]. In the last few

C. Kumari · J. Manam (✉) · S. K. Sharma
Department of Physics, Indian Institute of Technology (ISM), Dhanbad 826004, India

Z. H. Khan et al. (eds.), *Recent Advances in Nanomaterials*, Springer Proceedings in Materials 27, https://doi.org/10.1007/978-981-99-4878-9_33

years, trivalent rare-earth-doped antimonate-based phosphor has gained much attention due to its special properties such as sharp band emission, wide color tuning, large photoluminescence efficiency, low phonon energy, and relatively large decay time. Among all rare earth, Eu^{3+} is highly recommended because of its larger rate of radiative recombination and certain energy levels. These specialties of antimonates lead them to be easily applicable in the field of X-ray imaging scintillators, plasma display panels, light-emitting diodes, sensors, etc. [2]. Among these materials, Ca_2GdSbO_6 has been rated to be an excellent oxide host. Therefore, Eu^{3+}-doped Ca_2GdSbO_6 phosphor materials have been manufactured, and prime focus of this work is to explore the photoluminescence properties of the samples under consideration.

2 Materials and Methods

$Ca_2Gd_{(1-x)}SbO6$: xEu^{3+} ($x = 0, 0.005, 0.01, 0.02, 0.03$) phosphor materials were prepared using the traditional high-temperature solid-state method. The chemicals were ground for 2 h in the ethanol medium. Obtained mixture was annealed at 1400 °C for 6 h in a furnace, and finally, cooled samples were reground to fine powder to perform further characterizations.

3 Results and Discussion

3.1 Phase, Morphology, and X-Ray Photoelectron Spectroscopic Studies

Figure 1a represents the powder XRD of the CGSO: xEu^{3+} sample, according to the JCPDS no. 01-073-0085 of Ca_2GdTaO_6. Sharp diffraction peaks corresponding to each sample suggest the high crystallinity within the phosphor materials. The well-known equation of radius percentage difference $D_R = \frac{R_p(CN)-R_q(CN)}{R_p(CN)} * 100$ [3] verifies the successful incorporation of Eu^{3+} into the host ion Gd^{3+}. The prepared phosphor crystallized into the monoclinic phase with the $P2_1/n$ space group, with lattice constants $a = 5.489$ Å, $b = 5.763$ Å and $c = 8.083$ Å with $V = 255.69$ Å. Further, Debye–Scherrer equation [4]: $D = \frac{0.9\lambda}{\beta Cos\theta}$, provides average crystallite size as 42 nm. Further, Fig. 1b, c reveals FESEM images of the samples, showing particles of irregular shapes and sizes.

Figure 1d demonstrated the full stretched XPS survey scan of the phosphor, while Fig. 2e–h attributed the high-resolution XPS (HR-XPS) spectra of C 1s, Ca 2p, Gd 4d, Eu 4d, Sb 3d, and O 1s, respectively. Survey spectrum reveals the presence of all desired element without any impurities, except C 1s. Binding energies corresponding to each deconvoluted state of each element have been mentioned in the concerned

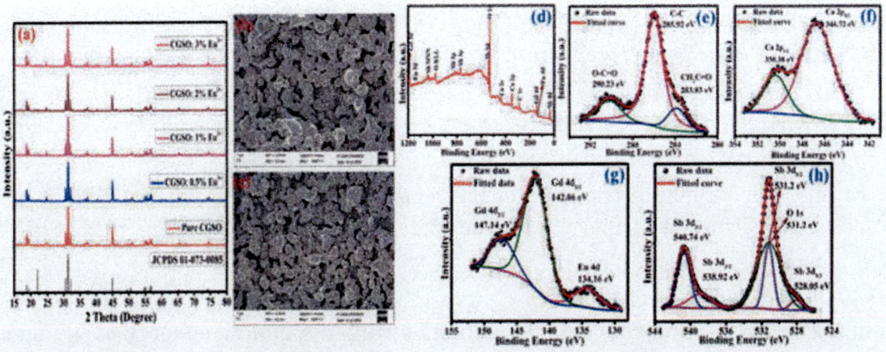

Fig. 1 **a** XRD pattern; **b**, **c** FESEM images of pure and Eu^{3+}-doped CGSO. **d** Survey spectra; **e**–**h** The XPS spectra of C (1s), Ca (2p), Gd (4d), Eu (4d), Sb (3d), and O (1s) of 2 mol% Eu^{3+}-doped CGSO phosphor

figures. These spectra have been deconvoluted via XPSPEAK 4.1 program with Shirley background.

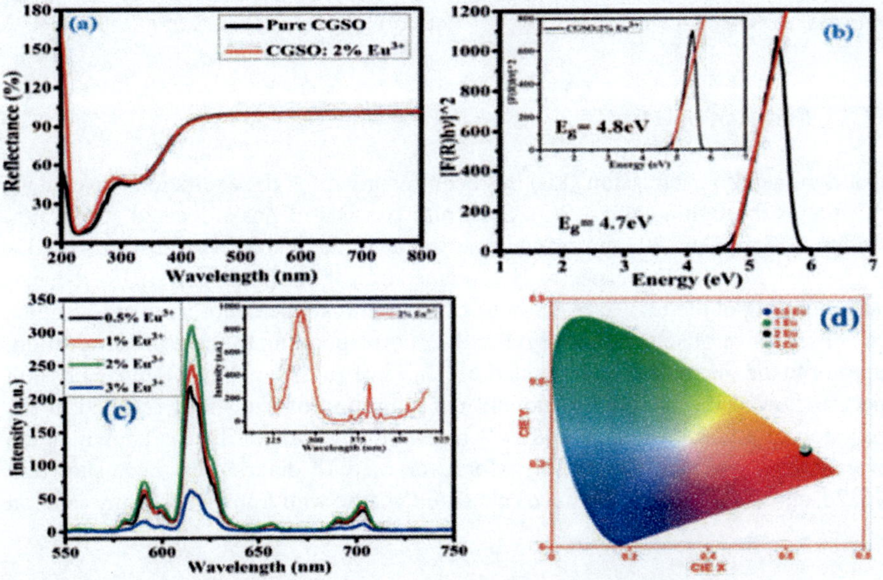

Fig. 2 **a** DRS spectra and **b** Tauc plot of CGSO and CGSO: 2 mol%Eu^{3+}. Photoluminescence **c** excitation spectrum of CGSO: 2 mol%Eu^{3+} **(inset)**, emission spectra and **d** CIE diagram of CGSO: xEu^{3+} ($x = 0.005, 0.01, 0.02, 0.03$)

3.2 Diffuse Reflectance Spectroscopy (DRS) and Photoluminescence Studies

3.2.1 DRS

K. M. and Tauc's equations determined the band gap energy of the material. Figure 2a, b shows the diffuse reflectance and Tauc plots of CGSO and CGSO: 2 mol%Eu^{3+}. KM equation is expressed as $F(R_\infty) = \frac{(1-R)^2}{2*R} = \frac{K}{S}$, and Tauc equation as $\alpha h\nu = B(h\nu - E_g)^n$ with α and B are constants and others with their usual meaning. The combined effect of the above two equations results in band gap energy as 4.7 and 4.8 eV corresponding to the pure CGSO and 2 mol% Eu^{3+}-doped CGSO.

3.2.2 Excitation Spectrum

Figure 2(c inset) represents the photoluminescence excitation (PLE) spectrum of 2 mol% Eu^{3+} activated CGSO at the fixed emission wavelength of 615 nm. The spectrum reveals the charge transfer band (CTB) from 228 to 310 nm, along with the 362 nm, 383 nm, 395 nm, 416 nm, and 466 nm corresponding to electronic transitions from $^7F_0$ to $^5D_4$, $^5L_7$, $^5L_6$, $^5D_3$, $^5D_2$, respectively.

3.2.3 Emission Spectra

Photoluminescence emission (PL) has been recorded for the excitation wavelength 395 nm, as the figure shows. 2(c). Spectrum consists of peaks located at 590 nm, 615 nm, 657 nm, and 703 nm could be attributed to the respective transitions as $^5D_0$ to $^7F_1$, $^7F_2$, $^7F_3$, and $^7F_4$, respectively. Obtained transitions have been emerged due to the 4f splitting of the Eu^{3+} rare earth ions. The dominant peak observed at 615 nm is the consequence of selection rule $\Delta J = \pm 2$, corresponding to the dipole transition, superior to the magnetic dipole transition $^5D_0 \rightarrow {}^7F_1$ ($\Delta J = 1$) at 590 nm. Emission spectra showed the increase of doping concentration of Eu^{3+} ions resulted in the reduction of emission intensity after 2 mol% of Eu^{3+}; this effect is known as the concentration quenching effect [5]. Moreover, the CIE diagram has been shown in Fig. 2d, and CCT values have been calculated via the well-known McGamy relation [6].

4 Conclusion

We have successfully prepared the series of CGSO: xEu^{3+} (0.005, 0.01, 0.02 and 0.03) phosphors. The phase, crystal structure, morphology, oxidation state, optical, and photoluminescence characteristics were studied in detail. Obtained optical band gaps were 4.7 and 4.8 eV for CGSO and CGSO: 2 mol% Eu^{3+}. In PL emission spectra, the phenomenon of concentration quenching was observed after the doping concentration of 2 mol% of Eu^{3+}. Moreover, the obtained CCT values were found to lie in the range of 2479–2770 K, i.e., less than 3000 K and CP was of 100%. These observations suggest that the proposed phosphor could be an auspicious element for warm red phosphors in W-LEDs for indoor and medical applications.

Acknowledgements The authors acknowledge the IIT(ISM) Dhanbad for providing a research fellowship funded by MHRD, Govt. of India. Also, grateful to the IIT(ISM) Dhanbad for XRD, FESEM, XPS and UV-Vis analyses.

Conflict of Interest The authors proclaim that they have no familiar conflicting interest.

References

1. Zhong J, Chen D, Yuan S, Liu M, Yuan Y, Zhu Y, Li X, Ji Z (2018) Tunable optical properties and enhanced thermal quenching of non-rare-Earth double-Perovskite $(Ba_{1-x}Sr_x)_2YSbO_6:Mn_4^+$ red phosphors based on composition modulation. Inorg Chem 57:8978–8987
2. Mondal K, Manam J (2018) Investigation of photoluminescence properties, thermal stability, energy transfer mechanisms and quantum efficiency of $Ca_2ZnSi_2O_7$: Dy^{3+}, Eu^{3+} phosphors. J Lumin 195:259–270
3. Wei C, Xu D, Li J, Geng A, Li X, Sun J (2019) Synthesis and luminescence properties of Eu^{3+}-doped a novel double perovskite Sr2YTaO6 phosphor. J Mater Sci Mater Electron 30:2864–2871
4. Zhang Z, Sun L, Devakumar B, Liang J, Wang S, Sun Q, Dhoble SJ, Huang X (2020) Novel highly luminescent double-perovskite $Ca_2GdSbO_6:Eu^{3+}$ red phosphors with high color purity for white LEDs: synthesis, crystal structure, and photoluminescence properties. J Lumin 221:117105
5. Geng X, Xie Y, Ma Y, Liu Y, Luo J, Wang J, Yu R, Deng B, Zhou W (2020) Abnormal thermal quenching and application for w-LEDs: double perovskite $Ca_2InSbO_6:Eu^{3+}$ red-emitting phosphor. J Alloys Compd 847:156249
6. Kumar Singh D, Mondal K, Manam J (2017) Improved photoluminescence, thermal stability and temperature sensing performances of K^+ incorporated perovskite $BaTiO_3:Eu^{3+}$ red emitting phosphors. Ceram Int 43:13602–13611

Synthesis of Nickel-Doped Zinc Selenide Nanoparticles and Study of Structural and Optical Properties

Sonia Sheokand, Dharmvir Singh Ahlawat, and Amrik Singh

Abstract Zinc selenide doped with nickel ($Zn_{1-x}Ni_xSe$) nanoparticles were synthesized using co-precipitation method with different compositions such as $x = 0.1$, 0.15, 0.2. Poly-vinyl-pyrrolidone was acted as a surfactant. The nature of as prepared samples was found crystalline and it was confirmed by using X-ray-diffraction (XRD) pattern. Along with it, the average crystallite size (D), dislocation density (δ) and micro-strain (ε) were calculated using XRD pattern. Ultra-violet visible (UV–visible) spectroscopy was employed for the calculation of optical energy band gap of the prepared nanoparticles. Nickel-doped zinc selenide nanoparticles are suitable choice for fabrication of nanoelectronic devices and also useful in optical storage devices.

Keywords Structural properties · UV–Visible and FTIR spectroscopy

1 Introduction

Owing to the novel properties, nanoparticles have earned the great interest in many applications such as light emitting diode, photo-detectors, photovoltaic devices, in biological applications, etc. [1–3]. It consists of fascinating properties like high quantum yield, higher molar extinction coefficient, high resistance to photo bleaching, etc. In binary semiconductors, cadmium-based nanoparticles have significant properties, but due to its toxic effect, these nanoparticles have limit on its many applications, e.g. in medical application, etc. To overcome this limitation of toxicity, zinc-based nanoparticles are favourable in medical applications, in vivo imaging, optoelectronic devices, in light emitting diodes, etc. [4–8]. Zinc selenide (ZnSe) lies in the II–VI semiconductor group and is widely useful due to its chemical stability, wide band gap (~2.8 eV), large exciton binding energy (21 meV) [9, 10].

As compared to the rare earth metals, transition metals are used as dopant because of their excellent and valuable optical, electrical and magnetic properties and successfully doped in the group II–VI semiconductor nanoparticles [11, 12]. A few work has

S. Sheokand (✉) · D. S. Ahlawat · A. Singh
Department of Physics, Chaudhary Devi Lal University, Sirsa, Haryana, India
e-mail: soniasheokand27@gmail.com

Z. H. Khan et al. (eds.), *Recent Advances in Nanomaterials*, Springer Proceedings in Materials 27, https://doi.org/10.1007/978-981-99-4878-9_34

been reported on the effect of nickel as a dopant on ZnSe nanoparticles. Among the various transition metals, Nickel (II) can be used as the dopant in Zn-based compound semiconductor due to 8.3% of difference in atomic radius of Ni^{2+} and Zn^{2+} [13]. The synthesis of ZnSe nanoparticles has been described using a variety of techniques including atomic-layer-deposition, electron-beam-evaporation, thermal-evaporation, chemical vapour deposition, electro-deposition and the sol–gel approach, among others [14–16]. The wet chemical approach is well accepted because to its many trustworthy qualities, including affordability, particle stability and ease of doping [17]. In this work, we have reported the synthesis of Ni-doped ZnSe nanoparticles and characterized samples for study of structural, optical properties using XRD pattern, FTIR spectra and UV–Visible absorption spectra, respectively.

2 Materials and Method

The chemicals such as zinc chloride ($ZnCl_2$) ($\geq$ 99% purity Merck), sodium selenite (Na_2SeO_3) (99% purity Merck), nickel chloride tetrahydrate ($NiCl_2.4H_2O$ Merck) of purity 99%, hydrazine hydrate ($N_2H_4.H_2O$ Merck), ethylene glycol ($C_2H_6O_2$ Merck), polyvinyl pyrrolidone (PVP Merck) are made available from sigma-Aldrich and no purification carried out before use.

Nickel-doped zinc selenide ($Zn_{1-x}Ni_xSe$, where $x = 0.1, 0.15$ and 0.2) was synthesized using wet chemical method, i.e. co-precipitation method. 0.2 M $ZnCl_2$ was prepared with continue stirring at 60 °C. Another solution was prepared by using 0.2 M sodium selenite in 150 ml distilled water and by adding hydrazine hydrate and ethylene glycol in 1:1 at 60 °C [17]. The as prepared solution was transferred into 0.2 M $ZnCl_2$ solution. 0.2 M $NiCl_2.4H_2O$ was added to the as prepared solution for the preparation of Ni was doped ZnSe nanoparticles and PVP was added to the solution which acted as surfactant. The solution was stirred at 60 °C (6–7 h) and left overnight to settle down the precipitates. Then after, the precipitates were processed for filtration and washed many times with distilled water, then dried out at 90 °C up to 24 h. The obtained sample was crushed and calcinated at 200 °C for 2 h and further used for the structural and optical study.

The prepared samples were characterized using a Rigaku Miniflex-II X-ray diffractometer (XRD) equipped with a Cu K_α (1.54 Å) source and the data was collected from 10° to 70° at a scanning rate of 4° per minute. Optical UV–visible absorption data was recorded by UV–VIS-NIR spectrometer (Varian-Cary-5000) in 200–800 nm wavelength range. Fourier transform infrared (FTIR) spectra were recorded in range of 300–4000 cm^{-1}.

3 Result and Discussion

3.1 XRD Study

XRD pattern of Ni-doped ZnSe (i.e. $Zn_{1-x}Ni_xSe$, where $x = 0.1, 0.15, 0.2$) is indicated in Fig. 1. The different observed diffraction-peaks (JCPDS card 80-0021 and JCPDS card 06-0362) at 2θ, ~ 29.7°, 31.8°, 34.5°, 36.3°, 45.4°, 47.6°, 56.4°, 62.9° correspond to (111), (100), (002), (101), (220), (110), (311), (400) planes, respectively [18, 19]. The left over peaks indicate the formation of metal oxides and also some correspond to the impurities induced in the sample during synthesis process. As the Ni concentration is increased from 0.1 to 0.2 in $Zn_{1-x}Ni_xSe$, the peaks shift towards higher diffraction which is due to the presence of strain in the sample after doping. Along with it, the intensity of (111) and (101) planes decreases and increases, respectively. Average crystallite size (D_{hkl}) for Ni-doped ZnSe nanoparticles was calculated using the Debye Scherrer formula [20]:

$$D_{hkl} = 0.9\lambda/(\beta_{hkl} \cos\theta),$$

where λ denotes the wavelength of X-ray, β_{hkl} is used for FWHM, and θ denotes Bragg angle. Average crystallite size decreases from 14.68 nm to13.09 nm as the value of x enhances from 0.1 to 0.15 in $Zn_{1-x}Ni_xSe$. As the value of x changes to 0.2, the value of D_{hkl} increases to 15.15 nm (Table 1).

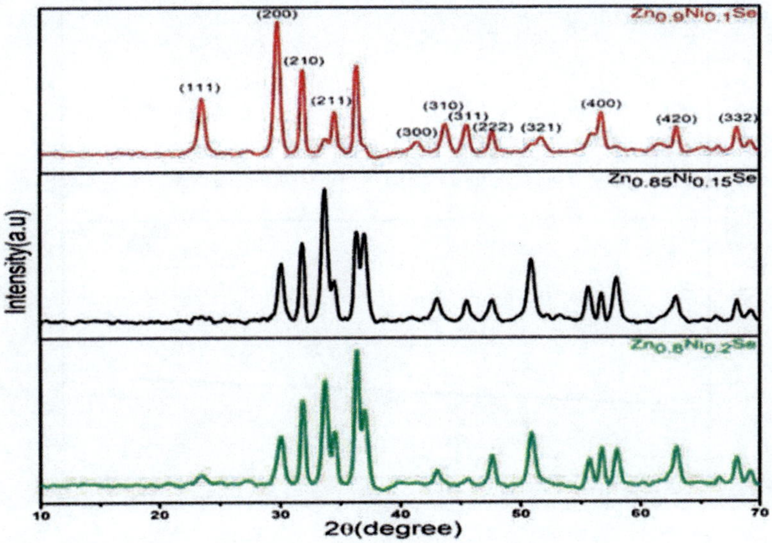

Fig. 1 X-ray diffraction pattern of $Zn_{1-x}Ni_xSe$ with $x = 0.1, 0.15, 0.2$

Table 1 Crystallite size, strain and the dislocation density of Ni-doped ZnSe

Samples	Crystallite size (nm)	Strain ε ($\times 10^{-3}$)	Dislocation density δ ($\times 10^{16}$ m^2)
$Zn_{0.9}Ni_{0.1}Se$	14.68	0.9	0.46
$Zn_{0.85}Ni_{0.15}Se$	13.09	2.3	0.58
$Zn_{0.8}Ni_{0.2}Se$	15.15	–	0.43

The induced strain in prepared samples was calculated using the W–H plot (Fig. 2).The decreased crystallite size for $x = 0.1$, 0.15 is due to increase in micro-strain in the crystal. The dislocation term represents the defects or irregularities in crystal and the dislocation-density (δ) is calculated by the following equation [21],

$$\delta = 1/D^2.$$

here, D denotes the crystallite size. The dislocation density increases with the increase in Ni concentration ($x = 0.1$–0.15) and further increase in the Ni ion concentration results a decreased value of δ (Table 1). From XRD analysis, it is found that size of prepared sample is in nanometre range which indicate the quantum confinement of particles due to which strain in the samples gets increased which can be due to incorporation of Ni ions in host lattice which further results in increase in irregularities, i.e. dislocation density in material.

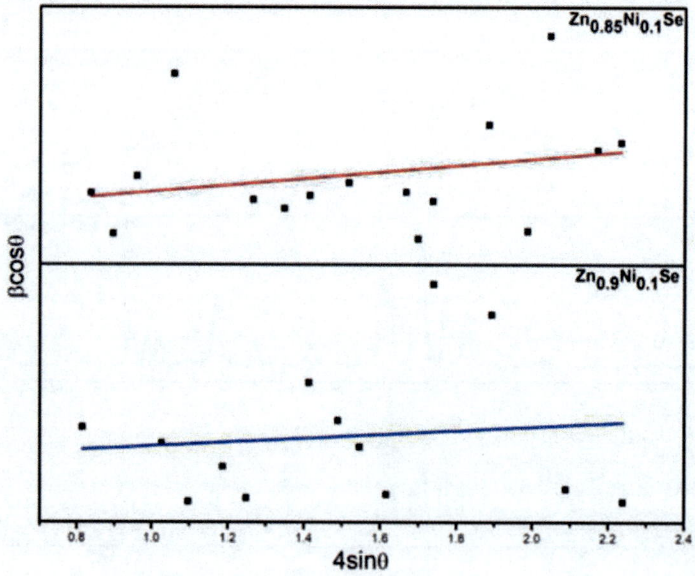

Fig. 2 W–H plot for Ni-doped ZnSe

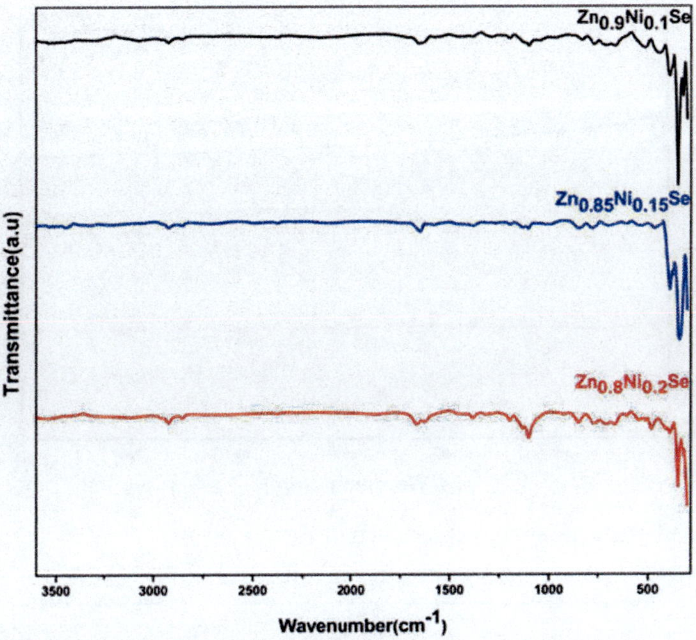

Fig. 3 FTIR spectra of Ni-doped ZnSe nanoparticles

3.2 FTIR Study

Using the Fourier Transform Infrared (FTIR) technique, the molecular vibrations of Ni-doped ZnSe have been investigated in the 300–4000 cm^{-1} range and are shown in Fig. 3. In Ni-doped ZnSe nanoparticles, the bands at 2854, 1656, 1101, and 1383 cm^{-1} have been attributed to C=H stretching, C=C stretching, C–H in plane stretching and C–H rock stretching, respectively. The band at 1195 cm^{-1} has been attributed to C–O stretching. There is slight shifting in the bands as the concentration of Ni increased in ZnSe nanoparticles. The ZnSe bending is allocated by the band present at 2924 cm^{-1}, while the bands observed at 387, 447, 543, 663 cm^{-1}are attributed as the ZnSe stretching [11].

3.3 UV–Visible-Spectroscopy Study

The plot for absorption spectra of synthesized samples (in ethanol) has been by shown in Fig. 4. At room temperature, the absorption spectra of the produced sample were captured in the 200–800 nm wavelength range.

Using the relationship given below, band gap was calculated [12],

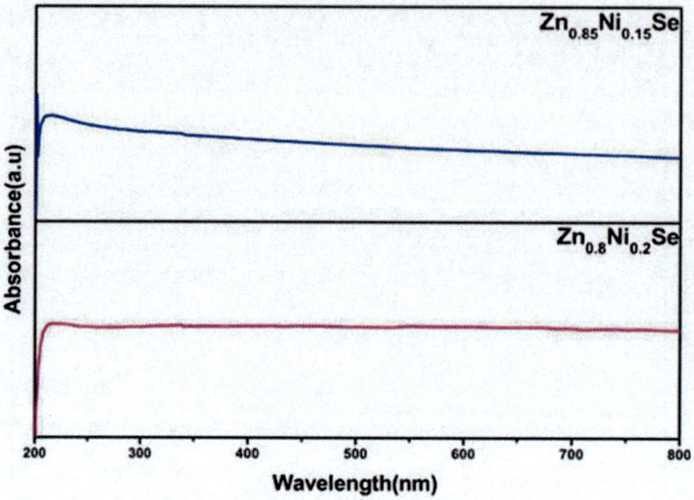

Fig. 4 Absorbance spectra of $Zn_{1-x}Ni_xSe$; $x = 0.15, 0.2$

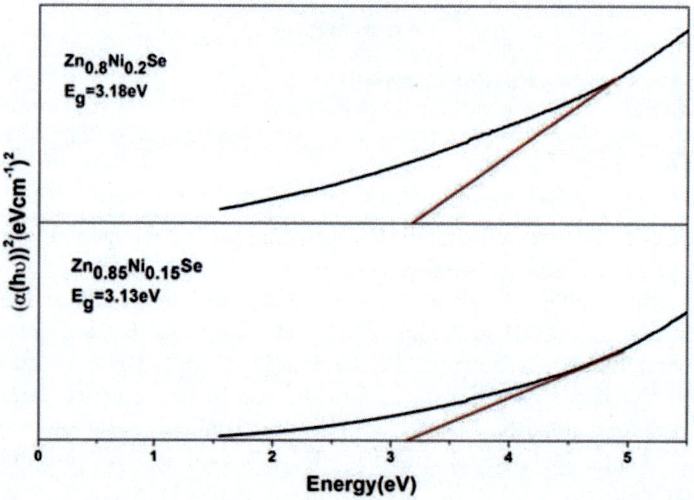

Fig. 5 $(\alpha h\upsilon)^2$ versus photon energy for $Zn_{1-x}Ni_xSe$; $x = 0.15, 0.2$

$$\alpha h\upsilon = A\left(h\upsilon - E_g\right)^n,$$

where α denotes absorption coefficient, $h\upsilon$ as the photon energy, E_g is taken as the energy band gap and A represents constant, and value of n is ½ and 2 for allowed direct transition and indirect transition, respectively [10].

The energy band gap was calculated using Tauc plot. Direct band gap value was calculated by extra plotting the straight line portion (Fig. 5). The calculated values of energy band gap ($Zn_{1-x}Ni_xSe$ nanoparticles) are 3.13 and 3.18 eV for $x = 0.15$ and 0.2, respectively.

As the concentration of Ni increased in ZnSe nanoparticles, the optical band gap increased due to quantum confinement of particles and obtained energy band gap which is tunable. Increase in the band gap is the favourable condition for the photocatalytic activities [17].

4 Conclusion

Ni-doped ZnSe ($Zn_{1-x}Ni_xSe$, where $x = 0.1$, 0.15, 0.2) nanoparticles were successfully synthesized using the co-precipitation method. Average crystallite size was calculated using XRD and was calculated to be about 13–15 nm due to quantum confinement, strain in the nanocrystals increases with increase in the nickel concentration and the dislocation density and irregularities in host material also increased with the crystallite size. For $x = 0.2$ in $Zn_{1-x}Ni_xSe$, the crystallite size increases, while strain and dislocation density decrease. FTIR study shows the presence of ZnSe stretching and also confirms presence of different functional groups. The optical energy band gap increases with the Ni concentration ($x = 0.15$, 0.2) from 3.13 to 3.18 eV which is the result of quantum confinement and favourable condition for using synthesized material for different application, e.g. in light emitting diodes, in photoelectronics, etc.

Acknowledgements CIL, GJUST, Hisar (Hry.) and JCDV, Sirsa (Haryana) are gratefully acknowledged for providing experimental facilities of XRD, UV-Visible and FTIR, respectively.

References

1. Xuan T, Wang X, Zhu G, Li H, Pan L, Sun Z (2013) One-step microwave-assisted synthesis of water soluble CdSe quantum dots for white light-emitting diodes with excellent color rendering. J Alloys Componds 558:105–108
2. Hu Z-S, Hung F-Y, Chen K-J, Chang S-J, Hsieh W-K, Liao T-Y, Chen T-P (2013) Recovery of thermal-degraded ZnO photodetector by embedding nano silver oxide nanoparticles. Appl Surf Sci 279:31–35
3. Pan Z, Zhao K, Wang J, Zhang H, Feng Y, Zhong X (2013) Near Infrared absorption of $CdSe_xTe_{1-x}$ alloyed quantum dot sensitized solar cells with more than 6% efficiency and high stability. ACS Nano 7:5215–5222
4. Lan GY, Lin YW, Huang YF, Chang HT (2007) Photo-assisted synthesis of highly fluorescent ZnSe (S) quantum dots in aqueous solution. J Mater Chem 17(25):2661–2666
5. Gaul DA, Rees WS Jr (2000) True blue inorganic optoelectronic devices. Adv Mater 12(13):935–946
6. Hamada Y, Gonda K, Takeda M, Sato A, Watanabe M, Yambe T, Satomi S. Ohuchi N (2011) In vivo imaging of the molecular distribution of the VEGF receptor during angiogenesis in a mouse model of ischemia. Blood J Am Soc Hematol 118(13):e93–e100

7. Trindade T, O'Brien P, Pickett NL (2001) Nanocrystalline semiconductors: synthesis, properties, and perspectives. Chem Mater 13(11):3843–3858
8. Chatterjee A, Mondal G, Guthikonda KK, Saha A (2021) Aqueous synthesis of mercaptopropionic acid capped ZnSe QDs and investigation of photoluminescence properties with metal doping. J Ind Chem Soc 98(12):100254
9. Hile DD, Swart HC, Motloung SV, Pawade VB, Kroon RE, Egbo KO, Koao LF (2020) Phase transformation on zinc selenide thin films deposited by photo-assisted chemical bath method: the effect of annealing temperature. Mater Sci Semicond Process 115:105118
10. Liu XD, Ma JM, Peng P, Meng WJ (2010) One-Pot hydrothermal synthesis of ZnSe hollow nanospheres from an ionic liquid precursor. Langmuir 26:9968
11. Ikhioya IL, Onah EO, Maaza M, Ezema FI (2022) Influence of precursor pH on the optical and electrical properties of electrochemically deposited cobalt-doped ZnSe thin films for photovoltaic application. Current Res Green and Sust Chem 5:100286
12. Nadeem MS, Munawar T, Mukhtar F, Rahman MN, Riaz M, Hussain A, Iqbal F (2020) Hydrothermally derived co, Ni co-doped ZnO nanorods; structural, optical, and morphological study. Opt Mater
13. Qi P, Dong YJ, Li YD (2003) ZnSe semiconductor hollow microspheres. Angew Chem Int Ed 42:3027–3030
14. Balu AR, Nagrethinam VS, Ahamed Syed Basheer MG, Thayumanavan A, Murali KR (2010) Influence of thickness on the microstructural, optoelectronic and morphological properties of nanocrystalline ZnSe thin films. Mater Sci Eng B, 171:93–98
15. Syed Basheer Ahamed MG, Nagarethinam VS, Balu AR, Thayumanavan A, Murali KR, Sanjeeviraja C, Jayachandran M (2010) Influence of substrate temperature on the properties of electron beam evaporated ZnSe films. Crystal Res Technol 45:421–426
16. Chaliha S, Borah MN, Sarmah PC, Rahman A (2008) Effect of substrate temperature on structural properties of thermally evaporated ZnSe thin films. J Optoelectron Adv Mater 10(2):427–433
17. Fu H, Li H, Jie W, Yang L (2006) The growth and characterization of ZnSe nanoneedles by a simple chemical vapor deposition method. J Cryst Growth 289:440
18. Priyadharsini N, Elango M, Vairam S, Venkatachalam T, Thamilselvan M (2016) Effect of temperature and pH on structural, optical and electrical properties of Ni doped ZnSe nanoparticles. Optik 127(19):7543–7549
19. Yadav K, Dwivedi Y, Jaggi N (2015) Structural and optical properties of Ni doped ZnSe nanoparticles. J Luminesc 158:181–187
20. Beena V, Ajitha S, Rayar SL, Parvathiraja C, Kannan K, Palan G (2021) Enhanced photocatalytic and antibacterial activities of ZnSe nanoparticles. J Inorg Organomet Polym Mater 31:4390–4401
21. Hassanien AS, Akl AA, Sáaedi H (2018) Synthesis, crystallography, microstructure, crystal defects, and morphology of Bi x Zn 1–x O nanoparticles prepared by sol–gel technique. Cryst Eng Comm 20:1716–1730

Structural and Photometrical Investigation of Bi_2O_3 Incorporated Polycarbazole Nanohybrid for Emissive Layer Application

Sanjeev Kumar, Debashish Nayak, Gobind Mandal, Sarfaraz Ansari, Jayanta Bauri, and Ram Bilash Choudhary

Abstract In recent times, polycarbazole (PCz) has attracted the scientists and researchers for its use in Flextronics industries due to its flexible and chemically stable nature. Herein, intension behind the incorporation of bismuth oxide (Bi_2O_3) in PCz to enhance its photoluminescence intensity. PCz/Bi_2O_3 nanocomposite was synthesized by using the oxidizing agent $FeCl_3$ in presence of the carbazole monomer and Bi_2O_3 powder. Structural study was done by X-ray diffraction (XRD). Morphological and elemental analysis of as-synthesized PCz/Bi_2O_3 nanocomposite has been characterized by the Field Emission Scanning Electron Microscope (FE-SEM) and Energy Dispersive X-ray spectroscope (EDX). PCz/Bi_2O_3 nanocomposite showed good photoluminescence intensity which suggested that it could be used as an efficient emissive layer material for optoelectronic devices.

Keywords Nanocomposites · Emissive layer · Conducting polymer · Optoelectronics

1 Introduction

Polymers were considered as insulating material before discoveries of conducting polymers in 1970. Conducting polymers (CPs) are identified as a class of polymeric materials having excellent optical and electrical properties like many inorganic semiconductors and metals. They have also advantages of the mechanical flexibility and of low cost than inorganic one [1]. Ordinary CPs having conductive networks needed high conducting filler metals/metal oxide to enhance the conductivity [2]. Conventional polymers consist of lakhs of monomer units, but conducting polymers only consist of thousands of monomer units, and conventional polymers are soluble in

S. Kumar · D. Nayak · G. Mandal · S. Ansari · J. Bauri · R. B. Choudhary (✉)
Nano Structured Composite Materials Laboratory (NCML), Department of Physics, Indian Institute of Technology (Indian School of Mines), Dhanbad, Jharkhand 826004, India
e-mail: rbcism@gmail.com

Z. H. Khan et al. (eds.), *Recent Advances in Nanomaterials*, Springer Proceedings in Materials 27, https://doi.org/10.1007/978-981-99-4878-9_35

many solvents, but CPs are stiff in nature are not easily soluble [3]. Polypyrrole (PPy), polyindole (PIN), polyaniline (PANI), and polycarbazole (PCz) are the CPs having high conductivity which are recently used in current research field. These CPs have wide applications in light emitting devices, sensors, supercapacitors, and transistors [4, 5]. Recently, metal oxide incorporated conducting polymer known as composite has attracted scientists and researchers due to their outstanding properties compared to polymeric one. Both polymeric and inorganic materials have different chemical, physical and mechanical properties which are going to form the composite. Some exceptional properties will be introduced by composite than single one. Optical and electrical properties are easily tuned by the inhibition of the metal oxide/sulfide into the CPs [6–10]. Among above mentioned fluorescent polymers, PCz's excellent optoelectronic properties have made them incredibly popular: a controllable optical band gap and good electrical conductivity, flexibility, and environmental friendliness than other CPs. PCz has wide technological applications in the field of optoelectronics [11]. Bismuth oxide (Bi_2O_3) is a semiconducting material having interesting chemico-physical properties like wide direct optical band gap, better photoconductive response and inflated refractive index. Talari et al. [12] have studied PCz/Fe_2O_3 composite for optoelectronic application. In our best of knowledge, PCz/Bi_2O_3 composite has not studied till now for emissive material application.

In this work, we report the synthesis of pure PCz and PCz/Bi_2O_3 composite by chemical oxidative polymerization and studied their surface morphologies and elemental composition. Inflated photoluminescence study was observed for PCz/Bi_2O_3 composite compare to pure PCz. These interesting results were suggested that it could be used as an emissive layer material application.

2 Experimental

2.1 Chemicals

Carbazole monomer ($C_{12}H_9N$) was procured from the Merck. Bi_2O_3 powder was purchased from the Alfa Aesar. $FeCl_3$ and acetonitrile were purchased from Sigma Aldrich company. Methanol and deionized (DI) water were used for washing purposes.

2.2 Synthesis of Polycarbazole/Bi_2O_3 Nanohybrid

0.168 gm of carbazole monomer was well mixed into the 30 ml of acetonitrile by stirring vigorously for 30 min at room temperature. $FeCl_3$ solution was prepared by dissolving 1.622 gm of anhydrous ferric chloride into the 20 ml of the acetonitrile. 0.0168 gm of the Bi_2O_3 powder was dissolved into the 15 ml of the acetonitrile

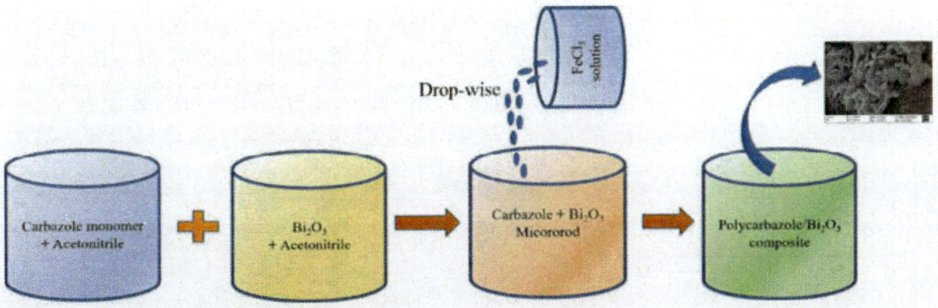

Fig. 1 Schematic representation of synthesis procedure of the PCz/Bi_2O_3 composite

and stirred it for 30 min. Then, this prepared solution was mixed dropwise into the carbazole solution with continuous stirring for 1 h. After that, $FeCl_3$ solution was added dropwise into the carbazole/Bi_2O_3 solution and solution's color changed from brown to light green. This solution was kept 26 h to complete the polymerization process. Finally, intense green solution was formed which was filtered and washed with methanol and DI water to remove the unwanted impurities present in the solution. Then, obtained precipitate was dried at 80 °C for 24 h in hot air oven and dried precipitate was grounded with the mortar-pestle which was stored for future experimental purposes. Figure 1 depicts the synthesis process schematically.

3 Results and Discussion

3.1 XRD Analysis

The characteristic XRD peaks of the polycarbazole exhibited at 9.9°, 19.2°, 20.26°, 25.12°, and 28.8° corresponding to the hkl planes (100), (20–1), (12–1), (32–1), and (210), respectively, as shown in Fig. 2a. Strong peak located at 9.9° represents the parallel planes to the polymer chain and small peaks reflect the periodicity is perpendicular to the polymer chain [13]. XRD peaks located at (−111), (020), (−102), (120), (−122), (133), (113), and (622) are the characteristic peaks of Bi_2O_3 which are well matched with JCPDS card no. 76–1730 as depicted in Fig. 2a [14]. Characteristic peaks of PCz and Bi_2O_3 are present in the XRD spectra of the PCz/Bi_2O_3 composite which confirms the successful inhibition of Bi_2O_3 into the PCz.

Debye–Scherrer's Eq. (1) was employed to calculate crystallite sizes (*D*) of the synthesized composites.

$$D = \frac{N\lambda}{\rho \mathrm{Cos}\theta}, \tag{1}$$

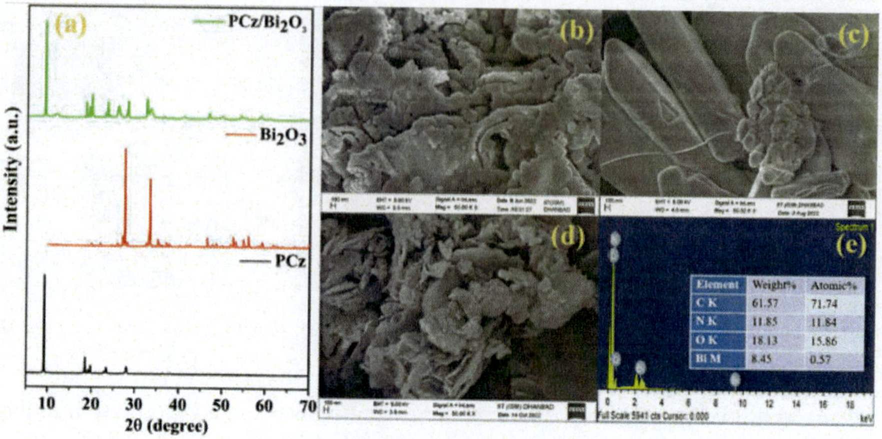

Fig. 2 XRD spectra of synthesized nanocomposites **a**; FE-SEM images of **b** PCz **c** Bi_2O_3 **d** PCz/Bi_2O_3 composite **e** EDS spectra of the synthesized samples

where N is denoted for Debye–Scherrer's constant, $\lambda = 1.543$ Å is defined as wavelength of X-ray used for the differaction, θ is Bragg's angle and ρ is defined as full width at half maximum (FWHM) for corresponding peak [15]. Crystallite sizes for PCz, Bi_2O_3 and PCz/Bi_2O_3 composite were evaluated as 73 ± 0.20, 87 ± 0.18, and 63 ± 0.11 nm, respectively.

3.2 FE-SEM and EDS Study

FE-SEM is an experimental technique which is used to take image of the surface of the sample in the micro- to nano-range [16]. Figure 2b explains the spherical agglomerated form of the PCz, and Fig. 2c clearly shows the rod-like structure of the Bi_2O_3 powder. Micro-rod-like structure and agglomerated spherical structure were both can be seen in Fig. 2d. Energy Dispersive X-ray spectroscopy (EDX) was used to confirm the presence of elements and their weight percentage in the selected area of the sample. EDX spectra of the PCz/Bi_2O_3 composite were shown in Fig. 2e and revealed the presence of elements C, N, O, and Bi into the sample.

3.3 Photoluminescence Analysis

Photoluminescence spectra of PCz/Bi_2O_3 are similar to the pure PCz. Characteristic photoluminescence emission peaks of PCz are found around ~ 400–420 nm which lie in the violet range shown in Fig. 3a [17, 18]. Violet emission peaks arise owing to the S_2–S_0 transition in the carbazole unit [19, 20]. Most intense peak of PCz/Bi_2O_3

is shifted to the red side in comparison with PCz due to the stronger π–π* interaction between PCz and Bi_2O_3 [13]. Photoluminescence intensity is enhanced after the insertion of Bi_2O_3 powder into the PCz due to the increment of electron–hole recombination rate into PCz/Bi_2O_3.There are three main reasons for the increase in the photoluminescence intensity: (i) surface stacking defects, (ii) oxygen vacancies, and (iii) self-trapped excitons. Herein, photoluminescence intensity of PCz/Bi_2O_3 composite enhanced may be due to the incorporation of Bi_2O_3 into PCz which increase oxygen vacancies and defect states into the composite. This is the main reason for enhancement in the photoluminescence intensity of the composite [21]. This enhanced photoluminescence intensity of PCz/Bi_2O_3 is suggesting for its use as emissive layer material application in optoelectronic devices.

3.4 Photometric Investigation

Color chromaticity coordinates gives insight of position of emission color of respective material. Herein, GO–CIE software is used to calculate color purity and color chromaticity coordinate of the composites [22]. From Fig. 3b, it is clearly depicted that emission color of PCz is found in the light violet region, and it is shifted to the intense blue region for the PCz/Bi_2O_3 composite. Color purity (CP) is calculated by the following Eq. (2).

$$\mathrm{CP} = \frac{\sqrt{(c_x - x_i)^2 + (d_x - y_i)^2}}{\sqrt{\left(c_y - x_i\right)^2 + \left(d_y - y_i\right)^2}} \times 100\%. \tag{2}$$

$(c_x, d_x) = (0.22, 0.23)$ is the coordinate of the PCz/Bi_2O_3, $(x_i, y_i) = (0.31, 0.316)$ is the coordinate of the white illumination point, and $(c_y, d_y) = (0.09, 0.12)$ is the coordinate of the dominant wavelength [23]. Color purity is found ~ 44.5% and 27.9% for PCz/Bi_2O_3 and PCz, respectively. Photometric investigation suggested that PCz/Bi_2O_3 composite may be used in the emissive layer material application for color emission in blue region in various optoelectronic device application.

4 Conclusions

Successful synthesis of these composites was confirmed by XRD and FE-SEM–EDX analysis. Crystallite sizes of the PCz and PCz/Bi_2O_3 composite were estimated by XRD analysis as 73 and 63 nm, respectively. Spherical and rod-like morphologies of PCz and Bi_2O_3 were determined by FE-SEM images. Elemental presence (C, N, Bi, and O) was confirmed by the EDX spectra analysis. Photoluminescence study revealed that the PCz/Bi_2O_3 composite shows the higher luminescence intensity comparison with pristine PCz. Photoluminescence emission of composite was found

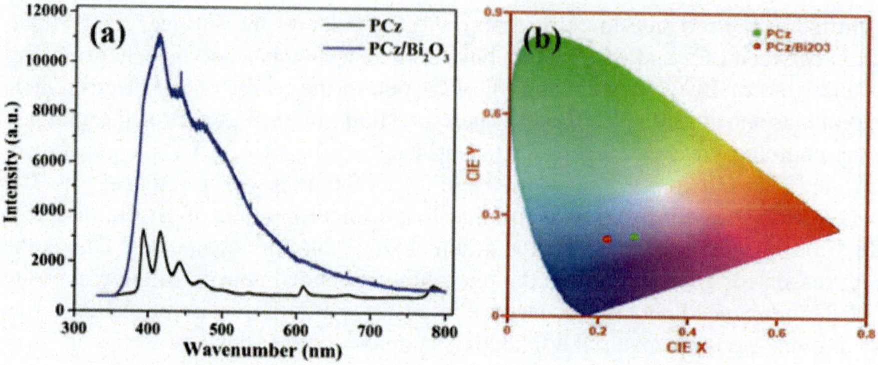

Fig. 3 **a** Depicts the photoluminescence spectra and **b** CIE plot of the composites

in the blue region. Photometric investigation unveiled that color purities of PCz and PCz/Bi_2O_3 composite were estimated as 27.9 and 44.5%. These results were suggested that PCz/Bi_2O_3 composite might be employed as an emissive layer material in optoelectronic devices.

Acknowledgements The authors would like to express their gratitude to the Indian Institute of Technology (Indian School of Mines), Dhanbad, India, for providing financial assistance.

Conflict of Interest There is no conflict of interest.

References

1. Nezakati T, Nezakati T, Seifalian A, Tan A, Seifalian AM (2018) Conductive polymers: opportunities and challenges in biomedical applications. Chem Rev 118(14):6766–6843
2. Huan P, Ling X, Ding-Xiang Y, Zhong-Ming L (2014) Conductive polymer composites with segregated structures. Progress Polymer Sci 39(11):1908–1933
3. Namsheer K, Rout CS (2021) Conducting polymers: a comprehensive review on recent advances in synthesis, properties and applications. RSC Adv 11(10):5659–5697
4. Bauri J, Choudhary RB, Mandal G (2021) Recent advances in efficient emissive materials-based OLED applications: a review. J Mater Sci 56(34):18837–18866
5. Chauhan M, Kumar Singh V (2021) Review on recent experimental SPR/LSPR based fiber optic analyte sensors. Optical Fiber Tech 64:102580
6. Choudhary RB, Ansari S (2022) Mesoporous complexion and multi-channeled charge storage action of PIn-rGO-TiO_2 ternary hybrid materials for supercapacitor applications. J Energy Storage 46:103912
7. Choudhary RB, Ansari S, Purty B (2020) Robust electrochemical performance of polypyrrole (PPy) and polyindole (PIn) based hybrid electrode materials for supercapacitor application: a review. J Energy Storage 29:101302
8. Choudhary RB, Nayak D (2021) Tailoring the properties of 2-D rGO-PPy-ZnS nanocomposite as emissive layer for OLEDs. Optik 231:166336

9. Nayak D, Choudhary RB (2022) Tuning the optical properties of high quantum-yield g-C3N4 with the inclusion of ZnS via FRET for high electron–hole recombination. Spectrochimica Acta Part A: Molecular and Biomol Spectrosc 122162
10. Nayak D, Choudhary RB (2022) Influence of ZnS on the structural, morphological, optical and thermal properties of Polyindole for an emissive layer. Inorg Chem Commun 144:109824
11. Gross S et al (2007) PMMA: A key macromolecular component for dielectric low-κ hybrid inorganic–organic polymer films. Eur Polymer J 43(3):673–696
12. Talari S, Basapathi R, Malge A (2021)Structural, optical and electrical conductivity studies in Polycarbazole and Its metal oxide nano composites
13. Horti NC et al (2021) Synthesis and photoluminescence properties of polycarbazole/tin oxide (PCz/SnO2) polymer nanocomposites. Polym Bull 78(11):6321–6336
14. Sindhu S, Mj J, CV N (2015) α-Bi_2O_3 photoanode in DSSC and study of the electrode-electrolyte interface. RSC Adv 5
15. Choudhary RB, Kumar S (2022) Optimum chemical states and localized electronic states of SnO_2 integrated PTh–SnO_2 nanocomposites as excelling emissive layer (EML). Opt Mater 131:112736
16. Mirovoi YA, Burlachenko AG, Buyakova SP, Sevostiyanova IN, Kulkov SN (2016) Producing composite materials based on ZrB_2, ZrB_2 -SiC. IOP Conf Ser: Mater Sci Eng 156:012035
17. Srivastava A, Chakrabarti P (2017) Experimental characterization of electrochemically polymerized polycarbazole film and study of its behavior with different metals contacts. Appl Phys A 123
18. Zhu X, Xu L, Wang M, Wang Z, Liu R, Zhao J (2011) Electrosyntheses and characterizations of a soluble blue fluorescent copolymer based on pyrene and Carbazole. Int J Electrochem Sci 6
19. Siraj N et al (2015) Enhanced S 2 emission in carbazole-based ionic liquids. RSC Adv 5(13):9939–9945
20. Zheng C, Xiang X (2019) Spectroscopic study of aggregation of carbazole units. J Fluoresc 29(6):1343–1348
21. Kandulna R, Choudhary RB, Singh R, Purty B (2018) PMMA–TiO2 based polymeric nanocomposite material for electron transport layer in OLED application. J Mater Sci: Mater Electron 29(7):5893–5907
22. Kumar S, Choudhary RB (2022) Influence of MnO2 nanoparticles on the optical properties of polypyrrole matrix. Mater Sci Semicond Process 139:106322
23. Nayak D, Choudhary RB (2019) Augmented optical and electrical properties of PMMA-ZnS nanocomposites as emissive layer for OLED applications. Opt Mater 91:470–481

NiFe-LDHs as an Effective Electrocatalyst for Electrooxidation of Urea

Sanjeeb Kumar Ojha, Archana Singh, Deepika Tavar, Kamlesh Goyre, and Satya Prakash

Abstract The development of an efficient and cost-effective electrocatalyst for urea oxidation reaction (UOR) not only help in countering energy crisis through hydrogen production but also in the treatment of toxic urea waste. Although the theoretical voltage required for H_2 generation from urea electrolysis is lower than water but the sluggish kinetics associated with UOR and stability of catalyst remains challengeable and various development were made to counter the same. In this work, we synthesized NiFe-LDHs catalyst via one step, low temperature co-precipitation method. In contrast to monometallic $Ni(OH)_2$, which takes 1.55 V versus RHE to obtain the same current density, the catalyst achieved it at 1.48 V versus RHE, demonstrating superior performance of bimetallic NiFe-LDHs over monometallic $Ni(OH)_2$.

Keywords UOR · LDHs · Bimetallic · Low temperature · Cost-effective

1 Introduction

To overcome the energy crisis and slow down the adverse impact of climatic change, the attention was shifted towards the development of sustainable fuel which will decrease our dependence on fossil fuels and generate low carbon. Hydrogen (H_2) is

A. Singh (✉) · D. Tavar · K. Goyre · S. Prakash
Academy of Scientific & Innovative Research (AcSIR), Ghaziabad 201002, India
e-mail: archanasingh@ampri.res.in

D. Tavar
e-mail: depineha@gmail.com

K. Goyre
e-mail: kamalgoyre@gmail.com

S. Prakash
e-mail: satyapra150592@gmail.com

S. K. Ojha · A. Singh · D. Tavar · K. Goyre · S. Prakash
CSIR-Advanced Material and Processes Research Institute (AMPRI), Bhopal 462026, India
e-mail: sanjeeb.sko@gmail.com

Z. H. Khan et al. (eds.), *Recent Advances in Nanomaterials*, Springer Proceedings in Materials 27, https://doi.org/10.1007/978-981-99-4878-9_36

an eco-friendly ideal fuel with the highest gravimetric energy density (142 MJ kg^{-1}) [1]. H_2 generation from water splitting is a green pathway but requires a theoretical voltage of 1.23 V, whereas it only entails a voltage of 0.37 V versus RHE in case of urea [2]. For the production of H_2, electrolysis of urea in alkaline media is thought to be a practical and affordable method, and the by-products are both desirable and safe. Hence, H_2 production from urea will help counter the energy crisis along with the remediation of toxic urea wastewater [3]. UOR, which goes through a 6e-transfer process, has relatively low reaction kinetics; thus, a highly effective catalyst is needed to speed up the reaction kinetics because UOR affects the effectiveness and efficiency of urea electrolysis [4]. Over the years, diverse catalysts are developed amid them noble metal catalysts are restricted due to their high cost and scarcity, attention has been directed towards non-noble metals and especially to nickel-based catalysts because nickel oxyhydroxide (NiOOH) has shown to be the active intermediate for UOR in alkaline medium [5]. However, monometallic Ni suffers from high onset potential, poor stability, susceptibility to CO poising and low activity. Under alkaline conditions, NiFe-based catalyst showed excellent promise for water splitting because Fe in Ni encourages the creation of NiOOH, which can significantly improve the kinetics of the reaction [6]. Since NiOOH is also the active intermediate for urea oxidation reaction so various NiFe-based catalysts are being explored for urea electrolysis. Layered Double Hydroxides (LDHs) are 2D nanomaterials which in electrocatalysis have been broadly used due to their simple synthesis methodology, easier composition regulation and functionalization [7]. In our work, using a one-pot co-precipitation approach at low temperature, we produced bimetallic NiFe-LDHs with a specific composition, and the catalytic activity of NiFe-LDHs and $Ni(OH)_2$ is investigated for UOR in alkaline electrolytes.

2 Materials and Methods

2.1 Materials

All the reagents were procured from SRL Pvt. Ltd., India, which were nickel nitrate hexahydrate ($Ni(NO_3)_2$), ferric nitrate nonahydrate ($Fe(NO_3)_3$), Sodium Hydroxide (NaOH), Ethanol (CH_3CH_2OH) and Polyvinylpyrrolidone (PVP). Nafion 117 was purchased from Sigma Aldrich.

2.2 Synthesis of NiFe-LDHs and Ni(OH)₂

NiFe-LDHs was synthesized by adding $Fe(NO_3)_3$ and $Ni(NO_3)_2$ with a molar ratio in 1:3 in 50 mL distilled water to which a definite amount PVP was added under stirring. 1 M NaOH solution was added dropwise, and after 3 h of stirring, the precipitate formed was filtered and dried in the oven. Similarly, $Ni(OH)_2$ was synthesized without the addition of $Fe(NO_3)_3$.

2.3 *Material Characterization and Electrochemical Measurements*

XRD analysis of the compound was performed by Bruker AXS D8 instrument equipped with CuKα radiation of wavelength 1.54 A°, and the morphology was determined by using JEM-F200 TEM. FT-IR spectra were recorded with ALPHA-II (Bruker) over the range of 400–4000 cm^{-1}. CHI604C workstation with a 3-electrode configuration set-up is used for electrochemical studies. Ag/AgCl (3 M KCl) as reference electrode, GCE as working electrode and Pt plate as counter electrode were used. 5 mg of the compound was dispersed in solvent (water and ethanol 3:1) with 20 μL Nafion and sonicated to generate catalyst ink which was then drop-casted onto GCE and dried before performing tests.

3 Results and Discussion

NiFe-LDHs XRD pattern was recorded in between the angle 2θ value of 10° and 80° shown in (Fig. 1a). The obtained pattern matches well with the reference NiFe-LDHs (JCPDS 40-0215). The typical diffraction patterns for 2θ values are 11.40°, 22.97°, 33.53°, 34.42°, 38.58°, 59.93°, 71.27° which correspond with the planes of (003), (006), (101), (012), (015), (110), and (119) respectively. The weak and broadened peak observed in the XRD pattern suggests weak crystallinity of the compound. From TEM images we observed nanosheets-like morphology for NiFe-LDHs as seen in Fig. 1b. The nanosheets diameter ranges from 20 to 70 nm.

FT-IR absorption spectra of the compound are shown in Fig. 1c. The strong and broad band present near 3421 cm^{-1} arises from the stretching vibration of hydroxy groups with hydrogen bonding and band observed at 1630 cm^{-1} ascribed to the bending vibration of absorbed H_2O molecules. The absorption band at 612 cm^{-1} ascribed to (M–O) stretching vibration and 491 cm^{-1} for (O–M–O) bending vibration [8]. The stretching vibration of the intercalated nitrate anion causes the peak at 1335 cm^{-1} [9]. The weak absorption peak at 2920 cm^{-1} corresponds to stretching vibration of CH_2 group of PVP.

Performance of UOR was assessed in 1 M KOH comprising 0.33 M urea, which is the typical concentration of urine. NiFe-LDHs reach a current density of 10 $mA.cm^{-2}$ at 1.48 V versus RHE, while $Ni(OH)_2$ needs a voltage of 1.55 V versus RHE for the same (Fig. 2a). It shows that incorporation of iron helps lowering the onset potential for UOR along with increasing the current density. Furthermore, in the absence of

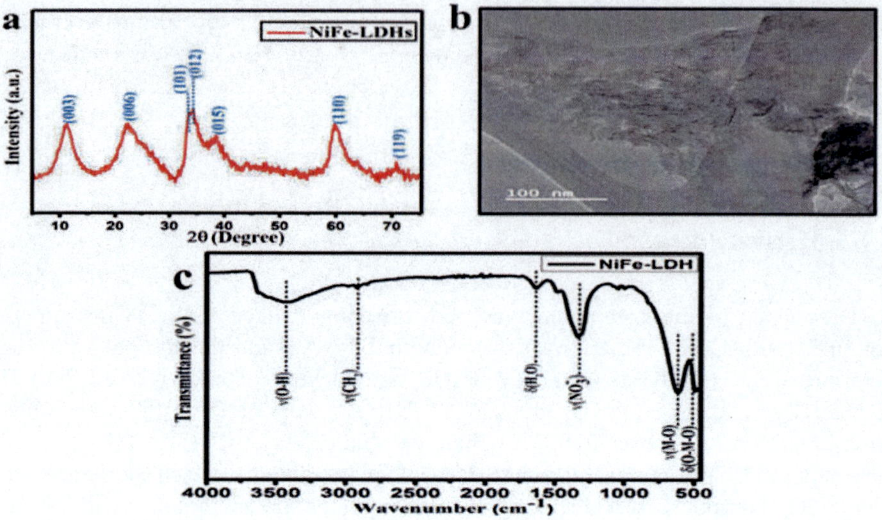

Fig. 1 **a** XRD of NiFe-LDHs, **b** TEM image of NiFe-LDHs, **c** FT-IR spectra of NiFe-LDHs

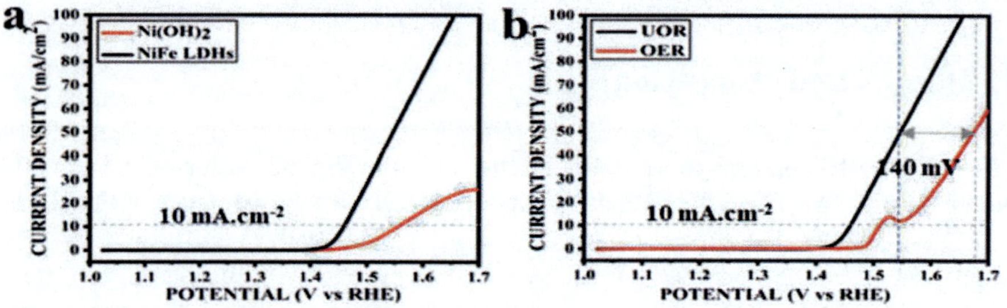

Fig. 2 **a** LSV curve of NiFe-LDHs and $Ni(OH)_2$ with urea, **b** LSV curve of NiFe-LDHs in 1 M KOH with and without urea (polarization curve were obtained at a scan rate of 5 mVs^{-1})

urea (OER) the onset potential shift towards higher positive potential and @in urea presence it accomplished 50 $mA.cm^{-2}$ current density at 1.58 V, which is 140 mV lower than that of urea free counterpart shown in Fig. 2b.

4 Conclusion

In summary, we hereby reported a cost-effective and facile strategy for the synthesis of NiFe-LDH nanosheets. The characteristic peaks obtained in XRD confirm the formation of NiFe-LDHs, while from TEM analysis the nanosheet morphology of catalyst was observed. Besides, NiFe-LDHs afford superior activity in comparison with $Ni(OH)_2$; since in order to attain a current density of 10 mA cm^{-2}, it needs a potential of 1.48 V, which is 70 mV below what required for $Ni(OH)_2$. Further, in presence of urea the catalyst achieved a current density of 50 mA cm^{-2} at a potential which is 140 mV lower than that of urea free counterpart which represent the UOR low energy usage attribute. The catalyst can be used to produce hydrogen, which will aid in addressing the current energy problem because H_2 can be utilized as a substitute fuel. Generating hydrogen from urea rich waste water will lower the investment cost in comparison to clean water and it is cost-effective way for the treatment of urea-rich waste water.

Acknowledgements Authors acknowledges the Department of Science and Technology, Govt., India, grant no. SERB/F/4835/2021-22 for the fellowship and for providing financial support to carry out the work.

Declaration of Interest Statement The authors declare that there is no conflict of interest.

References

1. He T et al (2016) Hydrogen carriers. Nat Rev Mater 1(12):16059
2. Boggs BK, King RL, Botte GG (2009) Urea electrolysis: direct hydrogen production from urine. Chem Commun (Camb) 32:4859–4861
3. Rollinson AN et al (2011) Urea as a hydrogen carrier: a perspective on its potential for safe, sustainable and long-term energy supply. Energy and Environ Sci 4(4)
4. Li J et al (2022) A review of hetero-structured Ni-based active catalysts for urea electrolysis. J Mater Chem A 10(17):9308–9326
5. Vedharathinam V, Botte GG (2012) Understanding the electro-catalytic oxidation mechanism of urea on nickel electrodes in alkaline medium. Electrochim Acta 81:292–300
6. Yun WH et al (2021) Ni–Fe phosphide deposited carbon felt as free-standing bifunctional catalyst electrode for urea electrolysis. Sci Rep 11(1):22003
7. Cai Z et al (2019) Recent advances in layered double hydroxide electrocatalysts for the oxygen evolution reaction. J Mater Chem A 7(10):5069–5089
8. Qu J et al (2022) One-pot synthesis of nitrate-intercalated NiFe layered double hydroxides with an 8.2 Å interlayer spacing. Adv Mater Interfaces 9(25)
9. Hunter BM et al (2016) Effect of interlayer anions on [NiFe]-LDH nanosheet water oxidation activity. Energy Environ Sci 9(5):1734–1743

Growth of WO_3–SnO_2 Composite Using Chemical Method for NO_2 Sensing

J. P. Singh, A. Sharma, R. Gupta, M. Tomar, and A. Chowdhuri

Abstract With the degrading air quality around the globe, the detection of air pollutants has attracted increasing interest. NO_2 is a major contributor towards air pollution in urban areas mainly due to emissions from automobiles and industries. Exposure to high NO_2 levels can lead to a range of human health complications and increased levels can have an adverse effect on the environment as well. In order to understand people's exposure to air pollution and take action when necessary, there is a need for fast, sensitive, and portable real-time sensing devices. The present work focuses on development of conductometric gas sensor based on Tungsten Trioxide-Tin Oxide (WO_3–SnO_2) composite films developed using chemical solution deposition (CSD) technique for the detection of NO_2 gas. The gas sensing parameters were studied and compared for WO_3–SnO_2 composite sensors as well as for pristine SnO_2 and WO_3 sensors in the temperature range of 30–180 °C towards 10 ppm of NO_2 gas. A high sensing response of ~ 286 was obtained at a low operating temperature of 110 °C for the WO_3–SnO_2 composite sensor.

Keywords Gas sensor · NO_2 sensor · Conductometric sensor · Environmental monitoring

---

J. P. Singh
Department of Physics and Astrophysics, University of Delhi, Delhi 110007, India

J. P. Singh · A. Chowdhuri (✉)
Department of Physics, Acharya Narendra Dev College (University of Delhi), Kalkaji, Govindpuri, New Delhi 110019, India
e-mail: arijitchowdhuri@andc.du.ac.in

A. Sharma
Department of Physics, Atma Ram Sanatan Dharma College (University of Delhi), Dhaula Kuan, New Delhi 110021, India

A. Sharma · M. Tomar
Institute of Eminence, University of Delhi, Delhi 110007, India

R. Gupta
Department of Physics, Hindu College, University of Delhi, Delhi 110007, India

M. Tomar
Department of Physics, Miranda House, University of Delhi, Delhi 110007, India

Z. H. Khan et al. (eds.), *Recent Advances in Nanomaterials*, Springer Proceedings in Materials 27, https://doi.org/10.1007/978-981-99-4878-9_37

1 Introduction

Semiconducting metal oxide-based gas sensors have grabbed the attention of researchers and environmentalists alike due to their efficient detection of toxic gases. Especially thin film-based conductometric metal oxide sensors are ideal for sensing gases including NO_2 due to their large surface area and comparatively simple sensing mechanism. One such promising n-type metal oxide is Tungsten Trioxide (WO_3), having a tunable bandgap, and has been used for a variety of applications, including gas sensing [1]. There are various ways to synthesize WO_3 thin films, nanostructures, and composites including physical vapour deposition, chemical vapour deposition, hydrothermal, solvothermal, sol–gel, spray pyrolysis, etc. [2]. Out of these various synthesis routes, the chemical solution deposition (CSD) technique is preferred due to its low cost, scalable, and simple process. Another semiconductor metal oxide that has been widely used for gas sensing applications is Tin Oxide (SnO_2). It is one of the most studied materials for gas sensing because of its high ability to adsorb surface oxygen and is very sensitive to most gases [3]. For the last couple of decades, the focus has been on tinkering with the properties of semiconducting metal oxide by some additives/catalysts in order to augment sensing response besides imparting partial selectivity towards a particular gas.

In the present work, conductometric gas sensors based on thin films of pristine WO_3, pristine SnO_2, and WO_3–SnO_2 composite have been developed using the chemical solution deposition (CSD) technique for the detection of NO_2 gas. The sensing response has been studied at different temperatures varying from 30 to 180 °C towards 10 ppm of NO_2.

2 Materials and Methods

All the chemicals used in the synthesis were used as obtained without further purification, Tungsten Trioxide powder (WO_3, extra pure) was obtained from Loba Chemie, Tin Tetrachloride Pentahydrate ($SnCl_4.5H_2O$) from Central Drug House (P) Ltd., Ethanol (absolute, 99.9% pure) was procured from Changshu Hongsheng Fine Chemical Co. Ltd., while Propan-2-ol and Propanol were purchased from Fisher Scientific.

Chemical Solution Deposition (CSD) technique has been employed for the synthesis of thin films of pristine, SnO_2, pristine WO_3, and WO_3–SnO_2 composite. 120 mg of the WO_3 powder was dispersed in 3 ml of ethanol [4], and ten microlitres of this solution were drop-casted on Platinum (Pt) interdigital electrodes (IDEs) patterned over a glass surface [3]. SnO_2 was also prepared via the chemical route wherein $SnCl_4.5H_2O$ was added to propan-2-ol to form $SnCl_x(OC_3H_7)_y$ type compound. A solution of propanol and water was prepared separately and then added to the above solution forming a colloidal solution of SnO_2 [5]. The solution of SnO_2 was dissolved in the previously prepared solution of WO_3 (5% v/v), and ten microlitres of this WO_3–SnO_2 composite were drop-casted on the Pt IDEs. Another configuration was also used where ten microlitres of the SnO_2 solution were drop-casted on the IDEs in order to determine the sensing response of pristine SnO_2. All the thin films were further annealed at 225 °C.

Using an X-ray diffractometer (make: Panalytical X'Pert Pro), the crystalline structures of all the deposited thin films (WO_3, SnO_2, and WO_3–SnO_2 composite) were examined. All the gas sensing measurements were carried out in a gas sensor test rig (GSTR) designed indigenously. The sensor response was calculated for all the sensors at various temperatures as

$$S = \frac{R_g - R_a}{R_a}. \tag{1}$$

Here, R_a and R_g are the resistance of the sensor in air and in the presence of the target gas, respectively.

3 Results and Discussions

The X-ray diffractogram (XRD) of the post-annealed WO_3–SnO_2 composite thin film in Fig. 1a agrees well with the XRD pattern of the monoclinic phase of WO_3 [6]. It is worth mentioning that the XRD pattern of the composite sample does not show any discernible SnO_2 diffraction peak due to its low concentration (5% v/v).

The sensing response towards 10 ppm NO_2 was recorded for all the three thin film sensor configurations at different temperatures (30–180 °C) and is exhibited in Fig. 1b. It is clear from the plot that the operating temperature for the pristine WO_3

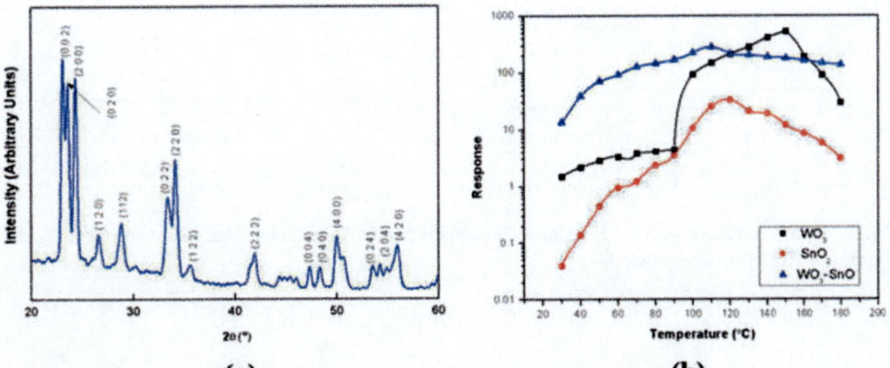

(a) (b)

Fig. 1 **a** XRD of WO_3–SnO_2 composite film and **b** temperature dependent sensing response for all the three sensor (WO_3, SnO_2, and WO_3–SnO_2 composite) configurations film sensor is 150 °C, whereas that of pristine SnO_2 film is 120 °C, while the WO_3–SnO_2 composite film-based sensor shows its maximum response at 110 °C. Table 1 summarizes the NO_2 sensing response characteristics for all the sensor structures.

Table 1 shows that pristine WO_3 showed the best sensing response of 534.97 at an operating temperature of 150 °C. However, it is exciting to observe that post incorporation of SnO_2 in the WO_3 matrix, the operating temperature got reduced to 110 °C with a sensing response of 286.22 and improvement in response speed and recovery times (39 s and 130 s, respectively). The sensing response of pristine SnO_2 thin film is noted to be quite ordinary compared to the other two configurations.

Apart from this, the films were also tested for multiple cycles of adsorption and desorption of 10 ppm of NO_2, and a similar response was observed, confirming good repeatability of the sensors towards sensing of NO_2 as Fig. 2 shows.

Table 1 Sensing response characteristics for all the sensors

Sensing structure	Operating temperature (°C)	Response	Response time(s)	Recovery time (s)
Pristine WO_3	150	534.97	15	121
WO_3–SnO_2	110	286.22	39	130
Pristine SnO_2	120	33.70	43	642

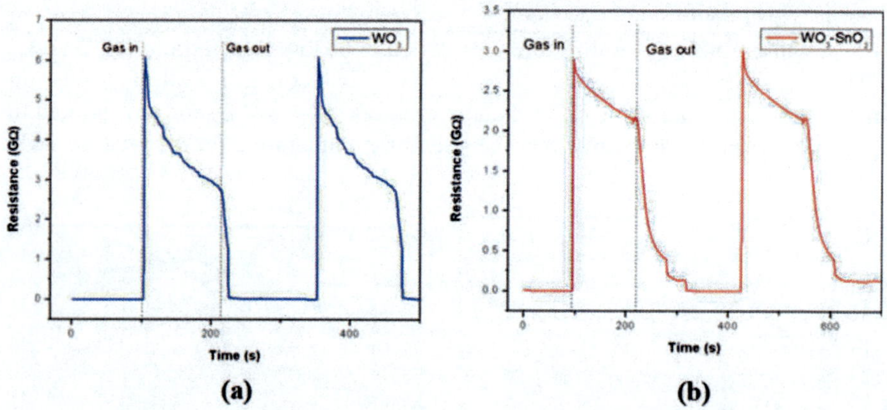

Fig. 2 Transient response for **a** Pristine WO_3 sensor and **b** WO_3–SnO_2 composite sensor towards 10 ppm of NO_2 for two cycles

4 Conclusion

In the temperature range of 30–180°C, NO_2 gas (10 ppm) sensing response characteristics of pristine WO_3, pristine SnO_2, and WO_3–SnO_2 composite films synthesized using chemical solution technique were investigated. It is interesting to note that at a temperature of 150 °C, pristine WO_3 film's sensitivity was 534.97; whereas post incorporation of SnO_2, the WO_3–SnO_2 film composite exhibited a sensing response of 286.22; however, the operating temperature could be seen to reduce substantially to 110 °C. Compared to the other two, the reaction of pristine SnO_2 film was seen to be minimal towards 10 ppm NO_2.

Acknowledgements Author JPS is thankful to the Council of Scientific and Industrial Research (CSIR), India, for JRF-NET-CSIR fellowship. Authors JPS and AC also express gratitude to Principal, Acharya Narendra Dev College (University of Delhi) for infrastructural support.

Declaration of Interest Statement The authors declare that they have no conflict of interest.

References

1. Buch VR, Chawla AK, Rawal SK (2016) Review on electrochromic property for WO_3 thin films using different deposition techniques. Mater Today: Proc 3(6):1429–1437. https://doi.org/10.1016/j.matpr.2016.04.025
2. Hassanzadeh A, Moazzez B, Haghgooie H, Nasseri M, Golzan M, Sedghi H (2008) Synthesis of SnO_2 nanopowders by a sol-gel process using propanol-isopropanol mixture. Open Chem 6(4):651–656. https://doi.org/10.2478/s11532-008-0072-x
3. Khan H, Zavabeti A, Ou JZ, Daeneke T, Li Y, Kalantar-zadeh K (2017) Two dimensional tungsten oxide nanosheets with unprecedented selectivity and sensitivity to NO_2. IEEE Sens 2017:1–3. https://doi.org/10.1109/ICSENS.2017.8234283
4. Ramkumar S, Rajarajan G (2017) A comparative study of humidity sensing and photocatalytic applications of pure and nickel (Ni)-doped WO_3 thin films. Appl Phys A 123(6):401. https://doi.org/10.1007/s00339-017-0983-5
5. Sharma A, Tomar M, Gupta V (2011) SnO_2 thin film sensor with enhanced response for NO_2 gas at lower temperatures. Sens Actuators, B Chem 156(2):743–752. https://doi.org/10.1016/j.snb.2011.02.033
6. Shendage SS, Patil VL, Vanalakar SA, Patil SP, Harale NS, Bhosale JL, Kim JH, Patil PS (2017) Sensitive and selective NO_2 gas sensor based on WO_3 nanoplates. Sens Actuators, B Chem 240:426–433. https://doi.org/10.1016/j.snb.2016.08.177

A Comparative Study of Antimony Telluride and Bismuth Telluride for Thermoelectric Generation

Jai shree Choudhary, Anisha, Aditya Gupta, Arijit Chowdhuri, Geeta Rani, Bilasni Devi, Mallika Verma, Monika Tomar, Ranjana Jha, and Anjali Sharma

Abstract The present work presents a comparative study of the thin film thermoelectric generators (TEG) followed by facile growth of nanoparticles of both materials by hydrothermal synthesis. The morphological and structural studies of the prepared nanomaterial for the TEG application are carried out using characterization techniques such as X-ray diffraction (XRD) and UV–Visible spectroscopy. The growth of thin film is carried out using the thermal evaporation technique. The thermal studies of the grown films are performed by an indigenously developed thermoelectric measurement setup creating the thermal gradient on both the ends of the thin films. Antimony telluride showed better results than bismuth telluride till 45 °C however at slightly higher temperatures Bismuth Telluride provides better thermoelectric voltage generation. Seebeck coefficient is calculated from the results obtained for Bismuth Telluride and Antimony Telluride and found to be 366.9 and 226.8 μV/K, respectively.

Keywords Thermoelectric generators · Bismuth telluride · Antimony telluride · Seebeck effect

---

J. Choudhary · R. Jha
Research Lab for Energy System, Department of Physics, Netaji Subhas University of Technology, New Delhi 110078, India

A. Sharma (✉)
Department of Physics, Atma Ram SantanDharm College, University of Delhi, Delhi 110021, India
e-mail: anjalisharma31@gmail.com

Anisha · A. Gupta · G. Rani · B. Devi · M. Verma · M. Tomar
Department of Physics, Miranda House, University of Delhi, Delhi 110007, India

M. Tomar · A. Sharma
Institute of Eminence, University of Delhi, Delhi 110007, India

A. Chowdhuri
Department of Physics, Acharya Narendra Dev College, University of Delhi, Delhi 110007, India

Z. H. Khan et al. (eds.), *Recent Advances in Nanomaterials*, Springer Proceedings in Materials 27, https://doi.org/10.1007/978-981-99-4878-9_38

1 Introduction

The rising cost and adverse effects of non-renewable energy sources have become a major concern for researchers worldwide. Thermoelectric power generation has significant potential to emerge as an alternative to fossil fuels and other non-renewable resources [1]. Thermoelectric (TE) materials have the ability to convert the temperature gradient from externally applied heat sources or wasted heat sources into electricity or to create a temperature difference by applying electric potential [2]. TE materials are efficiently described by their dimensionless Figure of Merit ($ZT = \alpha^2 \sigma T/\kappa$) and Power Factor ($PF = \alpha^2 \sigma$), where α, κ, T, and σ are Seebeck coefficient, thermal conductivity, absolute temperature and electrical conductivity, respectively. The mutual dependence of thermal and electrical conductivity and the Seebeck effect restrict to diminish the thermal conductivity because this will lead to a reduction of electrical conductivity also. Recent research has shown that it is possible to overcome the issue by nano-structuring thermoelectric materials. Because of phonon scattering nanostructures (e.g., thin films) will have low thermal conductivity [3]. So, researchers are focusing on thin film fabrication to enrich the TE performance of the materials Bismuth Telluride (Bi_2Te_3) and Antimony Telluride (Sb_2Te_3) are the two most efficient TE generators near room temperature. A high ZT value has attracted an enormous amount of research interest for decades for Bi_2Te_3 and Sb_2Te_3. They have potential applications in thermal detectors, nanoscale power generators, and solid-state thermoelectric cooling devices. Present work focuses on the growth of the Bi_2Te_3 and Sb_2Te_3, their thin films for thermoelectric applications.

2 Materials and Methods

2.1 Bismuth Telluride and Antimony Telluride Synthesis

Chemicals used:- Tellurium powder (Te), Bismuth Trichloride Anhydrous ($BiCl_3$, 99.0%) Ethylenediaminetetraacetic Acid Disodium Salt (Na_2-EDTA), Sodium Terahydridoborate ($NaBH_4$), Sodium Hydroxide (NaOH), Tellurium dioxide (TeO_2, 99.99%), Antimony Trichloride Anhydrous ($SbCl_3$, 99.0%), Polyvinyl Alcohol (PVA), Hydrazine Hydrate (($N_2H_4H_2O$), 85%(wt.%)) were taken as the starting materials for the growth of Bi_2Te_3 and Sb2Te3 thin films.

Experiment:- In a beaker, solution A is prepared by mixing 4 mmol $BiCl_3$, 1 g Na_2-EDTA, and 6 mmol Te powder to 60 mL DI water. Then after, the solution was stirred for 30 min. To the above solution mixture, 0.8 g NaOH and 0.8 g $NaBH_4$ were added and further stirred for 30 min [4] and in another beaker, solution B is prepared by mixing $SbCl_3$ (0.320 g), PVA (0.400 g), and thin films using the thermal evaporation technique. The Seebeck coefficients of films were calculated with a TeO_2 (0.319 g) powder to 50 mL DI water. Then, 5 mL of hydrazine hydrate ($N_2H_4H_2O$) is added to the mixture [5]. After that, both the solutions were stirred for 30 min at room temperature. Both solution A and solution B were poured into two different 75 ml Teflon-lined autoclaves. Once sealed, the autoclaves were placed inside a hot air oven for 24 h at 180 °C. Finally, the autoclaves were kept to cool up to ambient temperature. Afterward, the washing of the resultant mixtures was carried out several times with deionized water and ethanol followed by centrifuges. Then, the obtained powders were vacuum dried using a vacuum oven for 12 h at 60 °C.

Structural and optical properties, morphologies of the synthesized NPs of both Bi_2Te_3 and Sb_2Te_3 were studied using X-Ray diffraction and UV–visible spectroscopy respectively. A hydraulic press was used to mold the resulting powder into half-inch diameter pellets for the fabrication of Bismuth and Antimony Tellurides indigenously developed measurement setup [6].

3 Results and Discussion

Figure 1a shows the XRD patterns of synthesized nanoparticles of Bi_2Te_3 and Sb_2Te_3. All peaks observed in the XRD (PANalytical X'Pert Pro MRD, Cu *K-α radiation*) patterns of the synthesized Bi_2Te_3 and Sb_2Te_3 nanoparticles were comparable with the standard diffraction pattern as given in the JCPDS files numbers [JCPDS:00-015-0863] and [JCPDS:00-015–0874]. The peak obtained at (015) on both Bi_2Te_3 and Sb_2Te_3, represents NPs' most preferred orientation in the [015] direction. The diffraction peaks of powder XRD confirms rhombohedral phase along with R3 space group for both the samples. Hence, no trace of any other phase was found.

From UV-Visible spectroscopy, the optical band gap was calculated for both using Tauc's equation, $\alpha h\nu = A\ (h\nu - Eg)^n$, here n is equal to 2 and 1/2 for direct and indirect transitions, respectively, α is absorption coefficient. Energy independent constant is denoted with A. Due to the overlapping of orbital and energy levels in atoms and molecules in the particle, Eg is dependent on particle size. As a result, the energy gap between the valence and conduction bands will widen because of nanoeffect [7]. Using UV-Spectrometer (Shimadzu UV-2600i Spectrometer), the band gap calculated using a powder assembly by taking Barium Sulfate ($BaSO_4$) as reference and from Fig. 1b, c with the variations in $(\alpha h\nu)^2$ with energy (E) for both

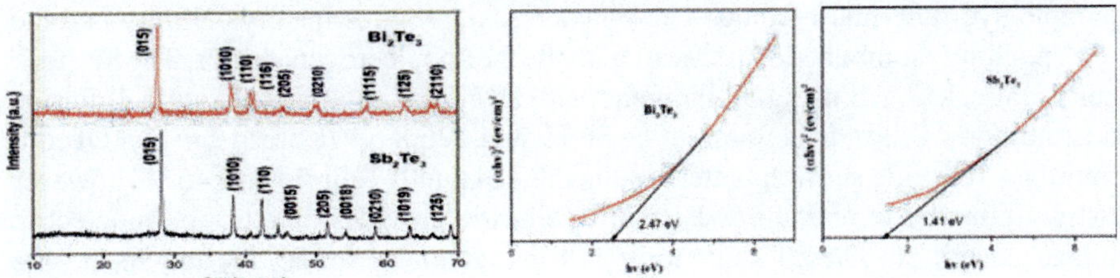

Fig. 1 **a** XRD of both Bi_2Te_3 and Sb_2Te_3 powder synthesized at 180 °C for 24 h by the hydrothermal method. **b** Tauc plot of Bi_2Te_3 and **c** Tauc plot for Sb_2Te_3

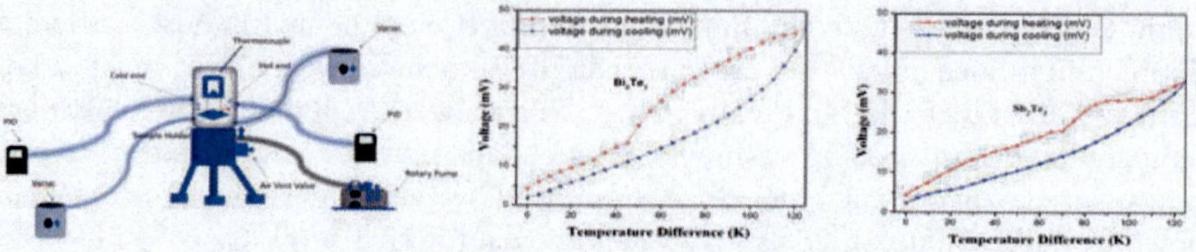

Fig. 2 **a** Schematic diagram of the indigenously developed setup for Seebeck coefficients measurement. **b** Seebeck coefficient measurements for thin films of (i) Bismuth Telluride and (ii) Antimony Telluride

Bi_2Te_3 and Sb_2Te_3. A straight line intersects the E-axis and gives the value of the band gap.

These characteristics confirm that the nanostructure has a remarkable effect on the band gap values and their electronic properties [8]. From the Tauc plot, the value of their band gaps for Bi_2Te_3 and Sb_2Te_3 is 2.47 and 1.4 eV, respectively, which is higher than that of the standard bulk sample.

It is observed that all Seebeck coefficient values show a direct correlation with temperature. As shown in Fig. 2a, the Seebeck coefficient was calculated using an indigenously developed setup, and the voltage generated by the temperature gradient was measured using a digital multimeter. The Seebeck coefficient was evaluated using the linear slope of the temperature gradient and the voltage generated data which was collected with a maximum temperature difference of $\pm$ 10 °C. From the obtained results during heating, Antimony Telluride shows better response up to 45 °C, then after Bismuth Telluride overcomes and during cooling Bismuth completely outcompeted Antimony Telluride. As shown in Fig. 2b(i), (ii) Seebeck coefficients calculated from the results obtained for Bismuth Telluride and Antimony Telluride were found to be 366.9 μV/K and 226.8 μV/K, respectively. The ZT value for thin films calculated for Bi_2Te_3 and Sb_2Te_3 are 7.0×10^{-3} and 5.6×10^{-3}, respectively.

4 Conclusion

In conclusion, the synthesis of both Bi_2Te_3 and Sb_2Te_3 has been demonstrated using a simple hydrothermal method. The powder XRD verifies the crystallinity and the corresponding rhombohedral phase of both the thermoelectric materials. The Seebeck coefficient calculated using an indigenously developed setup for Bismuth Telluride and Antimony Telluride is found to be 366.9 and 226.8 μV/K, respectively. Overall Antimony Telluride showed better results than Bismuth Telluride till 45 °C however at slightly higher temperatures Bismuth Telluride provides a better thermoelectric voltage generation. The ZT value for thin films calculated for Bi_2Te_3 and Sb_2Te_3 are 7.0×10^{-3} and 5.6×10^{-3}, respectively.

Acknowledgements One of the authors (JC) is grateful to the Vice Chancellor, NSUT, Delhi, for all the support and the funding given to the Research Lab for Energy System (RLES) for this research.

Declaration of Interest Statement There is no conflict of interest.

References

1. Siddique RM, Mahmud S, Van Heyst B (2017) A review of the state of the science on wearable thermoelectric power generators (TEGs) and their existing challenges. Renew Sustain Energy Rev 73:730–744. https://doi.org/10.1016/j.rser.2017.01.177
2. Haidar SA et al (2021) Deposition and fabrication of sputtered bismuth telluride and antimony telluride for microscale thermoelectric energy harvesters. Thin Solid Films 717:138444. https://doi.org/10.1016/j.tsf.2020.138444
3. Aronov D, Andalman AS, Fee MS (2008) A specialized forebrain circuit for vocal babbling in the juvenile songbird. Science 1979 320(5876):630–634. https://doi.org/10.1126/science.1155140
4. Kim HJ, Han M-K, Kim H-Y, Lee W, Kim S-J (2012) Morphology controlled synthesis of nanostructured Bi_2Te_3. Bull Korean Chem Soc 33(12):3977–3980. https://doi.org/10.5012/bkcs.2012.33.12.3977
5. Dong GH, Zhu YJ, Cheng GF, Ruan YJ (2013) Sb_2Te_3 nanobelts and nanosheets: hydrothermal synthesis, morphology evolution and thermoelectric properties. J Alloys Compd 550:164–168. https://doi.org/10.1016/j.jallcom.2012.09.061
6. Vieira E et al (2018) Bi_2Te_3 and Sb_2Te_3 thin films with enhanced thermoelectric properties for flexible thermal sensors. EUROSENSORS 2018, pp 815. https://doi.org/10.3390/proceedings2130815
7. Zhang Y et al (2020) Few-layer hexagonal bismuth telluride (Bi_2Te_3) nanoplates with high-performance UV-Vis photodetection. Nanoscale Adv 2(3):1333–1339. https://doi.org/10.1039/D0NA00006J
8. Ruhul Amin Bhuiyan M, Mamur H (2019) Accepted: 5.3.2019 Final Version: 1.06 (2019)

Scalable One-Step Template-Free Synthesis of Ultralight Edge-Multifunctionalized g-C_3N_4 Nanosheets with Enhanced Optical and Electrochemical Properties

Debashish Nayak, Gobind Mandal, Sanjeev Kumar, Sarfaraz Ansari, Jayanta Bauri, and Ram Bilash Choudhary

Abstract The multifunctional polymeric graphitic carbon nitride (g-C_3N_4) nanosheets were synthesized via a one-step template-free calcination method at various temperatures from melamine monomer. An X-ray diffraction investigation confirmed the formation of g-C_3N_4. Raman study corroborated the synthesis materials' chemical structure, phase, polymorphy, and molecular interactions. The surface morphological analyses of the as-synthesized g-C_3N_4 nanosheets were examined using FESEM. XPS and EDX analysis validated the purity and components included in the g-C_3N_4. UV–Vis analysis was used to study the optical absorbance and optical conductivity of the synthesized materials with an optical band gap of 2.79 eV. Photoluminescence spectra were used to investigate the enhanced rate of electron–hole recombination in g-C_3N_4 nanosheets that emit blue light. The color coordination, as well as the color purity of g-C_3N_4 calculated from the CIE diagram. The electrochemical performance of g-C_3N_4 was calculated using galvanostatic charge–discharge and cyclic voltammetry. All of these findings indicate that g-C_3N_4 are a potential material for multifunctional applications.

Keywords g-C_3N_4 · XPS · Optical properties · Electrochemical properties

1 Introduction

Two-dimensional graphene is employed in a variety of applications, including flexible materials, OLEDs, solar cells, optoelectronic devices, and more. Recent research has concentrated on two-dimensional materials other than graphene. Examples of more acceptable 2D materials include h-BN, GaN, MoS_2, phosphorene, and carbon

D. Nayak · G. Mandal · S. Kumar · S. Ansari · J. Bauri · R. B. Choudhary (✉)
Nanostructured Composite Materials Laboratory (NCML), Department of Physics, Indian Institute of Technology (Indian School of Mines), Dhanbad 826004, India
e-mail: rbcism@gmail.com

Z. H. Khan et al. (eds.), *Recent Advances in Nanomaterials*, Springer Proceedings in Materials 27, https://doi.org/10.1007/978-981-99-4878-9_39

nitride (C_3N_4) [1]. Because of their chemical and thermal stability, wear resistance, low density, water resistance, super hardness, and biocompatibility, carbon nitrides are suited for surface modification, light emitting equipment, and photocatalysis [2]. The most stable ambient allotrope is graphite carbon nitride (g-C_3N_4) with three triazine units [3]. Polymeric graphitic carbon nitride (g-C_3N_4) is (C_3N_3H)n, one of the first polymers. In 1834, first time introduced to the science society. A unique carbon-dependent material is polymeric graphite carbon nitride. Because of its simple synthesis, low band gap energy, visible light absorption, fast fictionalization, attractive electronic band structure, and high physicochemical stability, graphitic carbon nitride (g-C_3N_4), a semiconductor polymeric carbon-based material has gained interest in visible light-induced hydrogen evolution reactions (HER), light emitting diodes [4], and other applications [5]. The metal-free polymer n-type semiconductor g-C_3N_4 has many benefits. Because of their electric, optical, structural, optical, and physiochemical capabilities, g-C_3N_4-based materials are a one-of-a-kind family of multifunctional nanoplatforms for electronics, catalysis, and energy [6].

2 Materials and Synthesis Process

Melamine monomer from TCI Chemicals Pvt. Ltd. was used for this synthesis. For washing, we use ethanol and DI water. A muffle furnace and 10 g crucible with a lid are used for synthesis. As demonstrated in Fig. 1a, thermal polymerization generated g-C_3N_4. 5 g of melamine ($C_3N_6H_6$) was placed in a lidded 10 g alumina crucible. Wrapping the crucible in aluminum foil and placing it in a muffle furnace followed. The muffle's temperature was held constant at 450, 500, and 550 °C for 4 h at 6 °C /min. After cooling to ambient temperature, the light-yellow sample was recovered and suitably ground. Again, the sample was annealed for 2 h at the corresponding temperature. Sample was collected for further characterizations after cooling to ambient temperature. These samples are g-(T), where T stands for temperature.

3 Results and Discussion

The XRD pattern of pure g-C_3N_4 (at 550 °C) (Fig. 1b) revealed a larger diffraction peak at around 27.7° and a lesser one at about 13°, which was consistent with previous results and JCPDS data. The tri-s-triazine unit-containing in-plane structural packing motif that the diffraction peak at 13° corresponds to is indexed as (1 0 0). $d = 0.68$ nm is the estimated distance. The peak at 27.7° corresponds to inter-layer stacking of aromatic segments separated by 0.32 nm, and is indexed as the (0 0 2) peak of the conjugated aromatic system stacking. The estimated crystallite size of the g-C_3N_4 was ~ 11.60 nm with the micro-strain of 12.3×10^{-3}. Figure 1c shows Raman semiconductor quantum dots in PL. Melem unit dispersion affects PL. Melem unit PL in the blue zone links electron transitions from δ* and π* conduction bands to unshared electron pair states [9]. Thermal condensation and carbon nitride disorder microstructure enhance the FWHM of PL emission in the blue light region. Shape of cyclic voltammeter curve (CV) and galvanostatic charge–discharge (GCD) curve, as shown in Fig. 3g, h, elaborate the redox active behavior of the material. Further, it was observed that as the scan rate in CV increases the peak current density also increases due to rapidity in the redox process [10]. Moreover, as the current density in GCD increases discharging time decreases. Figure 3I, j showed the EIS and Bode plot for the synthesized g-C_3N_4.

All the above results shown that g-C_3N_4 is a good material for multifunctional applications in optoelectronics.

4 Conclusion

Polymeric graphitic carbon nitride was synthesized by thermal condensation method. XRD and Raman measurements verified the synthesis of the samples. The calculated crystallite size of g-C_3N_4 was 11.60 nm with a micro-strain of 12.3×10^{-3}. XPS and EDS analysis showed that the composition of the N and C in the g-C_3N_4. FESEM analysis showed that g-C_3N_4 has a well distributed morphology for the charge transfer properties. The band gap of the g-C_3N_4 was calculated to be 2.79 eV. g-C_3N_4 showed blue emission with a high PL intensity at 438 nm. Electrochemical analysis showed that g-C_3N_4 is the prominent materials for the energy storage devices. All the results showed that g-C_3N_4 is an attractive material for the multifunctional application.

Acknowledgements The authors would like to express their gratitude to the Indian Institute of Technology (Indian School of Mines), Dhanbad, India, for providing financing as well as access to experimental and characterization facilities.

Declaration of Interest Statement The authors are transparent about the fact that they do not have any financial conflicts of interest.

References

1. Zhao S, Zhang Y, Zhou Y, Wang Y, Qiu K, Zhang C, Fang J, Sheng X (2018) Facile one-step synthesis of hollow mesoporous g-C3N4 spheres with ultrathin nanosheets for photoredox water splitting. Carbon 126:247–256
2. Sharma G, Kumar A, Sharma S, Naushad M, Vo D, Ubaidullah M, Shaheen SM, Stadler FJ (2022) Visible-light driven dual heterojunction formed between g-C3N4/BiOCl@MXene-Ti3C2 for the effective degradation of tetracycline. Environ Pollut 308:119597
3. Bauri J, Choudhary RB (2023) Thermal and electronic states of exfoliated gC3N4-based nanocomposite with ZrO2 nanoparticles as a robust emissive layer. Mater Sci Semicond Process 154(2023)

4. Nayak D, Choudhary RB (2022) Influence of ZnS on the structural, morphological, optical and thermal properties of Polyindole for an emissive layer. Inorg Chem Commun 144:109824
5. Nayak D, Choudhary RB (2023) "Tuning the optical properties of high quantum-yield g-C3N4 with the inclusion of ZnS via FRET for high electron–hole recombination. Spectrochim Acta Part A Mol Biomol Spectrosc 289:122162
6. Vijayarangan R, Sakar M, Ilangovan R (2022) Stabilization of melon phase during the formation of g-C3N4 from melamine and its structure-property relationship towards photocatalytic degradation of dyes under sunlight. J Mater Sci Mater Electron 33(12):9057–9065
7. Kumar S, Choudhary RB (2022) Influence of MnO2 nanoparticles on the optical properties of polypyrrole matrix. Mater Sci Semicond Process 139:106322
8. Zeng X, Shu S, Meng Y, Wang H, Wang Y (2023) Enhanced photocatalytic degradation of sulfamethazine by g-C_3N_4/Cu, N-TiO_2 composites under simulated sunlight irradiation. J Chem Eng 456:141105
9. Binhong Q, Sun J, Li P, Jing L (2022) Current advances on g-C_3N_4-based fluorescence detection for environmental contaminants. J Hazard Mater 425:127990
10. Choudhary RB, Ansari S (2022) Mesoporous complexion and multi-channeled charge storage action of PIn-rGO-TiO2 ternary hybrid materials for supercapacitor applications. J Energy Storage 46:103912

Synthesis and Characterization of Thermoelectric Material Bi_2Te_3—A Potential Alternative for Power Generation

Avinash Kumar, Nirmal Manyani, and S. K. Tripathi

Abstract In recent years, thermoelectric materials attained great attention due to their capability of solving the problem of increasing energy demand. In this work, nanostructured thermoelectric material Bismuth Telluride (Bi_2Te_3) powder was synthesized using the one-step hydrothermal method with pre-reaction ultrasonication treatment in 10 h which is far less synthesis time than reported by other researchers. The structural characterizations of the product were studied using X-ray diffraction (XRD) and Fourier transforms infrared (FTIR) spectroscopy. The optical characterizations of the product were studied using Ultraviolet–visible (UV–Vis) and Photoluminescence (PL) spectroscopies, and the direct energy band gap was calculated by using the Tauc plot. Crystallite size, dislocation density, and lattice strain of synthesized powder were calculated by XRD analysis. Surface morphology was also studied using Field Emission Scanning Electron Microscopy (FESEM), and the elemental composition of synthesized Bi_2Te_3 was studied using Energy Dispersive X-Ray Spectroscopy (EDS). The synthesized Bi_2Te_3 can be used as potential thermoelectric material in a thermoelectric module for power generation and refrigeration applications.

Keywords Hydrothermal method · Thermoelectric materials · Renewable energy sources · Thermoelectric generators · Peltier coolers

1 Introduction

The increasing global population and decreasing fossil fuels have encouraged researchers to find an alternative for power generation to fulfill exponentially increased energy demand [1]. Renewable energy sources such as solar cells and

A. Kumar · N. Manyani · S. K. Tripathi (✉)
Department of Physics, Centre for Advanced Study in Physics, Panjab University, Chandigarh 160014, India
e-mail: surya@pu.ac.in

Z. H. Khan et al. (eds.), *Recent Advances in Nanomaterials*, Springer Proceedings in Materials 27, https://doi.org/10.1007/978-981-99-4878-9_40

fuel cells emerged as a possible solution to this problem. Due to low energy conversion efficiency, renewable energy sources waste a significant part of energy in form of heat [2]. To overcome this problem, in recent years, researchers focused on thermoelectric materials because of their capability to convert heat into electricity and vice versa and other advantages such as smaller size devices, low maintenance cost, high reliability, long life, and negligible emission of greenhouse gases [1, 2]. Bi_2Te_3-based alloys are presently the best efficient thermoelectric materials in the temperature range of 200–400 K. It has gained great attention due to its application in power generation devices (thermoelectric generators), chip sensing, chip cooling, Peltier coolers, television cameras, optoelectronic devices, and IR spectroscopy. Several physical and chemical processes have been discovered for the synthesis of nanostructured Bi_2Te_3 powder such as hydrothermal, solvothermal, electrochemical deposition, melting, ball milling, powder metallurgy, and wet chemical method [3]. The hydrothermal route has several advantages over other synthesis routes such as low reaction temperature, simple operation, fine crystallite size, low cost, and higher yield. In the hydrothermal route product purity, morphology, and crystallinity can be governed easily by varying reaction time, operating temperature, solvents, precursors, and surfactants [3, 4].

In this work, we synthesized nanostructured Bi_2Te_3 powder by the one-step hydrothermal method in 10 h with pre-reaction ultra-sonication treatment which is far less than earlier reported by Zhong et al. (48 h) [5], Saberi et al. (48 h) [6], Zhao et al. (24 h) [7], Zhang et al. (16 h) [8]. The results show that surfactant EDTA and pre-reaction ultrasonic treatment played a key role in the synthesis of material in low reaction time.

2 Materials and Methods

Synthesis of Bi_2Te_3 First of all, 2:3 molar ratio solution was prepared in DI water using Bi $(NO_3)_3.5H_2O$ (CDH Chemical)) and Te shots (Sigma Aldrich) as the Bi and Te precursors, respectively. NaOH (NICE Chemicals), $NaBH_4$ (SRL), and Ethylenediaminetetraacetic acid disodium salt (EDTA) (CDH Chemicals) were used as the PH controller, reducing agent, surfactant, respectively. The solution was given pre-reaction ultrasonic treatment at room temperature for 3 h at 40 kHz in an ultrasonic generator. The entire operation of the synthesis is explained in Fig. 1. Bismuth Telluride gray colored powder was obtained as the product of this synthesis process.

Characterization Techniques Structural and optical analysis of obtained product are done using XRD technique (Rigaku Miniflex 600 diffractometer) with a scanning range of $2\theta = 10°–80°$ and FTIR spectroscopy (Perkin Elmer-Spectrum RX-I) in the range of 4000–500 cm^{-1}, UV–Vis spectroscopy (Perkin Elmer LABINDIA3000 spectrophotometer) in the spectral range 200–800 nm and PL spectroscopy (Shimadzu spectro fluoro photometer RF-6000PC) with three different excitation wavelengths, i.e., 590, 595, and 600 nm are used to study the linear absorption

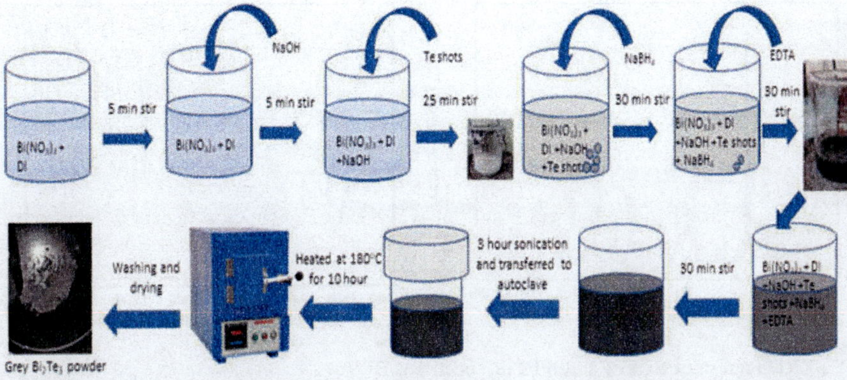

Fig. 1 Schematic representation of the synthesis procedure of Bi_2Te_3 by hydrothermal process

spectrum and emission spectrum, respectively. Morphologies and elemental distributions and mapping are studied using FESEM (SU-8000) and EDS (Brukers Xflash 6130) analysis, respectively.

3 Results and Discussion

As-prepared nanostructures of Bi_2Te_3 are studied with the help of XRD pattern and FTIR spectra. As depicted in Fig. 2a, the XRD pattern of the synthesized material is perfectly matched with the JCPDS Card no.15–0863 [9]. The average crystallite size (D) of prepared material has been calculated corresponding to the prominent planes of Bi_2Te_3 with the help of the Debye-Scherer's formula: $D = \frac{K\lambda}{\beta COS\theta}$, where K, λ, β, and θ are Scherer's constant, wavelength of X-ray, full-width half maxima (FWHM) of a diffraction peak, and diffraction angle, respectively [10]. Apart from this, the strain and dislocation density have been calculated correspond to prominent planes that are given in Table 1. The calculated crystallite size is in the range of 9–11 nm confirming the formation of the Bi_2Te_3 nanostructures.

In the XRD data, no peak of any oxides, impurities and unreacted Bi or Te are observed that confirming the important role played by surfactant EDTA and ultra-sonication. Presence of growth inducing agent EDTA may have promoted the formation of nanostructures by serving as bridging agent to create multinuclear Bi complexes and strong hydrogen bonding in EDTA led Bi complexes to self-associate to lamellar phases [5]. Ultrasonication at 40 kHz dispersed the reactants uniformly (Fig. 4b–d) and reduced the grain size. These nano-sized grains became the core of grain throughout the reaction. Reduced grain size and increased number of particles enhanced the rate of nucleation [1, 2, 5]. In FTIR spectra, as shown in Fig. 2b, the band that appeared at 3200–3925 cm^{-1} is due to vibration of O–H stretching that shows the presence of the coordinated water (H_2O) linked in the Bi_2Te_3 [11]. The

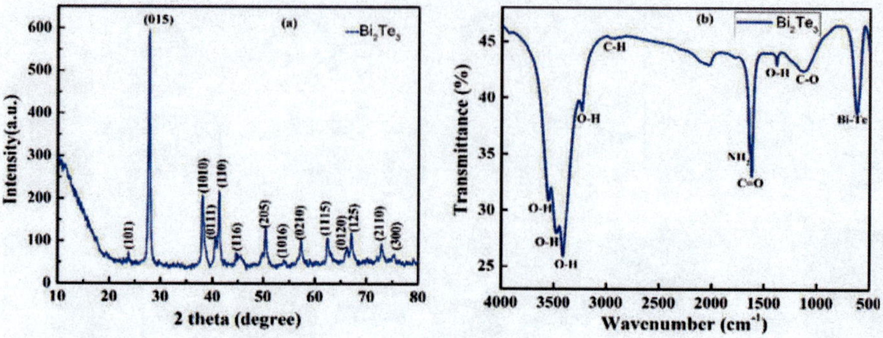

Fig. 2 a XRD Pattern of Bi_2Te_3 and b FTIR spectra of Bi_2Te_3

Table 1 Microstructural parameters of Bi_2Te_3 corresponding prominent peaks

2 theta (°)	FWHM (β) (°)	avg $D = k.\lambda/\beta.\cos\theta$ (nm)	$D = k.\lambda/\cos\theta$ (nm)	Strain (ε) = $\beta/4\tan\theta$ (10^{-3})	Dislocation density (δ) = $1/(D)^2$ (10^{14} lines cm^{-2})
27.764	0.1776	51.338	9.096	0.1796	0.0120
41.245	0.2935	32.220	9.435	0.1949	0.0113
50.376	0.4247	23.029	9.758	0.2257	0.0105
67.046	0.3503	30.309	10.592	0.1321	0.0089
75.147	0.3905	28.599	11.142	0.1268	0.0080

weak band appeared at 2926 cm^{-1} shows the presence of C–H band of the –CH_3 and strong band appeared at 636 cm^{-1} is due to the presence of Bi–Te band. The strong band appeared in the region of 1637 cm^{-1} shows the existence of N–H groups linked in EDTA, and the band appeared at 1616.08 cm^{-1} shows the asymmetric stretching vibration of (–COO–) group present in EDTA. The band at 1383 cm^{-1} is because of O–H bond, and the band at 1128 cm^{-1} is because of C–O groups stretching vibration present in Bi_2Te_3 absorbed CO_2 [11].

The band gap of the synthesized material is calculated by extrapolating the line up to the energy axis of the Tauc plot which is shown in Fig. 3a, and inset shows the absorbance spectra of Bi_2Te_3. The Tauc equation used for the calculation of the band gap is

$$\alpha hv = A\left(hv - E_g\right)^n \tag{1}$$

where A is a constant, hv is incident photon energy, and E_g is the band gap energy in Eq. (1). The calculated direct band gap of the Bi_2Te_3 powder is 0.92 eV which is in the optimum range for power generation and cooling application [1, 2].

The PL spectra of Bi_2Te_3 have been shown in Fig. 3b. Emission spectra of the Bi_2Te_3 have been obtained at the excitation wavelengths (λ_{ex}) of 590 nm, 595 nm, and 600 nm, respectively. In Bi_2Te_3, emission bands have been observed at 712, 718, and 721 nm corresponding to λ_{ex}. The PL emission decreases with the increase in λ_{ex} for the Bi_2Te, and the emission peaks exhibit a red shift with increasing λ_{ex}. The emission in the UV–Vis wavelength range is due to free excitons and may be due to surface defects, impurities, oxygen vacancies, etc., in the visible region [9].

Figure 4a depicts the FESEM analysis of Bi_2Te_3. As the size of these particles is in the range of 10–15 nm, so we may say that nanoparticles of Bi_2Te_3 have been formed which is also well matched with the crystallite size calculated by XRD analysis. Figure 4b–d shows the elemental mapping which reveal the homogeneous distribution and co-existence of Bi and Te in the nanocomposite. EDS analysis verifies the successful synthesis of nanostructured Bi_2Te_3 [11].

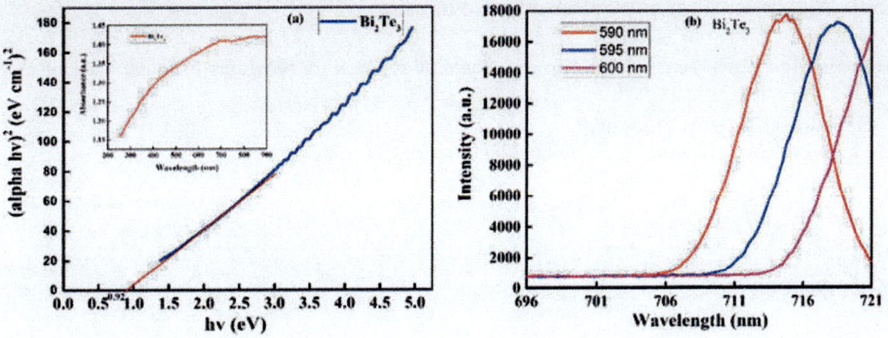

Fig. 3 **a** Tauc plot of Bi_2Te_3 and inset shows absorbance of Bi_2Te_3 and **b** PL spectra of Bi_2Te_3

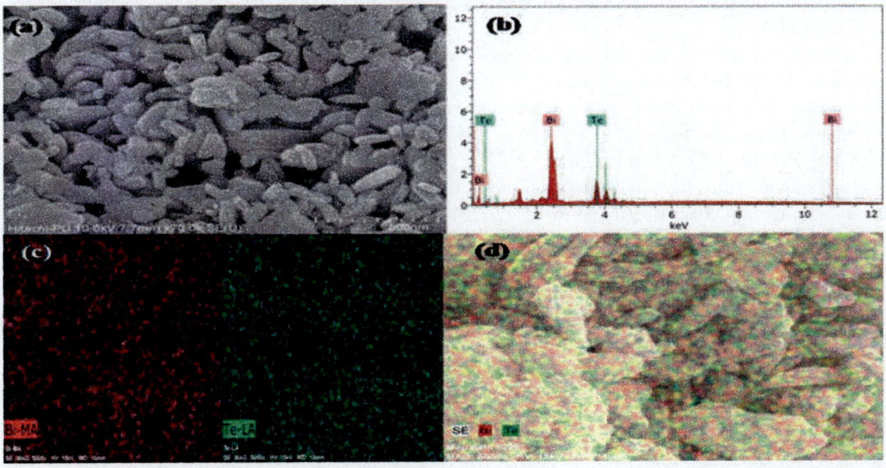

Fig. 4 FESEM and EDS mapping of Bi_2Te_3

4 Conclusion

Nanocrystalline Bi_2Te_3 powder was successfully synthesized in 10 h by single-step hydrothermal method with pre-reaction ultrasonication. No peak of any oxides, impurities or unreacted Bi or Te in XRD analysis and homogeneous distributions in EDS mapping confirmed the important role played by surfactant EDTA and ultrasonication. Crystallite size of 9–11 nm confirmed that fine Bi_2Te_3 can be synthesized by this approach without the need for any mechanical crushing and sieving. This simple and convenient approach may reduce time and cost in the large-scale synthesis of Bi_2Te_3. The calculated band gap of synthesized product was 0.92 eV that makes it suitable material for many applications in thermoelectric devices for power generation, Peltier coolers, chip sensing, and optoelectronic devices.

Acknowledgements Mr. Avinash Kumar is thankful to CSIR for providing financial assistance.

Declaration of Interest Statement
The authors declare that they have no conflict of interests.

References

1. Gayner C, Kar KK (2016) Recent advances in thermoelectric materials. Prog Mater Sci 83:330–382
2. Soleimani Z, Zoras S, Ceranic B, Shahzad S, Cui Y (2020) A review on recent developments of thermoelectric materials for room-temperature applications. Sustain Energy Technol Assess 37:100604
3. Mamur H, Bhuiyan MRA, Korkmaz F, Nil M (2018) A review on bismuth telluride (Bi_2Te_3) nanostructure for thermoelectric applications. Renew Sustain Energy Rev 82:4159–4169
4. Abishek NS, Naik KG (2022) Structural and morphological properties of bismuth telluride nanoparticles synthesized by two-step hydrothermal method. Mater Today: Proc 49:1319–1322
5. Wang Z, Wang FQ, Chen H, Zhu L, Yu HJ, Jian XY (2010) Synthesis and characterization of Bi_2Te_3 nanotubes by a hydrothermal method. J Alloy Compd 492(1–2):L50–L53
6. Saberi Y, Sajjadi SA, Mansouri H (2020) Comparison of characteristics of Bi_2Te_3 and $Bi_2Te_{2.7}Se_{0.3}$ thermoelectric materials synthesized by hydrothermal process. J Mater Sci: Mater Electron 31(21):18988–18995
7. Zhao XB, Ji XH, Zhang YH, Cao GS, Tu JP (2005) Hydrothermal synthesis and microstructure investigation of nanostructured bismuth telluride powder. Appl Phys A 80(7):1567–1571
8. Zhang HT, Luo XG, Wang CH, Xiong YM, Li SY, Chen XH (2004) Characterization of nanocrystalline bismuth telluride (Bi_2Te_3) synthesized by a hydrothermal method. J Cryst Growth 265(3–4):558–562
9. Suriwong T, Plirdpring T, Threrujirapapong T, Thongtem T, Thongtem S (2015) Thermoelectric properties of Bi_2Te_3 disk fabricated from rice kernel-like Bi_2Te_3 powder. Micro Nano Lett 1
10. Nirmal N, Sharma K, Poonam P, Tripathi SK (2020) Synthesis and characterization of Ni-BTC MOF for supercapacitor electrode. AIP Conf Proc 2265:030617. https://doi.org/10.1063/5.0017162
11. Mamur H, Bhuiyan MRA (2019) Bismuth telluride (Bi_2Te_3) nanostructure for thermoelectric applications. Int Sci Vocat Stud J 3(1):1–7

Enhanced Photocatalytic Performance of β-Bi_2O_3 Nanospheres Under Visible Light Irradiation

Aamir Sohail, Malik Aalim, Reyaz Ahmad, Arshid Mir, Asif Majeed, M. A. Shah, and Kowsar Majid

Abstract Beta-bismuth (III) oxide (β-Bi_2O_3) nanoparticles with a demonstrated diameter of roughly 200,350 nm were successfully fabricated using a hydrothermal approach under the low calcination temperature of 350 °C. Several characterization methods including powder X-ray diffraction (XRD), Field emission scanning electron microscopy (FESEM) outfitted with Energy-dispersive X-ray spectroscopy (EDXS), and UV–Vis spectroscopy were carried out in detail to study the crystal geometry, surface morphological, and optical properties of the prepared sample. UV–vis spectroscopy analysis demonstrated that the resulting β-Bi_2O_3 nanopowder exhibited broad absorption spectra with a narrow band gap of about 2.25 eV. Visible light stimulated photocatalytic activity of β-Bi_2O_3 nanoparticles was investigated thoroughly using a 1000 W Xu lamp as the light source outfitted with a 420 nm cutoff filter. Under visible range, the as-synthesized photocatalyst demonstrated excellent results for the degradation and decolorization of Methylene blue (MB). The superior photocatalytic activity is attributed to high oxidation power, low energy band gap, and enhanced effective surface area offered by β-Bi_2O_3 nanospheres.

Keywords Hydrothermal · Nanospheres · Photocatalysis · Methylene blue (MB)

1 Introduction

The removal of harmful pollutants from wastewater bodies, especially wastewater from dye industries, is one of the primary problems for building a clean environment owing to industrial growth and increasing population demand. A variety of treatment techniques, such as biological treatment, adsorption, coagulation, membrane filtering, ozonation, and photocatalytic degradation, have been used to remove

A. Sohail (✉) · M. Aalim · R. Ahmad · A. Mir · A. Majeed · M. A. Shah
Department of Physics, National Institute Technology, Hazratbal Srinagar, J&K 190006, India
e-mail: aamirphy111@gmail.com

K. Majid
Department of Chemistry, National Institute Technology, Hazratbal Srinagar, J&K 190006, India

Z. H. Khan et al. (eds.), *Recent Advances in Nanomaterials*, Springer Proceedings in Materials 27, https://doi.org/10.1007/978-981-99-4878-9_41

harmful chemical substances in water, such as dye molecules. Considering the disadvantages of the other techniques, such as limited efficiency, secondary pollutant generation, high operating costs, and so on, photocatalysis is the most promising way among them. In addition to this semiconductor, photocatalysis has various other benefits, including the utilization of sunlight, cost-effective operation, functioning in moderate circumstances, and photocatalyst reusability. For this purpose, several semiconductor materials have been reported so far; however, photocatalyst efficiency must be increased by lowering the energy band gap in order to better exploit the visible part of the sunlight [1, 2].

Recently, metal-based nanomaterials such as TiO_2, ZnO, NiO, Bi_2O_3 have been widely employed by researchers as potential photocatalysts owing to their higher efficiency and considerable physical and chemical stability. Nevertheless, because of larger energy band gap of TiO_2, ZnO, and NiO, these oxides are incompatible for the fabrication of visible light photocatalysts [3, 4]. On the other hand with the reputation of being optically active maintaining a narrow band gap of (2.75 eV for monoclinic and 2.25 for tetragonal Bi_2O_3 phase), high photostability, astonishing photoluminescence and formidable photoconductivity, bismuth oxide nanostructures find robust employability and are widely accepted nanomaterial in photocatalysis. Bi_2O_3 exists in five crystallographic polymorphs; however, owing to the narrow band gap and stronger optical absorption, β-Bi_2O_3 (tetragonal) exhibited superior photocatalytic activity under visible light irradiation compared to the other phases. Furthermore, as the beta phase represents a thermodynamically metastable state, it is challenging to prepare pure tetragonal Bi_2O_3, particularly at the nanoscale [5]. Therefore, it is crucial to establish a cost-effective facile method for the synthesis of pure Bi_2O_3 with nanostructures for prospective use in visible light photocatalytic activity.

In this study, we report a simple and efficient hydrothermal route to synthesize uniform and free-standing β-Bi_2O_3 nanospheres. The photocatalytic degradation of methylene blue (MB) under visible light has been studied systematically, and it was found that β-Bi_2O_3 nanospheres exhibited better degradation efficiency and a higher reaction rate constant. The superior photocatalytic activity is attributed to the narrower band gap and high surface area offered by free-standing nanospheres.

2 Materials and Methods

In the synthesis process, analytical grade precursors were employed without additional purification. Initially, 0.97 g of $Bi(NO_3)_3\cdot 5H_2O$ was dissolved in a mixture of 8 ml ethanol and 32 ml ethylene glycol solution. After that, 1.5 mmol of D-fructose was administered to the above mixture, which was then magnetically stirred until complete dissolution. The solution was poured in 50 ml Teflon-lined autoclave and kept in hydrothermal reactor at 180 °C for 12 h. After the reaction is completed, the precipitate formed was filtered, washed repeatedly with ethanol and deionized water, and subsequently dried in an oven at 80 °C. To get the required sample, the product was finally annealed at 350 °C for 3 h.

3 Results and Discussions

Figure 1c illustrates the powder X-ray diffraction (XRD) pattern of the prepared sample. The XRD profile of the product can be well indexed to the tetragonal β-Bi_2O_3 structure (JCPDS NO 27-0050) with the major peaks at 27.94°, 31.71°, 32.72°, 46.20°, 46.91°, 54.24°, 55.49°, 57.75°, and 74.48°values of 2θ corresponding to the diffractions from planes (201), (002), (220), (222), (400), (203), (421), (402), and (610) [6]. This indicates that heat treatment at 350 °C for 3 h led to the complete formation of β-Bi_2O_3. Moreover, no additional peak attributable to unwanted impurities was obtained, confirming the high purity of the prepared material. FESEM analysis was employed to study the surface morphological aspects of the sample. As revealed in Fig. 1a, b, the prepared β-Bi_2O_3 is composed of free-standing monodisperse nanospheres with an average diameter of 200–350 nm. This sort of morphology caters high specific surface area and is advantageous for photocatalytic degradation of MB. Further, the EDX spectra, Fig. 1d, demonstrate the presence of only Bi, O, elements in the sample.

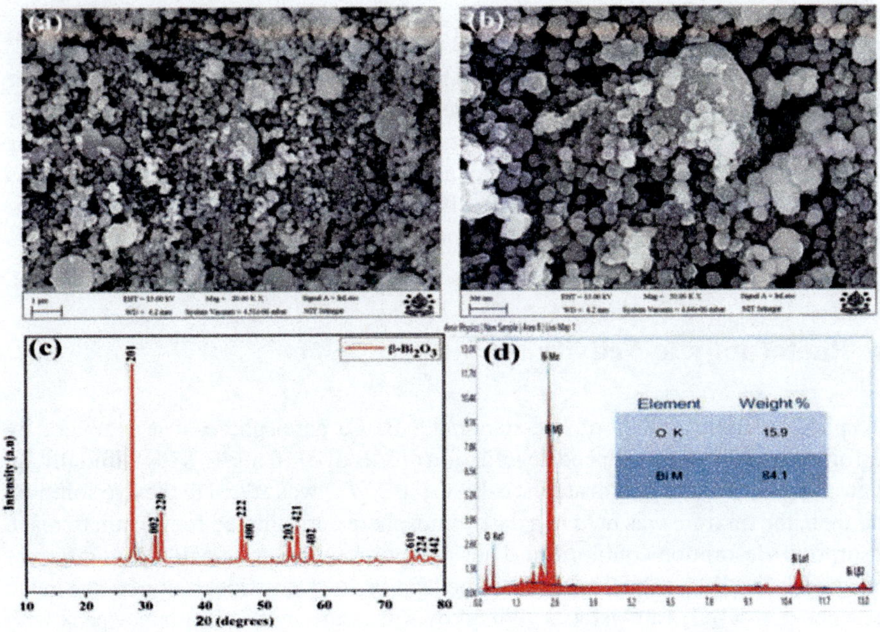

Fig. 1 **a**, **b** FESEM images and **c**, **d** XRD and EDX spectra of β-Bi_2O_3 nanospheres

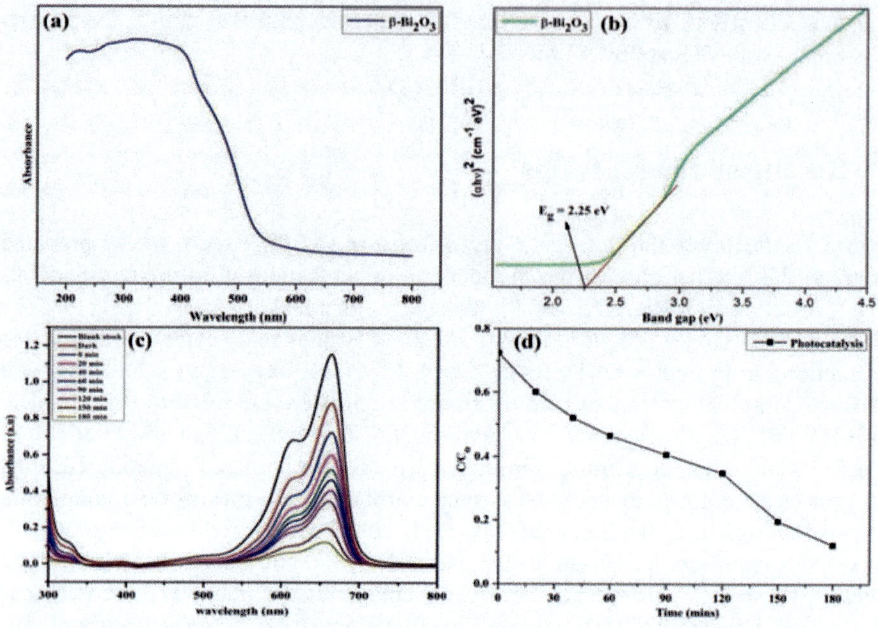

Fig. 2 **a**, **b** Absorbance spectra and Tauc plot of β-Bi_2O_3 nanospheres, **c** absorbance spectra and degradation rate of MB dye

The UV–vis DRS spectra of the prepared β-Bi_2O_3 nanospheres were recorded in the wavelength range from 200 to 800 nm. The absorption edge is located in the visible portion of the electromagnetic spectrum at a wavelength of about 500 nm, as shown in Fig. 2a. Using the Tauc plot method, in Fig. 2b, the energy band gap of the as-prepared photocatalyst was calculated to be 2.25 eV [4]. This narrow band gap and strong visible light absorption efficiency of the sample will accelerate the breakdown of MB dye.

4 Photocatalytic Activity

The photocatalytic activity of free-standing β-Bi_2O_3 nanospheres was estimated by performing visible light induced degradation of MB dye (10 mg/L, 8 PH). Initially, an appropriate amount of photocatalyst β-Bi_2O_3 (0.7 g/L) was added to the dye solution, and then, the mixture was held in a dark atmosphere and agitated for 30 min to reach adsorption–desorption equilibrium. The absorption spectrum of MB dye at different time intervals at $\lambda_{max} = 663$ nm is described in Fig. 2c. Degradation of MB dye in the presence of β-Bi_2O_3 catalyst was verified by a decrease in the absorbance peak with an increase in irradiation time which confirmed that degradation of MB dye took place in the presence of β-Bi_2O_3 catalyst. As illustrated in Fig. 2d, it was observed that about 85% MB was degraded in 180 min in presence of the catalyst. This superior photocatalytic activity is attributed to narrow band gap and large surface area offered by monodisperse β-Bi_2O_3 nanospheres.

5 Conclusion

In summary, monodisperse β-Bi_2O_3 nanospheres were prepared through facile hydrothermal method. The product was characterized via different techniques such as XRD, FESEM, and UV/Vis spectroscopy analysis. Further, the prepared sample was employed as an efficient photocatalyst for the degradation of MB dye and it was found that about 85% of MB dye (10 mg/L, 8 PH) was degraded into simpler non-toxic simpler compounds in the presence of β-Bi_2O_3 catalyst (0.7 g/L) under 180 min of visible light irradiation. Thus, in the view of excellent photocatalytic efficiency, β-Bi_2O_3 nanospheres may have potential applications in the field of environmental remediation for the treatment of industrial wastewater.

Acknowledgements The authors are grateful to the Department of Science and Technology (DST) for providing financial assistance.

References

1. Gupta G, Kaur M, Kansal SK, Umar A, Ibrahim AA (2022) α-Bi_2O_3 nanosheets: an efficient material for sunlight-driven photocatalytic degradation of Rhodamine B. Ceram Int 48(20):29580–29588
2. Jalalah M, Faisal M, Bouzid H, Park JG, Al-Sayari SA, Ismail AA (2015) Comparative study on photocatalytic performances of crystalline α-and β-Bi_2O_3 nanoparticles under visible light. J Ind Eng Chem 30:183–189
3. Ayekoe PY, Robert D, Goné DL (2015) TiO_2/Bi_2O_3 photocatalysts for elimination of water contaminants. Part 1: synthesis of α-and β-Bi_2O_3 nanoparticles. Environ Chem Letters 13(3):327–332
4. Ahmad R, Shah MA (2023) Hydrothermally synthesised nickel oxide nanostructures on nickel foam and nickel foil for supercapacitor application. Ceram Int 49(4):6470–6478
5. Ding S, Dong T, Peppel T, Steinfeldt N, Hu J, Strunk J (2022) Construction of amorphous SiO_2 modified β-Bi_2O_3 porous hierarchical microspheres for photocatalytic antibiotics degradation. J Colloid Interface Sci 607:1717–1729
6. Sudapalli AM, Shimpi NG (2022) Hierarchical self-assembly of 0D/2D β-Bi_2O_3 crossandra flower morphology exhibits excellent photocatalytic activity against bromophenol dyes. Opt Mater 132:112849

Thermal and Electrical Study of Polypyrrole and TiO_2/Polypyrrole Composite

Neha Luhakhra and Sanjiv Kumar Tiwari

Abstract The present study explores the thermal and electrical properties of polypyrrole (PPy) and TiO_2-doped PPy (TiO_2/PPy) composite. The chemical oxidative polymerization method was used to synthesize the PPy and TiO_2/PPy composite (in-situ approach was used for doping). The synthesized samples were characterized by experimental techniques, viz. differential scanning calorimetric (DSC) and IV characteristics to investigate the thermal and electrical properties. DSC revealed that the TiO_2/PPy composite has higher glass transition and melting temperature than bare PPy suggesting that the composite is more thermally stable. Additionally, TiO_2/PPy composite has enhanced electrical conduction compared to bare PPy. This signifies that the interfacial interaction between TiO_2 and PPy provides a smooth path for the transfer of charges from one site to another. Hence, the thermal and electrical properties significantly enhance with the formation of TiO_2/PPy composite compared to bare PPy.

Keywords Organic–inorganic composite · Conducting polymers · Thermal and electrical properties · Titanium dioxide and polypyrrole

1 Introduction

Conducting polymers belong to the new category of polymers and have been explored as organic materials in numerous fields [1]. Conducting polymers have alternate combinations of single and double bonds along the chain length of the polymers which impart novel features [2]. There is a planetary of research articles where the researchers have described the special properties of conducting polymers, having usage in various disciplines such as supercapacitors, storage, battery, photovoltaic, photocatalyst and sensors [3, 4]. Conducting polymers are also known as conjugated polymers and have special tunable characteristics due to the presence of

N. Luhakhra (✉) · S. K. Tiwari
Jaypee University of Information Technology, Waknaghat, Solan, H.P, India
e-mail: neha1996luhakhra@gmail.com

Z. H. Khan et al. (eds.), *Recent Advances in Nanomaterials*, Springer Proceedings in Materials 27, https://doi.org/10.1007/978-981-99-4878-9_42

charges (polaron and bipolarons) in it. Polypyrrole (PPy), poly-thiophene (PT), polyaniline (PANI) and poly-acetylene (PA) are the types of conducting polymers [4, 5]. Among above-mentioned conducting polymers, PPy has been extensively studied for different techniques. Easy synthesis, cost-effectiveness, good environmental stability and facile preparation procedure by chemical polymerization approach make PPy a demanding material [6, 7]. The polarons and bipolarons are the charge carriers responsible for altering the thermal, optical, magnetic and electrical properties of PPy [8]. The number of polarons and bipolarons can be controlled in PPy by simply varying the type and concentrations of an oxidizing agent. Therefore, PPy is a promising candidate for the researcher to explore due to its unique properties and vast applications.

The efficiency of these conducting polymers can be enhanced by tuning them with inorganic materials. One of the fascinating classifications of composite materials is organic–inorganic-based composites [9]. Titanium dioxide (TiO_2), zinc oxide (ZnO), aluminum oxide (Al_2O_3), magnesium oxide (MgO), manganese dioxide (MnO_2) and calcium oxide (CaO) are well-known metal oxide, which has been remarkably utilized for the manufacturing of semiconductor integrated circuits (IC), energy storage, drug delivery, pharmaceutical products, etc. [10, 11]. Out of all, TiO_2 is a well-known semiconductor which exists in three different basic crystal structures, viz. anatase, rutile and brookite having different properties [12].

In the present case, a composite of PPy (organic) and TiO_2 (inorganic) has been investigated thoroughly. The integrated study of thermal and electrical properties gives better insight to understand the utilization of TiO_2/PPy composite for most of the above-mentioned applications. TiO_2/PPy composites are emerging important materials for electronic devices, photocatalyst, photovoltaic, storage materials and energy conversion because of high charge carrier mobility, electrical conductivity and mechanical strength. Hence, in the present study, an attempt has been made on the synthesis and characterization of PPy and TiO_2/PPy composite. Differential scanning calorimetry (DSC) explains the thermal behavior and stability of the PPy and TiO_2 composite, and the electrical properties have been studied by the measurement of the I–V characteristic.

2 Experimental Details

2.1 Materials and Method

Pyrrole (C_4H_5N) ($\geq$98%) from Alfa Aesar and titanium dioxide (TiO_2) nanopowder (average particle size of 40 nm) from SRL have been purchased. Iron (III) chloride anhydrous ($FeCl_3$) and sodium dodecyl sulfate (SDS) were procured from Merck Chemicals, India. All the chemicals are of analytical grade and used without further purification. PPy was prepared by the chemical oxidative polymerization method (Fig. 1).

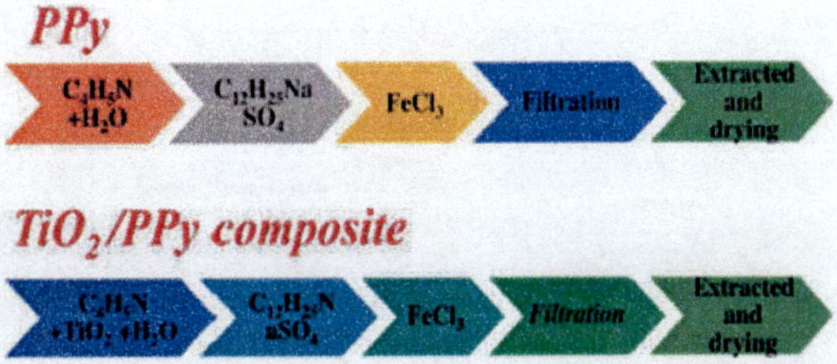

Fig. 1 Schematic illustration of synthesis method for the preparation PPy and TiO_2 by chemical oxidative polymerization approach

Preparation steps of PPy: 1 M pyrrole and 0.002 M SDS were mixed in the aqueous solution followed by 15 min stirring. This mixture is denoted as the "sol (a)". The aqueous solution of 0.5 M anhydrous iron (III) chloride was prepared simultaneously, viz. "sol (b)". After achieving homogenous solutions, both solutions were mixed and allow to react for 4 h. After the polymerization, reaction precipitates were obtained and separated by the process of filtration (Grade 602 H) followed by simultaneously washing with deionized water. The precipitates so obtained were dried in an oven at 40 °C overnight. Thereafter, the dried powder was crushed to obtain a fine powder.

Preparation steps of TiO_2/PPy: The aqueous solution of 1 M pyrrole and 0.002 M SDS was prepared with magnetic stirring for 15 min. About 0.5 M of TiO_2 was added to the above solution. The combination was given the term "sol (c)". In a separate beaker, 0.5 M anhydrous iron (III) chloride aqueous solution (sol (d)) was made concurrently. Both solutions (sol (c) and (d)) were combined and left to react for 4 h with continuous stirring. During the reaction, pyrrole polymerizes in the presence of TiO_2 nanoparticles and forms TiO_2/PPy composite, and this phenomenon is known as in situ polymerization. Precipitates were collected, separated and repeatedly rinsed with DI water, ethanol and acetone while being filtered using filter paper (Grade 602 H). The precipitates were then dried in an oven overnight at 40 °C. To get fine powder, the dried powder was further crushed.

2.2 *Characterization*

Thermal properties measurement: The thermal behavior of as-synthesized PPy and TiO_2/PPy composite was studied using DSC (PerkinElmer Pyris 1 setup). The sample ~ 10 mg was loaded in a Pt crucible under a nitrogen gas atmosphere and analyzed in the temperature range of 20–500 °C with a heating rate of 10 °C per minute.

I–V measurement: Fig. 2 shows the circuit and schematic diagram of the I–V measurement setup. I–V measurement graph was recorded using Keithley's Picoammeter setup on pellets at room temperature. The pellets of PPy and TiO_2/PPy having a thickness of 0.3 cm were prepared with the help of a hydraulic press by applying 60 MPa pressure. With the help of silver contact, the wires were connected for I–V measurements.

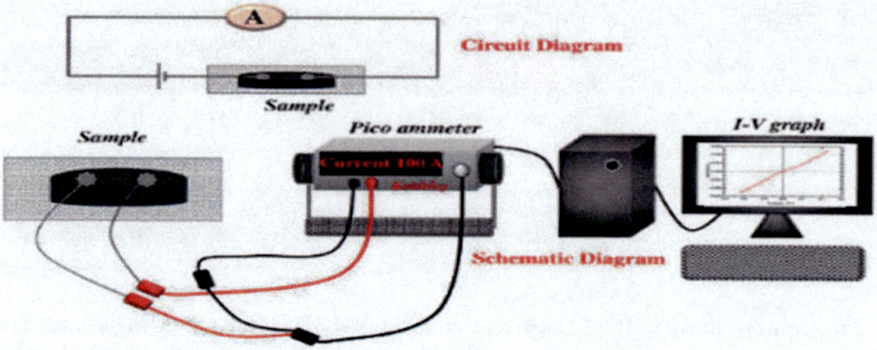

Fig. 2 Circuit diagram and schematic diagram of instrumental setup for I–V measurement

3 Result and Discussion

3.1 *Differential Scanning Calorimetry (DSC)*

Figure 3 shows the thermo graph of PPy and TiO_2/PPy composite. With the raising temperature, PPy and TiO_2/PPy experience thermal transformation, viz. glass transition temperature (T_g), melting temperature (T_m) and decomposition temperature (T_D). The peaks positioned at temperatures 88 °C, 270 °C and 430 °C correspond to T_g, T_m and T_D for PPy (Fig. 3a), Fig. 3b shows the T_g, T_m and T_D for TiO_2/PPy composite positioned at 155 °C, 312 °C and 470 °C, respectively. On comparing the thermograph, T_g, T_m and T_D values got enhanced for TiO_2/PPy composite, attributed to reducing in the free volume due to the loss of mobility of the intercalated polymer chain [13]. As TiO_2 is incorporated in the PPy, TiO_2 occupies the vacant space which results in a decrease in the free volume of the composite. Additionally, the intermolecular interaction between TiO_2 and PPy in the composite hold them strongly. The charge carriers polarons and bipolarons remarkably contribute to the intermolecular electrostatic interaction between PPy and TiO_2 [14]. Further, the electrostatic force between PPy and TiO_2 increases the binding strength. Hence, the thermal stability of TiO_2/PPy increases as compared to pure PPy.

3.2 *I–V Characteristics*

The electrical properties of the samples were studied on the I–V measurement set-up using PPy and TiO_2/PPy composite pellets. Figure 4 demonstrates the I–V characteristics' graph for PPy and TiO_2/PPy. In PPy, the charge carriers polarons and bipolarons cause conduction in it [15]. These charge carriers in PPy are highly delocalized parallel and perpendicular to the chain length. Conduction in the PPy has

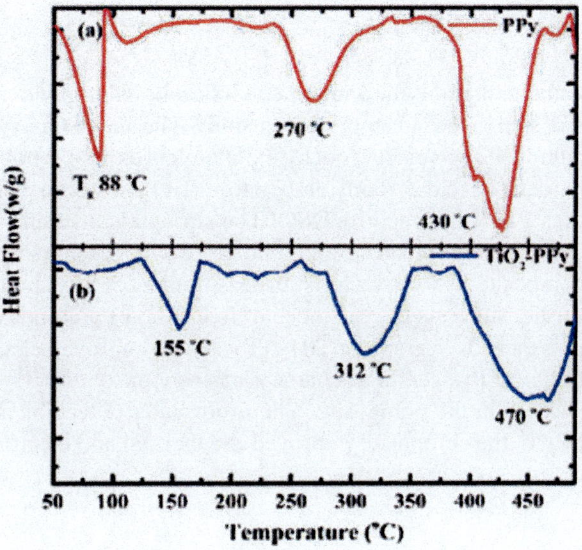

Fig. 3 DSC thermograph of **a** PPy and **b** TiO_2/PPy composite

been observed even applying a small potential, which signifies the presence of a sufficient number of polarons and bipolarons.

The blue curve in Fig. 4 demonstrated the I–V characterization for TiO_2/PPy composite. Since TiO_2 has high charge carrier mobility, TiO_2/PPy composite shows an improved conduction phenomenon [11]. The intermolecular interaction between TiO_2 and PPy provides the direction to the charges. The interaction between PPy and TiO_2 is majorly through polarons and bipolarons charges. Therefore, the charge carrier mobility enhances in TiO_2/PPy composite.

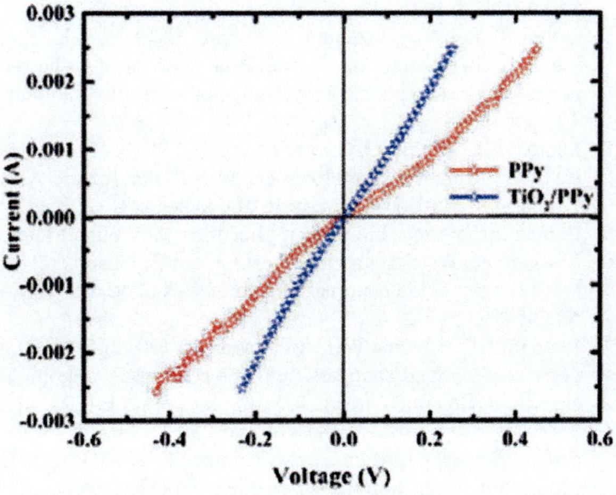

Fig. 4 I–V characteristics' measurement of PPy and TiO_2/PPy composite

3.3 Conclusion

Conducting polymers and metal oxide-based organic–inorganic composite has been invested in the current research study. The successful synthesis of PPy and TiO_2/PPy composite was carried out by the chemical oxidative polymerization using $FeCl_3$ as an oxidative agent at room temperature. TiO_2/PPy composite has better thermal stability than PPy confirmed by DSC. The strong electrostatic interaction between PPy and TiO_2 through polarons and bipolarons charges enhances the binding strength of the composite. I–V characteristic measurement of PPy shows its conductive nature which further suggests the existence of free charges polarons and bipolarons in conducting polymers. I–V graph of TiO_2/PPy shows improved electrical conductivity compared to pure PPy. The high charge carrier mobility of TiO_2 helps to manage the flow of charges in the composite. Therefore, intermolecular interaction between TiO_2 and PPy in the composite improved the thermal and electrical properties.

References

1. Nguyen DN, Yoon H (2016) Recent advances in nanostructured conducting polymers: from synthesis to practical applications. Polymers 8:118
2. Nezakati T, Seifalian A, Tan A, Seifalian AM (2018) Conductive polymers: opportunities and challenges in biomedical applications. Chem Rev 118:6766–6843
3. Das TK, Prusty S (2012) Review on conducting polymers and their applications. Polym-Plast Technol Eng 51:1487–1500
4. Goto H, Komaba K, Eguchi N, Toyofuku M, Nomura N (2022) A possibility of polaron vortex magnet of polypyrrole prepared in virus liquid crystal. J Polym Sci 60:90–96
5. Namsheer K, Sekhar Rout C (2021) Conducting polymers: a comprehensive review on recent advances in synthesis, properties and applications. RSC Adv 11:5659–5697
6. Tan Y, Ghandi K (2013) Kinetics and mechanism of pyrrole chemical polymerization. Synth Met 175:183–191
7. Gogoi R, Singh A, Moutam V, Sharma L, Sharma K, Halder A, Siril PF (2022) Revealing the unexplored effect of residual iron oxide on the photoreforming activities of polypyrrole nanostructures on plastic waste and photocatalytic pollutant degradation. J Environ Chem Eng 10:106649
8. Santos MJL, Brolo AG, Girotto EM (2007) Study of polaron and bipolaron states in polypyrrole by in situ Raman spectroelectrochemistry. Electrochim Acta 52:6141–6145
9. Parola S, Julián-López B, Carlos LD, Sanchez C (2016) Optical properties of hybrid organic-inorganic materials and their applications. Adv Func Mater 26:6506–6544
10. Siriwong C, Wetchakun N, Inceesungvorn B, Channei D, Samerjai T, Phanichphant S (2012) Doped-metal oxide nanoparticles for use as photocatalysts. Prog Cryst Growth Charact Mater 58:145–163
11. Leonard KC, Suyama WE, Anderson MA (2012) Evaluating the electrochemical capacitance of surface-charged nanoparticle oxide coatings. Langmuir 28:6476–6484
12. Bonetta S, Bonetta S, Motta F, Strini A, Carraro E (2013) Photocatalytic bacterial inactivation by TiO_2-coated surfaces. AMB Expr 3:59
13. Sultan A, Anwer T, Ahmad S, Mohammad F (2016) Preparation, characterization, and dynamic adsorption–desorption studies on polypyrrole encapsulated TiO_2 nanoparticles. J Appl Polymer Sci 133

14. Ullah H (2017) Inter-molecular interaction in polypyrrole/TiO_2: a DFT study. J Alloy Compd 692:140–148
15. Heydari Gharahcheshmeh M, Gleason KK (2020) Texture and nanostructural engineering of conjugated conducting and semiconducting polymers. Mater Today Adv 8:100086

Study of Optical Properties of 5-Fluorouracil-Conjugated Nanoparticles

Sharma Swati, Jain Shikshita, and S. K. Tripathi

Abstract With a view to creating specialized applications in biomedical physics, the utilization of nanoparticles (NPs) for anticancer therapies has now emerged as a high research priority. Over the past few decades, I-III-VI NPs have gained more interest than II-VI or IV-VI because these NPs are less toxic, cadmium-free and possess excellent optical properties. This paper reports the conjugation of the anticancer drug 5-fluorouracil (FU), which is a water-soluble drug, on glutathione (GSH)-capped copper indium sulfide core (CIS) NPs, which are formed using an aqueous synthesis method. Different concentrations of the drug, FU are taken into consideration for the conjugation with CIS NPs. This paper includes the study of change in the optical properties of CIS NPs after the drug FU is conjugated on them. The optical properties include Photoluminescence (PL) spectroscopy, Fourier Transform Infrared (FTIR) spectroscopy, and the UV–Visible (UV–Vis) spectroscopy. The PL spectroscopy of FU-conjugated CIS NPs shows that as the concentration of the drug is increased, there is quenching in the fluorescence intensity of the CIS NPs after the drug FU is conjugated with CIS NPs. The FTIR spectroscopy confirms the conjugation of the drug FU to CIS NPs and UV–Vis spectroscopy shows the absorbance of the drug on CIS NPs. The results strongly suggest the successful chemical conjugation of FU to CIS NPs and can be used in various biomedical applications in the future.

Keywords CIS nanoparticles · Fluorouracil · Drug conjugation · GSH

1 Introduction

The science of nanotechnology focuses on producing NPs with a size range of 1–100 nm and varying particle size and structure. These days, the application of NPs has increased unexpectedly in a variety of areas including physics, molecular biology, inorganic and organic chemistry, material science, and medicine [1].

S. Swati · J. Shikshita (✉) · S. K. Tripathi
Department of Physics, Centre of Advanced Study in Physics, Panjab University, Chandigarh 160014, India
e-mail: surya@pu.ac.in

Z. H. Khan et al. (eds.), *Recent Advances in Nanomaterials*, Springer Proceedings in Materials 27, https://doi.org/10.1007/978-981-99-4878-9_43

In the past years, II-VI, III-V, or IV-VI core–shell NPs are mostly used because of their distinct shape and size-dependent optical characteristics, but the presence of lead (Pb) and cadmium (Cd) made them toxic and less environment friendly. Therefore, it is necessary to investigate novel NPs that are less harmful and more environmentally benign [2].

Ternary QDs (I-III-VI) grab more attention in recent years as they are less toxic, environment friendly, and possess excellent optical properties. Additionally, because I-III-VI QDs include a variety of different atoms, their optical properties can be modified by changing their composition [3].

For the treatment of a multitude of cancers, such as breast, colorectal, head and neck, stomach, and pancreatic cancers, FU is employed as a first-line antineoplastic drug. However, the development of medication resistance significantly restricts the clinical application of FU, because it has a low bioavailability [4]. Therefore, it is crucial to create an effective drug delivery system for FU in order to enhance therapeutic benefit while minimizing adverse effects which can be overcome by using some biomolecules along with FUs to improve the short half-life, low efficacy, and have a good targeting impact [5]. So, in this study, ternary core NPs are used for the conjugation of the drug FU which can be further used in various biomedical applications.

Therefore, in this study, CIS NPs are taken as a delivery agent, and these NPs are hydrophilic in nature, also the drug FU is hydrophilic in nature, and hence, the drug conjugation on the CIS NPs becomes easy. Here only, the change in the optical properties of CIS NPs is studied after the drug FU is conjugated on them. The UV–Vis spectra, FTIR spectra and change in PL spectra of drug FU-conjugated CIS NPs show that the drug is successfully conjugated on the CIS NPs.

2 Materials and Synthesis Method

Materials Analytical-grade reagents are utilized during the synthesis process without additional purification. Cupric chloride dihydrate (SRL) sodium sulfide-fused flakes (Fisher Scientific), indium chloride, glutathione (CDH), sodium hydroxide (NICE), 5-Fluorouracil (SRL) have been used, respectively. The reaction is performed in standard glassware which is rinsed properly by using deionized (DI) water and then dried before they are used.

Method For the conjugation of the drug on CIS NPs, firstly the solution of the drug is prepared, and for that, some quantity of the drug FU (0.029 mM) has been taken and added it into a beaker having DI water, so that the drug is completely dissolved in DI water. Then, sonicate it for 15 min at room temperature, and after sonication, stir it for few minutes by using the magnetic stirrer.

After that, for the conjugation of the drug FU on CIS NPs, the drug FU is directly added in the CIS NPs because both are hydrophilic in nature. Firstly, the CIS NPs are taken into the round bottle flask and stirred by using the magnetic stirrer. After that the drug FU is added slowly and dropwise to the CIS NPs by using the pipette during the stirring at room temperature. Then, the solution containing the drug FU and CIS NPs is stirred in a round bottle flask at ambient temperature for about 2 h by using a magnetic stirrer.

For the conjugation of the drug to CIS NPs, different concentrations (300, 400, and 500 μl) of the drug are taken into consideration, and for all the concentrations of the drug, the process of conjugation of the drug on CIS NPs is same as mentioned above.

The synthesized samples have been characterized by UV–Vis spectroscopy performed by LABINDIA-UV 3000^{+} spectrometer, FTIR spectroscopy by employing Perkin Elmer (Spectrum-400 FTIR) Spectrometer, and PL spectroscopy using RF6000 Spectrofluorophotometer.

3 Results and Discussion

3.1 PL and UV–Vis Spectroscopy

Figure 1 represents the PL spectra of FU-conjugated CIS NPs and CIS NPs (black line), by using the different concentrations (300, 400, and 500 μL) of the drug FU. It is observed that on increasing the concentration of drug, the intensity of CIS NPs also increases [6], but after that, there is quenching in the intensity of CIS NPs as shown in Fig. 1 [7].

The decrease in the intensity of CIS NPs, and no shift in the peak position has been observed when the drug FU is conjugated with them which confirms the conjugation of the drug to the CIS NPs [8]. The decrease in the intensity is because there is π–π

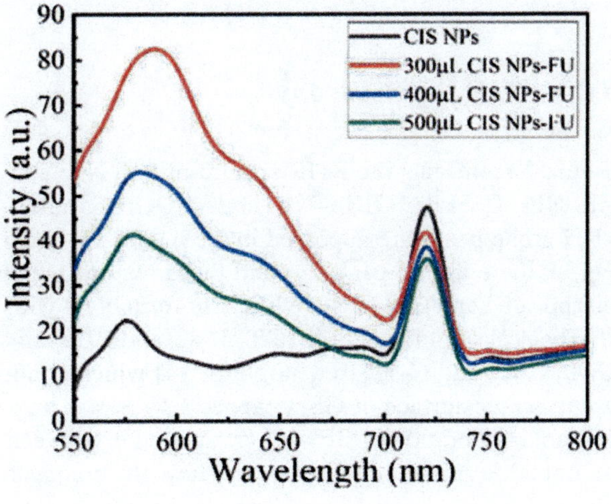

Fig. 1 Represents the PL spectra of CIS NPs and FU-conjugated CIS NPs at different concentrations (300, 400, and 500 μL) of FU

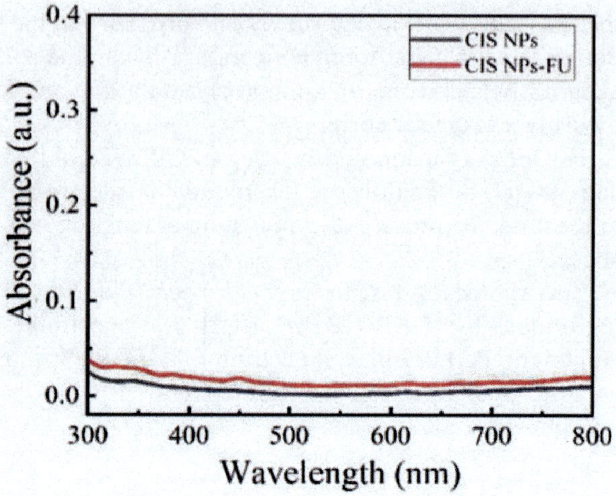

Fig. 2 Represents the UV–Vis spectra of CIS NPs and FU (300 μL) conjugated CIS NPs

interaction between the CIS NPs and FU due to which the aggregation of particles starts and their intensity starts decreasing and as the concentration of the drug starts increasing the aggregation is more and more, due to which there is more quenching in the intensity of CIS NPs at higher concentration [9]. From the result, it is confirmed that the drug FU is successfully conjugated on the CIS NPs.

Figure 2 represents the UV–Vis spectra of CIS NPs and the drug FU (300 μL) conjugated CIS NPs between the range 300–800 nm. The CIS NPs do not show any absorbance peak in the range 300–800 nm [10]. Also FU shows the characteristic absorption peak at 266 nm but has no absorbance peak in this region 300–800 nm [11]. Further, after conjugation of the drug FU on CIS NPs, there are no formation of new peak and no change in the UV–Vis spectra, because the drug FU itself has no peak in this region; therefore, results reveal that there is successful absorption of the drug FU on CIS NPs as shown in the figure below [12].

3.2 FTIR Spectroscopy

Figure 3 represents the FTIR spectra of CIS NPs and FU-conjugated GSH-capped CIS NPs. GSH has –SH, –NH, and –COOH groups on their surface. The –SH and –NH group peaks are appeared in the region 2525 and 1536 cm^{-1}, but as shown in Fig. 3, there are no peaks around these regions which means that these peaks are disappeared and CIS NPs are likely to form bond with –SH and –NH groups [13].

The peaks at 1128, 1214, 1678, 1682, and 2929 cm^{-1} represent the C–C, C–O–H, C=O, C=C, and C–H stretching [13, 14] which confirms the presence of carboxyl group on the surface of GSH-capped CIS NPs. The peaks observed (green line) at 778 cm^{-1} (vibration of CF=CH group) and in between 3000 and 3500 cm^{-1} (shows un-bound F groups of FU) greatly show the conjugation of the drug FU with CIS

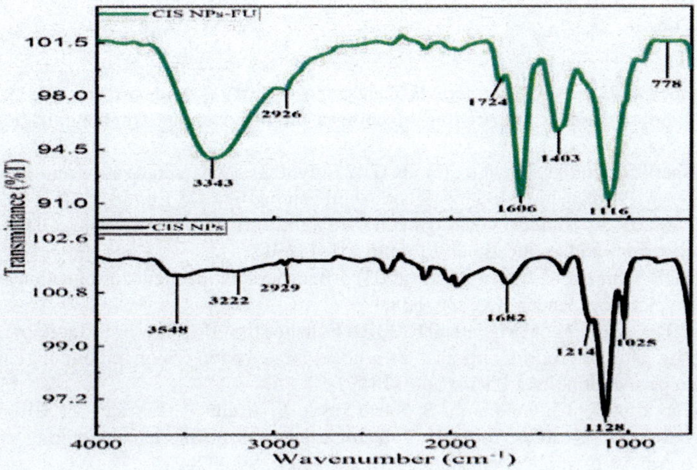

Fig. 3 Represents the FTIR spectra of CIS NPs and FU (300 μL) conjugated CIS NPs

NPs. Therefore, –NH and –COOH group of drug and CIS NPs interact with each other through hydrogen bonding [15, 16]. Hence, the results show that the drug FU is successfully conjugated with CIS NPs.

4 Conclusion

This study offers a straightforward method for the conjugation of the drug FU to GSH-capped CIS NPs. Also, the drug FU is hydrophilic in nature; therefore, it can be easily conjugated with the CIS NPs. The results of PL spectra show quenching in emission intensity of CIS NPs on increasing the concentration of FU. Absorption of drug FU on CIS NPs is confirmed by UV–Vis spectra and FTIR spectra shows the conjugation of FU with CIS NPs via hydrogen bonding. Hence, the drug FU is successfully conjugated on CIS NPs and can be further used for biomedical application.

Acknowledgements Ms. Swati Sharma is thankful to the CSIR for providing a scholarship to carry out this work.

Declaration of Interest Statement The authors declare that they have no conflict of interests

References

1. Jamkhande PG, Ghule NW, Bamer AH, Kalaskar MG (2019) Metal nanoparticles synthesis: an overview on methods of preparation, advantages and disadvantages, and applications. J Drug Delivery Sci Technol 53:101174
2. Jain S, Bharti S, Bhullar GK, Tripathi SK (2022) Synthesis, characterization and stability study of aqueous MPA capped $CuInS_2$/ZnS core/shell nanoparticles. J Lumin 252:119279
3. Jain S, Bharti S, Bhullar GK, Tripathi SK (2020) I-III-VI core/shell QDs: synthesis, characterizations and applications. J Lumin 219:116912
4. Longley DB, Harkin DP, Johnston PG (2003) 5-fluorouracil: mechanisms of action and clinical strategies. Nat Rev Cancer 3(5):330–338
5. Pasban S, Raissi H, Pakdel M, Farzad F (2019) Enhance the efficiency of 5-fluorouracil targeted delivery by using a prodrug approach as a novel strategy for prolonged circulation time and improved permeation. Int J Pharm 568:118491
6. Mangaiyarkarasi R, Chinnathambi S, Karthikeyan S, Aruna P, Ganesan S (2016) Paclitaxel conjugated Fe_3O_4@LaF_3: Ce^{3+}, Tb^{3+} nanoparticles as bifunctional targeting carriers for Cancer theranostics application. J Magn Magn Mater 399:207–215
7. Dai Y, Yang D, Kang X, Zhang X, Li C, Hou Z, Cheng Z, Lin J (2012) Doxorubicin conjugated $NaYF_4$: Yb^{3+}/Tm^{3+} nanoparticles for therapy and sensing of drug delivery by luminescence resonance energy transfer. Biomaterials 33(33):8704–8713
8. Duman FD, Akkoc Y, Demirci G, Bavili N, Kiraz A, Gozuacik D, Acar HY (2019) Bypassing pro-survival and resistance mechanisms of autophagy in EGFR-positive lung cancer cells by targeted delivery of 5FU using theranostic Ag_2S quantum dots. J Mater Chem B 7(46):7363–7376
9. Wang Y, Wang X, Deng F, Zheng N, Liang Y, Zhang H, He B, Dai W, Wang X, Zhang Q (2018) The effect of linkers on the self-assembling and anti-tumor efficacy of disulfide-linked doxorubicin drug-drug conjugate nanoparticles. J Control Rel 279:136–146
10. Jain S, Bharti S, Kaur G, Tripathi SK (2020) Synthesis and characterization of copper indium sulphide ternary chalcopyrite nanoparticles. In: AIP conference proceedings, vol 2220, no. 1. AIP Publishing LLC, p 020105
11. Salem DS, Sliem MA, El-Sesy M, Shouman SA, Badr Y (2018) Improved chemo-photothermal therapy of hepatocellular carcinoma using chitosan-coated gold nanoparticles. J Photochem Photobiol B 182:92–99
12. Bharti S, Jain S, Kaur G, Gupta S, Tripathi SK (2018) pH dependent conjugation of Ibuprofen to PEGylated nanoparticles. In: AIP conference proceedings, vol 1942, no 1. AIP Publishing LLC, p 050092
13. Zhang F, Ma P, Deng X, Sun Y, Wang X, Song D (2018) Enzymatic determination of uric acid using water-soluble CuInS/ZnS quantum dots as a fluorescent probe. Microchim Acta 185:1–8
14. Arocikia Jency D, Parimaladevi R, Vasant Sathe G, Umadevi M (2018) Glutathione functionalized gold nanoparticles as efficient surface enhanced Raman scattering substrate for poly chlorinated biphenyl detection. J Cluster Sci 29(2):281–287
15. Aydin RST, Pulat M (2012) 5-Fluorouracil encapsulated chitosan nanoparticles for pH-stimulated drug delivery: evaluation of controlled release kinetics. J Nanomater 2012:42–42
16. Olukman M, Sanli O, Solak EK (2012) Release of anticancer drug 5 Fluorouracil from different ionically crosslinked alginate beads. J Biomater Nanobiotechnol 2012(3):469–479

Facile Synthesis of Polymer Dot and Its Antibacterial Action Against *Staphylococcus aureus*

Aleena Ann Mathew, Neethu Joseph, Elcey C. Daniel, and Manoj Balachandran

Abstract Antimicrobial resistance (AMR) rising from nosocomial infections is an escalating threat to human life nowadays due to the overuse of drugs. The multidrug-resistant pathogenic bacteria have increased morbidity and mortality rates, becoming a crucial global clinical challenge. Gram-positive *Staphylococcus aureus* bacteria is one of the nosocomial pathogens that cause severe invasive diseases and skin infections to human health worldwide. Herein, a non-conjugated polymer dot (NCPD) was synthesized from less toxic and biocompatible polyvinyl alcohol (PVA) via hydrothermal treatment. The fluorescence of the polymer dots was enhanced by nitrogen doping. The as-synthesized nitrogen-doped polymer dots (PDs) exhibit excitation-dependent luminescence emission and show green color fluorescence under UV light. The average size of the synthesized functionalized non-conjugated polymer dot is obtained as 4.08 nm, and they exhibit an amorphous structure. No antibacterial property was observed for bulk polymer, but the doped polymer dots showed antibacterial activity against Gram-positive *Staphylococcus aureus* bacteria.

Keywords Polymer dot · Fluorescence · Nitrogen doping · Antibacterial

1 Introduction

Antimicrobial resistance (AMR) is a global epidemic that threatens the effectiveness of antibiotics and, as a result, the effectiveness of the medication provided. The recent developments in nanotechnology have given rise to a novel method for enhancing the effectiveness of antimicrobial therapies. Nanostructured antimicrobial polymers have fascinated more attention over the last few years due to their remarkable advantages, such as biocompatibility, surface permeation, low toxicity, biodegradability,

A. A. Mathew (✉) · N. Joseph · M. Balachandran
CHRIST (Deemed to be University), Bengaluru 560029, India
e-mail: aleena.mathew@res.christuniversity.in

E. C. Daniel
Kristu Jayanti College (Autonomous), Bengaluru 560077, India

Z. H. Khan et al. (eds.), *Recent Advances in Nanomaterials*, Springer Proceedings in Materials 27, https://doi.org/10.1007/978-981-99-4878-9_44

and synergistic therapy. Therefore, nanostructured polymer systems are emerging as an ideal approach for preventing AMR and establishing effectual therapeutic alternatives. Polymer dots (PDs) are polymer nanoparticles with a size of less than 10 nm that own a carbon core surrounded by polymer chains with a cross-linked structure. The modification by nitrogen-containing carbon is renowned for promoting the creation of reactive oxygen species (ROS), disrupting the cell's physiology, gradually leading to cell death [1, 2]

In this work, water-soluble and optically transparent non-conjugated polyvinyl alcohol (PVA) is used as the precursor to synthesize PDs by one-pot hydrothermal treatment. The PDs are modified by adding nitrogen via doping and studied morphological, structural, and optical properties compared with their pure form. The as-synthesized PDs are more semiconducting, amorphous, and luminescent, while the undoped PVA did not possess any fluorescence. Nitrogen-doped PDs show antibacterial action against Gram-positive *Staphylococcus aureus* bacteria, but the PVA did not show any inhibition zone. Compared to the polymer form, higher rates of antibacterial activity are observed for doped PDs. Therefore, the nitrogen-doped PVA dots can offer a better alternative for antibiotics as a fluorescent tag.

2 Materials and Methods

About 0.6 g of PVA (granular) from Sigma Aldrich was mixed with 40 μL deionized water and stirred (400 rpm). Nitric acid by Sigma Aldrich (20, 40, 60, and 80 μL) was added to the mixture and stirred in the solution for one hour (300 rpm). Then, pour the sample into a Teflon-lined autoclave and keep it for hydrothermal treatment at 200 °C for six hours. The brown-colored obtained residue was dried at 60 °C and powdered.

3 Results and Discussion

Spherical-shaped PDs are observed at an average diameter of 4.08 nm, and the particle size distribution of PDs is represented in Fig. 1a. The XRD peaks at 20.6° and 23.8° are associated with the semi-crystalline nature of PVA (Fig. 1b). Intermolecular and intramolecular hydrogen bonds in the polymer are responsible for the semi-crystalline nature of PVA. The interlayer spacing (d) of the most intense peak $2\theta = 20.6°$ is calculated as 0.43 nm. Doped PVA dots have a strong and broad peak at 18.8° in XRD, signifying the formation of polymer dot with a carbon core. The widening of the peak indicates the primary structural change, which is the formation of amorphous nanocarbon [3]. The semi-crystalline structure of PVA has transformed into an amorphous-structured PVA dot. The lattice spacing of the doped PVA dot was calculated for $2\theta = 18.8°$ which is 0.47 nm. No substantial change was observed in the FTIR spectrum of nitrogen-doped PDs by varying its doping concentrations (Fig. 1c).

The sharp transmission FTIR peak observed at 3386 cm^{-1} signifies the characteristic O–H stretching and confirms the occurrence of water. The C–H stretching, C–H bending, and C–H deformation bonds correspond at 2914, 1422, and 1327 cm^{-1} confirming the presence of carbon core in the nanoparticle formation. The functional group C=O stretching is assigned to the peak at 1692 cm^{-1}. The existence of nitrogen has confirmed the peak at 1065 cm^{-1}, which is attributed to the N–H group [4].

Pure PVA is entirely transparent in the visual spectrum and has one absorption peak which was around 257 nm. The peak can be ascribed to the π–π^* transition of the carbonyl group in the PVA material. The absorption peak of PVA dots shows a bathochromic shift toward the higher wavelength region, demonstrating the development of dots. The absorption of nitrogen-doped PVA dots has increased, and a bathochromic shift was also observed with increasing doping concentrations. The presence of graphitic nitrogen affects the electronic energy levels of the system and generates a significant red shift in their absorption spectra [5, 6]. The PVA has not shown any color under visible light and UV light. However, the doped PDs have shown brown color under visible light and green color fluorescence under UV light (Fig. 1d inset). The PVA shows the optical bandgap energy as 3.22 eV and the direct bandgap of PDs is found to be lower than that of PVA (Fig. 1e). It is observed that the bandgap energies of doped PDs are decreased with increasing doping concentration. The reduction of the optical energy bandgap can be explained by the formation of new molecular dipoles due to point defects generated within the bandgap. Furthermore, the shift in the optical energy bandgap reveals the presence of local cross-linking inside the polymer's amorphous phase, which increases the degree of ordering in these areas. The creation of energy states arising from nitrogen dopant consequently

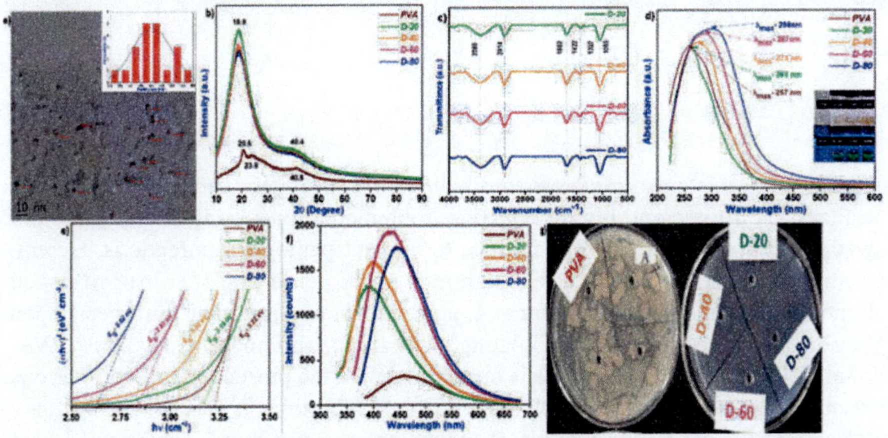

Fig. 1 **a** HR-TEM image of synthesized PDs (inset: particle size distribution of PDs), **b** XRD image of PVA and doped PDs, **c** FTIR spectra of nitrogen-doped PDs, **d** UV–visible absorption spectrum of PVA and the PDs (inset: PVA and PDs under visible light and UV lamp), **e** Direct optical bandgap energy of PVA and PVA dots, **f** Maximum excitation spectra of PVA and PVA dots, **g** Antimicrobial activity of samples on *Staphylococcus aureus*

shifts the conduction band edges and the bandgap decreases with increasing nitrogen doping in PDs [7, 8]. Hence, the addition of nitrogen doping makes the polymer into a more semiconducting material. In Fig. 1f, the maximum excitation wavelengths of each sample are plotted together for a comparative study and the PL emission spectra are plotted from 300 to 700 nm. The luminescence emission has widely distributed and exhibits excitation-dependent emission that indicates the presence of polymorphism in the synthesized PDs and the existence of multiple emission sites within the PDs. It is clear that the luminescence intensity of the PVA samples is increased with increasing doping concentration. The nitrogen content in the PDs augmented with increasing dopant concentration. Therefore, the enhancement of PL intensity may be ascribed to the oxygen-rich groups and the alteration of electronic and chemical features by the addition of nitrogen-doping concentrations [9].

The antimicrobial activity of the PVA and nitrogen-doped PVA dots against Gram-positive bacteria *Staphylococcus aureus* was tested by the agar well diffusion method (Fig. 1g). For the purpose of ensuring reproducibility, each test was done three times. Cultures were disseminated on sterile Mueller Hinton agar medium in sterile Petri plates to evaluate the antibacterial action. With PVA and D-20, *Staphylococcus aureus* did not show a zone of inhibition. While with D-40, D-60, and D-80, the *Staphylococcus aureus* shows a zone of inhibition of 12 mm, 14 mm, and 15 mm, respectively. These results also indicate that nitrogen-doped PDs performed as an improved antibacterial agent against *Staphylococcus aureus*. In addition, the antibacterial activity has increased with increasing nitrogen doping. Therefore, it can be concluded that the highly nitrogen-doped PVA dots possess an apparent deleterious effect on the tested organisms. The material's surface charge, size, and composition are the main key factors of the antimicrobial activity in polymer nanoparticles [10, 11].

4 Conclusion

Antimicrobial resistance from nosocomial infections is a rising threat to human lives nowadays. Non-toxic and highly effective implementations are incredibly relevant nowadays as a substitution of antibiotics to prevent pathogenic infections. Herein, an economical and facile synthesis method is proposed to prepare nitrogen-doped polymer nanoparticles. Semiconducting, amorphous, and luminous nitrogen-doped PVA dots were produced from insulating, crystalline, and non-emissive bulk PVA. Excitation-dependent fluorescence is observed due to the presence of PDs of various sizes at an average diameter of 4.08 nm. The green luminescent PDs possess antibacterial action against Gram-positive *Staphylococcus aureus* and can be used as a fluorescent antibacterial agent.

References

1. Travlou NA, Giannakoudakis DA, Algarra M, Labella AM, Rodríguez-Castellón E, Bandosz TJ (2018) S-and N-doped carbon quantum dots: surface chemistry dependent antibacterial activity. Carbon NY 135:104–111
2. Varghese M, Balachandran M (2021) Antibacterial efficiency of carbon dots against Gram-positive and Gram-negative bacteria: a review. J Environ Chem Eng 9(6):106821
3. Singh CP, Shukla PK, Agrawal SL (2020) Ion transport studies in PVA: NH_4CH_3COO gel polymer electrolytes. High Perform Polym 32(2):208–219
4. Sirotkin N et al (2022) Synthesis of chitosan/PVA/metal oxide nanocomposite using underwater discharge plasma: characterization and antibacterial activities. Polym Bull 1–20
5. Sarkar S et al (2016) Graphitic nitrogen doping in carbon dots causes red-shifted absorption. J Phys Chem C 120(2):1303–1308
6. Joseph N (2022) Green synthesized fluorescent nano-carbon derived from *Indigofera tinctora* (L.) leaf extract for sensing of Pb^{2+} Ions. ECS Trans 107(1):15255
7. Witjaksono G et al (2021) Effect of nitrogen doping on the optical bandgap and electrical conductivity of nitrogen-doped reduced graphene oxide. Molecules 26(21):6424
8. Saji M, Elsa Saji B, Joseph N, Mathew AA, Daniel EC, Balachandran M (2022) Investigation of fluorescence enhancement and antibacterial properties of nitrogen-doped carbonized polymer nanomaterials (N-CPNs). Int J Polym Anal Charact 1–13
9. Santiago SRM, Wong YA, Lin T-N, Chang C-H, Yuan C-T, Shen J-L (2017) Effect of nitrogen doping on the photoluminescence intensity of graphene quantum dots. Opt Lett 42(18):3642–3645
10. Mathew AA, Antony M, Thomas R, Sarojini S, Balachandran M (2022) Fluorescent PVDF dots: from synthesis to biocidal activity. Polym Bull 1–18
11. Joseph N, Mathew AA, Daniel EC, Balachandran M (2023) Polymer-carbon nanocomposite: synthesis, optical and biocidal properties. Results Chem 5:100826

Simple One-POT Hydrothermal Synthesis of CTAB-Assisted Spinel Manganese Ferrite Nanoparticles for Dye Removal: Kinetic and Isotherm Studies

Zafar Iqbal, Mohd Saquib Tanweer, and Masood Alam

Abstract Cetyltrimethylammonium bromide (CTAB)-assisted spinel manganese ferrite ($MnFe_2O_4$) (CTAB/MnF) nanoparticles have been developed via a hydrothermal route and applied against the removal of cationic dye. CTAB/MnF was studied by several characterization tools like X-ray diffractometer (XRD), field emission-scanning electron microscopy (FESEM), Fourier transformed infrared spectroscopy (FTIR), and ultraviolet–visible (UV–Vis) spectroscopy. Morphology of CTAB/MnF was found to be spherical with 85 nm particle size as an average size. Cationic dye such as crystal violet (CV) and methylene blue (MB) was used for selective dye removal onto CTAB/MnF via adsorption technique. The adsorption capacity of CV was measured by UV–Vis spectroscopy. The influence of parameters like pH, dosage, primary dye concentration, and adsorption time was considered for the optimization of the adsorption capacity of CTAB/MnF. The maximum adsorption capacity for CV by CTAB/MnF was better suited for the Langmuir isotherm model. The value of maximum adsorption capacity (q_{max}) of CTAB/MnF against CV was found 238 mg g^{-1}. The adsorption kinetic for CV dye was found well-tuned with the pseudo second order model. Thus, concludingly, modified $MnFe_2O_4$ nanoparticles (CTAB/MnF) have been proven an excellent material for the abatement of crystal violet.

Keywords CTAB · Adsorption · Hydrothermal · Manganese ferrite · Crystal violet

Z. Iqbal (✉) · M. S. Tanweer · M. Alam (✉)
Department of Applied Sciences & Humanities, Faculty of Engineering & Technology, Environmental Science Research Lab, Jamia Millia Islamia, New Delhi 110025, India
e-mail: zafariqbal.amu88@gmail.com

M. Alam
e-mail: malam@jmi.ac.in

Z. H. Khan et al. (eds.), *Recent Advances in Nanomaterials*, Springer Proceedings in Materials 27, https://doi.org/10.1007/978-981-99-4878-9_45

1 Introduction

Water is considered as a most important aspects of life on the earth. Rapid industrialization and urbanization have diminished water quality to the next level. Around 10,000 different kinds of dyes were produced annually [1]. Dye disposal in the environment is an important concern as it creates pollution of surface and ground water [2]. The previous studies show that cationic dyes are more dangerous and toxic than anionic dyes as these can easily interacts with cytoplasm [3]. Crystal violet is an industrial cationic dye of synthetic origin. It is one among the triaryl methane family. It is one among many dyes which severely used in textile and paper industries. Excess intake of crystal violet dye via respiration causes infuriation in the respiratory tract, vomiting, diarrhea, and head pain [4]. Adsorption is nowadays cost-effective popular process to remove wastewater pollutants. It is though moderated by some parameters like temperature, pH, dose, and concentration of adsorbate. Adsorption is thus contemplated to be one of the effective methods for their reusability, efficiency, operational suitability, and comparatively low cost [5]. A number of adsorbents including polymeric, silica-based, activated carbon, rGO-based, and metal oxides were used for cleaning of wastewater. Metal oxides specially chemically modified have good sorption capacity [6, 7]. For the modification of metal oxides, surfactant, polymers, carbonaceous material, etc., may be used. To avoid secondary pollution and separation inconvenience, metal ferrites with surfactant modification is a good option. Use of cationic surfactant CTAB acts as a capping agent and reduces the size of metal oxides [8]. Metal ferrites do not only contain good sorption capacity, but it has a good separability also. The strong concept of the present work is to fabricate a ferrite-based nanomaterials that is capable in removing organic wastewater pollutants even in the low concentrations. Herein, we synthesized CTAB assisted $MnFe_2O_4$ nanoparticles for the better adsorption result for the abatement of crystal violet.

2 Materials and Method

For the synthesis of manganese ferrite nanoparticles, manganese nitrate hexahydrate, Iron (III) nonahydrate, cetyltrimethylammonium bromide, hydrazine hydrate, and sodium carbonate were purchased from CDH chemicals. Ethyl alcohol is purchased from EMSURE. DDW water is used. All the reagents used in the synthesis are of AR grade and are used without further purification. CTAB/MnF nanoparticles were fabricated using the hydrothermal method. Typically, 3.1 g of $Mn\ (NO_3)_2.6H_2O$ and 10 g $Fe\ (NO_3)_3.9H_2O$ were dissolved in 110 mL of H_2O and stirred. 3.35 g of CTAB was separately dissolved in alcohol and finally poured into the metallic solution. 0.39 mL of hydrazine hydrate is also added to the final solution. pH is maintained using a 2 M Na_2CO_3 solution. A thick slurry containing metal hydroxide precipitate is transferred to the autoclave (Teflon lined) for 24 h under 180 °C. On cooling, the filtrate was washed with DDW until the pH is neutral. The collected material is then dehydrated in an oven at 100 °C. The obtained material finally calcined at 550 °C.

2.1 Adsorption Experiments

In the entire set of experiment, a 1000 mg L^{-1} of stock solution was made for all dyes. Adsorption experiments were done by batch method by taking 20 mL solution in several 100 mL Erlenmeyer flasks. The solution pH was maintained by taking 0.1 N solution of NaOH/HCl. Concentrations were taken in between 5 and 40 mg L^{-1}. The concentration in supernatant was measured using ultraviolet spectrophotometer in between 200 and 650 nm wavelength at λ_{max} of 590 nm.

The % removal of CV dye was calculated using Eq. 1. Adsorption capacity was computed from Eq. 2.

$$\%R = \frac{C_o - C_f}{C_o} \times 100 \tag{1}$$

$$q_e = \frac{(C_o - C_e)}{m} \times V \tag{2}$$

Here, C_f and C_o are the final and initial concentration, respectively. C_e and q_e represent equilibrium concentrations and adsorption capacity in mg L^{-1} and mg g^{-1}, respectively. '*V*' represents volume of dye solutions taken in liter, and m (in gram) is the adsorbent weight taken for adsorption experiments.

3 Results and Discussion

3.1 Characterization of CTAB/MnF

To know the crystallinity of the sample, diffraction is done with X-ray. Powder XRD pattern was obtained at 25 °C from Rigaku TTRAX-III coupled with copper source (Cu K_α; $\lambda = 1.5406$ Å). Figure 1a shows all the characteristic peaks of spinel phase of manganese ferrite in CTAB/MnF nanoparticle. All the peaks correspond to the FCC crystal structure of manganese ferrite nanoparticle. The average crystal size was calculated using Debye Scherrer equation [9] and found as 85 nm. The functional group determination for CTAB/MnF and CV loaded CTAB/MnF was done using FTIR spectroscopy. All the distinguished peak of metallic bond as well as hydroxyl linkage has been shown in Fig. 1b. The characteristic peaks for CV dye adsorption were found at 1587 and 1174 cm^{-1} which were ascribed for C=C and C–H stretching for aromatic ring. This shows the successful adsorption of CV dye onto CTAB/MnF

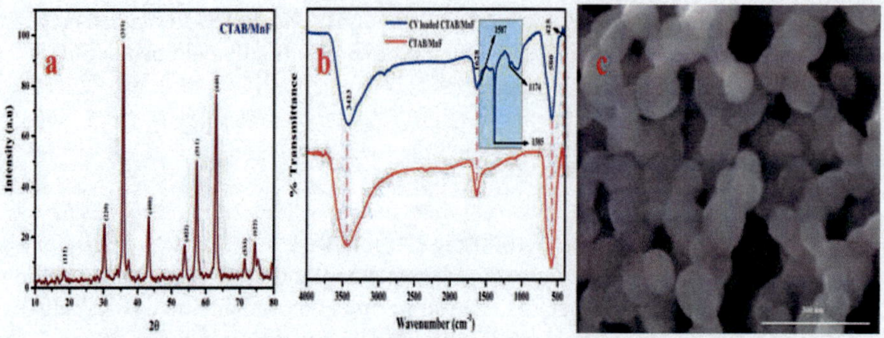

Fig.1 **a** XRD pattern, **b** FTIR graph and **c** SEM micrograph of CTAB/MnF

nanoparticle. The organizational and morphological survey of manganese ferrite was studied using the Scanning Electron Microscopy, Ziess Gemini instrument at an accelerating voltage of 15 kV. CTAB/MnF nanoparticle shows the perfectly spherical shape as seen from Fig. 1c. The homogeneity of the nanoparticle shows that they are free from any type of agglomeration.

3.2 *Effect of Parameters*

3.2.1 Time Effects and Kinetics of Adsorption

By keeping all other parameters constant, the influence of contact time was observed in between 10 and 70 min. Figure 2a shows a gradual increase in percentage removal until the equilibrium is achieved at 50 min, then after it becomes constant. The vacant surface-active sites of CTAB/MnF are responsible for increasing adsorption, and it decreases after some time with depletion of active sites. Kinetics of a reaction are usually explicated using pseudo first and pseudo second order (PFO and PSO) models. The value of q_e and q_t (Table 1) shows the adsorption of CV dye onto CTAB/MnF nanoparticles at equilibrium point and at time *t*, subsequently. The adsorption rate constants for PFO and PSO are k_1 and $k_{2,}$ respectively. The value of coefficient of regression (r^2) is 0.995 for PSO model (Fig. 2b), which is closer to 1 as compared to PFO in the adsorption of CV dye. Since the data are well matched by PSO model, hence, the rate determining step is chemisorption [10].

3.2.2 Concentration Effects and Isotherm of Adsorption

From the Fig. 2c, it is evident that percentage removal or capacity of adsorption is increasing with increasing CV concentration. This occurs because of rising driving force from the concentration gradient that compel the diffusion rate of crystal violet

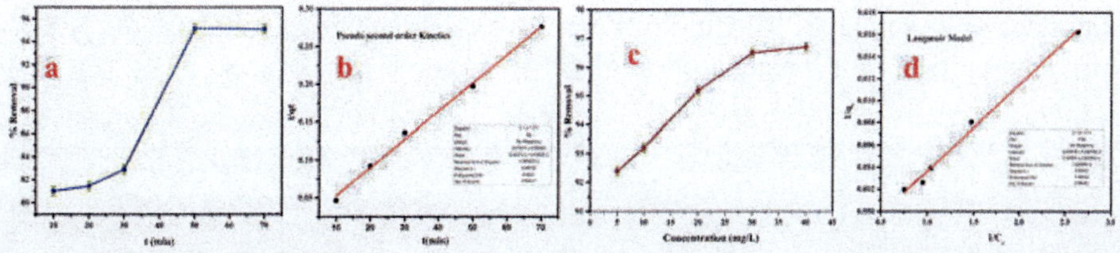

Fig. 2 **a** Influence of contact time, **b** PSO model for CV dye adsorption, **c** Effect of initial CV concentration, **d** Langmuir isotherm model onto CTAB/MnF

Table 1 Parameters for PFO and PSO kinetic model for the removal of CV dye by CTAB/MnF

Adsorbent	PFO				PSO			
	q_e^{exp}	q_e^{cal}	K_1 (min^{-1})	r^2	q_e^{exp}	q_e^{cal}	K_2 (g mg^{-1} min^{-1})	r^2
CTAB/MnF	221.87	253.8	0.1145	0.75	270.27	253.8	0.0009	0.99

Table 2 Isotherm version and its parameters for the adsorption of CV dye onto CTAB/MnF nanoparticles

Adsorbent	Langmuir model				Freundlich model		
	q^{max} (mg g^{-1})	K_l (L mg^{-1})	R_l	r^2	$1/n$	K_f (mg g^{-1}) (L mg^{-1})$^{1/n}$	r^2
CTAB/MnF	238.09	0.000032	0.999	0.99	1.75	11.94	0.96

toward CTAB/MnF [11]. It ultimately changes the concentration at equilibrium of the dye solution that increases the adsorbed dye per unit CTAB/MnF. Isotherm studies could be done using two main models known as Langmuir and Freundlich isotherm models. Langmuir model (Fig. 2d) deduces the monolayer broadcasting of adsorbate molecules on the uniform adsorbent surface that contain equivalent adsorption sites [12]. A multilayer adsorption could usually be surfaced by Freundlich isotherm model (Table 2).

4 Conclusion

From the present work, we can conclude that the use of surfactant in magnetic manganese ferrite nanoparticles not only increases the adsorption capacity but also enhances the morphology of material. X-ray diffraction study reveals the size of crystal calculated from Debye Scherrer equation to be 85 nm. CTAB/MnF was used for the adsorptive removal of CV dye. Kinetics of adsorption reveal that it follows PSO model while isotherm of the adsorption suits better with Langmuir model. The maximum value of adsorption capacity from Langmuir adsorption isotherm was calculated as 238 mg g^{-1}. The findings of this work show the potential use of CTAB/MnF nanoparticles for the adsorption of CV dye.

Acknowledgements Zafar Iqbal is pleased to CSIR HRDG-New Delhi for the research support (SRF-Direct fellowship).

Declaration of Interest
All the authors assert that there are not any financial or personal conflicts of interest.

References

1. Rahmat M, Rehman A, Rahmat S, Bhatti HN, Iqbal M, Khan WS, Bajwa SZ, Rahmat R, Nazir A (2019) Highly efficient removal of crystal violet dye from water by MnO_2 based nanofibrous mesh/photocatalytic process. J Mater Res Technol 8:5149–5159
2. Foroutan R, Mohammadi R, Ramavandi B (2019) Elimination performance of methylene blue, methyl violet, and Nile blue from aqueous media using AC/$CoFe_2O_4$ as a recyclable magnetic composite. Environ Sci Pollut Res 2019 2619. 26:19523–19539
3. Renita AA, Kumar PS, Jabasingh SA (2019) Redemption of acid fuchsin dye from wastewater using de-oiled biomass: kinetics and isotherm analysis. Bioresour Technol Rep 7:100300
4. Tanweer MS, Iqbal Z, Alam M (2022) Experimental insights into mesoporous polyaniline-based nanocomposites for anionic and cationic dye removal. Langmuir 38:8837–8853
5. Puri C, Sumana G (2018) Highly effective adsorption of crystal violet dye from contaminated water using graphene oxide intercalated montmorillonite nanocomposite. Appl Clay Sci 166:102–112
6. Iqbal Z, Tanweer MS, Alam M (2022) Recent advances in adsorptive removal of wastewater pollutants by chemically modified metal oxides: a review. J Water Process Eng 46:102641
7. Iqbal Z, Tanweer MS, Alam M (2023) Reduced graphene oxide-modified spinel cobalt ferrite nanocomposite: synthesis, characterization, and its superior adsorption performance for dyes and heavy metals. ACS Omega
8. Khoshnevisan K, Barkhi M, Zare D, Davoodi D, Tabatabaei M (2012) Preparation and characterization of CTAB-coated Fe_3O_4 nanoparticles. Synth React Inorganic Met Nano-Metal Chem 42:644–648
9. Iqbal Z, Siddiqui VU, Alam M, Siddiqi WA (2020) Synthesis of copper (II) oxide nanoparticles by pulsed sonoelectrochemical method and its characterization. In: AIP conference proceedings. American Institute of Physics Inc., p 020010
10. Ali S, Tanweer MS, Alam M (2020) Kinetic, isothermal, thermodynamic and adsorption studies on Mentha piperita using ICP-OES. Surfaces Interfaces 19:100516
11. Adel M, Ahmed MA, Mohamed AA (2021) A facile and rapid removal of cationic dyes using hierarchically porous reduced graphene oxide decorated with manganese ferrite. FlatChem 26:100233
12. Bir R, Tanweer MS, Singh M, Alam M (2022) Multifunctional ternary NLP/ZnO@ l- cysteine-grafted-PANI bionanocomposites for the selective removal of anionic and cationic dyes from synthetic and real water samples. ACS Omega 7:44850

Facile Synthesis of π Conjugated Heptazine PPy/gC_3N_4 Nanocomposite as an Emissive Layer Material for OLED Applications

Jayanta Bauri, Gobind Mandal, Debashish Nayak, Sanjeev Kumar, Sarfaraz Ansari, and Ram Bilash Choudhary

Abstract Heptazine-based π conjugated gC_3N_4 has represented itself as a novel metal-free compound due to its unusual electrical, thermal, and optical behavior. gC_3N_4 and Puppy/gC_3N_4 (5, 10, and 15 wt.%) nanocomposites were synthesized via thermal polycondensation and in situ polymerization methods, respectively[1]. PPy/gC_3N_4 nanocomposite was scrutinized and optimized using X-ray diffraction (XRD), field emission electron microscopy (FESEM), UV–visible and photoluminescence spectroscopic techniques. Average particle size and crystalline phase were determined using the XRD technique. Surface nature and elemental composition were investigated by FESEM and EDX techniques, respectively. Optical parameters such as optical band gap, refractive index, and optical conductivity were estimated using UV–visible spectroscopic technique. The photoluminescence spectroscopic method determined PL emission peaks, CIE coordinates, and color purity.

Keywords Conducting polymer · Heptazine · Optical band gap · Recombination

1 Introduction

Optoelectronic devices are a promising candidate in the electrically driven world. OLED has acquired much attention in front of researchers due to its commercial applications, such as small or large screen display devices, mobile phones, smartwatch display devices, and street or different lighting purposes in the modern world. OLED is an electroluminescence-based multilayers organic device. Advanced OLED has a combination of seven active layers such as an emissive layer, electron or hole injection, and transport layer and electron or hole blocking layers. In 1987, Tang and Van Slyke were introduced an organic electroluminescence material. Using active layer

J. Bauri · G. Mandal · D. Nayak · S. Kumar · S. Ansari · R. B. Choudhary (✉)
Department of Physics, Nanostructured Composite Materials Laboratory (NCML), Indian Institute of Technology (Indian School of Mines), Dhanbad 826004, India
e-mail: rbcism@gmail.com

Z. H. Khan et al. (eds.), *Recent Advances in Nanomaterials*, Springer Proceedings in Materials 27, https://doi.org/10.1007/978-981-99-4878-9_46

materials in OLED plays a vital role in enhancing the device's efficiency. Recently, conducting polymers have been extensively used as active layers in OLED because of their easy synthesis process, flexibility, low cost, and semiconducting nature [1, 2]. Polyaniline (PANI), polyindole (Pin), PEDOT PSS, polycarbazole (PCz), and polypyrrole (PPy) were successfully reported as a polymer in OLED applications. gC_3N_4 has unique optical and electrical behavior, such as high or broad PL emission and lower band gap energy (2.98 eV). Herein, we have synthesized gC_3N_4, PPy, and their nanocomposite with varying wt % of gC_3N_4 via hydrothermal and in situ polymerization methods. After that, we have gone through different cauterization techniques to check optical and electrical properties and conclude its OLED application.

2 Experimental Part

2.1 Chemicals

All required raw chemicals such as the highly oxidized ammonium persulfate $(NH_4)_2S_2O_4)$ (APS), hydrochloric acid (HCl) melamine ($C_3H_6N_6$), and pyrrole (C_4H_5N) monomer were bought from Merck. Ethanol and distilled water were utilized as a solvent in the synthesis processes. Chemicals were used as obtained from Merck.

2.2 Synthesis of gC_3N_4 Nanosheets

The gC_3N_4 nanosheet was synthesized by the hydrothermal method from the melamine unit. This process involves adding 0.3 M of the nitric acid solution to a mixture of 6 gm melamine with 40 ml of ethanol and constantly stirring for 30 min. The solution was then put into a 100 ml Teflon tube and heated to 80 °C for 12 h in a vacuum oven to undergo hydrothermal treatment. After the reaction was complete, the precipitate solution was centrifuged at 12,000 RPM for 30 min. The remaining product was dried in a 100 °C oven for 6 h. Finally, we got a fine yellow powder that had gone through different characterization techniques to check its physiochemical properties.

2.3 Synthesis of gC_3N_4/PPy Nanocomposite

Figure 1a depicts the schematic synthesis pathways for the PPy/gC_3N_4 nanocomposite. From the pure PPy, gC_3N_4 that was first made, PPy/gC_3N_4 nanocomposite was prepared in two phases utilizing in situ polymerization. Herein, 0.034 gm of previously synthesized gC$_3$N$_4$ was added to 0.67 gm of pyrrole monomer in 50 ml DI water, which was then constantly agitated for 30 min. Then, 0.2 M of the oxidizing agent APS was added dropwise into the solution and stirred with a magnetic stirrer for 4 h, producing a black solution. Then, for the following 24 h, this dark solution was placed in the freezer to finish the polymerization reaction. The residue was then repeatedly cleaned with ethanol and DI water on filter paper. The filter result was vacuum oven-dried overnight at 70 °C before being milled into a fine powder. A similar process was synthesizing the other two, which included 10 and 15% of PPy/gC$_3$N$_4$ binary nanocomposite.

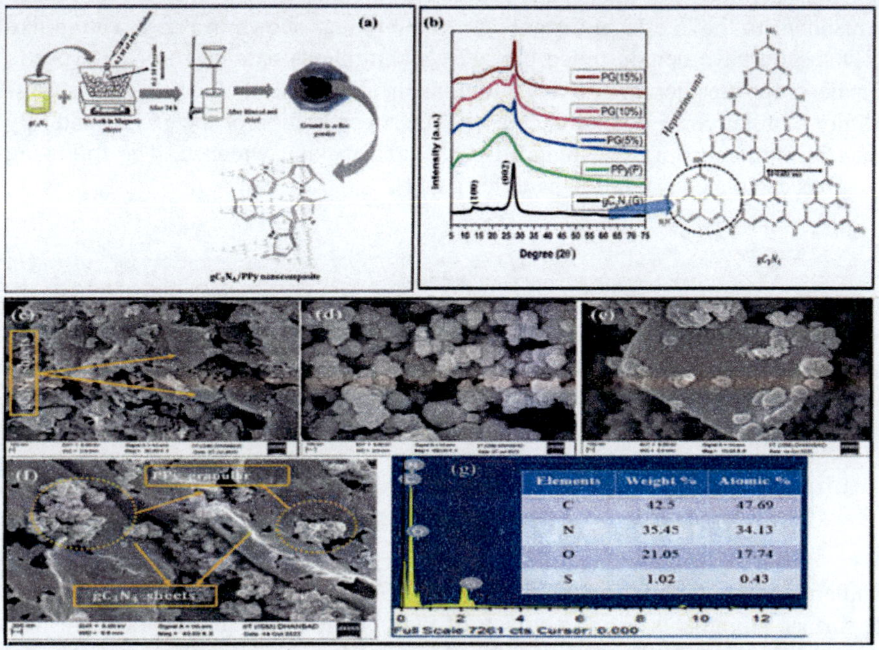

Fig. 1 **a** Schematic synthesis route of PPy/gC$_3$N$_4$ nanocomposite, **b** XRD spectra of pure gC$_3$N$_4$, PPy, and 5, 10, and 15 wt% of gC$_3$N$_4$/PPy nanocomposite. FESEM images of **c** pure gC$_3$N$_4$, **d** pure PPy and **e**, **f** gC$_3$N$_4$/PPy nanocomposite, **g** EDS elemental spectra of gC$_3$N$_4$/ZrO$_2$ nanocomposite

3 Results and Discussion

3.1 XRD Analysis

Crystallinity and average particle size were examined using the high-resolution X-ray diffractometer (Rigaku Smart Lab) through 1.541 CuK_α radiations. Figure 1b displays the XRD spectra of polypyrrole, graphitic carbon nitride (gC_3N_4), and their nanocomposites. When diffracted from the crystal planes *d* (100) and *d* (002), pure gC_3N_4 displayed two distinct, strong peaks at $2 = 13.07°$ and $2 = 27.06°$ [3]. gC_3N_4 is crystalline because of these strong peaks. The (100) plane corresponds to the regular arrangements of the heptazine unit at 0.681 nm, and the intense (002) plane corresponds to the interlayer distance $d = 0.33$ nm, as shown in Fig. 2. Hump-like XRD spectra have demonstrated pure PPy's amorphous nature. The XRD spectra of nanocomposites don't show any additional peaks, but as they grow, gC_3N_4 crystallinity by filler concentration increases. The superposition of the gC_3N_4 and PPy peaks in the nanocomposite reduced peak intensity and widened. The following equations calculated crystallite particle size [4],

$$D = \frac{k\lambda}{\beta \cos\theta} \tag{1}$$

Here, *D*-average crystallite size, *k*-Scherrer constant (= 0.9), β-full width and half maxima (FWHM), λ wavelength of the CuK_α line (= 0.154 Å). The average crystallite size of the gC_3N_4 nanosheets is 11.84 nm.

3.2 FESEM Study

Field emission scanning electron microscopy (FESEM) was performed to examine the surface alignments or morphology of the obtained pure and nanocomposites. Figure 1c–f shows FESEM images of the binary nanocomposite of pure gC_3N_4 and PPy. The disorganized configurations of gC_3N_4 nanosheets were confirmed by the FESEM pictures [5, 6]. By stacking 20 nm of an exfoliation unit of pure gC_3N_4, bulk nanosheets of 224 nm were formed. The surface morphology of PPy was shown by the FESEM image [7]. The FESEM picture of the binary gC3N4/PPy/ nanocomposite clearly demonstrated the presence of two phases. On the surface of the gC_3N_4 nanosheets, PPy was agglomerated. EDS spectra verified the elemental presence in the composite. Carbon, nitrogen, oxygen, and their atomic percentages were present in the gC_3N_4/PPy binary nanocomposite. The EDS spectrum is shown in Fig. 1g.

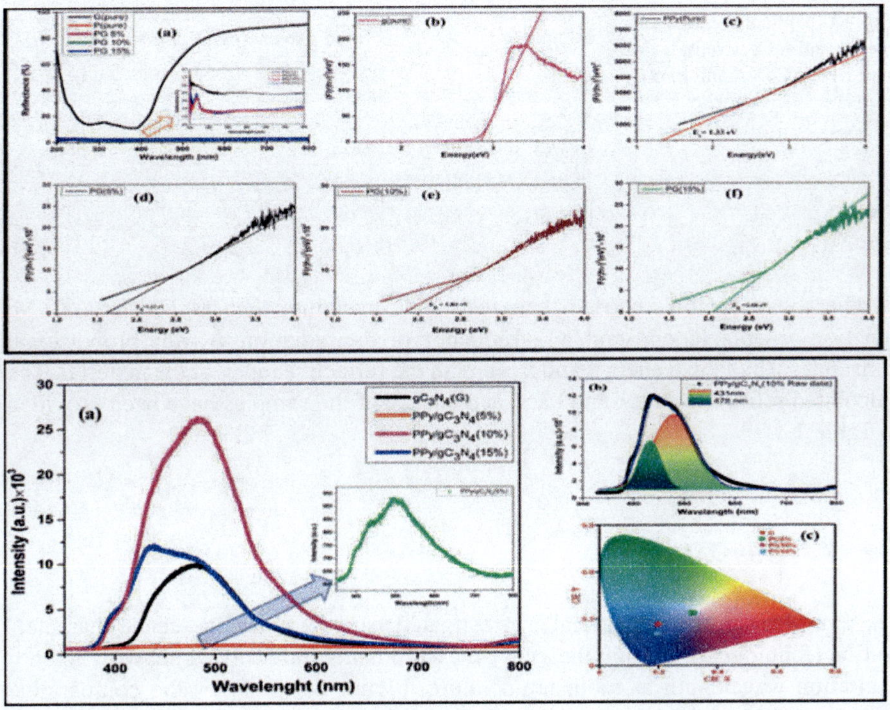

Fig. 2 **a** UV–visible spectra of pure and PPy/gC$_3$N$_4$ nanocomposites Tauc plots of **b** pure gC$_3$N$_4$, **c** pure PPy, **d–f** 5, 10, and 15 wt% of PPy/gC$_3$N$_4$ nanocomposites. PL spectra of **g** pure gC$_3$N$_4$ and nanocomposites, **h** deconvoluted PL spectra of pure gC$_3$N$_4$, **i** CIE diagram of pure and nanocomposite

3.3 *UV–Visible Analysis*

Figure 2a displays the diffusion reflectance spectra (DRS) of the pure gC$_3$N$_4$, PPy, and gC$_3$N$_4$/PPy binary nanocomposite as they were obtained. Pure PPy's stronger absorption characteristics in the visible range are consistent with its lower diffusion reflectance. Using Kubelka Munk's (K. M.) equations, the optical band gap energy of pure and ternary nanocomposites has been calculated. Tauc plots of the pure and nanocomposite have been shown in Fig. 2b–f. Equation (2) is well known as a Kubelka Munk equation, and the refractive index was determined using the Eqs. (2) and (3) [8, 9],

$$(F(r)h\nu)^{\frac{1}{\gamma}} = A\left(h\nu - E_g\right) \tag{2}$$

$$\frac{n^2 - 1}{n^2 + 1} = 1 - \surd\left(\frac{E_g}{20}\right)^{\circ} \tag{3}$$

Table 1 Optical bandgap energy and refractive index of pure PPY, gC_3N_4, and PPY/ gC_3N_4 nanocomposite

Sample	Band gap energy (E_g) (eV)	R. I. (n)
gC_3N_4	2.96	2.06
PPy	1.33	2.58
PPy/gC_3N_4 (PG5%)	1.62	2.38
PPy/gC_3N_4 (PG10%)	1.84	2.33
PPy/gC_3N_4 (PG20%)	2.06	2.24

The above symbols carried their respective meaning, such as $F(r)$ is a K. M. function, h—planks constant, ν—frequency of the radiation, A—the proportional constant, E_g band gap energy, and n refers to the refractive index of the material. The calculated refractive index and band gap energy of the samples have been provided in Table 1.

3.4 PL Analysis

The emissive nature of a material is investigated using photoluminescence characterization techniques. Photoluminescence measurements were carried out with 325 nm excitation wavelength at room temperature. Figure 2g, h shows the comparative emission spectra of the as-synthesized pure and nanocomposite. PL spectra reveal that emission intensity reached its highest for PG10% nanocomposite; after that, it decreased. Deconvoluted PL spectra of gC_3N_4 are exposed that it combines two peaks at 431 and 478 nm due to $n^* \to$ LP state and $n \to \sigma$ transition [10].

CIE Diagram CIE diagram is a 2D mapping system out of the three (red, green, and blue) fundamental colors of the light emitted from materials. GoCIE software drew the 2D CIE diagram from the PL data. CIE diagram of pure gC_3N_4 and PPy/gC_3N_4 nanocomposites has been seen in Fig. 2i. The color coordinate of the pure gC_3N_4 materials located in the green region and, in the case of nanocomposite location, shifted toward the blue to the white region in the 2D CIE diagram. Color purity and CCT of the materials were determined using the given equation [11],

$$\text{C.P.} = \frac{\sqrt{(a_x - a_i)^2 + (b_x - b_i)^2}}{\sqrt{(a_y - a_i)^2 + (b_y - b_i)^2}} \quad (4)$$

$$\text{CCT} = -437n^3 + 3601n^2 - 6861n + 5514.31 \quad (5)$$

where $n = \frac{a_x - a_e}{b_x - b_e}$, (a_x, b_x) is the CIE coordinate of the respective sample, (a_i, b_i) is the CIE coordinate of the white emission, (a_y, b_y) is the CIE coordinate of the intense PL emission, and CCT refers to the correlate color temperature, (a_e, b_e) epicenter of CIE coordinate. CCT value of the materials gives the knowledge about the cool and hot light emission. If the CCT value of 30,000–40,000 K signifies the cool emission, and above 40,000 K is called hot emission. The CCT value and color purity of the optimized sample PG10% were 242,813.02 K and 55%, respectively.

4 Conclusion

Heptazine-based metal-free gC_3N_4 nanosheet is synthesized using hydrothermal. Granular shape polypyrrole-based nanocomposites were successfully prepared in the lab via in situ polymerization method. The optimized PPy/gC_3N_4 (PG10%) nanocomposite reveals the highest optical properties compared to the other nanocomposites. Obtained structural and optical properties of the optimized PG10% nanocomposite are attributed to (I) gC_3N_4 and PPy looking like nanosheets and granular type surface morphology investigated by FESEM images, and elemental compositions are determined from the EDX spectra. (II) Optical band gap energy of 1.84 eV and refractive index of 2.33 of the PG10% nanocomposite were determined by the UV–visible technique. (III) Color purity of 55%, blue emission of optimized materials is clearly visualized from the CIE diagram and from the CCT value of 242,813.02 informed about warm emission. High PL emission intensity signifies the higher electron–hole recombination of the optimized material. Material has a higher recombination rate, optimized band gap (in the visible range), suitable refractive index, and narrow absorption that is suitable emissive layer material for OLED applications.

Acknowledgements The authors sincerely thank the IIT (ISM) Dhanbad for providing a research facility and continuous support in this communication.

Conflict of Interest The authors declare that they have no conflict of interest.

References

1. Hong G, Gan X, Leonhardt C, Zhang Z, Seibert J, Busch JM, Bräse S (2021) A brief history of OLEDs—Emitter development and industry milestones. Adv Mater 33
2. Wang Z, Wang C, Zhang H, Liu Z, Zhao B, Li W (2018) The application of charge transfer host based exciplex and thermally activated delayed fluorescence materials in organic light-emitting diodes. Org Electron 66:227–241
3. Peng L, Li ZW, Zheng RR, Yu H, Dong XT (2019) Preparation and characterization of mesoporous g-C_3N_4/SiO_2 material with enhanced photocatalytic activity. J Mater Res 34(10):1785–1794
4. Mote V, Purushotham Y, Dole B (2012) Williamson-Hall analysis in estimation of lattice strain in nanometer-sized ZnO particles. J Theor Appl Phys 6(1):2–9
5. Hong Y, Jiang Y, Li C, Fan W, Yan X, Yan M, Shi W (2016) In-situ synthesis of direct solid-state Z-scheme V_2O_5/g-C_3N_4 heterojunctions with enhanced visible light efficiency in photocatalytic degradation of pollutants. Appl Catal B Environ 180:663–673

6. Feng Z, Zeng L, Chen Y, Ma Y, Zhao C, Jin R, Lu Y, Wu Y, He Y (2017) In situ preparation of Z-scheme MoO_3/g-C_3N_4 composite with high performance in photocatalytic CO_2 reduction and RhB degradation. J Mater Res 19:3660–3668
7. Choudhary RB, Nayak D (2021) Tailoring the properties of 2-D rGO-PPy-ZnS nanocomposite as emissive layer for OLEDs. Optik 231
8. Bauri J, Choudhary RB (2023) Thermal and electronic states of exfoliated gC_3N_4-based nanocomposite with ZrO_2 nanoparticles as a robust emissive layer. Mater Sci Semicond Process 154:107205
9. Choudhary RB, Kumar S (2022) Optimum chemical states and localized electronic states of SnO_2 integrated PTh–SnO_2 nanocomposites as excelling emissive layer (EML). Opt Mater (Amst) 131:112736
10. Sun S, Fan E, Xu H, Cao W, Shao G, Fan B, Wang H, Zhang R (2019) Enhancement of photocatalytic activity of g-C_3N_4 by hydrochloric acid treatment of melamine. Nanotechnology 30(31)
11. Dey S, Kar AK (2021) Composition and excitation wavelength dependent photoluminescence color tuning in the nanocomposite of PMMA and ZnO nanorods for PLED. J Alloys Compd 879:160450

Structural, Morphological, Spectroscopic, and Magnetic Properties of Mg-Zn Nanoferrite for High-frequency Applications

Sonam Kumari, Neetu Dhanda, Saarthak Kharbanda, Atul Thakur, Satyendra Singh, and Preeti Thakur

Abstract Mg-Zn spinel ferrite of nominal composition $Mg_{0.4}Zn_{0.6}Fe_2O_4$ was synthesized by cost-effective citrate precursor approach. The prepared nanoparticles were characterized by an "X-rays diffractometer, scanning electron microscopy, diffused reflectance spectroscopy, and vibrating sample magnetometer". In the XRD pattern, the most intense reflection was observed from the (311) peak that corresponds to the cubic spinel phase of the produced ferrite. The average size of all crystallites was evaluated to be 38 nm. Uniformity in the size of the particles was detected with some agglomeration. The calculated energy band gap (1.88 eV) reveals the semiconducting nature of the prepared material. The low coercivity value (103 Oe) reveals that the synthesized material is a soft ferrite. The prepared sample of nanoferrite material was found to be suitable enough for high-frequency applications.

Keywords Spinel ferrite · X-rays diffractometer · Vibrating sample magnetometer · Diffused reflectance spectroscopy

1 Introduction

Ferrites are iron-based magnetic materials with highly intriguing magnetic characteristics. Due to their extremely small size and large surface area, ferrite nanoparticles exhibit distinct and improved characteristics over their bulk counterparts, making

S. Kumari · N. Dhanda · P. Thakur (✉)
Department of Physics, Amity School of Applied Sciences, Amity University Haryana, Gurugram, India
e-mail: pthakur@ggn.amity.edu

S. Kharbanda · A. Thakur
Amity Institute of Nanotechnology, Amity University Haryana, Gurugram 122413, India

S. Singh (✉)
Special Centre for Nanoscience, Jawaharlal Nehru University, New Delhi 110067, India
e-mail: satyambd@gmail.com

Z. H. Khan et al. (eds.), *Recent Advances in Nanomaterials*, Springer Proceedings in Materials 27, https://doi.org/10.1007/978-981-99-4878-9_47

them more applicable for a variety of commercial and industrial applications [1]. Regarding other ferrites, spinel ferrites are the most significant and frequently used material for a variety of applications due to their excellent qualities like strong magnetic properties, low eddy current losses, and relatively low conductivity [2]. These ferrites are used in a wide variety of applications, including telecommunications, optoelectronic and electromagnetic interference devices, power transformers, heterogeneous catalysis, and a host of other fields [3]. Mg-Zn spinel ferrite is regarded as a significant ferrite with a soft magnetic nature that is suitable for various electronic and magnetic devices [4, 5]. Mg–Zn ferrites possess a moderate saturation magnetization and minimal losses and are crucial factors for building deflection yokes [6]. The solution microwave-assisted technique was used by Gore [7] to produce $Mg_{1-x}Zn_xFe_2O_4$ (x = 0.0–0.5). With increasing content of Zn, the coercive force decreased and saturation magnetization increased, suggesting the feasibility of ferrite samples in applications such as read/write heads and an electromagnet. The citrate precursor synthesis approach, which produces nanoparticles, is a cost-effective and low-temperature process [8, 9]. Consequently, in this article, Mg–Zn ferrite with the formula $Mg_{0.4}Zn_{0.6}Fe_2O_4$ is made utilizing this method.

2 Experimental Procedure

2.1 Materials and Method

To produce $Mg_{0.4}Zn_{0.6}Fe_2O_4$ ferrite, the citrate precursor method was used. In this method, 250 ml of deionized water was taken, and the stoichiometric amounts of all precursors were mixed first. Next, an adequate quantity of citric acid ($C_6H_8O_7 \cdot H_2O$) was added, which was then kept at 70 °C on a magnetic hot plate. The heat was given to the solution until all of the solvents had evaporated. The dry residue was then ground and sintered for three hours at 800 °C.

3 Result and Discussion

3.1 Structural and Morphological Observations

The produced nanoferrites' XRD pattern is displayed in Fig. 1. The spinel phase of the produced nanoferrite was observed by peaks found at various angles. The ferrites were found to be nanometric in size due to the broad width of the peaks [10]. The highest intense peak (311) was used for all parameters' calculations. The average size of crystallite was calculated to be 38 nm by making use of Eq. (1) [11].

$$D = \frac{0.9\lambda}{\beta \mathrm{Cos}\theta} \quad (1)$$

Lattice constant and interplaner spacing were determined to be 8.357 Å and 2.535 Å, respectively. The sample's X-ray density and dislocation density were found to be 5.122 g/cm^3 and 0.70, respectively.

The Mg-Zn nanoferrite's SEM micrograph is displayed in Fig. 2. The particles are nearly spherical in form and are in the nanometric range. Some agglomeration of particles was also noticed because of the material's magnetic nature [12]. The average size of the particles is calculated to be ~93 nm.

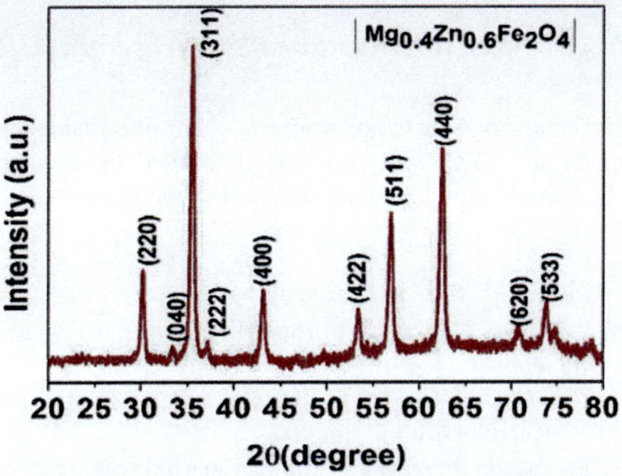

Fig. 1 XRD pattern of $Mg_{0.4}Zn_{0.6}Fe_2O_4$ nanoferrites

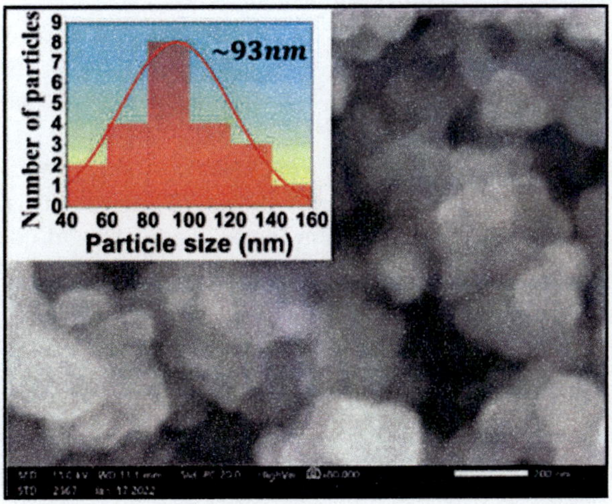

Fig. 2 SEM micrograph with histogram of Mg–Zn nanoferrite

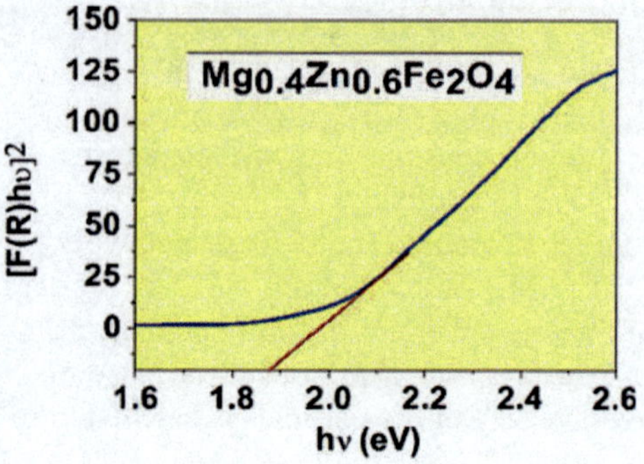

Fig. 3 Kubelka–Munk plot of synthesized Mg–Zn nanoferrite

3.2 *Diffused Reflectance Spectroscopy (DRS)*

The bandgap of the nanoparticles is determined using the UV–Vis "diffuse reflectance spectrum". DRS spectrum was employed to calculate the bandgap of using the "Kubelka–Munk (K-M) function described in Eq. (2) [13]

$$(R)h\nu = A\left(h\nu - E_g\right)^n \tag{2}$$

Here is A → Constant of proportionality, E_g → Energy band gap, n → Transition coefficient"

The value of $n = 1/2$ for direct allowed transition. Figure 3 shows the $[F(R)h\nu]^2$ v/s $h\nu$ Kubelka–Munk plot [14].

The bandgap energy of the nanoparticles is calculated by extrapolating of slop, and it was obtained to be 1.88 eV. The created sample is a semiconducting material, as shown by the measured band gap.

3.3 *Magnetic Properties*

The M-H curves for samples show a "S" shape. This demonstrated the material's soft ferrite nature as seen in Fig. 4. The values of coercivity (H_c) and "saturation magnetization" (σ_s) observed from the M–H graphs are 103 Oe and 34 emu/g, respectively. The magnetic moment (η_B) and the anisotropic constant (K) are determined by employing Eqs. (3) and (4) [15].

$$\eta_{B=\frac{M_w \times \sigma_s}{5585}} \tag{3}$$

where M_w is the nanoferrites' molecular.

$$K = \frac{H_{c \times \sigma_s}}{0.96} \tag{4}$$

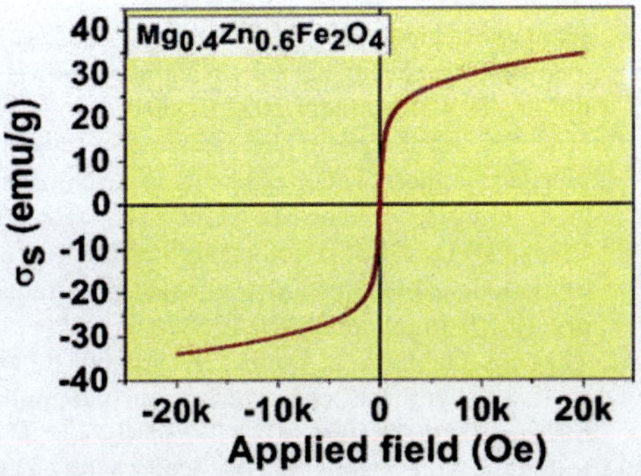

Fig. 4 M–H loop of produced $Mg_{0.4}Zn_{0.6}Fe_2O_4$ ferrites

The calculated value of magnetic moment (η_B) and anisotropic constant (K) is found to be 1.36 μB and 3647 erg/g, respectively. According to all the observed magnetic properties, the produced material is a soft ferrite and can be employed in high-frequency applications [16–18].

4 Conclusion

$Mg_{0.4}Zn_{0.6}Fe_2O_4$ nanoferrite were successfully prepared by citrate precursor method. Average crystallite size and particle size were found to be 38 nm and 93 nm, respectively. Band gap of the prepared ferrite was obtained to be 1.88 eV confirmed the semiconducting behaviour of the synthesized ferrites. The specific saturation magnetization value of 34 emu/g and low value of coercivity (103 Oe) revealed that prepared material is a soft ferrite and suitable for high-frequency applications.

References

1. Punia P et al (2021) Microstructural, optical and magnetic study of Ni–Zn nanoferrites. J Supercond Nov Magn 2131–2140. https://doi.org/10.1007/s10948-021-05967-y
2. Kumari S, Dhanda N, Thakur A, Gupta V, Singh S (2022) Nano Ca–Mg–Zn ferrites as tuneable photocatalyst for UV light-induced degradation of rhodamine B dye and antimicrobial behavior for water purification. Ceram Int. https://doi.org/10.1016/j.ceramint.2022.12.107
3. Sutka A, Mezinskis G (2012) Sol-gel auto-combustion synthesis of spinel-type ferrite nanomaterials. Front Mater Sci 6(2):128–141. https://doi.org/10.1007/s11706-012-0167-3
4. Mohseni H, Shokrollahi H, Sharifi I, Gheisari K (2012) Magnetic and structural studies of the Mn-doped Mg–Zn ferrite nanoparticles synthesized by the glycine nitrate process. J Magn Magn Mater 324(22):3741–3747. https://doi.org/10.1016/j.jmmm.2012.06.009
5. Dhanda N, Thakur P, Thakur A (2022) Materials today: proceedings green synthesis of cobalt ferrite: a study of structural and optical properties. Mater Today Proc. https://doi.org/10.1016/j.matpr.2022.07.202

6. Kumari S, Thakur P, Singh S, Thakur A (2022) Materials today: proceedings a detailed structural analysis, morphological and optical study of Mg–Zn nano ferrite. Mater Today Proc 3–6. https://doi.org/10.1016/j.matpr.2022.07.201
7. Gore SK, Tumberphale UB, Jadhav SS, Kawale RS, Naushad M, Mane RS (2017) Microwave-assisted synthesis and magneto-electrical properties of Mg–Zn ferrimagnetic oxide nanostructures. Phys B Condens Matter 530:177–182. https://doi.org/10.1016/j.physb.2017.11.044
8. Chahar D et al (2021) Photocatalytic activity of cobalt substituted zinc ferrite for the degradation of methylene blue dye under visible light irradiation. J Alloys Compd 851:156878. https://doi.org/10.1016/j.jallcom.2020.156878
9. Thakur P, Gahlawat N, Punia P, Kharbanda S, Ravelo B, Thakur A (2022) Cobalt nanoferrites: a review on synthesis, characterization, and applications, no. 0123456789. Springer US. https://doi.org/10.1007/s10948-022-06334-1
10. Rahaman MD, Dalim Mia M, Khan MNI, Akther Hossain AKM (2016) Study the effect of sintering temperature on structural, microstructural and electromagnetic properties of 10% Ca-doped $Mn_{0.6}Zn_{0.4}Fe_2O_4$. J Magn Magn Mater 404:238–249. https://doi.org/10.1016/j.jmmm.2015.12.029
11. Thakur P et al (2020) Manganese zinc ferrites: a short review on synthesis and characterization. J Supercond Nov Magn 33(6):1569–1584. https://doi.org/10.1007/s10948-020-05489-z
12. Abdel Maksoud MIA, El-Sayyad GS, Abokhadra A, Soliman LI, El-Bahnasawy HH, Ashour AH (2020)Influence of Mg2+ substitution on structural, optical, magnetic, and antimicrobial properties of Mn–Zn ferrite nanoparticles. J Mater Sci Mater Electron 31(3):2598–2616. https://doi.org/10.1007/s10854-019-02799-4
13. Rajivgandhi GN et al (2021) Effect of Ti and Cu doping on the structural, optical, morphological and anti-bacterial properties of nickel ferrite nanoparticles. Results Phys 23:104065. https://doi.org/10.1016/j.rinp.2021.104065
14. Heidari P, Masoudpanah SM (2021) Structural, magnetic and optical properties and photocatalytic activity of magnesium-calcium ferrite powders. J Phys Chem Solids 148:109681. https://doi.org/10.1016/j.jpcs.2020.109681
15. Punia P, Thakur P, Kumar R, Syal R, Dhar R, Thakur A (2022) Synthesis and characterization of Ca substituted Ni–Zn nanoferrites-microstructural, magnetic and dielectric analysis. J Alloys Compd 928:167248. https://doi.org/10.1016/j.jallcom.2022.167248
16. Taneja S, Chahar D, Thakur P, Thakur A (2021) Influence of bismuth doping on structural, electrical and dielectric properties of Ni–Zn nanoferrites. J Alloys Compd 859:157760. https://doi.org/10.1016/j.jallcom.2020.157760
17. Thakur P, Taneja S, Chahar D, Ravelo B, Thakur A (2021) Recent advances on synthesis, characterization and high frequency applications of Ni–Zn ferrite nanoparticles. J Magn Magn Mater 167925. https://doi.org/10.1016/j.jmmm.2021.167925
18. Kumari S, Dhanda N, Thakur A, Singh S, Thakur P (2023) Investigation of calcium substitution on magnetic and dielectric properties of Mg–Zn nano ferrites. Mater Chem Phys 297:127394. https://doi.org/10.1016/j.matchemphys.2023.127394

Dielectric and Ferroelectric Properties of $BaTiO_3$–$CoFe_2O_4$ Composites

Surbhi Sharma and Shakeel Khan

Abstract In this research work, the influence of ferroelectric phase has been examined on the various physical properties of the ferrite phase. Composites with general formula (x) $BaTiO_3-(1-x)$ $CoFe_2O_4$ ($x = 0$, 0.25, and 0.5) were synthesized by mixing of BTO and CFO nanocrystalline powders via ball milling. The XRD analysis reveals that the composite structure consists of both perovskite and cubic spinel phases. Average crystallite size and micro-strain induced within the system have been investigated using Williamson-Hall method. Dielectric measurements divulge that the dielectric constant increases with the incorporation of BTO in the CFO matrix and follows typical dispersion behaviour. The polarization-field measurement manifests increment in the ferroelectric properties with the increasing percentage of BTO phase in the composite. The improved dielectric properties suggest BTO-CFO composite powders as potential candidate for dielectric resonators and dynamic random-access memories (DRAMs).

Keywords Ball milling · Williamson-Hall method · Dielectric measurements · Ferroelectric properties

1 Introduction

Over the last few decades, multiferroics have played a significant role in the advancement of multifunctional materials that display ferromagnetic, ferroelectric, and piezoelastic orders simultaneously. Due to their potential use in various technological applications, including transducers, sensors, ferroelectric photovoltaics, spintronic devices, non-volatile memory elements, and terahertz radiation, scientific groups are paying attention to these multiferroic materials [1]. Ferroelectricity and ferromagnetic arrangement are mutually exclusive; hence, single-phase multiferroic materials

S. Sharma · S. Khan (✉)
Department of Applied Physics, Z.H. College of Engineering & Technology,
Aligarh Muslim University, Aligarh 202002, India
e-mail: skhanapd@gmail.com

Z. H. Khan et al. (eds.), *Recent Advances in Nanomaterials*, Springer Proceedings in Materials 27, https://doi.org/10.1007/978-981-99-4878-9_48

like $BiFeO_3$, $BiMnO_3$, and $YMnO_3$ are extremely rare [2]. Additionally, single-phase multiferroics magnetic response, dielectric constant, and magnetic permeability are constrained by phenomenological theory. Due to this, it is difficult to create single-phase multiferroic compounds with high magnetoelectric coupling for application in multifunctional devices [3]. To overcome the aforesaid drawback, various researchers around the globe are interested in synthesizing multiferroic materials with high extrinsic magnetoelectric effect using two-phase (ferromagnetic–ferroelectric) composites such as $BiFeO_3 + PbTiO_3$, NZ + PZT, and NCZ + BT reported earlier [4]. Getting motivated with above studies, the present investigation deals with the synthesis of pure $BaTiO_3$ and $CoFe_2O_4$ nanocrystalline powders using co-precipitation method followed by the composite powder preparation of (x) $BaTiO_3$–($1-x$) $CoFe_2O_4$ ($x = 0$, 0.25, and 0.5) via ball milling and exploration of the various physical properties of the composite samples as a function of $BaTiO_3$ content. To understand the influence of $BaTiO_3$ (ferroelectric) phases on the structural, dielectric and ferroelectric properties of $CoFe_2O_4$ (ferrite) phases above investigation have been carried out. So, that in the future, these properties can serve as basis to further extend our work on determining the magnetoelectric response for these BTO-CFO-based composites for their potential applications.

2 Materials and Method

2.1 $CoFe_2O_4$ Nanoparticles Preparation

Initially, the salts ($FeCl_3.6H_2O$, $CoCl_2.6H_2O$, and $MnCl_2.4H_2O$) were dissolved in double distilled water and kept at 80 °C for magnetic stirring. After that aqueous NaOH solution was added dropwise until the pH reaches 13. Finally, the precipitated products were obtained by washing alternately with ethanol and water and dried in vacuum oven for 8 h at 150 °C. Further, the samples were annealed at 800 °C for 4 h in the muffle furnace.

2.2 $BaTiO_3$ Nanoparticle Preparation

Initially, 0.12 molar barium nitrate aqueous solution was prepared using double distilled water. The above-prepared solution was then stirred continuously, while TiO_2 powder was added. The aforesaid combination was then gradually combined with 0.4 molar solution of oxalic acid. The pH of the solution was then raised by the addition of ammonia solution until it reached 7. The obtained precipitates were then thoroughly washed using ethanol and water. The resulting precipitate was then dried in vacuum oven and further calcined at 1000 °C for 5 h to obtain the resultant $BaTiO_3$ nanoparticles.

2.3 Composite Preparation

Both the powders were then mixed in the molar ratio (*x*) $BaTiO_3$−(1−*x*) $CoFe_2O_4$ (*x* = 0, 0.25, and 0.5) and grounded for 2 h each via ball milling and finally, calcined at 1000°C to obtain the resultant composite powders. The calcined composite powders were then characterized by using various techniques.

3 Results and Discussion

Phase purity and crystallinity of the XRD patterns have been determined by using Powder X software as shown in Fig. 1a. XRD patterns divulge that the composite powders consist of both perovskite and cubic spinel phases. In order to evaluate the average crystallite size (*D*) and micro-strain (ε) associated with all the composite powder samples, Williamson-Hall method has been used and can be expressed using the following equation [5]:

$$\beta\cos\theta = \frac{0.9\lambda}{D} + 4\varepsilon\sin\theta \tag{1}$$

where symbols have their usual meaning. As shown in Fig. 1b, value of both crystallite size and strain associated with the composite system was found to increase from 28 to 38 nm and 0.000424 to 0.000611, respectively.

Dielectric measurements were performed at room temperature in order to study the behaviour of dielectric constant and dielectric loss of the (*x*) $BaTiO_3$−(1−*x*) $CoFe_2O_4$ (*x* = 0, 0.25, and 0.5) samples under investigation in the frequency range 42 Hz-5 MHz as illustrated in Fig. 2a, b, respectively. It can be noticed that the dielectric constant for all the samples has higher values within the lower frequency

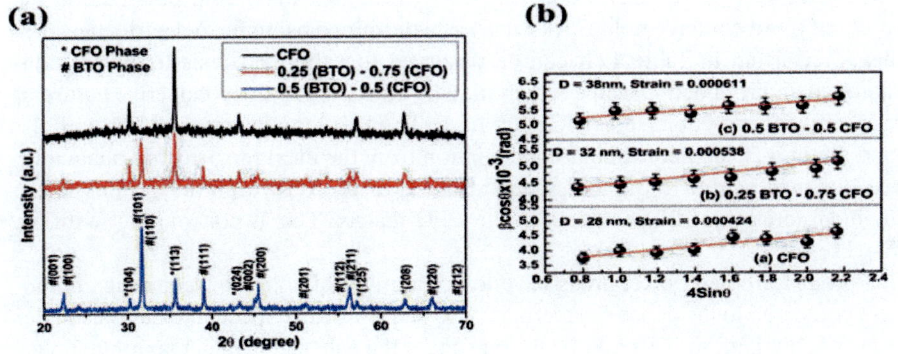

Fig. 1 **a** XRD patterns of (*x*) $BaTiO_3$−(1−*x*) $CoFe_2O_4$ (*x* = 0, 0.25, and 0.5) samples indexed using Powder X software and **b** Williamson-Hall plots for (*x*) $BaTiO_3$−(1−*x*) $CoFe_2O_4$ (*x* = 0, 0.25, and 0.5) samples

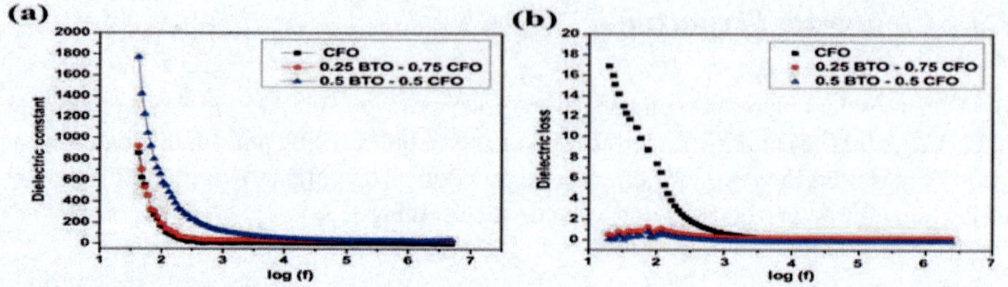

Fig. 2 **a** Variation of dielectric constant and **b** the variation of dielectric loss for (x) $BaTiO_3$–($1-x$) $CoFe_2O_4$ ($x = 0$, 0.25, and 0.5) samples

regime and decreases on increasing frequency and thereby attain saturation at higher frequencies like typical dielectrics. This can be understood by virtue of the Maxwell–Wagner space charge polarization in accordance with the Koop's phenomenological theory. According to this model, due to active participation of grain boundary and fast response to the applied field, greater polarization occurs at low frequencies. As a result, grain boundary interface experiences higher charge accumulation that results in higher value of dielectric constant. In low-frequency regime, electric dipoles are in phase with the applied field whereas in high-frequency regime, these dipoles are unable to follow the rapidly varying alternating field. Hence, display a decreasing trend in the dielectric constant. A significant enhancement can be observed in the dielectric constant with the incorporation of the BTO content in the CFO matrix, and also the dielectric losses are small as compared to the pristine CFO sample. The decrease in the value of dielectric losses may be attributed to the ferroelectric phase present in the cubic spinel matrix.

The room temperature hysteresis loops exhibiting the variation of polarization versus electric field for the (x) $BaTiO_3$–($1-x$) $CoFe_2O_4$ ($x = 0$, 0.25, and 0.5) samples are shown in Fig. 3. The appearance of distinct polarization loops suggests the ferroelectric behaviour in the composite samples under investigation. Electrical parameters such as remnant polarization P_r ($\mu C/cm^2$), maximum polarization P_m ($\mu C/cm^2$), and coercive field E_c (kV/cm) as determined using ferroelectric measurements are listed in Table 1. It can be observed that the hysteresis loops are non-saturated and rounded in shape which may be attributed to the conductive nature of the ferrimagnetic cubic spinel (CFO) phase. The P-E loop becomes wider with the increase in BTO content stipulating deviation from the ideal ferroelectric behaviour, and this may be due to the increased structural distortion within the system due to the incorporation of BTO content in the CFO matrix. This is concomitant with the W–H results.

Porous ferroelectric ceramics have lesser number of active ferroelectric component compared to the dense counterparts and therefore are expected to have a smaller value of polarization. This may be the reason for the enhancement in maximum polarization and remnant polarization in case of 0.5 BTO–0.5 CFO composite sample as compared to pristine CFO sample. The insensitivity of coercive field to the porosity might be due to accumulation of space charge on porous sites [6].

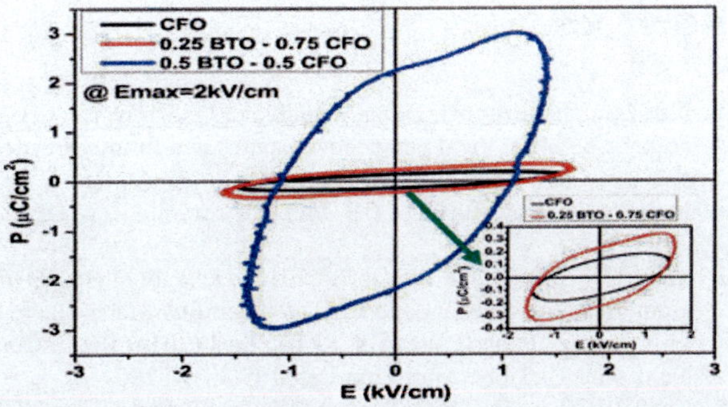

Fig. 3 P-E hysteresis loops of $(x)\ BaTiO_3-(1-x)\ CoFe_2O_4$ ($x = 0$, 0.25, and 0.5) samples at room temperature

Table 1 Ferroelectric parameters determined using P-E loop hysteresis measurements recorded at room temperature

Specimen samples	E (kV/cm)	P_r (μC/cm²)	P_m (μC/cm²)	E_c (kV/cm)
CFO	2	0.14	0.18	1.15
0.25 BTO–0.75 CFO	2	0.24	0.34	1.36
0.5 BTO–0.5 CFO	2	2.25	3.04	1.04

4 Conclusion

The XRD analysis reveals that synthesized composite powders consist of both perovskite and cubic spinel phases. Average crystallite size and micro-strain induced within the system increase with the increase in BTO content in the composite system. Dielectric constant decreases with the increase in the frequency of applied field. The polarization versus electric field measurement manifests enhancement in the ferroelectric properties with the increasing percentage of BTO phase in the composite system. The improved dielectric properties suggest BTO-CFO composite powders as potential candidate for dielectric resonators and dynamic random-access memories (DRAMs).

Acknowledgements One of the authors, Surbhi Sharma, gratefully acknowledges the Department of Chemistry, AMU, Aligarh for providing XRD facility.

Declaration of Interest Statement There is no competing interest of any sort. The authors declare that they have no conflict of interests.

References

1. Nan CW, Bichurin MI, Dong S, Viehland D, Srinivasan G (2008) Multiferroic magnetoelectric composites: Historical perspective, status, and future directions. J Appl Phys 103(3):031101
2. Kar BS, Goswami MN, Jana PC (2022) Enhancement of dielectric and multiferroic properties in Sr-modified 0.7 $BaTiO_3$–0.3 $ZnFe_2O_4$ ceramics. J Mater Sci: Materi Electron 33(31):23949–23963
3. Yang H, Wang H, He L, Yao X (2012) Hexagonal $BaTiO_3/Ni_{0.8}Zn_{0.2}Fe_2O_4$ composites with giant dielectric constant and high permeability. Mater Chem Phys 134(2–3):777–782
4. Li Y, Liao Z, Fang F, Wang X, Li L, Zhu J (2014) Significant increase of Curie temperature in nano-scale $BaTiO_3$. Appl Phys Lett 105(18):182901
5. Chen W, Wang ZH, Zhu W, Tan OK (2009) Ferromagnetic, ferroelectric and dielectric properties of $Pb(Zr_{0.53}Ti_{0.47})O_3/CoFe_2O_4$ multiferroic composite thick films. J Phys D: Appl Phys 42(7):075421
6. Zhang Y, Roscow J, Lewis R, Khanbareh H, Topolov VY, Xie M, Bowen CR (2018) Understanding the effect of porosity on the polarisation-field response of ferroelectric materials. Acta Mater 154:100–112

Photo and Piezocatalytic Behavior of Ag-NPs-Hybridized Barium Titanate

Moin Ali Siddiqui, Shahzad Ahmed, Arshiya Ansari, and Pranay Ranjan

Abstract Barium Titanate, a piezoelectric material having fascinating dye adsorbing properties, is forecasted as a promising material for the adsorption of cationic dyes. However, its efficiency remains a challenge that needs to be enhanced due to less surface charge. Herein, photocatalysis (PC) and piezocatalytic (PzC) induced degradation of Rhodamine B (RhB) in the wastewater using barium titanate and barium titanate hybridized with silver nanoparticles (Ag-NPs) metal particles were investigated and compared. The solid-state synthesis route was employed to synthesize Barium Titanate ceramics. Ag-NPs metal was hybridized with Barium Titanate particles (through a physical deposition technique) under ambient conditions at STP in the absence of light and heat. A broader hump was observed during UV–Vis spectrophotometry due to the surface plasmonic resonance on the Ag-NPs-hybridized Barium Titanate sample. Visible light source and vibrational energy were employed in the solution as an energy source in the PC and PzC, respectively. More than 99% of the rhodamine B (RhB) was degraded in an aqueous solution employing an Ag-NPs-hybridized Barium Titanate sample, its promising PC, and PzC activity. It was observed that the Ag-NPs-hybridized Barium Titanate powder gives more activity in RhB than pure Barium Titanate powder. It was observed that the PC and PzC performance was caused by the –OH radical species. The PC/PzC outcome was shown to be stable after five cycles, revealing the favoring property of the Ag-NPs-hybridized Barium Titanate.

Keywords Photo/piezocatalysis · RhB · Ferroelectric · Ag-NPs · Barium titanate

M. A. Siddiqui (✉) · S. Ahmed · A. Ansari · P. Ranjan (✉)
Department of Metallurgical and Materials Engineering, Indian Institute of Technology Jodhpur, Rajasthan 342030, India
e-mail: moinalisiddiqui@gmail.com

P. Ranjan
e-mail: pranay.ranjan@iitj.ac.in

Z. H. Khan et al. (eds.), *Recent Advances in Nanomaterials*, Springer Proceedings in Materials 27, https://doi.org/10.1007/978-981-99-4878-9_49

1 Introduction

Pure, clean, odorless, and colorless are some of the important characteristics of pure water. Indeed, it is the most necessary or essential requirement for the survival of mankind, animals (aquatic life), and for plants. However, the majority of the world's water supplies are contaminated by wastes in the form of liquid and solid that are created by human settlements and industrial operations. Interestingly, recently, it was found that there is an increase in the demand for beauty products and changing fashion. The chemical and textile industry, therefore, further fueled the dye-based products on a very large scale and contaminated water bodies (lakes, rivers, ponds, etc.) through the discharge of the industrial effluents. In the recent past, the growth of the textile industry was found to be dramatically increasing (exponentially) due to more demand and less supply. Around 7×10^7 tons of synthetic dyes are generated annually, in which over 10,000 tons of these are solely used by the textile industries. Moreover, these textile industries use more than one kind of dye (to give different shades to the fabric) for their end products which further increases the risk of polluting the water upon being discharged without treatment. It was found that the discharged dye water contains sulfur, nitrates, soaps, enzymes, heavy metals, and some auxiliary chemicals. This results in the primary source of water contamination and is also the major source of untreated effluents. It is therefore the need of the hour to treat the dyes before being consumed by aquatic animals or given to animals/plants as it contains toxic materials which can ultimately lead to the loss of marine and aquatic forests.

Literature suggests a few remedies or treatment methods for the filtration of dyes, some of them are membrane filtration, electrochemical oxidation, ultrasonication, sand filtration, etc. Although these techniques are effective, they are not cost-effective and consume a high amount of energy and intricate operational procedures. Thus, limiting their use in dye degradation. It was observed that dye wastewater can be treated by using a technique known as PzC which involves the use of ferroelectric materials, known for their inherent spontaneous polarization (the displacement of the positive and negative charges relative to its center). This spontaneous polarization causes the photogenerated holes and electrons to spatially separate from their space charge area and moves in the opposite polarity direction. As a result, there is reduced recombination of the charge carriers, which contributes to improving PzC efficacy. During the process, such materials demonstrate enhanced degradation of organic pollutants including color, medicines, etc. However, the fact that ferroelectric materials can only be triggered by UV light is one of their main limitations for water-cleaning purposes. Numerous methods, including doping, composite synthesis with another semiconductor, loading with noble metals, etc., have been proposed in the literature to produce visible light PC in ferroelectric materials [1].

The utilization of inverse opal/porous structures for photocatalytic activity [2], electrochemical sensing [3], and photodetector [4, 5] has been reported in other publications. Additionally, a structure based on MIP [6] might be used for photocatalytic activity [7]. On the other hand, magnetic nanoparticles [8] were also employed for dye degradation photocatalysis [9].

In this method, ferroelectric materials are triggered by ultrasonic vibrations rather than light. Moreover, it is been reported that adding noble metals (Ag-NPs, Au, etc.) to ferroelectric materials can enhance PC and trigger this phenomenon in visible light. Therefore, we did a comparative study and analyzed the PC and PzC performance in this work by loading Ag-NPs on Barium Titanate ferroelectric ceramic.

2 Methods

A solid-state reaction technique was used to synthesize the Barium Titanate powder. In this technique, stoichiometric molar ratios were used to measure the oxide powders of Barium Carbonate and Titanium Oxide. These powders were thoroughly amalgamated and processed for 50 min in agate mortar and pestle. The dry, homogeneous powder was calcined in the Nabertherm furnace for four hours at 1150 °C. The same procedure was followed for the Ag-NPs-hybridized Barium Titanate. In this technique, 90 ml of polyvinyl glycol was stirred with 1.9 g of each Barium Titanate and silver nitrate. The solution was stirred for 18 h, to achieve Ag-NPs loading on Barium Titanate and to prevent the oxidization of the Ag nanoparticles, a controlled ambient condition was created under nitrogen-sealed beakers. The powder was then washed, cleaned using double deionized water, and ethanol, and subjected to vacuum heating in a nitrogen atmosphere at 120 °C for 2 h.

The absorbance spectra using diffuse reflectance spectroscopy (DRS) were recorded for both Barium Titanate and Ag-NPs-hybridized Barium Titanate powder. However, the Beer-Lambart law was used for the study of bandgap tuning. Rhodamine B (RhB) was primarily investigated as a typical pollutant to study the PC and PzC activities of the developed catalyst. RhB dye was used as a model dye. The initial RhB dye concentration was taken at 7 mg/L. 18 ml dye solution was taken in a beaker, to which 0.12 g of the sample was added. The dye solution was placed under dark conditions for 24.2 h before performing the experiment for the dye adsorption equilibrium was achieved. During PC investigations, the beaker was exposed to visible light while being constantly stirred at 700 rpm. The catalyst was stimulated during the experiment using a visible light source (18 W; Havells LED). The sample was kept around 4 inches away from the light source. In the piezocatalysis experiments, an ultrasonicator (frequency—40 kHz) was used as an excitation source. During both catalytic experiments, a UV/Vis Spectrophotometer (Perkin Elmer Lambda-35) was used to record the dye degradation absorbance spectrum at various intervals. The reusability experiments were carried out for five cycles to examine the sample of Ag-NPs-hybridized Barium Titanate. After performing one catalysis cycle, the sample was centrifuged to settle in the dye solution. Once the sample had settled, a pipette was used to drain the dye solution and was further cleaned with ethanol and dried in an oven. The dried sample was used for the next catalysis cycle.

3 Results and Discussion

The result of PC experiments using Barium Titanate and Ag-NPs-hybridized Barium Titanate is represented in Fig. 1. Fig. 1a, b show the absorbance versus wavelength spectra obtained for Barium Titanate and Ag-NPs-hybridized Barium Titanate samples, respectively (during the PC of the RhB dye degradation in an aqueous solution). It was observed when the dye solution was subjected to a Barium Titanate powder sample, there is a slight decrease in the characteristic absorbance peak of RhB dye, as compared to the continuous and significant decrease in the case of Ag-NPs-hybridized Barium Titanate sample under visible light irradiation. Ag-NPs-hybridized Barium Titanate sample was found to be more effective in comparison with the initial 185 min. The graph of C/C_0 versus time plots indicates Barium Titanate, Ag-NPs-hybridized Barium Titanate, sample performance when loaded in a 3% (for pure dye), 18% (Barium Titanate), and 94% (Ag-NPs-hybridized nanoparticles) for the RhB dye (see Fig. 1c). It was observed that the performance activity for adsorption of the RhB dyes in the case of Ag-NPs-hybridized Barium Titanate is much enhanced and improved in comparison with Barium Titanate. Fig. 1d reveals the value of the kinetic rate constant (k) of dye degradation which was calculated to be 0.0002186 min^{-1}, 0.00111 min^{-1}, and 0.01559 min^{-1}, respectively for pure dye, Barium Titanate sample, and Ag-NPs-hybridized Barium Titanate sample.

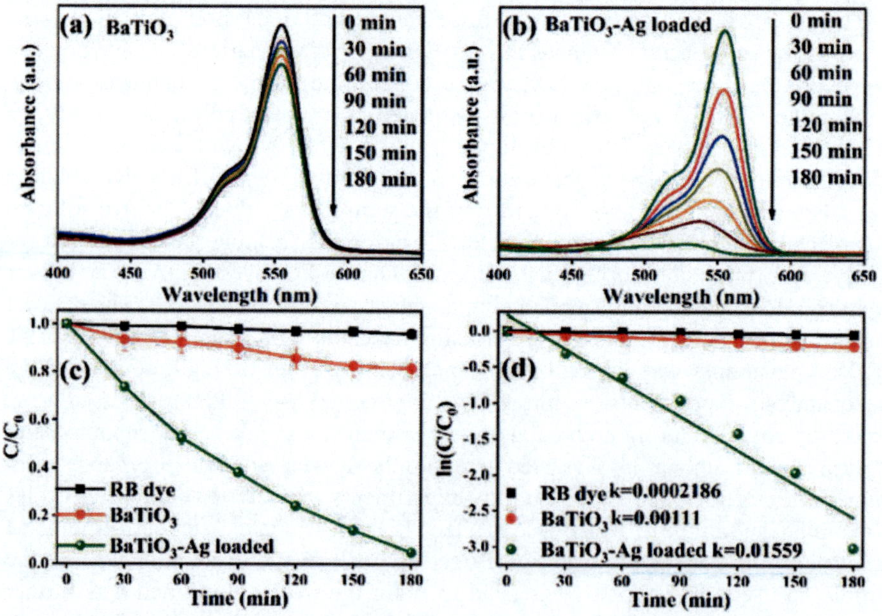

Fig. 1 PC activity of the as-synthesized Barium Titanate and Barium Titanate-Ag-NPs-hybridized sample, respectively

Fig. 2a, b show the absorbance versus wavelength spectra of the Barium Titanate and Ag-NPs-hybridized Barium Titanate samples when RhB dye was degraded through piezocatalytic. The characteristic RhB dye absorbance peak showed a consistent and significant decline in the Ag-NPs-hybridized Barium Titanate sample compared to a very modest decline in the pure Barium Titanate sample. This indicates that for the same time period of 152 min under ultrasonication, more dye was degraded using the Ag-NPs-hybridized Barium Titanate. According to the C/C_0 versus the time graphs in Fig. 2c, only 4, 23, and 98% of the RhB dye were degraded in the cases of no sample, Barium Titanate sample, and Ag-NPs-hybridized Barium Titanate sample, respectively. Thus, it can be taken as evidence that the piezocatalytic dye degradation activity of the Ag-NPs-hybridized Barium Titanate sample was higher than that of the Barium Titanate sample. Fig. 2d reveals the kinetic rate constant (k) of piezocatalytic dye degradation using the Ag-NPs-hybridized sample which was calculated to be 0.0178 min^{-1} as compared to 0.00217 min^{-1} for the Barium Titanate sample. In the PC process, the Ag-NPs-hybridized Barium Titanate sample demonstrated over 94–91% dye degradation even after 5 cycles. Similarly, the PzC process gives approximately 95.3% to 93.2 dye degradation. The Ag-NPs-hybridized Barium Titanate sample showed excellent and repeatable dye degradation which can be seen in Fig. 3.

The surface plasmon resonance (SPR) phenomenon was observed when Ag-NPs are being hybridized with Abrium Tiatatnate due to the size effect. SPR triggers the visible light and photogenerated charge carriers (e− /h^+ pairs) in the Ag-NPs metal

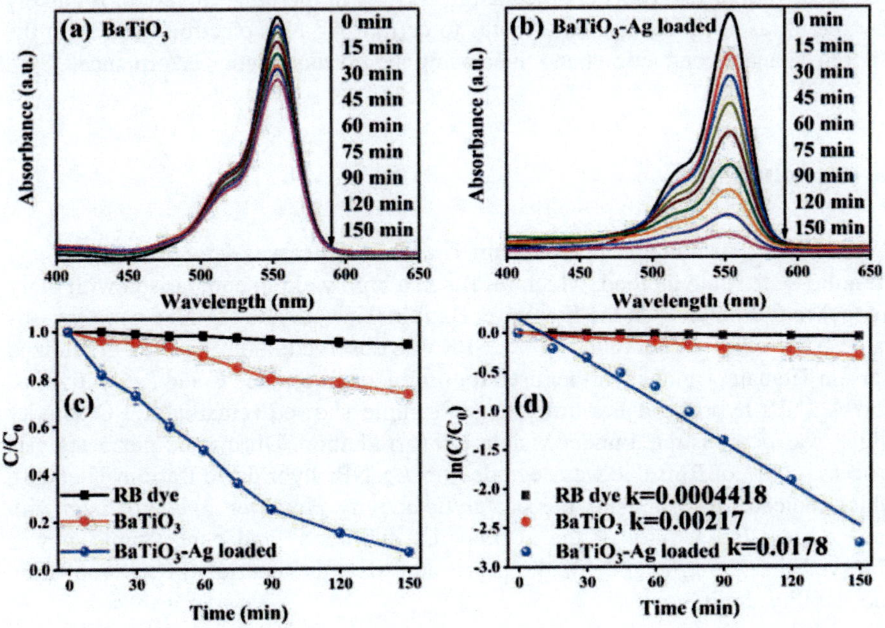

Fig. 2 PzC activity of the Barium Titanate and Barium Titanate-Ag-NPs-hybridized sample

Fig. 3 Brief comparison of the PC ad PzC reusability activity of Ag-NPs-hybridized Barium Titanate sample (5 cycles)

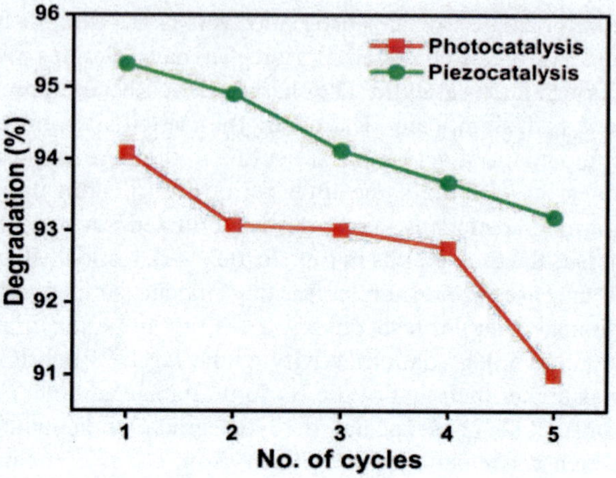

during PC, which has been widely reported in the case of Ag-NPs metal. The electron migrates from Ag-NPs metal to the conduction band of Barium Titanate ceramic, and subsequently, reduction of O_2 to $–O_2$ reactive species occurs. The $–O_2$ species then further convert into ·OH radical. In the process of piezocatalysis, ultrasonication results in a polarized electric field and band bending. The band bending makes it simple for these charge carriers to be transported to the surface, where they easily interact with O_2 and OH– to create ·OH species that ultimately destroy RhB dye. The polarization field transfers the charge carriers in the other direction. It's likely that the ultrasonication also contributes to certain Ag-NPs electron transfer to the Barium Titanate conduction band, increasing the piezocatalytic performance.

4 Conclusion

A facile synthesis route for pure Barium Titanate powder was demonstrated through the solid-state route method, which results in a high yield in comparison with other sol–gel routes existing in the literature. Barium Titanate and Ag-NPs were investigated as ferroelectric ceramics (STP). SPR was observed in the Ag-NPs-hybridized Barium Titanate, which is further used for promising reusable PC and PzC activities. The Ag-NPs-hybridized Barium Titanate sample showed remarkable PC activity which was demonstrated under visible light irradiation. During the piezocatalysis process, ~95% of RhB dye was degraded (in Ag-NPs-hybridized Barium Titanate), which indicates its promising piezocatalytic activity. However, in comparison with the PzC and PC activity of the as-prepared hybridized and pure sample, it was observed that the efficiency of the piezocatalysis was superior (in terms of time and stability) to PC.

Acknowledgements MAS would like to acknowledge MHRD for financial support and the Indian Institute of Technology Jodhpur for providing the research facility. Pranay Ranjan, Devendra Singh Negi, and Amitava Banerjee would like to thank SERB for SRG (Grant no. SRG/2022/000192, SRG/2022/000825, and SRG/2022/001377 respectively) for collaboration and SEED grant no I/SEED/PRJ/DSN/AB/20220044.

Declaration of Interest Statement The author declared no conflict of interest.

References

1. Siddiqui MA, Jaiswal P (2021) Photocatalytic behavior of ferroelectric materials: comparative study of $BaTiO_3$ and Ag-loaded $BaTiO_3$ for wastewater treatment. In: IOP conference series: materials science and engineering, vol 1166, no 1. IOP Publishing, p 012031
2. Pham K, Temerov F, Saarinen JJ (2020) Multicompound inverse opal structures with gold nanoparticles for visible light photocatalytic activity. Mater Des 194:108886
3. Ahmed S, Ansari A, Haidyrah AS, Chaudhary AA, Imran M, Khan A (2022) Hierarchical molecularly imprinted inverse opal-based platforms for highly selective and sensitive determination of histamine. ACS Appl Polym Mater 4(4):2783–2793
4. Ahmed S, Khatun S, Sallam S, Ansari A, Ansari ZA, Kumar RR, Hakami J, Khan A (2022) Photoresponse of porous silicon for potential optical sensing. Europhys Lett 139(3):36001
5. Ahmed S, Ansari A, Siddiqui MA, Khan A, Ranjan P (2023) A potential optical sensor based on nanostructured silicon. J Mater Sci: Mater Electron. https://doi.org/10.1007/s10854-023-10187-2
6. Imran M, Ahmed S, Abdullah AZ, Hakami J, Chaudhary AA, Rudayni HA, Khan SU, Khan A (2022) Nanostructured materials based optical and electrochemical detection of amoxicillin antibiotic. Luminescence https://doi.org/10.1002/bio.4408
7. Guo M, Hu Y, Wang R, Yu H, Sun L (2021) Molecularly imprinted polymer-based photocatalyst for highly selective degradation of methylene blue. Environ Res 194:110684
8. Imran M, Chaudhary AA, Ahmed S, Alam MM, Khan A, Zouli N, Hakami J, Rudayni HA, Khan SUD (2022) Iron oxide nanoparticle-based ferro-nanofluids for advanced technological applications. Molecules 27(22):7931
9. Fatimah I, Pratiwi EZ, Wicaksono WP (2020) Synthesis of magnetic nanoparticles using Parkia speciosa Hassk pod extract and photocatalytic activity for Bromophenol blue degradation. Egypt J Aquatic Res 46(1):35–40

Hydrothermal Synthesis of Pure and Cadmium-Doped MoS_2 for Comparative Study and Application in Humidity Sensing

Ravi Kant Verma and R. K. Shukla

Abstract In this work, MoS_2 nanosheets were produced using a hydrothermal technique for humidity sensing applications. Also, cadmium doping was done to study the effects on morphology and humidity sensing properties of MoS_2. To analyze the produced MoS_2 nanosheets, X-ray diffraction (XRD) and scanning electron microscopy (SEM), FTIR spectroscopy, and UV–visible spectroscopy were used. XRD data confirms the hexagonal phase of MoS_2, and from SEM images, flower-like spheres to wrapped nanosheet structures were observed. The comparative analysis of pure and cadmium-doped MoS_2 was discussed here. Also, the humidity sensing behavior of both pure and cadmium-doped MoS_2 was studied in this paper.

Keywords MoS_2 · Hydrothermal synthesis · Cadmium doping · Humidity sensing

1 Introduction

Atoms in molybdenum disulfide (MoS_2) connect through covalent interactions, as is typical of two-dimensional (2D) semiconducting materials, with layers interacting through van der Waals forces [1]. Such structural characteristics equip MoS_2 with unique qualities for a variety of applications, including photocatalysis, lubricants, electrocatalysts, supercapacitors, photodetectors, rechargeable batteries, hydrogen generation, and hydrogen storage [2]. MoS_2 with various morphologies, such as nanotubes, nanorods, nanowires, nanoribbons, nano-sphere, fungus-like, nanoflowers, rambutan-like, coral-like, and nanosheet, has been synthesized and used in a variety of applications [3, 4]. The nanoflowers of MoS_2 were synthesized with the help of a hydrothermal method using sodium molybdate and thiourea as precursors [5]. The reaction time and temperature play key roles in the morphology of MoS_2 [6]. It is known that in the hydrothermal synthesis technique, the precursor materials, molarity ratios, hydrothermal temperature, and reaction time will play

R. K. Verma (✉) · R. K. Shukla
Department of Physics, University of Lucknow, Lucknow 226007, India
e-mail: rviivarma@gmail.com

Z. H. Khan et al. (eds.), *Recent Advances in Nanomaterials*, Springer Proceedings in Materials 27, https://doi.org/10.1007/978-981-99-4878-9_50

important roles in the morphologies control, crystal phase, and performance [7]. Thus in this work, we used the hydrothermal method for the synthesis of MoS_2 nanosheets from sodium molybdate and thiourea precursors. The nanosheet structure of MoS_2 plays a key role in humidity sensing. In this paper, pure and cadmium-doped MoS_2 were synthesized using hydrothermal technique, and a comparative study of both pure and doped MoS_2 was done using XRD, FESEM, FTIR, and UV–visible characterization. Also, the humidity sensing behavior of both pure and cadmium-doped MoS_2 was done to study the effect of cadmium doping.

2 Materials and Method

2.1 Synthesis of Materials

The precursor for hydrothermal synthesis was made with sodium molybdate dihydrate, thiourea, and cadmium sulfide. All chemicals were purchased from ThermoFisher with 99.89% purity.

For the synthesis of pure $MoS_{2,}$ 0.008 mol of sodium molybdate and 0.016 mol of thiourea were dissolved in 30 ml deionized water. For the synthesis of cadmium-doped MoS_2, 0.008 mol of sodium molybdate and 0.016 mol of thiourea were dissolved in 30 ml of deionized water. 0.0004 mol, 0.0008 mol, and 0.001 mol of cadmium sulfide were dissolved in solution for 10% cadmium doping, respectively. After that solution was transferred into a 100 ml Teflon autoclave. For thin film deposition, a glass substrate was placed inside the autoclave making an angle of 45°. The autoclave was kept in a hot air oven for 24 h. After 24 h of reaction, the autoclave was removed. The black color synthesized material and the thin film were separated from the autoclave and washed several times with deionized water and with absolute ethanol. The washed material was transferred to a vacuum oven for drying at 80 °C for 5 h.

2.2 Characterization Techniques

X-ray diffraction, SEM, and UV–visible were used to characterize the powder and thin film of pure and cadmium-doped MoS_2. Rigaku Ultima IV X-ray diffractometer was used for XRD data. JEOL JSM- 7610F was used for FESEM analysis, and Shimadzu IRAffinity-1S Spectrometer was used to study the UV–Vis spectroscopy.

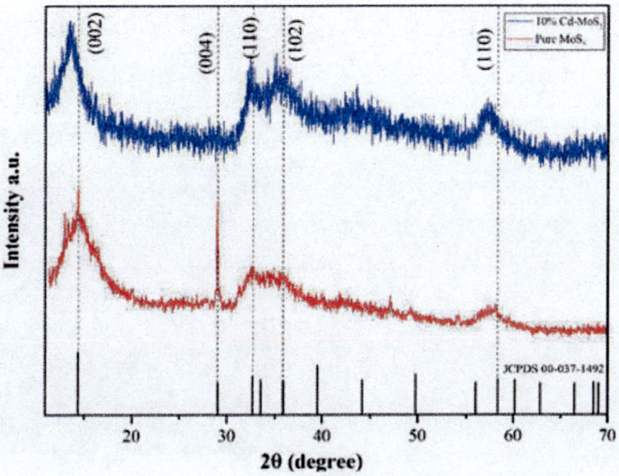

Fig. 1 XRD patterns of pure and cadmium-doped MoS_2

3 Results and Discussion

3.1 XRD Analysis

All samples' X-ray patterns are displayed in Fig. 1. The observed patterns closely resemble the hexagonal phase of MoS_2 (JCPDS 00-037-1492). The large peak at 002 in pure MoS_2 reflects the sample's nanostructure [8]. The Debye–Scherrer formula was used to compute the crystallite sizes of all samples [9]. The average crystallite size calculated with FWHM data was found to be 68 and 72 nm for pristine and cadmium-doped MoS_2. Lattice parameter was found to be $a = b = 2.89$ A° and $c = 12.52$ A° which agrees with the lattice parameter of MoS_2 [10]. Diffraction peaks at 002 and 110 shifted to lower angles in cadmium-doped MoS_2 which indicates changes in inter-planer distance and suggests the incorporation of cadmium in MoS_2 [11]. Also, the intensity of all peaks is increased in cadmium-doped MoS_2 which further indicates that the crystallinity of MoS_2 increased with doping.

3.2 SEM Analysis

The SEM figures are shown in Fig. 2. From SEM analysis, it was observed that the formation of a flower-like structure has formed. These flower-like structures have consisted of several nanosheets, and the growth of these sheets is due to lamellar structures of MoS_2 which occurs in the hydrothermal synthesis process [12]. These nanosheets further resemble and form a spherical flower-like structure due to van der Waals interaction [13]. The average diameter of each sphere was found to be 400 nm and 154 nm for pure and 10% cadmium-doped MoS_2, respectively. Also, the size of the sphere starts reducing with an increase in cadmium doping as shown in Fig. 3.

Fig. 2 SEM images of pure MoS_2

Fig. 3 SEM images of 10% cadmium-doped MoS_2

It is observed that the size of microsphere reduced rapidly with doping of cadmium but further increasing doping percentage of cadmium tends to decrease microsphere size linearly with increase in doping percentage of cadmium.

3.3 *UV–Visible Analysis*

The UV–visible was performed with Rigaku UV–Vis spectrometer. All samples are dissolved in nitric acid for absorption spectra analysis. The UV–visible graphs are shown in Fig. 4. The graphs contain absorbance versus wavelength plots of pure and cadmium-doped MoS_2. Peaks in the absorption spectra were observed at 328.5 nm and 407.8 nm for pure and 10% cadmium-doped MoS_2, respectively. This spectrum was used to calculate the optical band gap. The optical band gap was obtained using

Tauc's relation [14]. The optical band was calculated by extrapolating the intercept at $\alpha = 0$ from Tauc's plot as shown in Fig. 5. The direct band gap values found are 3.72 eV and 3.06 for pure and 10% cadmium-doped MoS_2. It was observed that the optical band gap decreased with an increase in the doping percentage of cadmium in MoS_2.

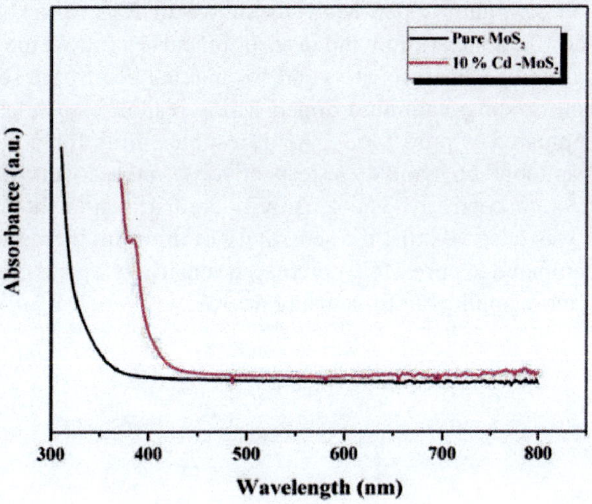

Fig. 4 UV–Vis spectra of pure and cadmium-doped MoS_2

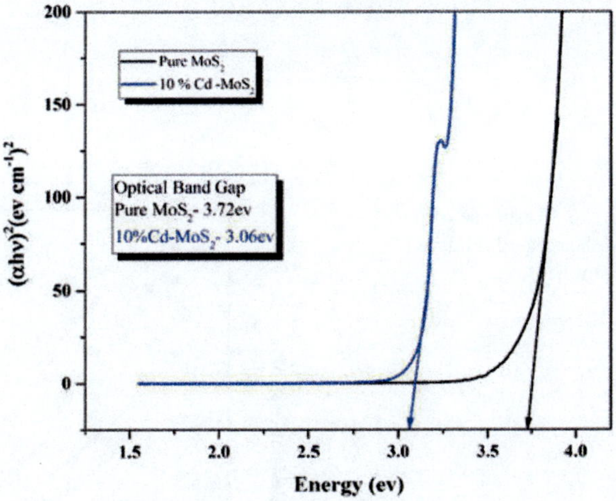

Fig. 5 Plot of $(\alpha h\nu)^2$ versus $(h\nu)$ for pure and cadmium-doped MoS_2

3.4 Humidity Sensing Performance

For humidity sensing, a thin film of MoS_2 was used. Humidity sensing was done with both pure MoS_2 and 10% cadmium-doped MoS_2 thin films. With the help of the humidity sensing setup, the response and recovery time were recorded. Recorded data were studied and plotted using Origin software. The humidity response of pure and cadmium-doped MoS_2 is shown in Figs. 6 and 7, respectively. Both graphs show that adsorption and desorption curves follow the nearly same path. The lower values of hysteresis are good for making electronic sensor devices. On comparing both graphs, cadmium-doped MoS_2 response to relative humidity is more linear compared to pure MoS_2. Also, response time for pure and cadmium-doped MoS_2 was found 56 s and 49 s, respectively. And recovery time was found to be 71 s and 59 s, respectively. The sensitivity comparison is shown in Fig. 8. From the figure, it was observed that the sensitivity of thin film increased for cadmium-doped MoS_2 compared to pure MoS_2. Increased sensitivity makes cadmium doping in MoS_2 make it more applicable for sensing devices.

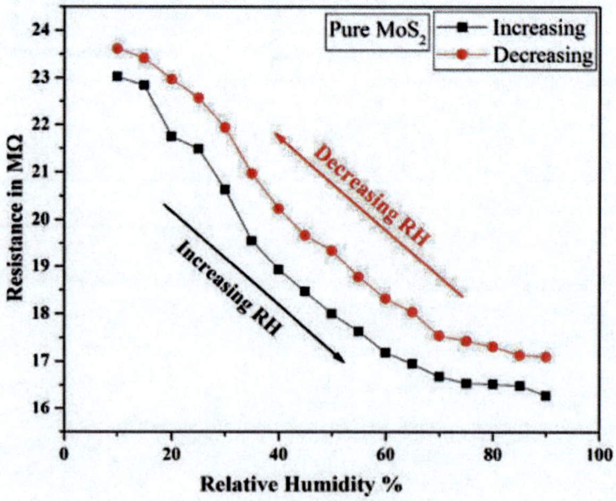

Fig. 6 Response of pure MoS_2 toward relative humidity

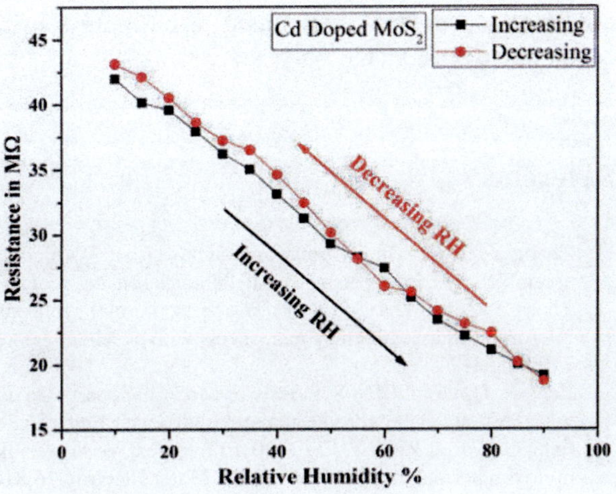

Fig. 7 Response of cadmium-doped MoS_2 toward relative humidity

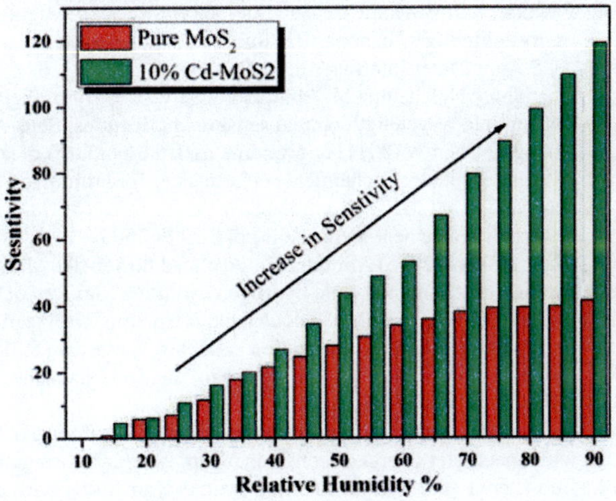

Fig. 8 Sensitivity comparison of pure and cadmium-doped MoS_2

4 Conclusion

In this work, comparative study has been done between pure and cadmium-doped MoS_2. Also, humidity sensing behavior was studied for both samples. It was observed that the doping of cadmium makes an effect on the crystallography of MoS_2. The increase in doping of cadmium tends to increase the crystallinity of MoS_2. Also, the crystallite size increases with cadmium doping. The microspheres formed reduce their size with increasing doping percentage of cadmium. Also, the distance between two nanosheets increased with an increase in the doping of cadmium. The optical band gap also decreases with an increase in the doping of cadmium in MoS_2. The cadmium-doped MoS_2 was found to be more sensitive toward relative humidity compared to pure MoS_2.

References

1. Zhang Y (2018) The hydrothermal synthesis of 3D hierarchical porous MoS_2 microspheres assembled by nanosheets with excellent gas sensing properties. J Alloys Compd 749:355–362
2. Zhang Y, Zeng W, Li Y (2018) The hydrothermal synthesis of 3D hierarchical porous MoS_2 microspheres assembled by nanosheets with excellent gas sensing properties. J Alloys Compd 749:355–362
3. Zeng X, Qin W (2016) Synthesis of MoS_2 nanoparticles using MoO_3 nanobelts as precursor via a PVP-assisted hydrothermal method. Mater Lett 182:347–350
4. Feng G, Wei A, Zhao Y, Liu J (2015) Synthesis of flower-like MoS_2 nanosheets microspheres by hydrothermal method. J Mater Sci Mater Electron 26:8160–8166
5. Kumar Y, Sharma A, Shirage PM (2017) Shape-controlled $CoFe_2O_4$ nanoparticles as an excellent material for humidity sensing. RSC Adv 7:55778–55785
6. Pawbake AS, Waykar RG, Late DJ, Jadkar SR (2016) Highly transparent wafer-scale synthesis of crystalline WS_2 nanoparticle thin film for photodetector and humidity-sensing applications. ACS Appl Mater Interfaces 8:3359–3365
7. Chaudhary N, Khanuja M, Abid, Islam SS (2018) Hydrothermal synthesis of MoS_2 nanosheets for multiple wavelength optical sensing applications. Sens Actuators A Phys 277:190–198
8. Chang K, Chen W (2011) L -Cysteine-assisted synthesis of layered MoS_2/graphene composites with excellent electrochemical performances for lithium ion batteries. ACS Nano 5:4720–4728
9. Joseph D et al (2020) Thermoelectric performance of Cu-doped MoS_2 layered nanosheets for low grade waste heat recovery. Appl Surf Sci 505
10. Wang FZ et al (2016) Ammonia intercalated flower-like MoS_2 nanosheet film as electrocatalyst for high efficient and stable hydrogen evolution. Sci Rep 6(1):31092
11. Lai W (2016) A NiMoS flower-like structure with self-assembled nanosheets as high-performance hydrodesulfurization catalysts. Nanoscale 8:3823–3833
12. Liang S (2013) PVP-assisted synthesis of MoS_2 nanosheets with improved lithium storage properties. CrystEngComm 15:4998–5002
13. Mahdavi M, Kimiagar S, Abrinaei F (2020) Preparation of few-layered wide bandgap MoS_2 with nanometer lateral dimensions by applying laser irradiation. Crystals 10:164
14. Luo L et al (2019) Hydrothermal synthesis of MoS_2 with controllable morphologies and its adsorption properties for bisphenol A. J Saudi Chem Soc 23:762–773

Effect of Different Synthesis Methods on the Optical Properties of Graphene/ SnO_2 Nanocrystals Composite Prepared via Chemical Reduction and Microwave Method

Farheen and Azra Parveen

Abstract In this study, we have examined the luminescence characteristics of a compound including tin oxide and graphene, and also variations in the synthesis process. To synthesize SnO_2 and graphene nano-composites, both chemical reduction and microwave synthesis were used. The SnO_2/graphene-nanosheets composite was morphologically and optically altered by modification of the Sn^{2+} and GO manufacturing processes. SnO_2 nanocrystals are randomly distributed on the surface of the graphene nanosheets (GN) and can be seen in the TEM image. The band gap for SnO_2 and graphene vary depending on the synthesis technique, from 3.95 to 3.16 for SnO_2: GNs-1 and SnO_2: GNs-2, respectively. This work includes testing changes in the aggregation process. TGA testing shows that the composite has high thermal stability below 500 °C, which is useful for applications in electrical and optical storage devices.

Keywords Photoluminescence · Nanocomposite · XRD Analysis · FTIR

1 Introduction

Graphene has unique and remarkable properties like chemical stability, electronic conductivity, and high surface area, etc. make it a promising material in the world of composite. For loading tin oxide nanoparticles on graphene nanosheets, several methods have been developed like wet chemical or atomic layer deposition method. However, graphene surfaces lack strong interactions between them [1]. In graphene, loading functional groups on its surfaces are not evenly distributed. The properties of graphene oxide can be altered by its functionalization and depending on the desired application various methods can opt. The chemically modified graphene obtained

Farheen · A. Parveen (✉)
Department of Applied Physics, Z.H. College of Engineering and Technology, Aligarh Muslim University, Aligarh 202002, India
e-mail: azrap2015@gmail.com

Z. H. Khan et al. (eds.), *Recent Advances in Nanomaterials*, Springer Proceedings in Materials 27, https://doi.org/10.1007/978-981-99-4878-9_51

could offer more applications that can be used in various fields of electrical energy storage devices such as electric vehicles, pulsed light generators, and backup power for computer memory. For preparing the composite of SnO_2: GNs, the key method to load the SnO_2 nanoparticles on both sides of single-layer graphene, graphene oxide (GO) single sheet was chosen to start with instead of a graphene sheet. The Sn^{2+} ions initially form electrostatic bonds with oxygen groups (probably hydroxyl and carbonyl groups) before forming SnO_2 nanocrystals with a size of about 3 nm. SnO_2 nanocrystals were uniformly charged on the in situ monolayer GO sheets and oxygen groups on the reduced graphene oxide layer. The high dispersion of the SnO_2 nanocrystals was maintained while GO was converted to graphene.

2 Experimental Methods

2.1 Method 1: In Situ Chemical Reduction

Firstly, the GO solution was prepared by combining 20 mg of GO, which was produced using the modified Hummer's technique, with 200 ml of distilled water and further the GO solution was sonicated for 1 h. $SnCl_2 \cdot 2H_2O$ (0.8 gm) was mixed with the GO solution and stirred for 24 h. The residual solution eventually washed with distilled water and dried at 100 °C. Resultant sample is known as "SnO_2: GNs-1."

2.2 Method 2: Microwave Method

In this procedure, 20 mg of GO in 100 ml of H_2O was sonicated for 1 h. 4.4106 g of $SnCl_2 \cdot 2H_2O$ was added to the mixture of ethylene glycol (EG) and H_2O(DI) with a ratio of $H_2O/EG = 0.1$. The mixture was heated for 4 min in a microwave oven (700 W) [2]. The synthesized products was filtered and washed with distilled water followed by drying at 70 °C for 24 h. The sample prepared from this method was denoted as "SnO_2: GNs-2."

3 Results and Discussion

Figure 1a depicts the XRD images of graphene oxide (GO). Facile exfoliation caused by weak van der Waals interactions between layers of graphene oxide, peak at 10.4° in the curve is typical for graphene oxide. According to Fig. 1b, the SnO_2 phase is responsible for the four dominating widened peaks (110), (101), (211), and (301), which reveal the creation of tetragonal nanocrystals of SnO_2. The TEM images show

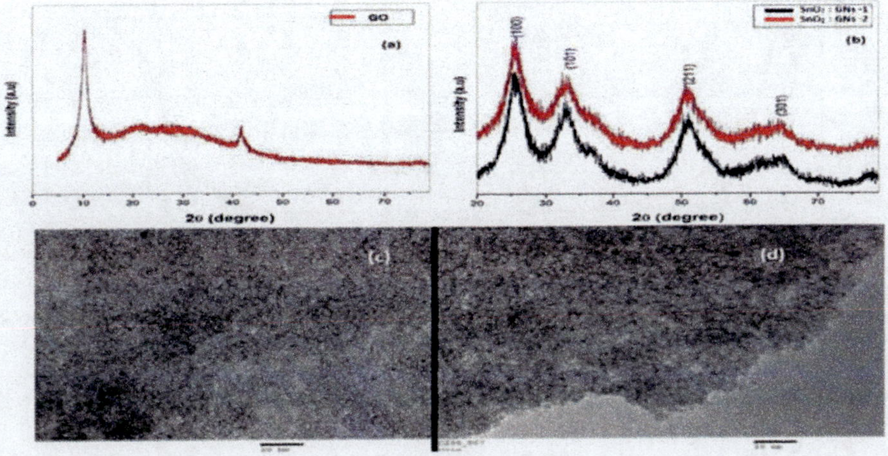

Fig. 1 X-ray diffraction pattern of **a** GO and **b** SnO_2: GNs-1 & 2 and TEM Images of **c**, **d** SnO_2: GNs-1 & 2

that SnO_2 nanoparticles are randomly arranged on the graphene nanosheets and the particle size of the SnO_2 nanoparticles is less than 3–4 nm. The SnO_2 nanocrystals are less than 3–4 nm in diameter and are randomly arranged on the graphene, according to the TEM pictures.

The UV–Vis spectrum of GO nanoparticles is shown in Fig. 2a 4.15 eV is the estimated band gap for GO. The energy band gap of the nanocomposite (E_g) is calculated by the Tauc relation [3].

$$\alpha h\nu = \left(h\nu - E_g\right)^n \quad (1)$$

where $n = ½$ for a direct band gap semiconductor and α is the absorption coefficient. Figure 2b shows the absorption peak at 290 nm and 374 nm, respectively. The energy band gap (E_g) of prepared sample from method 1 and 2 nanocomposite was found to be 3.26 eV and 3.24 eV, respectively.

One of the fascinating characteristics of SnO_2 nanomaterials is photoluminescence, which has its significance in optoelectronic devices including UV–Vis emitting diodes and laser diodes. PL spectra of the graphene and tin oxide nanocomposite are shown in Fig. 3a, b. The excitation points of SnO_2;GNs-1 & 2 were determined at 301.05 nm and 301.77 nm, respectively. The difference in the excitation peaks clearly shows the redshift that means on changing the synthesis method, the particle size changes. These results agree with our UV–Vis spectral observations and correspond well to the absorption spectrum. The emissions peaks of prepared sample 1 and 2 of graphene and tin oxide are observed at 332.7, 355.1, 375.5, 381, 410 nm and 301.4, 332.9, 354, 382, 387.2, 409.7 nm, respectively. Indicated by the TGA curves (Fig. 3c), an abrupt weight loss occurs from 360 to 560 °C indicating the oxidation of SnO_2: GNs. The spectrum pattern shows the weight loss of SnO_2, mainly caused

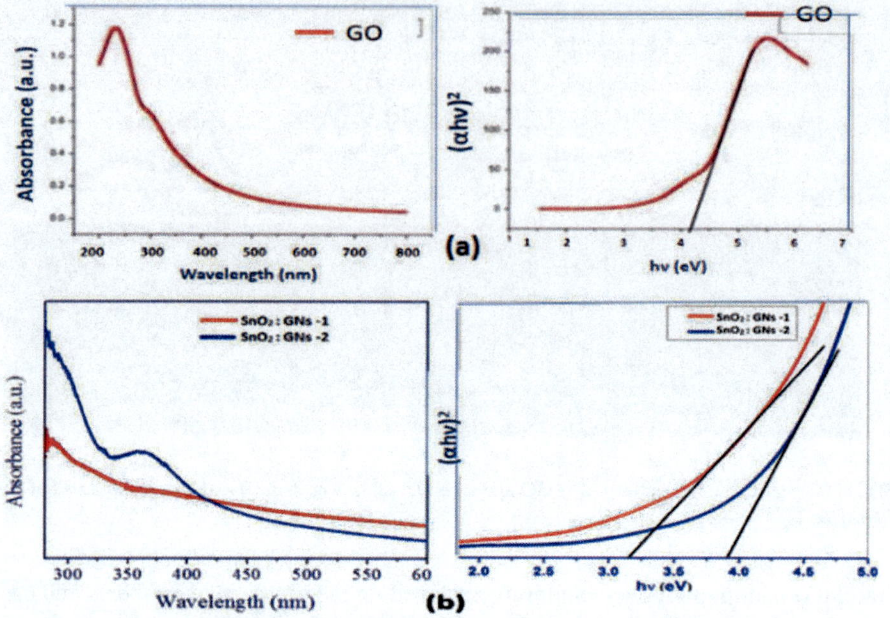

Fig. 2 Absorbance spectra and plot of $(\alpha h\nu)^2$ versus $h\nu$ of **a** GO and **b** SnO_2: GNs

by its dehydration. Weight loss in the temperature region 360 °C to 560 °C for SnO_2: GNs-1 & 2 Composite is about 56.7% and 58%, respectively. SnO_2 mass per cent is about 39%, 38% for SnO_2: GNs-1 & 2, respectively.

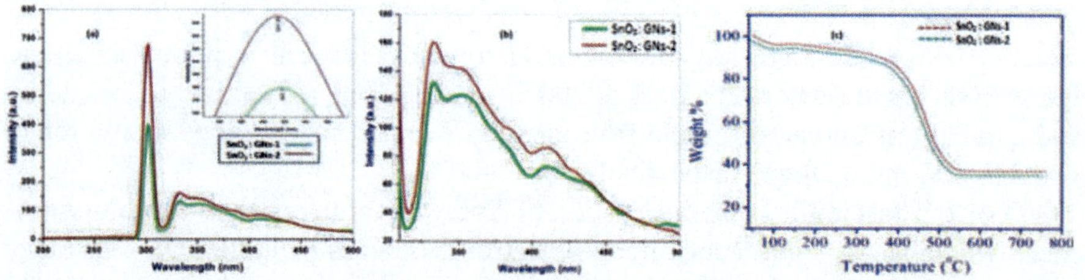

Fig. 3 **a**, **b** PL emission and excitation analysis of SnO_2;GNs-1 & 2 composite and **c** TGA analysis of SnO_2;GNs-1 & 2 composite

4 Conclusion

In method 1, the approach behind the synthesis was based on the chemical reduction in one step with no additional chemical used; therefore, it is a low-cost and efficient method for the synthesis of graphene nanosheets and SnO_2 nanocomposite. In method 2, the time taken for sample preparation was very less and gave high purity as compared to method 1. On changing the synthesis method, the band gap varies from 3.16 to 3.95 eV for prepared sample from method 1 and method 2 resp. Thermogravimetric Analysis (TGA) demonstrated that the composite had good thermal stability below 500 °C making it worth its application in the optical and electrical storage device.

References

1. Perreault F, De Faria AF, Elimelech M (2015) Environmental applications of graphene-based nanomaterials. Chem Soc Rev 44:5861–5896
2. Wang S, Jiang SP, Wang X (2011) Microwave-assisted one-pot synthesis of metal/metal oxide nanoparticles on graphene and their electrochemical applications. Electrochim Acta 56:3338–3344
3. Farheen, Parveen A (2022) Materials science in semiconductor processing enhanced visible light energy harvesting and efficient photocatalytic antibiotic drug degradation over egg albumen mediated Sr doped Fe_2O_3 nanoparticles. Mater Sci Semicond Process 148:106804

Studies on Nanochalcogenide $Se_{75}Te_{22}In_3$ and $Se_{75}Te_{19}In_6$ Thin Films Synthesized by Physical Vapor Condensation Technique

Imtiyaz H. Khan, Ravi P. Tripathi, and Shamshad A. Khan

Abstract Our purpose in this research work is to be synthesized $Se_{75}Te_{22}In_3$ and $Se_{75}Te_{19}In_6$ chalcogenide alloys which have been synthesized by using melt quenching method. Physical vapor condensation technique (PVCT) was used to prepared nano-thin films of synthesized $Se_{75}Te_{22}In_3$ and $Se_{75}Te_{19}In_6$ alloys. From HRXRD measurement clear that prepared thin films have amorphous texture. The morphological studies using FESEM suggest that the nanochalcogenide thin films contain particles of size ranges from 30–60 nm. Based on UV–Visible spectroscopy studies, absorption coefficient (α) is found to increase with photon energy and had significant change due to the change in Indium concentration. The optical absorption obeys rule of direct transition. Optical band gap increases from 2.18 to 2.22 eV when Indium content increases. These studies suggest that our prepared materials have possible application in different optical and storage devices.

Keywords Chalcogenide · HRXRD · UV–visible spectroscopy · Optical band gap

1 Introduction

Chalcogenide glasses (ChGs) have contributed in the emerging field of nano-sciences and nanotechnology. The physical, chemical, and thermal stability of ChGs have influenced the researchers of diverse fields for their implementation in practical applications for optoelectronic devices, photonics, photovoltaics, detectors, and in phase change memory devices [1]. Due to magnificent structural and optical properties, ChGs are acknowledged by the global communities for serving in medical and defense areas like infrared sources, thermal imaging, and chemical sensing applications [2]. ChGs consist low maximum phonon energy with high linear refractive

I. H. Khan (✉) · S. A. Khan
Materials Science Research Lab, Department of Physics, St. Andrew's College, Gorakhpur, U.P. 273001, India

R. P. Tripathi
Department of Physics, Mahatma Gandhi P. G. College, Gorakhpur, U.P. 273001, India

Z. H. Khan et al. (eds.), *Recent Advances in Nanomaterials*, Springer Proceedings in Materials 27, https://doi.org/10.1007/978-981-99-4878-9_52

index always available for possible modification. Such properties of ChGs are broadly used in many fields related to optical recording media, laser writer sensitivity and xerography. The photo-induced prospective of ChGs has electrographic applications like photon acceptors in laser fiber techniques, laser printing and in infrared spectroscopy [3]. Nowadays, innumerable amount of research is going on ChGs at the nanoscale regime. In current scenario, researcher has made attention to nanometric scale phase change recording materials [4, 5]. Andzanea et al. [6] have synthesized the thickness-dependent study Bi and Sb chalcogenide films synthesized by PVCT for their application in thermoelectric generators. Nur-E-Alam et al. [7] have prepared the added Ag-based metal-dielectric nanocomposite thin-film. Yavorskyi et al. [8] have synthesized the structural and optical characteristic of CdTe nanofilms. Alvi et al. [9] have synthesized Se–Te nanochalcogenide thin films. Benjamin et al. [10] have synthesized optical, and physical properties of electron-beam deposited a-Se-Te-M thin films. Khan et al. [11] have studies the structural and optical properties of $Se_{85}Te_{15-x}Bi_x$ nanofilms influenced by γ-irradiation. Alqahtani et al. [12] have studied a- Al-Se-Te thin films for application in optoelectronic. Mannu et al. [13] have studied Indium (In) based on chalcogenide thin films and their electrical transport properties for thermoelectric applications. Many research groups and scientists across the globe paid attention to develop various preparation techniques and characterizations of ChGs at this level [14–17].

2 Experimental Method

In the present work, Se has used as main content having exceptional uses in various fields of science and technology. The extensive used of selenium (Se) drag attention of researchers to advance its physical properties of thermal instability and low sensitivity by making their alloy with other elements. In their system, tellurium (Te) used as additive for reducing short coming of Se. Te is used to vulcanized rubber, tint glass and ceramics, in solar cell, DVD, etc. In is used because it has photovoltaic property, softness, and easily fusible which made them suitable for electronic devices and solar cells. The preparation of $Se_{75}Te_{22}In_3$ and $Se_{75}Te_{19}In_6$ ChGs was done by Melt-Quench Technique. PVCT applied for deposition of nanochalcogenide films having thickness 60 nm. Nano-thin films were synthesis liquid nitrogen on cooled glass substrates, and atmosphere in the chamber was ambient argon gas film thickness which was measured by Quartz crystal monitor. The structural studies of synthesized thin films by using Regaku X-ray diffract meter have been done. The studies of surface morphology of thin films by using field emission scanning electron microscopy (FESEM) have been done. A JASCO, spectrophotometer had used for optical studies of prepared films. Optical studies of nanostructured $Se_{75}Te_{22}In_3$ and $Se_{75}Te_{19}In_6$ thin films were done in the wavelength the region of 400–1000 nm.

3 Result and Discussion

3.1 Structural Studies

HRXRD measurements were done for the confirmation of structural properties of alloys and thin film at room temperature. X-ray diffraction was recorded from 10–80° range with having scanning 2°/min and chart preparing speed 1 cm/min. The HRXRD pattern is shown in Fig. 1 which consists no prominent peaks which was the evidence for amorphous nature of the synthesized thin films.

The FESEM images for $Se_{75}Te_{19}In_6$ (Fig. 2a) and $Se_{75}Te_{22}In_3$ (Fig. 2b) film show that films having signification of nano-particles with diameter in 30–60 nm range.

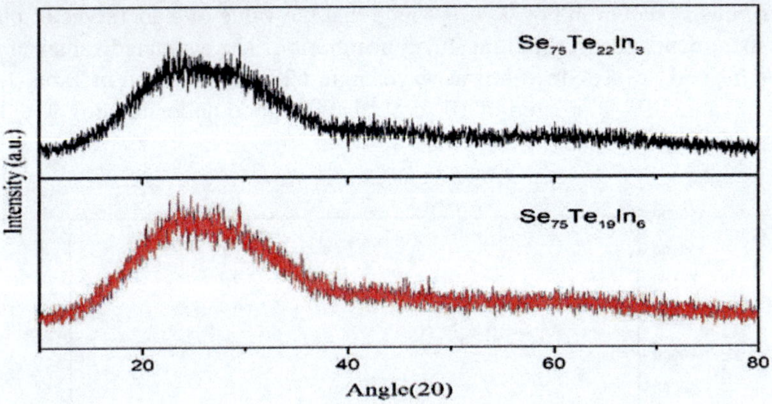

Fig. 1 HRXRD pattern of $Se_{75}Te_{19}In_6$ and $Se_{75}Te_{22}In_3$ nanochalcogenide thin films

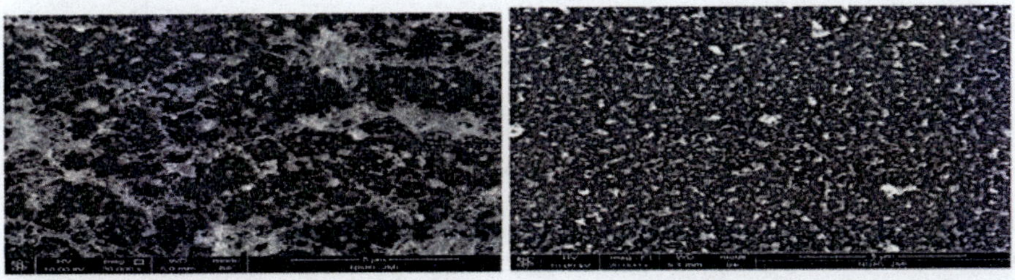

(a) FESEM image of $Se_{75}Te_{19}In_6$ Nano-thin Films

(b) FESEM image of $Se_{75}Te_{22}In_3$ Nano-thin Films

Fig. 2 **a** FESEM image of $Se_{75}Te_{19}In_6$ nano-thin films, **b** FESEM image of $Se_{75}Te_{22}In_3$ nano-thin films

3.2 Optical Studies

The optical absorption of nanochalcogenides gives basic idea for structural and optical energy gap. For $Se_{75}Te_{19}In_6$ and $Se_{75}Te_{22}In_3$ nanochalcogenide thin film using absorption data, we have calculated absorption coefficients (α) and optical band gap. An absorption process at absorption edge in photon having proper energy excited an electron from lower to upper level. This absorption edge can be assigned by one of the three process either by Urback tails, inter band absorption, or residual below-gap absorption. The value of α is estimated by the given equation [18, 19]

$$\text{Absorption coefficient } (\alpha) = \text{Optical density/thickness of the thin film} \quad (1)$$

Change in α with photon energy for $Se_{75}Te_{19}In_6$ and $Se_{75}Te_{22}In_3$ nanochalcogenides films is shown in Fig. 3. It is found that the value of α increases of photon energy increment and after Indium (In) concentration. The evaluated values of α for $Se_{75}Te_{19}In_6$ and $Se_{75}Te_{22}In_3$ films at wavelength 620 nm are listed in Table 1. The measured value of α is ordered of 10^4 cm^{-1} having good understanding with other workers [20, 21] results.

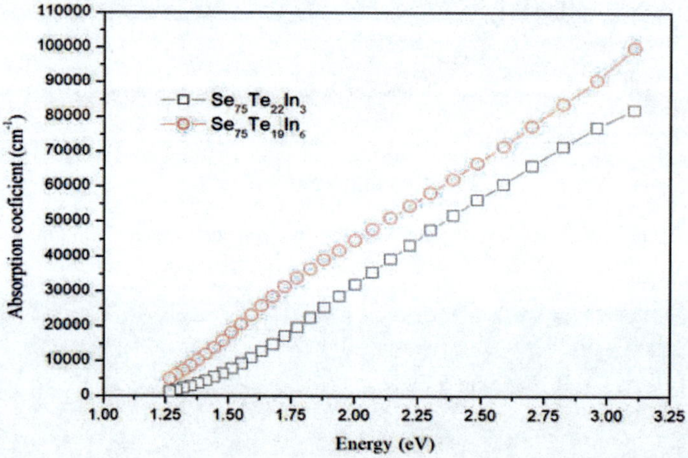

Fig. 3 Absorption coefficient (α) against photon energy ($h\nu$) of $Se_{75}Te_{22}In_3$ and $Se_{75}Te_{19}In_6$ nano-thin films

Table 1 Optical parameters of $Se_{75}Te_{22}In_3$ and $Se_{75}Te_{19}In_6$ nano-thin films

Optical parameters	Sample	
	$Se_{75}Te_{22}In_3$	$Se_{75}Te_{19}In_6$
Absorbance coefficient (α) at $\lambda = 620$ nm (10^4) cm^{-1}	3.6272	2.9693
Optical band gap (E_g) (eV)	2.22	2.18

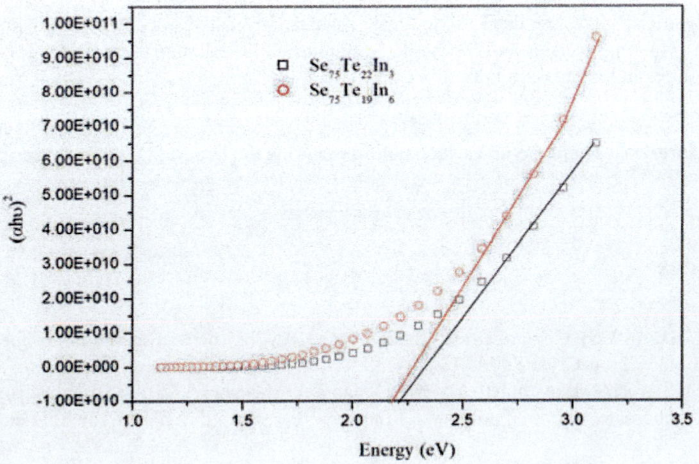

Fig. 4 $(\alpha\, h\nu)^2$ versus energy for $Se_{75}Te_{22}In_3$ and $Se_{75}Te_{19}In_6$ nano-thin films

The $Se_{75}Te_{22}In_3$ and $Se_{75}Te_{19}In_6$ nano-thin films follow the direct transition rule, and it can be estimated by equation

$$(\alpha\, h\nu)^2 = A(h\nu - E_g) \tag{2}$$

where A is dimensionless constant, E_g is the optical band gap. The graph of $(\alpha\, h\nu)^2$ with $(h\nu)$ of $Se_{75}Te_{19}In_6$ and $Se_{75}Te_{22}In_3$ nanochalcogenide thin films is presented in Fig. 4. The intersection of extrapolation of linear portion of graph on energy axis provides E_g values. Estimated values of E_g of $Se_{75}Te_{19}In_6$ and $Se_{75}Te_{22}In_3$ films are listed in Table 1. We observed that band gap increases (2.18–2.22 eV) with increase in Indium concentration. The change in band gap may be due to shift in Fermi level or increment in gain size. In our synthesized sample, the increment in gain size looks more prominent with small contribution of shift in Fermi level.

4 Conclusion

In this research work, our main focus is to analyze optical and structural characteristic of nanochalcogenide $Se_{75}Te_{19}In_6$ and $Se_{75}Te_{22}In_3$ thin films. HRXRD measurement suggests that films have amorphous nature. FESEM data shows high yield of nano-particles having size ranges from 30–60 nm. We observed (α) increases with increase in In content. E_g has sufficient change as In content changes. Further, following direct transition rule, the calculated value of E_g changes from 2.18 to 2.22 eV as In concentration increases. The estimated value of band gap suggests that our synthesized nano-thin films are handy for solar cell industries as well as many other optoelectronic and photonic devices.

Acknowledgements We are thankful to Prof. Zishan H. Khan Dept. of Applied Science and Humanities, F/O Engineering and Technology, Jamia Millia Islamia, New Delhi for providing research facilities in his research lab.

Conflict of Interest There are no as such conflicts for the declaration in this manuscript.

References

1. Seddon AB (1995) Chalcogenide glasses: a review of their preparation, properties, and applications. J Non-Cryst Solids 184:44–50
2. Khan SA, Tiwari G, Tripathi RP, Alvi MA, Khan ZH, Al-Agel FA (2014) Structural, optical, and structural characterization of polycrystalline $Ga_{15}Te_{85-X}Zn_X$ thin films. Adv Sci Lett 20:1715–1718
3. Khan SA, Al-Agel FA, Al-Ghamdi AA (2010) Optical characterization of nanocrystalline $Se_{85}Te_{10}Pb_5$ and $Se_{80}Te_{10}Pb_{10}$ chalcogenides. Superlattices Microstruct 47:695–704
4. Ni J, Bi X, Jiang Y, Li L, Lu J (2017) Bismuth chalcogenide compounds Bi_2X_3 (X=O, S, Se): Applications in electrochemical energy storage. Nano Energy 34:356–366
5. Bhatt VS, Yadav AK, Dixit D, Tomy CV (2022) High-temperature solution growth of large size chalcogenide FeT_XSe (T:Fe, Co) superconducting single crystals. Superconductivity 3:100016
6. Andzane J, Felsharuk A, Sarakovskis A, Malinovskis U, Kauranens E, Bechelany M, Niherysh KA, Komissarov IV, Erts D (2021) Thickness-dependent properties of ultrathin bismuth and antimony chalcogenide films and their application in thermoelectric generators. Mater Today Energy 19:100587
7. Nur-E-Alam M, Basher MK, Vasiliev M, Das N (2021) Physical vapor-deposited silver (Ag)-based metal-dielectric nanocomposites for thin-film and coating applications. Appl Sci 6746
8. Yavorskyi R, Nykyruy L, Wisz G (2019) Structural and optical properties of cadmium telluride obtained by physical vapor deposition technique. Appl Nanosci 9:715–724
9. Alvi MA, Al-Ghamdi AA, Madani JH (2014) Synthesis and characterization of Bi doped Se–Te nanostructured thin films. Measurement 58:325–329
10. Benjamin LK, Dube P, Tabi CB, Muiva CM (2021) Physical, linear and nonlinear optical properties of amorphous $Se_{90-x}Te_{10}M_x$ (M = Zn, In, Pb, x = 0, 5) chalcogenide thin films by electron-beam deposition. J Non-Cryst Solids 557:120646
11. Khan SA, Sahani RM, Tripathi RP, Shaheer Akhtar M, Srivastava A (2021) Influence of gamma-irradiation on the optical and structural properties of $Se_{85}Te_{15-x}Bi_x$ nano-thin chalcogenide films. Radiat Phys Chem 188:109659
12. Alqahtani A, Yakout HA, Shaaban ER, Qasem A (2022) Extended study into the prominence of Al content in controlling optical parameters, thermal properties and dielectric behavior of amorphous Al-Se-Te thin films for optoelectronic applications. Opt Laser Technol 156:108459
13. Mannu P, Palanisamy M, Bangaru G, Ramakrishnan S, Ramcharan M, Kandasami A (2022) Electrical transport properties of Indium chalcogenide thin films and their thermoelectric applications. Mater Today: Proc 48(2):115–118
14. Tripathi RP, Akhtar SM, Khan SA (2018) Thermally deposited $Se_{85}In_{15-x}Sb_x$ chalcogenide thin films: structural, electrical and optical properties. Mater Focus 7(2):251–258
15. Tripathi RP, Alvi MA, Khan SA (2020) Investigations of thermal, optical and electrical properties of $Se_{85}In_{15-x}Bi_x$ glasses and thin films. J Therm Anal Calorim 146:2261–2272
16. Srivastava A, Tiwari SN, Lal JK, Khan SA (2019) Phase transformation in $Se_{75}Te_{13}In_{12}$ chalcogenide thin films. Glass Phys Chem 45:111–118
17. Tripathi RP, Zulfequar M, Khan SA (2016) Structural, optical and electrical properties of $Se_{85}In_9Bi_6$ nanochalcogenide thin film. Curr Nanomater 1(3):176–182

18. Lal JK, Khan SA, Al-Gahamdi AA, Khan ZH (2009) Characterization of amorphous $Se_{97}Te_3$ nanoparticles prepared by ball milling. Int J Nanomanuf 4:208–218
19. Khan SA, Al-Agel FA, Faidah AS, Yaghmour SJ, Al-Gahamdi AA (2010) Characterization of $Se_{88}Te_{12}$ nanostructured chalcogenide prepared by ball milling. Mater Lett 64:1391–1393
20. Urbach F (1953) The long-wavelength edge of photographic sensitivity and of electronic absorption of solids. Phys Rev 92:1324
21. Dow JD, David R (1972) Toward a unified theory of urbach's rule and exponential absorption edges. Phys Rev B 5:594

Studying Effect of TiO_2 Nanoparticles on Soil Fertility and Plant Physiology Using IoT-Enabled Controlled Growth Chamber

Mridul Kumar, Khagendra Sharma, and Zeeshan Saifi

Abstract With the ever-increasing use of nanoparticles in many applications, the risk of entry of these particles into the natural food chain has also increased. As per recent reports, about 10^7 kilograms of TiO_2 nanoparticles are consumed annually by several industries like cosmetics, toothpastes, paints, and paper mills. Therefore, these nanoparticles are entering the natural ecosystem not only through industrial waste but also from the wastewater of every individual household. Currently, the scientific community has mixed opinions about the phytotoxicity of TiO_2 nanoparticles. There are many experiments in which the genotoxicity of TiO_2 and its adverse effects on the DNA and other proteins are reported; in contrast, numerous recent articles report that nanoprimed plants with TiO_2 may benefit in terms of crop yield. In this article, we present a systematic study of nanoprimed seeds with TiO_2 nanoparticles in the growth media of an IoT-enabled plant growth chamber. The method includes continuous monitoring of the electrical resistance of the growth media using our recently developed sensor. A healthy plant exhibits a rhythmic pattern in terms of the electrical resistance of the growth media around its root; however, this rhythm is disrupted when the plant is stressed. The method is generic and detects both biotic and abiotic stress. In this article, a controlled study of the induced stress by adding different concentrations of TiO_2 nanoparticles (size < 100 nm) in agarose growth media in which *Cicer arietinum* seedlings were germinated is reported.

Keywords TiO_2 nanoparticles · Plant stress · Soil fertility · Internet of Things

---

M. Kumar (✉) · K. Sharma · Z. Saifi
Department of Physics and Computer Science, Dayalbagh Educational Institute, Dayalbagh, Agra 282005, India
e-mail: mridulsharma@dei.ac.in

Z. Saifi
e-mail: zeeshan@dei.ac.in

Z. H. Khan et al. (eds.), *Recent Advances in Nanomaterials*, Springer Proceedings in Materials 27, https://doi.org/10.1007/978-981-99-4878-9_53

1 Introduction

Recent advancements in nanotechnology have increased the reach of the nanoparticles (NPs). TiO_2, one of the most manufactured NPs, is being consumed in a wide range of products including cosmetics and food. Hence, it is expected that these NPs would find their way into various ecosystems, and it becomes necessary to study their effect on living beings. Researchers are divided on how these NPs affect animals and plants. On one hand, some studies show that exposure to TiO_2 NPs can cause both genotoxicity and cytotoxicity to humans and animals [1, 2]; on the other hand, these NPs have shown to positively affect the root length, shoot length, germination percentage, germination rate, and fresh weight of the plants [3–5]. These positive effects on the plants are caused by the changes in seed metabolism and signaling pathways by TiO_2 NPs. When it is deliberately done, it is called nanopriming [6]. Nanopriming has been used to enhance germination, growth, and yield in different seeds in an eco-friendly and sustainable way [7, 8]. However, since NPs are still chemicals, which are sometimes toxic, it is yet unclear how they will affect the seed quality and soil quality over a long period of time. Therefore, methods are needed to evaluate the effect of nanopriming on the seed and soil quality which are non-destructive in nature. The present study focuses on studying nanoprimed *Cicer arietinum* seeds with the help of our recently developed sensor mechanism [9]. It focuses on the continuous measurements of electrical resistance of the growth media and surrounding temperature throughout the lifecycle of the plant. Plants show a rhythmic behavior in terms of electrical resistance of the growth media. However, this rhythmic behavior changes when plants are stressed because plants start to uptake more nutrients than usual because of the initiation of their defense mechanism [9–11]. Our goal is to study the change in rhythmic behavior of the plants with and without the effect of NPs on them. To validate our results, we have also studied seed germination using traditional approach.

2 Materials and Method

Seeds of chickpeas (*Cicer arietinum*) were chosen for the experiment because they have been well studied. We have used TiO_2 (Brookite size < 100 nm) NPs purchased from Sigma Aldrich for studying their effect on the plants. This paper reports two experiments. In the first experiment, chickpeas pre-exposed to TiO_2 NPs were grown in artificial growth chambers with agarose as the growth medium. In the second experiment, chickpeas were grown in beakers with both agarose (Run 1) and soil (Run 2) as the growth medium.

Experiment 1: An artificial growth chamber was created using 8 borosilicate bottles for growing chickpeas and measuring the electrical resistance characteristic of the growth media (see Fig. 1). Electrodes were made with copper and graphite tape and placed at the wall of the growth chamber. Agarose was used as growth medium and 100 mL of 8 g/L concentration of it was poured in all the bottles. Plants in bottles 1–3 were not exposed to TiO_2 NPs, plants in bottles 4–6 were exposed with TiO_2 NPs of concentration 1000 mg/L, 100 mg/L and 10 mg/L respectively, bottles 7 and 8 were without plants. Bottle 7 was poured with 200 µL of TiO_2 NPs of 100 mg/L concentration. The setup consists of a Raspberry Pi for IoT-enabled data collection, a relay module with 8 channels for turning on the circuit for only one bottle at a time, a multimeter for resistance measurement of the growth media.

Working: Raspberry Pi sends a signal to relay to turn the circuit ON for a bottle then sends a command to measure the electrical resistance of the growth media to the digital multimeter which then returns the measured value. Raspberry Pi also sends a signal to DHT11 module for the measurement of temperature and humidity and stores all these values (date, time, temperature, humidity, and electrical resistance) in different csv files for each plant.

Experiment 2: This experiment included two runs, in the first run agarose was used for growing chickpeas and in the second run soil was used. In both runs, three sets of seven 250 mL beakers were used to grow the chickpeas, to assess the impact of TiO_2 NPs on them. In first 3 beakers, chickpeas which were pre-exposed with TiO_2 NPs of 1000 mg/L, 100 mg/L, and 10 mg/L, respectively, were grown. In the next three beakers, chickpeas grown without NPs exposure were sprayed every 48 h with

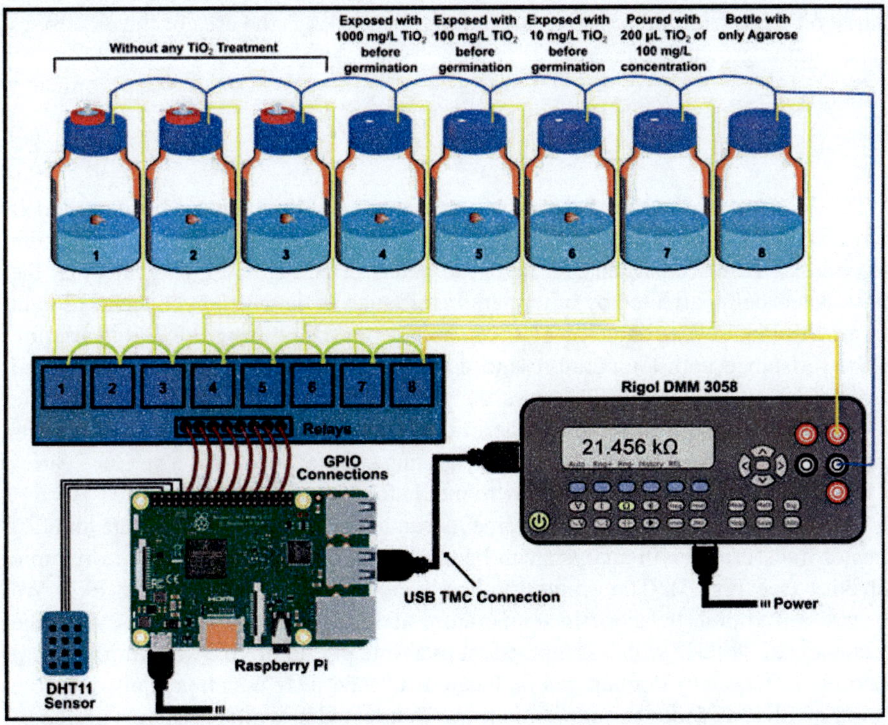

Fig. 1 This figure shows the experimental setup used for electrical resistance measurements [13]

concentrations of TiO_2 NPs of 1000 mg/L, 100 mg/L, and 10 mg/L, respectively. The natural growth of the chickpea was observed in the last beaker. The average weight of the seedling was calculated at the end of the experiment to assess the health of the plants.

3 Results and Discussion

Experiment 1: The data shows that as the temperature increases during daytime the resistance of the growth media decreases (see Fig. 2). This can be associated with the temperature compensation factor of the agarose media due to its ionic nature [12]. The variation of resistance for ionic solutions can be given by the following equation, where symbols have their usual meaning [13].

$$R = \frac{R_0}{1 + \alpha(T - T_0)} \tag{1}$$

Here, α is the temperature compensation factor (TCF) of agarose media, and R_0 can be given by the Drude's model as below, where m is the mass of the charge carriers, l is the distance between electrodes, n is the charge carrier concentration, q is the charge on the carrier ion, τ is the relaxation time, and A is the area of the electrodes dipped in the growth media,

$$R_0 = \frac{ml}{nq^2\tau A} \tag{2}$$

TCF leads to a rhythmic pattern in resistance (see Eq. 1) which can be identified as increased value in nighttime and decreased value in daytime because of temperature (see Fig. 1). However, when this rhythmic behavior is disturbed, it means that the resistance is being affected by other parameters such as the nutrient concentration in the growth media. The value of TCF can be calculated by using the daily variation of the resistance with temperature and using Eq. 1. In our case, it was calculated to be 0.0191/°C. Equation (2) shows that the charge carrier concentration and the electrical resistance are inversely proportional to each other, therefore, when plants initiate their self-defense mechanism while being stressed electrical resistance varies as the amount of nutrients in the growth media fluctuates [9–11].

The continuous electrical measurements can be taken as time series data and fast Fourier transform (FFT) analysis can be applied to it for finding out the rhythmic behavior (see Fig. 3). The amplitude has been normalized after taking FFT. We observe a rhythmic behavior in temperature and the electrical resistance with the frequency 26.78 μHz which is equivalent to a time period of 10.36 h. Another local maxima in frequency domain can be located at 19.64 μHz which is equivalent to a period of 14.13 h. Since the experiment was conducted in winter season, a period of 10.36 h corresponds to daytime, and a period of 14.13 h corresponds to nighttime

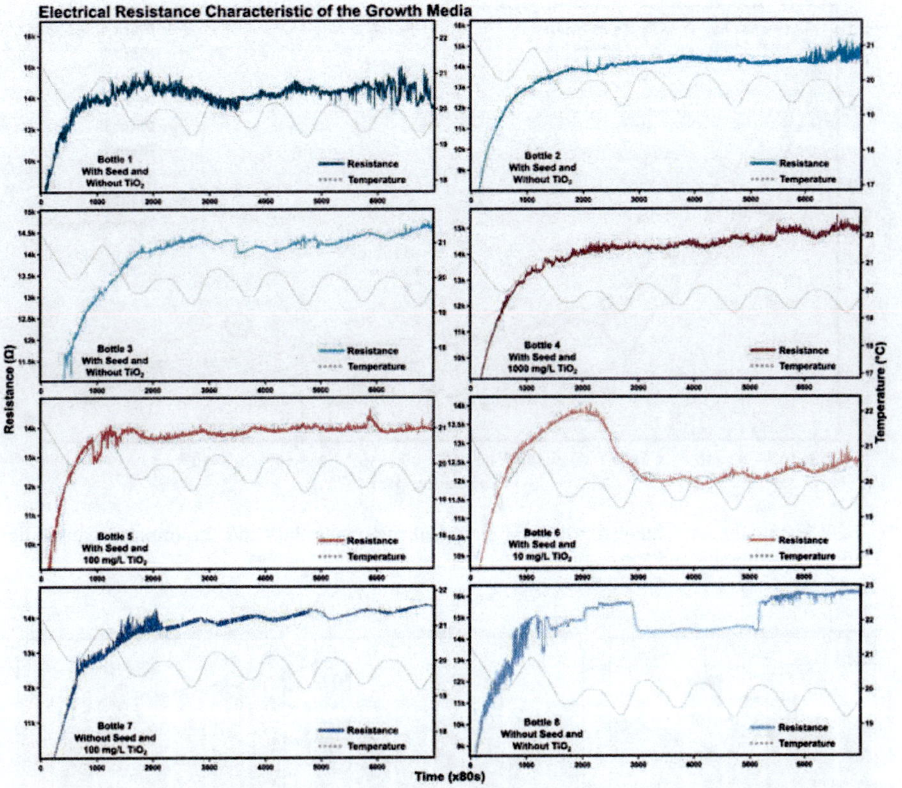

Fig. 2 This figure shows the variation of electrical resistance with time for the growth media in all the bottles

and adds up to ~ 24 h. Bottle 8 which is without any plant and TiO_2 NPs shows the lowest amplitude (0.286) compared to other bottles at 26.78 μHz, even bottle with TiO_2 in agarose shows an amplitude of 0.429. Plant 6 shows distinguishable variation in amplitude at frequencies 55.36 μHz and 76.79 μHz. Bottles 1, 6, and 8 show a periodic behavior at 8.93 μHz which is close to one quarter of a day.

Experiment 2: There is little or no effect on average shoot length and root length of the seedlings when chickpea seeds are pre-exposed with all the concentrations (1000, 100, and 10 mg/L) of TiO_2 NPs in both agarose and soil (see Fig. 4). However, there is a significant effect of spraying TiO_2 NPs on both average shoot and root length after the initial germination in both agarose and soil. It has been seen that the average shoot and root length decrease with the exposure to decreasing concentrations of TiO_2 NPs. However, the average root and shoot length remain less than the naturally grown seeds except for 1000 mg/L concentration in which for most of the cases, it is better than naturally grown seeds. The average weight of the seedlings in agarose media was unaffected by both the pre-exposure to NPs and spraying of them post-germination and remains close to naturally grown seedling weight (0.33 ± 0.04 gm). In the case

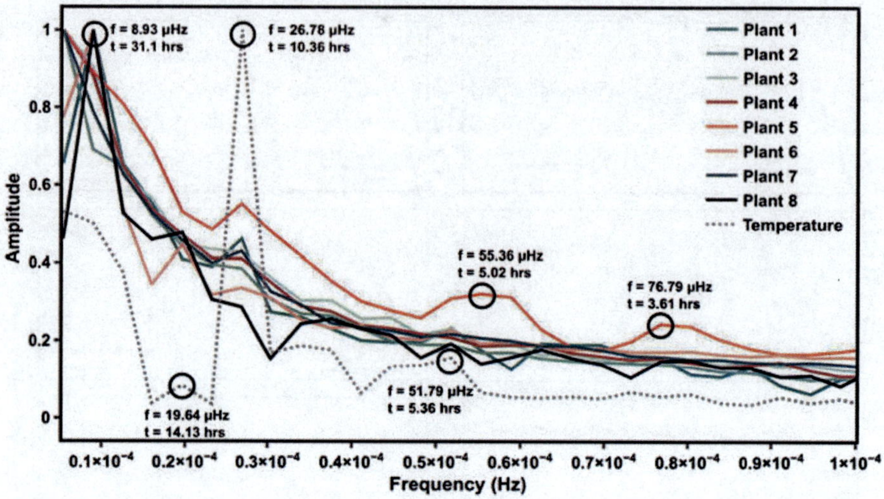

Fig. 3 It shows the FFT analysis of plant electrical resistance data and the important periodic behavior. Here, plant 7 and 8 are controls

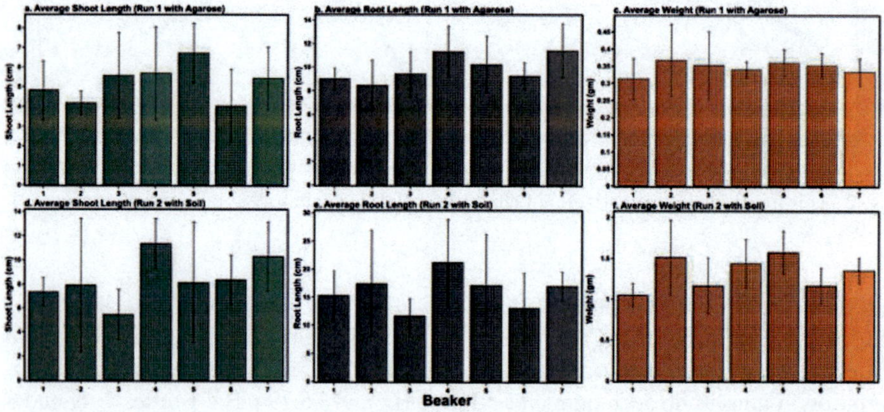

Fig. 4 This figure shows the variation of average shoot length, root length, and weight of the seedlings at the end of the Experiment 2. Beakers 1–3 are pre-exposed with the concentration 1000 mg/L, 100 mg/L, and 10 mg/L of TiO_2 NPs, respectively. Beakers 4–6 were sprayed with the concentration of TiO_2 NPs in the same manner after the seeds were germinated. Finally, beaker 7 shows the naturally grown seedling

of soil as growth media average weight is almost same for pre-exposed and sprayed seeds for concentration 100 and 10 mg/L. Since the experiment in soil ran for 13 days, the average weight of the seedlings is higher than the average weight in agarose media which only ran for 7 days. Even though the experiment was elaborate and tried to limit the possible variables which might have affected the overall results, some data points have high standard deviation which demands for more experimentation to verify these results.

4 Conclusion

In this paper, we studied the nanopriming of chickpeas by growing them in lab conditions. Our results show that the plants show a rhythmic behavior in electrical resistance of growth media, and when plants are stressed, the rhythm is changed. The stress may be caused by biotic or abiotic stressors (in our case, TiO_2 NPs). We also found that when chickpeas are pre-exposed to TiO_2 NPs, there is no significant change in the average seedling root or shoot length. However, in the case of spraying, the NPs on seedlings after germination exhibited positive effects. The average weight of the seedlings remains almost the same in the case of agarose growth media and varies by little margin in soil.

5 Declaration of Interest Statement

All authors declare no competing interests.

Acknowledgements Authors would like to thank Priyanshi, Prachi, Deepali, and Aditi for their help in the experiment.

References

1. Jalili P et al (2018) Investigation of the in vitro genotoxicity of two rutile TiO_2 nanomaterials in human intestinal and hepatic cells and evaluation of their interference with toxicity assays. NanoImpact 11:69–81
2. Wani MR, Shadab G (2020) Titanium dioxide nanoparticle genotoxicity: a review of recent in vivo and in vitro studies. Toxicol Ind Health 36(7):514–530
3. Dehkourdi EH, Mosavi M (2013) Effect of anatase nanoparticles (TiO_2) on parsley seed germination (*Petroselinum crispum*) in vitro. Biol Trace Elem Res 155(2):283–286
4. Mahmoodzadeh H, Aghili R (2014) Effect on germination and early growth characteristics in wheat plants (*Triticumaestivum l.*) seeds exposed to TiO_2 nanoparticles. J Chem Health Risks 4:29–36
5. Szymańska R et al (2016) Titanium dioxide nanoparticles (100–1000 mg/l) can affect vitamin E response in *Arabidopsis thaliana*. Environ Pollut 213:957–965
6. do Espirito Santo Pereira A, Caixeta Oliveira H, Fernandes Fraceto L, Santaella C (2021) Nanotechnology potential in seed priming for sustainable agriculture. Nanomaterials 11(2):267
7. Acharya P, Jayaprakasha GK, Crosby KM, Jifon JL, Patil BS (2020) Nanoparticle-mediated seed priming improves germination, growth, yield, and quality of watermelons (*Citrullus lanatus*) at multi-locations in Texas. Sci Rep 10

8. Mahakham W, Sarmah AK, Maensiri S, Theerakulpisut P (2017) Nanopriming technology for enhancing germination and starch metabolism of aged rice seeds using phytosynthesized silver nanoparticles. Sci Rep 7
9. Kumar M, Krishnananda SD, Zeeshan (2021) Apparatus and method for determining plant stress. Indian Patent Appl 202111033265, Published: 2021
10. Wang M, Zheng Q, Shen Q, Guo S (2013) The critical role of potassium in plant stress response. Int J Mol Sci 14(4), Art. no. 4
11. Batish DR, Lavanya K, Singh HP, Kohli RK (2007) Phenolic allelochemicals released by *Chenopodium murale* affect the growth, nodulation and macromolecule content in chickpea and pea. Plant Growth Regul 51(2), Art. no. 2
12. Kim J-S, Chun K-Y, Han C-S (2018) Ion channel-based flexible temperature sensor with humidity insensitivity. Sens Actuators, A 271:139–145
13. Kumar M, Saifi Z & Krishnananda SD (2023) Decoding the physiological response of plants to stress using deep learning for forecasting crop loss due to abiotic, biotic, and climatic variables. Sci Rep 13:8598. https://doi.org/10.1038/s41598-023-35285-3

Detection of Mercury Ions Using PVP-Capped AgNPs for Wastewater Analysis

Shailja Pandey, S. K. Sharma, and Shipra Mital Gupta

Abstract The present study reports comparative qualitative and quantitative detection of Hg^{2+} by using polyvinyl pyrrolidone-capped silver nanoparticle (PVP-AgNPs) from aqueous sample in a colorimetric detection system of PVP-AgNPs/aspartic acid/Hg^{2+} and PVP-AgNPs/L-arginine/Hg^{2+}. PVP-AgNPs were synthesized using chemical reduction method. The prepared PVP-AgNPs were characterized by UV–visible spectrophotometer. On addition of electron-rich molecules like aspartic acid and L-arginine, the as-synthesized yellow colored PVP-AgNPs form conjugate with them. However, on addition of Hg^{2+}, the solutions turn yellow to colorless. The variation in the intensity of yellow color which eventually turns colorless at higher concentration of Hg^{2+} can be utilized for the colorimetric sensing of Hg^{2+}. This assay shows immediate color change with a naked-eye detection limit of 200 μM each for PVP-AgNPs/aspartic acid/Hg^{2+} and PVP-AgNPs/L-arginine/Hg^{2+} detection systems, respectively. Hg^{2+} is quantitatively determined using a UV–Vis spectrophotometer. PVP-AgNPs/L-Arginine/Hg^{2+} had better detection limit of 11.402 μM compared to 13.784 μM for PVP-AgNPs/aspartic acid/Hg^{2+} system. This study concluded that PVP-AgNPs in the presence of a conjugating agents like L-arginine and aspartic acid can be helpful in development of colorimetric sensors for detection of Hg^{2+}.

Keywords Colorimetric detection · Silver nanoparticles · Mercury · Environmental monitoring · Heavy metals

S. Pandey · S. M. Gupta (✉)
USBAS, Guru Gobind Singh Indraprastha University, New Delhi 110078, India

S. K. Sharma
USCT, Guru Gobind Singh Indraprastha University, New Delhi 110078, India

Z. H. Khan et al. (eds.), *Recent Advances in Nanomaterials*, Springer Proceedings in Materials 27, https://doi.org/10.1007/978-981-99-4878-9_54

1 Introduction

Heavy metal ions are considered highly toxic for living beings. Mercury is one of the most toxic heavy metals. Mercury contamination in environment is due to three major sources namely, metallic, organic, and inorganic salts [1]. Once mercury is able to enter food chain and then the biochemical pathways, it hinders with the healthy functioning of several organ systems causing severe health-related issues like damage in kidney, lungs, neurological damage, and cardiovascular damage [2]. Therefore, detection of Hg^{2+} is need of the hour due to its highly toxic effect on environment and human health. For this purpose, several sophisticated instruments have been used. These instruments offer good sensitivity but they are expensive, tedious, require long hours of sample preparation process, and skilled technicians to perform and interpret data [3, 4]. An alternative method to develop a detection system for Hg^{2+} must be discovered that is not only rapid and cost-effective but also easy to analyze and interpret. Colorimetric detection systems have been developed for this purpose [5]. Plasmonic nanoparticles like gold nanoparticles (AuNPs) are popular with this colorimetric technique. But the problem with AuNPs is that their synthesis is expensive. Hence, researchers have aimed to replace AuNPs by more economical materials like silver nanoparticles (AgNPs) and copper nanoparticles [6]. The problem with copper nanoparticles is that they oxidize very easily, offering low stability. But AgNPs prove to be a good alternative to AuNPs. Here, we try to develop a detection system for Hg^{2+} using polyvinyl pyrrolidone-capped AgNPs (PVP-AgNPs) in presence of L-arginine and aspartic acid.

2 Materials and Method

2.1 Instrumentation

The UV–visible absorption spectrophotometer (U-2190, HITACHI) was used for the characterization of Ag NPs.

2.2 Materials

Double distilled water was used for synthesis and dilutions throughout the experiments. Silver nitrate ($AgNO_3$) was sourced from Rankem. Sodium borohydride ($NaBH_4$) was purchased from Fisher Scientific. L-arginine ($C_6H_{14}N_4O_2$) and aspartic acid ($C_4H_7NO_4$) were sourced from Thomas Baker. Mercuric sulfate ($HgSO_4$) was obtained from SRL Pvt Ltd. Polyvinyl pyrrolidone (PVP) $(C_6H_9NO)_n$ was bought from Qualikems. All chemicals used were of analytical grade. They were used as received without further purification.

2.3 *Methodology*

2.3.1 Synthesis of PVP-AgNPs

PVP-AgNPs were prepared as previously reported [7] with slight modification. 20 mL of 1 mM $AgNO_3$ was prepared. To that, 2 mL of 0.001 w/v% of PVP was added and stirred using magnetic stirrer. 30 mL of 1 mM of $NaBH_4$ was prepared and kept in ice-bath for chilling. $AgNO_3$ and PVP mixture were added drop by drop to $NaBH_4$ solution. The solution turned yellow, indicating the formation of PVP-AgNPs.

2.3.2 Addition of L-arginine and Aspartic Acid to PVP-AgNPs

1 mM of L-arginine solution was made. L-arginine was added in as-synthesized PVP-AgNPs in 3:1 volume ratio of PVP-AgNPs and L-arginine, respectively. The samples were kept for 30 min. Similar procedure was followed with aspartic acid and PVP-AgNPs.

2.3.3 Hg^{2+} Sensing Using PVP-AgNPs

To each of the above systems of PVP-AgNPs/L-arginine and PVP-AgNPs/aspartic acid, 1 mL of Hg^{2+} of varying concentrations (0, 10, 20, 30, 40, 50, 60, 70, 80, 90, 100, and 200 μM) was added. The volume ratio of the resultant solution of PVP-AgNPs, L-arginine/aspartic acid, and Hg^{2+} was 3:1:2. The samples were subjected to UV–Vis analysis immediately.

3 Results and Discussion

UV–Vis absorption spectra of as prepared PVP-AgNPs are shown in Fig. 1. The figure clearly depicts the surface plasmon resonance (SPR) peak of PVP-AgNPs at λ_{max} = 406 nm. Figures 2 and 4 show the photographs of PVP-AgNPs/L-arginine/Hg^{2+} and PVP-AgNPs/aspartic acid/Hg^{2+} detection systems, respectively. It is clearly visible in the photographs that yellow PVP-AgNPs turn colorless on addition of 200 μM Hg^{2+} to both the systems. Figures 3 and 5 show the change in the absorbance spectra of both the detection systems along with the calibration graph. The absorbance at 406 nm decreased gradually and the SPR band showed a blue shift with increase in the concentration of Hg^{2+}. There was a linear relationship ($R^2 = 0.97627$ for PVP-AgNPs/L-Arginine/Hg^{2+} and $R^2 = 0.94278$ for PVP-AgNPs/aspartic acid/Hg^{2+}) between the absorbance at λ_{max} = 406 nm versus Hg^{2+} concentration. The limit of detection (LOD) of the detection systems was calculated by the equation LOD = 3a/k, where a is the standard deviation and k is the slope of the calibration graph.

PVP-AgNPs/L-Arginine/Hg^{2+} shows better detection limit of 11.402 μM compared to 13.784 μM of PVP-AgNPs/aspartic acid/Hg^{2+} detection system. The detection parameters including linear range and detection limit are summarized in Table 1. Aspartic acid and L-arginine have carboxylic group that can link with the aliphatic group of PVP forming conjugates with PVP-AgNPs. This conjugate then attaches to Hg^{2+} from -NH_2 end similarly proposed by [7].

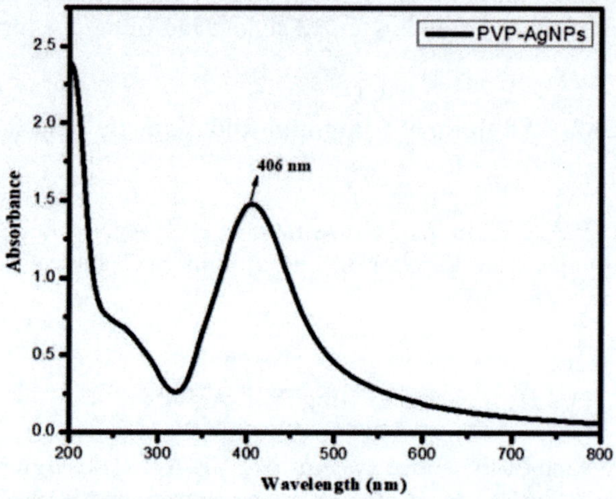

Fig. 1 UV–Vis absorption spectra of PVP-AgNPs

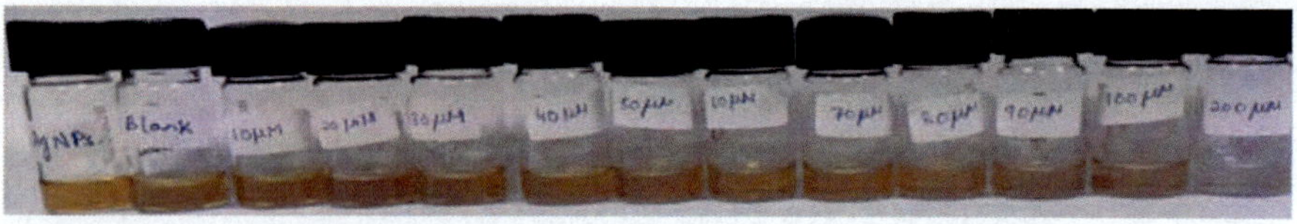

Fig. 2 Photograph of PVP-AgNPs/aspartic acid/Hg^{2+} system in presence of different concentrations of Hg^{2+} (0, 10, 20, 30, 40, 50, 60, 70, 80, 90, 100, and 200 μM)

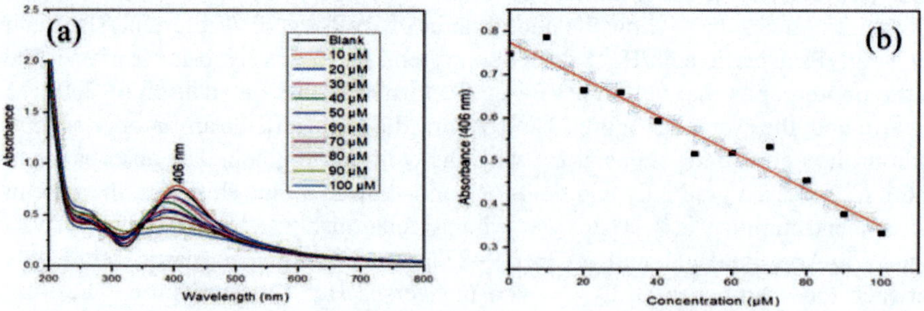

Fig. 3 **a** UV–Vis absorbance spectra, **b** calibration curve of PVP-AgNPs/aspartic acid/Hg^{2+} system as function of Hg^{2+} concentration (0, 10, 20, 30, 40, 50, 60, 70, 80, 90, and 100 μM)

Fig. 4 Photograph of PVP-AgNPs/L-arginine/Hg^{2+} system in presence of different concentrations of Hg^{2+} (0, 10, 20, 30, 40, 50, 60, 70, 80, 90, 100, and 200 μM)

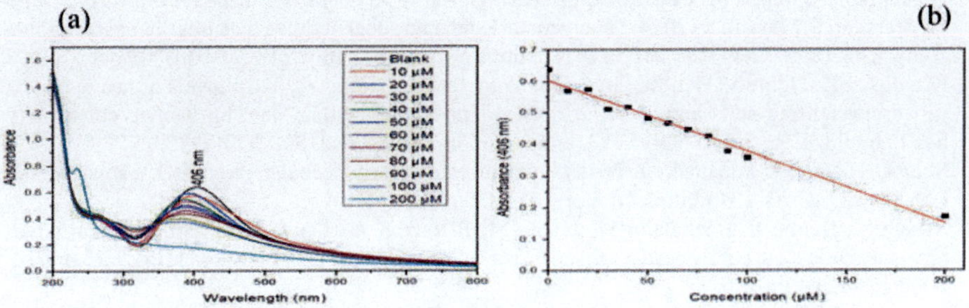

Fig. 5 **a** UV–Vis absorbance spectra, **b** Calibration curve of PVP-AgNPs/L-arginine/Hg^{2+} system as function of Hg^{2+} concentration (0, 10, 20, 30, 40, 50, 60, 70, 80, 90, and 100 μM)

Table 1 Comparison of developed detection systems

System	Limit of detection	Linear range
PVP-AgNPs/aspartic acid/Hg^{2+}	13.784 μM	0–100 μM
PVP-AgNPs/L-arginine/Hg^{2+}	11.402 μM	0–200 μM

4 Conclusion

The PVP-AgNPs/aspartic acid and PVP-AgNPs/L-arginine were sensitive toward the detection of Hg^{2+}. The developed method is rapid, cost-effective, and easy. The prepared detection systems can be applied in order to the sense Hg^{2+} in various wastewater samples in wastewater monitoring.

Acknowledgements This study is financially supported by GGSIP University, New Delhi, India, under FRGS.

Declaration of Interest Statement The authors declare that they have no conflict of interests.

References

1. Driscoll CT, Mason RP, Chan HM, Jacob DJ, Pirrone N (2013) Mercury as a global pollutant: sources, pathways, and effect. Environ Sci Technol 47:4967–4983. https://doi.org/10.1021/es305071v
2. Fernandes Azevedo B, Barros Furieri L, Peçanha FMI, Wiggers GA, Frizera Vassallo P, Ronacher Simões M, Fiorim J, Rossi De Batista P, Fioresi M, Rossoni L, Stefanon I, Alonso MJ, Salaices M, Valentim Vassallo D (2012) Toxic effects of mercury on the cardiovascular and central nervous systems. J Biomed Biotechnol 2012:1–11. https://doi.org/10.1155/2012/949048
3. Giacoppo S, Galuppo M, Calabrò RS, D'Aleo G, Marra A, Sessa E, Bua DG, Potortì AG, Dugo G, Bramanti P, Mazzon E (2014) Heavy metals and neurodegenerative diseases: an observational study. Biol Trace Elem Res 161:151–160. https://doi.org/10.1007/s12011-014-0094-5
4. Kinuthia GK, Ngure V, Beti D, Lugalia R, Wangila A, Kamau L (2020) Levels of heavy metals in wastewater and soil samples from open drainage channels in Nairobi, Kenya: community health implication. Sci Rep 10:1–13. https://doi.org/10.1038/s41598-020-65359-5
5. Shen L, Hagen JA, Papautsky I (2012) Point-of-care colorimetric detection with a smartphone. Lab Chip 12:4240–4243. https://doi.org/10.1039/c2lc40741h
6. Sabela M, Balme S, Bechelany M, Janot JM, Bisetty K (2017) A review of gold and silver nanoparticle-based colorimetric sensing assays. Adv Eng Mater 19:1–24. https://doi.org/10.1002/adem.201700270
7. Balasurya S, Syed A, Thomas AM, Marraiki N, Elgorban AM, Raju LL, Das A, Khan SS (2020) Rapid colorimetric detection of mercury using silver nanoparticles in the presence of methionine. Spectrochim Acta Part A Mol Biomol Spectrosc 228:117712. https://doi.org/10.1016/j.saa.2019.11771

An Analysis on the Effects of Metal Ion Dopant in the Structural, Optical, Morphological, and Magnetic Properties of Zinc Sulphide Nanoparticles

D. Sukanya, A. Antony Heartlin Sancta, and K. Shruti

Abstract ZnS is a versatile functional material and ZnS nanostructured assemblies of low-dimensional nanocrystals with controllable crystal phase and morphology which are highly desirable for exploiting novel properties and extending their applications. In the current research, a detailed study on the effect of metal ion dopant (Ni) in pure ZnS nanoparticles was investigated by varying their molar concentrations in pure ZnS. The simple, more reliable, and cost-effective co-precipitation route has been adopted for the synthesis of good quality Ni-doped ZnS nanoparticles. The effect of Nickel dopants on pure ZnS and the enhancement of the structural, optical, morphological, and magnetic properties was investigated employing X-ray diffraction, UV–Vis spectroscopy, photoluminescence, HRSEM, and VSM. The cubic zinc blende structural formation and the effect of Nickel substituents on pure ZnS nanoparticles are confirmed from the X-ray diffraction patterns. The band gap energy of pure ZnS nanoparticles was found to show a linear decrease with the addition of Ni dopants. Thus, the present report highlights the efficiency of Ni dopants and their concentration on pure ZnS in fine-tuning the internal properties of ZnS and transforming them into ideal candidates for photocatalytic, optoelectronic, and anti-microbial applications.

Keywords Transition metal ions · Photocatalysis · Photodegradation · Dopants

1 Introduction

The material science domain has been swiftly advancing their research in semiconductor nanoparticles for addressing the environmental hazards related to deadly and non-biodegradable waste, which are accountable for soil, water, and air pollution [1, 2]. Photocatalysis of semiconductor nanoparticles is an effective approach in sorting out various aqueous pollutions by decomposing organic contaminants into

D. Sukanya (✉) · A. Antony Heartlin Sancta · K. Shruti
Department of Physics, Stella Maris College, Chennai, Tamilnadu, India
e-mail: sukanderose@gmail.com

Z. H. Khan et al. (eds.), *Recent Advances in Nanomaterials*, Springer Proceedings in Materials 27, https://doi.org/10.1007/978-981-99-4878-9_55

simple inorganic compounds [2]. ZnS nanoparticles belonging to II–VI family have attained great attention due to its prominent structural, morphological, optical, electrical, magnetic, and photocatalytic properties which emerge from the property of quantum confinement [3]. ZnS nanoparticles promote high photocatalytic activities including CO_2 photoreduction, photoreduction of halogenated benzene derivatives by dehalogenation, organic pollutant degradation using photocatalysis, and photocatalytic splitting water of producing H_2 [4]. It is a polymorphous material that occurs in dual crystalline forms, viz. zinc blende (sphalerite) and wurtzite owning a wide band gap (E_g) of nearly 3.68 and 3.91 eV for zinc blende and wurtzite. Excitation binding energy of 40 meV is associated with a radius of 2.5 nm of the Bohr excitonic radius [5]. Zinc blende (ZnS) has a more lucid cubic form when compared to wurtzite which takes a hexagonal form. The coordination geometry at S and Zn is tetrahedral in these forms. A change in the electronic structure and transition probabilities of ZnS caused by doping with different transition metals (TM) such as nickel, copper, manganese, and cobalt can strongly influence its optical properties. Among these nickel ions have profuse electronic configurations, with a smaller ionic radius than Zn, implying that the nickel dopant can simply incorporated in ZnS crystal lattice [1, 3]. Also, as compared to the other TM, nickel dominates with intense magnetic moment [6]. It is widely known that large band gap semiconductors possess enriched efficiency for organic dye molecules through photocatalytic degradation, owing to the enhancement of the photogeneration electron–hole pair rate and a great negative reduction potential of the exciting electrons [7]. Many techniques such as solvothermal synthesis, thermal decomposition method, sol–gel, microwave irradiation, co-precipitation method, gas phases condensed, and solid-state reaction method have been widely utilized to prepare inorganic ZnS nanoparticles. The current research reports on the development of PVA capped pure and Ni-doped ZnS nanoparticles engaging a facile, cost-effective, high yield, efficient, and simple chemical co-precipitation method. The influences of Ni^{2+} doping content on the phase-structure, morphology, elemental composition, and optical properties on the ZnS nanostructures and their photocatalytic behaviour on the dye degradation of methylene blue were also investigated.

2 Methodology

2.1 *Experimental Procedure of Pure and Ni-Doped ZnS Nanoparticles*

In this typical synthesis procedure, polyvinyl alcohol (PVA) acts as a better stabilizer for preparing pure and metal (Ni) doped ZnS. 0.2 M of zinc chloride and sodium sulphide solutions each were prepared in double distilled water under constant stirring. The solutions of nickel nitrate for three different molar ratios (0.5 M, 0.1 M, 0.075 M) were prepared separately. After continuous stirring for certain hours to obtain homogenous solution, the solutions of $ZnCl_2$ (0.2 M) are mixed with an individual solution of nickel II nitrate (0.5 M, 1 M, 0.075 M). Then, the solution of Na_2S is added in drops to each solution and stirred continuously. The solution was enriched for the attainment of pH 11 by adding NaOH solution in drops. An aqueous solution of polyvinyl alcohol (PVA) was prepared and used as a stabilizer for preparing metal doped ZnS. The as-prepared solutions were allowed to form a homogenous mixture at 60 °C. The obtained samples were centrifuged with water and ethanol to remove impurities and dried overnight at 100 °C. The dried samples were calcined to obtain fine ZnS nano-powders. The final products are labelled as ZN1 (pure), ZN2 (0.05 M of Ni), ZN3 (0.1 M of Ni), and ZN4 (0.075 M of Ni).

3 Results and Discussion

The pure and nickel-doped ZnS nanoparticles were subjected to powder X-ray diffraction analysis for studying the crystallographic information. Figure 1 depicts the X-ray diffraction patterns of pure ZnS and Ni-doped ZnS nanoparticles at varied molar concentrations, respectively. The high crystalline nature of the sample was well revealed by the three sharp and prominent broad peaks conforming to the lattice planes (111) (311) and (220) at the specific angle of 29.07°, 49.54°, and 57.39°. These lattice planes are in good agreement with cubic geometry (JCPDS Card No. 05-0566). From the XRD patterns, it could be inferred that the synthesized pure and doped ZnS nanoparticles are of very high quality without the presence of any additional impurities. It is also observed that the diffraction peaks are broad in nature which attributes to the smaller dimension of the synthesized nanoparticles. The diffraction peaks also displayed a shift towards greater angles of 2θ with the addition of Ni impurity [8]. The obtained structural parameters are summarized in Table 1. It is affirmed from the tabulated values that nickel doping increases the crystallite size. As a result of XRD pattern, the average crystallite size was estimated to be 2 nm. The gradual increase in nickel concentration also corresponds to the increase in intensity of the Ni-doped ZnS peaks [9]. It is understood that typical lattice defects including dislocation and micro strain decreases with increasing nickel doping concentrations, possibly due to the improved crystalline nature and high orientation along (111) direction (Fig. 1). In addition, it is observed from the doped ZnS nanoparticles that the peak angle (111) shifts slightly towards the higher angles, indicating that nickel ions are integrated into the lattice structure of ZnS [8].

The morphological analysis using HRSEM accompanied by EDX depicted in Figs. 2 and 3 implies that the morphologies are noticeably dependent on the synthesis procedure and crystal composition. HRSEM images clearly affirm the microsphere morphology for pure ZnS which turns into agglomerated nanosheets in addition of Nickel dopant. The nanosheets of microspheres gradually crimped with increasing Nickel doping in ZnS nanoparticles [10]. The role of polyvinyl alcohol as a good stabilizer and Ni dopants in modulating the size and shape of the nanoparticles is obvious in the observed images. The EDX investigation of pure ZnS in Fig. 4a affirms

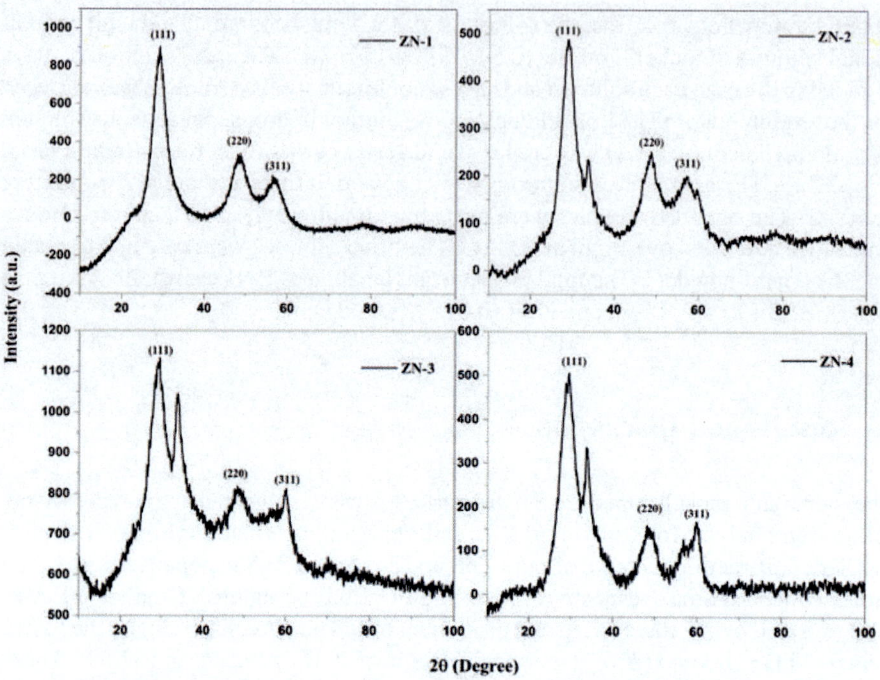

Fig. 1 XRD patterns of pure and Ni doped ZnS at varied molar concentration

Table 1 Structural and optical properties for pure and Ni-doped ZnS nanoparticles

Nickel doping (M)	Crystallite size D (nm)	Dislocation density (δ)	Micro strain (ε)	Absorbance peak (nm)	Emission peak (nm)	Bandgap energy (eV)
0	1.2876	0.6031	0.038154	315	371, 612	3.42
0.05	1.5654	0.4080	0.025966	314	371, 613	3.31
0.075	1.6846	0.3523	0.023996	308	611	3.38
0.1	1.7358	0.3318	0.042524	299	611	3.41

the existence of S and Zn in nearly stoichiometric ratios, and the atomic proportion of these elements is 58.21 and 41.79, individually. Figure 4b illustrates the occupation of the dopant nickel along with S and Zn with 22.13, 37.99, and 39.88%, respectively, which confirms the formation of good quality and highly pure product.

The absorbance spectra, band gap, and emission spectra at an excitation wavelength of 325 nm recorded for the pure and Ni-doped ZnS samples are shown in Fig. 5, and the values are shown in Table 1. From the absorbance spectra of Ni-doped ZnS nanoparticle, we can identify that for the lower content of Ni, a red-shift in the absorbance wavelength has been observed, whereas the wavelength is gradually blue-shifted with a higher amount of Ni doping which confirms the quantum

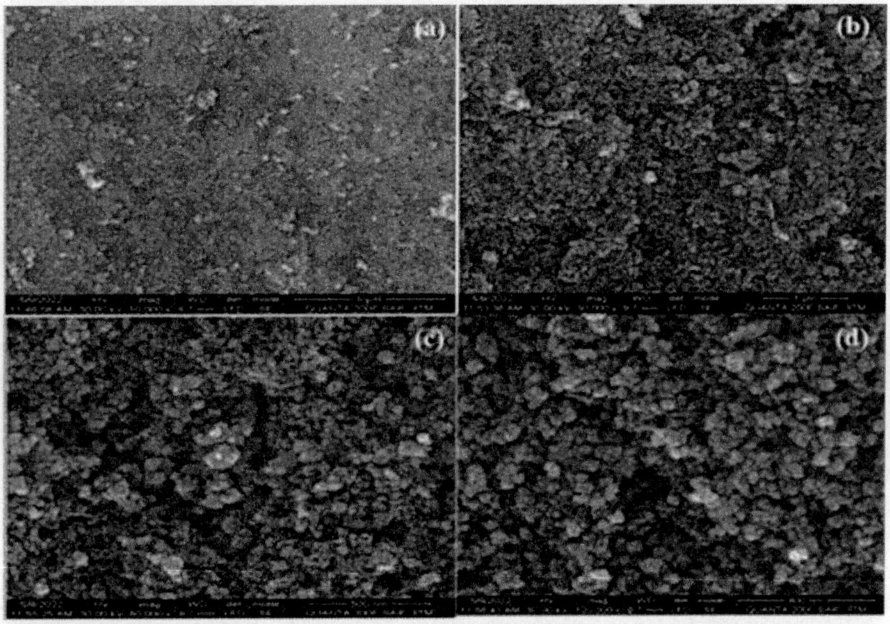

Fig. 2 HRSEM micrographs for pure ZnS

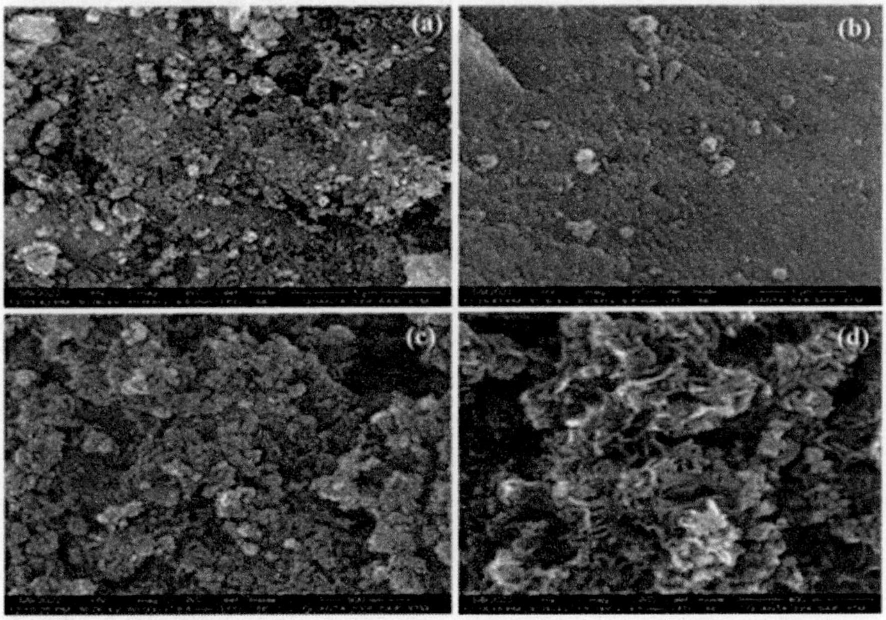

Fig. 3 HRSEM micrographs for Ni (0.1 M) doped ZnS

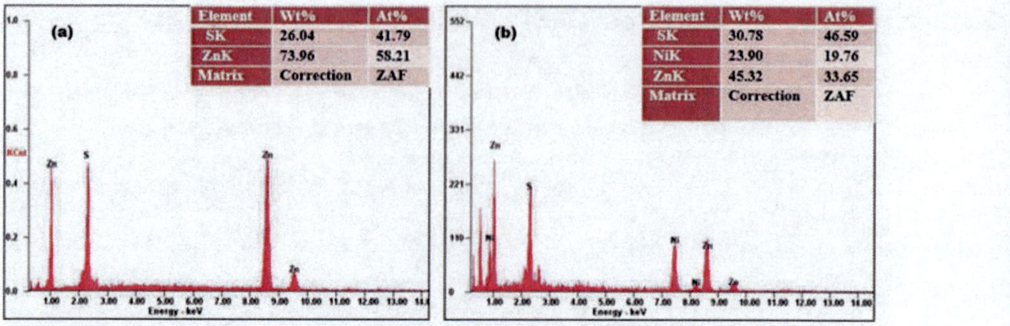

Fig. 4 EDAX spectrum of (**a**) pure ZnS, (**b**) Ni doped (0.1 M) doped ZnS nanoparticles

confinement effects of the Ni-doped samples due to the electron–hole pairs generated by the photons and the effect of positioning encumbrance exhibited by ZnS nanoparticles under the influence of Ni [11, 12]. Identically, Ni-doped ZnS samples synthesized by V. Kumar et al. also reported a similar trend of decreased band gap as in Fig. 6 [13]. In the PL spectra (Fig. 7), the excitation wavelength at 371 nm of pure ZnS nanoparticle may be attributed to the electron recombination at the donor sulphur sites with their holes confined to the acceptor zinc sites. Quenching in the intensity of photoemission with subsequent doping of Ni^{2+} on ZnS is consistent with the quenching on Ni-doped ZnS nanorods reported by Kumar and Verma [14]. Also, when Ni^{2+} is assimilated within the lattice structure of ZnS and replaced for cationic host sites, the hybrid between the d electrons of Ni^{2+} and s-p electrons of the host ZnS creates a forbidden transition from $^4T_1$ to $^6A_1$ partly allowed [15]. This results in yellow-orange photoemission nearly at 611 nm. This confirms the existence of nickel ions hosted within the lattice of ZnS. As Ni^{2+} has small ionic radii, it can be simply inserted with a T_d symmetry into Zn, creating a trapping electron centre that allows for electronic transitions by altering the energy states [16, 17]. Thus, it is also affirmed from the results of Lahariya and Ramrakhiani having observed the same for Mn^{2+}ion.

FTIR spectra are employed to discover the position and vibration of ions in a crystal. Transmittance FTIR spectrum of pure and Ni^{2+}-doped ZnS nanoparticles prepared at optimum temperature was documented (Fig. 8). The ZnS nanoparticles exhibited the characteristics of high-strength product formation, where the peaks at 3423 cm^{-1} correspond to O–H stretching due to the small amount of water absorption by the particles. The peak at 2935 cm^{-1}can be ascribed to N–H stretching. C–H stretching is identified at 2851 cm^{-1}, and the presence of nitrate groups is identified at 1384 cm^{-1}. Correspondingly, a peak at 1450 $cm^{-1}$is credited to the C–H bending. The peaks at 637, 640, 661, and 650 cm^{-1} are characteristics of ZnS cubic structure, and it is related to ZnS vibration.

Magnetic studies were performed by the vibrating sample magnetometer (VSM) to confirm the magnetic behaviour prompted in pure ZnS when doped with Ni at varied concentrations. The M–H curves observed at the ambient temperature are

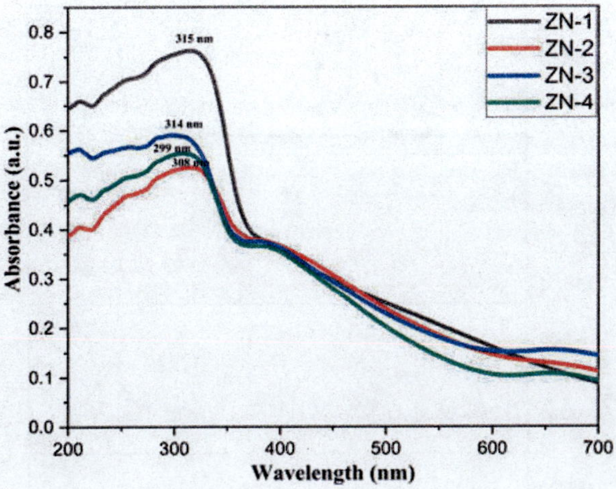

Fig. 5 Absorption spectra of pure and Ni doped ZnS nanoparticles

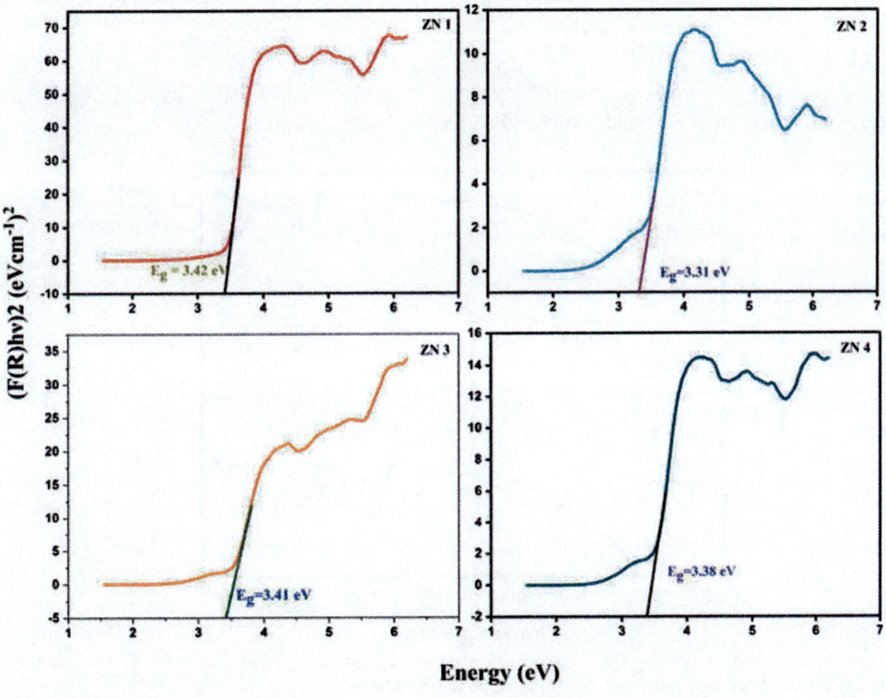

Fig. 6 Tauc plot and band gap energy estimation of pure and Ni doped ZnS nanoparticles

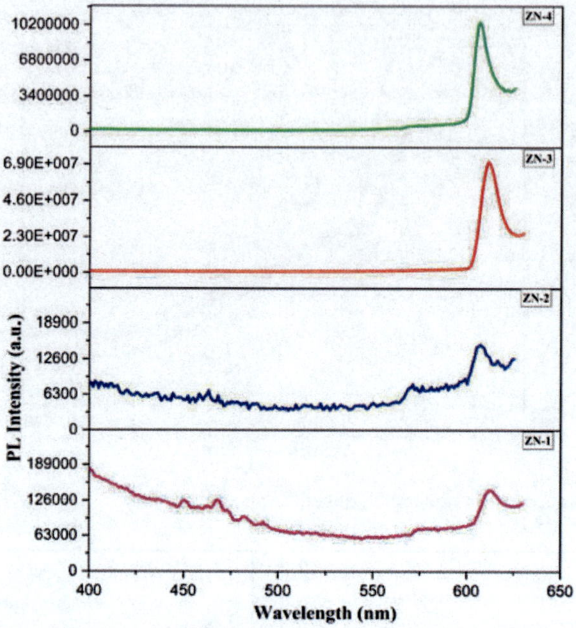

Fig. 7 PL spectrum of pure and Ni doped ZnS nanoparticles

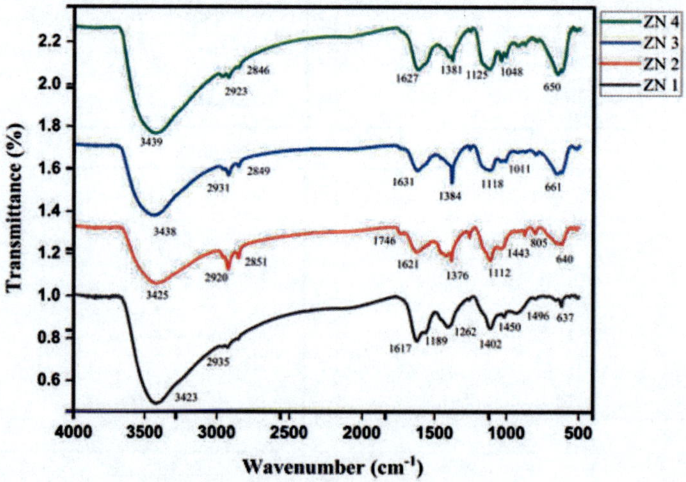

Fig. 8 FTIR spectra of pure and Ni doped ZnS nanoparticles

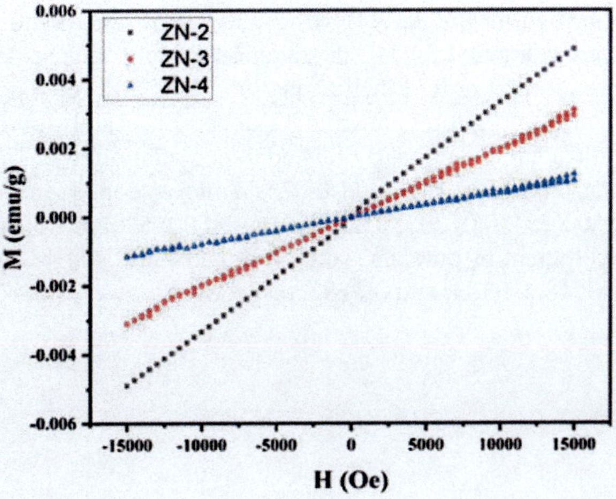

Fig. 9 M–H plot of Ni doped ZnS at different concentrations (0.05 M, 0.75 M, 0.1 M)

shown in Fig. 9. It is evident from the magnetization curves that Ni-doped ZnS nanoparticles exhibit superparamagnetic behaviour. This is due to the lower concentration of nanoparticles with small sizes [18]. The magnetization values for 0.05 M, 0.75 M, and 0.1 M of Ni doping are 0.005, 0.003, and 0.001 emu/g, respectively. Nortan et al. and Park et al. have reported superparamagnetism in cobalt-doped zinc oxide (ZnO) nanoparticles due to the smaller particle size nature [19, 20]. In addition, Kumar et al. have observed that at greater Ni-doped ZnS concentrations, there is a decrease in the distance of Ni–Ni ions which led to the development of an unsaturated M–H curve [21]. Thus, when the Ni ions are substituted in the lattice of ZnS, the concentration of Ni increases, from which it is understood that increasing Ni ion concentrations in the ZnS may result in ferromagnetic behaviour.

The photocatalytic activity can be controlled by light absorption property and the rate at which electron and hole recombines [4, 8]. In this regard, ZnS is an interesting catalytic entity in removing organic pollutants, as it is a material that owns the direct band gap with astounding chemical stability with resisting oxidation and hydrolysis [1, 12, 22]. As nickel present in the ZnS lattice, their PL spectra obtained a yellow-orange photoemission at 611 nm. Photo-catalytic responses of pure and 0.05 M, 0.1 M, and 0.075 M concentrations of Ni-doped ZnS were recorded for photodegradation under UV irradiation. The absorption spectra were noted for different intervals of time and the contaminant methylene blue (MB) used for degradation showed a maximal absorption wavelength at around 665 nm. In this context, photodegradation was performed on MB with pure and various concentrations of Ni-doped ZnS which were then calcined at 650 °C by treating the mixture of MB and photocatalyst to the UV light. Figure 10 displays the absorption spectra of MB during the photodegradation of pure and Ni-doped ZnS. The degradation efficiency per cent of the methylene blue is estimated and displayed in Fig. 11. It is inferred that 0.05 M Ni-doped ZnS catalyst degraded about 48.23% of MB in 60 min. In order to evaluate the degradation effect, the absorption peak of MB positioned at 665 nm was assessed by varying time breaks at 0, 10, 20, 30, 40, 50, and 60 min. The rate constant (k) values were calibrated for MB degradation by pure, and Ni-doped ZnS was modelled using pseudo-first order kinetic equation. Figure 11 represents the MB photodegradation under UV light by pure and Ni-doped ZnS catalysts. The graph evidences that time of irradiation is linearly related to degradation rate. The rate constant values of pure and 0.05 M, 0.1 M, and 0.075 M nickel-doped ZnS nanoparticles are determined as 0.0254, 0.01203, 0.000754, and 0.00452 min^{-1}. Over and above, the correlation coefficient of pure ZnS and Ni-doped ZnS catalysts was determined to be 0.9065, 0.8548, 0.0303, and 0.3782, respectively.

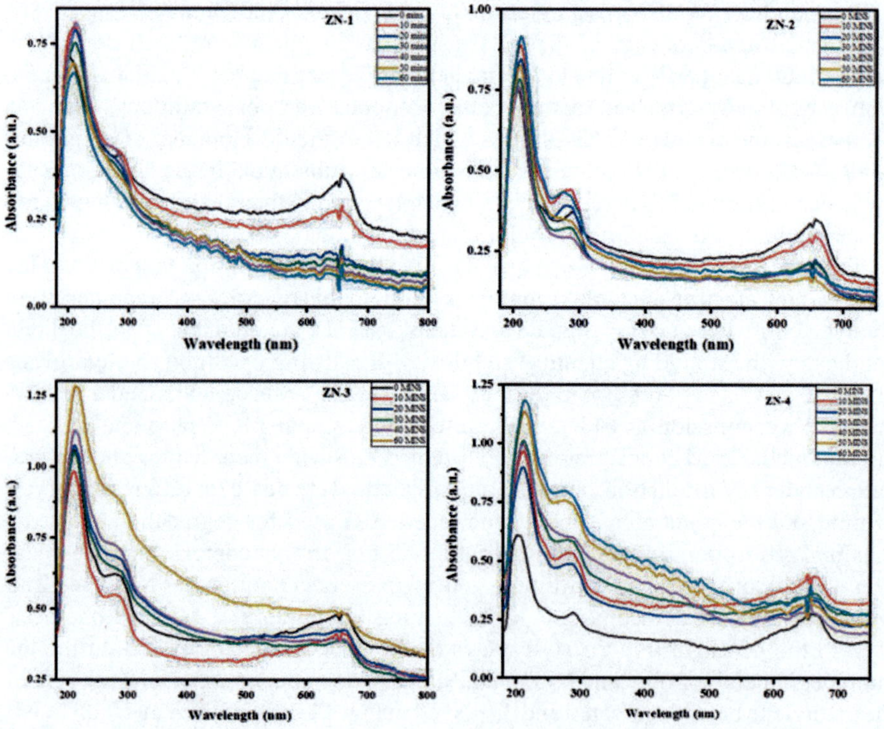

Fig. 10 Time dependent UV–Vis absorption spectra for degradation of methylene blue using pure ZnS and Ni doped ZnS at varied concentrations under visible light

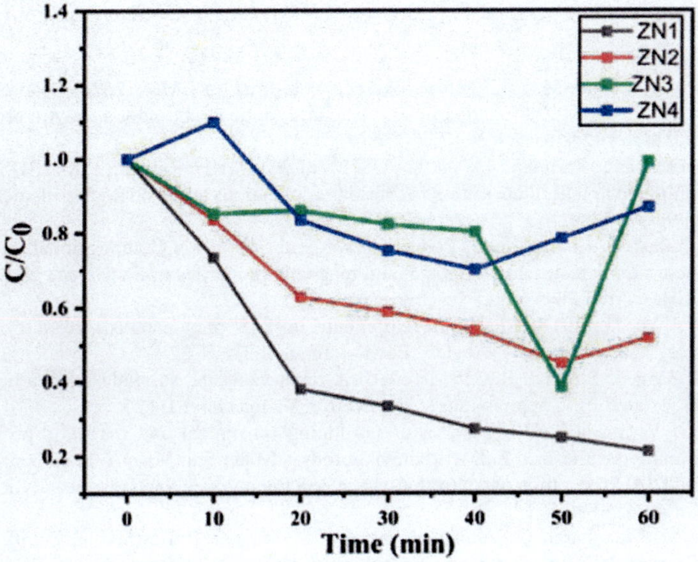

Fig. 11 Variation of C/C_0 against irradiation time for pure ZnS and Ni doped ZnS nanoparticles

4 Conclusion

Pure and nickel-doped zinc sulphide nanoparticles have been successfully synthesized by adopting a chemical co-precipitation route at varied molar concentrations, and it is proved to be an efficacious approach. XRD studies established the zinc cubic blende structure of ZnS with an average crystallite size of around 2 nm, and it is affirmed from the diffraction studies that Ni^{2+} ions doped with the Zn^{2+} ions in the lattice without any additional impurity formation. HRSEM images reveal the spherical microstructure and agglomerated nanosheet morphology of ZnS nanoparticles with reduced size caused due to excess Na_2S. Enhanced photoluminescence with increasing nickel concentration obtained yellow-orange photoemission at 611 nm, and beyond this, photoluminescence quenching was observed. Besides, the magnetic studies revealed that the Ni-doped ZnS nanoparticles exhibited a superparamagnetic behaviour due to the lower concentration of nanoparticles with smaller sizes. Photocatalytic activity of methylene blue under UV light substantiated pure and nickel-doped ZnS as a potential catalyst for the degradation of organic pollutants. Henceforth, the current research indicates that highly potential Ni-doped ZnS nano photocatalysts can be developed effectually at 0.05 M concentration for the degradation of organic pollutants.

References

1. Othman AA, Osman MA, Ali MA, Mohamed WS, Ibrahim EMM (2020) Sono-chemically synthesized Ni-doped ZnS nanoparticles: structural, optical, and photocatalytic properties. J Mater Sci: Mater Electron 31(2):1752–1767
2. Shakil MA, Das S, Rahman MA, Akther US, Majumdar MKH, Rahman MK (2018) A review on zinc sulphide thin film fabrication for various applications based on doping elements. Mater Sci Appl 9:751–778
3. Poornaprakash B, Chalapathi U, Prabhakar Vattikuti SV (2017) Compositional, morphological, structural, microstructural, optical, and magnetic properties of Fe Co, and Ni doped ZnS nanoparticles. Appl Phys Mater Sci Process 123:275
4. Lee G-J, Wu JJ (2017) Recent developments in ZnS photocatalysts from synthesis to photocatalytic applications—a review. Powder Technol 318:8–22
5. Kaur N, Kaur S, Singh J, Rawat M (2016) A review on zinc sulphide nanoparticles: from synthesis, properties to applications. J Bioelectron Nanotechnol 1(1):5
6. Kumar S, Verma NK (2014) Effect of Ni-doping on optical and magnetic properties of solvothermally synthesized ZnS wurtzite nanorods. J Mater Sci: Mater Electron 25:785–790
7. Zhang S (2014) Preparation of controlled-shape ZnS microcrystals and photocatalytic property. Ceram Int 40:4553–4557
8. Jothibas M, Manoharan C, Johnson Jeyakumar S, Praveen P, Kartharinal Punith-avathy I, Prince Richard J (2018) Synthesis and enhanced photocatalytic property of Ni doped ZnS nanoparticles. Solar Energy 159:434–443
9. Amaranatha Reddy D, Liu C, Vijayalakshmi RP, Reddy BK (2014) Effect of Al doping on the structural, optical and photoluminescence properties of ZnS nanoparticles. J Alloys Compd 582(5):257–264
10. Wang G, Huang B, Li Z, Lou Z, Wang Z, Dai Y, Whangbo M-H (2015) Synthesis and characterization of ZnS with controlled amount of S vacancies for photocatalytic H_2 production under visible light. Sci Rep 5:8544
11. Soni H, Chawda M, Bodas D (2009) Electrical and optical characteristics of Ni doped ZnS clusters. Mater Lett 63:767–769
12. Kaur J, Sharma M, Pandey OP (2015) Photoluminescence and photocatalytic studies of metal ions (Mn and Ni) doped ZnS nanoparticles. Opt Mater 47:7–17
13. Kumar V, Rawal I, Kumar V, Goyal PK (2019) Efficient UV photodetectors based on Ni-doped ZnS nanoparticles prepared by facial chemical reduction method. Physica B 575:411690
14. Kumar S, Verma NK (2015) Investigation of the magnetic and optical properties of Wurtzite Fe-doped ZnS nanorods. J Electron Mater 44:2829–2834
15. Pathak CS, Pathak PK, Kumar P, Mandal MK (2012) Characterization and optical properties of Ni^{2+} Doped Zns nanoparticles. J Ovonic Res 8(1):15–20
16. Lahariya V, Ramrakhiani M (2020) Luminescence study on Mn, Ni co-doped zinc sulfide nanocrystals. Luminescence 35:924–933
17. Borse PH, Deshmukh N, Shinde RF (1999) Luminescence quenching in ZnS nanoparticles due to Fe and Ni doping. J Mater Sci 34(24):6087–6093
18. Leena AMB, Raji K (2019) Magnetic, structural and optical properties of nickel doped zinc sulphide nanoparticles. Int J Sci Res Phys Appl Sci 7(3):17–21, 2348–3423
19. Norton DP, Overberg ME, Pearton SJ, Pruessner K (2003) Ferromagnetism in cobalt-implanted ZnO. Appl Phys Lett 83:5488
20. Park JH, Kim MG, Jang HM, Ryu S (2004) Co-metal clustering as the origin of ferromagnetism in Co-doped ZnO thin films. Appl Phys Lett 84:1338
21. Kumar S, Chen CL, Dong CL, Ho YK, Lee JF, Chan TS, Thangavel R, Chen TK, Mok BH, Rao SM, Wu MK (2013) Room temperature ferromagnetism in Ni doped ZnS nanoparticles. J Alloys Compd 554:357–362
22. Chauhan R, Kumar A, Chaudhary RP (2014) Photocatalytic degradation of methylene blue with Cu doped ZnS nanoparticles. J Lumin 145:6–12

Synthesis and Spectral Study of Ce^{3+} Activated $Sr_3(VO_4)_2$ Nanophosphor

Sajad Ahmad Bhat, Reyaz Ahmad, Pankaj Biswas, Pavneet Kour, Rozy Attri, and M. A. Mir

Abstract Ce^{3+} activated $Sr_3(VO_4)_2$ powders were prepared by combustion method. The structural and optical properties were determined by powder X-ray diffractometry (XRD), photoluminescence (PL) spectroscopy, and UV–Vis spectroscopy for diffuse reflectance (DR) studies. XRD was used to identify the phase of the material. Debye Scherrer equation was used to estimate the crystallite size of the material. From DR studies, optical band gap was found to be about 3.93 eV. On excitation with 325 nm near UV, the emission spectrum of the phosphor exhibited a broad band about 478 nm and a number of sharp peaks in the range 550–750 nm. The most prominent peaks about 574 and 618 nm were ascribed to the $^5d_1 \rightarrow {}^2f_{5/2}$ and $^5d_1 \rightarrow {}^2f_{7/2}$ transitions of Ce^{3+} ion. These experimental results reveal that the phosphor emits in yellow–red region with chromaticity coordinates as (0.40, 0.34) and thereby implying that it can be well pumped by 325 nm InGaN/GaN UV chips to give out emission near the white illuminant point for display applications.

Keywords Nanophosphors · WLEDs · Photoluminescence · Combustion

1 Introduction

In the past few decades, rare earth doped nanophosphors have been extensively studied for their use in optoelectronic industry, white light emitting diodes due to their efficient luminescent properties and possess low-power consumption [1]. Phosphor converted WLEDs have found a popular place in our daily life. PC-WLEDs can be obtained by combining blue light from blue led chip and yellow light from phosphor powder by exciting the phosphor with a suitable wavelength [2]. Second method consists of pumping green/red phosphors or green/yellow/red with blue LED

S. A. Bhat · P. Biswas · P. Kour · R. Attri · M. A. Mir (✉)
Shri Mata Vaishno Devi University (SMVDU), Katra, Jammu and Kashmir, India
e-mail: mudasir.mir@smvdu.ac.in

R. Ahmad
National Institute of Technology, Hazratbal, Srinagar, India

Z. H. Khan et al. (eds.), *Recent Advances in Nanomaterials*, Springer Proceedings in Materials 27, https://doi.org/10.1007/978-981-99-4878-9_56

chip [3]. But the main problem projected to us by the above-mentioned method is the strong reabsorption of blue or green color by yellow or red emitting phosphor which results in loss of efficiency of white light and light quality of PC-WLEDs gets decreased [4]. Therefore, multiphosphor mixtures due to above-mentioned disadvantages are weaker candidates for producing PC-WLEDs. After having a close study of existing literature on rare earth doped nanophosphors, different methods have been adopted to synthesize $Sr_3(VO_4)_2$. However, reports on Ce^{3+} doped strontium vanadate, employing combustion synthesis (CS), are quite scarce. In this article, we reported the synthesis and optical analysis of Ce^{3+} doped $Sr_3(VO_4)_2$ synthesized by CS route.

2 Material and Method

2.1 *Material Synthesis*

Combustion synthesis was carried out to prepare $Sr_3(VO_4)_2$: Ce^{3+} nanophosphor. Being time and energy saving synthesis, it involves nitrates as oxidizers and urea or glycine or citric acid as a fuel [5]. The enormous amount of heat that gets produced helps in carrying out the reaction in a very less time. The reagents used were strontium nitrate [$Sr(NO_3)_2$; 99.5%], ammonium metavanadate [NH_4VO_3; 99.5%], urea [CH_4N_2O; 99.5%], cerium nitrate [$Ce(NO_3)_2$; 99.5%]. All the reagents were taken in a stiochiometric ratio so that the oxidizer to fuel ratio turns out to be unity. Reaction that took place is given as

$$(3 - x)\mathrm{Sr}\ (\mathrm{NO_3})_2 + x\mathrm{Ce}\ (\mathrm{NO_3})_2 + 4\mathrm{CH_4N_2O} + 2\mathrm{NH_4VO_3} \\ \rightarrow \mathrm{Sr}_{3-x}\mathrm{Ce}_x(\mathrm{VO_4})_2 + \mathrm{N_2} + \mathrm{H_2O} + \mathrm{CO_2}$$

2.2 *Material Characterization*

XRD technique was used to identify the phase of a material. Diffuse reflectance spectra were recorded by using UV–Vis spectrophotometer equipped with ISR 2600 Plus assembly. Excitation and emission spectra were recorded using Cary Eclipse spectrofluorometer equipped with a 150 W Xenon lamp.

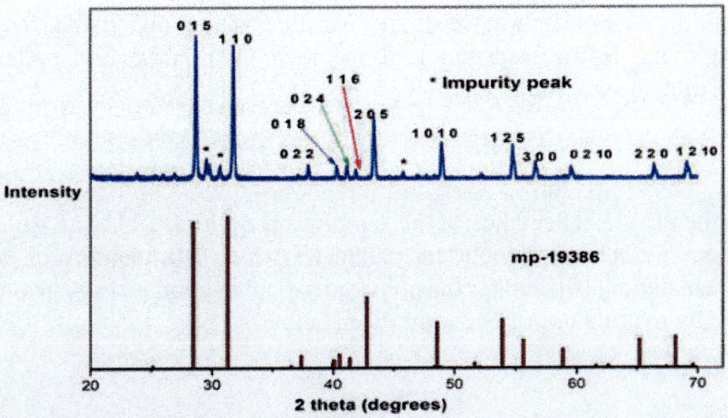

Fig. 1 XRD pattern of cerium doped strontium vanadate

3 Results and Discussion

3.1 *Structural Studies*

Figure 1 reflects the XRD pattern of cerium doped strontium vanadate, and it matches well with the reference XRD pattern taken from the materials project having card no MP-19386 [6]. XRD pattern so obtained projects its Rhombohedral structure lying in R3m space group. Impurity peaks (quoted as * in the figure) present in our XRD pattern may be due to moisture and/or presence of starting materials that we have used for sample preparation. Debye Scherrer equation, as given below, was used to calculate the crystallite size of a material [7].

$$D_c = \frac{k\lambda}{\beta \cos\theta} \tag{1}$$

where D_c represents the average crystallite size of a material, k is a constant often tagged by the name crystallite shape factor equal to 0.90, λ represents the wavelength of X-ray used for analyzing the sample, θ is the Bragg angle, and β represents peak width at the most intense diffraction peak. Crystallite size as calculated from above equation is 46 nm.

3.2 *DR Studies and Photoluminescence Analysis*

By using Kubelka–Munk theory, optical band gap from Tauc's plot with the help of Tauc's equation [8] given below was found to be equal to 3.93 eV. Further, on the basis of optical band gap, we can calculate the value of metallization constant

(M_c), and for the nanophosphor, we studied its value was estimated to be 0.44; which implies that it can also be used as nonlinear optical material having applications in sensors and optical switching [9].

$$[\mathrm{hv}F(R)]^2 = A(\mathrm{hv} - E_g) \tag{2}$$

where R is the diffuse reflectance of the sample and E_g is optical band gap.

Urbach energy being an important parameter which is a measure of degree of disorder of a material [10], and for the prepared nanophosphor, its value was obtained as 281 meV by making use of equation given below:

$$\alpha = \alpha_O \exp \frac{\mathrm{hv}}{E_u} \tag{3}$$

where α, α_O, and E_u represent absorption coefficient, absorption constant, and Urbach energy.

The excitation and emission spectra of 1.5 mol% concentration of Ce^{3+} in the optimized phosphor, as shown in Fig. 4a, were recorded in the range 400–700 nm at an excitation wavelength of 350 nm. Sharp peaks were observed at 547–618 nm. The excitation spectrum is a broad band centered about 340 nm and can be ascribed to $4f$-$5d$ transition of Ce^{3+} ion in the host. Excited state $5d$ can be divided into five energy levels by crystal field degeneracy. Hence, the electronic transitions from the ground state of Ce^{3+} ($^2F_{5/2}$ and $^2F_{7/2}$) reach the excited states of Ce^{3+} ($^5D_1$ and $^5D_2$) in the crystal field splitting environment [11]. The emission spectrum comprises of two components-one from self-activated $Sr_3(VO_4)_2$ host having a broad band about 475 nm and is ascribed to $^3T_1 \rightarrow {}^1A_1$ [12] transition and the other from the combined effect of activator-ligand interaction and low site symmetry giving away unprecedented yellow–red emission [13]. Figure 4b shows the chromaticity coordinates of cerium activated $Sr_3(VO_4)_2$ phosphor. CIE coordinates of prepared sample are (0.40, 0.34) which lie well closer to the white illuminant point (Figs. 2 and 3).

Fig. 2 Tauc's plot

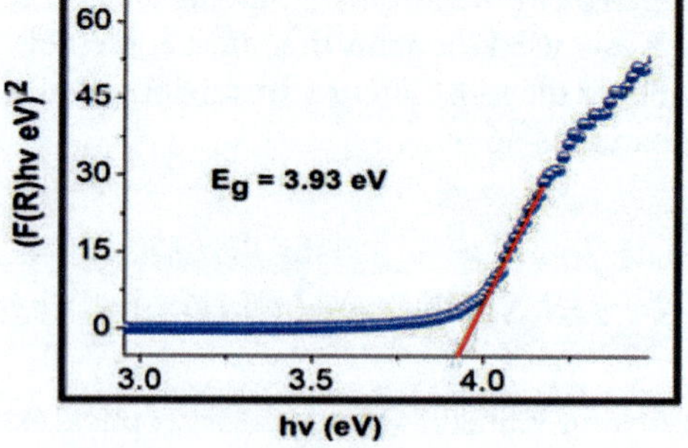

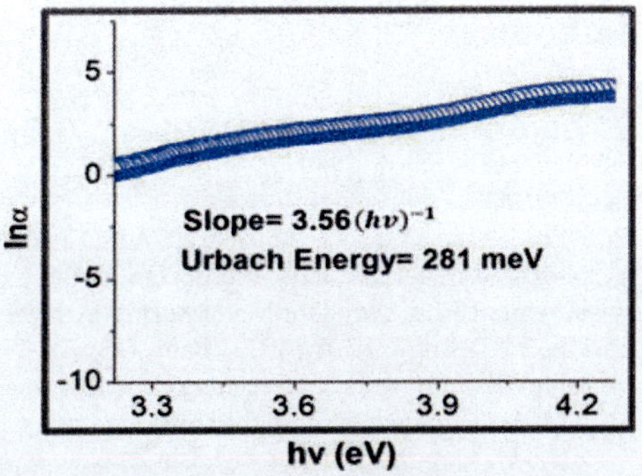

Fig. 3 Urbach energy estimation plot

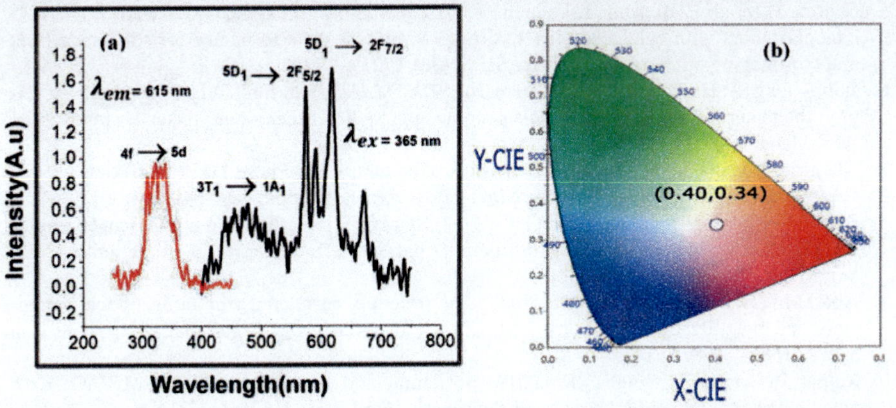

Fig. 4 **a** Excitation, emission spectra and **b** chromaticity coordinates of cerium doped strontium vanadate

4 Conclusions

Ce^{3+} ions were successfully incorporated into $Sr_3(VO_4)_2$ host by employing combustion synthesis method. Crystallite size was estimated using Debye Scherrer equation. Kubelka–Munk theory was used to calculate the optical band gap. On having an analysis of PL spectrum, we see that the prepared nanophosphor emits well closer to the white point; hence, it can be well pumped by 325 nm InGaN/GaN UV chips to give out emission about the white illuminant point.

Acknowledgements Authors would like to acknowledge SMVDU for providing the financial assistance in order to carry out this work.

References

1. Zhao M, Liao H, Molokeev MS, Zhou Y, Zhang Q, Liu Q, Xia Z (2019) Emerging ultra-narrow-band cyan-emitting phosphor for white LEDs with enhanced color rendition. Light: Sci Appl 8(1):38
2. Dang P, Liu D, Li G, Al Kheraif AA, Lin J (2020) Recent advances in bismuth ion-doped phosphor materials: structure design, tunable photoluminescence properties, and application in white LEDs. Adv Opt Mater 8(16):1901993
3. Xia YP, Wang CX, An LC, Zhang DS, Hu TL, Xu J, Chang Z, Bu XH (2018) Utilizing an effective framework to dye energy transfer in a carbazole-based metal–organic framework for high performance white light emission tuning. Inorg Chem Front 5(11):2868–2874
4. Zhou G, Jiang X, Molokeev M, Lin Z, Zhao J, Wang J, Xia Z (2019) Optically modulated ultra-broad-band warm white emission in Mn^{2+}-doped ($C_6H_{18}N_2O_2$) $PbBr_4$ hybrid metal halide phosphor. Chem Mater 31(15):5788–5795
5. Sharma P, Singh P, Bhushan I, Pathania K (2022) Combustion synthesis of NaSrVO4: Nd^{3+} nanophosphors with enhanced NIR 1.056 μm luminescent performance for solid-state laser and bio-imaging applications. J Mater Sci 57(36):17219–17233
6. Jain A, Ong SP, Hautier G, Chen W, Richards WD, Dacek S, Persson KA et al (2013) Commentary: the materials project: a materials genome approach to accelerating materials innovation. APL Mater 1(1):011002
7. Khajuria P, Mahajan R, Kumar S, Prakash R, Choudhary RJ, Phase DM (2020) Surface and spectral investigation of Sm^{3+} doped MgO-ZrO_2 phosphor. Optik 216:164909
8. Biswas P, Kumar V, Padha N, Swart HC (2017) Synthesis, structural and luminescence studies of $LiSrVO_4$: Sm^{3+} nanophosphor to fill amber gap in LEDs under n-UV excitation. J Mater Sci: Mater Electron 28:6159–6168
9. Abdo MA, Basfer NM, Sadeq MS (2021) The structure, correlated vibrations, optical parameters and metallization criterion of Mn–Zn–Cr nanoferrites. J Mater Sci: Mater Electron 32(12):15814–15825
10. Rajput P, Biswas P, Singh VK (2019) Structural and spectral studies of MgZnO: Cr^{3+} nanophosphors prepared by combustion synthesis. Appl Phys A 125(11):793
11. Que M, Que W, Zhou T, Shao J, Kong L (2016) Photoluminescence and energy transfer of YAG: Ce $^{3+}$, Gd^{3+}, Bi $^{3+}$. J Adv Dielectr 6(04):1650029
12. Zhu WL, Ma YQ, Zhai C, Yang K, Zhang X, Wu DD, Li G, Zheng GH (2011) Photoluminescence properties of Sr_3 (VO_4) $_2$, $Sr_2Y_{2/3}-yEu_y$ $(VO_4)_2$ and $Sr_2Y_{2/3}-zSm_z$ $(VO_4)_2$. Opt Mater 33(8):1162–1166
13. Maak C, Strobel P, Weiler V, Schmidt PJ, Schnick W (2018) Unprecedented deep-red Ce^{3+} luminescence of the nitridolithosilicates $Li_{38.7}RE_{3.3}Ca_{5.7}[Li_2Si_{30}N_{59}]O_2F$ (RE = La, Ce, Y). Chem Mater 30(15):5500–5506

Influence of the Decoration of Copper Metal Nanoparticles on the Structural and Electronic Properties of Carbon Nanotubes

Shah Masheerul Aalam, Mohammad Moeen Hasan Raza, Mohd Sarvar, Mohd Sadiq, Md. Faiz Akram, and Javid Ali

Abstract In this study, we have performed two properties (gas sensing and the field emission) of the MWCNTs simultaneously. We have first deposited the iron (Fe) catalyst on silicon (Si) substrate and have then used the low-pressure chemical vapour deposition (LPCVD) method to prepare the carbon nanotubes. We have decorated the CNTs by the metal (Cu) nanoparticles to enhance their electronic properties and to increase the surface area of the as prepared material. We have performed the gas sensing behaviour of the pristine/bare and decorated MWCNTs and compare the gas response behaviour of bare and decorated MWCNTs, a slightly high sensor response is obtained (Rg − Ra)/Ra × 100) of NH_3 gas in decorated 65% than in the bare 42% CNTs. Also we have improved the field emission behaviour of the CNTs by decorating with the Cu metal nanoparticles on the surface; in field emission behaviour, we have enhanced it by 1.82 V/μm in decorated than in pristine 2.5v/μm. The decoration of the CNT surface by the copper metal nanoparticles has been done by the RF-sputtering method for 6 min time interval at constant (non-variable) power of 120W by generating the nitrogen plasma inside the vacuum chamber. After that we have analyzed the structural and electrical properties of the prepared CNTs by FESEM and Raman spectroscopy, as they were being manufactured.

Keywords Carbon nanotubes · LPCVD · Metal decoration · Gas sensing · Field emission

S. M. Aalam · M. M. H. Raza · M. Sarvar · M. Sadiq · Md. F. Akram · J. Ali (✉)
Department of Physics, Jamia Millia Islamia, New Delhi 110025, India
e-mail: javi_reslab@rediffmail.com

M. Sadiq
A.R.S.D. College, University of Delhi, New Delhi 110021, India

Z. H. Khan et al. (eds.), *Recent Advances in Nanomaterials*, Springer Proceedings in Materials 27, https://doi.org/10.1007/978-981-99-4878-9_57

1 Introduction

As industrialization progresses and emission pollution worsens, several types of gas sensors have been thoroughly researched [1–4]. In the chemical, textile, fertilizer, paper goods, and food sectors, ammonia is one of the compounds that are frequently used as a coolant [5, 6]. Industries are discharging a significant amount of ammonia into the atmosphere as aerosols and haze. At concentrations of 16–28% by volume in air, ammonia is a highly flammable gas. Ammonia exposure that is too severe has a powerful irritant effect on our eyes, nose, mouth, lungs, and throat. A high ammonia dose can result in head-aches, nausea, vomiting, dyspnoea, pneumonedema, and even death [7, 8]. However, a variety of contemporary nano technological approaches can be used to increase the MWCNT sensor's sensitivity. MWCNTs can be decorated with various metal particles to increase the sensitivity of these sensors using a variety of techniques, such as spraying and RF-sputtering. MWCNTs are also being functionalized to get the highest sensitivity and maximum recovery possible [9–11]. According to a literature review, gold, silver, copper, and other metals are frequently utilized to decorate CNTs in order to increase their sensitivity. This enables a significant improvement in sensitivity, response/recovery characteristics, and low-temperature operation, however, compared to pure CNT sensors; it has been shown that CNT sensors decorated with noble metal nanoparticles exhibit higher-quality sensor response. According to a literature review, the CNT sensor response can be significantly improved by the decorating of Cu and other metal transition particles. In order to create adsorption sites for target gas molecules and maximize the amount of electron transport in the CNTs via charge transfer, the nanoscale size is required [11–15].

In this study, we have investigated about how Cu nanoparticles affected the sensor response of the MWCNT sensor after fabrication. FESEM and Raman spectroscopy were used to analyze the surface shape and structure. We have also discussed about how MWCNT sensors work to detect gases. Here, using Cu-decorated nanoparticles, we have achieved high quality enhancement in the field emission studies. RF-sputtering was used to adorn these Cu nanoparticles on the MWCNT surface. The self-assembled instrument coupled to a Keithley (2400) acquisition module, and a computer performed the resistive measurements. This effort resulted in a considerable and striking improvement in responsiveness and field emission.

2 Materials and Method

A silicon wafer is taken and put into the furnace for about 10 h at 900 °C for growing the SiO_2 layer on the surface. By using an RF-sputtering process, a thin layer of Fe catalyst was applied to an *n*-type (100) orientated Si substrate and then sliced into four 1×1 cm^2 pieces. Then, Fe-deposited substrates were put in the low-pressure CVD growth chamber, and the vacuum was pumped down to a level of 10^{-3} Torr.

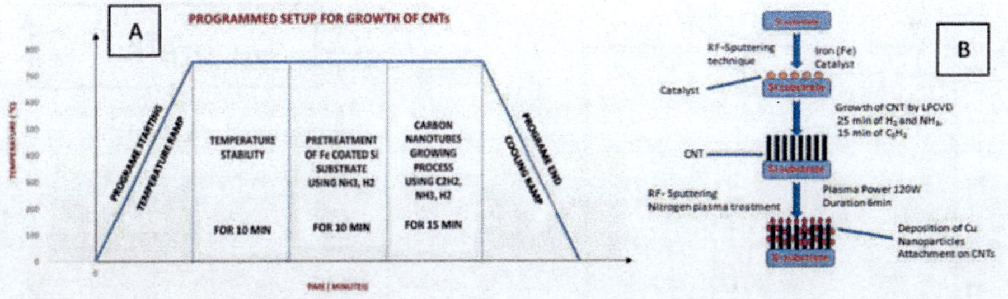

Fig. 1 **a** Represents the programmed setup for growth of CNT in LPCVD and **b** represents the schematic diagram of the growth and deposition of the material substances

Then, the surface of the Fe catalyst layer was uniformly etched from the H_2 and NH_3 gas at 750 °C for 25 min on the Si substrates on which Fe had been placed. Acetylene, a carbon source gas, was then supplied into the chamber for 15 min (growth time of MWCNTs). The ratio of the flowing gases is 30, 100, and 50 in sccm for C_2H_2, H_2, and NH_3, respectively. NH_3 is essential for the development mechanism of CNT because it lowers the activation energy and prevents the catalysts from being poisonous. The growth process mechanism and the setup for growth of the MWCNT are shown in Fig. 1.

3 Results and Discussion

The morphological studies have been confirmed by the FESEM micrographs shown in Fig. 1. The micrograph shows the dense, long, and high magnified MWCNT formed on the Si substrate. The Cu-decorated MWCNT is confirmed by the micrograph *B*, and this micrograph clearly shows the decoration of the NPs on the surface with diameter of few nanometers (Figs. 2 and 3).

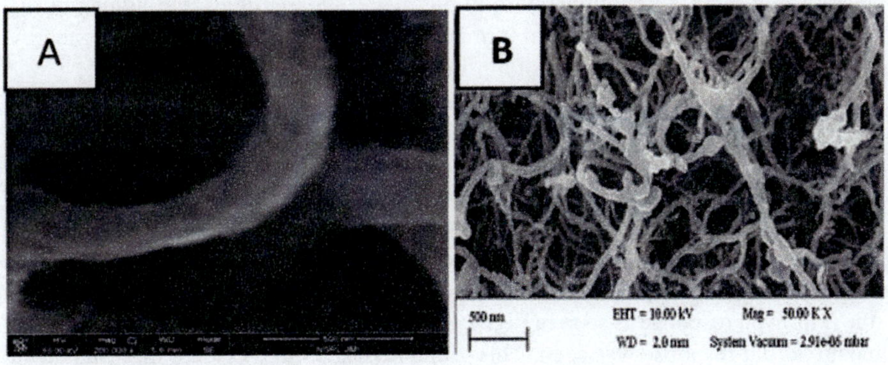

Fig. 2 FESEM micrographs, **a** Bare CNT and **b** Cu decorated CNT

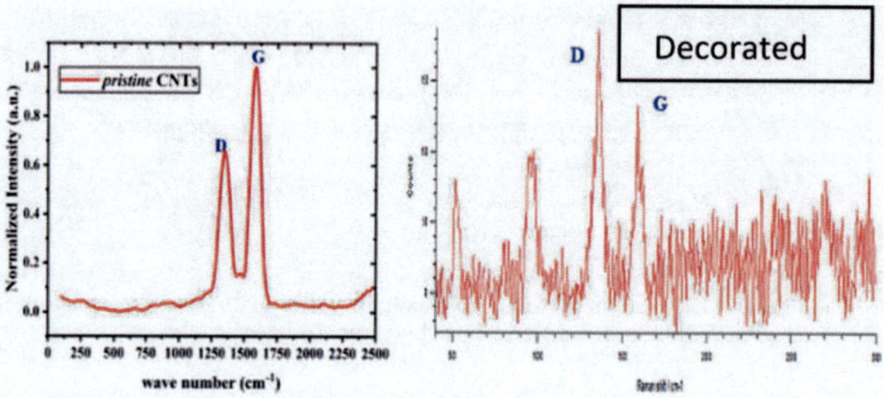

Fig. 3 Raman spectra of the pristine and decorated

The Renishaw Raman spectrometer at a wavelength of 532 nm confirmed the diameter distribution of the MWCNTs as grown. In our scenario, the RBM mode occurs at 232, whereas the *G*-band and *D*-band occur at 1587 and 1352, respectively. Due to its graphitic character, the graphitic *G* peak in our work occurs at 1587 cm^{-1} in the purest MWCNT field emitters and has a high intensity in contrast to the *D* peak at 1345 cm^{-1}. The stretching and splitting of E_{2g} by sp_2 hybridized C–C bonds in graphitic materials give birth to the *G*-band. The disorder of graphitic planes together with flaws in nanotubes and other impurities causes the *D*-band, which is typically associated to A_{1g}, to develop.

For plasma-treated CNT-Cu NPs decorating, the E_{to} and E_{th} fields substantially lowered see in Fig. 4, and the maximum J is seen at 1.82 V/μm. Due to the potential barrier dimension alteration in the coated CNT field emitters scenario, the emission starts even at low applied E. The facile tunnelling of electrons from the emitters into the vacuum was significantly improved by the lowering in the potential barrier's breadth and height. With the help of plasma-treated Cu NPs linked to CNT field emitters, the field enhancement factor was successfully increased. Additionally, by carefully adorning nanotubes with noble metal nanoparticles and adjusting the orientation of CNT in a way that minimizes the screening effect caused by nearby nanotubes, the electron field emission properties are improved.

The MWCNT sensor coated with Cu nanoparticles was put to the same NH_3 tests. In 2–4 s, an excellent improvement in sensor response magnitude of roughly 65% was made in comparison with pristine 42%. The target gas was taken out of the chamber after it reached saturation. When compared to the response and recovery properties of a pristine MWCNT sensor, high quality, swift response, and full recovery were seen in decorated sensors. This demonstrates unequivocally that the Cu nanoparticles produce locations on the MWCNT surface where gas molecules can interact, leading to the improved response as seen in Fig. 5. As compared to a bare MWCNT sensor, an almost twofold response was seen. This graph makes it quite evident that Cu coated nanoparticles were found to significantly improve responsiveness.

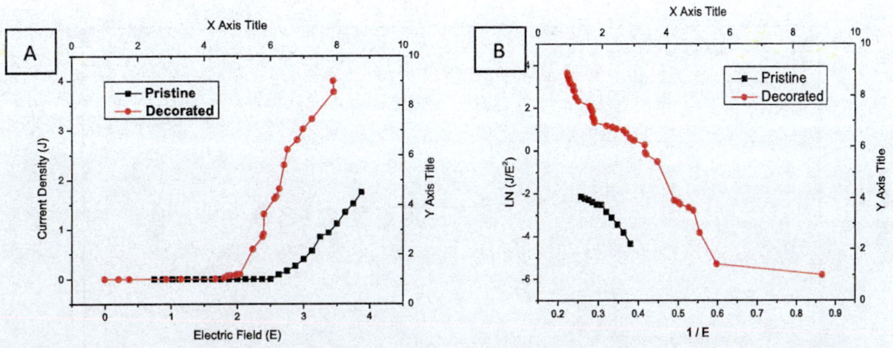

Fig. 4 **a** Field emission characteristics and **b** *F-N* plots of the pristine and decorated MWCNTs

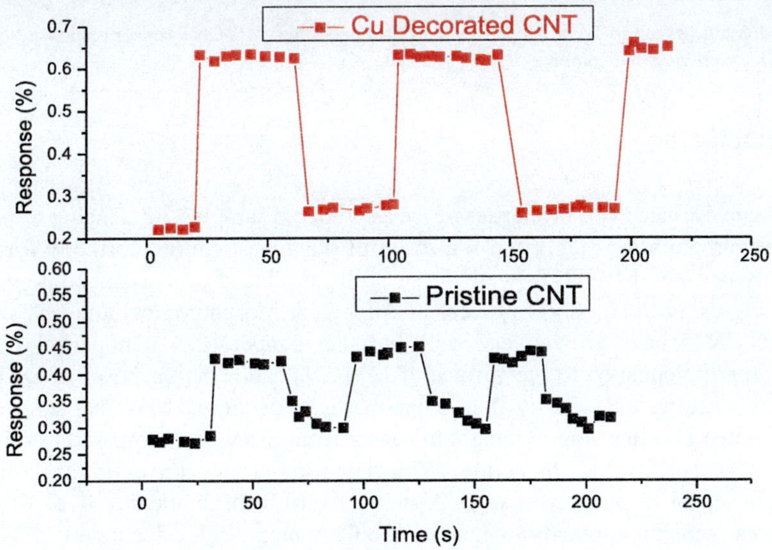

Fig. 5 Graph of pristine nanoparticles decorated (black line) and Cu-decorated (red line) MWCNT sensor response towards NH_3 with excellent enhancement

As shown in Fig. 6, the combined sensitivity graph from both sets of sensors has been compared and analyzed. This graph makes it quite evident that Cu coated nanoparticles showed a superb improvement in sensitivity.

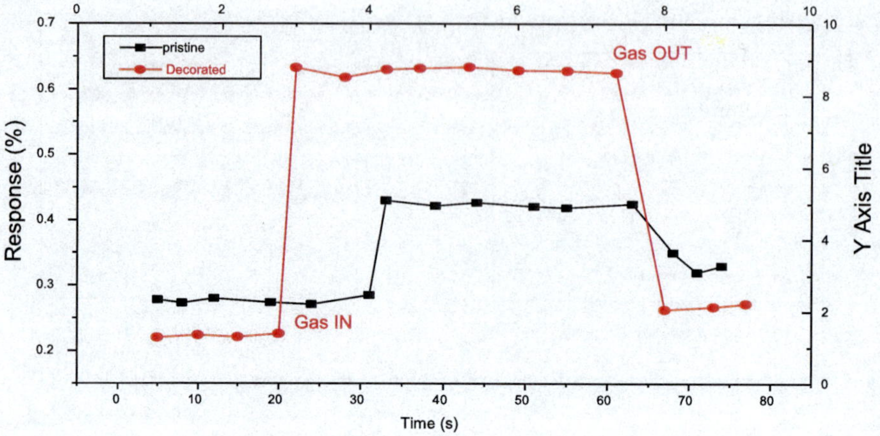

Fig. 6 Combine graph of pristine and nanoparticles decorated MWCNT sensor response towards NH_3 with excellent enhancement

4 Conclusion

We have investigated two properties of the as prepared samples, i.e. sensing response and the field emission stability. As a result of our investigation, correctly formed; randomly oriented MWCNTs were successfully produced on a Si substrate with a Fe coating using the LPCVD process at 750 °C. A morphological analysis shows that MWCNTs have grown very well and that copper (Cu) nanoparticles have been properly deposited to the surface of MWCNTs. For the purposes of environmental monitoring and security, the Cu nanoparticles decorated MWCNT sensor had demonstrated excellent improvement in sensor response when compared to pristine MWCNT sensor for NH_3 detection. With rising target gas concentration, a linear pattern in sensor response was seen. A step forward from the perspective of use in various gas sensing applications may be the Cu nanoparticles decorated MWCNT sensor. We have also successfully analyzed the field emission stability of the as prepared samples.

Acknowledgements We all are thankful to Centre for Nanoscience and Nanotechnology and Centre for Instrumentation Facility, JMI (New Delhi) for the characterization techniques such as FESEM and Raman. I am very grateful to ministry of higher education (Govt. of India) for providing me the Prime ministers research fellowship (PMRF) under lateral entry scheme in May cycle 2022 under PMRF-ID, 3302523. I am also thankful to my supervisor and my mentor Dr. Javid Ali for continuous support and supervision.

Declaration of Interest Statement
The authors affirm that they are free of any known financial conflicts of interest or close personal ties that might have looked to have affected the research presented in this study.

References

1. Yu K et al (2011) Growth of carbon nanowall at atmospheric pressure for one-step gas sensor fabrication. Nanoscale Res Lett 6(1):202–202
2. López-Mena ER et al (2017) Simple route to obtain nanostructured CeO2 microspheres and CO gas sensing performance. Nanoscale Res Lett 12(1):169
3. Choi HH et al (2011) noxious gas detection using carbon nanotubes with Pd nanoparticles. Nanoscale Res Lett 6(1):1–6
4. Chen T et al (2012) highly enhanced gas sensing in single-walled carbon nanotube-based thin-film transistor sensors by ultraviolet light irradiation. Nanoscale Res Lett 7(1):644–644
5. Timmer B, Olthuis W, Van Den Berg A (2005) Ammonia sensors and their applications—a review. Sens Actuators B Chem 107(2):666–677
6. Liu W, Liu Y, Do J, Li J (2016) Applied Surface Science Highly sensitive room temperature ammonia gas sensor based on Ir-doped Pt porous ceramic electrodes. Appl Surf Sci 390:929–935
7. Hou Y, Jayatissa AH (2014) Enhancement of gas sensor response of nanocrystalline zinc oxide for ammonia by plasma treatment. Appl Surf Sci 309:46–53
8. Pang Z, Fu J, Luo L, Huang F, Wei Q (2014) Fabrication of PA6/TiO2/PANI composite nanofibers by electrospinning-electrospraying for ammonia sensor. Colloids Surf A Physicochem Eng Asp 461(1):113–118
9. Choi SW, Kim J, Lee JH, Byun YT (2016) Remarkable improvement of CO-sensing performances in single walled carbon nanotubes due to modification of the conducting channel by functionalization of Au nanoparticles. Sens Actuators B 232:625–632
10. Dobrzanski LA, Pawlyta M, Krzton A, Liszka B, Tai CW, Kwasny W (2010) Synthesis and characterization of carbon nanotubes decorated with gold nanoparticles. Acta Phys Pol A 118:483–486
11. Nguyen LQ, Phan PQ, Duong HN, Nguyen CD, Nguyen LH (2013) Enhancement of NH3 gas sensitivity at room temperature by carbon nanotube-based sensor coated with Co nanoparticles. Sensors 13:1754–1762
12. Hou X, Wang L, Wang X, Li Z (2011) Coating multiwalled carbon nanotubes with gold nanoparticles derived from gold salt precursors. Diam Relat Mater 20:1329–1332
13. Joshi N, Silva LF, Jadhav HS, Shimizu FM, Suman PH, Peko JC, Orlandi MO, Seo JG, Mastelaro VR, Oliveira Jr ON (2018) Yolk-shelled ZnCo2O4microspheres: surface properties and gas sensing application. Sens Actuators B: Chem 257:906–915
14. Miller DR, Akbar SA, Morris PA (2014) Nanoscale metal oxide–based heterojunction for gas sensing: a review. Sens Actuators B 204:250–272
15. Bai H, Gaoquan S (2007) Gas sensors based on conducting polymers. Sensors 7(3):267–307

Fabrication of Silicon Nanowire Arrays by MACE for Effective Light Trapping

Sneha Rana, Anjali Saini, Manish K. Srivastava, and Sanjay K. Srivastava

Abstract Silicon nanowires, in particular, are being actively researched for solar-powered applications because they provide unique ways for converting solar radiation into electrical energy, resulting in highly efficient devices. Due to unique one-dimensional shape, solar cells-based on silicon nanowire (SiNW) arrays have the potential to be cost-effective and efficient sunlight trapping devices. Vertically aligned SiNW's arrays were fabricated across a large area using metal-assisted chemical etching (MACE) approach that involved etching the silicon substrate in the aqueous hydrofluoric acid (HF) and silver nitrate ($AgNO_3$) solution. This report describes the impact of various etching time at ambient temperature on the length of SiNWs. The time dependency of silicon nanowires growth rate was investigated for p-type substrates at 3 min, 6 min, and 9 min, respectively. In the 400–800 nm wavelength regions, reflectance was reduced to less than 5%. The results revealed that the silicon nanowires length might be readily regulated by varying the etching time. Our result demonstrates that the MACE technique provides a broad range of parameters for producing high-quality vertical silicon nanowires in a simple and controlled way. Light trapping capability is increased at the expense of surface imperfections. The decrease of surface imperfections has been shown to be more beneficial than the reduction of light trapping capability. Silicon nanowire arrays have high prospects for application as an effective radiation trapping material in the revolutionary photovoltaic and other optoelectronic devices due to its very low reflecting surface.

Keywords Silicon nanowire arrays · Light trapping · Metal-assisted chemical etching

S. Rana · M. K. Srivastava
Department of Physical Science, Banasthali Vidyapith, Rajasthan 304022, India

A. Saini · S. K. Srivastava (✉)
CSIR-National Physical Laboratory (NPL), New Delhi 110012, India
e-mail: srivassk@nplindia.org

Z. H. Khan et al. (eds.), *Recent Advances in Nanomaterials*, Springer Proceedings in Materials 27, https://doi.org/10.1007/978-981-99-4878-9_58

1 Introduction

Commercial silicon (Si) photovoltaic has been developed over the last 50 years to convert solar energy into electrical energy at efficiencies of roughly 20%, providing the most realistic carbon neutral route to substituting terawatts of non-renewable power consumed globally [1]. However, because of the high cost compared to conventional power sources, large-scale adoption is currently not economically practicable. Silicon has played an important part in solid-state electronics over centuries as an excellent semiconductor, and research on SiNWs has made remarkable progress due to its new electrical and optical features. Due to their large surface to bulk ratio, bio-compatibility, and quick response, small and long SiNWs are particularly highly suitable for the use in nanosensors. There are numerous optical and electronic systems, such as planar displays, solar cells (SCs), movies, camera displays, and light switches, benefiting from surface antireflection methods [2]. Perfectly flat Si surfaces, for example, have a high amount of reflectivity (35–40%) with a high spectrum dependency, limiting the energy conversion efficiency of Si-based SCs. Silicon has limited absorption in the visible and near-infrared regions of the solar-spectrum, efficient commercial silicon solar cells require comparatively large volumes of high purity silicon solar cells to fully absorb incoming sunlight. In this decade, a global effort has been launched to address these issues by producing silicon nanowires based solar cell. Several techniques to the growth of SiNWs have been widely examined in the past few years. They fall underneath the "bottom-up" or "top-down" approach categories. Several bottom-up approaches like vapor liquid solid (VLS) [3] method, molecular beam epitaxy [4], chemical vapor deposition, thermal evaporation [5], laser ablation [6], and others have been successfully produced to fabricate silicon nanowires. However, these methods generally require complicated and costly equipment, high temperature, templates, and dangerous Si precursors, which makes them time taking and costly, limiting their viability for cost efficient photovoltaic cells. Furthermore, bottom-up methods to SiNWs manufacture over broad areas are challenging to achieve. There are primarily three approaches in the top-down approach, namely reactive ion etching [7] (RIE), lithographic methods [8], and the MACE [9]. The first two methods are costly, as vast area and uniform silicon nanowires fabrication is difficult. The MACE approach is free of these constraints. It is a fairly simple solution-based beakers procedure that allows for manufacture of vertically aligned SiNW arrays on wafer size. Silicon nanowires could be produced on wafer scale using the MACE process, and their electrical properties are identical to those of the mother silicon wafer. This method can produce both irregular and regular silicon nanowire arrays. However, for periodic arrays, another lithographic method, such as colloidal or nanospheres lithography, is required prior to the MACE method [10]. In contrast, the standard MACE technique yields an aperiodic array of silicon nanowires. The single-step MACE is a self-controlled method that uses a solution of HF and $AgNO_3$ to etch silicon to fabricate silicon nanowire arrays [11]. It was also observed that the etching solution temperature, etching concentration (HF and $AgNO_3$), and crystallinity of silicon wafer all had a significant impact on the morphologies and lengths of silicon nanowires [12]. Many groups have recently demonstrated excellent antireflection or light harvesting capabilities of aperiodic silicon nanowires arrays fabricated using the MACE technique [13]. We report effect of etching time at room temperature on the surface morphologies and length of silicon nanowires by varying the etching time. This report is important for commercial silicon solar cells manufactures because silicon nanowire arrays could be employed as a light harvesting layer rather than additional antireflection coating (ARC) layer.

2 Materials and Method

2.1 Materials

Acetone and isopropyl (IPA) alcohol purchased from Thermo Fisher Scientific India Private Limited. Acetic acid (CH_3COOH), nitric acid (HNO_3), and hydrofluoric acid (HF) purchased from Merck Life Science Private Limited, and silver nitrate ($AgNO_3$) purchased from Sigma Aldrich.

2.2 Methods

Vertically aligned SiNW's arrays were fabricated using MACE technique on as-cut polished silicon substrate (100) oriented p- type wafers of thickness 190 μm, resistivity 1–5 Ω-cm. The sample size is fixed at 4.5×4.5 cm^2. The silicon substrates were cleaned with acetone (10 min) and IPA (10 min) in ultrasonic bath, then rinsed with de-ionized (DI) water for 2–3 times. In acidic etching, a solution of (HNO_3: CH_3COOH: HF = 5:1:1 by volume) for 30 s, then rinsed with DI water 2–3 times and dry it, then piranha cleaning, solution contains 3:1 mixture of H_2SO_4 and H_2O_2 by volume for 10 min on the hot plate at 150 °C. Then, wafer was copiously cleaned with DI water followed by dipping for 2 min in 2% HF solution to remove the surface oxide layer for 5 min. Then, cleaned silicon wafers were immediately transferred to a beaker (teflon) containing the etching solution of 0.04 M of $AgNO_3$ and 13 ml of HF for specific time, i.e., 3, 6, and 9 min, respectively, at room temperature to form vertically aligned silicon nanowire arrays. The Si substrate was rinsed thoroughly with DI water. The etched Si substrates were immersed in a solution of conc. HNO_3 for 5 min to remove silver dendrites deposited during the etching process. Again, rinsed with DI water for 2–3 times (Fig. 1).

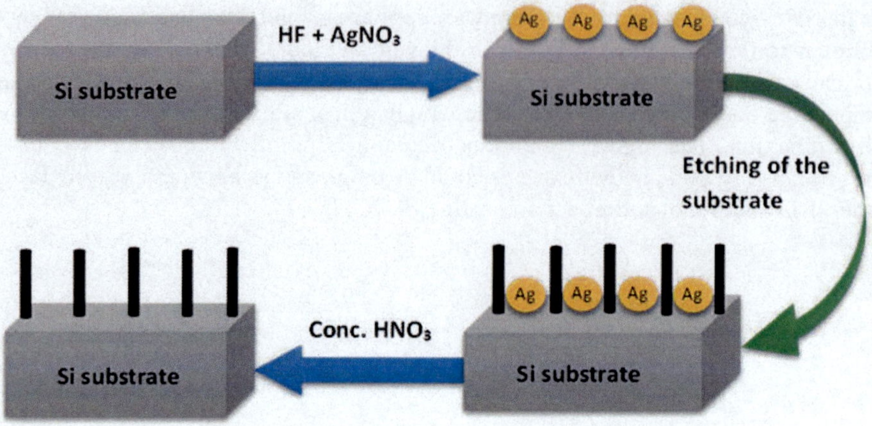

Fig. 1 Systematic diagram of MACE technique for the fabrication of SiNW arrays

2.3 Characterization

The surface morphology of SiNW arrays was inspected by scanning electron microscope (SEM–FESEM TESCAN Magna GHM), and reflectance of SiNW's arrays was measured by spectrophotometer (UV–Lambda 1050 Perkin Elmer).

3 Results and Discussion

Basically, when the Si substrate is dipped into aqeuous $AgNO_3/HF$ solution than the Ag ions in vicinity of the Si surface capture the electrons through the Si and deposit on the Si sample surface in the form of nuclei of metallic Ag; Si around the Ag nuclei is oxidized to silicon dioxide at the same time. There are two different reaction steps that make up the procedure.

(i) Cathodic reaction:

$$Ag^{+} + e^{-} = Ag \; E^{\theta} = 0.79\,V$$

(ii) Anodic reaction:

$$Si + 2H_2O = SiO_2 + 4H^{+} + 4e^{-} \; E^{\theta} = 0.91\,V$$

(iii) Overall etch reaction:

$$Si + 6HF + 4Ag^{+} = 4Ag + SiF_6^{2-} + 6H^{+}$$

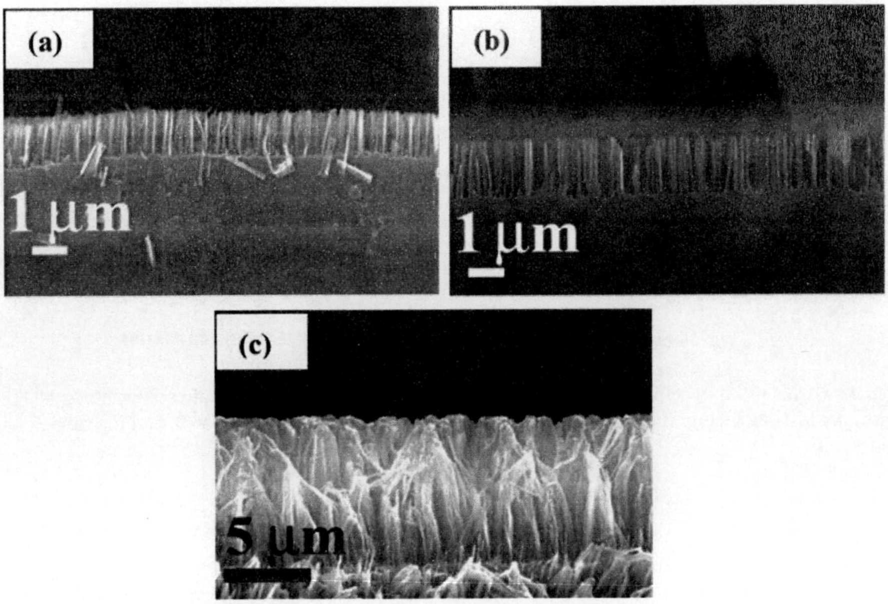

Fig. 2 Cross-sectional view SEM images of SiNW arrays in 0.04 M concentration of $AgNO_3$ for etching time **a** 3 min, **b** 6 min, and **c** 9 min

Due to their higher electronic reactivity than silicon atoms, the silver atoms linked to the Si wafer cause the cathode reaction to occur continuously, causing the silver atoms to eventually enlarge into silver nanoparticles. As a result of the Si atoms around Ag nanoparticles being oxidized to SiO_2 and then dissolving in HF as SiF_6^{2-}, the Ag nanoparticles fall into the holes/etched substrate at the same instant [14]. Figure 2 indicates the cross-sectional SEM views of the SiNW arrays fabricated for 3 min, 6 min, and 9 min etching time, respectively.

For these etching times, the nanowires length is 2 μm (Fig. 2a), 3.9 μm (Fig. 2b), and 5.8 μm (Fig. 2c), respectively. SEM images show that longer silicon nanowire arrays are complicated with one another close to the top and held together in bundles form. This is possible due to the springiness of silicon nanowires as well as surface tensional forces that act during the drying procedure. As a result, silicon nanowires are held together by van der Waals forces. These forces became dominant as etching time increases and thus nanowire length increase. SiNWs can collapse when their height becomes too long as shown in Fig. 2c.

These results represent that the length of silicon nanowire arrays rises with the etching time. Figure 2 clearly indicates that by judiciously adjusting the etching time duration for a given set of the etching conditions, SiNW arrays of a required length may be created at normal temperature. UV–visible reflectance measurements of the SiNW arrays demonstrate the spectral reflectivity of distinct lengths greatly minifies compared to that of silicon substrate over a broad spectral range from 400 to 1200 nm.

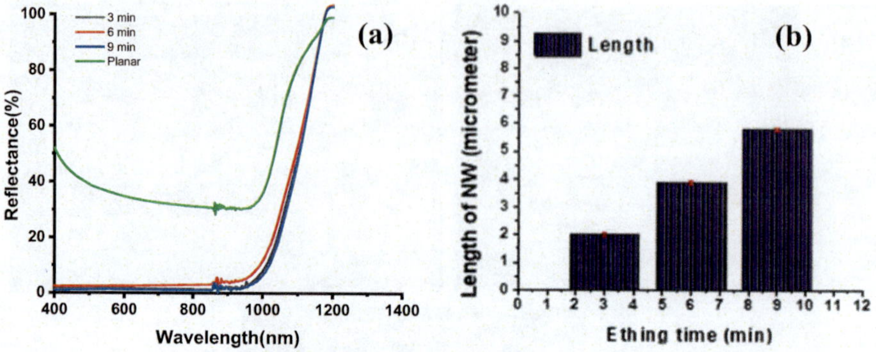

Fig. 3 **a** UV–Vis reflectance spectra of SiNW's for different etching time (keeping the $AgNO_3$ conc. 0.04 M). **b** Histogram curve represent the length variation of the NWs with etching time 3, 6 and 9 min

The sharp transition about 1000–1170 nm in reflectance spectrum be comparable to the band edge of silicon [15]. While the reflectance of SiNW arrays was dramatically repressed relative to that of the planar silicon substrate over a wide spectral wavelength range from 400 to 1000 nm. The reflectance of the planar silicon wafer varies from 50 to 30% in the range of 400–1000 nm. In case of 9 min, the minimum reflectivity is less than 5% at 400–800 nm range as shown in Fig. 3. Graph demonstrate that if we increase the etching time then increase the height of SiNW arrays and also decrease the reflectance. The SiNW arrays demonstrated lower reflection in short wavelength region than a planar silicon wafer.

4 Conclusion

A study of SiNW fabrication using the single-step MACE technique was carried out. This method for fabricating the vertically aligned SiNW arrays on Si substrate at ambient temperature is simple and scalable. This report shows the effect of variation of etching times such as 3, 6, and 9 min at room temperature on the length and surface morphology of silicon nanowires, respectively. Si substrate with SiNW arrays length of 2.0, 3.9, and 5.8 μm, respectively, that are vertically oriented, exhibits outstanding antireflection capabilities and can be prepared on large area directly on Si substrate at atmospheric temperature. The UV–visible reflectance spectra were investigated for different length of the SiNW arrays. Reflection was reduced to less than 5% over the broad spectrum range of 400–800 nm in comparison with the ~40% reflectance of the planar silicon sample in the same region. Reflection was reduced with increasing the length of SiNWs but after a certain value, no increment in the reflection property was observed. The current report could be useful in producing silicon nanowires with desired length and morphology for photovoltaic, optoelectronic, sensor, and photonic applications.

Acknowledgements Authors express gratitude to CSIR-NPL for providing instrument facility to this work (Grant code: 220432).

Declaration of Interest Statement The authors declare that they have no conflict of interests.

References

1. Garnet E, Yang P (2010) Light trapping in silicon nanowire solar cells. Nano Lett 10:1082–1087
2. Beyer T, Tacke M (1998) Antireflection coatings for PbSe diode lasers. Appl Phys Lett 73(9):1191
3. Latu RL, Mouchet C, Cayron C, Rouviere E, Simonato JP (2008) Growth parameters and shape specific synthesis of silicon nanowires by the VLS method. J Nanoparticle Res 10(8):1287–1291
4. Fuhrmann B, Leipner HS, Höche HR (2005) Ordered arrays of silicon nanowires produced by nanosphere lithography and molecular beam epitaxy. Nano Lett 5:2524
5. Pan H, Lim S, Poh C, Sun H, Wu X, Feng Y, Lin J (2005) Growth of Si nanowires by thermal evaporation. Nanotechnology 16:417
6. Yang YH, Wu SJ, Chiu SH, Lin P, Chen YT (2004) Catalytic growth of silicon nanowires assisted by laser ablation. J Phys Chem B 108(3):846–852
7. Fu YQ, Colli A, Fasoli A, Luo JK, Flewitt AJ, Ferrari AC, Milne WI (2009) Deep reactive ion etching as a tool for nanostructure fabrication. J Vacuum Sci Technol B: Microelectron Nanometre Struct Process Measurement Phenomena 27:1520
8. Mart J, Garcia R (2010) Silicon nanowire circuits fabricated by AFM oxidation nanolithography. Nanotechnology 21(24):245301
9. Qiu T, Wu XL, Siu GG, Chu PK (2006) Intergrowth mechanism of silicon nanowires and silver dendrites. J Electronics Mater 35:1879
10. Huang Z, Geyer N, Werner P, Boor JD, Gösele U (2011) Metal-assisted chemical etching of silicon: a review. Adv Mater 23(2):285–308
11. Kumar D, Srivastava SK, Singh PK, Sood KN, Singh VN, Dilawar N, Husain M (2010) Room temperature growth of wafer-scale silicon nanowire arrays and their Raman characteristics. J Nanoparticle Res 12:2267–2276
12. Srivastava SK, Kumar D, Schmitt SW, Sood KN, Christiansen SH, Singh PK (2014) Large area fabrication of vertical silicon nanowire arrays by silver-assisted single-step chemical etching and their formation kinetics. Nanotechnology 25(17):175601
13. Ozdemir B, Kulakci M, Turan R, Unalan HE (2011) Effect of electroless etching parameters on the growth and reflection properties of silicon nanowires. Nanotechnology 22:155606
14. Liu Y, Ji G, Wang J, Liang X, Zuo Z, Yi S (2012) Fabrication and photocatalytic properties of silicon nanowires by metal-assisted chemical etching: effect of H_2O_2 concentration. Nanoscale Res Lett 7:663
15. Dutta M, Fukata N (2015) Low-temperature UV ozone-treated high efficiency radial p-n junction solar cells: N-Si NW arrays embedded in a p-Si matrix. Nano Energy 11:219–225

Bifunctional Molybdenum Diselenide ($MoSe_2$) Nanosheets for High-Performance Symmetric Supercapacitor Device and Photocatalytic Dye Degradation Application

Ravi Pratap Singh, Pawanpreet Kour, Anu, Prashant S. Alegaonkar, and Kamlesh Yadav

Abstract In this study, $MoSe_2$ nanosheets (NSs) have been prepared via a hydrothermal route. The formation of the as-prepared sample is confirmed through X-ray diffraction (XRD) and field emission scanning electron microscope (FESEM) coupled with energy-dispersive spectroscope (EDS). Electrochemical testing of the formed $MoSe_2$NSs for a symmetric supercapacitor cell shows a specific capacitance (C_{sp}) of 300 Fg^{-1} @ 10 mVs^{-1}. It exhibits an energy density (E_D) of 30.4 $WhKg^{-1}$ @ power density (P_D) of 1620 WKg^{-1}. The photocatalytic performance of $MoSe_2$ NSs is examined by the degradation of rhodamine B (RhB) dye. The $MoSe_2$NSs exhibits superior catalytic performance with a 31% elimination efficiency for RhB after 70 min irradiation under natural sunlight. The results show that $MoSe_2$ is an effective bifunctional material for energy storage and photocatalytic applications.

Keywords $MoSe_2$ · Supercapacitor · Dye degradation

R. P. Singh · P. Kour · Anu · P. S. Alegaonkar (✉) · K. Yadav (✉)
Department of Physics, School of Basic Sciences, Central University of Punjab, Bathinda, Punjab 151401, India
e-mail: prashant.alegaonkar@cup.edu.in

K. Yadav
e-mail: kamlesh.yadav@cup.edu.in; kamleshyadav@allduniv.ac.in

K. Yadav
Department of Physics, University of Allahabad, Prayagraj, Uttar Pradesh 211002, India

Z. H. Khan et al. (eds.), *Recent Advances in Nanomaterials*, Springer Proceedings in Materials 27, https://doi.org/10.1007/978-981-99-4878-9_59

1 Introduction

The requirement of energy for daily needs gives rise to the energy crisis and environmental pollution. To overcome these issues, clean and renewable energy devices are crucial for the production and efficient storage of energy. Electrochemical capacitors are most promising candidates for energy converging and storage, due to high reliability, energy density, and fast charge discharge time [1]. On the other hand, industries such as textiles, lather, and plastics emit highly toxic nitrogen-containing organic pollutants, which significantly impact the environment and humans. The photocatalysts have enormous potential to degrade the organic and inorganic contaminants in water and air. $MoSe_2$ is a 2D material of type MX_2 (M = transition metals; X = chalcogens). The atoms in MX_2 are covalently coupled in the hexagonal structure and attracted to one another by the van der Waals interaction between adjacent planes. Owing to its layered 2D structure having high conductivity and specific surface area, $MoSe_2$ is developing as a strong candidate for energy storage devices and degradation of organic pollutants.

In the present work, $MoSe_2$ NSs are synthesized using a hydrothermal process and used further for bifunctional applications such as supercapacitors and photocatalytic dye degradation.

2 Experimental Procedure

$MoSe_2$ was synthesized using a hydrothermal method as mentioned in our previous report [2]. XRD (PAN analytical) study was performed in the 2θ range 10–70°. Fourier-transform infrared transmission (FTIR) spectra were recorded to identify different vibrational bands in the range of 600–3000 cm^{-1} with a Bruker Tensor 27 model. The morphological study of the prepared sample was carried out using FESEM (Carl Zeiss Merlin Compact) coupled with EDS. The electrochemical measurements and data analysis are done as per the reported work [2]. To confirm the photocatalytic activity of the prepared sample, RhB dye was degraded under dark and natural sunlight. In 100 ml of aqueous RhB dye solution, 20 mg of $MoSe_2$ (5 mg L^{-1}) was added. The solution was then agitated in the dark for 30 min to ensure that the dye molecules were in adsorption/desorption equilibrium with the photocatalyst surface. The solution was then exposed to natural sunlight. Approximately 3 ml of the solution was removed every 10-min interval and examined with a UV–visible spectrometer (Shimadzu UV-2450).

3 Results and Discussion

Figure 1a depicts the XRD of the $MoSe_2$ NS. The peaks at 11.7°, 29.7°, 41.4°, 46.5°, and 56.3° belong to the (002), (100), (006), (105), and (110) planes, respectively, corresponding to the hexagonal $MoSe_2$ (JCPDS no 77–1715) [3]. The average crystallite size (D) is ~ 26 nm, calculated using the Debye–Scherrer equation [2]. Figure 1b presents the FTIR spectra of the $MoSe_2$ NS. The vibrational bands at 647, 898, 1080, 1405, and 1640 cm^{-1}correspond to Se–O–Se, O–Mo–O, Se–O, C = O, and OH vibration bands, respectively [3, 4]. The FESEM micrographs of $MoSe_2$ consist of uniformly distributed nanosheets (Fig. 1c). The EDS spectra shown in inset confirm the existence of Mo and Se in the sample.

Figure 2a illustrates the CV plots of the symmetric device at different scan sweep from 10 to 100 mVs^{-1} in the operating window − 0.6 to 0.6 V. The redox peaks confirm the pseudocapacitive nature of the $MoSe_2$ NSs. Figure 2b shows the specific capacitance (C_{sp}) versus scan rates. The highest value of C_{sp} is 300 F g^{-1} @ 10 mV s^{-1}. Furthermore, the charge storage behavior of $MoSe_2$ is determined by a power-law equation, $i = av^b$, where i, v, and b are the peak current, scan rate, and dynamic variable, respectively [5]. The estimated value of b is 0.5 (see inset of Fig. 2b), which indicates the dominant diffusive charge behavior. Furthermore, the C_{sp} is also calculated from the GCD at diverse current densities (Fig. 2c). The C_{sp} value declines with an increment in current density because at larger current densities, charges do not get enough time for proper diffusion. The electrochemical impedance spectra are fitted according to the equivalent circuit R-QR-QR (inset of Fig. 2c). The high-frequency semicircle depicts the bulk contribution of electrode, and the straight line shows the dominance of capacitive behavior. The estimated value of bulk resistance (R_s) and charge transfer resistance (R_{ct}) is 1.35 Ω and 0.24 Ω, respectively. The energy (E_D) and power densities (P_D) are displayed in Ragone plot (Fig. 2d). The assembled device delivered E_D of 30.4 Wh kg^{-1} @ P_D of 1620 W kg^{-1}. The practical application of the fabricated symmetric SC cells is tested using three symmetric cells in series that can glow a standard red LED for 25 min (inset of Fig. 2d).

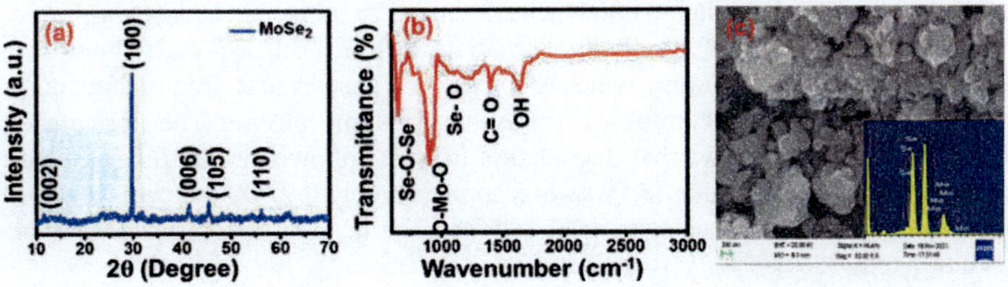

Fig. 1 **a** XRD of $MoSe_2$, **b** FTIR spectrum of $MoSe_2$, **c** FESEM image of $MoSe_2$ and inset show the EDS spectra of $MoSe_2$

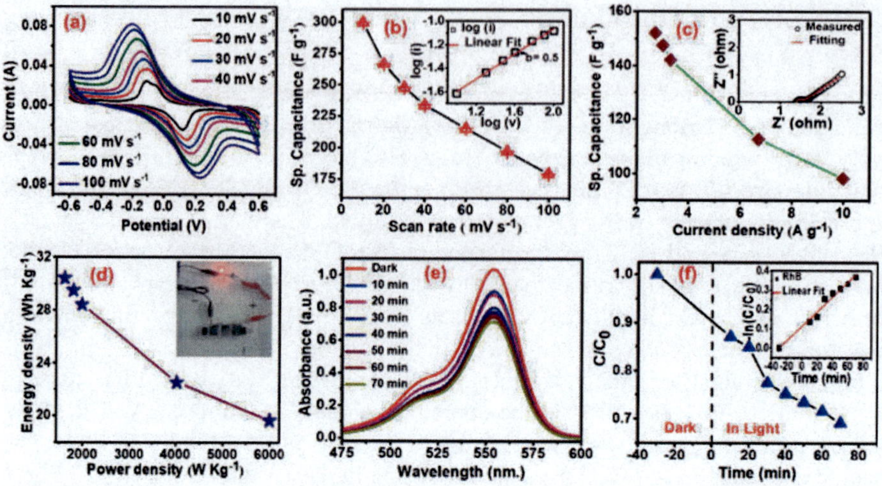

Fig. 2 **a** CV curves of $MoSe_2$ with diverse scan sweeps, **b** C_{sp} versus scan rates (inset shows log i versus log v), **c** C_{sp} versus current density (inset shows Nyquist plot with equivalent circuit), **d** Ragone plot (inset shows red LED glowing), **e** UV–vis absorption spectra of $MoSe_2$ for RhB dye under visible-light illumination, **f** C/C_0 versus time plot (inset shows the photocatalyst kinetic linear fitting curve)

Figure 2e shows the change in the UV–Vis spectrum of the RhB solution containing $MoSe_2$ NS as a catalyst over time. The intensity of the RhB peak (554 nm) in the presence of the $MoSe_2$ gradually decreases with increasing irradiation time, demonstrating the degradation of RhB dye. The photocatalytic efficiency of $MoSe_2$ for RhB dye is measured using the eqn: $\eta = (C_0 - C_t) \times \frac{100}{C_0}$, where C_0 and C_t are the dye concentrations at the starting time and after t secs, respectively. The rate of RhB dye degradation increases with the increasing irradiation time. After 70 min of illumination, the $MoSe_2$ sample exhibits catalytic performance, with a 31% RhB elimination efficiency. During the degradation of RhB dye, OH radicals and holes hit the carbon of the dye and break it into harmless products. Figure 2f shows the C/C_0 versus time plot. The absorption peaks are gradually declining with increasing exposure time. The degradation efficiency (C/C_0) of the prepared catalyst was found to be 0.69 after 90 min of visible-light irradiation. The inset of Fig. 2f shows the degradation kinetics of $MoSe_2$, which are fitted in a pseudo first-order using equation: $-\ln \frac{C_t}{C_0} = kt$, where k (min^{-1}) is the dye degradation constant. The close fitting of experimental data shows that degradation of dyes followed a first-order kinetic model. The correlation value (r), which is approximately 0.9748, indicates that the kinetic plot for catalysts is approximately linear. The results show that the value of k is 0.003 min^{-1}.

4 Conclusion

In this work, $MoSe_2$ has been synthesized via hydrothermal method. It exhibited outstanding electrochemical activity for SCs. The symmetric device delivered an E_D of 30.4 Wh kg^{-1} @ P_D of 1620 W kg^{-1}. Further, $MoSe_2$ also exhibited a remarkable catalytic performance, with a 31% RhB elimination efficiency after 70 min. of illumination. Thus, $MoSe_2$ can be used as a bifunctional material for both SCs and dye degradation.

Acknowledgements Ravi Pratap thanks the CSIR, India for its financial support and *CIL and DoP*, Central University of Punjab, Bathinda for the laboratory facilities.

Declaration of Interest Statement
The authors declare no competing financial interest.

References

1. Ramadoss A, Saravanakumar B, Lee SW, Kim Y-S, Kim SJ, Wang ZL (2015) Piezoelectric-driven self-charging supercapacitor power cell. ACS Nano 9(4):4337–4345. https://doi.org/10.1021/acsnano.5b00759
2. Kour P, Yadav K (2022) Electrochemical performance of mixed-phase 1T/2H MoS_2 synthesized by conventional hydrothermal v/s microwave-assisted hydrothermal method for supercapacitor applications. J Alloy Compd 922:166194. https://doi.org/10.1016/j.jallcom.2022.166194
3. Siddiqui I, Mittal H, Kohli VK, Gautam P, Ali M, Khanuja M (2018) Hydrothermally synthesized micron sized, broom-shaped $MoSe_2$ nanostructures for superior photocatalytic water purification. Mater Res Express 5(12):125020. https://doi.org/10.1088/2053-1591/aae241
4. Mishra A, Narang J, Pundir CS, Pilloton R, Khanuja M (2018) Morphology-preferable $MoSe_2$ nanobrooms as a sensing platform for highly selective apta-capturing of salmonella bacteria. ACS Omega 3(10):13020–13027. https://doi.org/10.1021/acsomega.8b01074
5. Karade SS, Nimbalkar AS, Eum J-H, Kim H (2020) Lichen-like anchoring of $MoSe_2$ on functionalized multiwalled carbon nanotubes: an efficient electrode for asymmetric supercapacitors. RSC Adv 10(66):40092–40105. https://doi.org/10.1039/D0RA06952C

Investigations of Impedance Properties of Calcium Copper Titanate ($CaCu_3Ti_4O_{12}$) Ceramic

Sukhanidhan Singh, Manisha Kumari, and P. M. Sarun

Abstract $CaCu_3Ti_4O_{12}$ (CCTO) is synthesized by solid-state route at a sintering temperature of 1100 °C for 4 h. Crystal structure and microstructural analysis are performed by X-ray diffraction (XRD) and field emission scanning electron microscopy (FE-SEM), respectively. XRD confirms the single phase CCTO with cubic crystal structure having *Im3* space group. FE-SEM reveals the distribution of dense grains with minimum porosity. The average grain size is obtained at around ~ 1.67 μm. The permittivity of synthesized CCTO ceramic is ~ 8099, and tangent loss is ~ 0.73 at 30 °C and 100 Hz. The activation energy of 0.11 eV is estimated from conductivity analysis. Positive temperature coefficient resistance (PTCR) behaviour is confirmed by impedance analysis. Contribution of grain and grain boundary on the impedance properties is explained by Nyquist analysis. Due to high dielectric constant and low dissipation factor along with good thermal stability, this material is a good candidate for energy storage devices. The CV plot confirms the capacitive behaviour of CCTO ceramic.

Keywords Microstructural · Porosity · Permittivity · PTCR coefficient · Impedance

1 Introduction

With the development and modernization of electronic industry, reduction in size and volume of electronic devices such as capacitive components becomes the significant and essential part of modern research community. Colossal permittivity (CP) and least magnitude of dissipation factor are the two most important parameters to fulfil the

S. Singh · P. M. Sarun (✉)
Functional Ceramics Laboratory, Department of Physics, Indian Institute of Technology (Indian School of Mines), Dhanbad 826004, Jharkhand, India
e-mail: sarun@iitism.ac.in

M. Kumari
Advanced Materials Design and Processing Group, Department of Metallurgical and Materials Engineering, Indian Institute of Technology, Jodhpur 342030, Rajasthan, India

Z. H. Khan et al. (eds.), *Recent Advances in Nanomaterials*, Springer Proceedings in Materials 27, https://doi.org/10.1007/978-981-99-4878-9_60

above mention requirements with thermal and frequency range stability. Normally, ferroelectric or relaxor type materials have giant dielectric constant [1, 2].

CCTO is a pseudo perovskite structure of $ACu_3Ti_4O_{12}$ family, where A = Ca, $Bi_{2/3}$, Cd, $Y_{2/3}$, etc. [1, 3]. CCTO maintains its magnitude of colossal permittivity in a wide range of temperature 100–400 K. Neither any kind of phase transition nor structural change is detected in CCTO ceramic by X-ray diffraction or Raman spectroscopy [4]. Internal barrier layer capacitor model (IBLC) is most trustable model. This model satisfactorily explains the origin of large value of dielectric constant in CCTO ceramics. According to this model, CCTO contains the large numbers of semi-conducting grains and insulating grain boundaries, which work as capacitor and connected in series combination, resulting the large magnitude of dielectric constant [5, 6]. This work inspects the phase structure, microstructure and detailed analysis of dielectric and electrical properties of CCTO ceramic at 1100 °C sintering temperature. This discussion also pointed out the practical applications of CCTO ceramics due to its large magnitude of dielectric constant and low dissipation factor, such as ceramic capacitor and sensors.

2 Materials and Methods

Commercially available $CaCO_3$, CuO and TiO_2 of *Alfa Aesar* with high purity (>99.9%) were taken as ingredients, and CCTO is synthesized by SSR route, which is explained in detail in previous literature. Calcination of synthesized material is done at 850 °C for 12 h, and sintering is done at 1100 °C for 4 h. For the measurement of electrical and dielectric properties, sintered pellet was coated with high-quality silver paste (Thermo Fisher Scientific) on both the surfaces. Using LCR metre (*HIOKI 3532 LCR HiTester*, Japan), electrical and dielectric measurements were performed from 303 to 353 K and 100 Hz to 1 MHz. *Bruker D8 Focus* model is used to perform the XRD of manufactured sample in Bragg's angle (2θ) from 10 to 90°. Surface morphology is analysed by FE-SEM (*Carl Zeiss Supra 55).*

3 Results and Discussion

XRD pattern of synthesized sample of CCTO ceramic after sintering at 1100 °C for 4 h is illustrated in Fig. 1a. XRD patterns show that all peaks are well matched with standard database of ICDD PDF NO-01-75-88 without any impurity phase with *Im3* space group. Obtained value of lattice parameters will be $a = b = c = 7.402$ Å. FE-SEM image of synthesized sample, shown by Fig. 1b, which shows the compactness of grains with minimum porosity and high grain density. The magnitude of average grain size is 1.65 μm, which is depicted in FE-SEM micrograph image. The values of green density, sintered density and relative density are 3.391 $g.cm^{-3}$, 5.005 $g.cm^{-3}$ and 92.

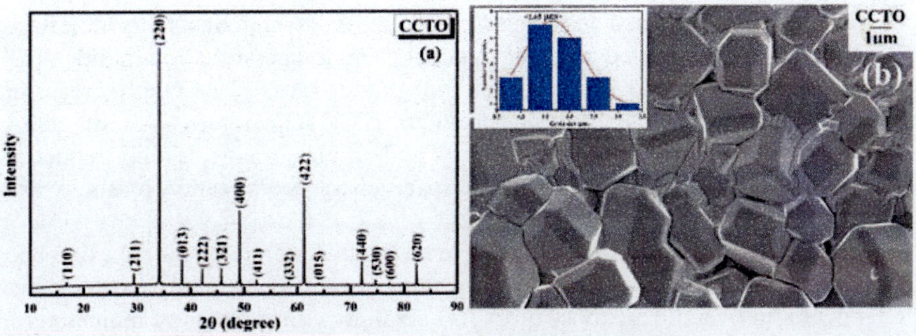

Fig. 1 (**a**) XRD pattern and **b** FE-SEM micrograph of CCTO ceramic

Figure 2a depicts the ε' from 303 to 353 K and 1 Hz to 1 MHz. Dispersion in the plots is more at low frequency, which overlap regardless of temperature at high frequency. At low frequency, dipoles are highly oriented, while at high-frequency lagging between the applied electric field and dipoles take place and space charge polarization is vanished at elevated frequency.

Figure 2b shows the tangent loss (tan δ) of prepared CCTO ceramic in above-mentioned temperature and frequency. Plots of tangent loss show peaks for each temperature in high frequency. In high frequency, dipoles do not follow the applied electric field and loss in energy is observed.

Figure 2c illustrates the inverse of temperature to DC conductivity. Using *Arrhenius* equation, the value of activation energy can be estimated. *Arrhenius.*

equation is as follows:

$$\sigma_{\mathrm{dc}} = \sigma_0 e^{\left(\frac{-E_a}{KT}\right)} \quad (1)$$

where E_a, K and T are the activation energy, *Boltzmann* constant and absolute temperature. At 100 Hz, the activation energy estimated by slope of σ_{dc} is 0.11 eV. The value of E_a is less than 1, which suggests that conduction mechanism present in the prepared ceramic is due to hopping of electrons or oxygen vacancies [2].

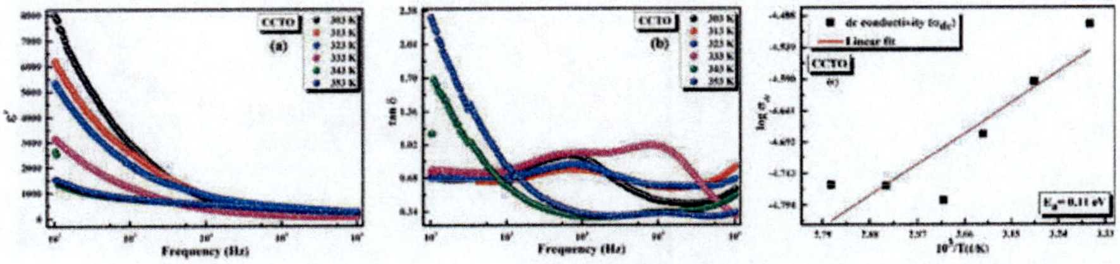

Fig. 2 Frequency dependent, **a** real, **b** imaginary part of permittivity and **c** inverse of temperature to DC conductivity

Figure 3a, b show the real and imaginary (Z' and Z'') components of synthesized CCTO ceramic in mentioned range of frequency and temperature. Magnitude of Z' at 303 K is minimum, and elevation in its magnitude takes place with increase of temperature and indicates the PTCR behaviour of CCTO ceramic [7]. Plots of Z'' have maximum dispersion at low frequency, and all plots marge at high frequency regardless of temperature indicating that space charge polarization exists in low frequency and vanish in high frequency [2].

Figure 3c shows Nyquist plot. For each temperature, a semicircular arc exists and diameter of semicircular arc decreases with increase of temperature and indicates the reduction in overall resistivity of CCTO ceramic with increase of temperature. Based on an IBLC model, interception in Z' axis towards high frequency represents grain contribution and large arc at low frequency corresponds to grain boundary contribution in CCTO ceramic. Each arc is associated with a circuit shown in inset of Fig. 2c, which is a series combination of parallelly connected R-Q circuit for grain and GB [2].

Cyclic voltammetry (CV) of the prepared sample was done for the evolution of the electrochemical performance as shown in Fig. 4. Symmetric nature other than the rectangular shape of CV curve indicates the rapid reversible Faradaic process along with truly capacitive behaviour of the prepared sample [8].

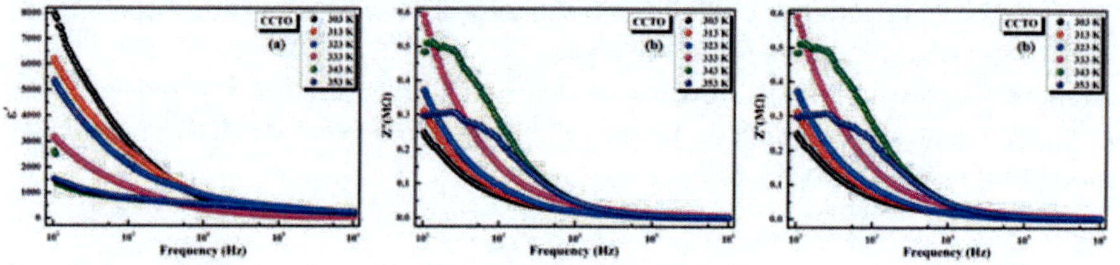

Fig. 3 Frequency dependent, **a** real part, **b** imaginary part of impedance and **c** Nyquist fitting of CCTO ceramic

Fig. 4 CV curves of CCTO at scan rate of 10 mV

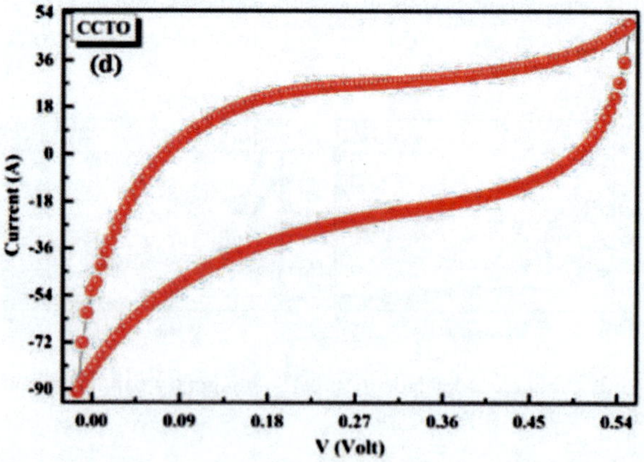

4 Conclusion

By SSR method, CCTO is synthesized at sintering temperature 1100 °C, successfully. XRD analysis comes out with the result of single phase formation of CCTO with space group *Im3* and cubic crystal structure. FE-SEM reveals the compactness of grains with minimum porosity, and obtained value of average grain size is ~1.67 μm. Dielectric constant is maximum (~8099) in low-frequency region at 30 °C and 100 Hz along with minimum tangent loss at the same frequency and temperature 0.73. Activation energy is 0.11 eV, estimated by the slope of plot between inverse of temperature and DC conductivity. This confirms that conduction is due to hopping of electrons or oxygen vacancies in synthesized CCTO ceramic. Impedance analysis confirms the positive temperature coefficient resistance (PTCR) behaviour in synthesized ceramic and Nyquist fitting carried out the result that grain and grain boundary contribution exist in synthesized CCTO ceramic, which is associated with an R-Q equivalent circuit. Capacitive behaviour is confirmed by the CV measurement.

Acknowledgements Sukhanidhan Singh and Manisha Kumari acknowledge the Indian Institute of Technology (Indian School of Mines), Dhanbad, India, for providing the Senior Research Fellowship. The author also acknowledges the CRF facilities of IIT(ISM), for the XRD and FE-SEM measurements.

Declaration of Interest Statement
The authors declare that they have no known competing financial interests or personal relationships that could have appeared to influence the work reported in this paper.

References

1. Ahmadipour M, Ain MF, Ahmad ZA (2016) A short review on copper calcium titanate (CCTO) electroceramic: synthesis, dielectric properties, film deposition, and sensing application. Nano-Micro Lett 8(4):291–311
2. Yadav A, Fahad M, Sarun PM (2021) Frequency dependent studies of dielectric and impedance properties of $NaNb_{0.92}V_{0.08}O_3$ ceramics. Mater Today: Proc 46:6286–6289
3. Subramanian MA, Li D, Duan N, Reisner BA, Sleight AW (2000) High dielectric constant in $ACu_3Ti_4O_{12}$ and $ACu_3Ti_3FeO_{12}$ phases. J Solid State Chem 151(2):323–325
4. Singh S, Yadav A, Kumari M, Sarun PM (2022) Analysis of giant dielectric permittivity and electrical properties for energy storage devices through impedance spectroscopy in $CaCu_3Ti_4O_{12}$. J Mater Sci: Mater Electron 33(12):9395–9402
5. Zhao J, Chen M, Tan Q (2021) Embedding nanostructure and colossal permittivity of TiO_2-covered CCTO perovskite materials by a hydrothermal route. J Alloy Compd 885:160948
6. Li F, Li Y, Wang S, Zhang J, Tang T, Liao Y, Lu Y, Liu X, Wen Q (2022) Improved co-substituted zinc vanadate ceramics based on LTCC for enhanced polarization converters. J Alloy Compd 921:166089

7. Chinnathambi M, Sakthisabarimoorthi A, Jose M, Robert R (2021) Study of the electrical and dielectric behaviour of selenium doped CCTO ceramics prepared by a facile sol-gel route. Mater Chem Phys 272:124970
8. Purty B, Choudhary RB, Biswas A, Udayabhanu G (2019) (2019), Chemically grown mesoporous f-CNT/α-MnO 2/PIn nanocomposites as electrode materials for supercapacitor application. Polym Bull 76:1619–1640

Effect of Rare Earth Lanthanides Doping on the Magnetic Properties of Magnesium Nanoferrite Prepared via Sol–gel Route

Umesh Chejara, Anamika Prajapati, and Rupesh Kumar Basniwal

Abstract This study investigates the synthesis and magnetic properties of rare earth (RE) lanthanides-doped magnesium nanoferrite. Specifically, the chemical composition $MgCe_xEr_yFe_{2-x-y}O_4$ (where $x = 0.4, 0.6, 0.8$ and $y = 0.6, 0.4, 0.2$) was synthesized by sol–gel method. The series that was synthesized underwent characterization through X-ray diffractometry (XRD), Fourier transform infrared spectroscopy (FTIR), and scanning electron microscopy (SEM). The samples were examined by means of a superconducting quantum interference device (SQUID) magnetometer. *M-H* curves plotted at 300, 100, 20, and 5 K. The results demonstrate that doping with RE ions (specifically Ce^{3+} and Er^{3+}) effectively controls their properties of samples. These research works are significant for the advancement of communication technology in the future.

Keywords Sol–gel method · Rare earth elements · Magnesium nanoferrites

1 Introduction

The quest for materials with desired properties has been a long-standing pursuit of scientists. Today, there is a growing need for materials that are small in size, energy-efficient, and capable of self-adjustment to meet specific requirements. This has led to a surge in nanotechnology, where metal oxide nanoparticles play a vital role. However, improving the properties of these particles is crucial for their successful application in various fields, particularly in electromagnetic applications such as memory devices, microwave units, transformers, cleaners, and antennas, among

U. Chejara (✉)
Department of Physics, University of Rajasthan, Jaipur 302004, India
e-mail: umeshchej@gmail.com

A. Prajapati
Department of Chemistry, University of Rajasthan, Jaipur 302004, India

R. K. Basniwal
AIARS (M&D), Amity University Uttar Pradesh, Noida, India

Z. H. Khan et al. (eds.), *Recent Advances in Nanomaterials*, Springer Proceedings in Materials 27, https://doi.org/10.1007/978-981-99-4878-9_61

others. Among the different synthesis methods available, the sol–gel route has shown results in enhancing the properties of magnesium ferrite nanoparticles. Here our aim is to synthesize Mg-ferrite NPs using the sol–gel route as well as investigate the impact of doping rare earth lanthanides, namely cerium and erbium, on their electromagnetic properties. By improving the properties of metal oxide nanoparticles, we can enhance their applications in several fields, both in daily use and research [1], $MgFe_2O_4$ is a cubic-structured, *n*-type semiconductor, and soft magnetic material with various applications, such as memory devices, microwave devices, and sensors even cancer treatment. Ferrite, particularly spinal structure, is a versatile material with exceptional properties that find applications in various fields, such as storage, MRI, radar absorbing materials, microwave antennas, and high-frequency power devices. Its high stability makes it an ideal material for applications requiring reliable and stable magnetic properties [2], our research required a material with high resistivity, good thermal and chemical stability, and ferrite was a critical consideration. Furthermore, by replacing the cations in the spinel structure with RE lanthanides, we enhanced the material's electromagnetic properties [3, 4], The XFe_2O_4 spinel structure has a face-centered cubic arrangement composed of metal cations such as Mg^{2+}, Co^{2+}, Mn^{2+}, and others. By altering the composition, substitution, and preparation processes, the structure's optical, magnetic, physical, chemical, and dielectric properties can be modified. Thus, ferrite, and especially spinal ferrite, has a significant impact on various technological advancements, making it an essential material for modern technology [5]. Consequently, our interest in the electromagnetic properties of this material lies in its potential use as a microwave antenna material with a negative refractive index and low energy loss, as well as minimal heat dissipation during communication processes. Our research focused on the synthesis and investigation of $MgCe_xEr_yFe_{2-x-y}O_4$nanoscale ferrite particles, with a particular emphasis on their size-dependent behavior, for practical planar antenna applications in design and construction. The findings from these studies will be beneficial for the development of practical planar antenna applications.

2 Materials and Methods

In this study, we utilized magnesium ferrite doped with rare earth elements such as cerium and erbium for our sample series. The nanoferrite materials were synthesized using the sol–gel technique, with magnesium, iron, cerium, and erbium nitrates, citric acid, and ammonia as starting materials. We employed X-ray diffraction (XRD) and scanning electron microscopy (SEM) to determine particle size (nanomaterials), and Fourier transform infrared spectroscopy (FTIR) to analyze the elements present in the sample. In addition, we determined the permeability values of the nanoferrite materials using a SQUID under controlled magnetic fields to evaluate their magnetic properties. The combination of these techniques provided a comprehensive analysis of the synthesized nanoferrite materials.

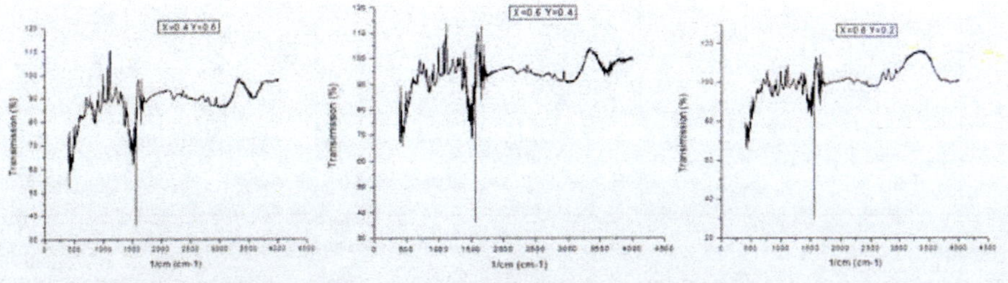

Fig. 1 FTIR spectra of $MgCe_xEr_yFe_{2-x-y}O_4$ nano-ferrites powders for, **a** $X = 0.4$ $Y = 0.6$, **b** $X = 0.6$ $Y = 0.4$ and **c** $X = 0.8$ $Y = 0.2$

2.1 FTIR Analysis

FTIR spectroscopy is a widely used experimental technique for characterizing a wide variety of materials due to its accessibility, non-destructiveness, and fast analysis. In this study, a Shimadzu spectrometer with a single reflection horizontal attenuated total reflection accessory was used to conduct measurements. The FTIR absorption spectrum of the prepared sample pellet was recorded at 300 K in the regime of 400–4000 cm^{-1}. To investigate the elements present in the prepared series $MgCe_xEr_yFe_{2-x-y}O_4$, FTIR spectroscopy was performed. The resulting FTIR spectra for the prepared sample pellets, recorded in KBr, were analyzed between 400 and 4000 cm^{-1} and are shown in Fig. 1a–c.

The prepared sample was analyzed using spectroscopy, which revealed several absorption bands. The presence of tetrahedral and octahedral vibration complexes was identified by absorption bands at 560 cm^{-1} (υ_1) and 435 cm^{-1} (υ_2), respectively. This suggests that the sample has a spinel structure of ferrite [6], and another absorption band was observed between 1544 and 1549 cm^{-1} (υ_3), which corresponds to O–H stretching vibration. The presence of water is indicated by this band, and it suggests the existence of hydroxyl groups in the nanoferrite material. Additionally, a band at 1404 cm^{-1} (υ_4) was observed, which corresponds to the $C = O$ stretching vibration, it is caused by the presence of carboxyl groups. The spectrum also indicated the presence of a nitrate group with a band (υ_5) between 1107 and 1147 cm^{-1}. Finally, a band at approximately 400 cm^{-1} (υ_6) was observed, which is attributed to metal–oxygen tetrahedral and octahedral vibration in the lattice. This metal–oxygen band confirms the formation of the prepared sample series, $MgCe_xEr_yFe_{2-x-y}O_4$ [7].

2.2 XRD Analysis

The $MgCe_xEr_yFe_{2-x-y}O_4$ sample series was prepared using the sol–gel method with concentrations of x and y ($x = 0.8$, 0.6, and 0.4, and $y = 0.2$, 0.4, and 0.6, respectively), and the resulting data was recorded and plotted using a Rigaku MiniFlex-600 diffraction spectrometer. Cu K-radiation with a wavelength of 1.5406 Å was used to

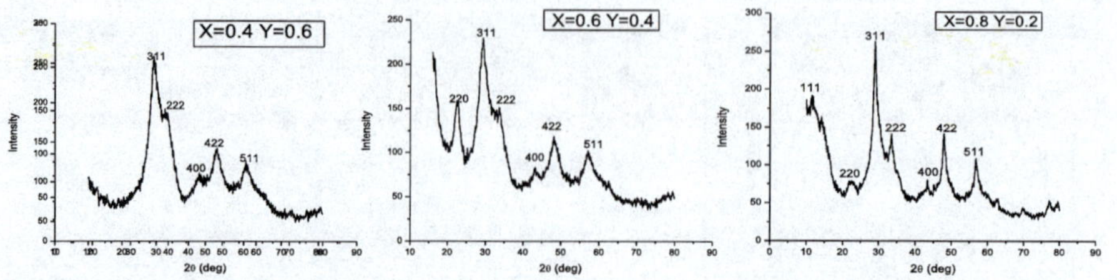

Fig. 2 XRD pattern of $MgCe_xEr_yFe_{2-x-y}O_4$nano-ferrites powders for, **a** $X = 0.4$ $Y = 0.6$, **b** $X = 0.6$ $Y = 0.4$ and **c** $X = 0.8$ $Y = 0.2$

obtain XRD patterns of the nanopowders. The recorded XRD patterns are presented in Fig. 2a–c, and the peaks observed around the (311), (220), (400), and (422) planes confirm the presence of a spinel structure of ferrite [8]. The analysis of the XRD spectra provides information about the FCC structure and the respective planes of the nanoferrite. The spinel unit cell structure consists of eight formula units, with 64 A-sites (tetrahedral sites) and 32 B-sites (octahedral sites) in each unit cell. Half of the site's cations are occupied at octahedral sites, and 1/8th part of the sites are occupied by cations at the tetrahedral site.

$$\left[Mg^{2+}Fe^{3+}\right]A\left(Ce_x^{3+}Er_y^{3+}Fe_{1-x-y}^{3+}\right)BO_4^{2-}$$

In the above relation, the Ce^{3+} and Er^{3+} ions exhibit inverse spinel structures. The non-magnetic Mg^{2+} ion tends to occupy the tetrahedral site as well. It is well-known that lanthanide substitution requires more energy, resulting in a more thermally stable material compared to un-substituted ferrite materials. Previous research work has reported that without doping, the material has a larger lattice constant, but with Ce^{3+} and Er^{3+} doping, the Mg ferrite exhibits a smaller lattice constant due to their occupation of the octahedral site [9].

The XRD patterns in Fig. 2a–c exhibit narrow reflections, indicating a narrow size distribution of crystallites. The nanomaterial size of the samples lies within the range of 1.31 nm ($X = 0.4$ $Y = 0.6$) to 2.75 nm ($X = 0.8$ $Y = 0.2$). Additionally, the lattice parameter increases with increasing Ce^{3+} doping concentration.

2.3 *Morphological Analysis*

The scanning electron microscope (SEM) analysis of the prepared $MgCe_xEr_yFe_{2-x-y}O_4$ sample series was performed using a Zeiss evo 18 microscope, and the micrographs are presented in Fig. 3a–c. The images reveal that the particles are agglomerated together, exhibit cubic faces, and have a uniform distribution. It is observed that when the concentration of Er^{3+} increases (or Ce^{3+} decreases), the average particle size decreases, which can be attributed to the larger radius of Ce^{3+}

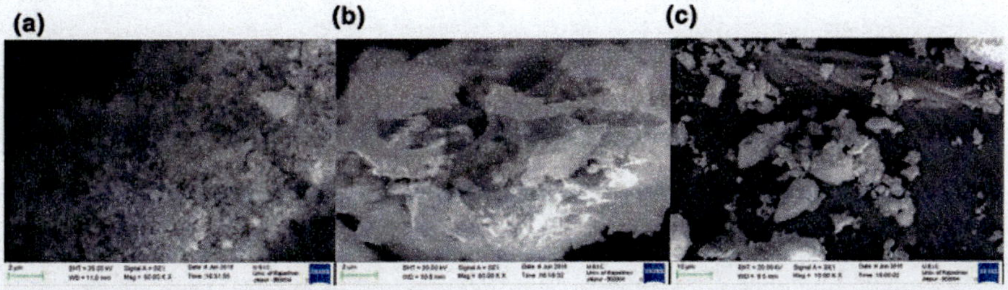

Fig. 3 SEM images of $MgCe_xEr_yFe_{2-x-y}O_4$ nano-ferrites powders for, **a** $X = 0.4$ $Y = 0.6$, **b** $X = 0.6$ $Y = 0.4$ and **c** $X = 0.8$ $Y = 0.2$

compared to the Er^{3+} atom. The particle size was discovered to measure less than 50 nm which is in agreement with the grain size predicted by Scherrer's formula [10].

2.4 *Magnetic Measurements*

The data recorded and plotted for the prepared series as.

Figure 4a–c show the saturation magnetization value for each sample. The sample with $X = 0.4$ $Y = 0.6$ exhibits the highest saturation magnetization value of 100 emu/gm at 5 K temperature, corresponding to a magnetization value of 55,000 G. The magnetic susceptibility (χ) was calculated using the formula:

$$\chi = \frac{B}{H} \tag{1}$$

The value of χ is found to be 0.00023057, and the corresponding magnetic permeability (μ) is calculated to be 1.002898. An increase in doping concentration and temperature has been found to result in an increase in μ, as observed. The different parameters, such as maximum magnetic moment, are shown in Table 1. This study

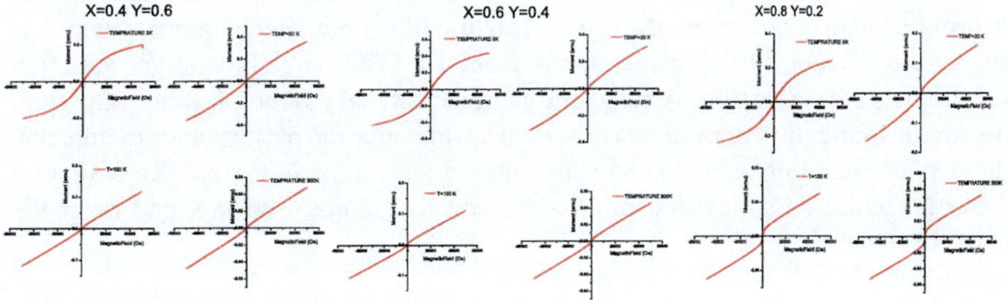

Fig. 4 Magnetization as a function of magnetic field for $MgCe_xEr_yFe_{2-x-y}O_4$ at, **a** $T = 5$ K, **b** $T = 20$ K, **c** $T = 100$ K and **d** $T = 300$ K for prepared series

Table 1 Magnetic moment values with increasing temperature in emu/gm units

S. No.	Temperature (K)	$X = 0.4$ $Y = 0.6$ magnetic moment (μ_B) (emu/gm)	$X = 0.6$ $Y = 0.4$ magnetic moment (μ_B) (emu/gm)	$X = 0.8$ $Y = 0.2$ magnetic moment (μ_B) (emu/gm)
1	5	100	84	40
2	20	70	60	30
3	100	24	22	12
4	300	8	7.6	10

shows that the magnetic properties are altered by doping Ce^{3+} and Er^{3+} in Mg ferrite. Observations have indicated that the inclusion of Ce^{3+} and Er^{3+} leads to an increase in saturation magnetization as reported by Rezlescu et al. [11], Ce^{3+} and Er^{3+} doped Mg ferrite exhibit higher values of saturation magnetization compared to un-doped ferrite, as presented in the similar results reported by Liu et al. [12]. Figure 4 shows that the prepared sample materials exhibit ferromagnetic properties, and with an increase in temperature, paramagnetic behavior is observed. Additionally, increasing Ce^{3+} concentration leads to the conversion of ferromagnetic to paramagnetic properties. The graphs in Fig. 4 are plotted at constant temperatures of 5, 20, 100, and 300 K. These curves show that the slope of the curve decreases with an increase in temperature, indicating lower values for the prepared series $X = 0.8$, $Y = 0.2$, which means that the ferromagnetic properties of the material are converting into paramagnetic properties, and the growth of domain is also decreasing.

3 Conclusions

The sol–gel method was used to successfully prepare a series of magnesium nanoferrites with various doping concentrations of Ce^{3+} and Er^{3+} ($X = 0.4, 0.6, 0.8$ and $Y = 0.6, 0.4, 0.2$). The X-ray diffraction and SEM results provide strong evidence for the nanomaterial structure and indicate that the lattice parameter decreases with decreasing concentration of Ce^{3+} and Er^{3+} ions due to their larger ionic radii compared to the host cations, Mg^{2+} and Fe^{3+}. The FTIR spectrum data exhibits expected absorption bands, confirming the spinel structure of the prepared nanomaterials. The magnetic permeability (μ) values range from 1.002898 and show slight variation with doping concentration. A magnetic permeability (μ) value plays an important role for the fabrication of microwave antenna to enhance the performance of antenna. These synthesized nanoferrites have potential applications in various areas such as planar antennas, soft magnets, microwave devices, remote sensing, and magnetic fluids for hyperthermia.

References

1. Yuan X, Chen M, Yao Y, Guo X, Huang Y (2021) Recent progress in the design and fabrication of multifunctional structures based on metamaterials. Curr Opin Solid State Mater Sci 25(1):100883–100895
2. Kaur N, Sivia J, Rajni (2021) Artificial neural network based metasurface inspired planar frequency reconfigurable antenna for wireless applications. RF Microw Comput Aided Eng 1–13
3. Junaid M, Khan M, Iqbal F, Murtaza G, Akhtar M, Ahmad M, Shakir I, Warsi M (2016) Structural, spectral, dielectric and magnetic properties of Tb–Dy doped Li-Ni nano-ferrites synthesized via micro-emulsion route. J Magn Magn Mater 419:338–344
4. Srivastava R, Yadav B (2012) Ferrite materials: introduction, synthesis techniques, and applications as sensors. Int J Green Nanotechnol 4:141–154
5. Akhtara M, Khan M (2018) Effect of rare earth doping on the structural and magnetic features of nanocrystalline spinel ferrites prepared via sol gel route. J Magn Magn Mater 460:268–277
6. Pradeep A, Priyadharsini P, Chandrasekaran G (2008) Sol–gel route of synthesis of nanoparticles of MgFe2O4 and XRD, FTIR and VSM study. J Magn Magn Mater 320:2774–2779. Elsevier
7. Ahmed A, Siddig M, Mirghni A, Omer M, Elbadawi A (2015) Structural and optical properties of Mg1-xZnxFe2O4 nano-ferrites synthesized using co-precipitation method. Adv Nanopart 4:45–52. Scientific Research Publishing
8. Muthuraman K, Naidu V, Ahamed S, Sahib K (2013) Study of electrical and magnetic properties of cerium doped nano smart magnesium ferrite material. Int J Comput Appl 65(23):18–24
9. Gateshki M, Petkov V, Pradhan S, Vogt T (2005) Structure of nanocrystallineMgFe2O4 from X-ray diffraction, Rietveldand atomic pair distribution function analysis. J Appl Crystallogr 38:772–779
10. Mulushoa Y, Wegayehu M, Aregai G, Murali N, Reddi M, Arunamani T, Samatha K (2017) Synthesis of spinel $MgFe_2O_4$ ferrite material and studying its structural and morphological properties using solid state method. Chem Sci Trans 6(3):1401–1411
11. Rezlescu N, Rezlescu E, Pasnicu C, Crau M (1994) Effect of the rare-earth ions on some properties of nickel-zinc ferrite. J Phys Condens Matter 6:5707–5719
12. Liu X, Zhong W, Yang S, Yu Z, Gu B (2002) Influence of La3+ substitution on the structural and magnetic properties of M-type strontium ferrites. J Magn Magn Mater 238(2):207–214

Synthesis and Characterization of Thin Film Nanocomposites of PEO/PMMA Blend Using SnO_2 Filler

S. Amudha, A. N. Nousheen Nisha, and R. Pooja

Abstract A new approach involving the prologue of tin oxide (SnO_2) nanofiller into nanocomposite solid polymer electrolyte system [NSPE] of polyethylene oxide (PEO)/polymethyl methacrylate (PMMA)–cadmium bromide ($CdBr_2$) has been adopted, intending to improve the electrochemical performance of thin film. Thin film specimens of nanocomposite solid polymer electrolytes [NSPE] were synthesized using solution casting technique and typified by diverse portrayal techniques such as X-ray diffractogram (XRD), Fourier transforms infrared spectroscopic analysis (FTIR) and UV–visible spectrophotometry. The XRD studies show the structural changes occurring within the chosen nanocomposite solid polymer electrolyte system. The coordinative interactions taking place within the polymer electrolyte system were revealed by FTIR studies and optical analysis was carried out by UV–visible spectroscopy. Furthermore, all these studies enabled the identification of the interaction of nanofiller SnO_2 within these blended polymer electrolytes exhibiting an improved amorphicity, and hence, the present system shows an advanced performance in the case of ion-conducting polymer electrolytes.

Keywords Solid polymer electrolyte · Tin oxide · Cadmium bromide

1 Introduction

In existing years, exploration on nanocomposite solid polymer electrolyte [NSPE] system serves as an unconventional to energy sources and serves as prospective candidate in electrochemical device applications, fuel cells, supercapacitors, sensors, high-performance solid-state batteries and smart windows. In comparison with liquid polymer electrolytes, nanocomposite solid polymer electrolyte [NSPE] displays major advantages such as high energy density, leakage resistant, lightweight and excellent mechanical stability [1]. Generally, the most commonly used polymer electrolytes include polyacrylonitrile (PAN) [2], polyvinylidene

S. Amudha (✉) · A. N. Nousheen Nisha · R. Pooja
Department of Physics, S.D.N.B. Vaishnav College for Women, Chrompet, Chennai, India
e-mail: amudha.sbs@gmail.com

Z. H. Khan et al. (eds.), *Recent Advances in Nanomaterials*, Springer Proceedings in Materials 27, https://doi.org/10.1007/978-981-99-4878-9_62

fluoride-co-hexafluoropropylene (PVDF-co-HFP) [3], polyvinylchloride (PVC) [4], PEO [5] and polymethyl methacrylate (PMMA) [6]. Among these, PEO is a partially crystalline and partially amorphous polymer having single helical structure, and it acts as a good polymer electrolyte due to its high salvation capacity, high thermal stability, good conductivity and adequate corrosion resistance [7] but it also exhibits some practical intricacy due to its crystalline structure. In order to overcome this difficulty blending polymer method has been adopted in which PEO is blended with PMMA since van der Waals type of bonding taking place between PEO and PMMA polymer electrolyte [8]. Lately, many researchers have reported that the addition of nanofillers such as Al_2O_3 [9], TiO_2 [10], Fe_2O_3 [11] enhances the conductivity of polymer matrix. With the aim of reducing crystallinity, to improve thermal stability, ionic conductivity and to obtain mechanically stable thin films an experimental endeavor on the development of suitably blended polymer electrolytes of PEO/PMMA with cadmium bromide ($CdBr_2$) salt in conjunction with further incorporation of ceramic filler such as tin oxide (SnO_2) has been undertaken.

2 Materials Required

2.1 *Starting Materials*

Aldrich-make chemicals of polyethylene oxide (PEO) (Mw $\approx$ 500,000) and polymethyl methacrylate (Mn $\approx$ 996,000) blend with cadmium bromide ($CdBr_2$) by the dispersion of x wt% of tin oxide (SnO_2) (size < 100 nm) nanofiller as plasticizer were purchased, whereas acetone was employed as the solvent medium.

2.2 *Composite Formation of Thin Film Polymer Electrolyte Samples*

Thin layers of blended nanocomposite solid polymer electrolyte [NSPE] films were prepared by complexing polyethylene oxide (PEO) (0.2 g) and polymethyl methacrylate (PMMA) (0.1 g) with cadmium bromide ($CdBr_2$) (0.024 g) by the dispersion of 3 wt% of tin oxide (SnO_2) (size < 100 nm) nanofiller as plasticizer. Initially, PMMA polymer was dried in vacuum oven at 373 K for 15 min. Then PEO and PMMA polymer were dissolved together using acetone as the common solvent followed by constant stirring for 1 h so as to form the consistent solution. The $CdBr_2$ salt was afterward dried in a vacuum oven at 373 K. It was then directly dissolved into the blended polymeric matrix and then stirred well for another 1 h. Subsequently, 3 wt% SnO_2 nanofiller is added into polymer-salt complex followed by stirring for another 2 h and the same was poured onto a clean Petri dishes and dried under free air at normal temperature for 24 h. Finally, mechanically stable and solvent-free samples of the PEO-PMMA-$CdBr_2$ −3 wt% SnO_2 blended nanocomposite solid polymer electrolyte [NSPE] thin films were obtained. These films are prepared based on the procedure reported earlier [12, 13].

2.3 Characterization Techniques

The X-ray diffraction measurement of thin film nanocomposite solid polymer electrolyte [NSPE] system was scrutinized using X-ray Diffractometer of Bruker D8 Advance Model with Copper $K\alpha$ as the radiation source with wavelength $\lambda = 1.541$ Å at normal temperature (298 K). The Fourier transform infrared (FTIR) spectra of samples were recorded by KBr pellet method using Bruker Tensor Model FTIR spectrophotometer at the wave number resolution of 1 cm^{-1}, whereas the ultraviolet–visible spectra were confirmed using Perkin Elmer UV spectrophotometer.

3 Results and Discussion

3.1 X-ray Diffractogram Analysis

X-ray diffractogram is a scheme involved in the area of science in order to analysis and evaluates the structure, particle size and phase identification of the crystalline samples. Figure 1 confirms the X-ray diffractograms obtained for PEO-PMMA-$CdBr_2$ and PEO-PMMA-$CdBr_2$-SnO_2 thin film nanocomposite solid polymer electrolyte [NSPE] system. The maximum peak observed at $2\theta = 19.14°$, 23.3° and 26.06° corresponds to crystalline nature of pure PEO [14] which may be assigned to (112), (120) and (222) planes. These diffraction peaks are found to get shifted to $2\theta =$ 19.33°and 23.46° and slight dwindling in height of the peak due to the incorporation of SnO_2 which increases the amorphous phase of PEO. The new peaks observed at 2θ = 26.92°, 34.03° and 52.05°corresponds to (110), (101) and (211) planes which indicates the existence of tetragonal rutile structure of SnO_2 [15]. Hence, the prepared nanocomposite thin film blended solid polymer electrolytes endures considerable miscibility and structural reorganization when SnO_2 filler is added into it, whereas the shifting of peaks, diminution in the height of prominent peak and emergence of innovative peaks tend to show an increase in the amorphous phase as well as the conductivity of polymer electrolyte.

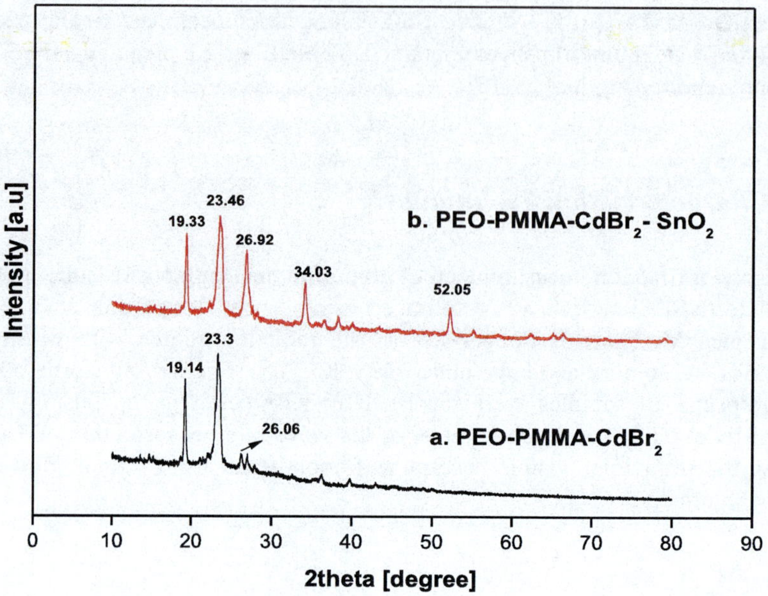

Fig. 1 X-ray diffraction patterns of **a** PEO-PMMA-$CdBr_2$ and **b** PEO-PMMA-$CdBr_2$-SnO_2 thin film nanocomposite solid polymer electrolyte system

3.2 *FTIR Results*

Fourier transform infrared (FTIR) spectroscopy recognizes the molecular bonds and interactions occurring in the polymer electrolyte sample with the help of infrared radiation. The FTIR spectrum obtained for PEO-PMMA-$CdBr_2$ and PEO-PMMA-$CdBr_2$-SnO_2 thin film nanocomposite solid polymer electrolyte [NSPE] system is exposed in Fig. 2. The vibrational spectrum observed at 2887 cm^{-1} corresponding to carbon-hydrogen stretching of pure PEO disappeared in the complexes, whereas the intensity of the spectrum at 1097 cm^{-1} corresponding to carbon–oxygen-carbon stretching of linkage of PEO gets reduced, and also, it gets shifted to 1056 cm^{-1} in the polymer matrix [16]. Besides the formation of doublet peak has been examined at 1129 and 1146 cm^{-1} which may be attributed to the coordination of PEO-PMMA-$CdBr_2$ with SnO_2 nanofiller. The spectra observed on 836 and 957 cm^{-1} corresponding to CH_2 wagging and CH_2 twisting of pure PEO get shifted to 843 and 965 cm^{-1} in the complex, whereas the band at 1236, 1275 and 1338 cm^{-1} corresponds to CH_2 twisting, wagging and scissoring form of PEO [17]. The intensity of the strong peak 1726 cm^{-1} which may be accredited to C-O stretching vibration of PMMA gets diminished after introducing SnO_2 filler into the polymer matrix, whereas the spectrum at 1465 cm^{-1} corresponding to asymmetrical meandering vibration of methyl group of PMMA also gets shifted to 1473 cm^{-1} in the complexes [18]. This sort of shifting, vanishing and diminishing in the intensity of peaks is probably due to

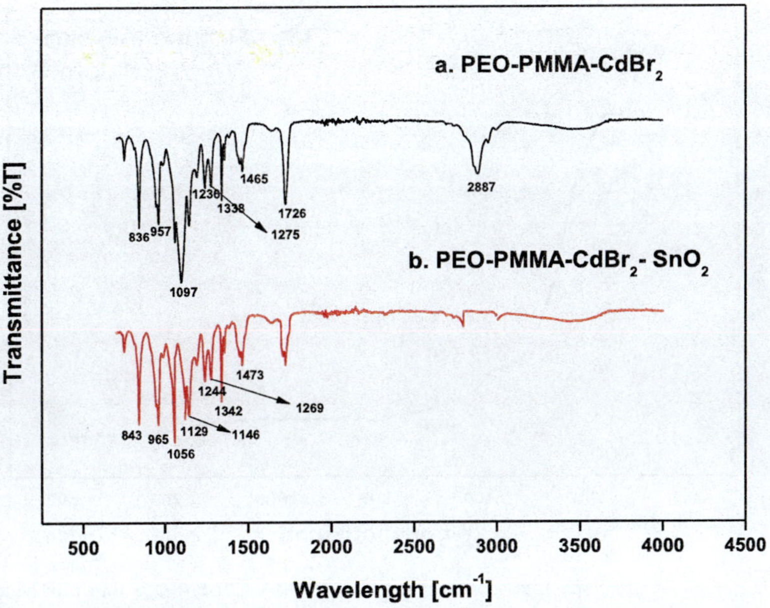

Fig. 2 FTIR spectra obtained for PEO-PMMA-$CdBr_2$-SnO_2 thin film nanocomposite solid polymer electrolyte system

most promising situation for coordinative interaction taking place between blended polymer, salt and filler which is related to amplify the mobility of movement of ions, and hence, the conducting nature also gets enhanced.

3.3 UV–Visible Optical Analysis

Akin to FTIR spectroscopy, UV–visible spectroscopy is another versatile optical technique used to examine the interactions occurring in the compounds where change in absorbance is measured for change in wavelength and using this band gap energy of the material is also calculated. Figure 3 shows the ultraviolet–visible absorption spectrum connecting the range 200–800 nm which is used to determine optically stimulated transitions and band gap energy of thin film polymer electrolyte specimen. Generally, pure PEO reveals no absorption peak in the range 200–380 nm [19], whereas in PEO-PMMA-$CdBr_2$-SnO_2 thin film nanocomposite solid polymer electrolyte [NSPE] system a new absorption band has been observed at 239 nm which reveals the $\pi \rightarrow \pi^*$ transition occurring within $C = O$ of PMMA polymer [20], and also the absorption band at 374 nm indicates the presence of SnO_2 nanofiller [21], and the energy gap between the band value is 3.31 electron volts.

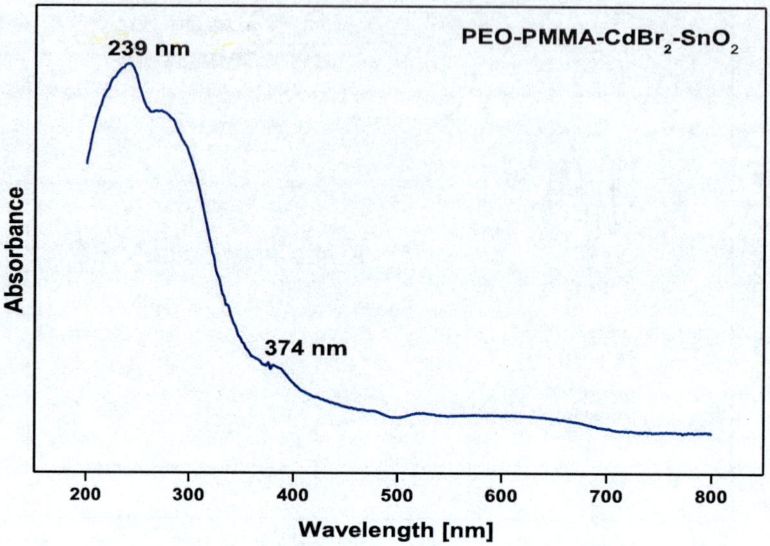

Fig. 3 Ultraviolet–visible spectrum obtained for PEO-PMMA-$CdBr_2$-SnO_2 thin film nanocomposite solid polymer electrolyte system

4 Conclusions

A new type of PEO and PMMA-based blended polymer electrolyte with $CdBr_2$ as salt and SnO_2 as filler was prepared by solution casting technique from which free standing, mechanically stable thin films were obtained and their structural, spectral and optical studies were investigated analytically. Here Cd^{2+} ions interact with carbon and oxygen present in polymer and SnO_2 filler since cadmium bromide salt has wide applications in rechargeable batteries, whereas tin oxide (SnO_2) is a spacious n-type partially conducting material used in optoelectronic devices. All the compounds used for preparing this thin film polymer electrolyte samples are peculiar one which may not be attempted earlier. From the X-ray diffractogram studies, it is understandable that the prologue of nanofiller into the solid polymer electrolyte system illustrated decrease in degree of crystallinity whereas the FTIR spectrum of thin films confirmed the complexation taking place between PEO-PMMA blended polymer, $CdBr_2$ salt and SnO_2 nanofiller. The UV–visible spectra clearly indicate the band gap energy value of 3.31 eV for 374 nm absorption band, and this low band gap energy value may enhance the electrical conductivity and make it as a promising candidate for batteries, fuel cells, LEDs and optoelectronic device applications.

References

1. Money BK, Hariharan K, Swenson J (2012) Glass transition and relaxation processes of nanocomposite polymer electrolytes. J Phys Chem B 116(26):7762–7770
2. Yu B, Zhou F, Wang C, Liu W (2007) A novel gel polymer electrolyte based on poly ionic liquid 1-ethyl 3-(2- methacryloyloxy ethyl) imidazolium iodide. Eur Polym J 43(6):2699–2707
3. Lee C, Kim JH, Bae JY (2003) Polymer gel electrolytes prepared by thermal curing of poly(vinylidene fluoride)–hexafluoropropene/poly(ethylene glycol)/propylene carbonate/ lithium perchlorate blends. Polymer 44(23):7143–7155
4. Jung HR, Lee WJ (2011) Electrochemical characteristics of electrospun poly(methyl methacrylate)/polyvinyl chloride as gel polymer electrolytes for lithium ion battery. Electrochim Acta 58:674–680
5. Song C, Li Z, Peng J, Wu X, Peng H, Zhou S, Qiao Y, Sun H, Huang L, Sun S (2022) Enhancing Li ion transfer efficacy in PEO-based solid polymer electrolytes to promote cycling stability of Li-metal batteries. J Mater Chem A 10(30):16087–16094
6. Sharma JP, Sekhon SS (2007) Nanodispersed polymer gel electrolytes: conductivity modification with the addition of PMMA and fumed silica. Solid State Ionics 178(5–6):439–445
7. Aziz SB, Nofal MM, Brza MA, Hussein SA, Mahmoud KH, El-Bahy M, Dannoun EMA, Kareem WO, Hussein AM (2021) Characteristics of PEO incorporated with $CaTiO_3$ nanoparticles: structural and optical properties. Polymers 13(3484):1–18
8. Jaipal Reddy M, Ravindar Reddy M, Subrahmanyam AR, Maheshwar Reddy M, Samba Siva Rao A (2018) Effect of $LiClO_4$ concentration on structural, morphological and thermal properties of PMMA and PEO polymer blends. Asian J Res Chem 11(2):463–466
9. Agrawal SL, Rai N, Chand N (2013) Dynamic mechanical, DSC, and electrical investigations on Nano Al_2O_3 filled PVA:NH_4SCN:DMSO polymer composite dried gel electrolytes. Int J Polym Mater Polym Biomater 62(2):61–67
10. Nimah YL, Cheng MY, Cheng JH, Rick J, Hwang BJ (2015) Solid-state polymer nanocomposite electrolyte of TiO_2/PEO/$NaClO_4$ for sodium ion batteries. J Power Sources 278:375–381
11. Rajasudha G, Durgalakshmi D, Thangadurai P, Nikos B, Narayanan V, Stephen A (2012) Preparation and characterization of polyindole-iron oxide composite polymer electrolyte containing $LiClO_4$. Polym Plastics Tech Eng 51(3):225–230
12. Reddy MR, Reddy MJ, Subrahmanyam AR (2017) Structural, thermal and optical properties of PMMA, PEO and PMMA/PEO/$LiClO_4$ polymer electrolyte blends. Mat Sci Res India 14(2):123–127
13. Sharma P, Kanchan D, Gondaliya N, Jayswal M, Joge P (2013) Influence of nano filler on conductivity in PEO-PMMA-$AgNO_3$ polymer blend. Indian J Pure Appl Phys 51(5):346–349
14. Abdelrazek EM, Abdelghany AM, Badr SI, Morsi MA (2018) Structural, optical, morphological and thermal properties of PEO/PVP blend containing different concentrations of biosynthesized Au nanoparticles. J Mat Res Tech 7(4):419–431
15. Debatarajaa A, Zulhendria DW, Yuliartoa B, Nugrahaa, Hiskia, Sunendar B (2017) Investigation of nanostructured SnO_2 synthesized with polyol technique for CO gas sensor applications. Proc Eng 170:60–64
16. Lim YS, Jung HA, Hwang H (2018) Fabrication of PEO-PMMA-$LiClO_4$-based solid polymer electrolytes containing silica aerogel particles for all-solid-state lithium batteries. Energies 11(2559):1–10
17. Hamisu A, Abubakar AM, Aminu KS (2019) PEO-hBN-$NaClO_4$ polymer composite electrolyte for sodium ion batteries. Int J Sci Basic Appl Res 45(2):104–117
18. Kanimozhi G, Vinoth S, Harish Kumar, Srinadhu ES, Satyanarayana N (2018) Conductivity and dielectric permittivity studies of KI-based nanocomposite (PEO/PMMA/KI/I_2/ZnO nanorods) polymer solid electrolytes. Polym Compos 40(7):1–10
19. Muhammed DS, Brza MA, Nofal MM, Aziz SB, Hussen SA, Abdulwahid RT (2020) Optical dielectric loss as a novel approach to specify the types of electron transition: XRD and UV-vis as a non-destructive techniques for structural and optical characterization of PEO based nanocomposites. Materials 13(2979):1–15

20. Rai VN, Mukherjee C (2017) Beena Jain: UV-Vis and FTIR spectroscopy of gamma irradiated polymethyl methacrylate. Indian J Pure Appl Phys 55(11):775–785
21. Mayandi J, Marikkannan M, Ragavendran V, Jayabal P (2014) Hydrothermally synthesized Sb and Zn doped SnO_2 nanoparticles. J Nanosci Nanotech 2(6):707–710

Nanocomposites of Reduced Graphene Oxide (RGO) with CdS Nanoparticles for Enhanced Photocatalytic Behaviour

Mansi Malik, Poonam Mahendia, O. P. Sinha, and Suman Mahendia

Abstract Semiconductors having their band gap in visible region are preferably used as photocatalytic applications. Cadmium sulphide (CdS) is one such suitable semiconductor with a band gap of 2.42 eV. However, direct use of CdS nanoparticles (NPs) for photocatalytic applications is limited because of their self-oxidizing property and less stability. Therefore, composites can be formed to overcome such limitation. Reduced graphene oxide (RGO) can be suitable matrix because of its larger specific surface area and facilitates the separation of charges for better photocatalytic (PC) performance. Thus, we endeavour the synthesis of CdS-RGO nanocomposites (NCs) via in situ solvothermal reduction method. The UV–Visible absorption spectroscopy of NCs shows slight red shift in absorption edge compared to virgin CdS nanoparticles and thus promotes the charge transfer from CB of CdS to fermi level of RGO sheets. Such observations are also confirmed by observed quenching in emission peak of photoluminescence (PL) spectra of CdS-RGO NCs. The enhancement in photocurrent observed through I-V characteristics confirms the better photocatalytic behaviour of CdS-RGO NCs.

Keywords Nanocomposites (NCs) · Photocatalytic · Cadmium sulphide (CdS) · Reduced graphene oxide (RGO)

1 Introduction

Cadmium sulphide (CdS) is widely used semiconductor in various optical applications including optoelectronic devices, photocatalyst, photoelectrochemical cells, etc. A good photocatalyst should have high absorption coefficient, large number of surface active sites, lesser defects in crystal structure, high crystallinity and good

M. Malik · S. Mahendia (✉)
Department of Physics, Kurukshetra University, Kurukshetra, Haryana, India
e-mail: smahendia@kuk.ac.in

P. Mahendia · O. P. Sinha
Amity Institute of Nanotechnology, Amity University UP, Noida, India

Z. H. Khan et al. (eds.), *Recent Advances in Nanomaterials*, Springer Proceedings in Materials 27, https://doi.org/10.1007/978-981-99-4878-9_63

stability. CdS NPs acquire all such properties but narrower band gap of CdS introduces limitation of higher recombination rate and self-oxidation. Thus, the lack of stability hinders the use of CdS as photocatalyst in optoelectronic devices as well as water-splitting application [1, 2]. Therefore, in order to improve the catalytic action of CdS, we endeavor the synthesis of nanocomposites of CdS with reduced graphene oxide (RGO). RGO offers high conductivity, extraordinary mechanical strength, large specific surface area and the fastest mobility of charge carriers [3]. Therefore, it acts as a supportive matrix to decorate CdS NPs on a larger area which enhances the surface active sites and helps in enhanced charge carrier separation and transportation. In the present study, the photocatalytic activity of CdS-RGO NCs deposited on Ni foam is analyzed by measuring the I-V characteristics in dark and under light illumination.

2 Experimental Section

Cadmium acetate dihydrate ($Cd(Ac)_2.2H_2O$), sodium sulphide (Na_2S), Ethylenediamine (EDA) and ethanol were purchased from Rankem Pvt. Ltd. and used as procured without further purification. Graphite powder of size $< 20\ \mu m$ and Nickel foam with purity > 99%, cell density - 110 PPI and thickness 1.5 mm are purchased from Sigma-Aldrich.

2.1 Synthesis of CdS NPs and CdS-RGO NCs

In order to synthesize CdS NPs, a solution of 0.05M of $Cd\,(Ac)_2.2H_2O$ and 0.05M of Na_2S was prepared in 80mL of EDA. Thereafter, resultant solution was transferred to a 100mL of Teflon-lined autoclave and heated at 120°C for 12 h [4]. Further in order to synthesize CdS-RGO NCs, firstly GO was synthesized using modified Hummer's method [5, 6]. Thereafter, dried powder of GO was mixed with 80ml of EDA and reduced using hydrazine hydrate during ultrasonication followed by magnetic stirring. Then, following the same procedure as above, in situ CdS-RGO NCs were synthesized. I-V measurements were done using two-probe set-up supported with Keithley 2400 source meter with light illumination by 100W light source. For this purpose, prepared samples were dissolved in DMSO followed by drop cast on Ni foam.

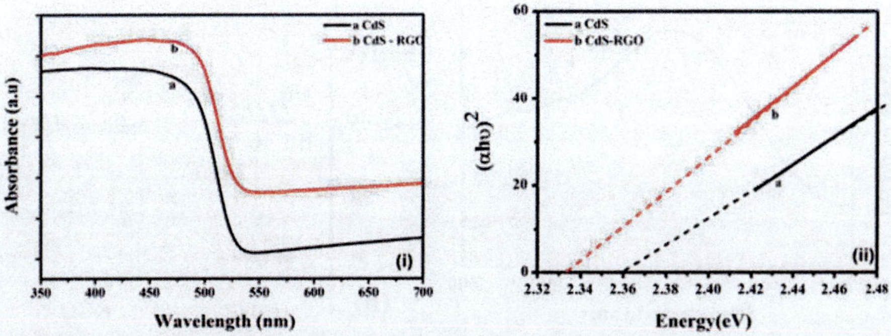

Fig.1 (i) UV–Vis absorption spectra and (ii) Tauc's plot (with $R^2 = 0.999$) for CdS NPs and CdS-RGO NCs

3 Results and Discussion

3.1 UV–Visible Absorption Spectroscopy

The absorption spectra for CdS NPs and their composite with RGO are shown in Fig. 1(i). The higher absorption observed in case of CdS-RGO NCs indicates the existence of more number of surface active sites due to the larger surface area of RGO sheet. The red shift is also observed in fundamental absorption band edge from 525nm (for CdS NPs) to 532nm in case of composite. This corresponds to the formation of trapping states because of RGO surface-active sites between valence band (VB) and conduction band (CB). The value of band gap (E_g) for direct band gap semiconductor like CdS can be determined from absorption spectra by extrapolating the intercept on X-axis in $(\alpha h\upsilon)^2$ vs energy ($h\upsilon$) plot (shown in Fig. 1(ii)) obeying Tauc's relation: $(\alpha h\upsilon)^2 = B(h\upsilon - E_g)$ (with all symbols have their usual meaning). The obtained values are tabulated in Table 1. The decrease in band gap energy from 2.36 $\pm$ 0.01eV for CdS NPs to 2.33 $\pm$ 0.02eV for CdS-RGO NCs is observed and may attribute to be due to the formation of additional trapping sites as an effect of RGO base matrix.

3.2 Photoluminescence (PL) Spectroscopy

PL spectrum for pure CdS NPs and their nanocomposites with RGO are shown in Fig. 2. As observed for CdS NPs (curve 'a') show good PL peak at 547 nm. However, for CdS-RGO NCs this emission gets quenched with slight blue shifting to 540 nm with small hump at 547 nm (curve 'b'). This decrease in intensity in case of composites signifies less recombination of e^--hole pairs due to efficient separation of charges. This supports facile transfer of electrons from CdS to RGO in case of

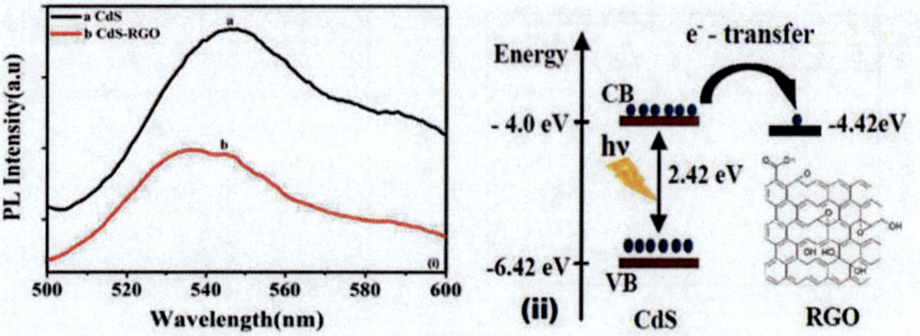

Fig. 2 (i) PL spectra of CdS NPs and CdS-RGO NCs at an excitation wavelength of 420nm. (ii) Schematic illustration showing band diagram for CdS-RGO NCs and charge transportation

CdS-RGO NCs, thus depicting good photocatalytic properties of NCs. Schematic of charge transfer is also illustrated in Fig. 2(ii).

3.3 *Electrical Measurements*

Figure 3 represents the current (I) vs voltage (V) characteristics measured for CdS NPs and CdS-RGO NCs in dark as well as under light illumination. The electrical conduction in semiconductors is quite low at room temperature. The dark current for CdS NPs (curve 'a') is of small order (also shown in inset in Fig. 3), whereas enhancement in current with light can be seen depicting its photoactive behavior. For CdS-RGO NCs (curve 'b') noticeable enhancement in dark current is observed signifying the improvement in charge conduction electrons due to RGO matrix. This might be due to the less recombination of e^{-}-hole pairs in the presence of enhanced surface active sites of RGO matrix. This facilitates the good separation of e^{-}-hole pairs thereby reducing their recombination rate leading to enhanced charge transportation from the CB of CdS (higher potential) to that of RGO having lower work function of −4.42eV. Further, enhancement in photocurrent after light illumination (curve 'e') for CdS-RGO is also observed attributing to the improved photocatalytic behavior of NCs.

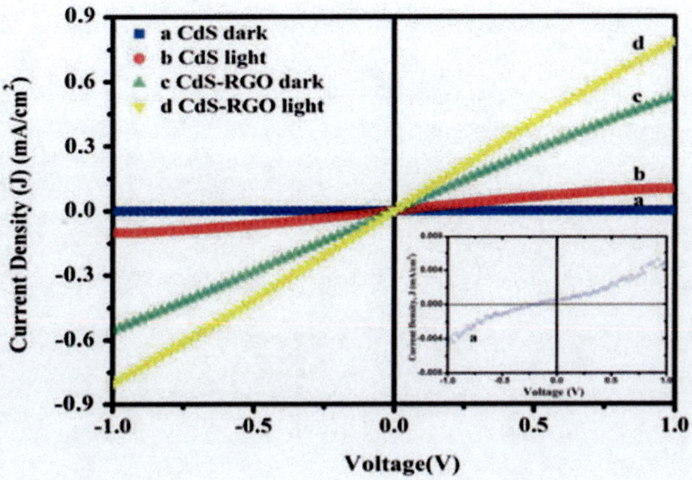

Fig. 3 Dark and light I-V curves for CdS NPs ('a' and 'b') and CdS-RGO NCs ('c' and 'd')

4 Conclusion

For the prepared CdS-RGO NCs decrease in band gap and PL quenching is observed. Further, appreciable enhancement in photocurrent is observed for these NCs. These observations suggested the formation of surface active sites in NCs due to the presence of RGO matrix. These sites act trapping states between CB and VB, thus help in improved charge transfer due to charge separation and their reduced recombination rate and thus help in acquiring good photocatalytic properties for CdS-RGO NCs.

Acknowledgements Authors from KUK are highly thankful for the Seed Money Grant, under RUSA 2.0, KUK.

References

1. Nasir J (2020) A: Recent developments and perspectives in CdS-based photocatalysts for water splitting. J Mater Chem A 8(40):20752–20780
2. Hullavarad N (2008) Cadmium sulphide (CdS) nanotechnology. J Nanosci Nanotechnol 8(7):3272–3299
3. Mondal A (2021) Boosting photocatalytic activity using reduced graphene oxide (RGO)/semiconductor nanocomposites. ACS Omega 6(13):8734–8743
4. Kumar S (2016) In-situ synthesis of reduced graphene oxide decorated with highly dispersed ferromagnetic CdS nanoparticles. Mater Chem Phys 171:126–136
5. Hummers WS Jr (1958) Preparation of graphitic oxide. J Am Chem Soc 80(6):1339–1339
6. Guan H (2019) CdS@ Ni_3S_2 core–shell nanorod arrays on nickel foam. J Mater Chem A 7(6):2560–2574
7. Khatter J (2020) Effect of temperature on properties of cadmium sulfide nanostructures. J Mater Sci: Mater Electron 31(3):2676–2685

Fabrication of Electrospun PVA-Aloe Vera Hybrid Nanofibers: Dye Removal Ability from Wastewater

Mohd Saquib Tanweer, Zafar Iqbal, and Masood Alam

Abstract Hybrid of polyvinyl alcohol and aloe vera gel (PVA@AVG) electrospun nanofiber composite has been developed by the electrospinning technique. AVG was used in 5% and 10% concentrations (*w/v*) to develop two adsorbent materials, named PVA@AVG5 and PVA@AVG10, respectively, with a fixed concentration of PVA, and glutaraldehyde (GA) as a crosslinker. The morphological structure and properties of developed adsorbent materials were characterized using several techniques, including SEM, EDAX, FT-IR, and XRD. PVA@AVG10 exhibited a high removal efficiency of 100% toward RhB dye in comparison with PVA@AVG5 under similar conditions. Sorption studies support its competency for pseudo-second-order (PSO) kinetics and Langmuir isotherm model with $q_{\max}$ of 119.76 and 277.7 mg.g^{-1} for PVA@AVG5 and PVA@AVG10, respectively. These results confirm the good candidacy and effectiveness of the prepared membrane, offering an encouraging solution to wastewater treatment.

Keywords Polyvinyl alcohol membrane · Electrospinning · Nanofiber · Adsorbent · Textile wastewater treatment

1 Introduction

Globally, dyes are present in aqueous environments because of untreated water discharge from anthropogenic activities. It resulted in groundwater and surface water pollution [1]. Dye effluents are poisonous and carcinogenic, thus posing a grave risk to the ecosystem and being dangerous for both humans and the environment [2]. Also, dyes limits photosynthetic activity in aquatic ecosystems, raising concerns about the oxygen production [3]. Hence, the abatement of such toxic organic dyes from water is an urgent need.

M. S. Tanweer (✉) · Z. Iqbal · M. Alam
Environmental Science Research Lab, Department of Applied Sciences and Humanities, Faculty of Engineering and Technology, Jamia Millia Islamia, New Delhi 110025, India
e-mail: saquibtanweer701@gmail.com

Z. H. Khan et al. (eds.), *Recent Advances in Nanomaterials*, Springer Proceedings in Materials 27, https://doi.org/10.1007/978-981-99-4878-9_64

Adsorption is the most efficient technique used to mitigate toxic dyes in aqueous solutions [3, 4]. Traditional adsorbents like activated carbon (AC) are widely used as adsorbents for the dyes' removal from wastewater. However, the use of AC for large-scale removal of dyes is quite expensive. Several researchers have developed different electrospun nanofibrous membranes for the abatement of water pollutants [5, 6].

A review of the literature revealed that previous studies did not utilize modified polyvinyl alcohol (PVA) nanofibers from aloe vera gel (AVG) for the abatement of dye. The current paper describes the preparation of PVA@AVG nanofibers for the purpose of sequestering RhB dye from wastewater.

2 Materials and Methods

2.1 Fabrication and Crosslinking of PVA@AVG Nanofibers

Two separate solutions were prepared by changing the amount of AVG, i.e., 5% and 10%. Previously prepared PVA (50 mL) solutions were taken in two separate Erlenmeyer flasks and mixed with AVG (5% and 10%) using an electromagnetic stirrer for 4 h at 26 ± 2 °C. Fabrication of PVA@AVG nanofibers was achieved by a vertical electrospinning process. In a typical electrospinning setup, the prepared PVA@AVG solutions were filled in 10 mL plastic, disposable syringe (0.5 mm inner diameter) with needles attached to the positive electrode. The applied voltage was 16.1 kV. A syringe pump (kd Scientific) was used to regulate the flow rate of the spinning mixture sample at 0.01 mL h^{-1}, while a 30 × 30 cm plate was covered with aluminum foil (as a collector) and conjoined to the negative electrode of a power supply. The collector and the syringe tip was maintained at 150 mm of distance. Once the voltage was applied, the electrospinning machine began generating and collecting the PVA@AVG fibers. The pump was controlled with a Pump-term code.

After generation, the PVA@AVG5 and PVA@AVG10 nanofibers were subjected to cross-linking in glutaraldehyde vapor at 70 °C for 12 h. They were then heated at 70 °C for an additional 10 h to eliminate any excess glutaraldehyde present.

2.2 Evaluation of RhB Dye Removal

Batch studies were used to assess the cationic RhB dye's adsorption onto PVA@AVG nanocomposites with continuous agitation at room temperature. Typically, 0.12 g of adsorbent material (PVA@AVG) was added to numerous Erlenmeyer flasks (100 mL) containing 25 mL of RhB dye (concentration: 10 to 350 mg.L^{-1}) for 120 min. pH was in between 2 to 10 adjusted by 0.1 M NaOH/HCl solutions. Once equilibrium was achieved, the PVA@AVG composite samples were subjected to centrifugation employing a Remi RM-12 C centrifuge to isolate the dye aliquots. The RhB remaining in the aqueous phase consequently analyzed by a UV–visible spectrophotometer at a wavelength of λ_{max} 554 nm.

To determine the percentage of dye uptake (R%) and sorption capacity (q_e in $mg.g^{-1}$) of the PVA@AVG composite, the preliminary RhB amount (C_i in $mg.L^{-1}$) and equilibrium RhB amount (C_e in $mg.L^{-1}$) were used to calculate the values using the following expressions.

$$R(\%) = \frac{(C_i - C_e)}{C_i} \times 100 \quad (1)$$

$$q_e = \frac{(C_i - C_e)V}{M}, \quad (2)$$

where M (g) is the amount of the PVA@AVG material, and 'V' (L) is the volume of the RhB dye solution.

3 Results and Discussion

3.1 Characterization of PVA@AVG Nanofibers

3.1.1 FE-SEM with EDAX

Figure 1 shows the SEM micrographs (recorded by microscope model JSM 6510LV, JEOL, Japan) of fine fiber formation of PVA@AVG5, PVA@AVG10, and RhB loaded PVA@AVG10. The diameter of PVA@AVG5 nanofiber in Fig. 1a is thicker and contains a greater number of beads than PVA@AVG10 nanofiber (Fig. 1c). Increase in amount of aloe vera gel causes a considerable reduction in the fabricated nanofibers' diameter and a smaller number of beads (Fig. 1c). The observed behavior may be attributed to the manifestation of hydrogen groups in both PVA and AVG. As the amount of aloe vera augmented, the electrostatic forces and coulombic repulsion between the molecules become stronger, resulting in greater molecular compactness [7]. It is observed from the EDS spectrum that the nanofiber contains, besides gold (Au), four elements (C, O, Al, Cl), which are coming from the PVA and the AVG. The appearance of a gold peak in the spectra is due to the thin layer employed for the EDS measurement. In Fig. 1e, a significant morphological change can be observed after RhB adsorption onto the PVA@AVG nanofiber composite.

3.1.2 FT-IR Spectroscopy

The infrared spectrometer (Perkin-Elmer 1600) was used to measure the FT-IR of PVA@AVG across a wave number range of 4000–400 cm^{-1}. The results of the

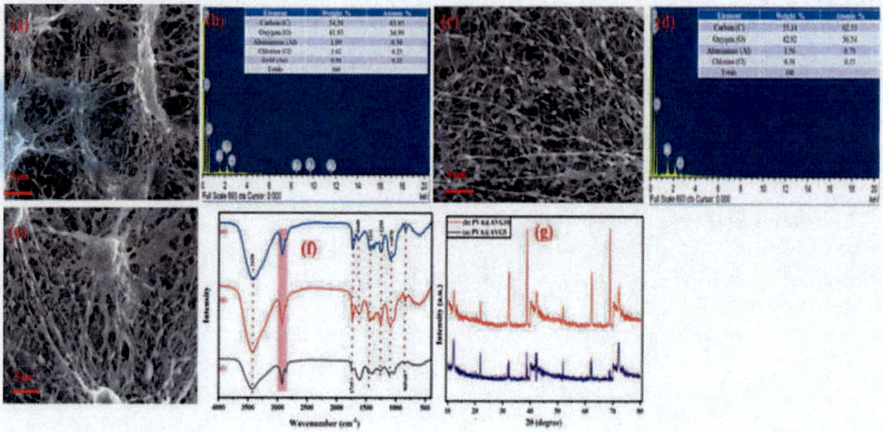

Fig. 1 Surface morphology of the fabricated electrospun nanofibers. **a** PVA@AVG5, **c** PVA@AVG10. EDX of **b** PVA@AVG5 and **d** PVA@AVG10. **e** RhB loaded PVA@AVG; FT-IR spectra of **f** PVA@AVG5, PVA@AVG10, and RhB loaded PVA@AVG10; **g** XRD spectra of PVA@AVG5 and PVA@AVG10

FT-IR spectroscopic studies for PVA@AVG are presented in Fig. 1f. The hydroxyl group, which is present in both PVA and aloe vera, is represented by the strong peak at 3420 cm^{-1}, while the band at 1070 cm^{-1} depicts the ether signals (C–O–C) in sugar units. The band between 2900 and 2850 cm^{-1} is ascribed to symmetrical/asymmetrical C–H stretching of the methylene groups, while the peak at 1066 cm^{-1} appears from the mannopyranose component [8]. Furthermore, the bands at 1735 and 1255 cm^{-1} are due to o-acetyl ester in aloe polysaccharide [8]. The peak at 849.48 cm^{-1} in PVA@AVG5 and PVA@AVG10 is attributed to mannose of aloe polysaccharide. In Fig. 1f, it can be observed that the peaks at 849.48, 1640, and 1431 cm^{-1} (due to—NH_2 group) have disappeared, indicating the successful adsorption of RhB dye.

3.1.3 XRD Study

X-ray diffraction spectra (recorded using X-ray diffractometer (MiniFlex, Rigaku) characteristic for PVA@AVG5 and PVA@AVG10 are presented in Fig. 1g. The diffraction pattern from $2\theta = 10–80°$ illustrated the semicrystalline nature of both PVA@AVG5 and PVA@AVG10. From Fig. 1g, it was observed that on increasing the aloe concentration, crystalline behavior of PVA@AVG10 nanofibers greatly improved. Ultimately, it improves the stability of nanofiber in water. The broad peaks observed at $2\theta = 20°$ and 40° are indicative of the crystalline and semicrystalline behavior of pristine PVA, respectively. The improvement in the crystalline peak of PVA in PVA@AVG10 can be attributed to the augmentation of H-bonding resulting from the increased concentration of aloe vera gel [9].

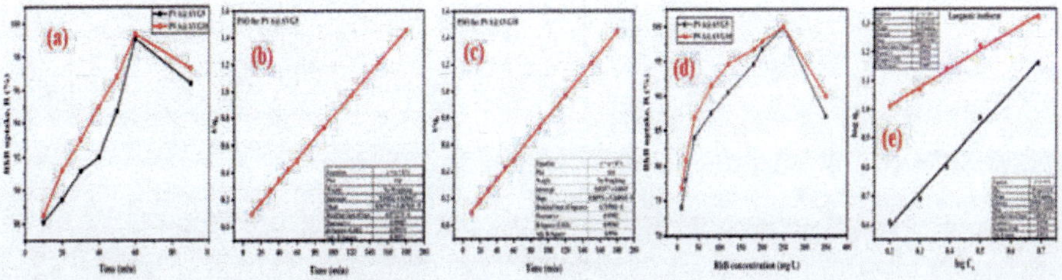

Fig. 2 **a** Impact of contact time; pseudo-second-order kinetic model for the adsorption of RhB **b** PVA@AVG5 and **c** PVA@AVG10; **d** Effect of initial RhB concentration; **e** Langmuir isotherm onto the PVA@AVG nanofiber composite

3.2 *Batch Study*

3.2.1 Impact of Time and Kinetic Studies

Figure 2a illustrates that the RhB uptake percentage of the PVA@AVG adsorbent increased on increasing contact time, and then adsorption process reached to equilibrium after 60 min with a removal percentage of 99%. Such adsorption behavior was due to accessibility of functional adsorptive sites onto PVA@AVG at the preliminary phase of the adsorption process

Two models, i.e., PFO and PSO, were applied to the experimental data of RhB adsorption from the aqueous phase to ascertain the mechanism and kinetics of the adsorption method. As presented in Fig. 2b–c and Table 1, it is apparent that RhB adsorption onto PVA@AVG is well suited to PSO kinetic model. For both PVA@AVG5 and PVA@AVG10, calculated $q_{e,RhB}$ values obtained from PSO model are near the experimental $q_{e,RhB}$ values which means that the chemisorption is rate-determining step [3].

3.2.2 Effect of RhB Concentration and Isotherm Study

The study also investigated the impact of the initial RhB concentration, ranging from 10 mg L^{-1} to 350 mg L^{-1}, on the adsorption capacity of PVA@AVG nanofiber composites. The findings in Fig. 2d indicated that the adsorption efficiency of both PVA@AVG5 and PVA@AVG10 increased from 74 to 99.97% and 76.8 to 100%, respectively, as the initial RhB concentration increased. Equilibrium was accomplished at RhB concentration greater than 250 mg L^{-1}. With higher RhB concentrations, a stronger driving force was generated, allowing dye molecules to prevail over their resistance to mass shift from the liquid media to the solid material. As a result, the uptake of dye increased. Figure 2e shows the Langmuir isotherm models for the RhB dyes on the PVA@AVG composite. Table 2 shows the parameters of each model. Langmuir models fit the best. As per Langmuir model, the maximum

Table 1 Parameters of PFO and PSO model for the sorption of RhB dye on the PVA@AVG nanofiber composites

Adsorbent	PFO				PSO			
	$q_{e,RhB}$ (exp) ($mg.g^{-1}$)	$q_{e,RhB}$ (cal) ($mg.g^{-1}$)	K_1 (min^{-1})	R^2	$q_{e,RhB}$ (exp) ($mg.g^{-1}$)	q_{eRhB} (cal) ($mg.g^{-1}$)	K_2 ($g.mg^{-1}.min^{-1}$)	R^2
PVA@AVG5	123.85	0.991	−0.00014	0.87	123.85	126.58	0.0035	0.99
PVA@AVG10	124.2	0.993	−0.00010	0.96	124.2	126.58	0.0031	0.99

Table 2 Parameters of Freundlich and Langmuir model for the sorption of RhB dye on the PVA@AVG nanofiber composites

Adsorbent	Langmuir model				Freundlich model		
	$q_{\max}$ (mg.g^{-1})	K_l (L.mg^{-1})	R_L	R^2	1/n	K_f (mg.g^{-1})(L.mg^{-1})1/n	R^2
PVA@AVG5	119.76	0.012	0.755	0.99	1.24	2.20	0.80
PVA@AVG10	277.7	0.003	0.538	0.99	0.98	0.74	0.96

adsorption capacities, q_m, for PVA@AVG5 and PVA@AVG10, respectively, were 119.76 and 277.7 mg g^{-1}.

In Table 2 we can see that for both adsorbent materials, PVA@AVG5 (0.75) and PVA@AVG10 (0.53), all intended separation factor (R_L) values are between 0 and 1, validating both the favorable adsorption of RhB dye and the suitability of the Langmuir isotherm model [10].

4 Conclusion

In conclusion, we have developed low cost, biodegradable, and highly efficient toxic organic dye removal membrane based on PVA loaded with aloe vera gel for wastewater treatment. The improvement in properties of PVA membranes depends on the concentration of AVG. Higher the AVG content, the better the performance of the membranes.

Acknowledgements One of the authors, M.S.T., is thankful to the University Grants Commission (UGC) for the Non-NET Fellowship.

Declaration of Interest Statement The authors declare that they have no conflict of interests.

References

1. Ahmad R, Ansari K (2022) Novel in-situ fabrication of L-methionine functionalized bionanocomposite for adsorption of Amido Black 10B dye. Process Biochem
2. Tanweer MS, Alam M (2022) Novel 2D nanomaterial composites photocatalysts: application in degradation of water contaminants 75–96
3. Tanweer MS, Iqbal Z, Alam M (2022) Experimental insights into mesoporous polyaniline-based nanocomposites for anionic and cationic dye removal. Langmuir
4. Iqbal Z, Tanweer MS, Alam M (2022) Recent advances in adsorptive removal of wastewater pollutants by chemically modified metal oxides: a review. J Water Process Eng 46:102641
5. Hosseini SA, Vossoughi M, Mahmoodi NM, Sadrzadeh M (2018) Efficient dye removal from aqueous solution by high-performance electrospun nanofibrous membranes through incorporation of SiO_2 nanoparticles. J Clean Prod 183:1197–1206

6. Patel S, Hota G (2014) Adsorptive removal of malachite green dye by functionalized electrospun PAN nanofibers membrane. Fibers Polym 15:2272–2282
7. Khanzada H, Salam A, Qadir MB, Phan DN, Hassan T, Munir MU, Pasha K, Hassan N, Khan MQ, Kim IS (2020) Fabrication of promising antimicrobial aloe Vera/PVA electrospun nanofibers for protective clothing. Materials 13:3884
8. Nejatzadeh-Barandozi F, Enferadi ST (2012) FT-IR study of the polysaccharides isolated from the skin juice, gel juice, and flower of Aloe vera tissues affected by fertilizer treatment. Org Med Chem Lett 21(2):1–9
9. Aziz SB, Kadir MFZ, Hamsan MH, Woo HJ, Brza MA (2019) Development of polymer blends based on PVA: POZ with low dielectric constant for microelectronic applications. Sci Rep 91(9):1–12
10. Iqbal Z, Tanweer MS, Alam M (2022) Reduced graphene oxide-modified spinel cobalt ferrite nanocomposite: synthesis, characterization, and its superior adsorption performance for dyes and heavy metals. ACS Omega

Structural and Electrical Characterization of Gadolinium and Sodium Co-Doped Barium Strontium Titanate

Sahil, Sahil Kumar, Anshu Gaur, Md. Ahamad Mohiddon, and Preeti

Abstract The effect of Gadolinium (Gd) and Sodium (Na) co-doping on the structural and electrical properties of $Ba_{0.8}Sr_{0.2}TiO_3$, i.e. $(Gd_{0.5}Na_{0.5})_{0.1}Ba_{0.7}Sr_{0.2}$ TiO_3 (GNBST) is reported. (Gd, Na) doping is expected to tune the dielectric characteristics of BST by ionic size effect. GNBST was synthesized by the solid-state mixed oxide method. The wet mixed powder was calcined at 1400°C for 12 hrs. A single-crystallographic phase formation was confirmed by the X-ray diffraction technique and analysed through Rietveld refinement using the X'Pert Highscore plus software. The calcined powder was pressed into pellets and sintered at 1450°C for 2 hrs. The dielectric and electrical characteristics was investigated by measuring the dielectric parameters at different temperatures starting from room temperature to 300°C.

Keywords Barium strontium titanate · Electro-caloric · Ferroelectric · X-ray diffraction

1 Introduction

The gradual restriction on the lead-based electroceramics has opened the opportunities for the lead-free ferroelectric materials to extend its usage in various novel applications such as electrocaloric effect (ECE)-based solid state refrigeration (SSR) [1]. In general, a large EC effect will be observed in the vicinity of ferroelectric (FE) to paraelectric (PE) phase transition; a large relative cooling power will be achieved

---

Sahil · S. Kumar · Md. Ahamad Mohiddon · Preeti (✉)
Center for Nanoscience and Nanotechnology, Department of Physics, Faculty of Science and Humanities, SRM University Delhi-NCR, Sonepat, Haryana 131029, India
e-mail: dahiyapreety1@gmail.com

Md. Ahamad Mohiddon
e-mail: mahamad.ahamad@srmuniversity.ac.in

A. Gaur
School of Physics, University of Hyderabad, Gachibowli, Hyderabad 500046, India

Z. H. Khan et al. (eds.), *Recent Advances in Nanomaterials*, Springer Proceedings in Materials 27, https://doi.org/10.1007/978-981-99-4878-9_65

across a diffuse phase transition. Doped $BaTiO_3$ (BST) is one of the highly investigated materials for room temperature SSR application due to its 'appropriate' and tunable dielectric-ferroelectric properties. Compared to sharp FE-PE transition of $BaTiO_3$ around $T_C = 120°C$, the phase transition in the 20 mol% Sr doped $BaTiO_3$ (BST) is broad, diffuse and characterized with relaxor nature phase transition at 63°C [2]. The relaxor profile is due to disordered arrangement of Sr^{2+} and Ba^{2+} cations at A site of ABO_3 perovskite structure and decrease in the temperature of maximum is due to change in the tolerance factor [3]. Ionic size effect of dopants also plays a crucial role in modifying the transition properties of perovskite ferroelectric materials. It is reported that in $Ba_{1-x}La_xTi_{1-x/4}O_3$, aliovalent La^{3+} with ionic radius smaller than Ba^{2+} decreases T_C in a faster manner (−24°C/at.% La) than what Sr^{2+} does (−3°C/at.%Sr) [3], which is due to the combined effect of difference in ionic radii and distortion in TiO_6 octahedral network due to charge imbalance. In order to check if there is room for improvement, the dielectric characteristics of BST is further tuned by doping with various impurities at both Ba and Ti sites [4, 5]. In the same line, we have modified BST with aliovalent Gd^{3+} along with Na^+ to overcome the charge imbalance of the BST unit cell which will lead to the sizable conduction and dielectric loss. In the present work, we have synthesized the (Gd, Na) modified BST by solid state oxide route and investigated the effect of same on the structural, dielectric and electrical properties of the BST.

2 Experimentation

Gd and Na modified BST, i.e. $(Gd_{0.5}Na_{0.5})_{0.1}Ba_{0.7}Sr_{0.2}TiO_3$ (will be called GNBST) was synthesized by conventional method based on solid-state reaction of mixed oxides and carbonates. The stochiometrically weighed constituents $BaCO_3$, SrO, TiO_2, Gd_2O_3 and Na_2CO_3 of GNBST composite were wet mixed in acetone media for 2 hrs. This powder was calcined at 1400°C for 12 hrs. The crushed powder was used for identifying the crystallographic phase confirmation. Malvern Panalytical X'pert pro X-ray diffractometer was employed for crystal phase confirmation of the powdered sample in θ–2θ configuration in the range of 20° to 70° at a scan step of 0.016°. Cu K_{alpha} radiation of wavelength 1.54 Å was used for X-ray diffraction. The calcined powder was pressed under uni-axial hydraulic press at a pressure of 2 GPa into cylindrical disc of approximate 1 mm thickness and 8 mm diameter. These cylindrical pellets were sintered at 1450°C for 2 hrs. Dielectric measurements were carried out using impedance analyser (Novocontrol Technologies GmbH) in the temperature range of room temperature to 200°C at a heating rate of 1°C/min.

3 Results and Discussion

Figure 1 displays the X-ray diffraction pattern of calcinated $(Gd_{0.5}Na_{0.5})_{0.1}Ba_{0.7}Sr_{0.2}TiO_3$ powder. The XRD pattern has seven sharp diffraction peaks without any impure or faint peaks in the background, which confirms the formation of single-phase crystal structure. These XRD peaks are identified as belongs to the tetragonal perovskite structure of $Ba_{0.8}Sr_{0.2}TiO_3$ with P4mm space group and are indexed according to the JCPDS file no 000-044-0093. The close observation of the peak centred around $2\theta = 46°$ reveals the asymmetric shape with a small shoulder in the right side (shown in the inset of Fig. 1). The extent of peak splitting indicates the degree of tetragonality. In case of pure $Ba_{0.8}Sr_{0.2}TiO_3$ this peak is well separated [2]. However, in the present investigation, due to replacement of Ba^{2+} (1.61 Å) with different ionic size Gd^{3+} (1.107 Å) and Na^+ (1.39 Å), the degree of tetragonality is reduced and the structure modified to pseudo cubic. Rietveld refinement was carried out using the X'pert high score software, the refined data and the residue are included in Fig. 1. The low intensity of residue presented in figure confirms the reasonably good fit. Low goodness of fit and low-profile R-parameters (R_p and R_{wp}) are obtained, which are the indications of the best fit achieved. The refined lattice parameters are $a = b = 3.97$ Å and $c = 3.99$ Å. To gain further insight of crystallographic information, the average crystallite size is calculated from the strongest (101) XRD peak full width half maxima ($\beta_{1/2}$) following the Scherrer's formula. average crystallite size $= k\lambda/\beta_{1/2}\cos\theta$, where λ is the wavelength of the X-ray used, $k = 0.89$ and θ is Bragg's diffraction angle. The calculated crystallite size of GNBST is 44.4 nm. This value is higher than the crystallite size of $Ba_{0.75}Sr_{0.25}TiO_3$ (sintered at 1300°C) which was reported to be 16.4 nm [6].

Figure 2a–b shows the real part of dielectric constant (ϵ') and dielectric loss (tanδ) of GNBST and pure BST as a function of temperature at four different frequencies. ϵ' of GNBST at 1 MHz shows a faint hump around 41 °C which may be the signature

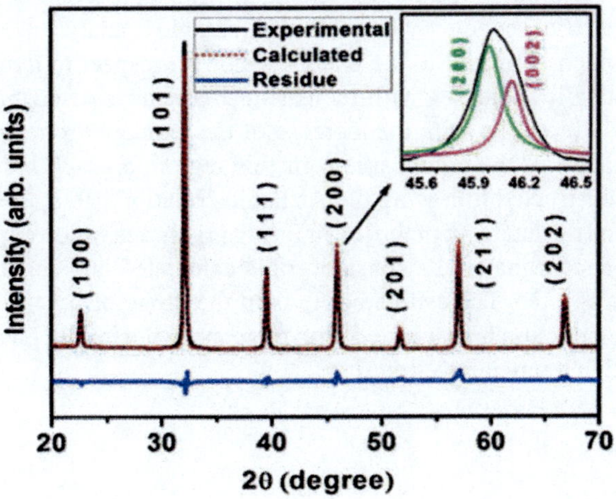

Fig. 1 XRD pattern of $(Gd_{0.5}Na_{0.5})_{0.1}Ba_{0.7}Sr_{0.2}TiO_3$ composite powder calcined at 1400°C for 12 hrs.

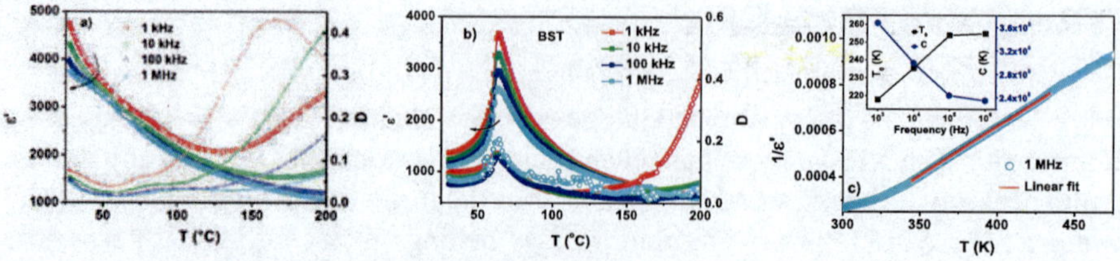

Fig. 2 Dielectric constant and dielectric loss of **a** $(Gd_{0.5}Na_{0.5})_{0.1}Ba_{0.7}Sr_{0.2}TiO_3$ composite, **b** BST and **c** inverse of dielectric constant versus absolute temperature. Inset shows the curie constant and critical temperature T_0 as a function of frequency

of FE to PE phase transition. The phase transition peaks at lower frequencies are expected to move below the 25°C, which is the minimum temperature that can be measured with our instrument. Corresponding to this, a relaxation peak in dielectric loss is noticed at 101°C for 1 MHz and at 74°C for 1 kHz applied a.c. frequency. The shift of the peak position as function of frequency is the indication of the relaxer characteristics of the GNBST [7]. In pure BST, this phase transition is recorded at 63°C which does not vary much with the change of frequency; this temperature is well matched with the literature [8].

In order to further analyse the data, Curie–Weiss law [9] was employed in the paraelectric region, $\varepsilon' = C/(T - T_0)$, where C is Curie constant, a material-dependent property and T_0 is the critical temperature which is theoretically the stability limit for the paraelectric phase. A graph between $1/\epsilon\prime$ and T is plotted and from the slope and intercept of the linear portion C and T_0 are calculated at the chosen frequencies and is presented in the inset of Fig. 2c 1. The constant C of GNBST is of the order of 10^5K and decreases with the increase of the frequency. These values are well compared with C of BST thin film (9.2×10^4 K) [10] and $BaTiO_3$ ceramic (1.05×10^5 K) [11]. T_0 systematically increases with the increase of the frequency showing that the dipolar motion is thermally activated.

In order to further investigate the transport properties, temperature dependent a.c. electrical conductivity of GNBST sample is analysed (Fig. 3a). The a.c. conductivity which is related to the energy loss with respect to dynamic characteristics systematically increases with increasing frequency. After the FE-PE transition peak, the σ_{ac} increases with the increase of the temperature representing the semiconducting nature of the composition. In this region, a graph between $\log_e\sigma_{\text{ac}}$ and 1000/T is constructed following the Arrhenius relation [9], $\sigma_{\text{ac}} = \sigma_0 e^{-E_a/k_BT}$. From the slope and the intercept of the linear fitting graph, activation energy E_a and temperature independent material's constant σ_0 is calculated and shown as a function of frequency in Fig. 3b. The estimated E_a is in the range of 0.34–0.37 eV and therefore the a.c. conduction is associated with oxygen vacancies [12], which might be created during high temperature sintering.

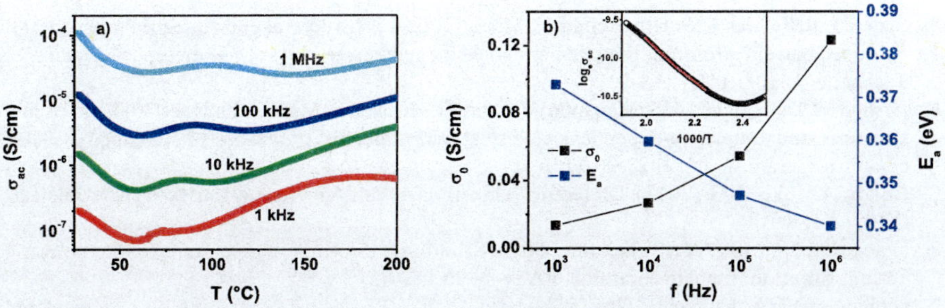

Fig. 3 **a** A.c conductivity (σ_{ac}) versus temperature and **b** σ_0 and activation energy (E_a) as a function of frequency for $(Gd_{0.5}Na_{0.5})_{0.1}Ba_{0.7}Sr_{0.2}TiO_3$ composite along with Arrhenius plot in the inset

4 Conclusion

Gd and Na modified Barium Strontium Titanate $(Gd_{0.5}Na_{0.5})_{0.1}Ba_{0.7}Sr_{0.2}TiO_3$ is synthesized by following the traditional mixed oxide solid-state reaction technique. The XRD investigation carried out on the calcined powder confirms the formation of single-phase perovskite pseudo cubic structure with lattice parameters $a = b =$ 3.97 Å and $c =$ 3.99 Å. With the modification of the Gd and Na, the ferroelectric to paraelectric phase transition of $Ba_{0.8}Sr_{0.2}TiO_3$ is shifted below 25 °C with large dielectric constant of 4800 at 1 kHz. The activation energy estimated from the a.c. electric conductivity is found to the range of 0.34–0.37 eV and is proposed to be associated with oxygen vacancies.

Acknowledgements The Authors acknowledges school of Physics, University of Hyderabad for providing the XRD characterization facility and impedance data. Dr. Anshu acknowledge DST (Ref. No.: SR/WOS-A/ET-38/2018) for providing the fund during this work. Dr. Md. A. Mohiddon and Dr. Preeti acknowledge Honourable Vice-Chancellor of SRM University Delhi-NCR, Sonepat, Prof. P. Prakash and Dean of Academic Affairs, Prof. Samuel Raj for their kind support in establishing the Centre for Nanoscience and Nanotechnology.

References

1. Srikanth KS, Patel S, Vaish R (2017) Electrocaloric behavior and temperature dependent scaling of dynamic hysteresis of $Ba_xSr_{1-x}TiO_3$ (x=0.7, 0.8 and 0.9) bulk ceramics. J Aust Ceram Soc 54:439–450
2. Şanlı K, Adem U (2020) Electrocaloric properties of $Ba_{0.8}Sr_{0.2}Ti_{1-x}Zr_xO_3$ $(0 \leq x \leq 0.1)$ system: the balance between the nature of the phase transition and phase coexistence. Ceram Int 46(2):2213–2219
3. Ben L, Sinclair DC (2011) Anomalous Curie temperature behavior of A-site Gd-doped $BaTiO_3$ ceramics: the influence of strain. Appl Phys Lett 98(9):092907
4. Jacob R, Nair HG, Isac J (2015) Impedance spectroscopy and dielectric studies of nanocrystalline iron doped barium strontium titanate ceramics. Process Appl Ceram 9(2):73–79

5. Sadeghzadeh Attar A, Salehi Sichani E, Sharafi S (2017) Structural and dielectric properties of Bi-doped barium strontium titanate nanopowders synthesized by sol-gel method. J Mater Res Technol 6(2):108–115
6. Arshad M, Du H, Javed MS, Maqsood A, Ashraf I, Hussain S, Ma W, Ran H (2020) Fabrication, structure, and frequency-dependent electrical and dielectric properties of Sr-doped $BaTiO_3$ ceramics. Ceram Int 46(2):2238–2246
7. Bokov AA, Ye Z-G (2012) Dielectric relaxation in relax or ferroelectrics. J Adv Dielectr 02(02):1241010
8. Gatea HA, Naji IS (2020) The effect of Ba/Sr ratio on the Curie temperature for ferroelectric barium strontium titanate ceramics. J Adv Dielectr 10(5)
9. Mohiddon MdA, Kumar R, Goel P, Yadav KL (2007) Effect of Nb doping on structural and electric properties of PZT (65/35) ceramic. IEEE Trans Dielectr Electr Insul 14(1):204–211
10. Lahiry S, Mansingh A (2008) Dielectric properties of sol-gel derived barium strontium titanate thin films. Thin Solid Films 516(8):1656–1662
11. Flores-Ramirez R, Huanosta A, Amano E, Valenzuela R, Westi AR (1989) Curie-Weiss behavior in polycrystalline barium titanate from ac measurements. Ferroelectrics 99(1):195–201
12. Li M, Pietrowski MJ, De Souza RA, Zhang H, Reaney IM, Cook SN, Kilner JA, Sinclair DC (2014) A family of oxide ion conductors based on the ferroelectric perovskite $Na_{0.5}Bi_{0.5}TiO_3$. Nature Mater 13(1):31–35

Sol–Gel Synthesis of Spinel-Structured Pure and Manganese-Activated Zinc Aluminate Nanoparticles

Bindiya Goswami, Neelam Rani, and Rachna Ahlawat

Abstract Pure and Mn^{2+}-doped $ZnAl_2O_4$ nanoparticles have been synthesized using the citrate sol–gel method. X-ray diffraction was used to acquire pure spinel structure, while the crystallinity has been assessed by HR-TEM at the nanoscale. The nanocrystalline size was calculated from the peak width of the most prominent reflection plane. The *W–H* formula was used to assess the micro-stress present in the samples. FTIR reveals the corresponding functional group and emphasized the characteristic fingerprint region of the spinel. The data of absorbance manifests the band gap energies that have remarkably lower values in the doped sample. The electrical characterization has inferred the enhanced conductivity in the doped sample where Mn^{2+} ions act as donors of electrons. The modified properties strengthened the usability of prepared aluminates in the semiconductor industry.

Keywords Pure and Mn^{2+}-doped $ZnAl_2O_4$ · XRD · Absorbance · Resistivity

1 Introduction

From the category of well-known host matrices, the important host for luminescent material is fascinatedly the Metal Aluminates with the general formula AB_2O_4 which crystallizes in cubic phase having space group $Fd\bar{3}m$ [1]. Nowadays, $ZnAl_2O_4$ has been investigated at length due to its less surface acidity, high thermal stability, and hydrophobicity [2]. The transition metals and rare earth ions are frequently doped in $ZnAl_2O_4$ owing to the simple band gap structure of $ZnAl_2O_4$. In this context, Motloung et al. [3] have prepared Cu^{2+}-doped $ZnAl_2O_4$ nanophosphor using the famous citrate sol–gel technique. Ravikumar et al. [4] synthesized the Dy^{3+}:$ZnAl_2O_4$ nanoparticles using the plant latex-mediated green synthesis. Tshabalala et al. [5] successfully prepared the Ce^{3+}/Tb^{3+} co-doped zinc aluminate by combustion process where urea was used as fuel. The normal and inverse spinel structure of $ZnAl_2O_4$ has

B. Goswami (✉) · N. Rani · R. Ahlawat
Material Science Lab, Department of Physics, Chaudhary Devi Lal University, Sirsa, Haryana, India
e-mail: ranibindiya92@gmail.com

Z. H. Khan et al. (eds.), *Recent Advances in Nanomaterials*, Springer Proceedings in Materials 27, https://doi.org/10.1007/978-981-99-4878-9_66

also been used in the catalyst and ceramic industry [6]. The spinel $ZnAl_2O_4$ is considered as a wide band gap semiconductor that can be altered significantly via suitable substitutional elements. Not only in optical phosphor nanomaterials but $ZnAl_2O_4$ should also be used in the optoelectronic industry owing to its semiconductor nature. Only a few studies exist in the literature about electrical conduction either in pure or doped $ZnAl_2O_4$ [7]. Therefore, it is imperative to analyze the conductivity and resistivity of pure and doped $ZnAl_2O_4$ also.

2 Materials and Methodology

The precursor materials like zinc nitrate, aluminum nitrate, ethylene glycol, and citric acid were purchased from Sigma-Aldrich with high purity. Pure and 0.5 mol % Mn^{2+}-doped spinel nanostructures are fabricated using a known sol–gel method in which citric acid acted as a chelator and further annealed the samples at 800 °C. Citric acid is the most common organic molecule used in sol–gel chemistry. Further, the detailed mechanism has been reported in our earlier publication [8].

3 Results and Discussion

3.1 XRD Analysis

The XRD patterns for prepared samples are shown in Fig. 1a with extremely pure spinel structure. The peaks in pure $ZnAl_2O_4$ are appeared at 2θ ~ 31.36°, 36.80°, 44.93°, 49.19°, 55.78°, 59.47°, 65.36°, 74.20°, and 77.45°. In Mn^{2+}-doped $ZnAl_2O_4$, the peaks are centered at 2θ ~ 31.06°(220), 36.68°(311), 44.63°(400), 48.86°(311), 55.42°(422), 59.13°(511), 65.00°(440), 73.96°(620), and 77.08°(533) with corresponding reflection planes that are correlated with the cubic spinel structure with $F\bar{d}3m$ space group [1, 2]. The obtained data of the XRD study is also matched with JCPDS No. 05-0669 [9]. Figure 1b shows the (311) diffraction peaks shift toward a lower angle in the Mn^{2+}-doped $ZnAl_2O_4$ sample. Due to this shift, the lattice constant has slightly changed to 8.10 Å from 8.06 Å in the doped sample. The large size Mn^{2+} ions ($r_{Mn} = 0.08$ nm) have partially replaced the smaller Zn^{2+}ions ($r_{Zn} = 0.07$ nm) at the '*A*' site. However, at the 'B' site when Al^{3+} ions are exchanged with Mn^{2+} ions, even then the lattice parameter has been altered due to size difference [10].

The average crystalline size (*D*) has been obtained by the famous Debye–Scherrer formula [5, 6]. It is revealed that the size for Mn-doped sample is lower (23 ± 0.5 nm) than the pure $ZnAl_2O_4$ (26 ± 0.5 nm). The calculated values show the well-mannered doping of Mn^{2+} at the lattice site of zinc aluminate. One may notice that the peak broadening is caused by nanosize and also the microstrain present in the crystalline structure. The induced microstrain in nanoparticles can be elaborated by

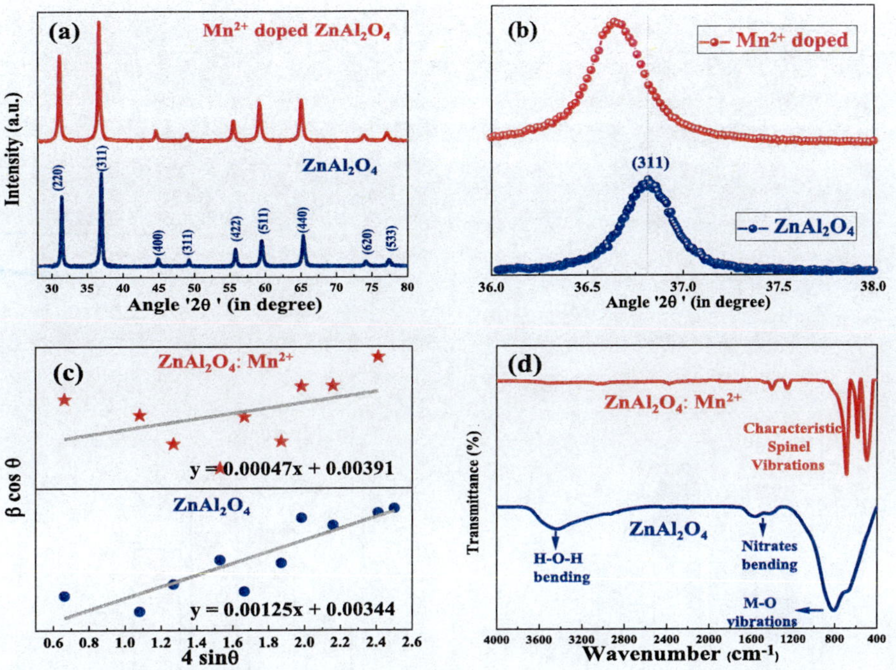

Fig. 1 **a** XRD pattern, **b** peak shift, **c** *W–H* plot, and **d** FTIR spectra for prepared samples

Williamson-Hall formula:

$$\beta_{\mathrm{hkl}} \cos\theta_{\mathrm{hkl}} = K\lambda/D + 4\varepsilon \sin\theta_{\mathrm{hkl}}, \tag{1}$$

where the symbols have their usual meanings [8]. Figure 1c depicts the *W–H* graph plotted between 4sinθ_{hkl} and β_{hkl}cosθ_{hkl} for the prepared samples. The calculated microstrain is smaller ($\varepsilon = 0.00047$) than that of pure $ZnAl_2O_4$ sample ($\varepsilon = 0.00125$). The observed results suggest the exact doping of Mn^{2+} at the lattice site of $ZnAl_2O_4$ spinel.

3.2 FTIR Analysis

In Fig. 1d, the FTIR spectroscopy investigates the bonding environment of pure and Mn^{2+}-doped $ZnAl_2O_4$ spinel nanoparticles. The wide band that occurs at 3437 cm^{-1} can be assigned to the –OH group due to the water's existence [3, 4]. Further, two bands present near 1567 cm^{-1} and 1416 cm^{-1} are assigned to the nitrates. Small peaks around 815 cm^{-1} and 808 cm^{-1} are related to vibrations modes of Al^{3+} ion occupied at octahedral and tetrahedral sites [10]. The bands lie in 450–700 cm^{-1} range

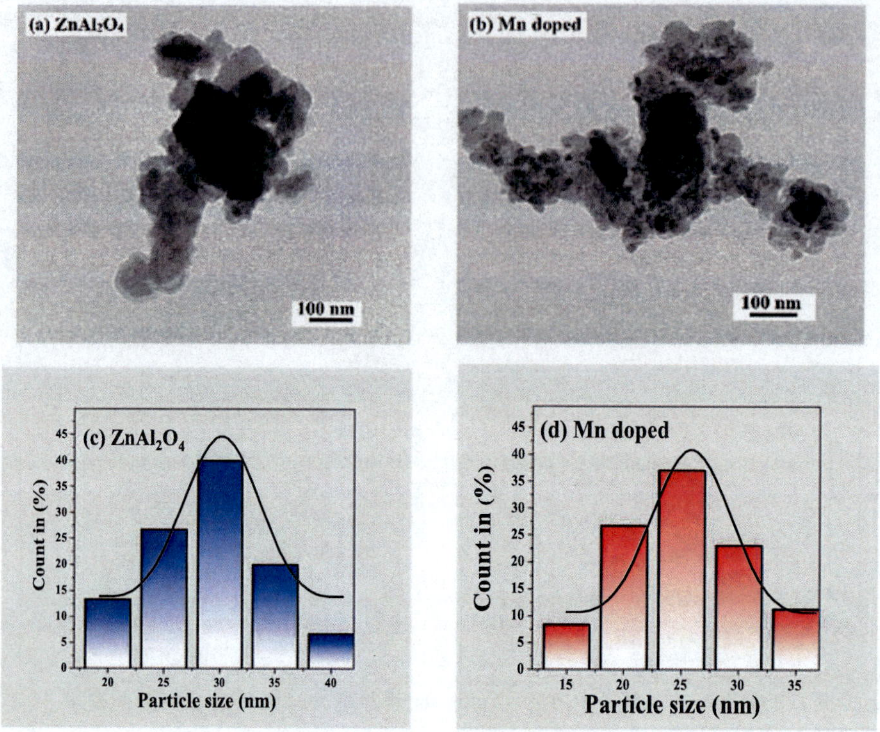

Fig. 2 **a** TEM micrograph of pure, **b** Mn^{2+} doped, and **c–d** histograms for prepared samples

ensuring the formation of spinel crystallites of $ZnAl_2O_4$ [5, 6]. The peak centered at 681 cm^{-1} may correspond to the Al-O vibrations in a tetrahedral coordinate state [11]. FTIR study infers that Mn^{2+} doped sample has improved crystallinity and high densification than the pure sample.

3.3 TEM Analysis

To explore the nanocrystallinity of the prepared samples, HR-TEM has been performed on the pure and Mn^{2+}-doped $ZnAl_2O_4$ samples in Fig. 2.

3.4 UV–Vis Spectroscopy

In Fig. 3a, the band found with the highest peak at 225 nm in absorption spectra of pure sample is ascribed due to defects states while a hump at 320 nm may attribute

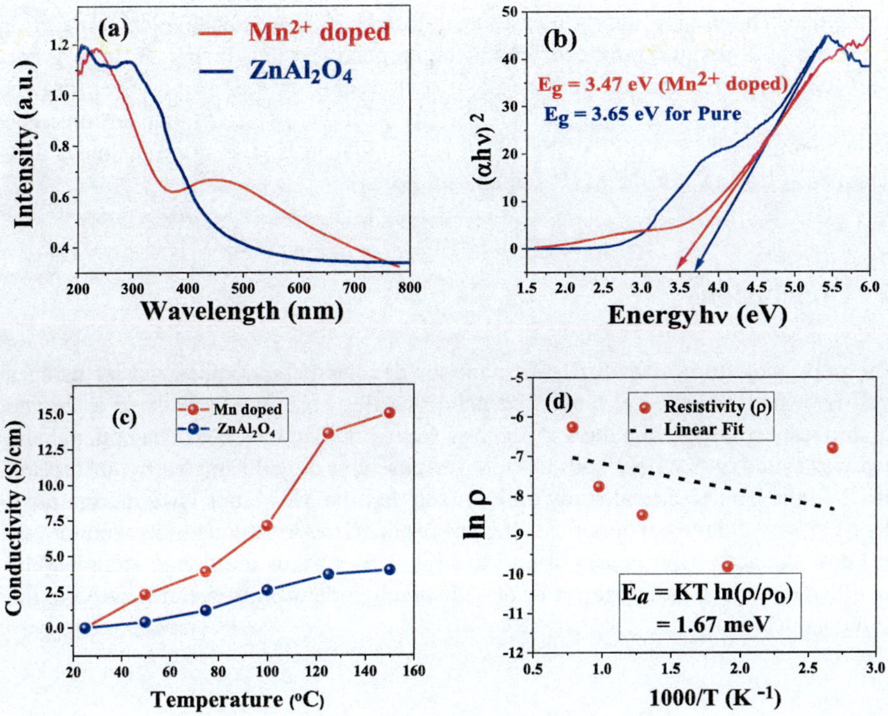

Fig. 3 **a** Absorbance, **b** Tauc's plot, **c** conductivity, and **d** Arrhenius plot for prepared samples

due to the excitonic transition of AlO_6 assembled in the pure sample, i.e., $ZnAl_2O_4$ [2]. Further, in the Mn^{2+}-doped sample, two bands occur at 235 nm and 465 nm, respectively. The maximum band at 465 nm arises due to the absorbance of Mn^{2+} ions that exist very close to the bottom of the conduction band [8]. In the doped sample, it is noticed that the absorbance has shifted to the higher wavelength side which signifies the band gap variation. However, the exact energy band gap has been intended from Tauc's plot and found ~ 3.65 eV and 3.47 eV for pure and Mn^{2+}-doped samples. In Fig. 3b, one may notice lower band gap values for the doped sample because the Mn^{2+} ions have situated themselves in between the VB and CB of zinc aluminate, so decrease the value [9, 10].

3.5 Electrical Studies

The analyzed data of the current and Arrhenius plot for conductivity is depicted in Fig. 3c and d. The Mn^{2+}-doped sample has a small band gap as one may notice from UV–Visible spectra. Therefore, the conduction in the doped sample is higher than in pure $ZnAl_2O_4$. In Fig. 3c, the current has linearly increased with the applied voltage difference. The doping and proper annealing induces more negative charge carrier ions, and as a result, significant conductance occurred [7]. In Fig. 3d, the graph between resistivity and the reciprocal of temperature in Kelvin has been drawn. One may notice a linear decrease in resistivity with the inverse of temperature. The activation energy 'E_a' (in eV) has been accomplished by the slope of linear data obtained ~ 1.67 eV for the Mn^{2+} doped sample.

4 Conclusions

The pure and Mn-doped $ZnAl_2O_4$, nanopowder has been synthesized by using a well-known citrate sol–gel route. The nanocrystalline size was calculated as 26 and 23 nm, respectively, while the morphology was examined by TEM. The optical band gap was found as 3.65 eV and 3.47 eV for pure and doped samples by absorption of UV light. The electrical study emphasized that the Mn^{2+} ions have incorporated the necessary charge carriers for sufficient conduction. Arrhenius plots were drawn to know the activation energy found to be ~ 1.67 eV for the doped sample. The modified optoelectronic features of doped spinel oxide promise its utilization in the semiconductor industry.

References

1. Tangcharoen T, Thienprasert JT, Kongmark C (2019) Effect of calcination temperature on structural and optical properties of MAl_2O_4 (M = Ni, Cu, Zn) aluminate spinel nanoparticles. J Adv Ceram 8:352–366
2. Sameera S, Vidyadharan V, Sasidharan S, Gopchandran KG (2019) Nanostructured zinc aluminates: a promising material for cool roof coating. J Sci Adv Mater Device 4:524–530
3. Motloung SV, Dejene FB, Koao LF, Ntwaeaborwa OM, Swart HC, Motaung TE, Ndwandwe OM (2017) Structural and optical studies of $ZnAl_2O_4$:x%Cu^{2+} ($0<x\leq1.25$) nanophosphors synthesized via citrate sol-gel route. Opt Mater 64:26–32
4. Ravikumar BS, Nagabhushana H, Sunitha DV, Sharma SC, Nagabhushana BM, Shivakumara C (2014) Plant latex mediated green synthesis of $ZnAl_2O_4$:Dy^{3+} (1–9 mol%) nanophosphor for white light generation. J Alloy Compd 585:561–571
5. Tshabalala KG, Cho SH, Park JK, Shreyas PS, Nagpure IM, Swart HC, Ntwaeaborwa OM (2011) Luminescent properties and X-ray photoelectron spectroscopy study of $ZnAl_2O_4$:Ce^{3+}, Tb^{3+} phosphor. J Alloy Compd 509:10115–10120
6. Cornu L, Duttine M, Gaudon M, Jubera V (2014) Luminescence switch of Mn-doped $ZnAl_2O_4$ powder with temperature. J Mater Chem C 2:9512–9522
7. Gupta P, Sahni M, Chauhan S (2021) Manifestation of multifunction capabilities by stabilizing cadmium together with zinc and aluminum in spinel oxide. J Mater Sci Mater Electron 32:15317–15330
8. Goswami B, Ahlawat R (2022) Impact of annealing on optimization of various thermal, structural, and optical parameters of spinel 'gahnite' for device fabrication. Phys Scr 97:105808
9. Maity G, Maji P, Sain S, Das S, Kar T, Pradhan SK (2019) Microstructure, optical and electrical characterizations of nanocrystalline $ZnAl_2O_4$ spinel synthesized by mechanical alloying: effect of sintering on microstructure and properties. Phys E 108:411–420

10. Trung DQ, Tu N, Quang NV, Tran MT, Du NV, Huy PT (2020) Non-rare-earth dual green and red-emitting Mn-doped $ZnAl_2O_4$ phosphors for potential application in plan- growth LEDs. J Alloy Compd 845:156326
11. Huang S, Wei Z, Wu X, Shi J (2020) Optical properties and theoretical study of Mn-doped $ZnAl_2O_4$ nanoparticles with spinel structure. J Alloys Compd 825:154004

Structural, Dielectric and Electrocaloric Characteristics of the Barium-Modified Bismuth Sodium Strontium Titanate

Ranjeet Ughade Sukachari, Preeti, Md. Ahamad Mohiddon, and Anshu Gaur

Abstract The structural, dielectric and electrocaloric characteristics of Ba-modified bismuth sodium strontium titanate, i.e., $(Bi_{0.35}Na_{0.35})_{0.98}Ba_{0.02}Sr_{0.3}TiO_3$ (BNBST) is reported. The BNBST was synthesized by the solid-state mixed oxide method. The wet mixed powder was calcined at 800 °C for 6 h and was used for crystal-phase confirmation. The powder was pelletized and sintered at 1250 °C for 2 h. X-ray diffraction confirmed the formation of a single-phase perovskite structure. The dielectric characteristics were investigated in the temperature range of 30 °C–350 °C at different frequencies (100 Hz–1 MHz). The electrocaloric investigations are also included.

Keywords Bismuth sodium titanate (BNT) · Electrocaloric · Ferroelectric · X-ray diffraction · Dielectric

1 Introduction

Solid-state refrigeration (SSR) technology based on the electrocaloric effect (ECE) has drawn the incredible attention of researchers, as a propitious alternative to substitute the vapor-cooling approach [1]. Positive (negative) ECE is the phenomenon of reversible positive (negative) temperature change in ferroelectric specimen upon application (removal) of electric field. Although the ECE phenomenon is familiar for the past 50 years, the high ECE of $\Delta T = 12$ K reported in $Pb(Zr,Ti)O_3$ thin film has rejuvenated a new hope for SSR applications [2]. Due to the toxic nature of Pb, which is considered as a serious threat to the balance of ecosystem, it is necessary to find environmentally friendly materials as an alternative. $Bi_{0.5}Na_{0.5}TiO_3$ (BNT) is

R. U. Sukachari · Preeti · Md. A. Mohiddon
Center for Nanoscience and Nanotechnology, Department of Physics, Faculty of Science and Humanities, SRM University Delhi-NCR, Sonepat, Haryana 131029, India
e-mail: mahamad.ahamad@srmuniversity.ac.in

A. Gaur (✉)
School of Physics, University of Hyderabad, Gachibowli, Hyderabad, Telangana 500046, India
e-mail: gauranshu20@gmail.com

Z. H. Khan et al. (eds.), *Recent Advances in Nanomaterials*, Springer Proceedings in Materials 27, https://doi.org/10.1007/978-981-99-4878-9_67

an environment-safe ferroelectric material. It undergoes structural phase transitions from rhombohedral to a tetragonal ($T_{R\text{-}T}$) at 320 °C and tetragonal to cubic ($T_{T\text{-}C}$) at 520 °C and exhibits a large polarization ($P = 38\ \mu C/cm^2$) at room temperature [3]. BNT's large polarization is the result of off-centering of Bi^{3+} ion due to the hybridization of Bi $6s^2$ and O 2p orbital [4].

Sr doping in perovskite ferroelectric materials like $BaTiO_3$ and $(Pb,La)(Zr,Ti)O_3$ [5, 6] is reported to bring down the phase transition following different mechanisms. In BTO, addition of Sr causes shifting of the crystal structure toward cubic symmetry causing diminished polarization [7]. On the other hand, in PLZT, substituting Pb by non-lone-pair ion Sr^{2+} decreases the stereochemical activity of lone-pair electrons, overall polarization and T_C [6]. In the present work, we have chosen Ba and Sr-modified BNT, i.e., $(Bi_{0.35}Na_{0.35})_{0.98}Ba_{0.02}Sr_{0.3}TiO_3$ and investigated the effect of Ba and Sr modification on the structural, electrical and electrocaloric properties of BNT materials.

2 Experimentation

Ba and Sr-modified BNT, i.e., $(Bi_{0.35}Na_{0.35})_{0.98}Ba_{0.02}Sr_{0.3}TiO_3$ (BNBST) was prepared by solid-state reaction method. Raw materials metal oxide and carbonates Bi_2CO_5 (Alfa Aesar), Na_2CO_3 (Sigma Aldrich), $SrCO_3$ (Alfa Aesar), $BaCO_3$ (Sigma Aldrich) and TiO_2 (Sigma Aldrich) were weighed according to their stoichiometric ratio and mixed in acetone medium for 2 h. The mixed powder was calcined at 800 °C for 6 h and used for the confirmation of the crystallographic phase. The X-ray diffraction (XRD) data were collected using X-ray Diffractometer (Malvern Panalytical X'pert Pro) in a wide range of 2θ (20°–70°) with CuK_α radiation ($\lambda = 1.54$ Å). The calcined powder was compacted into cylindrical pellets by using an uni-axial hydraulic press with 2GPa pressure with Polyvinyl Alcohol as a binder. Thickness and diameter of the pellets were 1 mm and 10 mm, respectively. These pellets were sintered at an optimized temperature of 1250 °C for 2 h. Dielectric measurements were carried out using an impedance analyzer (Novocontrol GmbH Technologies) in the temperature range of RT–300 °C. Polarization–electric field (*P–E*) loops are recorded at different temperatures from RT to 200 °C using Sawyer–Tower circuit-based *P–E* loop analyzer (Radiant Technologies).

3 Results and Discussion

The XRD pattern of BNBST ceramics measured at room temperature is shown in Fig. 1. From the figure, it is evident that the XRD pattern belongs to a pure perovskite phase without any secondary phases and impurities. The XRD peaks are identified as belongs to the rhombohedral structure with the R3c space group and are indexed according to JCPDS file number 98-028-0983. The XRD data are further refined by

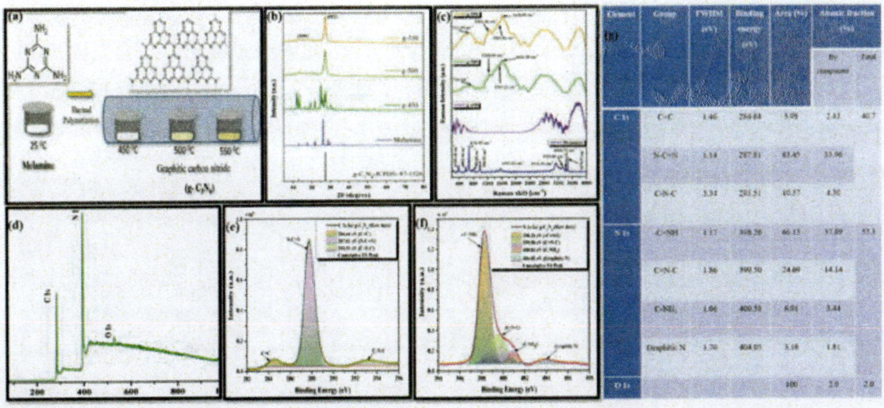

Fig. 1 (**a**) Schematic diagram for synthesis of g-C_3N_4, (**b**) XRD, (**c**) Raman, (**d**) XPS survey scan of g-C_3N_4, (**e**) C 1s, (**f**) N1s, (**g**) table for XPS convoluted data

spectrum analysis of synthesized materials. NH2 wagging causes the 389.48 cm^{-1} peak, whereas C–N–C bending and NH_2 twisting provide the 583.94 cm^{-1} peak. The spectrum's greatest peak, at 676.99 cm^{-1}, is caused by the ring-breathing II mode's in-plane triazine ring deformation. Ring-induced out-of-plane deformation peaks at 783.82 cm^{-1}. Experimental spectra showed another neutral melamine band at 983 cm^{-1}. C–N–C bending vibration occupied this band. This study found the NH2 bending vibration linked with melamine's C–N stretching in the Raman spectrum at 1557 cm^{-1}. As materials temperatures rise, noticeable peaks for synthesized g-C_3N_4 appear. This picture shows that hexagonal rings and carbon and nitrogen atom graphitization occur at 1628.99 cm^{-1}. N–C–N stretching vibrations, bending vibrations = C (sp2), and heterocycle C–N bond vibrations in the carbon nitride structure correspond to bands at 1503.56 cm^{-1}, 1266.44 cm^{-1}, and 565.36 cm^{-1}, respectively. Raman measurements matched the data base, confirming that g-C_3N_4 well synthesized. X-ray photoelectron spectra were able to disclose the elemental composition as well as the chemical states of the samples that were synthesized (XPS) [7]. The survey scan for g-C_3N_4 was shown in Fig. 1d. The table presented in Fig. 1g displays all the deconvolved peaks for g-C_3N_4.

FESEM analysis examined composite morphology and suggested improvements. Melamine resembles stone pebbles of various sizes (Fig. 2a). Thermal polymerization for synthesis modifies g-C_3N_4 at 450 °C to have a bulk fragmented shatters-like structure (Fig. 2b). At 500 °C (Fig. 2c), sheets with a fiber-like structure develop. (Fig. 2d), there are fibers and wrinkles. Figure 2e–h shows melamine, g-450, g-500, and g-550 elemental composite EDX spectra regions, respectively. Melamine's EDX spectrum shows 27.51% carbon and 70.80% nitrogen (Fig. 2e). As temperatures rise, synthesis reduces nitrogen and boosts carbon. The composites have 33.2% carbon and 63.98% nitrogen, according to the g-500. Characterization of the g-C_3N_4 validated its development. Morphology of g-C_3N_4 showed well growth structure which escalated the charge transfer properties.

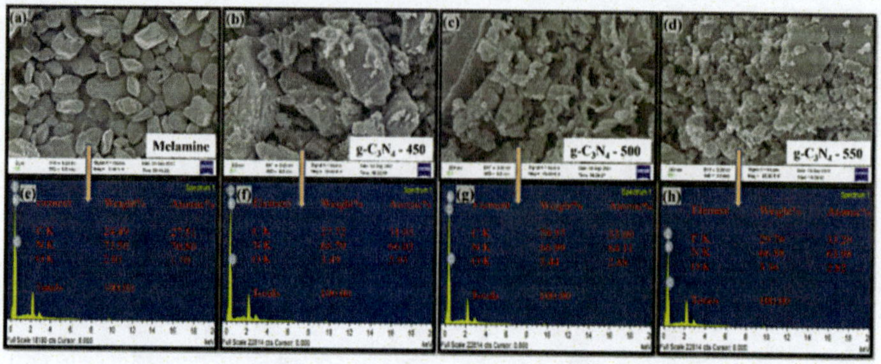

Fig. 2 (**a–d**) FESEM images of synthesized sample, (**e–h**) EDS spectra for synthesized materials

Figure 3a showed all synthetic samples' absorbance spectra. When melamine absorbed heat, electron production from carbon nitride's VB to CB induced typical semiconductor adsorption and strong absorption between 200 and 500 nm in all samples. Heat-induced structural flaws enhance optical absorption due to weak absorption tails. Photons that stimulate electrons provide optical conductivity. Figure 3b demonstrates how photon energy influences material optical conductivity. Optical conductivity rises with increasing temperature. Materials preserve optical conductivity at decreasing photon energies. g-C_3N_4's optical band gap conductivity peaks at ~ 2.79 eV. Localized tail states in the prohibited band gap transfer electrons from the valence band to the tail and conduction band. Localized states reduce band gap and boost optical conductivity [8].

Photon-excited electrons study synthetic photoluminescence. PL spectra optimize g-C_3N_4 (Fig. 3c). Powder-produced materials' emission spectra had an excitation wavelength of 365 nm. Optimized g-500 peak is 441 nm. Melamine had a low PL spectra at 438 nm. Melamine's redshift creates g-C_3N_4. g-C_3N_4 nanosheets vary from

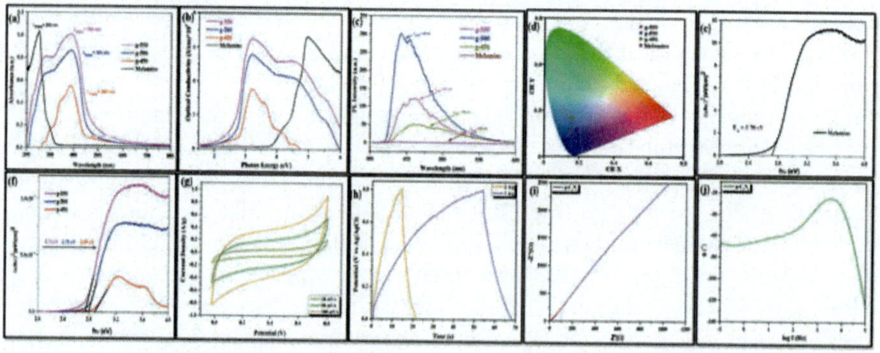

Fig. 3 (**a**) Absorbance, (**b**) Optical conductivity, (**c**) PL spectra, (**d**) CIE spectra, (**e–f**) Band gap of materials, (**g**) CV curve, (**h**) GCD plot, (**i**) EIS graph, (**j**) Bode plot

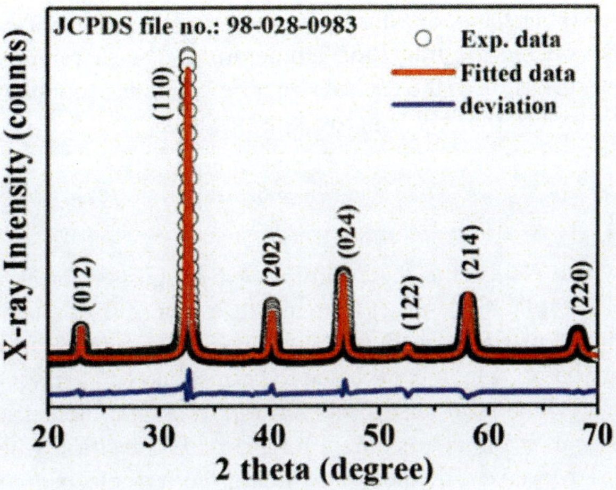

Fig. 1 XRD pattern of $(Bi_{0.35}Na_{0.35})_{0.98}Ba_{0.02}Sr_{0.3}TiO_3$-calcined powder at 800 °C for 6 h

the X'pert high score plus software, and the refined lattice parameter in the hexagonal notation are $a = b = 5.51$ Å and $c = 13.54$ Å. From the XRD pattern, average crystal size was calculated using the Debye–Scherrer's equation,

$$\text{average crystallite size} = k\lambda/\beta_{1/2}\cos\theta,$$

where $k = 0.89$, $\beta_{1/2}$ is full width at half maxima of the strong peak of BNBST, λ is the wavelength of the X-ray (λ = 1.54 Å for Cu Kα radiation) and θ is the Bragg's angle. We have used the strongest peak of BNBST (110) at $2\theta = 32.7°$ and the calculated average crystallite size is 54.4 nm.

The temperature-dependent dielectric constant and dielectric loss of the BNBST sample at five different frequencies of 100 Hz, 1 kHz, 10 kHz, 100 kHz and 1 MHz are presented in Figs. 2 and 3, respectively. Dielectric constant versus temperature graph has two characteristic peaks around 105 °C and 190 °C at 1 kHz, which are the indications of the phase transitions. It is reported in the literature that BNT undergoes a phase transition from rhombohedral ferroelectric (FE) phase to anti-ferroelectric (AFE) tetragonal phase at 200 °C, which is called as depolarization temperature (T_d) [3]. Crystal structure of BNT beyond 200 °C is under extensive investigation for the past 10 years. Structural transformation into a paraelectric tetragonal phase at 320 °C is widely accepted, which finally transforms to a paraelectric cubic phase at 520 °C [3]. Since the present dielectric investigations are carried out up to 300 °C, we could not observe features of the tetragonal to cubic phase transformation. Above-mentioned two transitions in 30% Sr-doped BNT are reported to occur at 50 °C and 160 °C [8]. Co-doping of 2% Ba^{2+} and 30% Sr^{2+} at the A-site of BNT in the present case triggers the transitions at intermediate temperatures of 105 °C and 190 °C. These observations indicate that the two ions are playing different roles in modifying the transition. Substitution of Sr^{2+} ion at Bi^{3+}-site decreases the stereochemical activity and the overall polarization by decreasing the number of lone-pair electrons. On the other hand, doping of 2% Ba^{2+} in BNT (or 30% Sr-doped BNT) is responsible for raising the transition temperature. The increment may be associated with the modification in the crystal structure which is accounted with change in the tolerance factor t of the ABO_3 perovskite structure.

$$t = (r_A + r_O) \Big/ \sqrt{2}(r_B + r_O),$$

where r's are the ionic radii. $t = 1$ suggests cubic symmetry and Ba^{2+} with ionic radius of 1.49 Å serves as to increase the t of BNT (having an r_A average of ~ 1.17 Å) more away from 1. Further, we have observed a clear peak in the dielectric constant at 105 °C, which is expected to correspond to T_d having shifted from 200 °C of pure BNT. Cao et al. also observed a peak in the dielectric loss and claimed that such a peak is the characteristic of FE-AFE transition whose T_d shifts with Sr doping [5]. Moreover, the peak position of the ferroelectric–paraelectric phase transition is

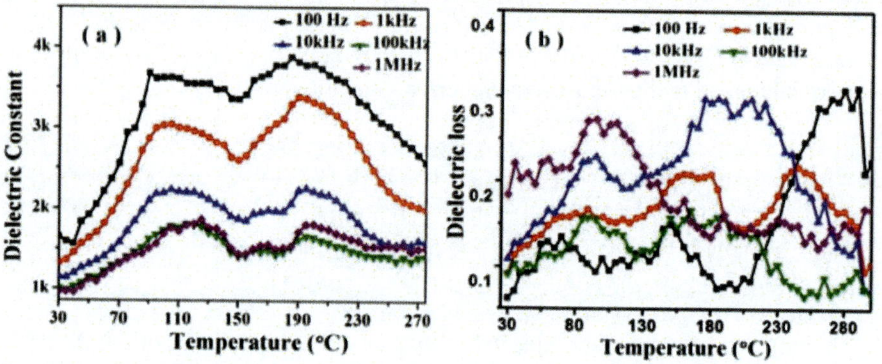

Fig. 2 Temperature dependence of **a** dielectric constant and **b** dielectric loss of $(Bi_{0.35}Na_{0.35})_{0.98}Ba_{0.02}Sr_{0.3}TiO_3$ measured at different frequencies

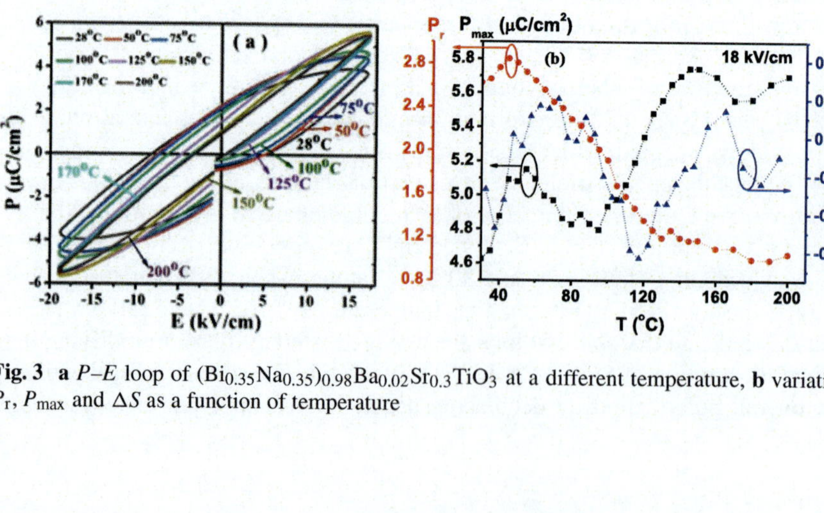

Fig. 3 **a** P–E loop of $(Bi_{0.35}Na_{0.35})_{0.98}Ba_{0.02}Sr_{0.3}TiO_3$ at a different temperature, **b** variation of P_r, P_{max} and ΔS as a function of temperature

shifted toward high temperature as the frequency increases, demonstrating a typical relaxor behavior. In addition, a widespread dielectric peak is observed which indicates the diffuse behavior of phase transition. This may be because of the multiple ion co-occupation at A-sites of the BNT perovskite structure.

The polarization–electric field (*P*–*E*) hysteresis loop measurements were carried out for BNBST ceramic at different temperatures starting from RT to 200 °C at each 5 °C step by applying a field of 18 kV/cm and a few selected temperature PE loops are presented in Fig. 3a. The RT PE loop shows a typical hysteresis curve which is a characteristic feature of ferroelectric behavior. With increasing the temperature up to 100 °C, the hysteresis loop slightly changes its shape, but the ferroelectric nature still persists. The remnant polarization (P_r) and maximum polarization (P_{max}) are extracted from each *P*–*E* loop and are plotted as a function of temperature in Fig. 3b. A small increase in the P_r at an early stage up to 55 °C and then followed by a continuous fall with temperature is observed. A change in the slope of P_r observed at 105 °C may be the signature of the phase transition from FE to AFE state (T_d). Compared to the P_r variation, the P_{max} as a function of temperature shows different trends after 55 °C, but at each point of inflection of P_r change, a counterpart anomaly in the P_{max} is noticed, i.e., P_{max} remained constant till 100 °C and then sharply rises up to 130 °C and finally shows a small dip. These all changes support the arguments drawn from the dielectric data investigation that BNBST is in FE state up to 105 °C where it transforms to AFE state and finally to paraelectric state at 190 °C.

From the temperature dependence of the *P*–*E* loop, the isothermal entropy change ΔS of the BNBST ceramic can be determined using Maxwell's relations $\Delta S = -\int_0^E \left(\frac{\partial P}{\partial T}\right)_E \mathrm{d}E$. The area under the curve of *P*–*E* loop is calculated for two different temperature and from that ΔS is estimated.

Following the present calculations, polarization changes and we obtained the ΔS value between 0.02 and −0.06 J/kg. *K* in the temperature range of 28 °C–200 °C. This value is comparable with the reported ΔS value for BNT ceramic [5, 9].

4 Conclusions

Sr and Ba-modified BNT ceramic $(Bi_{0.35}Na_{0.35})_{0.98}Ba_{0.02}Sr_{0.3}TiO_3$ (BNBST) is synthesized by solid-state mixed oxide. XRD analysis confirms that the BNBST composite crystallizes into a single-phase rhombohedral structure with lattice parameters in the hexagonal notation as $a = b = 5.51$ Å and $c = 13.54$ Å. BNBST is characterized by the two dielectric maxima which are closely spaced on the temperature scale mimicking a broad transition peak covering the 70–270 °C temperature range. The temperature-dependent *P*–*E* loop investigation confirms the phase transition from a ferroelectric state to an anti-ferroelectric state at 105 °C followed by a paraelectric phase transition at 190 °C. The electrocaloric ability of the BNBST is estimated by calculating the change in the entropy at a field of 18 kV/cm using the standard thermodynamic relation.

Acknowledgements Authors acknowledge Prof S. Srinath and UGC Network Resource Centers (UGC-NRC), School of Physics, the University of Hyderabad, for providing the characterization facilities used in the present work. Dr. Anshu acknowledges the Department of Science and Technology (DST) India for providing the fund through the Women Scientist A, project with reference number: SR/WOS-A/ET-38/2018. Dr. Md. A. Mohiddon and Dr. Preeti acknowledges Honorable Vice-Chancellor of SRM University Delhi-NCR, Sonepat, Prof. P. Prakash and Dean of Academic Affairs, Prof. V. Samuel Raj for their kind support in establishing the Center for Nanoscience and Nanotechnology.

References

1. Weyland F, Acosta M, Koruza J, Breckner P, Rödel J, Novak N (2016) Criticality: concept to enhance the piezoelectric and electrocaloric properties of ferroelectrics. Adv Funct Mater 26(40):7326–7333
2. Mischenko AS, Zhang Q, Scott JF, Whatmore RW, Mathur ND (2006) Giant electrocaloric effect in thin-film $PbZr_{0.95}Ti_{0.05}O_3$. Science 311(5765):1270–1271
3. Zhou X, Xue G, Luo H, Bowen CR, Zhang D (2021) Phase structure and properties of sodium bismuth titanate lead-free piezoelectric ceramics. Prog Mater Sci 122:100836
4. Schütz D, Deluca M, Krauss W, Feteira A, Jackson T, Reichmann K (2012) Lone-pair-induced covalency as the cause of temperature- and field-induced instabilities in bismuth sodium titanate. Adv Func Mater 22(11):2285–2294
5. Cao WP, Li WL, Dai XF, Zhang TD, Sheng J, Hou YF, Fei WD (2016) Large electrocaloric response and high energy-storage properties over a broad temperature range in lead-free NBT-ST ceramics. J Eur Ceram Soc 36(3):593–600
6. Kour P, Pradhan SK, Kumar P, Sinha SK, Kar M (2017) Effect of Sr doping on electrical properties of lead zirconate titanate nanoceraics. Ferroelectrics 517(1):104–112
7. Ben L, Sinclair DC (2011) Anomalous curie temperature behavior of A-site Gd doped $BaTiO_3$ ceraic: the influence of stain. Appl Phys Lett 98(9):092907
8. Sakata K, Masuda Y (1974) Ferroelectric and antiferroelectric properties of $(Na_{0.5}Bi_{0.5})TiO_3$-$SrTiO_3$ solid solution ceramics. Ferroelectrics 7:347–349
9. Bai Y, Zheng GP, Shi SQ (2011) Abnormal electrocaloric effect of $Na_{0.5}Bi_{0.5}TiO_3$–$BaTiO_3$ lead-free ferroelectric ceramics above room temperature. Mater Res Bull 46(11):1866–1869

Optimization of Process Parameter of Silver Nanoparticles Synthesis by the Leaf of *Paederia foetida* Using Green Technology

Sreyashi Sarkar

Abstract Silver nanoparticles are synthesized using the method that is facile, low cost and eco-friendly green technology by the aqueous leaf extract of raw *Paederia foetida* in a sealed Teflon container (18 mL capacity). The process operating parameters are then studied using a Response Surface Methodology wherein silver nitrate concentration, temperature and rotational speed are considered as factors, each of all the three factors are studied at three different levels viz. high, medium and low—1:9, 1:7 and 1:5 for silver nitrate concentration; 50 °C, 40 °C, 30 °C for temperature and 160 rpm, 140 rpm and 120 rpm for rotational speed. The total volume of the solution is maintained at 10 mL and incubated for 120 h. The surface plasmon resonance (SPR) peak centered at a wavelength of 430 nm for colloidal silver nanoparticles was considered as a response. The optimum values of process parameters for silver nanoparticles synthesis are determined (1:5, 50 °C, 140 rpm). The green synthesized silver nanoparticles are found to be very stable and not precipitated even after 1 year of storage. The model predicted the optimum AgNPs yield value of 2.0924 at optimal value of process parameters (1:5 $AgNO_3$ conc., 50 °C and 142.52 rpm). The experiment conducted with the same optimum conditions and 1.993 AgNPs yield value with 4.98% error is obtained.

Keywords Silver nanoparticles · Green synthesis · Process · Parameter · Optimization

S. Sarkar (✉)
Department of Food Technology and Biochemical Engineering, Jadavpur University, Kolkata, India
e-mail: sreyashisarkar4@gmail.com

Z. H. Khan et al. (eds.), *Recent Advances in Nanomaterials*, Springer Proceedings in Materials 27, https://doi.org/10.1007/978-981-99-4878-9_68

1 Introduction

Silver nanoparticles (AgNPs) have huge applications as an antimicrobial agent in the biomedical field and also, in other areas, e.g., cloths, fabrics, water purification, tooth paste, deodorants, filters, washing machines, kitchen utensils, wound dressings, toys, etc. [1].

Nowadays, scientists are very much engaged in synthesizing metal nanoparticles employing "green chemistry" which is one the major focuses in the context of today's environmental crisis. Different biological sources, e.g., microorganisms and extracts obtained from the plant portions, e.g., roots, fruit, seeds, leaves, etc. that contain both the reducing agent as well as the stabilizing agent are used for the synthesis of metal nanoparticles as reported earlier [2].

There are few studies reported earlier on the green synthesis of silver nanoparticles from the leaf extract of *Paederia foetida*, that are mainly focused on the synthesis, characterization and antimicrobial activity of silver nanoparticles [3, 4]. However, to the best of my knowledge, no report has been published yet focusing on statistical optimization of the process parameters that represented mathematical model to obtain maximum yield of silver nanoparticles synthesis by the *P. foetida* aqueous leaf extract.

In the present study, the synthesis of AgNPs is studied using the aqueous leaf extract of *P. foetida*. Effect of different process operating parameters in the shaken Teflon container is also studied using the Response Surface Methodology (RSM). Silver nitrate concentration, temperature and rotational speed are considered as factors and the surface plasmon resonance (SPR) peak found within the range, i.e., 420 nm–450 nm for silver nanoparticles represented as a response.

2 Experimental

2.1 *Preparation of* Paederia foetida *Aqueous Leaf Extract*

P. foetida raw leaves (Fig. 1) were collected from Garfa, Kolkata-700075, West Bengal. The fresh leaves (5 g) were thoroughly washed, then finely cut leaves were soaked in a 100 ml of sterile double distilled water for about 10 min and the aqueous plant extract was prepared using mortar pestle. It was then filtered through Whatman No. 1 filter paper, stored in a 100 mL volumetric flask at 4 °C and used within a week.

2.2 *Synthesis and Characterization of AgNPs*

Silver nitrate (99.99%) used in this study was purchased from Sigma. Now, 2 ml of the leaf extract was taken in a sealed Teflon container and, then, it was thoroughly mixed

Fig. 1 *P. foetida* raw leaves

with a 10 ml of freshly prepared 1 mM aqueous Silver nitrate ($AgNO_3$) solution. The mixture was run in an incubator shaker at 30 °C for 120 h. "Silver nanoparticle content" is represented as the maximum optical density obtained and, is expressed as Arbitrary Unit (AU) around 400 to 430 nm of the spectral curve.

2.3 *Optimization of Process Parameter of AgNPs Synthesis by the* **Paederia foetida** *Aqueous Leaf Extract and Its Statistical Analysis*

Design of Experiments (DOE) was used to optimize the process parameters of AgNPs synthesis and, then the data analysis was made using the Design-Expert (Version 12) statistical software. Subsequently, statistical analysis are investigated using a Response Surface Methodology wherein silver nitrate concentration, temperature and rotational speed are considered as factors, each of three factors are considered at three levels viz. high, medium and low—1:9, 1:7 and 1:5 for silver nitrate concentration; 50 °C, 40 °C, 30 °C for temperature; 160 rpm, 140 rpm and 120 rpm for rotational speed (while maintaining the total volume at 10 mL and 120 h of incubation period) and the SPR peak at 430 nm for Silver nanoparticle content (AgNP content) as a response, i.e., yield. The optimum process conditions for AgNPs synthesis were determined (1:5, 50 °C, 140 rpm). Statistical analysis is further performed for the *P. foetida* leaf aqueous extract. Coded values 1, 0, −1 are assigned to the high, medium and low levels of each factor.

3 Results and Discussion

3.1 *Statistical Analysis of AgNPs Synthesis by* Paederia foetida *Leaf Aqueous Extract Using Response Surface Methodology (RSM)*

The highest response, 2.572 was found at run 14, whereas the lowest response (0.159) was found at run 1 (Table 1). The quadratic model was found to be significant from the analysis of variance (ANOVA) study, as p-value was less than 0.05 (Table 2 and Eq. 1).

Figures 2 and 3 represent the 3D surface plot of AgNPs yield considering all the three variables studied. From Fig. 2, it can be observed that, although AgNP content decreases with the increase in concentration of $AgNO_3$ but it increases strongly with the increase in temperature. Again, from Fig. 3, it is observed that the AgNP content increases initially with the increase in rotational speed upto a certain level and then it decreases but for the temperature, AgNP content increases with the increase in temperature. The perturbation plot (Fig. 4) also indicates that the optimal condition of AgNP content synthesis that strongly depends on the temperature and the rotational speed compared to the $AgNO_3$ concentration.

Based on the statistical analysis, the optimum condition for AgNP content synthesis are 1:5 $AgNO_3$ concentration, 50 °C temperature and 142.52 rpm rotational speed. The model predicted the optimum AgNPs yield value of 2.0924 at optimal value of process parameters (1:5 $AgNO_3$ conc., 50 °C and 142.52 rpm). The experiment conducted with the same optimum conditions and 1.993 AgNPs yield value with 4.98% error is obtained.

$$\begin{aligned}&\text{Yield expressed as, AgNP content}\\&= +1.09 - 0.1836A + 0.3059B + 0.1211C - 0.0795AB\\&\quad - 0.0243AC + 0.0455BC + 0.0477A^2 + 0.3782B^2 - 0.7382C^2 \qquad (1)\end{aligned}$$

4 Conclusions

RSM successfully validates the optimum process parameters for the synthesis of AgNPs by the *Paederia foetida* leaf aqueous extract.

Table 1 The 3^3 design of experiments for *Paederia foetida* leaf aqueous extract

Run	Factor 1(A) $AgNO_3$ Conc	Factor 2(B) temp (°C)	Factor 3(C) rotational speed (rpm)	Response as AgNP content (AU)
1	1	0	−1	0.159
2	0	1	0	1.891
3	0	−1	0	1.151
4	0	0	1	0.432
5	−1	−1	−1	0.412
6	−1	−1	0	0.919
7	−1	0	1	0.762
8	1	0	1	0.321
9	−1	1	1	1.329
10	0	1	1	0.983
11	1	−1	1	0.589
12	0	0	0	0.781
13	0	−1	−1	0.584
14	−1	1	0	2.572
15	1	0	0	0.561
16	1	−1	0	0.681
17	0	0	−1	0.369
18	−1	1	−1	0.863
19	0	−1	1	0.892
20	−1	0	−1	0.584
21	1	1	−1	0.632
22	1	1	0	2.111
23	−1	0	0	1.663
24	1	1	1	0.876
25	0	1	−1	0.532
26	1	−1	−1	0.462
27	-1	−1	1	0.592

Table 2 ANOVA for the quadratic model of *Paederia foetida* leaf aqueous extract

Response as AgNP content					
Source	Sum of squares	df	Mean square	*F*-value	*p*-value
Model	6.80	9	0.7560	6.85	0.0004
A-$AgNO_3$ Conc	0.6065	1	0.6065	5.50	0.0314
B-Temp	1.68	1	1.68	15.27	0.0011
C-Rotational speed	0.2638	1	0.2638	2.39	0.1405
AB	0.0758	1	0.0758	0.6875	0.4185
AC	0.0071	1	0.0071	0.0640	0.8034
BC	0.0248	1	0.0248	0.2252	0.6412
A^2	0.0136	1	0.0136	0.1236	0.7295
B^2	0.8581	1	0.8581	7.78	0.0126
C^2	3.27	1	3.27	29.63	< 0.0001
Residual	1.88	17	0.1103		

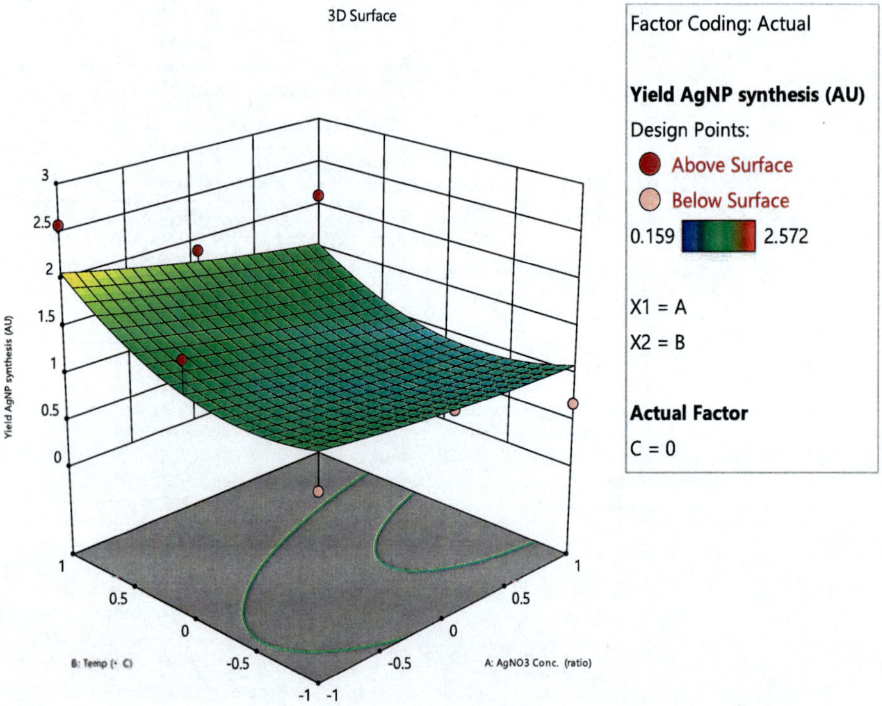

Fig. 2 3D surface plot of AgNPs yield by *Paederia foetida* leaf aqueous extract, interaction of concentration of $AgNO_3$ and temp

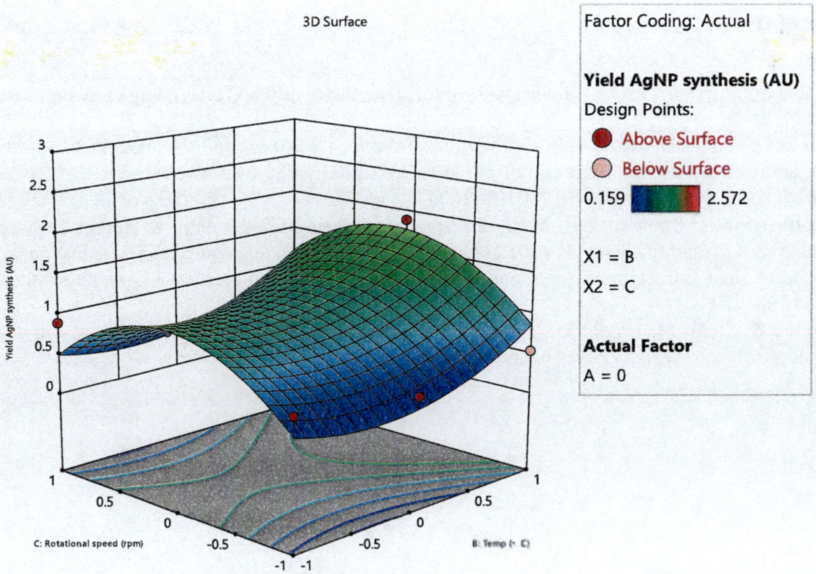

Fig. 3 3D surface plot AgNPs yield by *Paederia foetida* leaf, aqueous extract, interaction of temperature and rot, speed interaction

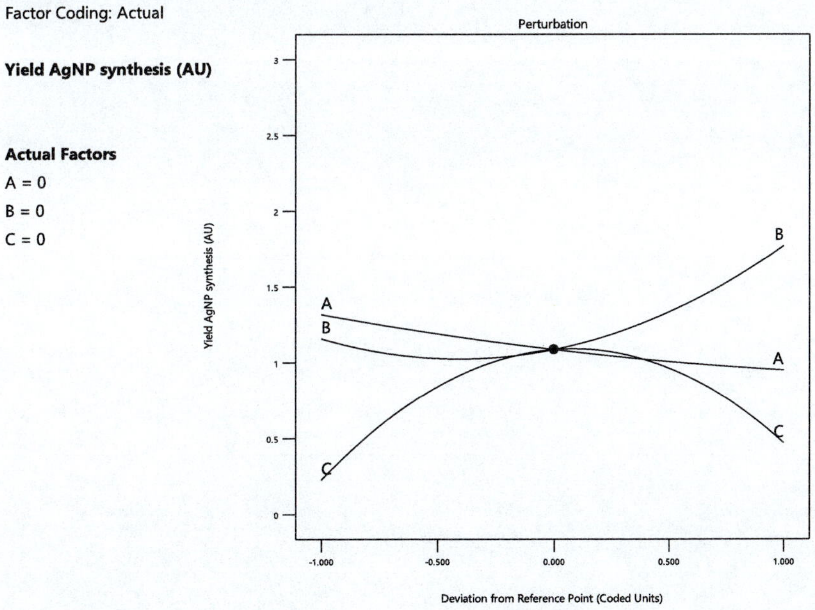

Fig. 4 Perturbation plot of optimization of AgNP content synthesis by the *Paederia foetida* leaf aqueous extract

References

1. Dubas ST, Pimpan V (2008) Green synthesis of silver nanoparticles for ammonia sensing. Talanta 76(1):29–33
2. Dinda G, Halder D, Mitra A, Pal N, Vázquez-Vázquez C, López-Quintela MA (2017) Study of the antibacterial and catalytic activity of silver colloids synthesized using the fruit of *Sapindus mukorossi*. New J Chem 41(19):10703–10711
3. Mollick MMR, Bhowmick B, Maity D, Mondal D, Bain MK, Bankura K, Sarkar J, Rana D, Acharya K, Chattopadhyay D (2012) Green synthesis of silver nanoparticles using *Paederia foetida* L. leaf extract and assessment of their antimicrobial activities. Int J Green Nanotechnol 4(3):230–239
4. Kumar S, Singh M, Halder D, Mitra A (2014) Mechanistic study of antibacterial activity of biologically synthesized silver nanocolloids. Colloids Surf A 449:82–86

Study of the Optical and Current Variation of Polymer Material in the Semiflexible Module Application

Ajit Singh, Sanjai Kumar, Neeraj Tyagi, and Arun Singh

Abstract Glass-based conventional crystalline silicon photovoltaic modules dominate the market due to their high efficiency and low cost. But these modules cannot be used for special applications due to their increased weight and rigidity. Thin film modules have the advantage of polymer but are less efficient and prone to degradation. Nowadays, people are trying to make flexible/bendable modules using substrates instead of traditional silicon-based modules. Because of these properties for solar photovoltaic modules, several candidate polymer films qualify as front sheets, such as Tedlar Polyethylene Terephthalate (TPT), PolyVinylidene Fluoride (PVDF), Ethylene Tetra Fluoro Ethylene (ETFE), Ethylene Chloro Tri Fluoro Ethylene (ECTFE), Fluorinated Ethylene Propylene (FEP). The transparency of the front sheet material in the visible range (300–1100) nm should be comparable to that of traditional glass. Above all cost of the material also plays a prime role in the commercialization of the polymer front sheets. This study will include selecting suitable materials for polymer films based on availability and physical properties data. This will also have an experimental investigation of materials based on quality and the electrical parameters of photovoltaic modules, including indoor and outdoor experiments, to get the best results for real-time applications. This experimental study helps to develop solar modules for low curvature applications such as facades and solar-powered transport systems where low weight and flexible surfaces with high efficiency are the keys to success.

Keywords ETFE · PET · TPT · ECTFE · Spectral response

A. Singh (✉) · N. Tyagi
Deen Dayal Upadhyaya College, University of Delhi, New Delhi 110078, India
e-mail: ajitsingh107@gmail.com

S. Kumar
Technology Mission Division (Energy, Water &Others) Department of Science and Technology, New Delhi, India

A. Singh
Department of Physics Jamia, Milia Islamia University, New Delhi, India

Z. H. Khan et al. (eds.), *Recent Advances in Nanomaterials*, Springer Proceedings in Materials 27, https://doi.org/10.1007/978-981-99-4878-9_69

1 Introduction

Energy security, climate change, and the need for sustainable development are all challenges that can be addressed using solar photovoltaic (PV) energy for electricity generation. It is also a technological advancement with the potential for long-term global growth. Due to government-supported programs and subsidies from all over the world, as well as considerable cost reductions, solar PV power generation is now developing quickly. From 2010 to 2021, the PV market grew at a compound annual growth rate of 32% [1]. Various types of PV solar cells are available in the market, like crystalline Si-based solar cells, thin film-based solar cells (CIGS, CdTe, amorphous silicon), and organic solar cells. Among these types of solar cells, CdTe and CIGS are manufactured with toxic materials that are hazardous to the environment. The organic type of solar cell has a maximum lab-reported efficiency of about 15%. The lifetime and stability of these types of cells are also important issues. Si is an abundantly available material on earth, and the silicon-based solar cell manufacturing processes are mature. Si-based solar cells cover over 90% of the market [1]. The crystalline-based solar cell's reported efficiency in mono and multi-cell cases is 26.7% and 24.4%, respectively. The modules based on crystalline Si have a reported efficiency of more than 20% [2].

The Silicon PV Module currently available in the market uses soda lime silica (SLS) float glass as its front sheet material. The glass used in the front sheet material has a thickness of more than 3 mm, and the weight of the standalone glass is more than 20–25 kg. Since the density of soda lime glass is near 2520 kg/m^3 [3]. Due to weight, a proper structure is required for the glass-based module to stall these panels, which cannot stall in the curved form. Several researchers did some work related to this. Gorteretal. [4] investigated a comparison of 15 polymer materials for application in photovoltaic modules in PV power boats; he found some polymer materials reinforced with glass fibers, such as fluorides, show potential to be used as photovoltaic materials. In their research, Thomson et al. [5] conclude that it is possible to create flexible modules that are more than 13% efficient, have a power-to-weight ratio of at least 150 W/Kg, and stretch to a radius of 5 cm for more than 1000 cycles. Kajisa et al. [6] use thin acrylic film instead of glass as front sheet material. The module based on the acrylic film as a front sheet was tested for long-term reliability against thermal changes and humidity. The results indicate that an acrylic sheet may be used for a lightweight solar PV module.

A systematic approach has been carried out with polymer materials like TPT and thin glass 0.7 mm, thin glass 0.5 mm, PVDF, FEP, ETFE, glass 3.2 mm, and ECTFE. Their optical characterization has been compared to that of glass. A comparative study on the current variation of different commercial crystalline silicon solar cells using the above front sheet material has been conducted for Mono, Multi, and PERC cells.

2 Experimental Procedure

Each polymer material's optical transmittance is measured between 300 and 1100 nm wavelengths. The Bentham PVE 300 device is used to test transmittance. A polymer sample measuring 2 cm by 2 cm is employed in this measurement. The polymer materials were cleaned with the help of an isopropyl alcohol solution. To calculate the total transmittance of the transparent film, the clean polymer material is mounted on the integrating sphere (DTR 6). Monochromatic light is transmitted through the polymer material film, while the Bentham PVE 300 is configured in the transmittance mode. The proportion of light transmitted at each wavelength is recorded for the intended wavelength range (Fig. 1).

For the current variation study, we have used full industrial crystalline silicon solar cells of standard size (15.6 × 15.6 cm) of Mono, Multi, and PERC type cells. And then, different polymer materials are put on the front side of the different types of crystalline cells. Then the spectral response (SR) of other cells with varying combinations of the cell is measured. This spectral response is converted into the cell's current using the standard AM 1.5 data of the irradiance. By this, we find the variation in the different cell currents.

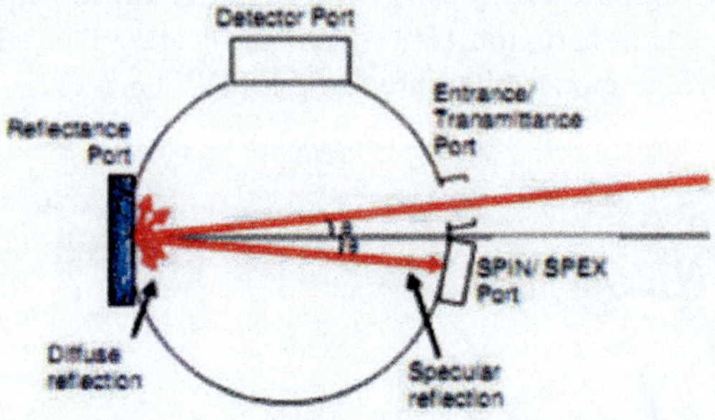

(a)

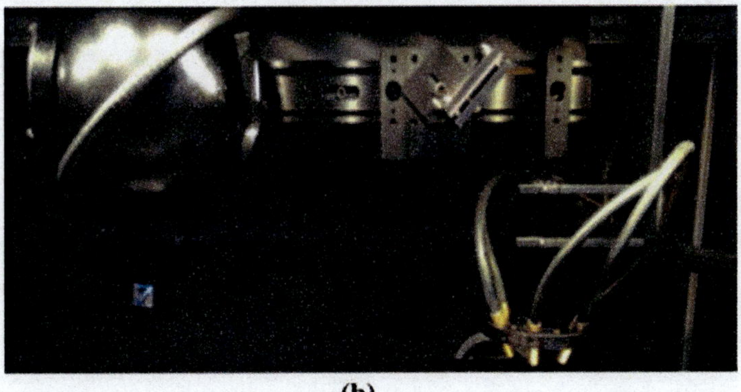

(b)

Fig. 1 **a** Integrated sphere **b** Bentham (PVE300) optical system

3 Result and Discussion

(1) **Transmittance Measurement:** The transmittance of different polymer materials' is measured in the wavelength range of (300–1100 nm). This range is used because silicon solar cells work well within this range. The transmittance of several polymer materials in the specified wavelength range is shown in Fig. 2.

Figure 2 indicates that the thick glass of 3.2 mm has the highest transmittance. The FEP polymer shows good agreement with glass and has nearly a 3% variation in transmittance compared to the glass after 400 nm wavelength. The polymer materials like ETFE, ECTFE, and thin glass (0.7 mm) and thin glass (0.5 mm) show a 5% variation in the transmittance compared to the glass after 400 nm. Polymer materials like PMMA and TPT show 7% variations compared to glass. PVDF shows the lowest transmittance in the lower wavelength range below 700 nm; after 700 nm, it offers a 7% variation in the transmittance compared to the glass (Table 1).

(2) **Spectral Response Measurement**

Figure 3 represents a different polymer material with a monocrystalline cell spectral response (SR) curve. A standard silicon solar cell is used for calibration purposes. The figure indicates that the FEP-based front material shows superior performance throughout the wavelength region (300–1100 nm); after that, ETFE shows some change in the spectral response compared to the FEP. Thin glass (0.7 mm) and thin

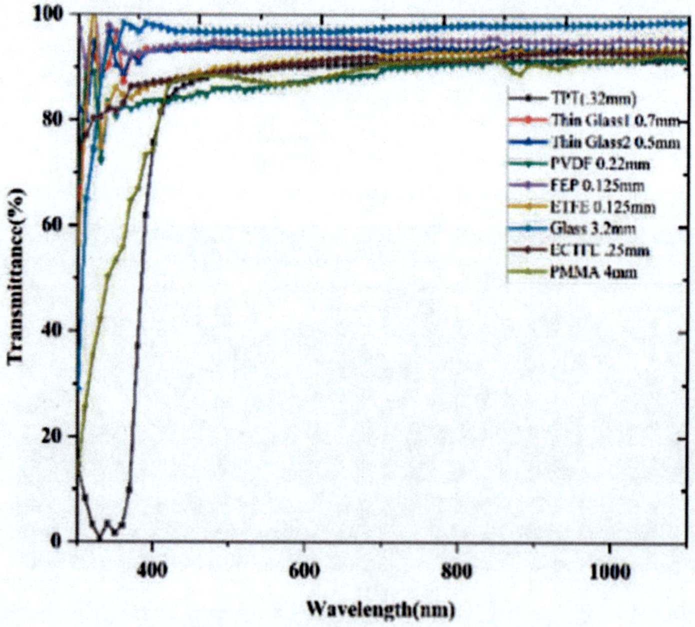

Fig. 2 Transmittance of different polymer materials

Table 1 Maximum transmittance of different polymer materials

Sr. No	Material	Maximum transmittance (%)
1	TPT	91
2	Thin Glass 0.7 mm	93
3	Thin Glass, 0.5 mm	93
4	PVDF	88
5	FEP	95
6	ETFE	93
7	Glass	98
8	ECTFE	93
9	PMMA	91

glass (0.5 mm) show similar spectral responses throughout the region but less than FEP and ETFE. ECTFE-based front material offers a similar spectral response as thin glass. PVDF-based material shows the lowest spectral response.

Figure 4 represents different polymer material's multi-crystalline spectral response (SR) in the 300–1100 nm wavelength range. The figure indicates that FEP shows the highest spectral response throughout the wavelength region. ETFE shows a slight variation in comparison to FEP. ECTFE, thin glass (0.7 mm, 0.5 mm) performance remains the same in spectral response, while PVDF and TPT show the lowest spectral response throughout the wavelength region. Figure 5 represents the PERC

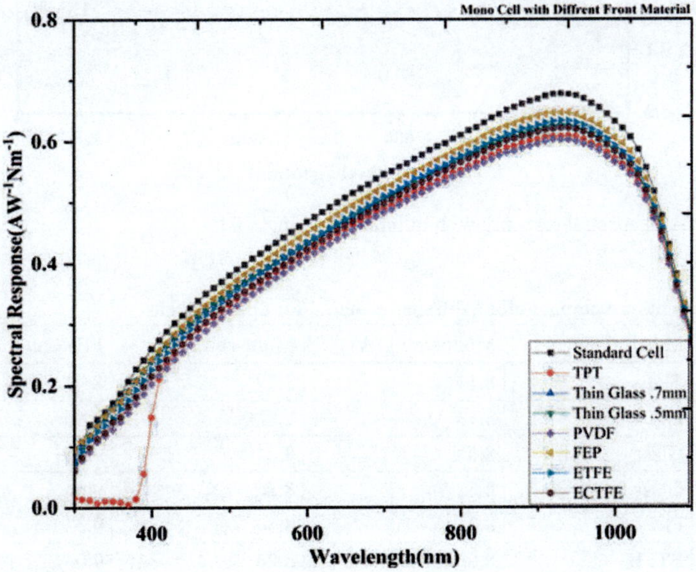

Fig. 3 Mono-cell spectral response with different front material

cell with different materials in spectral response terms in the 300–1100 nm wavelength range. FEP-based material shows the highest spectral response (SR). TPT offers the lowest performance in terms of spectral response.

(3) **Current measurement:** The current measurement of Mono, Multi, and PERC cells is given below in the table.

Table 2 indicates that FEP material performs better in all three combinations of cells, and TPT and PVDF-based material show an 8–9% reduction compared to FEP-based material in all combinations of the cell. Compared to thin glass, ETFE, ECTFE shows a 3–4% reduction compared to FEP.

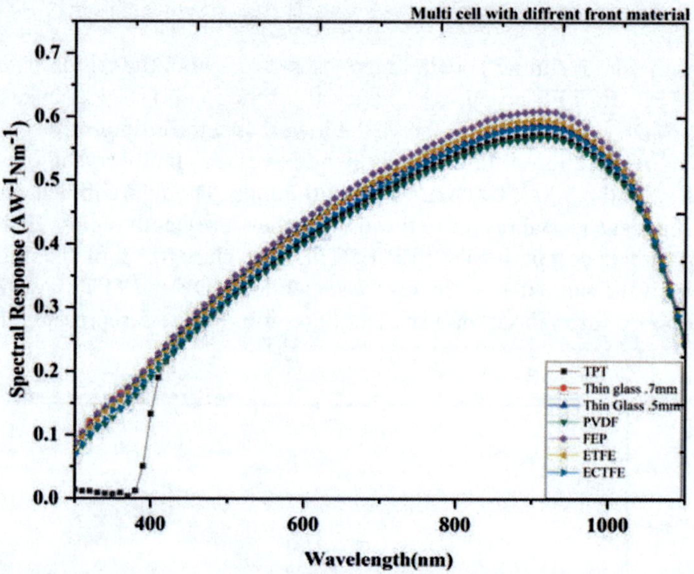

Fig. 4 Multi-cell spectral response with different front material

Table 2 Current measurement for a different material for different cell

Sr. No	Material name	Mono-cell I (A)	Multi-cell I(A)	PERC cell I (A)
1	TPT	8.11	7.77	8.47
2	PVDF	8.13	7.82	8.66
3	Glass 0.7 mm	8.61	8.24	9.05
4	Glass 0.5 mm	8.62	8.26	9.04
5	FEP	8.91	8.5	9.33
6	ETFE	8.63	8.26	9.04
7	ECTFE	8.42	8.06	8.80

4 Conclusion

The finding of this study is that polymer materials like TPT, PVDF, FEP, ETFE, and ECTFE may be used as the front sheet material in place of glass. In terms of transmittance, FEP shows good agreement with the glass. Other polymer materials also show promising results in terms of transmittance. Regarding electrical parameters, the PERC cell shows the maximum current with the above combination of front materials.

Acknowledgements The author would like to thank the Electronics Science Department, University of Delhi, DDU College, DST Delhi, and JMI for their support.

Declaration of Interest Statement There is no conflict of interest by any author.

References

1. International Technology Roadmap for Photovoltaic (ITRPV) (2022) September 2022, VDMAPhotovoltaicEquipment. https://www.ise.fraunhofer.de/en/publications/studies/photovoltaics-report.html
2. Green MA, Dunlop ED, Hohl-Ebinger J et al (2022) Solar cell efficiency tables (Version 60). Prog Photovolt Res Appl 30(7):687–701. https://doi.org/10.1002/pip.3595
3. https://onlinelibrary.wiley.com/action/showCitFormats?doi=10.1002%2Fpip.3595
4. Allsopp BL, Orman R, Johnson SR, Baistow I, Sanderson G, Sundberg P, Stålhandske C, Grund L, Andersson A, Booth J, Bingham PA, Karlsson S (2020) Towards improved cover glasses for photovoltaic devices. Progress in Photovoltaics. https://doi.org/10.1002/pip.3334
5. Gorter T, Reinders AHME (2012) A comparison of 15 polymers for application in photovoltaic modules in PV-powered boats. Appl Energy 92:286–297. ISSN 0306–2619, https://doi.org/10.1016/j.apenergy.2011.10.042
6. Thomsen EA, Muric-Nesic J, Xia L, Brauers M, Ratcliff T, Davies E, Skryabin I, Everett V, Blakers A (2010) Analysis of components and structures in flexible high-efficiency PV modules.. https://openresearch-repository.anu.edu.au/handle/1885/61939
7. Kajisa T, Miyauchi H, Mizuhara K, Hayashi K, Tokimitsu T, Inoue M, Hara K, Masuda A (2014) Novel lighter weight crystalline silicon photovoltaic module using acrylic-film as a cover sheet. Japanese J Appl Phys 53:092302. https://iopscience.iop.org/article/https://doi.org/10.7567/JJAP.53.092302

Dielectric, Optical, and Structural Analysis of Nanostructured Mn-Doped ZnO Using Phoenix Dactylifera

Azam Raza, Jhalak Gupta, Tanveer Ahamad, S. k. Najrul Islam, Mohammad Muaz, Syed Mohd Adnan Naqvi, and Absar Ahmad

Abstract This study report the fabrication of pristine and Mn-doped ZnO nanoparticles (in the proportion 0% (pristine) and 4%) using biological (Phoenix dactylifera Pulp) route. Structural and dielectric analyses were carried out to understand their properties. The XRD analysis exhibited a crystalline hexagonal wurtzite structure. The crystallite sizes were found to be 23 and 21.9 nm for pristine and 4% Mn-doped ZnO, respectively. For pure and 4% Mn-doped ZnO, the dielectric constant exhibits a declining trend with an increase in frequency. The fact that conductivity results from the same mechanism of switching between metal ions explains this. Higher frequencies cause the hopping to lose its ability to keep up with the field's rapid fluctuation and start to diminish because of the chaotic dipolar distribution and random orientation. At lower frequencies, the hopping is in sync with the applied ac field.

Keywords Dielectric · TEM · XRD

1 Introduction

One of the most promising metal oxide materials is zinc oxide (ZnO) due to its potential applications such as solar cells, light emitters, and optical sensors [1]. ZnO is receiving much interest due to direct bandgap (3.37 eV) and 60 meV binding energy. These characteristics make ZnO an ongoing study topic. As a result of its attractiveness as a research topic, transition metal-doped ZnO has been the subject of research. It was observed that the sort of transition metals used to replace Zn affected the physical characteristics of ZnO. For instance, as the Co concentration in $Zn_{1-x}Co_xO$ grew, the optical bandgap E_g, dielectric constant, and magnetic characteristics of ZnO also increased [2–4]. By adding Fe ions to ZnO, both its optical and

A. Raza · S. Najrul Islam · M. Muaz · S. M. A. Naqvi · A. Ahmad (✉)
Interdisciplinary Nanotechnology Centre, Aligarh Muslim University, Aligarh, India
e-mail: aahmad786in@gmail.com

J. Gupta · T. Ahamad
Department of Physics, Aligarh Muslim University, Aligarh, Uttar Pradesh 202002, India

Z. H. Khan et al. (eds.), *Recent Advances in Nanomaterials*, Springer Proceedings in Materials 27, https://doi.org/10.1007/978-981-99-4878-9_70

magnetic characteristics were enhanced [5]. Although the crystal structure remained unchanged up to 5-wt.% Fe, a ferromagnetic transition was noticed at roughly room temperature (RT).

Additionally, for Mg-doped ZnO, a lower E_g, as well as a higher photocatalytic activity, were discovered [6]. The physical characteristics of Mn-doped ZnO have attracted the attention of many groups [7–9]. The combination of Mn-doped ZnO and poly(3-hexythiophene) is advantageous for solar energy production [10], the Eg is blue-shifted in comparison to that of pure ZnO [7], and a magnetic transition was detected at room temperature [7, 8]. Here is a summary of the findings: The amount of Mn ions determines the physical characteristics of ZnO, and the crystal structure is hexagonal wurtzite [7, 8]. The structure, morphology, and optical characteristics of $Zn_{1-x}Mn_xO$ nanoparticles (NPs), were discussed in a study by Mote et al. [9]. The outcomes demonstrated that the structure is still hexagonal and wurtzite. In contrast to the optical bandgap, the lattice parameters and packing fraction enhanced with the rise in Mn concentration. Additionally, the optical characteristics of nanoscale ZnO doped with various ions, including Cd, Mg, Mn, and Fe, were investigated [11, 12]. It was discovered that depending on the type of dopants, the E_g could vary from 2.9 to 3.8 eV. Mn-doped ZnO also exhibits semiconducting–like electrical conductivity that is influenced by the sintering temperature and the type of doping ions [11].

2 Experimental Details

All the chemicals were of analytical grade. The green synthesis-assisted sol–gel precipitation technique was used to produce pure and Mn-doped ZnO nanoparticles. Acetone, Whatman filter paper, magnesium acetate ($Mg (CH_3COO)_2$), zinc acetate ($Zn (CH_3COO)_2.2H_2O$), and Ajwa date pulp (Phoenix dactylifera L.) were utilised as the starting materials. During the synthesis, desired molar concentrations of zinc acetate and magnesium acetate (0% and 4%) were used, and each were dissolved in 100 ml of double distilled water; thereby, we added 20 ml of Ajwa plant extract for 30 min in the solution with precursor. The addition of plant extract acted as a capping, stabilising, and reducing agent. Three hours of continuous stirring produced a fine precipitate, which was subsequently calcined for three hours at 350 °C to produce ZnO and Mn-doped ZnO nanoparticles.

3 Characterisation

Using a Rigaku X-ray diffractometer (Cu anode material, wavelength of (CuK_α), i.e. 1.541838 Å), the crystallinity, phase, and unit cell dimension were determined. The particle size, shape, and crystalline structure of nanoparticles were examined using a transmission electron microscope (TEM) with a model number of JEM 2100 from JEOL Japan.

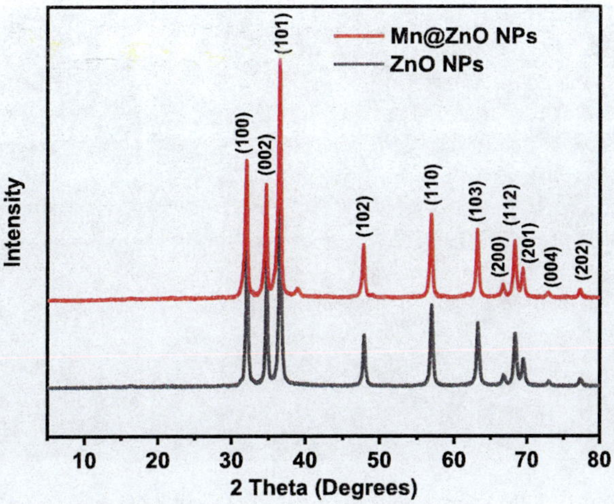

Fig. 1 XRD pattern of pure and Mn (4%) doped ZnO NPs

Table 1 Illustrating all parameters of $Zn_{1-x}Mn_xO$ ($x = 0.0, 0.04$) samples

Conc. of Mn in ZnO (%)	Lattice constant (Å)		Crystallite Size (nm)
	$a = b$	c	
0	3.215	5.199	23
4	3.208	5.196	21.9

4 Results and Discussion

4.1 Structural Analysis

The changes in phase, structure, and crystallite size of the undoped and Mn (4%) doped ZnO NPs were examined using X-ray diffraction patterns (Fig. 1). Regardless of the dopant concentrations, the XRD peaks were matched using Pan Analytical Software with the reference data (JCPDS 96–210-7060) without the presence of any additional impure phases. The diffraction peaks at 32.05°, 34.57°, 36.58°, 47.56°, 56.52°, 62.87°, 66.55°, 67.78°, 69.22°, 72.87°, and 77.02° were indexed as (100), (002), (101), (102), (110), (103), (200), (112), (201), (004), and (204) planes of hexagonal wurtzite structure of ZnO nanoparticles (Table 1).

4.2 TEM Analysis

Transmission electron microscopy measurements were used to examine the microstructures of the Mn ($x = 0.04$) doped ZnO nanoparticles and the findings

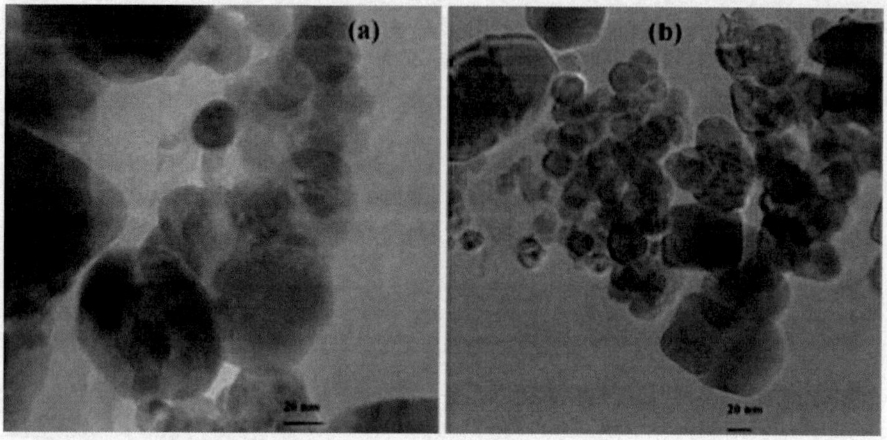

Fig. 2 TEM micrographs for **a** Pure ZnO and **b** 4% Mn-doped ZnO nanoparticles

are shown in Fig. 2a and b for pristine and 4% Mn-doped ZnO, respectively. Particle agglomeration is visible in both micrographs, although it is more pronounced in the 4% Mn-doped ZnO than in the pure one.

4.3 LCR Analysis

At room temperature, the real and imaginary parts of the complex dielectric permittivity of pure and 4%Mn-doped ZnO NPs were measured as functions of frequency. The real part of the complex dielectric permittivity describe the energy that is stored, whereas the imaginary part describe the energy that is dissipated. The impacts of "dipolar" and "charge carrier" behaviour are typically present in the dielectric response. While the latter is linked to partial recovery on discharge but always leaves a limited polarisation in the system, the former is in charge of restoring the zero residual polarisation after discharging. Figure 3a and b illustrate the dielectric constant as a function of frequency for all compositions.

4.4 Dielectric Loss

According to the relationship, loss tangent tan is produced from the phase difference resulting from the energy loss within the sample at a specific frequency is as follows:

$$\tan\,\delta = \frac{1}{\tan\,\theta},$$

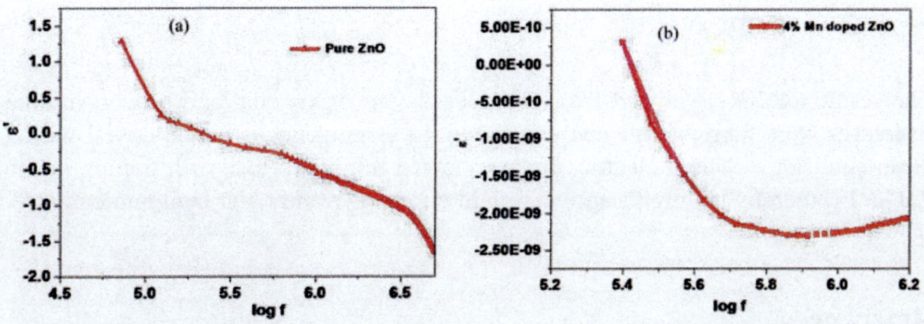

Fig. 3 Variation of dielectric constant with frequency for **a** Pure ZnO, **b** 4% Mn-doped ZnO

where θ is the phase difference. At low frequencies, dipoles often follow the applied ac field with a minor phase difference, which causes the loss tangent to have a significant value. Dipoles start to fall behind the field as frequency rises, increasing the phase difference and diminishing the tan. Dipoles do not follow the applied ac field at high frequencies, when tan exhibits frequency-independent behaviour. Figure 4a and b depicts the variation of loss factor with frequency for all compositions.

As frequency rises, the dielectric constant (ε') gradually decreases. This decrease is due to the space charge polarisation effect being lessened. Following that, it was nearly constant but gradually got worse as the temperature rose periodically. At low frequencies, the dielectric constant was substantial, nevertheless.

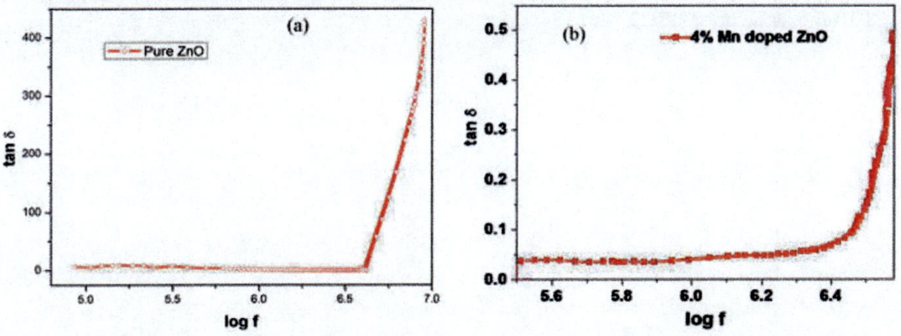

Fig. 4 Dielectric loss variation with frequency for **a** Pure ZnO & **b** 4% Mn-doped ZnO

5 Conclusion

The results of XRD showed that ZnO NPs having hexagonal and polycrystalline structures were fabricated. TEM confirmed the homogeneous nanomaterial with a hexagonal shape. The dielectric studies showed minimum loss with doping in the ZnO NPs, thereby having its application in microwave and radar communications.

References

1. Chen KJ, Fang TH, Hung FY, Ji LW, Chang SJ, Young SJ, Hsiao YJ (2008) Appl Surf Sci 254:5791
2. Franco A, Pessoni HVS (2015) Phys B 476:12
3. Nam G, Yoon H, Kim B, Lee DY, Kim JS, Leem JY (2014) J Nanosci Nanotechnol 14:8544
4. Hammad TM, Griesing S, Wotocek M, Kuhn S, Hempelmann R, Hartmann U, Salem JK (2013) Int J Nanoparticles 6:324
5. Raj KP, Sadaiyandi K, Kennedy A, Sagadevan S, Chowdhury ZZ, Johan MRB, Abdul-Aziz F, Rafique RF, Selvi RT, Bala RR (2018) Nanoscale Res Lett 13:229
6. Sharma RK, Patel S, Pargaien KC (2012) Adv Nat Sci Nanosci Nanotechnol 3:035005
7. Sundar SA, John NJ (2016) Int J Eng Appl Sci (IJEAS) 3:26
8. Han J, Shen M, Cao W, Senos AMR, Mantas PQ (2003) Appl Phys Lett 82:67
9. Mote V, Dargad J, Dole B (2013) Nanosci. Nanoeng. 1:116
10. Wang YS, Thomas PJ, O'Brien P (2006) J Phys Chem B 110:21413
11. Han J, Mantas PQ, Senos AMR (2001) J Eur Ceram Soc 21:1883
12. Raza A, Shoeb M, Mashkoor F, Rahaman S, Mobin M, Jeong C, Yusuf Ansari M, Ahmad A (2022) Materials chemistry, and physics. 286:126173. ISSN 0254–0584. https://doi.org/10.1016/j.matchemphys.2022.126173

Effect of Zinc Doping on the Structural, Dielectric and Electrical Properties of the Modified Bismuth Sodium Strontium Titanate

Preeti Saroha, Anshu Gaur, Preeti, and Md.Ahamad Mohiddon

Abstract The structural identification of the lead free $Bi_{0.35}Na_{0.27}Sr_{0.3}Zn_{0.08}TiO_3$ electro-ceramic along with its dielectric and electrical characterization is reported. The solid solution was synthesized using carbonates and oxides via a solid-state mixed oxide technique at 1000 °C for 12 h. The hexagonal perovskite phase formation was confirmed from XRD investigation. Crystal parameters are estimated from the Rietveld refinement. Dielectric measurements as a function of temperature at different frequencies were carried out and confirm the diffuse nature of phase transition. Activation energy for ionic conductivity is also estimated.

Keywords Bismuth sodium titanate (BNT) · X-ray diffraction · Dielectric · Electrical characterization

1 Introduction

Lead free ferroelectric materials are the promising candidate to be used as an alternate refrigerant in the electrocaloric (EC) effect based cooling devices due to their eco-friendly nature. ECE is basically the phenomenon of reversible temperature change of a polar material when an external electric field is applied [1]. Mischenko et al. in 2006 noted a giant electrocaloric responsivity in $PbZr_{0.95}Ti_{0.05}O_3$ (PZT) thin film with a large ΔT of 12K. Pb based materials exhibit high ferroelectric, piezoelectric, pyroelectric properties [2] but due to its toxic nature, one has to switch to alternate material for use in cooling devices. From the last few decades, researchers found $Bi_{0.5}Na_{0.5}TiO_3$ (BNT) as a promising leadfree piezoelectric material [3]. BNT has

P. Saroha · Preeti · M. Mohiddon (✉)
Center for Nanoscience and Nanotechnology, Department of Physics, Faculty of Science and Humanities, SRM University Delhi-NCR, Sonepat, Haryana 131029, India
e-mail: mahamad.ahamad@srmuniversity.ac.in

A. Gaur
School of Physics, University of Hyderabad, Prof. C.R. Rao Road, Gachibowli, Hyderabad, Telangana 500046, India

Z. H. Khan et al. (eds.), *Recent Advances in Nanomaterials*, Springer Proceedings in Materials 27, https://doi.org/10.1007/978-981-99-4878-9_71

an ABO_3 perovskite structure and happen to be ferroelectric at room temperature (RT). BNT has the Curie temperature (T_C) of 320 °C and large remnant polarization of 38 μC/cm^2 and large coercive field of 73 kV/cm with high dielectric loss, high conductivity and low depolarization temperature. Doping with appropriate chemical constituents may lead to intensify the results obtained. It is reported that Sr doping helps to bring the phase transition temperatures towards RT [4] and Zn helps to densify the structure and lowers the T_c [5]. In the present work, we have modified BNT with Sr and Zn and investigated the effect of modification on the structural and electrical properties of BNT.

2 Experimentation

Conventional solid-state method was employed to synthesize $Bi_{0.35}Na_{0.27}Sr_{0.3}$ $Zn_{0.08}TiO_3$ (BNSZT) solid solution. Stoichiometric amounts of Bi_2CO_5, Na_2CO_3, $SrCO_3$, ZnO, and TiO_2 were homogeneously mixed in an acetone media for 2h. The mixed powders were kept for calcination at 1000 °C for 12h in an alumina crucible. Calcined powders were pressed into cylindrical pellets using a uniaxial hydraulic press at the applied pressure of 2GPa. These pellets are then sintered at a temperature of 1250 °C for 2h. Identification of phases and crystal structure was carried out using X-ray Diffractometer (PANalytical X'pert Pro) using Cu Kα radiation of wavelength of 0.154 nm. Unit cell parameters are calculated with the help of X'pert highscore plus software. Before taking the electrical measurements, sintered pellets were polished and electrodes was achieved by applying high purity silver paste followed by drying at 150 for 30 min. Dielectric measurements were carried out using an Impedance Analyser (Novo Control Technologies) at five discrete frequencies of 100 Hz, 1 kHz, 10 kHz, 100 kHz and 1 MHz in the temperature range of RT-350 °C.

3 Results and Discussion

The XRD pattern of calcined $Bi_{0.35}Na_{0.27}Sr_{0.3}$ $Zn_{0.08}TiO_3$ powder is presented in Fig. 1. The diffraction pattern has sharp six peaks along with few small peaks in the background. This strong peak pattern is identified as belonging to hexagonal BNT perovskite structure with R3C space group and are indexed according to JCPDS file number: 98–002-0786. The tiny small peaks in the background are identified belonging to the $Bi_4Ti_3O_{12}$ (JCPDS file number: 98–019-5919) and Na_2CO_3 (JCPDS file number: 98–016-8130). The growth of such impurity is expected to be due to doping of Zn^{2+} at the Na^+ site. The difference in the oxidation states and the ionic radii of the dopant atoms and the atoms of the host lattice may result to the segregation of such impure phases. The ionic radii of the dopant, i.e. Zn^{2+} (${r_8}^{2+} = 0.9$ nm) is smaller than the ionic radii of Na^+(${r_{12}}^{+} = 1.32$ nm) and Bi^{3+} (${r_8}^{3+} = 1.17$ nm). The

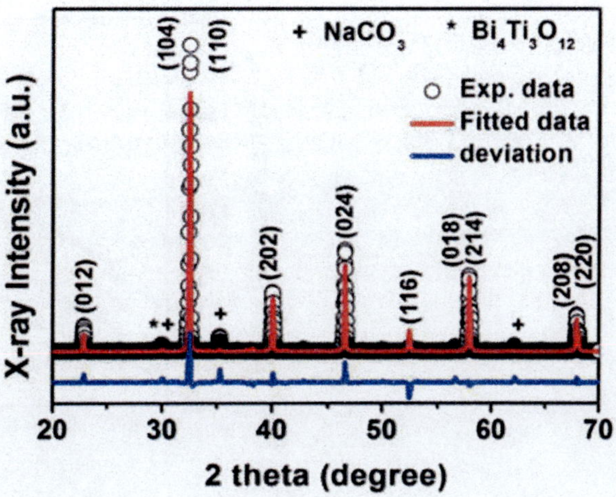

Fig. 1 XRD pattern of $Bi_{0.35}Na_{0.27}Sr_{0.3}Zn_{0.08}TiO_3$ powder calcined at 1000 °C for 12h

experimental XRD data is refined using X'pert high score software keeping all above three phases together. After the refinement the obtained lattice parameters are $a = b = 5.52$Å; $c = 13.54$Å. Average crystal size of BNSZT sample was calculated using the full width at half maximum ($\beta_{1/2}$) of the strongest XRD peak at $2\theta = 32.5°$ (110) by employing the famous Debye Scherrer's Equation. Average Crystal Size, $D = k\lambda/\beta_{\frac{1}{2}} \cos\theta$, where k = 0.89, θ is the Bragg's angle, λ is the wavelength of the X-Rays ($\lambda = 0.154$ nm). The calculated average crystallite size is 47.4 nm.

Figure 2 represents the temperature response of the dielectric constant measured at five different frequencies. It is observed from the figure that there are two phase transition peaks exists, first phase transition at a temperature around 150 °C and the second phase transition at 230 °C. In the literature it is reported that the BNT undergo following three phase transitions. The first phase transition is from rhombohedral ferroelectric (R-F) to orthorhombic anti-ferroelectric (O-AF) at 200 °C (T_{c1}) followed by second phase transition from O-AF to tetragonal paraelectric (T-PE), ferroelastic (T-FEla) at 320 °C (T_{c2}) and third phase transition from T-PE, T-FEla to Cubic paraelectric (C-PE), paraeleastic (C-PEla) at 520 °C [6].

In the present dielectric characterization from RT to 320 °C, we have observed that the R-F to O-AF phase transition is shifted to 150 °C and O-AF to T-PE phase transition is shifted to 230 °C. This shift in the phase transition temperature is expected to be due to the modification of BNT by Zn and Sr. Phase transition temperature and the dielectric constant at phase transition for different frequencies are reported in the Table 1. From the table it is evident that both phase transition temperature of the BNSZT sample shifted to higher temperature with increasing the frequency. A large shift of 20 °C in R-F to O-AF phase is observed compared to 5 °C shift in the O-AF to T-PE phase transition for a change of frequency from 100 Hz to 1 MHz. This behaviour is the typical of relaxor characteristics of BNT.

The maximum dielectric constant at transition temperature decreases with the increase in the frequency. It is due to the fact that when the frequency is high the

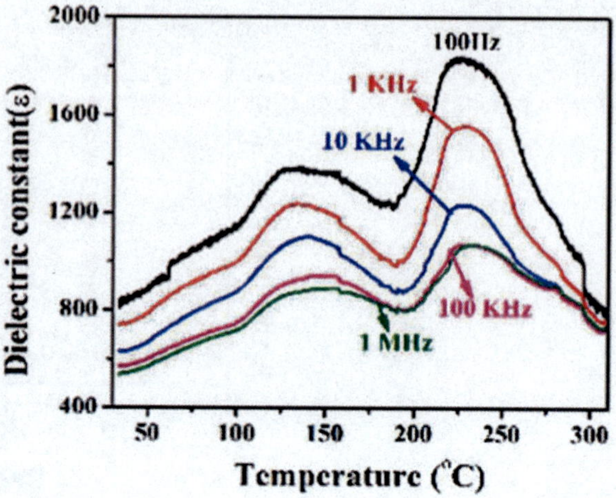

Fig. 2 Variation of dielectric constant of $Bi_{0.35}Na_{0.27}Sr_{0.3}Zn_{0.08}TiO_3$ as a function of temperature measured at different frequencies

Table 1 The phase transition temperature, maximum dielectric constant, diffusion coefficient and activation energy of $Bi_{0.35}Na_{0.27}Sr_{0.3}Zn_{0.08}TiO_3$ at different frequencies

Frequency (Hz)	Tc_1, °C (ε_{max1})	Tc_2, °C (ε_{max2})	Diffusion parameter (ϒ)	Activation energy (eV)
100	135(1387)	227 (1843)	2.46	0.46
1000	134 (1236)	229 (1555)	2.6	0.69
10,000	140 (1102)	228 (1235)	1.81	0.99
100,000	150 (945)	230 (1073)	2.44	0.72
1,000,000	155 (891)	232 (1071)	1.91	0.60

contribution from orientational polarization of dipoles drops down. The dielectric constant observed in our samples is relatively low compared to the one reported by Zhou et al. for BNT ceramics [4]. This indicates that doping with Zn and Sr decreases the dielectric constant of BNT along with decreasing the phase transition temperature. Further the close observation of the dielectric peak shows a diffuse nature of phase transition and also it is frequency dependent. At lower frequency the phase transition is relatively sharp compared to the high frequency transition. The diffusion parameter is calculated following modified Curie–Weiss law for the AFE to PE phase transition.

$$\frac{1}{\varepsilon} - \frac{1}{\varepsilon_{max}} = \frac{(T - T_c)^{\gamma}}{2\delta^2 \varepsilon_{max}}$$

where γ and δ are called diffusion coefficients. The value of γ is calculated from the slope of logarithmic plot log ($1/\varepsilon$-$1/\varepsilon_{max}$) versus log (T-T_C). The value of γ between 1 and 2 indicates the partially diffuse phase transition and $\gamma = 2$ indicates completely diffuse phase transition and strong relaxor ferroelectric behaviour. The calculated value of the γ, i.e. the degree of diffuseness for the different frequencies is given in Table 1 and it indicates the pure diffuse phase transition. The activation energy of BNSZT a.c. electrical conductivity is calculated using the Arrhenius equation, $\sigma = \sigma_0 e^{E_a/k_B T}$,

where σ is the conductivity, σ_0 is the temperature independent material's constant, E_a is the activation energy, k_B is the Boltzmann constant, T is the temperature in Kelvin. From the slope of the linear graph between ln(σ) and 1/T, E_a is calculated and is included in Table 1. The E_a calculated in the temperature range of 250–300 °C is found to be in the range of 0.5–1eV for all measured frequencies. These values closely matched with that observed for vacancy conduction in the BNT system [6]. The conductivity in the BNSZT ceramic is associated with the diffusion of vacancies formed due to the substitution of Na by Zn^{2+} followed by partial segregation of BNSZT as Bismuth sodium titanate supported from XRD result.

4 Conclusions

Sr and Zn modified Bismuth sodium titanate, i.e. $Bi_{0.35}Na_{0.27}Sr_{0.3}Zn_{0.08}TiO_3$ (BNSZT) ceramic is synthesized via solid-oxide mixed route. XRD analysis confirms the formation of hexagonal perovskite phase with R3C space group for the powder calcined at 1000 °C for 12 h. The substitution of Sr and Zn in BNT has strongly influenced the dielectric properties: it reduced the phase transition temperature towards RT; the phase transition temperature and dielectric constant are frequency dependent, which confirms the relaxor nature. The diffuseness parameter is estimated from the modified Curie–Weiss law and confirms the complete diffuse phase transition. The activation energy of a.c. conductivity is estimated to range from 0.5 to 1 eV for different frequency range from 100 Hz to 1 MHz.

Acknowledgements The Authors acknowledges school of Physics, University of Hyderabad for providing the XRD characterization facility and impedance data. Dr. Anshu acknowledge DST (Ref. No.: SR/WOS-A/ET-38/2018) for providing the fund during this work. Dr. Md. A. Mohiddon acknowledge Honourable Vice-Chancellor of SRM University Delhi-NCR, Sonepat, Prof. P. Prakash and Dean of Academic Affairs, Prof. Samuel Raj for their kind support in establishing the Centre for Nanoscience and Nanotechnology.

References

1. Crossley S, Mathur ND, Moya X (2015) New developments in caloric materials for cooling applications new developments in caloric materials for cooling applications. In: AIP advances, vol 5. 067153 0–8
2. Fu H, Cohen RE (2000) Polarization rotation mechanism for ultrahigh electromechanical response in single-crystal piezoelectric. Nature 403:281–283

3. Reichmann K, Feteira A, Li M (2015) Bismuth sodium titanate based materials for piezoelectric actuators. Materials 8(12):8467–8495
4. Zhou L, Vilarinho PM, Baptista JL (2001) Dielectric properties of bismuth doped $Ba_{1-x}Sr_xTiO_3$ ceramics. **21**(4):531–534
5. Jain A, Panwar AK, Jha AK (2016) Improvement in dielectric, ferroelectric and ferromagnetic characteristics of $Ba_{0.9}Sr_{0.1}Zr_{0.1}Ti_{0.9}O_3$-$NiFe_2O_4$ composites. Ceramics Int 43(13):10253–10262
6. Zhou X, Xue G, Luo H, Bowen CR, Zhang D (2021) Phase structure and properties of sodium bismuth titanate lead-free piezoelectric ceramics. Progress in Mater Sci 122:100836

Fabrication and Characterization of rGO-Doped TiO_2 Nanocomposite for Temperature Variable I–V Measurements

Mohd Azharuddin and Rana Tabassum

Abstract The measurements of the electronic characteristics of a device at elevated temperatures are found to be very useful in various applications. Therefore, researchers have made a handful of efforts to evaluate the electronic characteristics of a device at elevated temperatures which may be achieved by careful optimization of the governing factors. One of the prominent governing factors may be the type as well as the dimensionality of the sensing materials. This work reports the evaluation of the temperature variable electronic characteristics of the fabrication of the device via prepared reduced graphene oxide (rGO) doped titanium-di-oxide (TiO_2) nanocomposite. This electronic device is found to be accurate, simple, cost-effective, and capable of operating temperature in the range (30–120) °C. The improved performance of the fabricated device is found to be due to the extraordinary structural, morphological, thermal, and electronic properties of rGO-doped TiO_2 nanostructures. To measure the output characteristics of the device operating at elevated temperatures, i.e., at 90 °C or above may be a crucial issue because it may overheat the device and there would be a risk of burning the device. Therefore there would be a sleek safe window of operating temperatures of the fabricated electronic device. Finally, the temperature variable electronic phenomenon was discussed well via suitable band diagrams where it may be concluded that the type of nanostructures its dimensionality and the operating temperature, significantly affect the overall performance of the device.

Keywords Titanium-di-oxide (TiO_2) · Reduced graphene oxide (rGO) · Nanostructures · Temperature-dependent IV measurement

---

M. Azharuddin · R. Tabassum (✉)
Centre for Nanoscience and Nanotechnology, Jamia Millia Islamia (A Central University), New Delhi 110025, India
e-mail: rtabassum1@jmi.ac.in

Z. H. Khan et al. (eds.), *Recent Advances in Nanomaterials*, Springer Proceedings in Materials 27, https://doi.org/10.1007/978-981-99-4878-9_72

1 Introduction

Over the past few decades, researchers have indulged in carrying out thermal, electronic, and electrical measurements at ambient or at slightly lower/higher values of temperature. The electrical/electronic measurements at a higher temperature may lead to certain disadvantages/limitations which may ultimately hampers the device operation or its performance. The limited operation of the fabricated device at a certain elevated temperature may be observed in the form of its poor performance parameters like the poor figure of merit (FOM) and hence subsequent decrease in the overall performance of the device. Therefore, nowadays, several combinational materials have been fabricated which can further be used in devices to work in higher temperature regimes. In these detecting materials, several modifications have also been performed to further improve the performance of the device. The sensing materials may be anything from a single material to combinational material which can be used in various device applications. The researches have already been done to obtain the temperature-dependent properties of various single and combinational nanostructures to further fabricate/design devices for several diversified applications [1–6]. The motive of using nanostructured materials for device application is to get the improved performance of the device such as high sensitivity, ultrafast speed, and excellent chemical and thermal stability. Nowadays, researchers indulge in carrying out device applications using combinational nanostructures for several applications such as solar cells, various types of sensors, photodetectors, photocatalytic activities, biological tissue engineering [7–14]. Carbon-based nanomaterials such as carbon nanotubes and fullerenes are also used for the enhancement of the performance of devices [15]. Sometimes carbon-based nanomaterials in coordination with several oxides materials are also found to possess favorable properties for device applications [16].

Here, we report temperature variable input–output characteristics for device applications. The experimental work reports the fabrication and characterization of the rGO-doped TiO_2 nanostructures to use it further for the measurement of temperature-dependent I–V characteristics. The work is accomplished using the thermal transport behavior of electrons in nano dimension rGO-doped TiO_2 nanostructures at elevated temperatures because size-dependent nanostructures promote a comparatively higher rate of charge carrier transportation. Since the performance of the device can further be improved by using combinational nanostructures as detecting materials in comparison to their bare nanostructures therefore we researchers have employed combinational nanostructures as detecting materials for several electrical and electronic device applications. Here, pure rGO, pure TiO_2, and rGO-doped TiO_2 nanostructures were fabricated employing well-known hydrothermal method where three types of nanostructured samples were prepared; they are pure rGO, pure TiO_2, and rGO-doped TiO_2 nanostructures. After preparing such three samples, morphological characterization was performed using the scanning electron microscopic technique. After successful fabrication and characterization of pure rGO, pure TiO_2, and rGO-doped TiO_2 nanostructures, a device was prepared onto a glass substrate having silver electrodes on its two extreme ends. In between these Ag electrodes the synthesized samples were drop-casted separately by which three separate electrical devices were obtained for temperature variable current–voltage (I–V) measurements employing a two-probe experimental setup as discussed in our previous publications [6].

2 Materials and Methods

2.1 *Materials Used*

For the experimentation purpose, the bulk powder of titanium-di-oxide and N-Methyl-2-pyrrolidoneidone (NMP) were procured commercially from Sigma Aldrich. Both the materials were used in their pure form as obtained by the company many. For the experimental purpose, we have used de-ionized (DI) water which was obtained from an in-house Millipore system (~18. Planted intended our university.

2.2 *Fabrication of Titanium-Di-Oxide (TiO_2) Doped Reduced Graphene Oxide (rGO) Nanostructures Using the Hydrothermal Approach*

There are various methods of fabrication of nanocomposite of carbon-based material with oxide materials. Carbon-based nanomaterials such as reduced graphene oxides in coordination with TiO_2 can be fabricated using varied experimental techniques such as using chemical methods, ultrasound-assisted mechanical mixing [7, 14], chemical vapor deposition technique. These fabrication methods possess their own merits, demerits and limitations, one of the easy and cost-effective methods of fabrication of rGO and TiO_2 nanocomposite may be using a hydrothermal approach [17]. The schematic of the experimental method for the synthesis of pure rGO, pure TiO_2, and rGO-doped TiO_2 nanostructures is shown in Fig. 1. These three nanostructures were synthesized using the well-known hydrothermal method and then using these three nanostructures, a device for the measurement of temperature variable IV characteristics was also fabricated. It involves four-step process depicted well in Fig. 1. In a typical method, first of all, dispersion of a TiO_2, rGo-doped TiO_2 was prepared in NMP by continuous magnetic stirring ultrasonication. This completes step-1. Now, in step-II, three prepared dispersion solutions, viz. TiO_2, rGO, and rGO-doped TiO_2 dispersion, were put separately in the hydrothermal autoclave at 240 °C for 12 h. Then, we get three separate final suspensions of TiO_2 nanostructures, rGO nanostructures, and rGO-doped TiO_2 nanostructures. This completes step-II. Then, in step-III, these prepared three dispersions were drop-casted separately onto three separate glass slides to do the morphological characterizations. In the final step-IV, we have deposited silver electrodes on two ends of the already TiO_2 nanostructures,

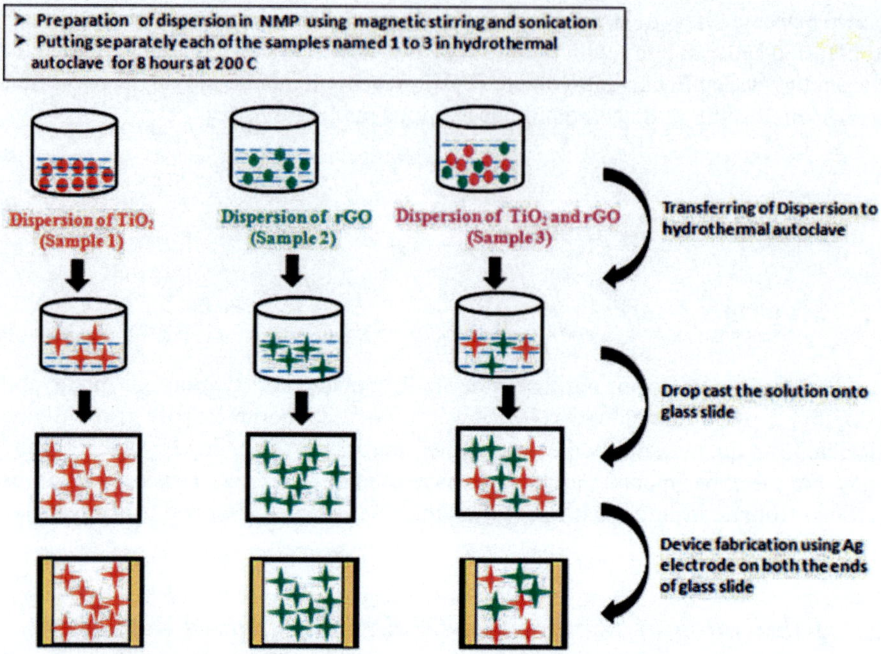

Fig. 1 Schematic of the synthesis of TiO_2 and rGo and rGO-doped TiO_2 nanoflower structures using the hydrothermal approach

rGO nanostructures, and rGO-doped TiO_2 nanostructures deposited glass slides. This is the final fabricated device to perform temperature-dependent current–voltage measurements by which we get the temperature variable electronic response of the fabricated device.

2.3 Assembling the Two-Probe Experimental Setup to Characterize the Device Having Pure TiO_2, Pure rGO, and RGO-Doped TiO_2 Nanostructures

Since the fabricated device having TiO_2 nanostructures, rGO nanostructures, and rGO-doped TiO_2 nanostructures were aimed to use them to measure the thermal transportation of the charge carriers within these detecting nanostructures, therefore we performed experiments on these three devices to obtain the current–voltage characteristics (V–I) on changing the temperature in the range (30–120) °C. The detailed representation of the experimental setup for the measurement of the temperature controlled I–V measurements shown in Fig. 2. By performing such experiments on I–V characteristics at varying temperatures, we have the evaluated the performance of the device on varying temperature regimes.

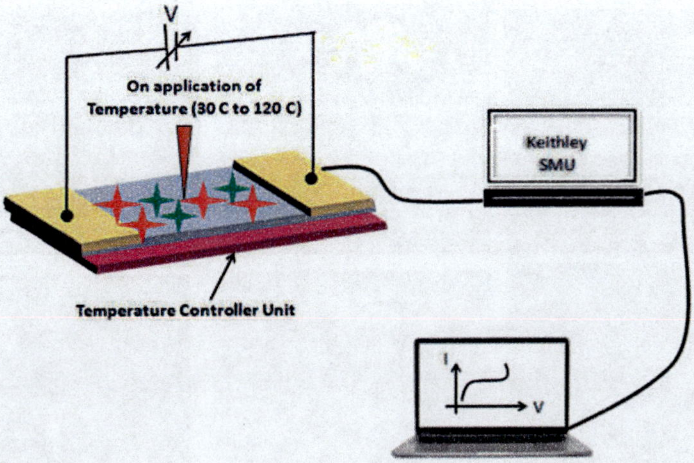

Fig. 2 Schematic of the experimental setup to obtain temperature variable IV characteristics of the device having TiO_2, rGO, and rGO-doped TiO_2 nanoflower structures

3 Results and Discussion

3.1 Scanning Electron Microscopy (SEM) Analysis

The crux of the temperature variable I–V measurements of the electrical device is based on the tailored/favorable properties of the synthesized nanostructures, therefore it is important to make a detailed discussion on the obtained morphology of TiO_2, rGO, and rGO-doped TiO_2 nanostructures. In this respect, first of all, we have observed the scanning electron microscopic image of the fabricated three samples discussed in Fig. 2. In Fig. 3a–c, we have displayed SEM images of the fabricated rGO nanostructures at three different magnifications where it may be noticed that a uniform morphology of the fabricated rGO nanostructures. Similarly, Fig. 3d–f shows the morphology of the fabricated TiO_2 where bunches of the flowers-like structure of TiO_2 at three different magnifications are clearly depicted. As magnifications increase from Fig. 3a–c changes, bunches of nanoflower get more clearly visible, and it gets converted into petal-like nanostructures. Now, as we study the rGO-doped TiO_2 samples, we can see the mixture of the morphologies of rGO and TiO_2 which are depicted well in Fig. 3g–i.

3.2 Temperature Variable Current–Voltage Characteristics

The main application of the fabricated rGO nanostructures, TiO_2 nanostructures, and rGO-doped TiO_2 nanostructures is to obtain their electrical characteristics [18].

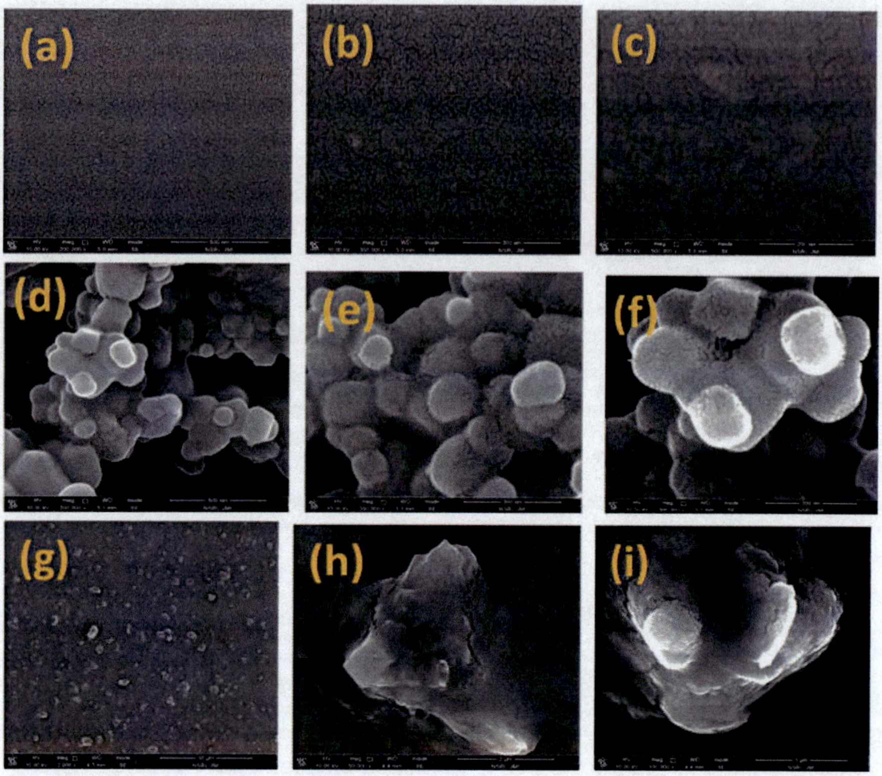

Fig. 3 SEM micrograph with different magnifications of the fabricated **a–c** pure rGO **d–f** pure TiO_2 and **g–i** rGO-doped TiO_2 nanostructures

Therefore to observe the temperature variable electrical characteristics of the fabricated device by the aforesaid these nanomaterials, we performed experiments to obtain characteristics for the I–V measurements under room temperature and above the room temperature. In this direction, here we performed experiments to obtain the current–voltage (I–V) characteristics on changing the temperature of the fabricated device. Here, an experimental setup for temperature variable I–V measurements is shown in Fig. 4. Figure 4a–c shows the temperature variable current–voltage characteristics of the fabricated device having pure rGO nanostructures, pure TiO_2 nanostructures, and rGO-doped TiO_2 nanomaterials. From the experimentally observed curves displayed in Fig. 4, it may be seen that for all the fabricated nanomaterials as the voltage varies − 12 to + 12, the current increases starting from minimum temperature 30 °C and as the temperature varies from 30 °C to 120 °C, there is an increase in output current for all three devices fabricated with pure rGO nanostructures, pure TiO_2 nanostructures, and rGO-doped TiO_2 nanostructures. There is a similarity in the trends of the I–V curves for changing temperature. The similarity in the trend is such that for the entire device as the voltage increases from 0 to

30 °C, the current increases linearly and the rate of increase in the output current is also low for lower values of voltage. As the voltage increases from 9 to 12 V, the current starts saturating and acquired a certain higher value of current. The reason behind such a trend in the V–I characteristics for all three fabricated devices is that when the temperature low says 30 °C and while keeping the voltage low the charge carriers within the detecting nanostructures were found to be limited consequently the output current was low. For such a lower temperature, i.e., 30 °C as the voltage increased the corresponding output current also gets increased accordingly. Now, as the temperature of the detecting nanostructures of the device increases from 30 °C to 120 °C, then in addition to the charge carriers via biasing, charge carriers generated due increased temperature also contributed significantly which ultimately increases the output current; this process continues but at higher voltage, the charge carriers due to biasing and the thermally generated carriers both get saturated as a result the current also gets saturated. This phenomenon is found to be in all the devices fabricated with pure rGO nanostructures, pure TiO_2 nanostructures, and rGO-doped TiO_2 nanostructures.

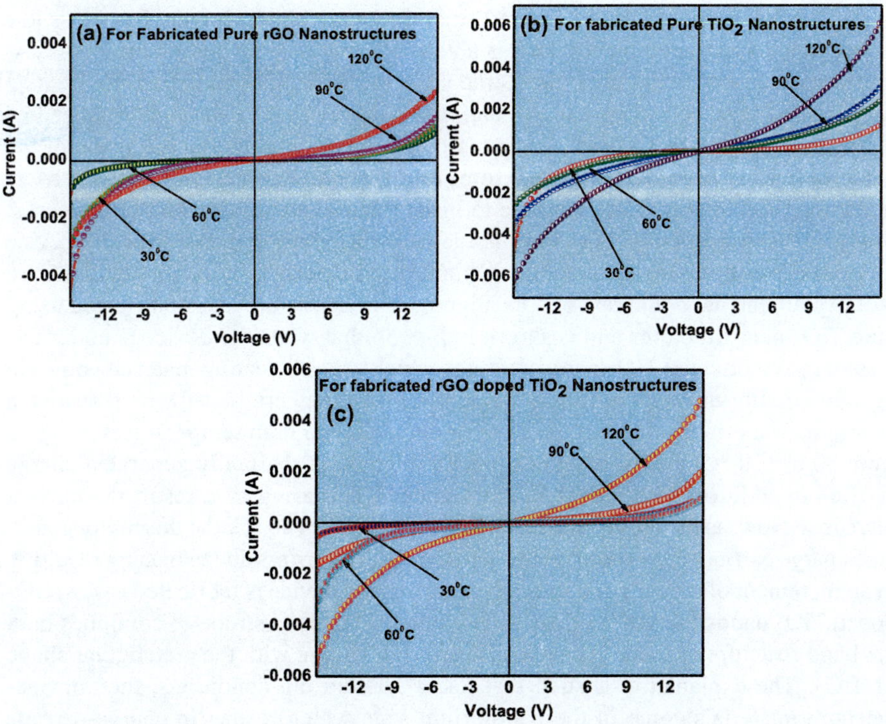

Fig. 4 The temperature-dependent IV measurements of the device fabricated with TiO_2, rGO, and rGO-doped TiO_2nanoflower structures

3.3 Band Diagram Representation to Explain the Mechanism of Temperature Variable Current–Voltage Characteristics

The working of this device to obtain temperature variable current–voltage characteristics can be discussed using the important physical parameters of the synthesized detecting nanostructures. The three main parameters of the synthesized pure rGO nanostructures, pure TiO_2 nanostructures, and rGO-doped TiO_2 nanostructures are Fermi energy (EF), work function (W), and electron affinity (χ). These are the main parameters on which the band bending and the direction of charge-carriers (e and h) transportation depend. Figure 5a–c shows the energy band diagram of the three fabricated nanostructures and the direction of flow of electrons on the application of temperatures are depicted in detail. Since it is well-known that there is a drastic difference in the shape of the energy bands of TiO_2 and rGO. Therefore to explain one by one, we consider first the band diagram and its alignment when the device is fabricated using pure rGO nanostructures where it may be understood from Fig. 5a that under the DC biasing condition of the device, the nanostructures come in the non-equilibrium state which results in splitting of the Fermi level. Since, in the hydrothermal technique, the rGO nanostructures were fabricated which possess lots of graphene layers and possess defects levels [19] that help in charge-carriers transportation at low temperatures, the Dirac cone shape of the band structure of rGO supports the flow of charge carriers and, the output current is obtained under low temperatures upto 30 °C. Now, as the temperature varies, although the shape of the band is same as Dirac cone, there are some more thermally generated charge carriers in addition to the charge carriers due to biasing consequently the output increases. This is the explanation of Fig. 5a of the I-V characteristics obtained for the device fabricated using rGO nanostructures. Similarly, for pure TiO_2 nanostructures, we will explain the band diagram and its alignment when the device is fabricated using pure TiO_2 nanostructures and is shown in Fig. 5b that when the device is under DC biasing, the fabricated TiO_2 nanostructures which act as detecting material comes in the non-equilibrium state resulting in the splitting of the Fermi level which results in charge-carriers transportation. On application of low to high temperatures, ranging from 30 to 120 °C, the band again supports the flow of thermally generated charge carriers in addition with charge carriers under DC biasing as a result the current increases. Now, as the temperature increases from 30 to 120 °C, the thermally generated charge carriers increase in addition to the charge carriers due to biasing resulting in an increment of the output current. Now, when the device is fabricated using rGO-doped TiO_2 nanostructures as detecting materials, the band structure combines both the band structure of the Dirac cone shape of rGO along with the rectangular shape of TiO_2. The explanation is such that under DC biasing conditions, then device-detecting materials come in the equilibrium state which results in charge-carriers transportation. Since the rGO contains a large number of graphene layers and due to which the defects levels increase to increase the introduction of rGO in TiO_2, the defects levels of overall rGO-doped TiO_2 nanostructures, the defects levels again increase which hugely supports the flow of charge carriers from the valence band to the conduction band, as a result, the output current increases under low temperature. Now, on increasing the temperature, there is a huge number of thermally generated charge carriers in addition to the charge carriers due to biasing consequently the output increases. This is the explanation of Fig. 5c of the I–V characteristics obtained for the device fabricated using rGO-doped TiO_2nanostructures.

4 Conclusion

In summary, in this experimental report, we have fabricated and characterized pure rGO nanostructures, pure TiO_2 nanostructures, and rGO-doped TiO_2 nanostructures for device application. All three single and combinational nanostructures were fabricated using the same hydrothermal experimental technique. The morphological characterizations of all three nanostructures have been performed and extensive studies have been performed to justify their morphology to use them further in temperature-dependent device applications. The application of the synthesized pure rGO nanostructures, pure TiO_2 nanostructures, and rGO-doped TiO_2 nanostructures was evaluated from temperature-dependent current–voltage measurements. From morphological characterization, it was found that nanostructures of rGO, TiO_2, and rGO-doped TiO_2 can be fabricated employing well-known hydrothermal techniques under controlled experimental conditions. First, the morphology of the synthesized aforesaid nanomaterials were justified, and the measurement of current–voltage (V–I) characteristics under changing temperature varies from 30 °C to 120 °C performed. From temperature-dependent current–voltage (V–I) characteristics, it has been found that for all the temperatures in the current–voltage (V–I) characteristics, the current varies non-linearly with voltage from 0 to 12 V. For lower values of temperature, i.e., 30 °C, as the voltage increases, the rate of increase in the output current is low but for higher values of temperature from 60 to 120 °C, the rate of increase of the output current was found to be high. The increment in the output current is obtained to be low for the device fabricated using pure rGO nanostructures and pure TiO_2 nanostructures pure but is found to be very high for the device fabricated using rGO-doped TiO_2 nanostructures. The reason for such a trend in the current–voltage (V–I) characteristics under varying temperatures was explained using a suitable band diagram. In this way, from the findings/analysis of the performed experimentations, it may be concluded that the fabricated pure rGO nanostructures, pure TiO_2 nanostructures, and rGO-doped TiO_2 nanostructures possess all the favorable properties which can further be successfully used in the study of temperature-dependent I–V characteristics.

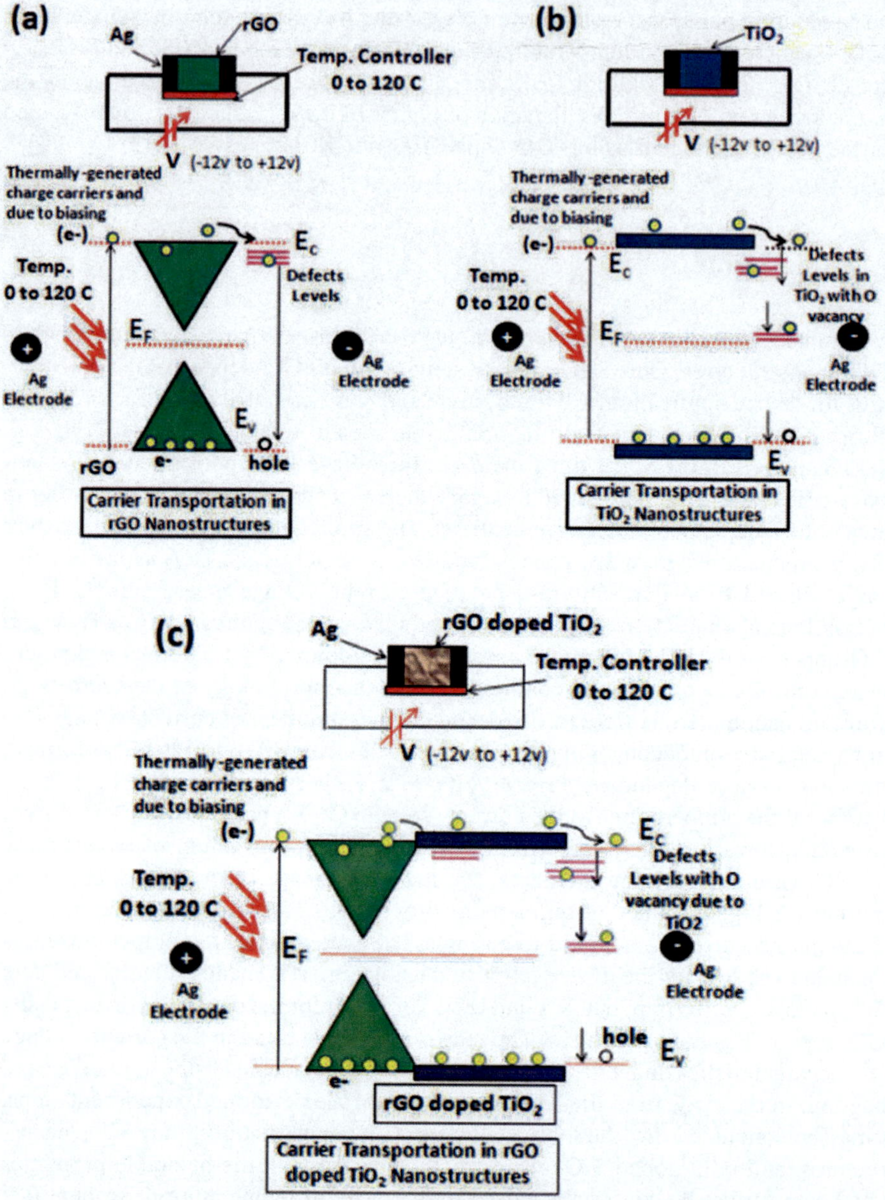

Fig. 5 Pictorial representation of the band diagram explaining the temperature-dependent IV phenomenon in the device fabricated with TiO_2, rGO, and rGO-doped TiO_2 nanostructures

Acknowledgements Dr. Rana Tabassum is grateful to the Department of Science and Technology (DST) for INSPIRE Faculty scheme and Science and Engineering Research Board for SERB-Research Scientist fellowships and Research grants.

Declaration of Interest Statement The authors declare that they have no conflict of interest.

References

1. Haque A, Abdullah-Al Mamun M, Taufique MFN, Karnati P, Ghosh K (2018) TempTemperature-dependent trial transport properties of high carrier mobility reduced graphene oxide thin film devices. IEEE Trans Semicond Manuf 31(4):535–544
2. Haque A, Abdullah-Al Mamun M, Taufique MFN, Karnati P, Ghosh K (2018) Temperature-dependent electrical transport properties of high carrier mobility reduced graphene oxide thin film devices. IEEE Trans Semicond Manuf 31(4):535–544
3. Nam WH, Lim YS, Kim W, Seo HK, Dae KS, Lee S, Seo W-S, Lee JY (2017) A gigantically increased ratio of electrical to thermal conductivity and synergistically enhanced thermoelectric properties in interface-controlled TiO_2–RGO nanocomposites. Nanoscale 9(23):7830–7838
4. Ahmed MAM, Meyer W, Nel J (n.d.) Investigation of the structural and temperature dependent electrical properties of Mzno (M= CE and Sm) Schottky diode devices fabricated using the sol-gel spin coating technique. Available at SSRN 4141027
5. Ahmad F, Azharuddin M, Bansal J, Tabassum R, Hafiz AK (2022) Understanding of temperature-dependent photoluminescence in graphite and SixZnO(1–x) tri-composite nanostructure. Opt Mater 134:113165
6. Rahman S, Azharuddin M, Bansal J, Bilal M, Tabassum R, Hafiz AK (2023) Role of temperature on CdS and MoS_2 doped SnO_2 nanostructures: potential applications in photodetection and temperature dependent current-voltage characteristics. J Alloy Compd 941:168901
7. AlShammari AS, Halim MM, Yam FK, Kaus NHM (2020) Effect of precursor concentration on the performance of UV photodetector using TiO_2/reduced graphene oxide (rGO) nanocomposite. Results Phys 19:103630
8. Aqeel M, Anjum S, Imran M, Ikram M, Majeed H, Naz M, Ali S, Ahmad MA (2019) TiO_2@ RGO (reduced graphene oxide) doped nanoparticles demonstrated improved photocatalytic activity. Mater Res Express 6(8):086215
9. Rostami M (2019) Photodecomposition and adsorption of hazardous organic pollutants by Ce-doped ZnO@ Ce-doped TiO_2-N/S-dual doped RGO ternary nano-composites photocatalyst for water remediation. J Mol Struct 1185:191–199
10. Ghosal K, Mondal P, Bera S, Ghosh S (2021) Graphene family nanomaterials-opportunities and challenges in tissue engineering applications. FlatChem 30:100315
11. Daneshnazar M, Jaleh B, Eslamipanah M, Varma RS (2022) Optical and gas sensing properties of TiO_2/RGO for methanol, ethanol and acetone vapors. Inorg Chem Commun 145:110014
12. Liu H, Zou X, Chen Q, Fan W, Gong Z (2022) Pumice-loaded rGO@ MnO_2 nanomesh photocatalyst with visible light response for rapid degradation of ciprofloxacin. Sep Purif Technol 297:121502
13. Benyakhlef M, Bensouici F, Bououdina M, Dakhel AA, Tala-Ighil R, Toubane M (2020) Surface, structural and optical properties dependence of Fe-doped TiO_2 films deposited onto soda–lime–glass. Surf Interfaces 21:100682
14. Marichy C, Russo PA, Latino M, Tessonnier J-P, Willinger M-G, Donato N, Neri G, Pinna N (2013) Tin dioxide–carbon heterostructures applied to gas sensing: structure-dependent properties and general sensing mechanism. J Phys Chem C 117(38):19729–19739
15. Galindo Lorente S (2015) Studies on organic solar cells based on small-molecules: tetraphenyldibenzoperiflanthene and fullerene C70

16. Cuong DLT, Trinh DN, Linh NTT, Hai ND, Chau PTL, Hoang NT, Phong MT, Hieu NH (2022) Synthesis of titanium dioxide/reduced graphene oxide nanocomposite by ultrasound-assisted mechanical mixing method for fabricating photo anode to upgrade the performance and stability of dye-sensitized solar cell. Mater Res Bull 156:112000
17. Dang CLT, van Le C, Le NTT, Nguyen MTT, Tran DC, Pham KT, Nguyen HT, Pham CTL, Hoang NM, Mai PT (2021) Synthesis of titanium dioxide/reduced graphene oxide nanocomposite material via the incorporated hydrothermal co-precipitation method for fabricating photoanode in dye-sensitized solar cell. Synth Met 281:116919
18. Raval B, Mahapatra S, Banerjee I (2019) Processing of graphene oxide for enhanced electrical properties. In: Advanced battery materials, pp 613–644
19. Terrones H, Lv R, Terrones M, Dresselhaus MS (2012) The role of defects and doping in 2D graphene sheets and 1D nanoribbons. Rep Prog Phys 75(6):062501

Synthesis, Characterization and Photocatalytic Application of GO/Fe_3O_4 Nanocomposite

S. Shreya, S. Rahul, Si. Rahul, J. Noor, and K. Jasmeet

Abstract Magnetic nanocomposites are being widely used to increase the recovery rate of the photocatalysts. Magnetite (Fe_3O_4) and Graphene Oxide (GO) nanocomposites can prove to be promising in degrading the organic dyes and reduce the water pollution. Photocatalysis has been used for the treatment of waste water and has proved to be promising for solving environmental remediation effectively. To improve the recovery rate of photocatalysts and increase the reusability of photocatalysts, magnetic nanocomposite of GO/Fe_3O_4 was used. Modified hummer's method was employed to prepare Graphene Oxide (GO) and the nanocomposite of GO/Fe_3O_4 was prepared using hydrothermal process. The structural and morphological characteristics were studied using XRD, FTIR, FESEM and UV–VIS spectrophotometer. XRD results demonstrated that nanocomposite had formed and the formation of spinel structure of Fe_3O_4. The presence of functional groups was examined using FTIR Spectroscopy. FESEM gave the confirmation of cubic structure of Fe_3O_4 at different magnifications. Bandgap of synthesized nanocomposite was calculated using absorbance through UV–VIS spectrophotometer. Finally, photocatalytic degradation of organic methyl orange dye was studied.

Keywords Photocatalysis · Graphene oxide (GO) · Magnetite (Fe_3O_4) · Methyl orange (MO)

1 Introduction

Industries like textile and paper industry uses harmful dyes which affect the marine life as well as human health. Using such dyes on regular basis and disposing them into water results in some serious issues. These toxic chemicals need to be reduced into

S. Shreya · S. Rahul · Si. Rahul · J. Noor (✉)
Department of Physics and Astronomical Sciences, Central University of Himachal Pradesh, Shahpur, Kangra, H.P 176206, India
e-mail: noor.jahan@hpcu.ac.in

K. Jasmeet
Department of Physics, Sri Guru Gobind Singh College, Sector 26, Chandigarh 160019, India

Z. H. Khan et al. (eds.), *Recent Advances in Nanomaterials*, Springer Proceedings in Materials 27, https://doi.org/10.1007/978-981-99-4878-9_73

biodegradable ones [1]. The wastewater can be treated using photocatalysts which can decompose these toxic chemicals into mineralized ones.

Before discussing about photocatalysis one must know about AOP, i.e. Advanced Oxidation Processes–one used to remove toxic chemicals from water [2]. Among all the AOPs, photocatalysis has gained much attention. In this process, when natural sunlight is irradiated on to the surface of photocatalysts there is electron–hole pairs generation. The electrons present in valence band absorb the energy $\geq$ to the band gap of material and get excited into the conduction band. Creation of hydroxyl and superoxide radicals decompose dye molecules and degrade them into less toxic minerals. Main hindrance in use of these photocatalysts is their recovery which can be improved by using magnetically separable nanoparticles.

The present work will focus on the synthesis of magnetically separable nanoparticles which can be easily recovered and reused. This will ultimately lower the cost of their production and will degrade toxic chemicals into biodegradable ones. The photocatalytic activity will be studied under natural sun light irradiation.

2 Materials and Methods

2.1 *Chemicals and Reagents*

Potassium Permanganate ($KMnO_4$), Graphite Powder, Sulphuric Acid (H_2SO_4), Hydrochloric Acid (HCl), Hydrogen Peroxide (H_2O_2), Iron Sulphate ($FeSO_4$), Distilled Water were used without any further purification.

2.2 *Preparation of Graphene Oxide (GO) Using Modified Hummer's Method*

In this process, 90 mL of H_2SO_4 was added into 3 g of graphite powder. Ice bath setup was prepared and 12 g of $KMnO_4$ was added. After getting green colour, slowly 100 mL of water was added and stirred for 30 min. After this, 30 mL of H_2O_2 was added and stirred for an hour. Then, 100 mL of water was added instantly and solution was left for overnight. Next day, 10 mL of HCl was added and stirred. Final solution was filtered. Prepared GO was washed with 5 mL of H_2O_2 and 200 mL of distilled water along with 200 mL of distilled water and 10 mL of diluted HCl several times.

2.3 Preparation of Graphene Oxide/Fe_3O_4 Nanocomposite by Hydrothermal Process

2 g of $FeSO_4$ was added into 50 mL of distilled water and stirred on magnetic stirrer. 70 mg of prepared GO was added to this solution and mixed with magnetic stirrer for 10 min. The prepared solution was sonicated for 30 min and transferred to Teflon-lined autoclave. The sample was heated for 10 h at 180 °C. Obtained magnetic GO/ Fe_3O_4 nanocomposite was washed, dried and grinded.

3 Results and Discussion

X-ray Diffraction-

X-ray diffraction peaks observed at 2theta values of 30.04°, 35.4°, 37°, 43°, 57°, 62.54° can be assigned to diffraction of (111), (311), (222), (400), (511), (440) planes as shown in Fig. 1. The observed peaks were found to be matching with the standard XRD data of spinel structure of Fe_3O_4 (JCPDS card no.- 89–4319) [3]. Absence of characteristic peak of graphite at 26.4° confirmed the successful deposition of Fe_3O_4 on graphene surface. Scherrer equation was used to calculate the crystallite size of 33 nm given as $D = k\lambda/\beta_{hkl}Cos\theta$, where β_{hkl} is full width half maximum, k is 0.9, θ is the angle of diffraction and λ is the X-rays wavelength [4].

FTIR Spectroscopy

FTIR spectra given below in Fig. 2 depicts the presence of functional groups like O–H bond, C = C bond, O = C-O bond and Fe–O bond. Observed peaks at 3600 cm^{-1} indicate the stretching and bending bands of -OH groups adsorbed on the surface. Vibrations present within the range of 540–600 cm^{-1} confirming the presence of iron (Fe) in the prepared nanocomposite [5].

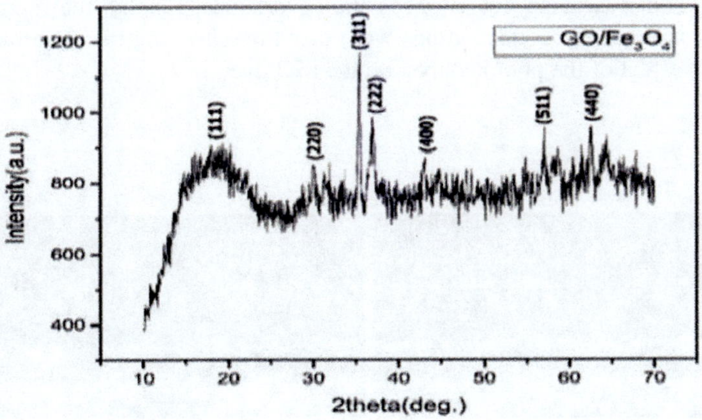

Fig. 1 XRD spectrum of GO/Fe_3O_4

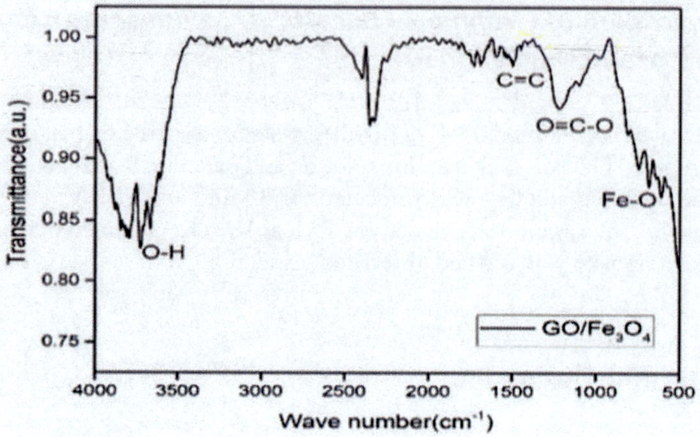

Fig. 2 FTIR Spectra of GO/Fe_3O_4

FESEM Characterization

FESEM gave the topographical and morphological information of the nanocomposite formed at magnifications of 20 and 50 kx. The image shows that Fe_3O_4 nanoparticles with cubic structures are uniformly distributed on the GO sheets (Fig. 3).

Photocatalytic Degradation

Photocatalytic activity of GO/Fe_3O_4 was used to check the degradation of MO dye under natural sunlight illumination. In this experiment, few drops of MO dye (10^{-5} M) were added into 80 mL of distilled water. The solution was sonicated and 3 mL of this solution was taken out by checking the absorbance using double beam UV–VIS spectrophotometer (Fig. 4).

Then, photocatalyst was added into the dye solution and kept in dark for 120 min under continuous stirring to maintain adsorption–desorption equilibrium. Then 3 mL of this solution was taken out to check the absorbance. Finally, the mixture was kept under sunlight and four readings were taken by checking the absorbance and degradation effect of the photocatalyst on the MO dye.

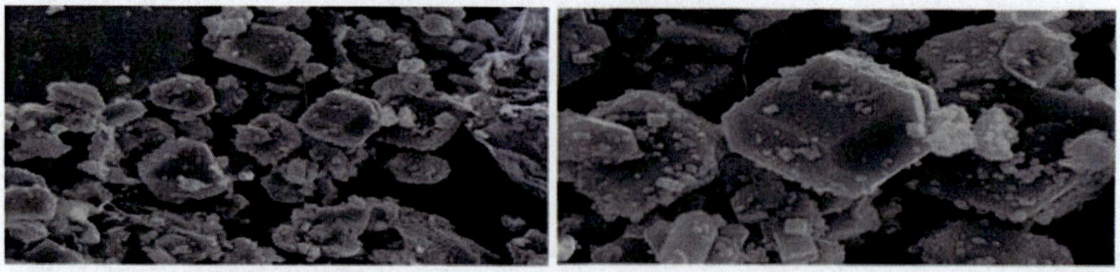

Fig. 3 FESEM micrograph of GO/Fe_3O_4

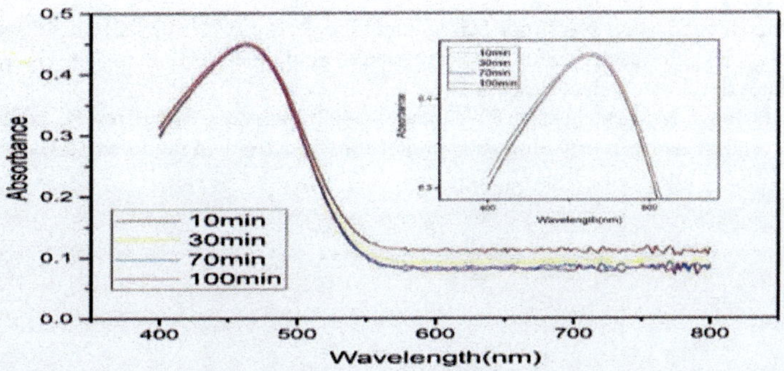

Fig. 4 UV–VIS spectra of methyl orange dye

4 Conclusion

In this paper, we have synthesized GO using modified hummer's method. The nanocomposite of GO/Fe_3O_4 has been prepared using hydrothermal method. The characterization techniques like XRD, FTIR, FESEM and UV–VIS spectroscopy were performed. XRD spectrum and FTIR confirmed the formation of GO/Fe_3O_4 nanocomposite. The photocatalytic activity showed the degradation of MO dye to some extent which may be due to the evaporation of water content from the solution. Large surface area of GO helps in removing dye molecules through adsorption and functional groups containing oxygen adhered to the surface of GO prevents the aggregation of nanoparticles.

For future work, the degradation efficiency of MO dye can be improved by changing the operational parameters in order to mineralize the pollutants completely. The synthesized magnetic nanocomposite of GO/Fe_3O_4 can be used for many cycles and their degradation efficiencies can be compared.

Acknowledgements Authors truly acknowledge the research facility provided by the Central University of Himachal Pradesh, Shahpur, H.P. and UGC BSR grant.

Declaration of Interest Statement Authors affirm that they are clear of any known conflicts of interest.

References

1. Khatri J et al. (2018) Advanced oxidation processes based on zero-valent aluminium for treatingtextile wastewater. Chem Eng J Elsevier B.V. 348:67–73
2. Legrini O, Oliveros E, Braun AM (1993) Photochemical processes for water treatment. Chem Rev 93:671–698
3. Cheng M, Zeng G, Huang D, Lai C, Xu P, Zhang C, Liu Y (2016) Hydroxyl radicals based advanced oxidation processes (AOPs) for remediation of soils contaminated with organic compounds: a review. Chem Eng J 284:582–598

4. Yilmaz M, Mengelizadeh N, Saloot MK, Shahbaksh S, Balarak D (2022) Facile synthesis of Fe_3O_4/ZnO/GO photocatalysts for decolorization of acid blue 113 under solar, visible and UV lights. Mater Sci Semiconduct Process 144
5. Uzunova-Bujnova M, Kralchevska R, Milanova M, Todorovska R, Hristov D, Todorovsky D (2010) Crystal structure, morphology and photocatalytic activity of modified TiO_2 and of spray-deposited TiO_2 films. Catal Today 151:14–20

Structural and Optical Properties of Reduced Graphene Oxide

Ankita, Umang Berwal, and Vinod Singh

Abstract A wide range of doped materials, including fluorescence, optical fibres, lasers, and heterojunction PV devices, have many uses for carbon-based materials due to their intriguing structural and optical characteristics. Graphene has a special character of high electron transfer rate, this attribute leads to structural modifications, enhancing the optical properties of various substances. Here, using a variety of characterisation approaches, we have evaluated the methods used to create graphene-based nanoparticles and emphasised their structural (X-ray diffraction (XRD), Raman spectroscopy and scanning electron microscopy (SEM)) and optical properties (UV–Visible spectroscopy). The structural and optical qualities of diverse materials are improved with an increase in the concentration of reduced graphene oxide (rGO). The diffraction peak of rGO in the XRD diffraction pattern was virtually centred at $2\theta = 26°$, indicating the crystalline character of rGO. Additionally, as rGO concentration grows, existing grain boundaries in FESEM characterisation increases as a result of electron–hole recombination, and these GO nanoparticles also intensify the G and D bands in Raman spectra. It is obvious that altering the oxidation level of graphene, which can be demonstrated using the UV–Visible spectroscopic characterisation method, can result in significant control over the band gap. The increase in surface roughness of rGO doped material leads to an increase in surface area of rGO doped samples, making photon absorption easier and increasing the rate of light absorption, which is why the UV–Vis absorbance spectrum peak's intensity rises at 255 nm. Because rGO can readily tune the band gap of the material, there have been a wide range of futuristic research opportunities due to the multiple applications based on its optical features.

Keywords Reduced graphene oxide · XRD · Raman spectroscopy · UV–visible spectroscopy

---

Ankita · U. Berwal · V. Singh (✉)
Departmentof Applied Physics, Delhi Technological University, New Delhi 110042, India
e-mail: vinodsingh@dtu.ac.in

Z. H. Khan et al. (eds.), *Recent Advances in Nanomaterials*, Springer Proceedings in Materials 27, https://doi.org/10.1007/978-981-99-4878-9_74

1 Introduction

Phillip Russel Wallace created a theoretical idea in the 1940s that would allow us to build single-atom carbon structures. This substance has been the focus of arduous research for more than 60 years and is regarded as having the potential to completely transform various sectors. But for many years, this theory had been disproven by numerous scientists due to its low electrical and thermal conductivity [10]. Graphene is essentially a monolayer of securely bound carbon atoms in a two-dimensional hexagonal honeycomb lattice, which is an allotrope of carbon atoms. Strong σ-bonds help connect each atom in the graphene sheet to its closest neighbours. Carbon nanotubes also exhibit this type of bonding [44]. In essence, graphene is a superior thermal and electrical conductor with intriguing optical and light-absorbing capabilities. It is lightweight, flexible, chemically inert, and it conducts electrical and thermal energy. There are several photonics uses for graphene, ranging from transparent conductors in photonic devices to high-bandwidth photodetectors, due to its special optical properties that allow it to absorb a wide range of electromagnetic radiation [19].

In essence, graphene oxide is an oxidised form of the material that has been infused with oxygen-containing groups. It is thought to be simple to process because it disperses in water and even in other solvents. GO is commonly available in powered or coated form for substrates [5]. Any substrate may be used for the deposition of graphene oxide layers, which will eventually be transformed into conductors. Thus, GO is employed in the fabrication of transparent conductive films, such as those utilised in sensors [9], solar cells [21], and a variety of other devices. Batteries [46], capacitors [38], and solar panels [40] all use graphene oxide due to its large surface area. Since graphene oxide is more accessible and less expensive to produce than graphene, it may eventually be put into mass production. GO can be blended with other polymers and substances to improve the tensile strength, elasticity, conductivity, and other qualities of substances [14]. Although graphene oxide is not a good conductor, there are two methods to improve its characteristics by doping some other element or by reducing it.

So, for the present study we are focusing on the reduced graphene oxide to improve its conductivity by tuning the band gap. Reduced graphene oxide might be a good option if graphene is going to be used for large-scale applications like energy storage because it is easier to make in big quantities than ideal single-layer or few-layer graphene and the quality is good enough for these bulk material applications [8].

There are several techniques for reducing graphene oxide including chemical, thermal, electrochemical, IR or UV irradiation to obtain a powdered form. It mainly has a dark colour, has no smell, and is insoluble [27]. Applications involving bulk materials, like energy storage, are well suited to it. Graphene research, batteries [49], biomedicine [45], supercapacitors [25], and printable graphene electronics [50] are some of its application areas. Research institutions all over the world are seeking for new approaches to producing superior graphene sheets. Graphene oxide (GO) to create reduced graphene oxide can accomplish this (rGO). How closely a structure resembles flawless graphene is produced depends greatly on the precise process parameters that are employed to reduce GO into rGO.

Chemical reduction is a scalable way for reducing GO, however the rGO produced by this technique frequently has poor electrical conductivity and surface area [32]. On the other hand, thermally decreasing GO at temperatures exceeding 1000 °C results in rGO, which has a large surface area and resembles pure graphene. The heating procedure damages the structure of the graphene platelets, which has a considerable negative influence on mass loss and could weaken the mechanical strength of the rGO [13]. According to some reports, oxygen plasma treatment can make graphene luminous. Similar to this, enhanced electronic structures in chemically produced graphene oxide (GO) display a wide photoluminescence [31]. GO is far more chaotic and insulating than crystalline grapheme [34]. However, the accessibility of diverse oxygen-containing functional groups with both sp^2 and sp^3 carbon sites makes them desirable for the synthesis of large-scale graphene manufacture using appropriate reduction kinetics. By choosing an appropriate reduction procedure for GO, it is possible to change the properties of graphene, causing it to go from an insulator to a semiconductor and then to a metal. As a result, it is anticipated that by reducing graphene oxide (GO) into reduced graphene oxide sheets (rGO), the optical properties of graphene sheets can be controlled by adjusting the sp2 to sp3 ratio [18].

In this paper, we are reviewing modified Hummer's method in order to produce graphite oxide in which we firstly treated graphite powder with concentrated sulfuric acid and potassium permanganate, to produce graphite oxide which alters the structural and optical properties that leads to changes in the band gap. The process is quite easy and have low risk, so that it can be done in college laboratory. Graphite oxide was then sonicated and reduced into reduced graphene oxide (rGO) using hydrazine hydrate [12]. The obtained samples were then studied using various characterisation techniques to study their structural (X-ray diffraction (XRD), Raman spectroscopy and scanning electron microscopy (SEM)) and optical properties (UV-Visible spectroscopy). The main objective of this work is to synthesize reduced graphene oxide and study its various properties.

2 Materials and Methodology

2.1 Materials

Graphite Powder, Potassium Permanganate, Sulphuric Acid, Hydrazine Hydrate, Orthophosphoric Acid, Hydrochloric Acid, Hydrogen Peroxide, Cobalt (III)Nitrate Hexahydrate, Magnesium Nitrate, Ammonium Hepta Molybdate, Aluminium Oxide.

2.2 Methodology

GO Preparation. GO was created using a modified version of Hummer's approach from graphite flakes. To begin with, 50 mL of concentrated sulfuric acid and 2 gm of graphite flakes were combined. The liquid was chilled using an ice-water bath while being continuously stirred, and the suspension was formed. Potassium permanganate ($KMnO_4$), an oxidising agent, was moderately introduced into the suspension at a regulated temperature of below 10 °C. The suspension was then agitated for 25 min at room temperature before being sonicated for 8 min in an ultrasonic bath. After swirling and sonicating the mixture ten times, 250 mL of distilled water was added. Then, after being subjected to ultrasonication for one hour to exfoliate the graphite oxide, 30 mL of hydrogen peroxide (H_2O_2) was added. After centrifuging the mixture at 9000 rpm, GO was repeatedly cleaned using distilled water and 2 M HCl (to eliminate excess metal ions) until the pH reached 7. The GO precipitates were dried for 24 h at room temperature [35].

rGO Preparation. The above suspension was once more sonicated for two hours to produce rGO. At room temperature, hydrazine hydrate was gradually added up drop by drop to the exfoliated GO suspension. The reduction process was carried out for one hour at 100 C. For this sample, hydrazine hydrate and GO were maintained at a weight ratio of 9:7. The resulting black precipitates were removed through cellulose filter paper and rinsed with distilled water and a 1 M HCl solution to achieve a pH of zero. The filtrated product was then dried for 24 h at ambient temperature to produce rGO [35] (Fig. 1).

Fig. 1 Flowchart illustrating the synthesis of rGO

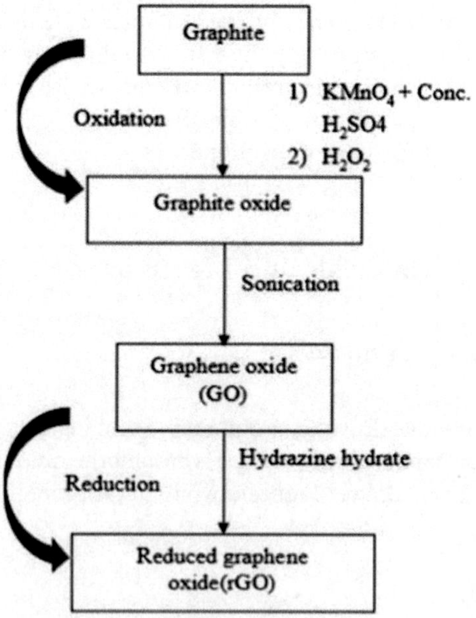

3 Results and Discussion

In this work, we have reviewed the synthesis of rGO and analyse its structural and optical properties. The colour change from brown to black, served as a visual indicator of the reduction reaction converting graphene oxide into graphene sheets. The rGO material's dark colour implied that the exfoliated GO had been re-graphitized by the removal of oxygenated functional groups.

3.1 *Structural Properties*

X-Ray Diffraction (XRD) Analysis. Figure 2 displays the GO and rGO XRD patterns. XRD data were obtained using a copper rotating anode and an incident beam with a wavelength of 1.54. Data acquisition employed the reflection mode. Using Bragg's relation, the d-spacing of the peak with the highest intensity was determined [35].

$$2d \sin \theta = n\lambda \quad (1)$$

where n is an integer, λ is the copper target's X-ray wavelength, and θ is the angle formed by the incident and reflected rays. 1.45 nm is the computed d-spacing from the primary peak of GO in the (001) plane, higher than the d-spacing produced in formation of rGO (0.379 nm). The hydroxyl, epoxy, and carbonyl groups in GO are to blame for this. When oxygen-containing functional groups are removed from rGO, the d-spacing decreases, indicating that the sp^2 network has been restored after reduction. The broad rGO peak at 21419.6 could be the result of insufficient oxidation.

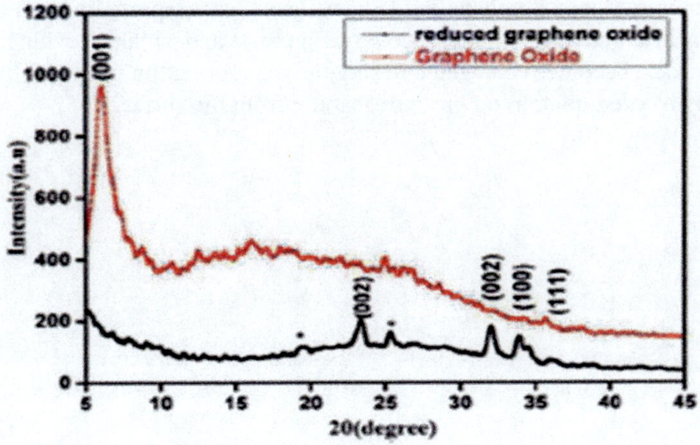

Fig. 2 XRD spectra of rGO and GO [35]

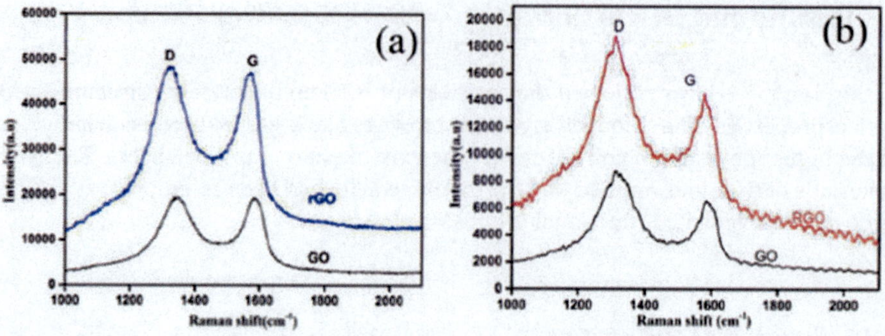

Fig. 3 Raman spectra of GO and rGO at **a** 532 nm and **b** 785 nm [35]

Due to the oxidation of graphite, the diffraction peak of GO shifted towards $2\theta = 10°$. The XRD pattern of rGO, which exhibits a large diffraction peak at $2\theta = 26°$, further supports the reduction of graphene oxide into rGO. The graphene oxide's total reduction into rGO is further supported by the peak of the material at $2\theta = 10°$ completely disappearing and the creation of a new broad peak at $2\theta = 26°$ [35].

Raman Spectroscopy Analysis. The Raman spectra of GO and rGO at two laser wavelengths, 532 and 785 nm, are shown in Fig. 3a and b. The oxidation and reduction processes caused structural changes in the materials, which were seen using Raman spectroscopy. Two distinct characteristic bands for GO can be seen in spectrum 5(a) at 1347 cm^{-1} (D band) and 1587 cm^{-1} (G band), which are shifted to 1324 cm^{-1} (D band) and 1573 cm^{-1} (G band) following reduction. The increase in flaws causes the relative intensity ratio ID/IG to rise from 1.80 to 1.84 after reduction [35].

Field Emission Scanning Electron Microscope (FESEM) Analysis. The surface morphology of GO and rGO was examined using FESEM to see how the corrugation changed. In contrast to GO, the extremely wrinkly structure of rGO is seen in Fig. 4b. Highly wrinkly reduced graphene oxide nanosheets have reportedly been found to be more sensitive and exhibit better recovery. Figure 4a and b make it evident that the sample's surface becomes corrugated during the transformation from GO to rGO. It is caused by oxygen-containing groups escaping from the surface [35].

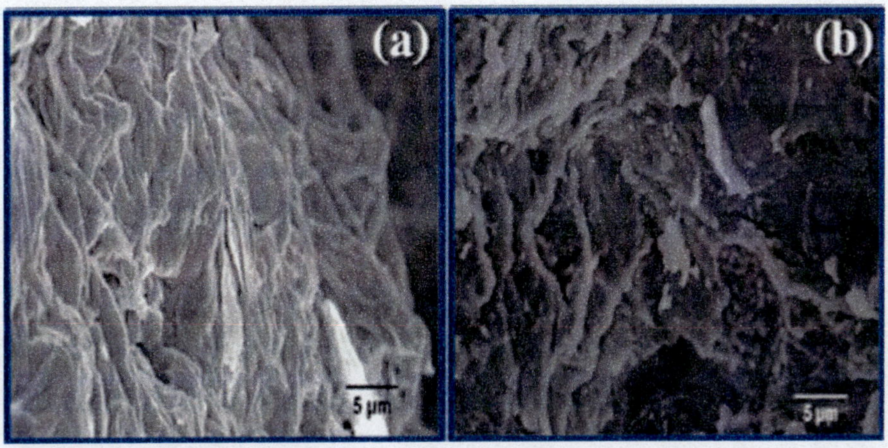

Fig. 4 FESEM image of **a** GO and **b** rGO [35]

3.2 *Optical Properties*

3.2.1 UV-Vis Spectroscopy

Figure 5 displays the UV-Visible spectra of XGO and RGO. The primary spectrum of XGO exhibits an absorption peak at 230 nm, which is attributable to the π-π* transition of the aromatic C-C ring, and weak absorption at 303 nm because of the n-π* transition of the C=O bond which can be observed according to the absorbance spectra. However, because there are fewer oxygen functional groups and more aromatic rings, the peak of RGO is relocated to 255 nm, where electrons can easily be excited at lower energies [23].

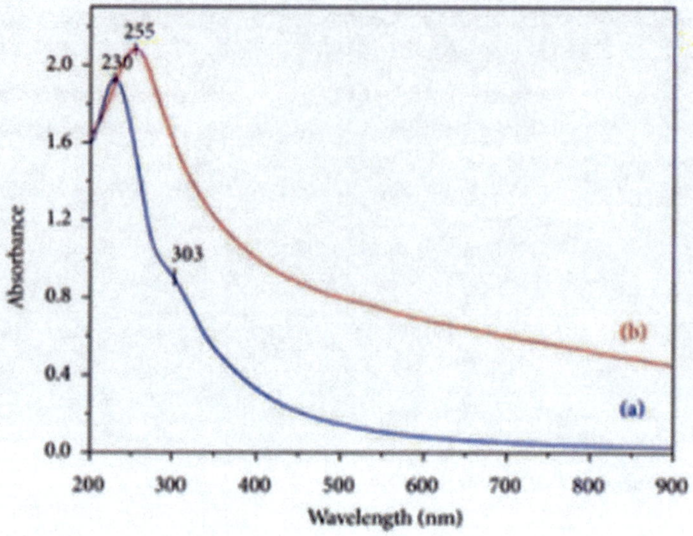

Fig. 5 UV–VIS Spectrum of rGO [23]

4 Conclusion

In this study, Hummer's method is used to create reduced graphene oxide (rGO). The diffraction peak of rGO in the XRD diffraction pattern was almost perfectly centred about 2θ~26°, indicating the crystalline structure of rGO. The elimination of oxygen functional groups during the reduction process was the primary cause of this peak's creation. Due to electron-hole recombination, existing grain boundaries in FESEM grow as rGO concentration rises. It is obvious that altering graphene's amount of oxidation, which can be demonstrated using the UV-Visible spectroscopic characterisation method, can result in significant control over the band gap. As the surface roughness of the rGO doped material grows, so does the intensity of the UV-Vis absorbance spectrum peak.

Acknowledgements We are thankful to DTU for the financial support and obliged to DTU family for encouraging us to carry out the research smoothly.

References

1. Abdelrazik AS, Tan KH, Aslfattahi N, Saidur R, Al-Sulaiman FA (2020) Optical properties and stability of water-based nanofluids mixed with reduced graphene oxide decorated with silver and energy performance investigation in hybrid photovoltaic/thermal solar systems. Int J Energy Res 44(14):11487–11508

2. Ahmed SH, Bakiro M, Aljasmi FIA, Albreiki AMO, Bayane S, Alzamly A (2020) Investigation of the band gap and photocatalytic properties of ceo2/rgo composites. Molecul Catalysis 486:110874
3. AlShammari AS, Halim MM, Yam FK, Mohd Kaus NH (2020) Synthesis of titanium dioxide (tio2)/reduced graphene oxide (rgo) thin film composite by spray pyrolysis technique and its physical properties. Mater Sci Semiconduct Process 116:105140
4. Anand A, Meena D, Dey KK, Bhatnagar MC (2020) Enhanced piezo-electricity properties of reduced graphene oxide (rgo) loaded polyvinylidene fluoride (pvdf) nanocomposite films for nanogenerator application. J Polym Res 27(12):1–11
5. Austria HFM, Subrahmanya TM, Setiawan O, Widakdo J, Chiao Y-H, Hung W-S, Wang C-F, Hu C-C, Lee K-R, Lai J-Y (2021) A review on the recent advancements in graphene-based membranes and their applications as stimuli-responsive separation materials. J Mater Chem A 9(38):21510–21531
6. Azarang M, Shuhaimi A, Yousefi R, Sookhakian M (2014) Effects of graphene oxide concentration on optical properties of zno/rgo nanocomposites and their application to photocurrent generation. J Appl Phys 116(8):084307
7. Baghchesara MA, Azimi HR, Shiravizadeh AG, Teridi MAM, Yousefi R (2019) Improving the intrinsic properties of rgo sheets by s-doping and the effects of rgo improvements on the photocatalytic performance of cu3se2/rgo nanocomposites. Appl Surf Sci 466:401–410
8. Bo Z, Shuai X, Mao S, Yang H, Qian J, Chen J, Yan J, Cen K (2014) Green preparation of reduced graphene oxide for sensing and energy storage applications. Sci Rep 4(1):1–8
9. Borini S, White R, Wei D, Astley M, Haque S, Spigone E, Harris N, Kivioja J, Ryhanen T (2013) Ultrafast graphene oxide humidity sensors. ACS Nano 7(12):11166–11173
10. Brush SG, Segal A (2015) In: Making 20th Century science: how theories became knowledge. Oxford University Press, USA
11. Chakraborty A, Agresti A, Pizzoferrato R, Matteis FD, Orsini A, Medaglia PG (2017) Study of structural and optical properties of low temperature photo-activated zno-rgo composite thin film. Mater Res Bullet 91:227–231
12. Chen J, Yao B, Li C, Shi G (2013) An improved hummers method for eco-friendly synthesis of graphene oxide. Carbon 64:225–229
13. Silva KKHD, Huang H-H, Joshi R, Yoshimura M (2020) Restoration of the graphitic structure by defect repair during the thermal reduction of graphene oxide. Carbon 166:74–90
14. Jinhong D, Cheng H-M (2012) The fabrication, properties, and uses of graphene/polymer composites. Macromol Chem Phys 213(10–11):1060–1077
15. Hajimazdarani M, Naderi N, Yarmand B, Kolahi A, Sangpour P (2018) Enhanced optical properties of zns–rgo nanocomposites for ultraviolet detection applications. Ceramics Int 44(15):17878–17884
16. Hasson DA, Al-Mammory MH (2022) Investigation structural and optical properties for pbs/rgo nano composite synthesis by pulse laser ablation. In: AIP Conference proceedings, vol 2450. AIP Publishing LLC, pp 020035
17. Jamil A, Mustafa F, Aslam S, Arshad U, Ahmad MA (2017) Structural and optical properties of thermally reduced graphene oxide for energy devices. Chinese Phys B 26(8):086501
18. Johari P, Shenoy VB (2012) Tuning the electronic properties of semiconducting transition metal dichalcogenides by applying mechanical strains. ACS Nano 6(6):5449–5456
19. Karanjikar SR, Sena AS, Manekar P, Mudagi S, Juneja AS (2021) Utilization of graphene and its derivatives for air and water filtration: A
20. Krishnamoorthy K, Veerapandian M, Mohan R, Kim S-J (2012) Investigation of raman and photoluminescence studies of reduced graphene oxide sheets. Appl Phys A 106(3):501–506
21. Li S-S, Kun-Hua T, Lin C-C, Chen C-W, Chhowalla M (2010) Solution-processable graphene oxide as an efficient hole transport layer in polymer solar cells. ACS Nano 4(6):3169–3174
22. Li X, Wei Y, Ma C, Jiang H, Gao M, Zhang S, Liu W, Huo P, Wang H, Wang L (2021) Multichannel electron transmission and fluorescence resonance energy transfer in in2s3/au/rgo composite for co_2 photoreduction. ACS Appl Mater Interfaces 13(10):11755–11764

23. Loryuenyong V, Totepvimarn K, Eimburanapravat P, Boonchompoo W, Buasri A (2013) Preparation and characterization of reduced graphene oxide sheets via water-based exfo-liation and reduction methods. Adv Mater Sci Eng
24. Ma L, Li S, Wang H (2021) Enhanced nonlinear optical properties of rgo via au modification: application for q-switched waveguide laser. Opt Mater Expr 11(6):1583–1591
25. Malik MTU, Sarker A, Rahat SMSM, Shuchi SB (2021) Performance enhancement of graphene/go/rgo based supercapacitors: a comparative review. Mater Today Commun 28:102685
26. Nugraheni AY, Jayanti DN, Soontaranon S, Putra EGR, et al. (2017) Structural analysis on reduced graphene oxide prepared from old coconut shell by synchrotron x-ray scattering. In: IOP Conference series: materials science and engineering, vol 196. IOP Publishing, pp 012007
27. Oliveira AEF, Braga GB, Tarley CRT, Pereira AC (2018) Thermally reduced graphene oxide: synthesis, studies, and characterization. J Mater Sci 53(17):12005–12015
28. Priyadharsan A, Shanavas S, Vidya C, Kalyana Sundar J, Acevedo R, Anbarasan PM (2020) Structural and optical properties of sn doped zno-rgo nanostructures using hydrothermal technique. Mater Today: Proc 26:3522–3525
29. Hanifah MFR, Jaafar J, Aziz M, Ismail AF, Othman MHD, Rahman MA (2015) Effects of reduction time on the structural, electrical, and thermal properties of synthesized reduced graphene oxide nanosheets. Bullet Mater Sci 38(6):1569–1576
30. Rahimi A, Kazeminezhad I, Ghahfarokhi SEM (2018) Synthesis and investigation of sns2/rgo nanocomposites with different go concentrations: structure and optical properties, photocatalytic performance. J Mater Sci: Mater Electron 29(6):4449–4456
31. Rani JR, Lim J, Oh J, Kim J-W, Shin HS, Kim JH, Lee S, Jun SC (2012) Epoxy to carbonyl group conversion in graphene oxide thin films: effect on structural and luminescent characteristics. The J Phys Chem C 116(35):19010–19017
32. Ray SC (2015) Application and uses of graphene oxide and reduced graphene oxide. In: Applications of graphene and graphene-oxide based nanomaterials, vol 1
33. Revathi B, Krishna Chandar N (2022) Magnetic field sensing characteristics of rgo/bimno$_3$ nanocomposites loaded clad-modified optical fiber sensor. J Sci: Adv Mater Dev 7(4):100488
34. Saleki Z, Majarshin AJ, Luo Y-A, Zhang D-L (2021) Spectral statistics of light in one-dimensional graphene-based photonic crystals. Phys Rev E 104(1):014116
35. Sharma N, Sharma V, Jain Y, Kumari M, Gupta R, Sharma SK, Sachdev K (2017) Synthesis and characterization of graphene oxide (go) and reduced graphene oxide (rgo) for gas sensing application. In: Macromolecular symposia, vol 376. Wiley Online Library, pp 1700006
36. Shen H, Zhao X, Duan L, Liu R, Hongjing W, Hou T, Jiang X, Gao H (2017) Influence of interface combination of rgo-photosensitized sno2@ rgo core-shell structures on their photocatalytic performance. Appl Surf Sci 391:627–634
37. Smith AT, LaChance AM, Zeng S, Liu B, Sun L (2019) Synthesis, properties, and applications of graphene oxide/reduced graphene oxide and their nanocomposites. Nano Mater Sci 1(1):31–47
38. Sohail M, Saleem M, Ullah S, Saeed N, Afridi A, Khan M, Arif M (2017) Modified and improved hummer's synthesis of graphene oxide for capacitors applications. Modern Electron Mater 3(3):110–116
39. Sookhakian M, Amin YM, Zakaria R, Basirun WJ, Mahmoudian MR, Nasiri-Tabrizi B, Baradaran S, Azarang M (2015) Significantly improved photocurrent response of zns-reduced graphene oxide composites. J Alloys and Compound 632:201–207
40. Tiwari SK, Mishra RK, Ha SK, Huczko A (2018) Evolution of graphene oxide and graphene: from imagination to industrialization. ChemNanoMat 4(7):598–620
41. Tuan PV, Hoang CM, Tuong HB, Tan VT, Hoa TTQ, Xuan Sang N, Khiem TN, et al. (2020) Effects of annealing temperature on the structure, morphology, and photo- catalytic properties of sno2/rgo nanocomposites. Nanotechnology 32(1):015201
42. Venkatesh G, Geerthana M, Prabhu S, Ramesh R, Prabu KM (2020) Enhanced photocatalytic activity of re- duced graphene oxide/srsno3 nanocomposite for aqueous organic pollutant degradation. Optik 206:164055

43. Wang C-I, Periasamy AP, Chang H-T (2013) Photoluminescent c-dots@ rgo probe for sensitive and selective detection of acetylcholine. Anal Chem 85(6):3263–3270
44. Weiss NO, Zhou H, Liao L, Liu Y, Jiang S, Huang Y, Duan X (2012) Graphene: an emerging electronic material. Adv Mater 24(43):5782–5825
45. Yang K, Feng L, Shi X, Liu Z (2013) Nano-graphene in biomedicine: theranostic applications. Chem Soc Rev 42(2):530–547
46. Ye M, Gao J, Xiao Y, Tong X, Zhao Y, Liangti Q (2017) Metal/graphene oxide batteries. Carbon 125:299–307
47. Yuan Y, Zhu B, Cao F, Jiawen W, Hao Y, Yuzong G (2021) Enhanced nonlinear optical properties of the cu2se/rgo composites. Results in Phys 27:104568
48. Zainuddin MF, Nik Raikhan NH, Othman NH, Abdullah WFH (2018) Synthesis of reduced graphene oxide (rgo) using different treatments of graphene oxide (go). In: IOP conference series: materials science and engineering, vol 358. IOP Publishing, pp 012046
49. Zhang E, Jia X, Wang B, Wang J, Xinzhi Y, Bingan L (2020) Carbon dots@ rgo paper as freestanding and flexible potassium-ion batteries anode. Adv Sci 7(15):2000470
50. Zhang W, Bi E, Li M, Gao L (2016) Synthesis of ag/rgo composite as effective conductive ink filler for flexible inkjet printing electronics. Colloids Surf A 490:232–240
51. Zhong WenWu, AiJiao X, Guo RenQing, Wang WenJun (2017) Role of graphene on the structure and optical properties of rgo/bipbo2cl nanosheet composites with enhanced photocatalytic activity. J Alloy Compd 696:246–250
52. Yanli Z, He C, Liu D, Chen L, Li W, Zhang W (2021) Nonlinear optical properties of poly (vinyl alcohol) thin films doped with in-situ wse2/rgo composite. Opt Laser Technol 142:107198

Synthesization of Mn_4Si_7 by High-Energy Ball Milling Technique

Anjali Saini, R. Gowrishankar, Mukesh Kumar Bairwa, and S. Neeleshwar

Abstract Mn_4Si_7 is earth-abundant, environment friendly, non-toxic cost-effective material, non-degenerating semiconductor material having many applications such as thermoelectric, lithium-ion batteries, electrical, optical, and photoelectrical. Mn_4Si_7 was synthesized by high-energy ball milling which is scalable and eco-friendly process. The Mn_4Si_7 tetragonal phase was obtained at 600 RPM for 50H which was confirmed by X-ray diffraction (XRD). From the XRD spectrum, the average crystallite size was estimated ~ 19 nm. Subsequently, non-uniform different shape was observed by scanning electron microscopy (SEM).

Keywords Phase purification · Ball milling · Manganese silicide · Earth-abundant

1 Introduction

Manganese silicide consists of economically earth-abundant and non-toxic elements and has been explored due to their potential applications in thermoelectric, lithium-ion batteries [1], electrical [2], photodiodes [3], optical [2], etc. Manganese silicides are known as Higher Manganese Silicides (HMS) because it exists in different phases such as Mn_4Si_7, Mn_5Si_3, Mn_7Si_{12}, $Mn_{11}Si_{19}$, $Mn_{15}Si_{26}$, $Mn_{19}Si_{33}$, $Mn_{26}Si_{45}$, $Mn_{27}Si_{47}$, and $Mn_{39}Si_{68}$, $Mn_{81.50}Si_{18.50}$ [4–6]. Among the HMS material family, Mn_4Si_7 is a non-degenerating semiconductor with an indirect band gap of 0.77 eV, while all other phases are degenerate semiconductors [7, 8]. The Mn_4Si_7 is a Nowotny chimney-ladder (NCL) structure and crystallizes into the tetragonal phase where the Mn atoms form the chimney frame wall, and Si atoms are arranged as spiral ladders [9–11] as shown in Fig. 1.

A. Saini · M. K. Bairwa · S. Neeleshwar (✉)
University School of Basic and Applied Sciences, Guru Gobind Singh Indraprastha University, Delhi 110078, India
e-mail: sn@ipu.ac.in

R. Gowrishankar
Department of Physics, Sri Sathya Sai Institute of Higher Learning, Prasanthinilayam, A.P. 515134, India

Z. H. Khan et al. (eds.), *Recent Advances in Nanomaterials*, Springer Proceedings in Materials 27, https://doi.org/10.1007/978-981-99-4878-9_75

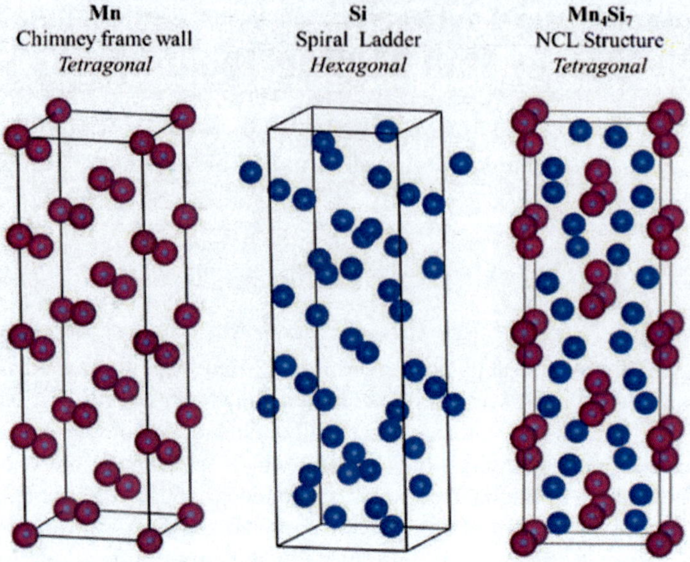

Fig. 1 The Nowotny chimney-ladder structure of Mn_4Si_7 composed of Mn (chimney frame wall) atoms and superimposed Si (spiral ladder) atoms

The Mn_4Si_7 synthesizing process is one of the challenging and difficult tasks due to the multiple phases of HMS. So far, many research groups synthesized Mn_4Si_7 by different methods such as Bridgman, Czocharalski, Arc-melting, high-energy planetary ball milling etc. [12–17]. However, in this paper, we need to discuss the synthetization of high-energy ball milling which is a simple, fast, easy, scalable, and eco-friendly technique. Because of the multiple domain applications of Mn_4Si_7, it is very important to synthesize and understand the intrinsic nature. This paper addresses the synthetization challenges of a single-phase tetragonal Mn_4Si_7 system by high-energy ball milling technique. The details of optimizations of Mn_4Si_7 were discussed in detail.

2 Experimental

Mn_4Si_7 was synthesized by planetary high-energy ball milling technique at room temperature. The high-purity elements such as manganese pieces (Mn, 99.9%, Alfa Aesar) and silicon lump (Si, 99.999%, Alfa Aesar) were homogenously mixed in tungsten carbide jar and balls. Samples were synthesized by optimizing ball milling parameters such as RPM, milling time, ball-to-powder weight ratio, and on: off milling time. Then the phase and crystallinity study were examined by X-ray diffraction (XRD) utilizing PANalytical X-pert diffractometer with Cu Kα-rays ($\lambda =$

1.54 Å). The morphology of samples was evaluated by scanning electron microscope (SEM, ZEISS EVO/18).

3 Results and Discussion

High-energy ball milling process is an eco-friendly and scalable technique which have limited tuning parameters such as RPM, milling time, ball-to-powder weight ratio, and on: off milling time. Based on the phase diagram [18], Mn_4Si_7 was synthesized at RPM 400, 450 and 500 but no desired phase was observed. Due to the non-formation of Mn_4Si_7, the RPM was increased to 600, and milling time also subsequently increased by 20H, 30H, 40H, and 50H. For 20H, 30H, and 40H, the HMS and MnSi phase was presented as shown in Fig. 1a. Below 40H, no pure phase was found. However, some other phases were presented such as MnSi and $Mn_{81.5}Si_{18.5}$. To eliminate the other phases, the milling time was increased by 50H. In 50H only Mn_4Si_7 was presented and no other phases were found as shown in Fig. 2a (A4). The clear view one can observed in Fig. 2b and compared with a standard database (JCPDS# 01–072-2069) no other phase was presented.

As, Fig. 2b indicates the nanostructure formation, need to understand the crystallite size. The crystalline size was estimated by Debye–Scherrer equation given:

$$D = \frac{k\lambda}{\beta \cos \theta} \tag{1}$$

where k is Scherrer constant ($k = 0.9$), λ the wavelength of the incident X-rays (1.5418 Å), θ the diffraction angle, and β stands for FWHM (full width at half

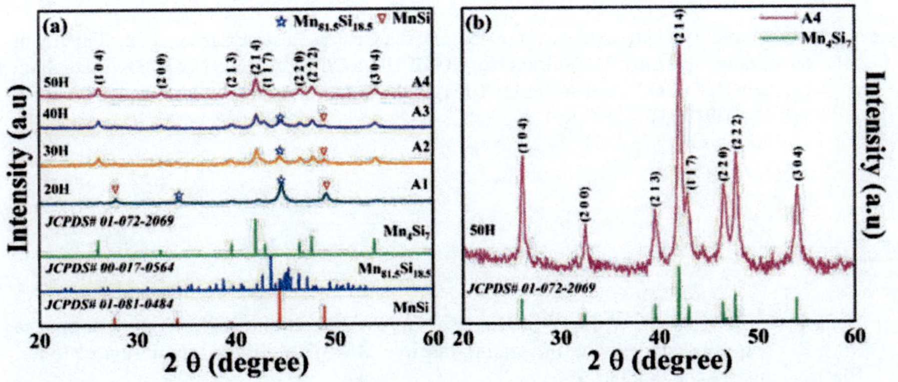

Fig. 2 X-ray diffraction patterns of **a** samples compared with standard data MnSi and Mn_4Si_7 **b** pure sample compared with standard data of HMS family. The peaks are well matched with MnSi (JCPDS # 01–081-0484), $Mn_{81.5}Si_{18.5}$ (JCPDS # 01–017-0564), and Mn_4Si_7 (JCPDS # 01–072-2069)

Fig. 3 Secondary scattered SEM micrographs of Mn_4Si_7 sample synthesized via high-energy ball milling thru WC vial and balls for 50H at **a** 1 μm **b** 2 μm **c** 5 μm **d** 10 μm scale

maximum) for peaks. The average crystallite size for the Mn_4Si_7 sample was ~ 19 nm. From XRD patterns the lattice parameters were calculated $a = 5.524$ Å and $c = 17.478$ Å for tetragonal Mn_4Si_7. The unit cell volume for Mn_4Si_7 was ~ 533 Å^3.

High-energy ball milling synthetization is a scalable process, one can need to understand the morphology of the synthesized samples. The morphology of the synthesized sample has shown in Fig. 3 with different magnificent. Figure 3 indicates non-uniform particles of different shapes and sizes with random agglomeration was observed. The minimum average particle size was ~ 0.52 μm and maximum average particle size was ~ 0.67 μm.

4 Conclusion

The tetragonal phase Mn_4Si_7 was successfully synthesized by high-energy ball milling technique. From the XRD data, the average crystallite size was estimated ~ 19 nm, lattice parameters $a = 5.524$ Å and $c = 17.478$ Å and unit cell volume was ~ 533 Å^3. A non-uniform, different size and shape morphology was observed from scanning electron microscopy.

Acknowledgements The authors are acknowledging the Central Research Instruments Facility of SSSIHL. SN acknowledge the FRGS (Grant No. GGSIPU/DRC/FRGS/2022/1223/13). Anjali Saini acknowledge the STRF (GGSIPU/DRC/2021/675). Mukesh Kumar Bairwa acknowledge the IPRF (GGSIPU/DRC/2019/1453).

References

1. Deng L, Wu Z-Y, Yin Z-W, Lu Y-Q, Huang Z-G, You J-H, et al. (2018) High-performance SiMn/C composite anodes with integrating inactive Mn4Si7 alloy for lithium-ion batteries. Electrochimica Acta 260:830–7
2. Gao Y, Chen R, Zhou M, Lourenco M, Homewood K, Shao G (2008) Structural, electronic, and optical properties of Mn4Si7. In: Sixth international conference on thin film physics and applications: SPIE. pp 74–7
3. Shukurova D, Orekhov A, Sharipov B, Klechkovskaya V, Kamilov T (2011) Mn 4 Si 7-Si-Mn 4 Si 7 and Mn 4 Si 7-Si-M photodiodes. Tech Phys 56:1423–1428

4. Li Z, Dong J-F, Sun F-H, Hirono S, Li J-F (2017) Significant enhancement of the thermoelectric performance of higher manganese silicide by incorporating MnTe nanophase derived from Te nanowire. Chem Mater 29(17):7378–7389
5. Liu L, Oda H, Onda T, Yodoshi N, Wada T, Chen Z-C (2020) Microstructure and thermoelectric properties of higher manganese silicides fabricated via gas atomization and spark plasma sintering. Mater Chem Phys 249:122990
6. Boldrini S, Famengo A, Montagner F, Battiston S, Fiameni S, Fabrizio M, et al. (2013) Test rig for high-temperature thermopower and electrical conductivity measurements. J Electron Mater 42(7):1319–23
7. Caprara S, Kulatov E, Tugushev V (2012) Half-metallic spin polarized electron states in the chimney-ladder higher manganese silicides MnSi1− x (x= 1.75− 1.73) with silicon vacancies. The Europ Phys J B 85(5):1–12
8. Migas D, Shaposhnikov V, Filonov A, Borisenko V, Dorozhkin N (2008) Ab initio study of the band structures of different phases of higher manganese silicides. Phys Rev B 77(7):075205
9. Gascoin F, Kyratsi T (2022) 3.11. Silicides silicide thermoelectrics. Key properties of inorganic thermoelectric materials—tables (version 1). 4:102
10. Miyazaki Y (2021) Higher manganese silicides 2.15. In: Thermoelectric energy conversion: theories and mechanisms, materials, devices, and applications. vol 375
11. Neeleshwar S, Saini A, Bairwa MK, Bisht N, Katre A, Narsinga RG (2022) Major challenges toward the development of efficient thermoelectric materials: from high figure-of-merit (zT) materials to devices. In: Khan ZH (ed) Nanomaterials for innovative energy systems and devices. Springer Nature Singapore, Singapore, pp 103–141
12. Kawasumi I, Sakata M, Nishida I, Masumoto K (1980) Crystal growth and characterization of MnSi∼ 1.73. J Crystal Growth 49(4):651–8
13. Rao SP, Saw AK, Chotia C, Okram G, Dayal V (2021) Structural and thermoelectric properties of Mn15Si26, Mn4Si7 and MnSi2, synthesized using arc-melting method. Appl Phys A 127(8):1–6
14. Guo Q, Zhang W, Liu Z, Fu X, Le Tonquesse S, Sato N et al (2021) Thermoelectric performance of Cr doped and Cr–Fe double-doped higher manganese silicides with adjusted carrier concentration and significant electron-phonon interaction. ACS Appl Mater Interfaces 13(7):8574–8583
15. Aoyama I, Fedorov MI, Zaitsev VK, Solomkin FY, Eremin IS, Samunin AY, et al. (2005) Effects of Ge doping on micromorphology of MnSi in MnSi∼ 1.7 and on their thermoelectric transport properties. Japanese J Appl Phys 44(12R):8562
16. Ponnambalam V, Morelli DT (2012) Effect of Cr and Fe substitution on the transport properties of the Nowotny Chimney–Ladder MnSi δ (1.73< δ< 1.75) compounds. J Electron Mater 41(6):1389–94
17. Binti Hamzan N, Ng CYB, Sadri R, Lee MK, Chang L-J, Tripathi M, et al. (2021) Controlled physical properties and growth mechanism of manganese silicide nanorods. J Alloys and Compounds 851:156693
18. Perumal S, Gorsse S, Ail U, Prakasam M, Rajasekar P, Umarji AM (2019) Enhanced thermoelectric figure of merit in nano-structured Si dispersed higher manganese silicide. Mater Sci Semiconduct Process 104:104649. https://doi.org/10.1016/j.mssp.2019.104649

Synthesization of SnSe by High-Energy Ball Milling Technique

Mukesh Kumar Bairwa, R. Gowrishankar, Anjali Saini, and S. Neeleshwar

Abstract Tin selenide (SnSe) are the materials consisting nontoxic and economical earth-abundant elements. SnSe was synthesized by high-energy ball milling technique. The pure phase, orthorhombic phase was obtained at 300 RPM for 18H which was confirmed by X-ray diffraction (XRD). The average crystalline size was estimated ~ 10 nm. Surface morphology was carried out by scanning electron microscopy (SEM). The distribution of non-uniform particle size from few nanometers to micrometer range.

Keywords Phase purification · Ball milling · SnSe

1 Introduction

The binary compound SnSe is a long-known semiconductor for various potential applications such as thermoelectrics [1], photovoltaic [2], solar cells [3], electronic [4], optical [5], infrared optoelectronic devices [6], phase change memory alloys [7], and flexible systems [8]. SnSe is a p-type material with a hole concentration in the range of 10^{17}–10^{18} cm^{-3} at room temperature [9, 10]. And has an indirect bandgap of 0.829 eV at 300 K [11]. SnSe compound is typically a layered orthorhombic crystal structure [12, 13]. The phase transition from α-SnSe to β-SnSe for space group *Pnma* and *Cmcm* respectively, above 750 K [12, 14]. In the crystal structure of SnSe, the atom connected with strong hetero-polar bonds, which containing zigzag plan like chain [15, 16] as shown in Fig. 1. The adjacent layers are bound by van der Waals forces and a long-range electrostatic attraction [15, 17], which make it easy to crack

M. K. Bairwa · A. Saini · S. Neeleshwar (✉)
University School of Basic and Applied Sciences, Guru Gobind Singh Indraprastha University, Delhi 110078, India
e-mail: sn@ipu.ac.in

R. Gowrishankar
Department of Physics, Sri Sathya Sai Institute of Higher Learning, Prasanthinilayam, A.P 515134, India

Z. H. Khan et al. (eds.), *Recent Advances in Nanomaterials*, Springer Proceedings in Materials 27, https://doi.org/10.1007/978-981-99-4878-9_76

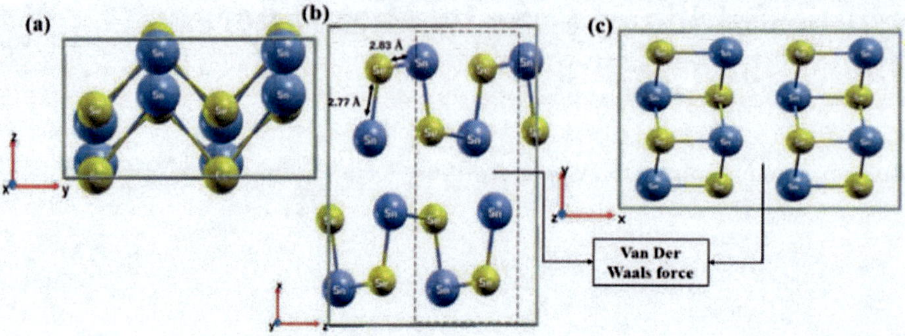

Fig. 1 The crystal structure of SnSe at room temperature with Pnma space group. **a** Structure along x-axis **b** Structure along y-axis **c** Structure along z-axis

when the shear force is applied [18]. SnSe is a strong anharmonic and anisotropic bonding [19].

Despite SnSe has become more and more popular recently, there are some challenges during the synthetization of SnSe, due to the formation of the oxides phase. SnSe was synthesized by various techniques such as Bridgman [20], Rocking furnace [21], Melt growth [22], Arc melting [23], Hydrothermal [24], Vertical vapor deposition [25], Rapid induction [26], and High-energy ball milling technique [27]. The main objective of this paper is to synthesize SnSe by high-energy ball milling which is an economical and viable method. This paper addresses the synthetization challenges of single phase orthorhombic *Pnma* space group SnSe system by high ball milling technique. The details of optimizations of SnSe were discussed in detail.

2 Experimental

The samples were synthesized by high-energy ball milling technique. High-purity elements such as tin pieces (Sn, 99.999%) and selenium powder (Se, 99.999%), obtained from Alfa Aesar were loaded in stainless steel (SS) and tungsten carbide (WC) jar with respective types of balls. Samples were synthesized by varying the RPM, milling time, on: off milling cycle, and ball-to-powder weight ratio. After the synthesis, fine powder was collected. The synthesized samples were characterized by X-ray diffraction (XRD) measurement on a PANalytical x-ray Diffractometer with Cu Kα-rays ($\lambda = 1.54$ Å) for the phase and structural studies. The morphology was studied by Scanning electron microscope (SEM, ZEISS EVO/18).

3 Results and Discussion

High-energy ball milling process is an environment-friendly and scalable process, where the samples were synthesized in stainless steel and tungsten carbide jar at different RPM and milling time. In stainless steel jar, on the basis of the phase diagram [28], SnSe was synthesized at RPM 200, 250, and 300 for 8H, 10H and 12H, but no desired phase was observed as shown in Fig. 2a. Further, increasing the RPM 300 for 10H, interestingly the SnSe phase was observed with no other impurity. We carried out 300 RPM for 18H using tungsten carbide jar. In this condition, we also get pure of SnSe phase without any other phases. The details are given in Fig. 2b (A5). The clear view one can observed in Fig. 2b and compared with a standard database (JCPDS# 014–0159), no other phase was presented.

As, Fig. 2b indicates the nanostructure formation, need to understand the crystallite size. The crystalline size was estimated by Debye–Scherrer equation given:

$$D = \frac{k\lambda}{\beta \cos \theta} \tag{1}$$

where k is Scherrer constant ($k = 0.9$), λ the wavelength of the incident X-rays (1.5418 Å), θ the diffraction angle, and β stands for full width at half maximum (FWHM) for peaks. From the XRD data, the average crystallite size for SnSe was ~ 10 nm. From XRD patterns the lattice parameters were calculated $a = 11.48$ Å, $b = 4.16$ Å, and $c = 4.39$ Å for the orthorhombic SnSe. The unit cell volume for SnSe was ~ 210 Å^3.

SEM image indicates the formation of particle size varies from nanometer to micrometer. The minimum particle size was ~ 176 nm and the maximum particle

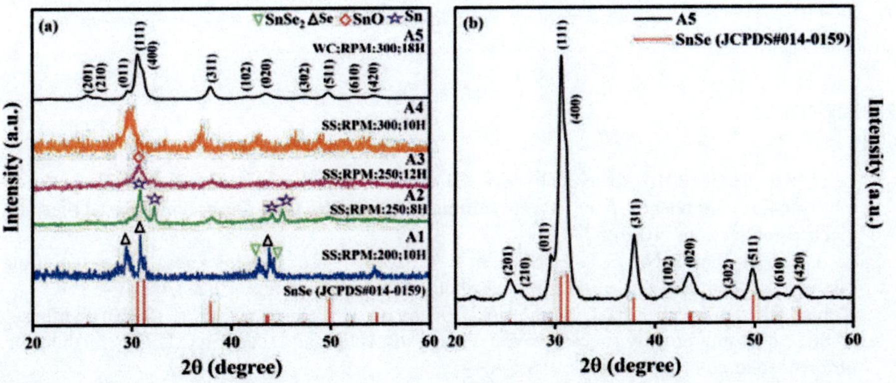

Fig. 2 The XRD of synthesized samples SnSe in stainless steel jar (A1, A2, A3, and A4) and tungsten carbide jar (A5) and compared with standard data base SnSe (JCPDS# 014–0159), $SnSe_2$ (JCPDS# 038–1055), Sn (JCPDS# 001–0926), Se (JCPDS# 001–0848) & SnO (JCPDS# 01–085-0423)

Fig. 3 Secondary scattered SEM micrographs of SnSe samples synthesized via high-energy ball milling thru WC vial and balls for 18H **a** 2 μm **b** 5 μm **c** 10 μm scale

size was ~ 10 μm. In this synthetization process, one can fine-tune the particle size from nanometer to micrometer scale by optimizing ball milling conditions (Fig. 3).

4 Conclusion

The orthorhombic phase SnSe was successfully synthesized by a high-energy ball milling technique. From XRD data the average crystallite size was estimated ~ 10 nm, the lattice parameter $a = 11.48$ Å, $b = 4.16$ Å and $c = 4.39$ Å and unit cell volume ~ 210 Å^3. Asymmetric non-uniform morphology was observed from scanning electron microscopy.

Acknowledgements The authors are grateful to Sri Sathya Sai Institute of Higher Learning Prasanthinilayam, A.P., India. SN acknowledge the FRGS (Grant No. GGSIPU/DRC/FRGS/2022/1223/13). Mukesh Kumar Bairwa acknowledge the IPRF (GGSIPU/DRC/2019/1453). Anjali Saini acknowledge the STRF (GGSIPU/DRC/2021/675).

References

1. Ju H, Kim JJAn (2016) Chemically exfoliated SnSe nanosheets and their SnSe/poly (3, 4-ethylenedioxythiophene): poly (styrenesulfonate) composite films for polymer based thermoelectric applications. 10(6):5730–9
2. Li Z, Guo Y, Zhao F, Nie C, Li H, Shi J, et al. (2020) Effect of film thickness and evaporation rate on co-evaporated SnSe thin films for photovoltaic applications. 10(28):16749–55
3. Kumar DK, Loskot J, Kříž J, Bennett N, Upadhyaya HM, Sadhu V, et al. (2020) Synthesis of SnSe quantum dots by successive ionic layer adsorption and reaction (SILAR) method for efficient solar cells applications. 199:570–4
4. Ju H, Park D, Kim JJJoMCA (2018) Solution-processable flexible thermoelectric composite films based on conductive polymer/SnSe 0.8 S 0.2 nanosheets/carbon nanotubes for wearable electronic applications. 6(14):5627–34
5. Haq BU, AlFaify S, Ahmed R, Butt FK, Laref A, Shkir MJPRB (2018) Exploring single-layered SnSe honeycomb polymorphs for optoelectronic and photovoltaic applications. 97(7):075438

6. ur Rehman S, Butt FK, Tariq Z, Hayat F, Gilani R, Aleem FJJoA, et al. (2017) Pressure induced structural and optical properties of cubic phase SnSe: an investigation for the infrared/mid-infrared optoelectronic devices. 695:194–201
7. Chung K-M, Wamwangi D, Woda M, Wuttig M, Bensch WJJoap (2008) Investigation of SnSe, Sn Se 2, and Sn 2 Se 3 alloys for phase change memory applications. 103(8):083523
8. Fan J, Huang X, Liu F, Deng L, Chen GJCC (2021) Feasibility of using chemically exfoliated SnSe nanobelts in constructing flexible SWCNTs-based composite films for high-performance thermoelectric applications. 24:100612
9. Li S, Li X, Ren Z, Zhang QJJoMCA (2018) Recent progress towards high performance of tin chalcogenide thermoelectric materials. 6(6):2432–48
10. Zhao L-D, Chang C, Tan G, Kanatzidis MGJE, Science E (2016) SnSe: a remarkable new thermoelectric material. 9(10):3044–60
11. Shi G, Kioupakis EJJoAP (2015) Quasiparticle band structures and thermoelectric transport properties of p-type SnSe 117(6):065103
12. Kumar M, Rani S, Singh Y, Gour KS, Singh VN (2021) Tin-selenide as a futuristic material: properties and applications. RSC Adv 11(12):6477–503
13. Neeleshwar S, Saini A, Bairwa MK, Bisht N, Katre A, Narsinga RG (2022) Major challenges toward the development of efficient thermoelectric materials: from high figure-of-merit (zT) materials to devices. In: Khan ZH (ed) Nanomaterials for innovative energy systems and devices. Springer Nature Singapore, Singapore, pp 103–141
14. Shi W, Gao M, Wei J, Gao J, Fan C, Ashalley E et al (2018) Tin selenide (SnSe): growth, properties, and applications. Adv Sci 5(4):1700602
15. Erdemir A (1994) Crystal chemistry and solid lubricating properties of the monochalcogenides gallium selenide and tin selenide. Tribol Trans 37(3):471–8
16. Abrikosov NK (2013) Semiconducting II–VI, IV–VI, and V–VI compounds. Springer
17. Kim S-u, Duong A-T, Cho S, Rhim S, Kim J (2016) A microscopic study investigating the structure of SnSe surfaces. Surface Sci 651:5–9
18. Carter R, Suyetin M, Lister S, Dyson MA, Trewhitt H, Goel S et al (2014) Band gap expansion, shear inversion phase change behaviour and low-voltage induced crystal oscillation in low-dimensional tin selenide crystals. Dalton Trans 43(20):7391–7399
19. Zhao L-D, Chang C, Tan G, Kanatzidis MG (2016) SnSe: a remarkable new thermoelectric material. Energy Environ Sci 9(10):3044–3060
20. Jin M, Shao H, Hu H, Li D, Shen H, Xu J et al (2017) Growth and characterization of large size undoped p-type SnSe single crystal by horizontal bridgman method. J Alloy Compd 712:857–862
21. Jin M, Shi X-L, Feng T, Liu W, Feng H, Pantelides ST, et al. (2019) Super large Sn1−x Se single crystals with excellent thermoelectric performance. ACS Appl Mater Interfaces 11(8):8051–9
22. Zhang M, Wang D, Chang C, Lin T, Wang K, Zhao L-D (2019) Oxygen adsorption and its influence on the thermoelectric performance of polycrystalline SnSe. J Mater Chem C 7(34):10507–10513
23. Serrano-Sánchez F, Gharsallah M, Nemes N, Mompean F, Martínez J, Alonso JJAPL (2015) Record Seebeck coefficient and extremely low thermal conductivity in nanostructured SnSe. 106(8):083902
24. Xu R, Huang L, Zhang J, Li D, Liu J, Liu J et al (2019) Nanostructured SnSe integrated with Se quantum dots with ultrahigh power factor and thermoelectric performance from magnetic field-assisted hydrothermal synthesis. J Mater Chem A 7(26):15757–15765
25. Jin M, Tang Z, Jiang J, Zhang R, Zhou L, Zhao S et al (2020) Growth of SnSe single crystal via vertical vapor deposition method and characterization of its thermoelectric performance. Mater Res Bull 126:110819
26. Yang S, Si J, Su Q, Wu H (2017) Enhanced thermoelectric performance of SnSe doped with layered MoS2/graphene. Mater Lett 193:146–149

27. Sidharth D, Nedunchezhian AA, Rajkumar R, Devi NY, Rajasekaran P, Arivanandhan M et al (2019) Enhancing effects of Te substitution on the thermoelectric power factor of nanostructured SnSe 1–x Te x. Phys Chem Chem Phys 21(28):15725–15733
28. Shi XL, Tao X, Zou J, Chen ZG (2020) High-performance thermoelectric SnSe: aqueous synthesis, innovations, and challenges. Adv Sci 7(7):1902923

Investigation of Static and Dynamic Magnetization in Ni_2FeAl Full Heusler Alloy Nanomaterials

Aarzoo Dhull, Vipul Sharma, Monika Sharma, Pawan S. Rana, and Bijoy K. Kuanr

Abstract Full Heusler alloys (FHA) offer excellent possibilities due to their high Curie temperature, spin polarization, and predicted half metallicity. Among various FHA, Ni_2FeAl is the least studied alloy that can offer vast possibilities in exploring spin-based devices. By keeping this aspect in mind, Ni_2FeAl nanoparticles have been prepared by the cost-effective ball milling process. As-prepared dried samples were vacuum annealed at 900 °C in the presence of the mixture of H_2 and N_2 gas to avoid oxidation for different time durations. XRD pattern revealed the B2 type atomic ordering with a crystallite size of approximately 21 nm. Morphological properties were investigated using SEM which revealed the formation of uniform nanoparticle flakes. Nearly perfect stoichiometry was obtained in EDS elemental mapping. The ferromagnetic properties were investigated by static magnetic measurements conducted by vibrating sample magnetometry. The dynamic magnetic response was measured from Vector Network Analyzer ferromagnetic resonance (VNA-FMR) technique in a broad range of microwave frequencies (5–20 GHz). A broad FMR linewidth was obtained from the FMR experiment due to the presence of magnetic inhomogeneities in the bulk sample. The Gilbert damping parameter which describes the magnetization relaxation was obtained from the fitting of FMR data with the Landau-Lifshitz-Gilbert equation. We obtained the lowest value of the Gilbert damping for the sample annealed at 900 °C temperature for 4 h. These investigations can open a new pathway in designing and fabricating Ni_2FeAl based ultra-fast, energy efficient spintronics devices.

Keywords Heusler alloys · Nanoparticles · Ferromagnetic resonance · Gilbert damping · Spintronics

---

A. Dhull · P. S. Rana
Deenbandhu Chhotu Ram University of Science and Technology, Murthal, Sonepat, Haryana, India

V. Sharma · M. Sharma · B. K. Kuanr (✉)
Special Centre for Nanoscience, Jawaharlal Nehru University, New Delhi 110057, India
e-mail: bkkuanr@gmail.com

M. Sharma
Department of Physics, Deshbandhu College, University of Delhi, New Delhi 110019, India

Z. H. Khan et al. (eds.), *Recent Advances in Nanomaterials*, Springer Proceedings in Materials 27, https://doi.org/10.1007/978-981-99-4878-9_77

1 Introduction

Magnetic nanoparticles exhibit prodigious properties which considerably deviate from bulk counterparts owing to their high surface to volume ratio. Therefore, they find applications in various fields such as spintronics, magnetic hyperthermia and topological insulators. Over the past few years, Heusler based intermetallic compounds have drawn the attention of researchers due to their unusual magnetic behaviour [1–3]. Heusler alloys are theoretically predicted to be half-metallic ferromagnets (HMFs). They are preferential candidates for achieving half-metallicity at room temperature owing to their Curie temperature values much higher than room temperature, large band gap in spin density of states (DOS) at Fermi level (E_F) and high spin-polarisation. Moreover, Heusler compounds exhibit high magnetic moment, low Gilbert damping and good lattice-match with most substrates, which makes them suitable for spintronics applications [3–6]. Heusler alloys are ternary intermetallic compounds comprising of three metallic elements—X, Y and Z. They are divided into two families: full Heusler (X_2YZ) and half-Heusler compounds (XYZ). Full Heusler alloys have X_2YZ stoichiometry where X and Y are transition metals and Z is a non-magnetic metal (from groups III-V) [6, 7]. They usually crystallize in cubic $L2_1$ structure, with space group Fm $\bar{3}$ m. Heusler alloys may deviate from the ideal $L2_1$ structure and crystallize into distinct disordered phases namely, B2 (Y–Z site disorder), D03 (X–Y site disorder) and A2 (X–Y–Z site disorder). The degree of disorder in Heusler nanostructures largely affects their magnetic properties. However, the nature of this dependency is not explored in detail. The properties of Heusler alloys can be easily tuned by varying combinations of constituent elements [5, 6]. Majority of the studies explore Heusler alloys in bulk and thin film geometries [8, 9]. The synthesis of Heusler alloy nanoparticles is a challenging task and poses difficulties in optimizing their properties. Ball milling is a cost-effective, environment-friendly method that can be used for synthesis of metallic alloy nanoparticles. The crystal structure can be tuned by varying process parameters like milling time, milling speed and ball to powder weight ratio [10]. Lee et al. [11] prepared Co_2MnSi nanoparticles by ball milling method and explored the effect of milling process variables on the crystal structure and magnetic behaviour of the alloy. Popa et al. [12] synthesized Ni_2MnSn nanoparticles by mechanical alloying and studied the process variables affecting phase formation. Ahmad et al. [13] prepared Co_2FeAl nanoparticles by co-precipitation method and reported the formation of disordered A2-phase in the material. Sarvanan et al. [14] synthesized Fe_2CrSi nanoparticles via ball milling method and studied the variation in magnetic properties with change in annealing temperature. So far, Ni-based Heusler alloy nanoparticles are the least investigated for their magnetic behaviour [15, 16]. Theoretically, Ni_2FeAl full Heusler alloys are predicted to have a total magnetic moment about 2.99 μB and a high spin polarization value of 46% for highly ordered $L2_1$ phase [17–19]. These properties make Ni_2FeAl Heusler alloy a promising candidate for spin-based devices. In this work, we have investigated the static and dynamic magnetic properties of Ni_2FeAl full Heusler alloy nanoparticles, and have determined Gilbert damping constant of the material for the first time.

2 Experimental Details

The Full Heusler Ni_2FeAl nanoparticles were prepared by the cost-effective ball milling method using a planetary ball mill (NST, Japan). Pure metal powders (Sigma, purity > 99%) of Ni, Fe, Al were taken in 2:1:1 stoichiometric proportion. These powders were thoroughly hand ground to get a uniform mixture. A total of 12 gm precursor mixture was transferred to Zirconia jars. Zirconia balls (diameters—8 mm and 10 mm) were used for grinding and Ball to Powder Weight ratio (BPWR) was kept to be 10:1. Ethanol was used as solvent for making a wet mixture of precursors. This is done in order to dissipate the heat produced during milling process and thereby prevent oxidation of metal particles. The milling was performed at a rotational speed of 550 rpm for 24 h. During the milling process, after each 1.5 h milling cycle was followed by 20 minutes cooling time to prevent overheating. The milled mixture was dried in vacuum and ground to a fine powder in mortar pestle. The X-ray diffraction pattern of as-prepared powder signified the need for annealing the sample further. So, the as-prepared sample was annealed in the presence of N_2 and H_2 gas mixture to prevent oxidation of sample during heating. Annealing was carried out at optimized temperature of 900 °C for different stay time periods. In the present work, the samples annealed at 2 h and 4 h duration were investigated and labelled as S1 and S2 respectively. X-ray diffractometer (XRD) technique [Rigaku, Miniflex 600 X-ray diffractometer] was used to study the phases and crystal structure of the samples. XRD analysis was done with Cu kα ($\lambda = 1.5418$ A°) radiation, keeping a step size of 0.05° at room temperature. The average crystallite sizes of the samples were calculated using Debye–Scherrer formula. Scanning electron microscope (SEM—ZEISS EV040) was used to investigate the surface morphology of the samples. EDS was carried out to assert the stoichiometric composition of Ni_2FeAl samples. Static magnetic measurements of samples were done using vibrating sample magnetometry (VSM—Cryogenics UK) by applying an external magnetic field upto 3 T. The dynamic magnetic response of the full Heusler samples was recorded from VNA-FMR technique pertaining to field sweep mode for a broad range of microwave frequencies 1–20 GHz at room temperature. Further, the Gilbert damping parameter was calculated by fitting the FMR data with the Landau-Lifshitz-Gilbert (LLG) equation.

3 Results and Discussion

The phase and crystal structure of Ni_2FeAl Heusler alloy nanoparticles were investigated using XRD technique. Figure 1 shows the XRD pattern of as-prepared, 2 h annealed and 4 h annealed samples. From the plot, it is evident that the prominent peak at $2\theta = 44°$ which corresponds to (220) plane is observed in all the samples. The superlattice peak at $2\theta = 30°$ corresponding to (200) plane arises in sample S1 and becomes more prominent in sample S2. But the (111) superlattice peak is absent in all the samples indicating that L_{21} phase is not achieved in any of the samples. However, the presence of (200) superlattice peak confirms the ordering of Ni sublattice and crystallization of Ni_2FeAl nanoparticles in B2-type atomic ordering [5–7] and usually appears to be of much low intensity in full Heusler materials. Comparing XRD patterns samples S1 and S2, we observe that the diffraction peak corresponding to (200) plane intensifies in sample S2 and a diffraction peak arises at Bragg's angle $2\theta = 64^{o}$ (which also corresponds to B2 phase) [15]. With the increase of stay time to 4 h we observe that some additional impurity peaks of pure metals are present. These impurities can further be removed by considering appropriate N_2 and H_2 gas pressures. So, both the samples depict B2 phase of Heusler alloys, which confirms the ordering of Ni sublattice but Fe-Al site disorder is quite high.

The average crystallite size (D) of nanoparticles was calculated using Debye–Scherrer formula [20] given by Eq. (1):

$$D = \frac{k\lambda}{\beta \mathrm{Cos}\theta} \quad (1)$$

where k is the shape factor (~ 0.9), λ is wavelength of Cu K_α radiation (~ 1.5418 A°), β is full width at half maximum intensity (FWHM), and θ is half the Bragg's angle. The average crystallite size of Ni_2FeAl Heusler alloy nanoparticles is calculated to be 20.52 nm for sample S1 while it is 27.58 nm for sample S2. This shows that crystallite size of nanoparticles increases with the increase in annealing stay time. The surface morphology of Ni_2FeAl nanoparticles is shown in Fig. 2. The SEM image showed uniform formation of flakes due to the agglomeration and the size of flakes is in μm range. The elemental composition of the nanoparticles was observed from EDS

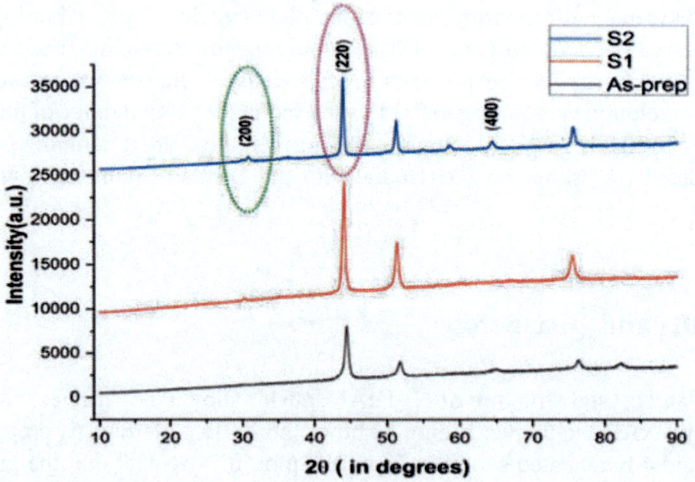

Fig. 1 Diffraction pattern obtained for Ni_2FeAl full Heusler alloy nanoparticles: as-prepared, S1 (annealed for 2 h) and S2 (annealed for 4 h) respectively

analysis. Figure 3a, b confirm the EDS spectra of samples S1 and S2. It clearly depicts the appropriate stoichiometric ratio of Ni, Fe and Al elements as 2:1:1. Further, it is clear that no other elements are present in the Ni_2FeAl nanoparticles. Figure 4 shows the colour mapping of the EDS spectra indicating uniformity of elements in the sample.

The room temperature static magnetic measurement for sample S1 is shown in Fig. 5. From the M-H curve, it is evident that Ni_2FeAl Heusler alloy nanoparticles has soft ferromagnetic nature. The saturation magnetization of 10 emu/gm and coercivity of 70 Oe was obtained for the Ni_2FeAl Heusler alloy nanoparticles.

Ferromagnetic Resonance (FMR) measurements is an established method to probe the dynamic magnetic behaviour of magnetic materials. It is based on the excitation of magnetic spins of the material by application of microwave field. It measures the interaction of ferromagnetic material (sample) with microwave (*rf*) energy before the energy dissipates to lattice as heat. An external dc magnetic field is applied to allow for precession of magnetic material, thereby causing FMR phenomena. Coupling of

Fig. 2 Surface view of full Heusler Ni_2FeAl nanoparticles

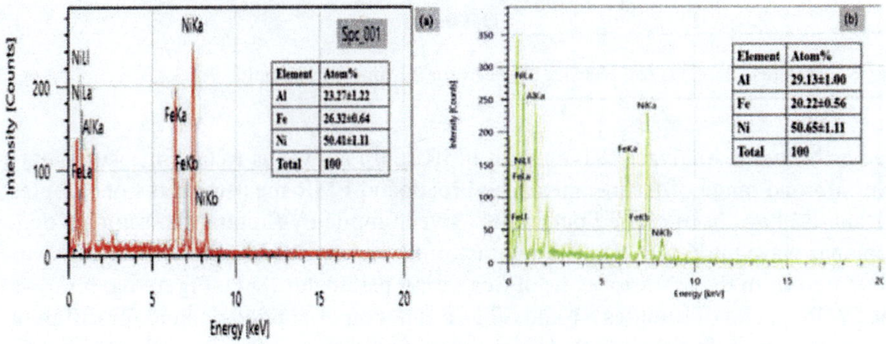

Fig. 3 **a** Element composition of sample S1 **b** elemental composition of sample S2 via element mapping

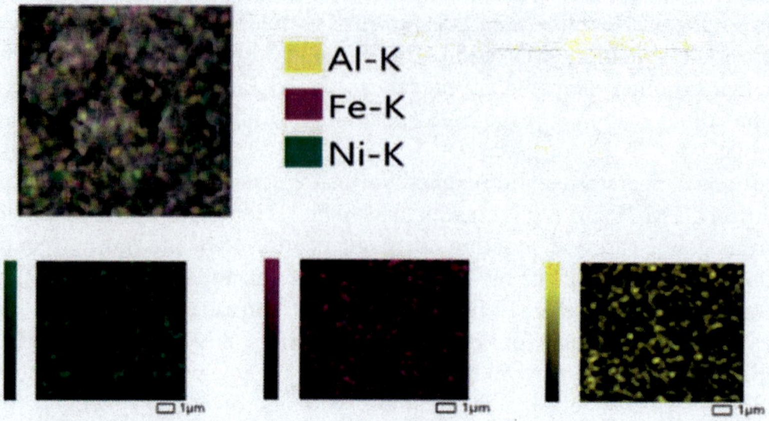

Fig. 4 Elemental colour mapping of Ni_2FeAl nanoparticles

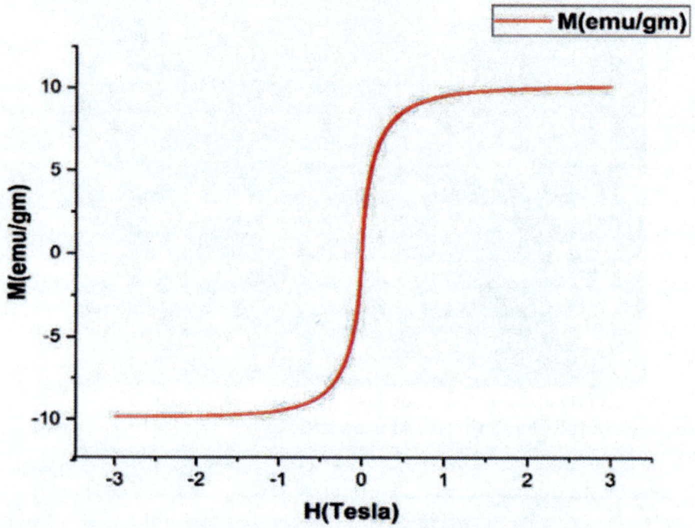

Fig. 5 Hysteresis loop for full Heusler Ni_2FeAl for 2 h annealed sample

Vector Network Analyzer (VNA) with FMR setup allows us to take measurements over a broad range of frequencies. The broadband FMR measurements of samples S1 and S2 have been carried out in field sweep mode by flipping the samples on a coplanar waveguide (CPW). The interactions between the sample and rf field was measured from the VNA in terms of Scattering parameter (S_{21}). Figure 6a, b shows the FMR spectra of samples S1 and S2 as a function of applied dc field for different frequency range. It revealed that microwave absorption occurs at higher magnetic fields for a given frequency in case of sample S2. Also, the energy absorption is more intense for sample S2 in comparison to sample S1. The Lorentzian fitting of FMR absorption spectra retrieved the resonance field (H_r) and field linewidth (ΔH) as functions of applied frequency. The Kittel expression given by Eq. (2) was used to determine intrinsic material parameters [21–23]

$$f = \gamma'(H_r + H_{\text{eff}}) \tag{2}$$

Here, f is microwave frequency, $\gamma = 2\pi\gamma\prime = g.\mu_B/\hbar$ is gyromagnetic ratio, H_r is resonance field, H_{eff} is effective magnetic field.

The gyromagnetic ratios for Ni_2FeAl samples were obtained by fitting of Kittel equation. The gyromagnetic ratios obtained for samples S1 and S2 are respectively 3.04 GHz/kOe and 3.125 GHz/kOe Figs. 7 and 8 compare the resonance fields (H_r) and linewidths (ΔH) of samples S1 and S2 at different frequencies. The material's intrinsic and extrinsic factors collectively contribute to the FMR linewidth. The intrinsic contribution is attributable to nature of the material whereas extrinsic contribution is attributable to defects and inhomogeneous broadening. The linear fitting of observed data was done using LLG model [24, 25] given by Eq. (3)

$$\Delta H = \Delta H_0 + \frac{2\alpha f}{\gamma\prime} \tag{3}$$

where α is the Gilbert Damping constant. ΔH_0 is the frequency independent contribution to ΔH and is known as extrinsic contribution to linewidth.

The fitting of ΔH versus f curve to LLG equation gives ΔH_0 from intercept and α from the slope. For sample S1, the gilbert damping constant was calculated to be 0.205 whereas for sample S2 it was 0.1377 (depicted in Table 1). However, damping constants for 2-d Heusler nanostructures are usually of the order of 10^{-2}. But our

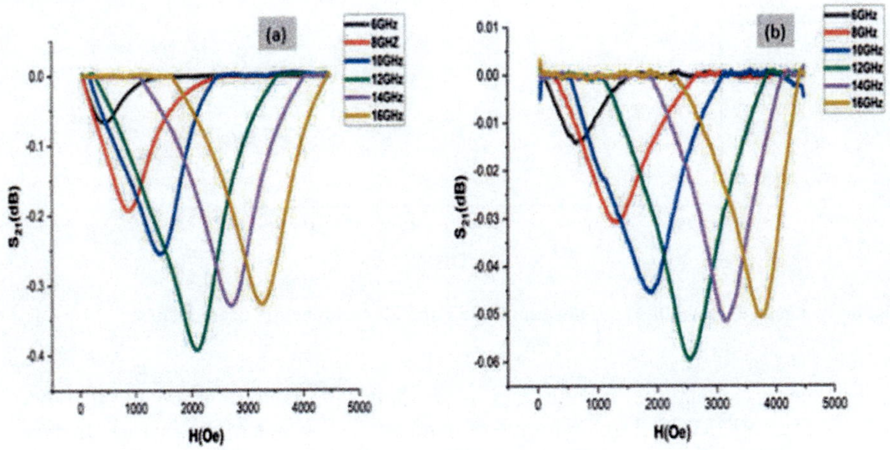

Fig. 6 **a** and **b** shows microwave absorption for samples S1 and S2 at different frequencies

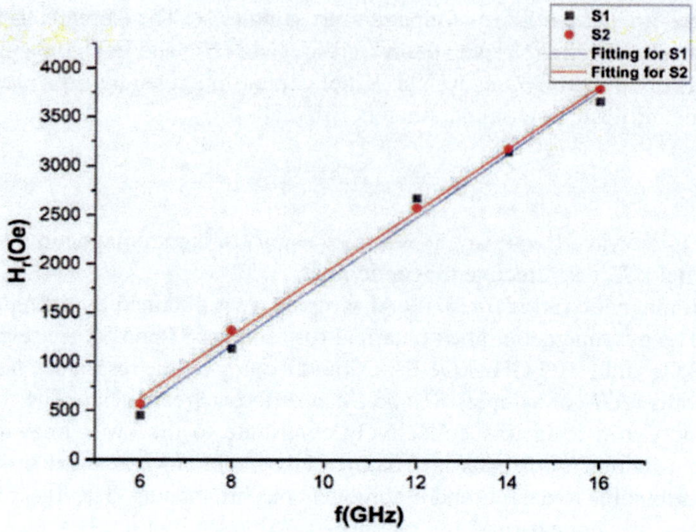

Fig. 7 Comparison of the resonance fields (H_r) for samples S1 and S2 as a function of frequency

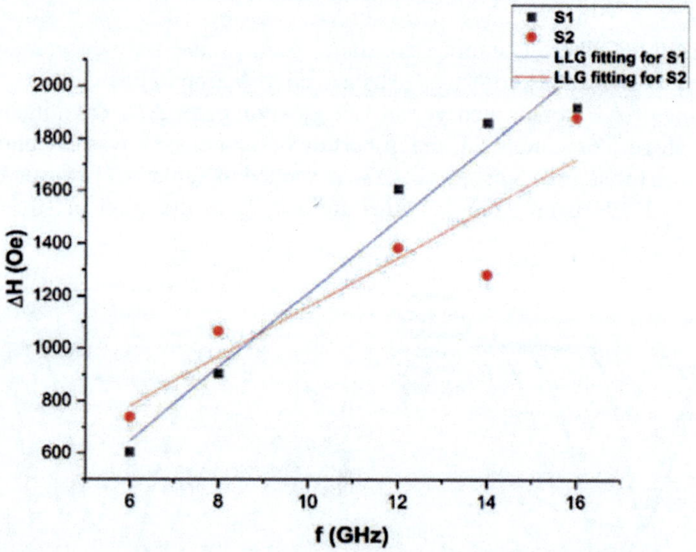

Fig. 8 Resonance linewidths (ΔH) of samples S1 and S2 as a function of frequency

material possesses a higher order damping constant. This behaviour may be attributed to several factors. Multi-domain nature of nanoparticles is one of the factors, as these multiple domains generate individual FMR spectra which superimpose and thereby enhance damping. Another reason could be inhomogeneous magnetization in the

Table 1 Gyromagnetic ratios and Gilbert damping constants obtained for samples S1 and S2

Sample	Gyromagnetic ratio (GHz/kOe)	Damping constant
S1	3.04	0.2055
S2	3.125	0.1377

sample due to high tendency of agglomeration of metal nanoparticles. The presence of defects in the material can also elevate damping. Thus, such nanoparticles with low Gilbert damping constant finds suitable place for spin-based device applications.

4 Conclusion

Ni_2FeAl full Heusler alloy nanoparticles have been successfully synthesized by ball milling technique. These nanoparticles depict B2-phase of full Heusler class. Ni_2FeAl nanoparticles have low value of coercive field indicating soft ferromagnetic nature of the material. The nanoparticles annealed for larger time duration exhibit a lower value of gilbert damping parameter. Further, ordering of Heusler alloy crystals largely affects their static and dynamic magnetic properties. We find appreciable scope of improvement in the synthesis of Ni_2FeAl Full Heusler nanoparticles to yield enhanced magnetic behaviour. To the best of our knowledge, Ni_2FeAl full Heusler material in nanoparticle regime has been investigated first in this work for its magnetic behaviour.

References

1. Galanakis I (2015) Theory of Heusler and full-Heusler compounds. Heusler alloys: properties, growth, applications. Springer International Publishing, Cham, pp 3–36
2. Felser C, Wollmann L, Chadov S, Fecher GH, Parkin SS (2015) Basics and prospective of magnetic Heusler compounds. APL Mater 3(4):041518
3. Elphick K, Frost W, Samiepour M, Kubota T, Takanashi K, Sukegawa H, ..., Hirohata A (2021) Heusler alloys for spintronic devices: review on recent development and future perspectives. Sci Technol Adv Mater 22(1):235–271
4. de Paula VG, Reis MS (2021) All-d-metal full Heusler alloys: a novel class of functional materials. Chem Mater 33(14):5483–5495
5. Graf T, Parkin SS, Felser C (2010) Heusler compounds—A material class with exceptional properties. IEEE Trans Magn 47(2):367–373
6. Webster PJ (1969) Heusler alloys. Contemp Phys 10(6):559–577
7. Palmstrøm CJ (2016) Heusler compounds and spintronics. Prog Cryst Growth Charact Mater 62(2):371–397
8. Bonda A, Uba S, Uba L (2022) Laser-induced two-step demagnetization process study in Ni–Mn–Sn Heusler alloy film. J Magn Magn Mater 546:168805
9. Kurdi S, Sakuraba Y, Masuda K, Tajiri H, Nair B, Nataf GF, Vickers ME, Reiss G, Meinert M, Dhesi SS, Ghidini M, Barber ZH (2022) Quantitative atomic order characterization of a Mn_2FeAl Heusler epitaxial thin film. J Phys D Appl Phys 55(18):185305

10. Ahamed R, Ghomashchi R, Xie Z, Chen L (2019) Powder metallurgy synthesis of Heusler alloys: effects of process parameters. Materials 12(10):1596
11. Lee CH (2020) Preparation and structural observation of Co_2MnSi Heusler alloys by mechanical alloying. J Nanosci Nanotechnol 20(9):5502–5505
12. Popa F, Chicinaş HF, Marinca TF, Chicinaş I (2017) Influence of mechanical alloying and heat treatment processing on the Ni_2MnSn Heusler alloy structure. J Alloy Compd 716:137–143
13. Ahmad A, Mitra S, Srivastava SK, Das AK (2019) Size-dependent structural and magnetic properties of disordered Co_2FeAl Heusler alloy nanoparticles. J Magn Magn Mater 474:599–604
14. Saravanan G, Asvini V, Kalaiezhily RK, Ravichandran K (2020) Effect on annealing temperature (Ta) of Ternary Full Fe_2CrSi Heusler alloy nanoparticles for spin-based device applications. J Supercond Novel Magn 33(12):3957–3962
15. Amudhavalli A, Rajeswarapalanichamy R, Iyakutti K (2018) Half metallic ferromagnetism in Ni based half Heusler alloys. Comput Mater Sci 148:87–103
16. Zhang W, Qian Z, Tang J, Zhao L, Sui Y, Wang H, Li Y, Su W, Zhang M, Liu Z, Liu G, Wu G (2007) Superparamagnetic behaviour in melt-spun Ni2FeAl ribbons. J Phys Condens Matter 19(9):096214
17. Saito T, Nishio-Hamane D (2018) Magnetic and thermoelectric properties of melt-spun ribbons of Fe_2XAl (X= Co, Ni) Heusler compounds. J Appl Phys 124(7):075105
18. Qawasmeh Y, Hamad B (2012) Investigation of the structural, electronic, and magnetic properties of Ni-based Heusler alloys from first principles. J Appl Phys 111(3):033905
19. Benhizia NE, Zaoui Y, Amari S, Beldi L, Bouhafs B (2020) Theoretical study of structural, electronic, dynamic and thermodynamic properties of Ni_2FeAl and Ni2CoAl alloys. Comput Condens Matter 24:e00480
20. Stokes AR, Wilson AJC (1944) A method of calculating the integral breadths of Debye-Scherrer lines: generalization to non-cubic crystals. Math Proc Cambridge Philos Soc 40(2):197–198. Cambridge University Press
21. Montoya E, Sebastian T, Schultheiss H, Heinrich B, Camley RE, Celinski Z (2016) Handbook of surface science 5
22. Polder DVIII (1949) VIII. On the theory of ferromagnetic resonance. London Edinb Dublin Philos Mag J Sci 40(300):99–115
23. Kittel C (1948) On the theory of ferromagnetic resonance absorption. Phys Rev 73(2):155
24. Van Vleck JH (1950) Concerning the theory of ferromagnetic resonance absorption. Phys Rev 78(3):266
25. Heinrich B, Cochran JF, Hasegawa R (1985) FMR line broadening in metals due to two-magnon scattering. J Appl Phys 57(8):3690–3692

Annealing-Induced Multilayer Formation of C8-BTBT Films for Better Electrical Performance

Nargis Khatun, Hasanur Jaman, Jayanta K. Bal, and A. K. M. Maidul Islam

Abstract The crystallinity, ordering, and large-area uniformity in the organic semiconducting thin films are crucial for their technological applications. However, preparing such films by solution process is challenging since the involved processes are far from thermodynamic equilibrium. C8-BTBT has emerged as the p-type key material for the elaboration of OTFT due to its proven high mobility, facile processability, environmental stability, and good thermal responses. In this study, to understand the dynamics of stable bilayer/multilayer formation, the thermal responses of C8-BTBT films prepared by Langmuir–Blodgett techniques were investigated with two complementary techniques, namely XRR and AFM. C8-BTBT LB films showed a potent temperature sensitivity over a range of temperatures, and XRR data shows clear Bragg peaks, indicating a strong alignment of crystallites parallel to the substrate but in the form of a discrete island as observed by AFM. After annealing and prolonged heating (220 min) at 70 °C, Kiessig fringes appeared, indicating that the initial disordered film transformed into a well-ordered bilayer/multilayer structure, maintaining the crystallinity of the C8-BTBT islands. In our study the detailed structural evolution may determine the optimal conditions for forming large-area uniform crystalline films, which are essential for improving device performance.

Keywords Organic small molecules · Langmuir–Blodgett film · XRR · AFM · Structural properties

N. Khatun · H. Jaman · A. K. M. Maidul Islam (✉)
Department of Physics, Aliah University, Kolkata, India
e-mail: maidul79@gmail.com

J. K. Bal
Department of Physics, Abhedananda Mahavidyalaya, Burdwan University, Sainthia, India

Z. H. Khan et al. (eds.), *Recent Advances in Nanomaterials*, Springer Proceedings in Materials 27, https://doi.org/10.1007/978-981-99-4878-9_78

1 Introduction

Small-molecule organic semiconducting materials having high carrier mobility have recently gained significant attention for their use in flexible, low-cost electronic devices, like organic photovoltaics (OPVs), organic light-emitting devices (OLEDs), and organic field-effect transistors (OFETs) [1–3]. In all such devices, particularly for OFET, it has been observed that charge transport primarily happens through a very thin layer, only a few nanometers thick, near the semiconductor-substrate interface. Therefore, using monolayer OFETs can significantly reduce the amount of organic semiconducting material required and may facilitate the development of a novel category of organic thin film electronic devices. However, the charge transport process in organic semiconductors is complex and influenced by factors such as chain conformation, π-stacking, and the uniformity of crystalline structures [4, 5]. Thus, preparing ultra-thin, large-area uniform films is essential for practical applications. The current research aims to prepare uniform films of small organic molecules with higher charge carrier mobility. Within the diverse range of organic materials, molecules based on thiophene, including 2, 7-dioctyl [1] benzothieno [3,2-b] benzothiophene (C8-BTBT), have gathered significant attention as prospective candidates for OFET applications. This is attributed to their high mobility, ease of processing, and enhanced environmental stability compared to other organic semiconductors [6]. However, the formation of homogeneous, large-area ultra-thin films from solutions of small molecules is a complicated task [3, 4]. The small size of such molecules results in grain boundaries and a high likelihood of self-assembling into polycrystalline aggregates with rough surfaces, making it difficult to deposit a continuous layer [3, 7, 8]. As a result, devices fabricated using these methods often exhibit poor performance compared to traditional thick film devices. This leads to significant problems in producing organic devices, where layer uniformity is essential for device performance and repeatability in manufacturing. Langmuir–Blodgett (LB) methods are recognized for producing highly organized films with a controllable number of monomolecular structures [9]. This approach involves forming monolayers, known as Langmuir monolayers, at the air/water interface, and subsequently transferred them to solid supports suitable for constructing organic devices [10, 11]. However, thiophene-based small molecules, C8-BTBT, exhibit low amphiphilicity, resulting in the formation of aggregated molecular layers, which leads to another challenge in forming a stable monolayer on the subphase. As reported in the literature [12], one way to overcome this issue is by adding a small amount of fatty acid to assist in the formation of a uniform monolayer. However, it should be noted that using such amorphous material may reduce the crystallinity of the semiconducting molecules and, therefore, may negatively impact the electrical performance. In contrast, pure C8-BTBT showed a pronounced temperature response over temperature ranges [13]. With increasing temperature, an initial disordered film undergoes a transition to a well-organized multilayer or bilayer structure characterized by a molecular herringbone packing arrangement. However, detailed systematic structural studies are essential for an optimum condition for the large-area uniform film. In the present work, to understand the dynamics of stable multilayer/multilayer formation, thermal responses of C8-BTBT films prepared by Langmuir–Blodgett techniques were investigated with two complementary techniques, namely XRR and AFM. The investigation revealed that an initial disordered film transforms a bilayer or multilayer structure upon annealing at elevated temperatures, which is crucial for enhancing electrical performance from the practical device perspective.

2 Materials and Methods

2.1 Material

In our study, 2,7-dioctyl [1] benzothieno [3,2-b] benzothiophene (C8-BTBT, MW = 464.77) was employed as the organic semiconductor. Chloroform was used as the solvent for C8-BTBT, and both the C8-BTBT and solvent were obtained from Sigma Aldrich and used as received. To prepare the thin films, 100 nm thermally grown silicon dioxide substrates were utilized, supplied by University Wafer Inc.

2.2 Methods

C8-BTBT LB films were prepared by a commercial LB instrument (Model No.: LBXD-NT, Apex Instruments). To prepare a C8-BTBT monolayer, a 0.5 g/L concentration of C8-BTBT was evenly distributed over the water (Milli-Q water) subphase. Subsequently, using two movable barriers, the monolayer was compressed on the water surface at a 5 mm/min speed. This compression enabled the recording of the isotherms, the surface pressure (π) versus area per molecule (A), as depicted in Fig. 1. The fabrication of LB films onto RCA cleaned [11] Si-wafers was performed by vertical dipping and lifting a substrate through the subphase monolayer. Once the monolayer had been formed, it was then transferred to a solid substrate at a surface pressure of 30 mN/m. This transfer process allowed the monolayer to be deposited onto the substrate, which then be further studied. X-ray reflectivity (XRR) measurements were performed at Elettra Synchrotron, Trieste [14], Italy, with an X-ray of wavelength $\lambda = 1.54$ Å. The angular scans (2θ) have been recalculated to scattering vector using the formula $q_z = (4\pi/\lambda)\sin\theta$. XRR technique essentially provides average in-plane electron density (ρ) as a function of depth (z), known as an electron density profile. By utilizing the electron density profile (EDP), it becomes feasible to make estimations regarding the structural characteristics of the films, such as out-of-plane crystallinity, film thickness and roughness [11, 15]. The investigation of film morphologies was conducted by atomic force microscopy, AFM (NT-MDT). Non-contact mode with a silicon cantilever was employed to acquire AFM images, and Gwyddion software was utilized for image processing and analysis.

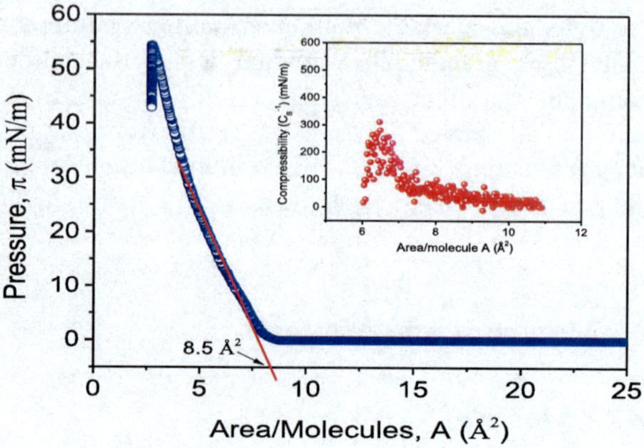

Fig. 1 Surface pressure (mN/m) vs area/molecule ($Å^2$) of C8-BTBT, inset shows the compressibility (Cs^{-1}) versus area/molecule

3 Results and Discussion

3.1 Isotherm Study

The π-A isotherm for pure C8-BTBT is depicted in Fig. 1, which illustrates a linear increase in surface pressure (π) against the decreasing area/molecule up to $\pi = 55$ mN/m. Beyond this point, the isotherm collapses. By extrapolating the linear portion of the isotherm towards the area per molecule axis (*X*-axis), we can calculate an area of 8.5 $Å^2$ per molecule. This value is lower than the expected area of 14 $Å^2$ per molecule if all the thiophene rings were oriented perpendicular to the surface, as reported in the literature [12]. However, in our case, unlike typical thiophene-based small molecules, C8-BTBT exhibits a considerably high collapse pressure of 55 mN/m, which is a significant value for such molecules. Increasing pressure in a typical Langmuir isotherm indicates that molecules are more compact, and depositing a Langmuir–Blodgett (LB) film at a relatively higher pressure is necessary. Additionally, the high collapse pressure monolayer film suggests a higher resistance to compression, likely due to increased compressibility. Furthermore, we also calculated the in-plane elasticity or compressibility (C_s^{-1}) of the isotherm, as shown in the inset of Fig. 1. A significantly high value of C_s^{-1} indicating a greater ability for rearrangement at the air–water interface and potentially attributed to the flexibility of the molecules to avoid the formation of stacked multilayers. This suggests that C8-BTBT molecules have a higher compressibility than typical thiophene-based small molecules, resulting in a higher compression resistance and an enhanced capacity for rearrangement at the air–water interfaces. Such compressed monolayers of C8-BTBT may be useful in fabricating thin films for electronic devices.

3.2 X-Ray Reflectivity (XRR) and AFM Study

Specular X-ray reflectivity (XRR) was performed to gain structural information on the C8-BTBT films. XRR spectra, presented in Fig. 2, were obtained under an elevated temperature of 70 °C over an extended period; different times were indicated in the figure. The bottom spectra (black curve) show a standard XRR curve, with a total external reflection plateau that extends up to $q_z = 0.031$ Å^{-1}. Notably, there is a significant reduction in the reflected intensity, and no visible Kiessig fringes are present. The absence of Kiessig fringes can be ascribed to the high surface roughness of the film, which is likewise evident in the AFM images (refer to Fig. 3). However, prominent diffraction peaks at $q_z = 0.22$ Å^{-1} and 0.43 Å^{-1} are identified as the [001] and [002] Bragg peaks, respectively, which indicates a strong preferred orientation of C8-BTBT crystallites parallel to the substrate surface. To investigate the temperature response of the film, in-situ X-ray reflectivity measurements were conducted at temperatures ranging from room temperature to 120 °C in steps of 10 deg, which shows notable changes in the film structure starting at 70 °C (data not shown). These changes are similar to those observed in the literature [13], where reversible structural changes were reported. Our study also focused on time-evolution measurements of structural changes at an elevated temperature of 70 °C. Figure 2 illustrates the kinetics data taken in-situ over a prolonged period (220 min) until no further changes occurred. It is noteworthy that after 40 min, clear Kiessig fringes start to appear with a minimum of 0.12 Å^{-1}, which indicates initial island-type rough film begins to convert a monolayer, which then progresses to a bilayer structure by heating for about 90 min. The minima observed around 0.01 Å^{-1} and 0.17 Å^{-1} confirm the formation of a bilayer arrangement, wherein a second monolayer is formed on the initial layer. It is to be noted that Bragg peaks ([001] and [002]) remained present after heating, with a slight increase in the width of the Bragg's peak indicating a reduction in sharpness of vertically grown islands, accompanied by monolayer and bilayer formation. Despite this, a uniform film was achieved without significant loss of crystallinity.

AFM height images of C8-BTBT films were obtained before and after heating at 70°C for 220 min. Figure 3a–c show that the X-ray reflectivity (XRR) data is consistent with the AFM results. The film prepared at room temperature exhibited a distribution of islands with micrometre size and limited coverage (Fig. 3a). The height of the islands, measuring approximately 100 nm, suggests the presence of aggregated C8-BTBT on the substrate. Upon annealing, the large islands appear to remain relatively unchanged but are more connected, resulting in an increase in the crystalline-rich area, as seen in Fig. 3b. Upon closer analysis, it becomes apparent that the gaps between the islands are now occupied by a monolayer or bilayer of C8-BTBT, as illustrated in Fig. 3c. This is indicative of re-wetting and subsequent filling of previously unoccupied regions. While a more detailed structural analysis would further reveal the dynamics of the structural evolution of the C8-BTBT films.

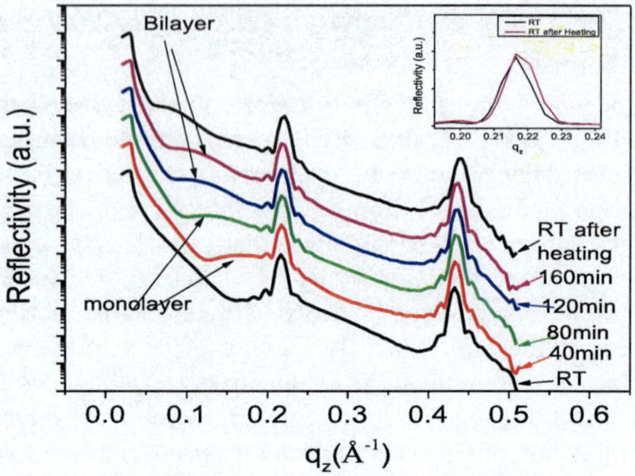

Fig. 2 Time evolution of XRR data of C8-BTBT films 70 °C. Inset shows the comparison of [001] Bragg's peak

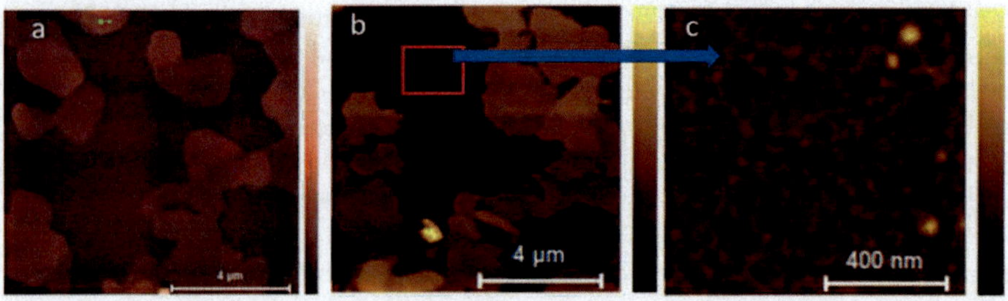

Fig. 3 A height image of a C8-BTBT film was captured using an Atomic Force Microscope (AFM), shown in **a**. The film was then annealed at a temperature of 70 °C for 220 min. The resulting AFM height images of the annealed film are shown in **b** and **c**

4 Conclusion

The structural evolution of C8-BTBT films, prepared using Langmuir–Blodgett techniques, was investigated by X-ray reflectivity (XRR), and results were complemented by atomic force microscopy (AFM) analyses. Results from the Langmuir isotherm study showed that the collapse pressure reached up to 55 mN/m, a reasonably high value, which led to the successful deposition of Langmuir–Blodgett (LB) thin films. XRR data revealed peaks at $q_z = 0.22$ Å^{-1} and 0.43 Å^{-1}, representing the [001] and [002] Bragg peaks, indicating a significant alignment preference of crystallites parallel to the surface of the substrate. The appearance of Kiessig fringes upon heating confirms that the initial disordered film transforms into a well-ordered bilayer/multilayer structure between the C8-BTBT islands, which remain crystalline. Further kinetic studies may provide optimal conditions for preparing large-area uniform, crystalline films, which are essential for improved device performance.

Acknowledgements The author acknowledges Indo-Italian-POC financial support (DST, India) to perform measurements at the MCX beamline, Elettra synchrotron, Trieste, Italy. Also, Cordial thanks to Dr. Jasper R. Plaisier(Beamline Scientist, Elettra) for his kind cooperation in helping us use the laboratory facility. Also, thanks to Pietro Parisse, Elettra, Italy, for AFM measurements.

Declaration of Interest Statement Authors have no conflict of interest to declare.

References

1. Yang F, Shtein M, Forrest SR (2005) Controlled growth of a molecular bulk heterojunction photovoltaic cell. Nat Mater 4(1):37–41
2. Sirringhaus H (2014) 25th-anniversary article field-effect transistors: the path beyond amorphous silicon. Adv Mater 26(9):1319–1335
3. Facchetti A (2007) Semiconductors for organic transistors. Mater Today 10(3):28–37
4. Virkar AA, Mannsfeld S, Bao Z, Stingelin N (2010) Organic semiconductor growth and morphology considerations for organic thin-film transistors. Adv Mater 22(34):3857–3875
5. Matsushima T, Sandanayaka AS, Esaki Y, Adachi C (2015) Vacuum-and-solvent-free fabrication of organic semiconductor layers for field-effect transistors. Sci Rep 5(14547):1–9
6. Yuan Y, Giri G, Ayzner AL, Zoombelt AP, Mannsfeld SC, Chen J, Bao Z (2014) Ultra-high mobility transparent organic thin film transistors grown by an off-centre spin-coating method. Nat Commun 5(3005):1–9
7. Yamashita Y (2009) Organic semiconductors for organic field-effect transistors. Sci Technol Adv Mater 10(2):024313
8. Ward JW, Lamport ZA, Jurchescu OD (2015) Versatile organic transistors by solution processing. Chem Phys Chem 16:1118–11321
9. Tanese MC, Farinola GM, Pignataro B, Valli L, Giotta L, Conoci S, Torsi L (2006) Poly (alkoxyphenylene-thienylene) Langmuir-Schäfer thin films for advanced performance transistors. Chem Mater 18(3):778–784
10. Xu G, Bao Z, Groves JT (2000) Langmuir-Blodgett films of regioregular poly (3-hexylthiophene) as field-effect transistors. Langmuir 16(4):1834–1841
11. Maidul Islam AKM, Mukherjee M (2008) Characterization of Langmuir–Blodgett film using differential charging in X-ray photoelectron spectroscopy. J Phys Chem B 112(29):8523–8529
12. Silva EAD, Oliveira VJRD, Braunger ML, Constantino CJL, Olivati CDA (2014) Poly (3-octylthiophene)/stearic acid Langmuir and Langmuir-Blodgett films: preparation and characterization. Mater Res 17:1442–1448
13. Dohr M, Ehmann HMA, Jones AOF, Salzmann I, Shen Q, Teichert C, Ruzié C, Schweicher G, Geerts YH, Resel R, Sferrazza M, Werzer B (2017) Reversibility of temperature driven discrete layer-by-layer formation of dioctyl-benzothieno-benzothiophene films. Soft Matter 13(12):2322
14. Rebuffi L, Plaisier JR, Abdellatief M, Lausi A, Scardi P (2014) MCX: a synchrotron radiation beamline for X-ray diffraction line profile analysis. Z Anorg Allg Chem 640(15):3100–3106
15. Daillant J, Gibaud A (2009) X-ray and neutron reflectivity: principles and applications, vol 770. Springer

The Magnetic Properties of MnX_2 (X = Br, I): First Principles and Monte Carlo Study

Lahiri Saurav and R. Thangavel

Abstract Herein, magnetic properties of the emerging class of transition metal halides MnX2 (X = Br, I) have been studied using first principles calculation. The electronic band structure and spin polarized density of states (DOS) were plotted for both the systems to analyze the contribution of both the spin channels toward band formation in these systems. The magnetic ground state is revealed from exchange interaction computation using Green's function formulation. The Goodenough–Kanamori–Anderson rule revealed the competing interaction between the direct and super exchange interactions. The magnetic phase transition temperature was obtained from Monte Carlo simulation on the proposed Heisenberg model incorporating the anisotropy as evaluated providing a practicality of these systems for potential spintronics applications.

Keywords Magnetic system · Spintronics · Green's function formalism · Wannierization

1 Introduction

Two-dimensional materials are garnering extensive interest in the field of spintronics. The discovery of graphene has paved the way for exploring the two-dimensional materials for different applications like water splitting for hydrogen generation, flexible electronics, and recently the spintronics devices. Experimental realization of $Cr_2Ge_2Te_6$ [1] and CrI_3 has added to this research. Interest in transition metal halides arose after the successful exfoliation of CrI_3 [2]. Signatures of quantum spin hall insulators were detected in $GaBiCl_2$, ZrBr [3], and HfCl [4] at room temperature. Half metallicity was observed in $FeCl_2$ [5] along with ferromagnetism. While VCl_2, $CrBr_2$, and CoI_2 are antiferromagnetic (AFM) having a potential nano spintronics application [6].

L. Saurav · R. Thangavel (✉)
Condensed Matter Physics Laboratory, Department of Physics, Indian Institute of Technology (Indian School of Mines), Dhanbad, Jharkhand, India
e-mail: rthangavel@iitism.ac.in

Z. H. Khan et al. (eds.), *Recent Advances in Nanomaterials*, Springer Proceedings in Materials 27, https://doi.org/10.1007/978-981-99-4878-9_79

Among this class of compounds, the present research focusses on emerging class of transition metal dihalides MnX_2 (X = Br, I). These are dynamically as well as thermally stable systems. This system has been reported previously for electric, optical, and magnetic properties [7]. The feasibility of exfoliation from its bulk is reported. But exchange interaction and phase transition temperature of these compounds still is not comprehensively studied.

Motivated by these, present article focusses on the magnetic properties and exchange interaction evaluation by Green' function formalism. Monte Carlo simulation using Metropolis algorithm predicts the phase transition temperature of these systems.

2 Computational Details

Quantum ESPRESSO [8] is used to compute the electronic structure and total density of states. To account for the electron–electron correlation and exchange, Perdew Burke Ernzerhof (PBE) functional along with generalized gradient approximation (GGA) is implemented. Projector augmented wave (PAW) pseudopotential is used with a vacuum of 15 Å applied along the c axis for avoiding interaction of the periodic copies of the monolayer. Wannier90 software [9] is used to plot the Wannier interpolated bands. TB2J [10], a Python package is used to compute the exchange interaction using the Green's function formulation. Monte Carlo simulation within Metropolis algorithm is used to compute the magnetic phase transition temperature [11].

3 Results and Discussion

The lattice parameter of the $MnBr_2$ and MnI_2 is 3.89 Å and 4.16 Å, respectively, which are close to its experimental bulk counterpart's result [7]. The band gap were found to be indirect in both the cases with a band gap value of 2.78 eV and 2.54 eV. These systems shown in Fig. 1 belong to space group of $P\overline{3}m1$ which is trigonal crystal system.

The spin polarized density functional theory (DFT) calculation was performed on both the systems, and their respective band structures for both the spin channels that is spin up and spin down were plotted in Fig. 2. The net magnetic moment is 5 μ_B per unit cell that is consistent with the Hund's rule. For a detailed analysis of the magnetic ground state of these systems, Green's function formalism is used in order to obtain analytical expression for evaluating the exchange integral using small spin rotation as perturbation to the Hamiltonian. For implementing this formalism, mapping of the DFT Hamiltonian in plane wave basis set to the Wannier Hamiltonian is done which is a localized basis set using the Wannier90 package. The corresponding Wannier

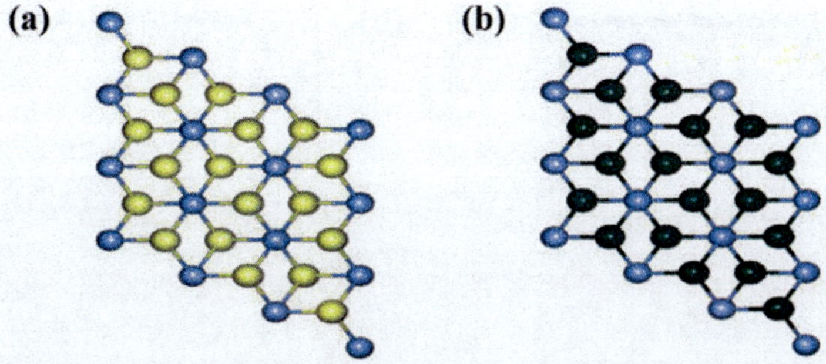

Fig. 1 **a** $MnBr_2$ and **b** MnI_2 monolayers where blue ball represents Mn atoms, yellow ball represents Br atoms, and green ball represents I atoms

interpolated bands are being plotted along with the DFT bands for depicting a match between them near about the Fermi level in Fig. 2.

A Python package TB2J is used to perform the exchange integral calculation employing the Green's function formalism. The resulting plots as shown in Fig. 3 depict an antiferromagnetic type behavior for both the systems with the nearest

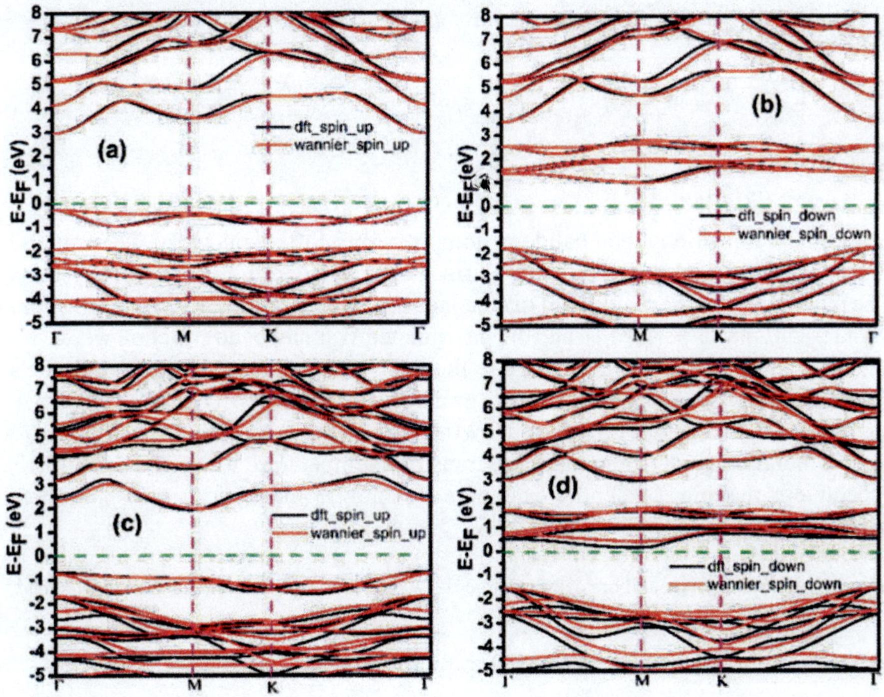

Fig. 2 Wannier interpolated and DFT band structure for both spin channels **a** and **b** of $MnBr_2$ as well as **c** and **d** of MnI_2

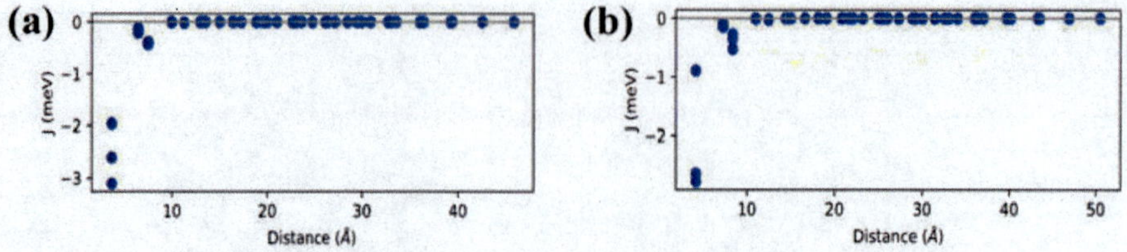

Fig. 3 Exchange integral versus nearest neighbor distance for **a** $MnBr_2$ and **b** MnI_2

neighbor exchange integral value of − 3.1 meV and − 2.64 meV for $MnBr_2$ and MnI_2, respectively.

The Monte Carlo simulations based on Metropolis algorithm were used to determine magnetic phase transition temperature. The Heisenberg Hamiltonian incorporating anisotropy is

$$H = -\sum_{i \neq j} J_{ij} S_i . S_j + S_i . J_{\text{anisotropy}} . S_j$$

where the magnetic anisotropy is 6.08 μeV as reported by Luo [7]. The resulting temperature was 25 K and 15 K for $MnBr_2$ and MnI_2, respectively. This study opens up the possibility for exploring antiferromagnetic spintronics application in this class of systems.

4 Conclusion

In summary, transition metal halide system, namely $MnBr_2$ and MnI_2, were studied within the DFT framework. The spin polarized DFT calculation showed these systems to be magnetic with the majority of the magnetic moment contribution from Mn atom. Wannierization was performed and the interpolated Wannier band structure were well matched within the fermi level range to the DFT bands. Subsequently, the Green's function method gave an AFM magnetic ground state with nearest neighbor exchange integral − 3.1 meV and − 2.64 meV for $MnBr_2$ and MnI_2, respectively, while Monte Carlo simulation gave magnetic phase transition temperature to be 25 K and 15 K, respectively, making it potential candidate for antiferromagnetic-based spintronics application.

Acknowledgements The authors acknowledge the Aryabhata HPC cluster, IIT (ISM) Dhanbad, for providing computational facility.

Declaration of Interest Statement The authors declare that they have no conflict of interests.

References

1. Gong C (2017) Discovery of intrinsic ferromagnetism in two-dimensional van der Waals crystals. Nature 546:265–269
2. Huang B (2017) Layer-dependent ferromagnetism in a van der Waals crystal down to the monolayer limit. Nature 546(7657):270–273
3. Li L (2017) Gallium bismuth halide GaBi-X_2 (X = I, Br, Cl) monolayers with distorted hexagonal framework: novel room-temperature quantum spin Hall insulators. Nano Res 10:2168–2180
4. Zhou L (2015) New family of quantum spin hall insulators in two-dimensional transition-metal halide with large nontrivial band gaps. Nano Lett 15(12):7867–7872
5. Torun E (2015) Stable half-metallic monolayers of $FeCl_2$. Appl Phys Lett 106:192404
6. Kulish VV (2017) Single-layer metal halides MX_2 (X = Cl, Br, I): stability and tunable magnetism from first principles and Monte Carlo simulations. J Mater Chem C (5):8734
7. Luo J (2020) The electric and magnetic properties of novel two-dimensional MnBr 2 and MnI2 from first-principles calculations. J Appl Phys 128:113901
8. Giannozzi P (2009) QUANTUM ESPRESSO: a modular and open-source software project for quantum simulations of materials. J Phys Condens Matter 21:395502
9. Pizzi G (2020) Wannier90 as a community code: new features and applications. J Phys Condens Matter 32:165902
10. He X (2021) TB2J: a python package for computing magnetic interaction parameters. Comput Phys Commun 264:107938
11. Evans RFL (2014) Atomistic spin model simulations of magnetic nanomaterials. J Phys Condens Matter 26:103202

Particle Size Effects on Magnetic Properties of $ZnFe_2O_4$ Ferrite

Mukesh C. Dimri, R. Stern, and H. Khanduri

Abstract Magnetic nanoparticles have gained a lot of interest in the last two decades due to their interesting properties and application in various devices. In this paper, we report the magnetic studies and low-temperature transitions in $ZnFe_2O_4$ nanoparticles and bulk samples synthesized by the citrate combustion method. XRD results confirm the spinel cubic phase in both samples treated at 600 °C and 1200 °C. Magnetic measurements show that nanopowders have hysteresis behavior without saturation due to their paramagnetic nature, whereas the sintered sample shows highly paramagnetic behavior. Temperature dependence of magnetization curves shows the blocking temperature of 17 K for the $ZnFe_2O_4$ nanopowders, higher than 11 K as observed for the corresponding bulk sample.

Keywords Spinel ferrite · Magnetic nanoparticles · Citrate combustion method

1 Introduction

Spinel ferrites are widely applicable materials due to their interesting magnetic, electronic, optical, and dielectric properties. The general formula is MFe_2O_4 (M = Mn^{2+}, Zn^{2+}, Fe^{2+}, Ni^{2+}, Co^{2+}, etc.), which has a lot of technological and medical applications [1–3]. The nanoparticles of these ferrites are quite important due to their modified properties as well as their application in medical, magnetic recording, and spintronic devices [4–6]. The properties of these ferrites strongly depend on the ratio of Fe^{2+} and Fe^{3+} ions at A (tetrahedral) and B (octahedral) lattice sites,

M. C. Dimri (✉)
Jaypee University of Engineering and Technology, Guna, Madhya Pradesh 473226, India
e-mail: mukesh.chandra@juet.ac.in

R. Stern
National Institute of Chemical Physics and Biophysics, 12618 Tallinn, Estonia

H. Khanduri
Indian Reference Materials Division, CSIR-National Physical Laboratory, New Delhi 110012, India

Z. H. Khan et al. (eds.), *Recent Advances in Nanomaterials*, Springer Proceedings in Materials 27, https://doi.org/10.1007/978-981-99-4878-9_80

which varies with the processing parameters of the synthesis method. In the present paper, we have synthesized the nano and bulk samples of $ZnFe_2O_4$ ferrite by the citrate combustion method, which is an eco-friendly method. The particle size effect on structural and magnetic studies of zinc ferrite is discussed and reported in this paper, spinel ferrites are widely applicable materials due to their interesting magnetic, electronic, optical, and dielectric properties. The general formula is Mfe_2O_4 (M = Mn^{2+}, Zn^{2+}, Fe^{2+}, Ni^{2+}, Co^{2+}, etc.), which have a lot of technological and medical applications [1–3]. The nanoparticles of these ferrites are quite important due to their modified properties as well as their application in medical, magnetic recording, and spintronic devices [4–6]. The properties of these ferrites strongly depend on the ratio of Fe^{2+} and Fe^{3+} ions at A (tetrahedral) and B (octahedral) lattice sites, which varies with the processing parameters of synthesis method. In the present paper, we have synthesized the nano and bulk samples of $ZnFe_2O_4$ ferrite by citrate combustion method, which is an eco-friendly method. The particle size effect on the structural and magnetic studies of zinc ferrite is discussed and reported in this paper.

2 Experimental

Polycrystalline powders of $ZnFe_2O_4$ ferrite were prepared by the citrate combustion method [7, 8]. Analytical grade zinc and iron nitrates were used as starting materials. Deionized (DI) water was used as solvent for preparing solutions, and citric acid acts as a combustion agent. As prepared fluffy powders were calcined at 600 °C/3 h and sintered at 1200 °C/5 h in the air for phase formation and particle size variation. Powder X-ray diffraction patterns were carried out with Rigaku X-ray diffractometer. Scanning electron micrographs (SEM) for microstructure were done by a Zeiss EVO-MA15 apparatus. Measurement of magnetic hysteresis and temperature dependence of magnetization were carried out using the vibrating sample magnetometer option of 14 T—PPMS (Quantum Design).

3 Results and Discussion

The X-ray diffraction patterns measured at room temperature for $ZnFe_2O_4$ nanopowder and bulk sample are shown in Fig. 1. Single-phase spinel ferrite formation is confirmed in both the samples, which reflects all desired peaks of $ZnFe_2O_4$. The peaks are broader for powders treated at 600 °C which can be related to lower grain sizes. Using the Scherrer formula the average crystallite size was 20 nm and 60 nm for the powders calcined at 600 and 1200 °C respectively. The particle sizes were found to be in a nanometre range which is confirmed by the SEM image shown in the inset of Fig. 1. The average grain size was obtained below 30 nm for the calcined powders estimated from SEM micrograph.

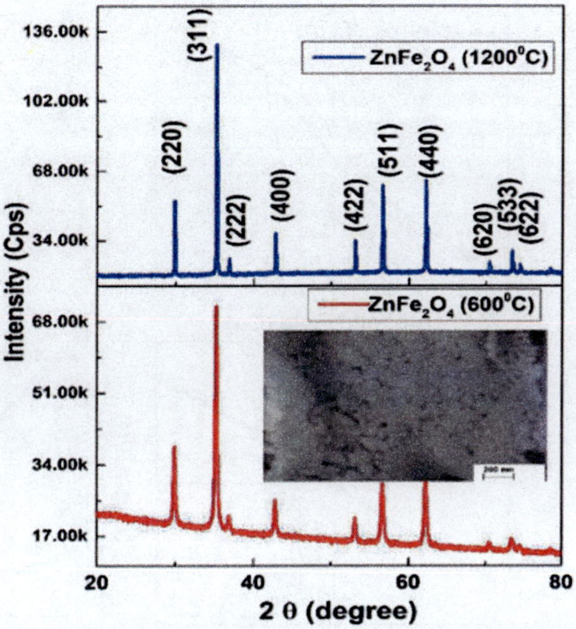

Fig. 1 XRD patterns for $ZnFe_2O_4$ powders heat treated at 600 °C/3 h and 120 °C/5 h (inset image shows the SEM for powder treated at 600 °C/3 h)

The magnetic hysteresis curves measured at 300 K for calcined powder and sintered samples are shown in Fig. 2a. We can see that powders calcined at 600 °C show some ferromagnetic nature in addition to the paramagnetic behavior, whereas the powders treated at 1200 °C show characteristics of paramagnetic M-H curve. This can be related to the nanoparticle nature of $ZnFe_2O_4$ powders treated at 600 °C, as nanomaterials have some change in structure with changes the positions of Fe^{3+} and Zn^{2+} ions causing enhancement in ferromagnetism [9].

Magnetization versus temperature curves (ZFC and FC, magnetic field = 1000 Oe) for these samples are shown in Fig. 2b. These curves exhibit characteristic low-temperature peak at 11 K in bulk $ZnFe_2O_4$ (expected around 10 K), whereas it shifted to 17 K in case of nanopowders. The high-temperature (300–850 K) dependence of magnetization was also measured (figures not included), showing the Curie temperature was around 640 K for bulk $ZnFe_2O_4$, whereas it was observed 750 K for the nanopowders. It is again because of the change in Fe^{3+} and Fe^{2+}/Zn^{2+} lattice sites, which modifies the magnetic interactions among different lattice sites.

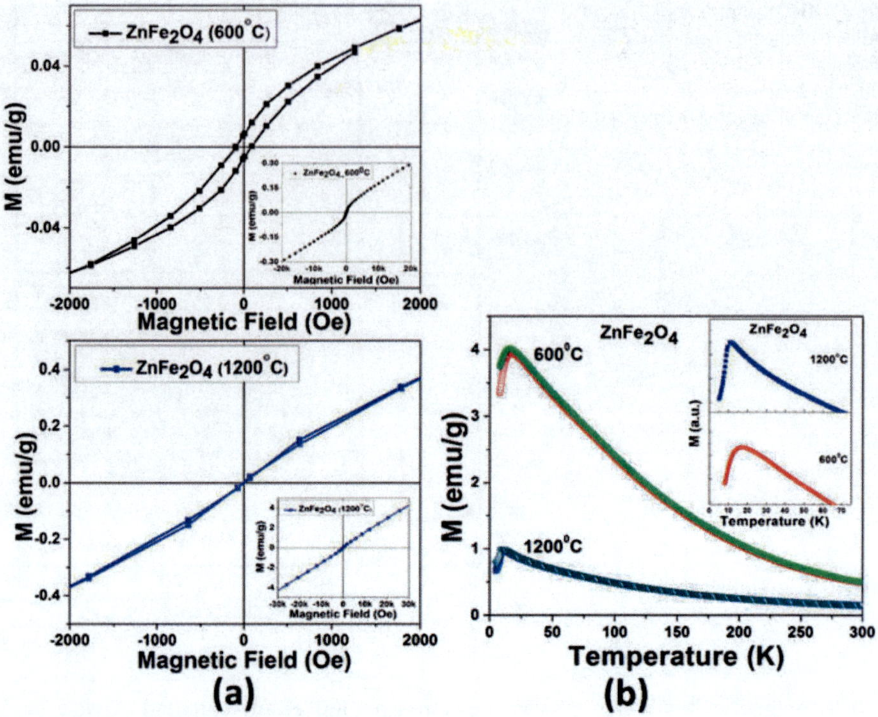

Fig. 2 **a** M-H curves measured at room temperature (300 K) for $ZnFe_2O_4$ powders heat treated at 600 °C/3 h and 1200 °C/5 h. **b** magnetization versus temperature curves measured for $ZnFe_2O_4$ samples (magnetic field = 1000 Oe)

4 Conclusions

Nano powders and polycrystalline bulk samples of $ZnFe_2O_4$ were prepared by the citrate combustion method. The XRD pattern confirms the spinel phase formation with all characteristic peaks of $ZnFe_2O_4$. Powders calcined at 600°C have nanoparticle nature (average grains ~ 30 nm), confirmed from SEM image. Bulk sample shows paramagnetic behavior, whereas the nano powders show some ferromagnetic nature in M-H curves measured at room temperature. Low-temperature magnetization measurement shows that the blocking temperature around 11 K and 17 K for the bulk and nano powders, respectively. The Curie temperature was higher for the nanoparticles as compared to the bulk samples due to change in the octahedral and tetrahedral lattice sites occupied by iron ions. These modified or tailored properties due to nanoparticle nature may be used for some electronic, magnetic, or medical applications.

Acknowledgements One of the authors H. Khanduri acknowledges the Department of Science and Technology, Government of India, for the DST-INSPIRE Faculty Award (DST/INSPIRE/04/

2017/002826). R. Stern is supported by the European Regional Development Fund Project (TK134) and the Estonian Research Agency Project (PRG4 and IUT23-9).

References

1. Dippong T (2021) Erika Andrea Levei. Oana Cadar Nanomaterials 11:1560
2. Dimri MC, Verma A, Kashyap SC, Dube DC, Thakur OP, Prakash C (2006) Mater Sci Eng B 133:42–48
3. Ansari SM, Ghosh KC, Devan RS, Sen D, Sastry PU, Kolekar YD, Ramana CV (2020) ACS Omega 5(31):19315–19330
4. Kefeni KK, Msagati TA, Mamba BB (2017) Mater Sci Eng B 215:37–55
5. Zhang J,Song JM, Niu HL, Mao CJ, Zhang SY, Shen YH (2015) Sens Actuators B Chem 211:55–62
6. Lal G, Punia K, Bhoi H, Dolia SN, Choudhary BL, Alvi PA, Dalela S, Barbar SK, Kumar S (2021) J Alloy Compd 886:161190
7. Dimri MC, Khanduri H, Agarwal P, Pahapill J, Stern R (2019) Magn Magn Mater 486:165278
8. Dimri MC, Khanduri H, Agarwal P, Garg V, Mere A, Stern R (2020) AIP Conf Proc 2265:030517
9. Milanović M, Moshopoulou EG, Stamopoulos D, Devlin E, Giannakopoulos KP, Kontos AG et al (2013) Ceram Int 39(3):3235–3242

Bandgap Engineering and Visible Luminescence in Cubic Yttrium Oxide via Lanthanides Doping/Co-Doping

R. Vats, C. Bhukkal, B. Goswami, and R. Ahlawat

Abstract We report synthesis of pure and lanthanide doped Y_2O_3 by a facile citrate mediated sol–gel approach. The prepared samples are investigated for their optical properties via UV–Vis and fluorescence spectroscopy. The Tauc's plot is used for the determination of band gap of the prepared samples. Effect of co-doping is also emphasized for the enhancement of luminescence in the entire visible region. In this paper, an effort is made to optimize the absorbance and photoluminescence (PL) in pure, Tb^{3+} doped and Tb^{3+}/Dy^{3+} co-doped samples of Yttria. The obtained emission results are emphasized via chromaticity curves.

Keywords UV–vis · Photoluminescence · Y_2O_3 · Doping · Co-doping · Tauc's plot

1 Introduction

The luminescence efficiency of Yttria (Y_2O_3) can largely be modified by the addition of a suitable Lanthanide [1]. In addition, to their unique spectral properties, Lanthanide ions have low sensitivity for the rare earth oxide hosts. The host and trivalent Lanthanide ions weak interactions favor their extensive use in luminescence applications [2, 3]. Among Lanthanide ions, trivalent Dysprosium and Terbium ions are technologically very useful due to their emission in visible as well as IR region [4, 5]. In previous studies, miraculous changes in the structural and thermal behavior are noticed on suitably doping the Lanthanides in the Y_2O_3 hosts. There are many reports that claim intense emission in the visible region due to Lanthanide dopants. Wang et al. [6] have reported improved photoluminescence emission as well as optical temperature sensing behavior of Er^{3+}: Y_2O_3 microtubes co-doped

R. Vats (✉) · C. Bhukkal · B. Goswami · R. Ahlawat
Ch. Devi Lal University, Sirsa, Haryana 125055, India
e-mail: sharmariddhi11@gmail.com

C. Bhukkal
Government College, Adampur, Hisar, Haryana 125052, India

Z. H. Khan et al. (eds.), *Recent Advances in Nanomaterials*, Springer Proceedings in Materials 27, https://doi.org/10.1007/978-981-99-4878-9_81

with Tm^{3+}, Ho^{3+} ions. They have found green and red emission in prepared microtubes. Prasanna Kumar et al. [4] have reported blue, red, and prominent yellow emission in Dy^{3+} doped cubic crystalline Y_2O_3. Therefore, we report synthesis of Pure, Tb^{3+} doped and Tb^{3+}/Dy^{3+} co-doped sample of Yttria nanopowder by 'Pechini-type sol–gel technique' and investigation of their structural and luminescent properties. The detailed synthesis scheme is described in our previous publication [7]. All the prepared samples are heated to a fairly high temperature around 650 °C for 4 h. The Tb^{3+} concentration is 2.50 mol% in doped and co-doped samples of Y_2O_3 along with 3.00 mol% of Dy^{3+} ion.

2 Results and Discussions

2.1 *Structural Analysis Using XRD*

Cubic crystalline phase is found to establish in XRD patterns of pure, doped and co-doped samples of Yttria. Figure 1(a) represents a comparison of XRD pattern of pure Yttria with standard JCPDF card—00-041-1105. The observed diffraction peak positions and (*hkl*) values in pure Yttria lie at (2θ) ~ 20.3438°, (211); 29.0266°, (222); 33.6388°, (400); 48.3939°, (440) and 57.5022°, (622) [5]. On doping and co-doping with different dopants, a meager shift in the main peak is also observed and is shown in the inset of Fig. 1(a). The crystallite size Fig. 1(b) as evaluated from the intercept of the W–H plot Eq. (1) are 18.28, 19.47 and 22.91 nm, respectively [8].

$$\beta_{hkl} \cos\theta = k\lambda/D + 4\epsilon \sin\theta \tag{1}$$

where D = crystallite size, β_{hkl} = FWHM, λ = X-ray wavelength, θ = diffraction angle, ϵ = microstrain and $k = 0.9$ for spherical particles.

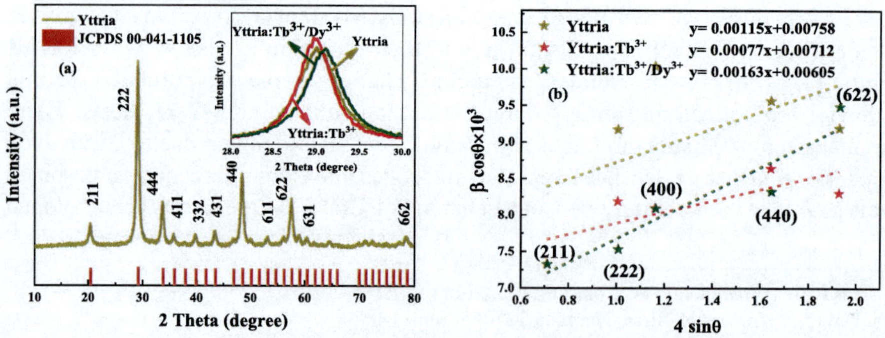

Fig. 1 (a) XRD spectra of pure Yttria, (b) W–H plots of pure, doped and co-doped samples of Yttria

2.2 UV–Vis and Photoluminescence Spectroscopy

UV–Vis spectra of pure, doped and co-doped samples are depicted in Fig. 2(a). Absorption spectra of pure Yttria comprise a band at 225 nm and a hump at 300 nm. These bands correspond to 2p level of oxygen and 4d level of Yttrium [7]. Further, there is a band at 270 nm in doped and co-doped Yttria sample, probably due to dopant ion. Tauc's plot are also plotted for determination of energy gap of prepared nanopowders [5]. The observed band gap of pure, doped and co-doped sample is 4.63, 3.42 and 3.89 eV, respectively. The observed band gaps are somewhat smaller than that of the bulk Yttria, which suggests the presence of defect states in the samples [7]. Figure 3(a) represents the PLE spectra of all the three samples (pure, doped and co-doped; $\lambda_{ex} = 270$ nm). The PLE spectra of pure Yttria consist of few weak peaks with low intensities. These peaks may be due to presence of defect states in the sample. On doping pure Yttria with Tb ions, strong emission peaks appear in the visible region of spectra with highest emission in the green region at 544 nm. This luminescence peak corresponds to $^5D_4 \rightarrow {}^7F_5$ transition of Tb^{3+} ion [9].

However, in the co-doped sample (Tb^{3+}/Dy^{3+} doped Yttria), the characteristic emission peak due to Tb^{3+} ion is suppressed, instead an intense peak at 486 nm, a small peak at 572 nm along with few weak peaks in visible region of spectra are present due to characteristic emission of Dy ions. These peaks are attributed to $^4F_{9/2} \rightarrow {}^6H_{15/2}$ and $^4F_{9/2} \rightarrow {}^6H_{13/2}$ transition of Dy^{3+} ion, respectively [10, 11]. The chromaticity curve of the samples is represented in Fig. 3(b). The color coordinates of pure, doped and co-doped samples of Yttria are (0.3258, 0.3816), (0.2890, 0.5575)

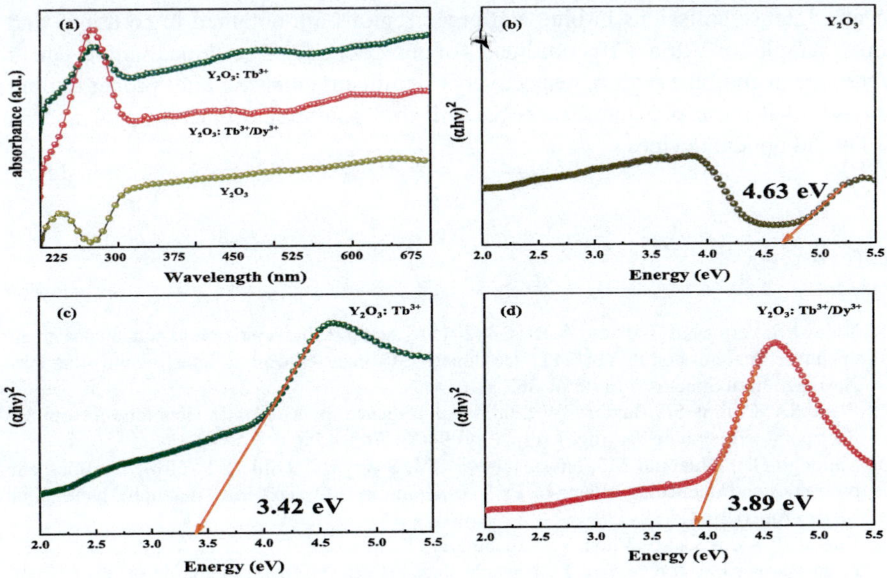

Fig. 2 (a) UV–Vis spectra, (b-d) Tauc's plot of pure, doped and co-doped samples of Yttria

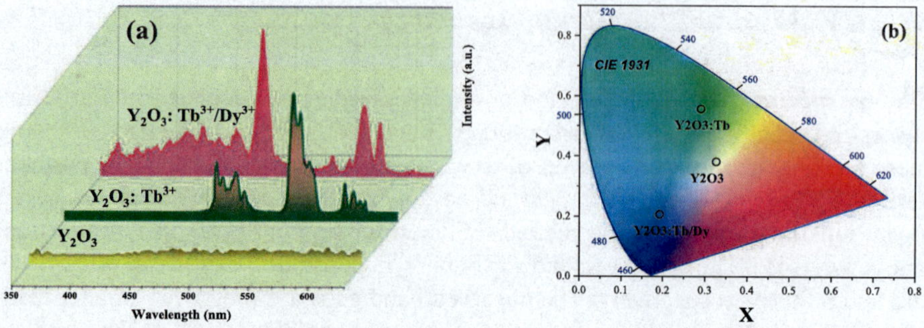

Fig. 3 (a) The PL spectra, (b) chromaticity curve of pure, doped and co-doped sample of Yttria

and (0.1924, 0.2065), respectively. These color coordinates lie in white, green and blue region, respectively, The PL emission in visible region reveals that the prepared nanopowders can be potentially used in phosphors, solid state lasers and modern LEDs.

3 Conclusion

In conclusion, we synthesized pure, doped and co-doped samples of Yttria via a facile and cost-cutting approach. All samples possess cubic crystalline phase at moderate annealing conditions. Crystallite sizes as calculated by W–H method are about 15–25 nm. Intense emissions in blue and green regions are obtained in co-doped and doped samples of Yttria. CIE coordinates of pure, doped and co-doped samples lie in white, green and blue region, respectively. Significant emission after proper doping suggests that these phosphor materials will find potential applications in modern LEDs and optical devices.

References

1. Yadav RS, Verma RK, Bahadur A, Rai SB (2015) Structural characterizations and intense green upconversion emission in Yb^{3+}, Pr^{3+} co-doped Y_2O_3 nano-phosphor. Spectrochim Acta Part A Mol Biomol Spectrosc 137:357–362
2. Wade SA, Collins SF, Baxter GW (2003) Fluorescence intensity ratio technique for optical fiber point temperature sensing. J Appl Phys 94(8):4743–4756
3. Mhlongo GH, Dhlamini MS, Ntwaeaborwa OM, Swart HC, Hillie KT (2014) Luminescent properties and quenching effects of Pr^{3+} co-doping in SiO_2:Tb^{3+}/Eu^{3+} nanophosphors. Opt Mater (Amst) 36(4):732–739
4. Kumar JP, Ramgopal G, Vidya YS, Anantharaju KS, Prasad BD, Sharma SC, Prashantha SC, Nagaswarupa HP, Kavyashree D, Nagabhushana H (2015) Green synthesis of Y_2O_3:Dy^{3+} nanophosphor with enhanced photocatalytic activity. Spectrochim Acta Part A Mol Biomol Spectrosc 149:687–697

5. Vats R, Ahlawat R (2022) Rietveld refinement, luminescence and catalytic study of as—synthesized and—Dy^{3+} -doped cubicnanopowder prepared by citrate mediated sol–gel technique. J Nanoparticle Res 24:188
6. Wang X, Wang Y, Marques-Hueso J, Yan X (2017) Improving optical temperature sensing performance of Er^{3+} doped Y_2O_3 microtubes via co-doping and controlling excitation power. Sci Rep 7(1):1–13
7. Vats R, Ahlawat R (2021) Impact of annealing time on structural evolution of pure and Dy^{3+} -doped CeO_2 nanopowder, rietveld refinement and optical behavior. Int J Nanosci 20(4):1–15
8. Malleshappa J, Nagabhushana H, Kavyashree D, Prashantha SC, Sharma SC, Premkumar HB, Shivakumara (2015) Shape tailored green synthesis of CeO_2:Ho^{3+} nanopowders, its structural, photoluminescence and gamma radiation sensing properties. Spectrochim Acta Part A Mol Biomol Spectrosc 145:63–75
9. Back M, Massari A, Boffelli M, Gonella F, Riello P, Cristofori D, Ricco R, Enrichi F (2012) Optical investigation of Tb^{3+}-doped Y_2O_3 nanocrystals prepared by Pechini-type sol-gel process. J Nanoparticle Res 14(4)
10. Mariscal-Bbecerra L, Acosta-Najarro D, Falcony-Gguajardo C (2018) Luminescent and structural analysis of yttrium oxide doped with different percentages of terbium and dysprosium to obtain different shades of green to yellow
11. Mishra K, Singh SK, Singh AK, Rai SB (2012) Optical characteristics and charge transfer band excitation of Dy^{3+} doped Y_2O_3 phosphor. Mater Res Bull 47(6):1339–1344

Effect of Impulse Pressure on Diyl Diphenol Cross-Linked Polymer

Navin S. Mathew, Raja Devangan, Navin Kumar, and Prashant S. Alegaonkar

Abstract Impulse pressure response of epoxy resin (Diyl Diphenol cross-linked polymer) is reported by subjecting it to the shock test by split-Hopkinson pressure bar (SHPB) at the strain rate of 1650 s^{-1}. In particular, phase-space curves such as pressure–volume, pressure–particle velocity and shock velocity–particle velocity have been investigated. These recorded shock properties can later lay foundation to the shock loading of other materials, whose dynamic response is to be investigated, can be embedded in the polymer specimen. The data analyzed from the experiment shows the yield strength and HEL which is in good agreement.

Keywords Impulse response · Shock · Bisphenol A · Split-Hopkinson pressure bar · Hugoniot elastic limit · Phase-space curves

1 Introduction

Mechanical waves having finite amplitudes resulting from rapid compression are called shock waves. They can be characterized as velocity of propagation (pressure dependent), a steep wave front formation, a strong decline in the velocity of propagation with increase in separation from the origin (nonplanar shock waves) and interaction and reflection properties. In gas, liquid, solid, plasma and multiple phases media, shock wave effects are observed. Shock waves range can scale from microscopic (nano) to macroscopic (cosmic) proportions which subsequently caused the emergence of its fields in various areas like physics, medicine, material science, military technology, engineering, etc. Perhaps World War II opened the biggest window of shock wave physics which produced new disciplines such as nuclear

N. S. Mathew · P. S. Alegaonkar (✉)
Department of Physics, School of Basic Sciences, Central University of Punjab, Bathinda 151401, India
e-mail: prashant.alegaonkar@cup.edu.in

R. Devangan · N. Kumar
Department of Mechanical Engineering, Indian Institute of Technology Ropar, Rupnagar, Punjab 140001, India

Z. H. Khan et al. (eds.), *Recent Advances in Nanomaterials*, Springer Proceedings in Materials 27, https://doi.org/10.1007/978-981-99-4878-9_82

explosions, implosions, exploding wires, hypersonic aerodynamics, magnetofluid dynamics, laser-supported detonation, rarefied gas dynamics, super aerodynamics, impact physics, fracture mechanics, cosmic gas dynamics, laser fusion and high-rate materials dynamics. These vast quantities of disciplines which fall under the category of shock waves emerged from complex interactions its disciplines or formed unconventionally from further branches. Most popular natural shock wave include thunder and meteor fall [1]. Pressure which generally exceeds the magnitude of GPa is considered to be an explosion. It is a mechanical perturbation that propagates at high shear velocities in a medium and is categorized as blast, shock, plastic, elastic and explosion waves with nonlinear amplitudes. An explosion is an uncontrollable event, but it can be achieved at the laboratory scale by utilizing split-Hopkinson pressure bar (SHPB), stand-off/contact explosion and shock wave tube techniques. SHPB is one among them, as it is a quick, low-cost and safe way to generate shock waves [2].

Various types of waves can travel in solids depending on the mobility of the solid particles in relation to the wave propagation direction and the boundary conditions. Classic elastic waves in solids include longitudinal (irrotational, primary and dilatational), transverse (shear, distortional or secondary), Rayleigh (Surface), Stoneley (interfacial), love and bending (flexural) waves [3]. The magnitudes of longitudinal and shear waves fall as r^{-1} in the region furthest from the free surface. The longitudinal and shear wave amplitudes decline as r^{-1} and quicker along the surface as r^{-2}. At $r^{-1/2}$, the Rayleigh waves decay more slowly, allowing it to be measured across greater distances. Rayleigh wave carries most of the energy (67% for $v = 0.25$). Only 7% of the energy is carried by longitudinal waves. Love waves particle displacements are aligned to the wave front due to changes in elastic constants near to the surface. Stoneley waves are a broader sort of wave that arises at the interface between layers that are experiencing discrete impedances of peak amplitude around the interface. Shock waves contain sharp front, necessitating a uniaxial strain state that allows for the accumulation of high amounts of hydrostatic factor of stress. When the hydrostatic aspect of the material surpasses the dynamic flow stress over many degrees, the material is believed to have no impedance to shear. The Rankine–Hugoniot conservation equations are used to compute the shock wave parameters [4].

Real shock waves have idiosyncrasies that are material and pressure dependent. An ideal shock wave profile constitutes regions of discontinuity, plateau and a steady return to zero pressure as shown in Fig. 1a while b depicts a general profile of interface velocity. Due to the deviatoric component of stress, the pressure–volume curve for an actual material varies from the Hugoniot (hydrostatic) curve. In the range of elastic region, the rate of increase in stress with volume is substantially faster.

The pressure–volume curve changes slope once the elastic limit beneath the applied stress and strain rate circumstances HEL (Hugoniot elastic limit) is achieved. The elastic and plastic portions of the wave are separated in Fig. 1b. Below the HEL, the elastic component travels at a faster rate than that of the plastic wave. As a result, the effects shown in Fig. 1 can be described. The HEL is attained after a sharp increase in pressure (or U_p). The pressure continues to climb to the top beyond the HEL (no discontinuity). The rate at which this pressure rises is determined by

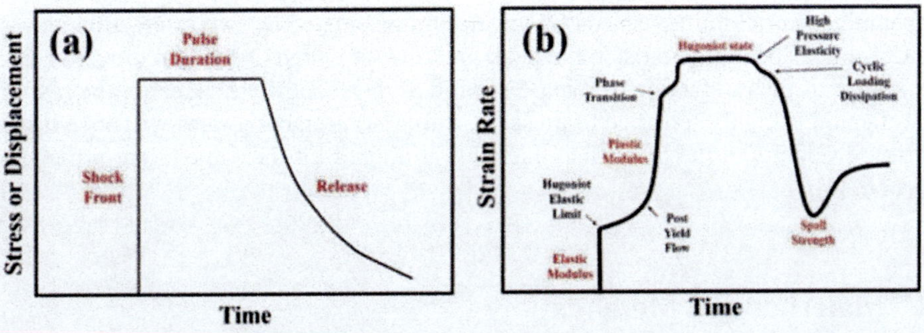

Fig. 1 Shock profile **a** idealized; **b** realistic

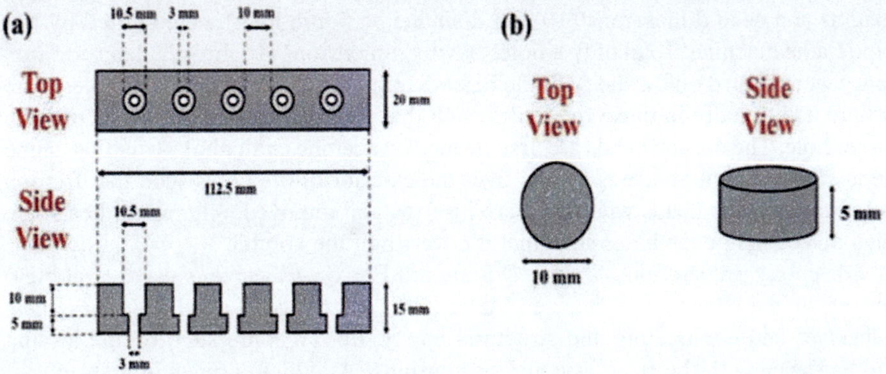

Fig. 2 Design of aluminum die showing both top and side views of **a** aluminum die; **b** aluminum bead

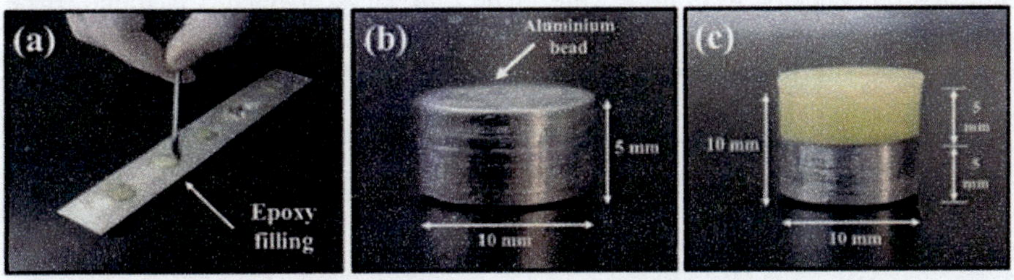

Fig. 3 **a** Specimen preparation for SHPB test; **b** aluminum bead as base; **c** epoxy

the material's constitutive behavior. If there is a phase change, then there can be a visible indication in the wave profile. It is possible that the wave will split into two. Pulse duration plateau can be found near the top of Fig. 1. When unloading begins, it happens in two stages: elastic and plastic. In a similar way as the HEL on loading, this elastoplastic transition leaves a signal on unloading phase. Spalling is a type of dynamic material failure caused by tensile strains created by two rarefaction waves interacting. The region outside the peak pressure, at which pressure returns to zero, is known as the rarefaction or release section of the shockwave. Because the rate of rarefaction is highly sensitive to pressure, different rarefaction rates will arise if the striking velocity of the flyer plate thickness is altered for the same target-projectile system [4].

2 Materials and Methods

Aluminum 6061 alloy standard grade solid was used for creating die and bead having density of 2.7 g cm^{-3}. The die dimensions (112.5 mm length × 20 mm width × 15 mm height) and bead dimensions (10 mm diameter × 5 mm long) were crafted by M3 type Lathe machine. Total of five holes having dimensions (10 mm diameter × 5 mm long) were drilled on the die to fit the bead. 5 mm long × 3 mm diameter holes were drilled additionally in those five holes such that the bead can be retrieved from the larger hole. The die and beads are first cleaned by acetone or alcohol. After that, a fine deposit of vacuum grease is spread over the exterior of the holes such that friction is reduced for the beads when they are inserted and removed. Aluminum beads are then placed inside the holes such that it covers half the volume.

For epoxy preparation, Araldite® Standard Epoxy Adhesive is used. Araldite® belongs to the Huntsman Advanced Materials having range of acrylic, polyurethane adhesives and engineering and structural epoxy. Epoxy is prepared by the mixing the hardener with the resin at equal proportions. Araldite® comes in a variety of packaging options, the most typical of which includes two hardener and resin tubes. Araldite® Standard Epoxy Adhesive of 180 g box containing (1 × 100 g) tube of resin and (1 × 80 g) tube of hardener was used. Equal quantities (by volume) of resin and hardener of Araldite® were taken out. Mass of resin was found to be 168.8 mg which appeared to be transparent, and the mass of hardener was 156.4 mg which appeared yellow. Resin and hardener were mixed well until a uniform color (creamy) was formed. Then, mixture was deposited inside the holes of aluminum die cast, where an Aluminum bead is resting by covering half the volume of the hole. Mixture is then allowed to fill the cavity completely and is at the same surface level of the die. The epoxy was then allowed to rest for 24 h such that it achieves full strength. The epoxy was removed from the die with the aluminum bead attached to it and packed for experimentation.

A commercial chemical known as bisphenol A (BPA) is used to make polycarbonate, a tough, clear plastic that can be found in a variety of consumer goods. Epoxy resins, which act as a protective coating on the inside of numerous beverage cans and metal-based products, also contain BPA. The chemical formula of BPA is $C_{15}H_{16}O_2$, and its molecular mass is 228.291 g mol^{-1}. It appears white solid, and its density is 1.20 g cm^{-3}. It is an inert material that dissolves in organic solvents but not in water. BPA's most common use is as a co-monomer in the manufacture of polycarbonates, which amounts for 65–70% of total BPA production, while 25–30% of BPA is used in the manufacture of vinyl ester resins and epoxy resins [5, 6]. The residual 5% is used as a primary construct in several high-performance plastics as well as a minor addition in thermal paper, PVC, polyurethane and many other products. Although it is frequently mislabeled as a plasticizer, it is not one. A powerful acid, such as concentrated sulfuric acid, hydrochloric acid or a solid acid resin such as polystyrene sulfonate, catalyzes the condensation of acetone (thus it gets the suffix "A" in BPA) [7] with two counterparts of phenol [8]. To achieve complete condensation and avoid by-products like Dianin's compound, a surplus of phenol is utilized. A reaction with phosgene is carried out under biphasic circumstances, with the hydrochloric acid being scavenged with aqueous base [9]. Individual BPA molecules are converted into massive polymer chains during this process, effectively trapping them. Epoxy resin is transformed to its diglycide ether first [10]. This is accomplished through a basic reaction with epichlorohydrin. Bis-GMA, that is used to manufacture vinyl ester resins, is formed by reacting some of this with methacrylic acid. BPA can also be ethoxylated and then transformed to diacrylate and dimethacrylate derivatives, though to a much smaller extent. To modify the physical properties of vinyl ester resins, they can be introduced at low amounts [11].

SHPB or Kolsky bar generally comprises a striker bar (projectile) which collides with the incident bar and gives rise to a pulse that has a larger dimension compared to the specimen. The incident bar carries the elastic wave, and it arrives at the specimen which is bounded amid the incident and transmitted bars. This wave amplitude is large enough such that in the specimen plastic deformations can be observed. The strain gages are joined to both incident and transmitter bars which take the measurements of incident, reflected and transmitted pulse, whose amplitudes are ε_i, ε_r and ε_t. Strain rates of the scale 10^2–10^4 s^{-1} are attained [4]. Generally, Kolsky bar apparatus contains three major mechanisms as shown in Fig. 4 [12]. The detonation method developed by Kolsky is of dynamic loading type [13]. The most frequent method for dynamic loading is hitting the incident bar with a striker. For Kolsky compression bars, gas pistols have been found to be effective, controlled and safe. A quick release of pressurized air or a mild gas from a pressure storage reservoir launches the striker, which accelerates in the gun barrel until it collides with the incident bar's end. The striker hits the incident bar at fixed speed thanks to gas venting holes carved into the side of the gun barrel near the exit. Just before the collision, the striking velocities are optically or magnetically monitored. The impact on the incident bar is controlled and repeatable with this type of striker launching mechanism. The striking speed can easily be changed by altering the level of the striker inside the gun barrel or the pressure of the pressurized gas in the tank. The length of the striker influences the loading time. An incident bar, a transmission bar, a momentum trap device and an optional extension bar are the components of a conventional Kolsky compression bar. Most of the time, all bars have the same diameter and are constructed from the same material. The material for the bars should be linearly elastic and have a high yield strength since surface strains are required to calculate the stress waves inside the bars. In Kolsky bar studies, strain gauges have become a typical tool for measuring bar strains. Across a bar diameter, two strain gauges are normally connected symmetrically on the bar surface. A Wheatstone

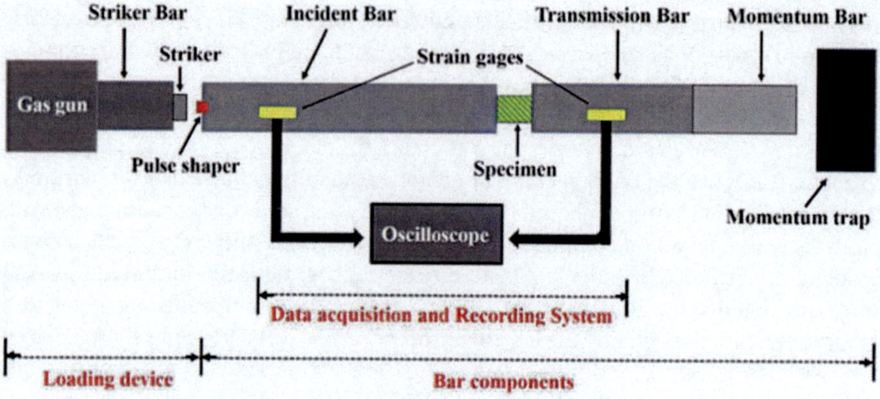

Fig. 4 General Kolsky compression bar apparatus

bridge is used to condition the signals from the strain gauges. The oscilloscope and amplifier in a typical Kolsky bar experiment need to have high enough frequency response to capture the signal, which typically lasts less than one millisecond. A data-acquisition system's components should all, in general, have a minimum frequency response of 100 kHz. Although the oscilloscope has a suitable frequency response, using the 3 kHz and 100 Hz filters causes the recorded signals to be significantly distorted [12].

The most frequent means of producing shock waves are planar, normal and parallel impact. Two plane surfaces are engaged in a planar impact. The term "parallel impact" refers to when two surfaces remain parallel and make contact at the same moment, i.e., once all points of the contacting surfaces connect at the same time. The most frequent means of producing shock waves are planar, normal and parallel impact. Two plane surfaces are engaged in a planar impact. The term "parallel impact" refers to when two surfaces remain parallel and make contact at the same moment, i.e., once all points of the contacting surfaces connect at the same time [4]. The path of motion of the projectile is orthogonal to its surface in a normal hit as shown in Fig. 5. Epoxy resin's response to planar shock waves produced by plate impact was examined [14]. In particular, the Hugoniot has been studied, as well as the relationship between shear strength and impact stress in the stress-particle velocity and shock velocity-particle velocity spaces. It is explored how shear strength changes as longitudinal load increases. It has been demonstrated that lateral stresses drop beyond the shock front, suggesting that the material's strength increases under shock loading. It has been suggested that the viscoplastic properties of epoxy-based resins may manifest. The behavior of a carbon-fiber epoxy composite to one-dimensional shock loading is discussed [15]. According to the findings, transit distance has no impact on either the shock velocity or the shock stress. The relationship between U_s and U_p over the range of measurements indicated a larger value of C_o and a lower value of S, which was caused by the microstructure's somewhat lower compressibility and slightly higher proportion of carbon fibers. Hugoniot stress and predicted hydrodynamic pressure

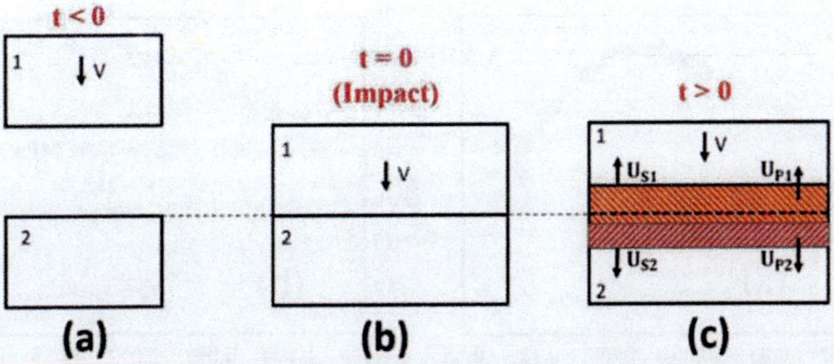

Fig. 5 Events of impact **a** before impact, projectile 1 moving at velocity V; **b** at instant of impact; **c** after impact, wave propagation stage

discrepancies indicate that this material's shear strength rises as shock stress does. The investigation of the strain rate response of PMMA under uniaxial compression at room temperature at strain rates ranging from 0.0001 s^{-1} to almost 4300 s^{-1} is reported [16]. Additionally, at strain rates of 1 s^{-1} and 0.001 s^{-1} with temperatures ranging from 0 °C to 115 °C, the temperature response of PMMA was studied. Using a split-Hopkinson pressure bar (SHPB) with pulse-shaping, high strain rate experiments at room temperature (more than 1 s^{-1} rates) were carried out. Under the influence of one-dimensional shock loading, the behavior of polymer polyether ether ketone (PEEK) has been researched [17]. In line with many other materials, a simple linear response can be seen when the relationship between shock velocity and particle velocity is examined. It has also been demonstrated that shear strength rises with shock stress, with a break in the slope occurring at 1.0 GPa. The material appears to operate simply elastically below this stress, which points to a Hugoniot elastic limit of 1 GPa. Additionally, it has been noted that shear strength both above and below 1 GPa increases dramatically after the shock front. Other polymeric materials have displayed a similar behavior, and it has been hypothesized that these materials were responding via a viscoplastic mechanism.

3 Results and Discussion

The data from Kolsky bar includes stress, time, strain and strain rate. Total time duration was 307 μs. The graphs of these parameters were plotted, and the curve of stress versus strain is represented as a bilinear function, where the elastic region is followed by region of plastic deformation as shown in Fig. 6a. This curve is also known as the dynamic work hardening of the material. An equation:

$$\sigma_o = \sigma - k^n \tag{1}$$

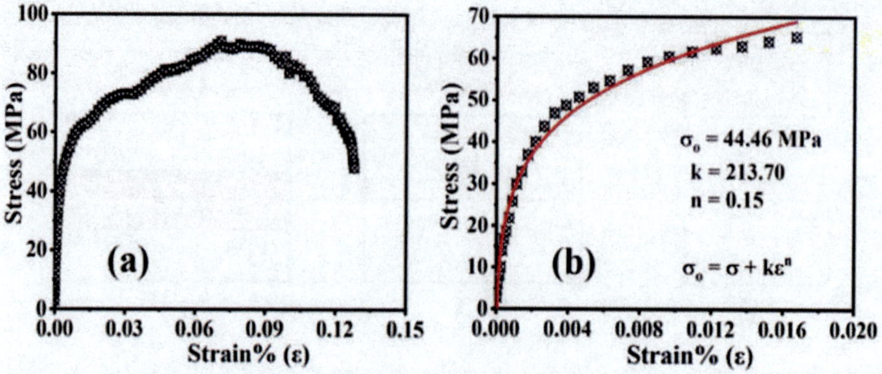

Fig. 6 Stress versus strain curve **a** complete curve; **b** elastic region showing the power law

where the material's maximal stress tolerance in the elastic regime is σ_o, n is the exponent having value less than one, and k is a pre-exponential factor.

Figure 6b depicts the system's observed behavior in the elastic area. A steeper graph would indicate a better elastic yield strength of the material. Higher value of n indicates a better stress accumulation ability of the material [18], where the value of n for epoxy was found to be 0.15. If the value is higher, say near one, then it indicates that the stress is almost independent of the applied strain [19]. Pre-exponential factor (k) indicates how swiftly the stress gets build up, where the value was found to be 213.70. Such materials could have significant impact on the static strength. The value of σ_o was estimated to be 44.46 MPa. The HEL of the epoxy was estimated from the graph to be to be 61.97 MPa. The elastic phase is followed by a plastic phase after HEL, from the graph, it is evident that when the curve crosses the 61.97 MPa stress mark, and there is an abrupt change in the slope of the graph. Maximum stress obtained is 90.87 MPa.

The strain vs time graph (Fig. 7b) is shown with curve which is found to mathematically obey the equation:

$$\varepsilon = A_0 + A_1t^1 + A_2t^2 + A_3t^3 + A_4t^4 + A_5t^5 \tag{2}$$

where A_{is} are the polynomial coefficients linked to the dynamic structure variables of the matrix experiencing deformations. Dynamic shear factors such as shear (A_1), compression (A_3) and work hardening (A_5), and these are interrelated to evaluate the overall reaction. The linear increase of strain against time is observed, and saturation is achieved beyond 90 μs. Shock damping characteristics can also be analyzed [20]. Figure 7c shows the shock profile of the epoxy where the shock loading and loading parts were smooth, and no phase transition was observed. The shock holding time duration was found to be approximately 60 μs. Also, there was no evidence of spall feature signifying the instant dissipation of the material without formation of spall. Unloading has taken place identical to loading phase without spall. Figure 7d also shows the behavior of strain rate against strain which appears identical to Fig. 7c.

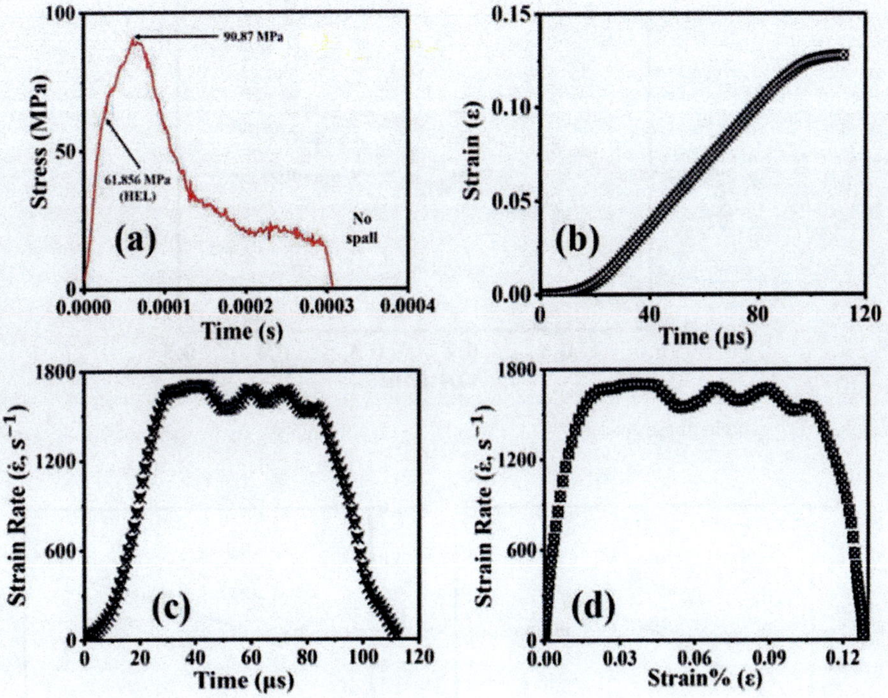

Fig. 7 Rankine plots

The phase-space curves were plotted in accordance with the pressure, volume, particle velocity and shock velocity. Volume was calculated by using stain percent and finding the subsequent volumetric changes of the epoxy sample by incorporating those values in the equation of volume for a cylinder. Shock velocity was found by using equation of states:

$$P = \frac{\rho_o}{S}\left(U_s^2 - C_o U_s\right) \tag{3}$$

where the value of S for epoxy is 1.52, and the particle velocity was found using equation:

$$P = \rho_o\left(C_o + SU_p\right)U_p \tag{4}$$

In Fig. 8, pressure versus volume graph is plotted showcasing the loading and unloading path, where the loading curve shows almost a linear behavior, whereas the unloading curve shows exponential decay. Figure 9 shows particle velocity versus pressure (a) and shock velocity versus pressure (b) where the unloading curve is higher than loading because of dissipation of particles at unloading phase

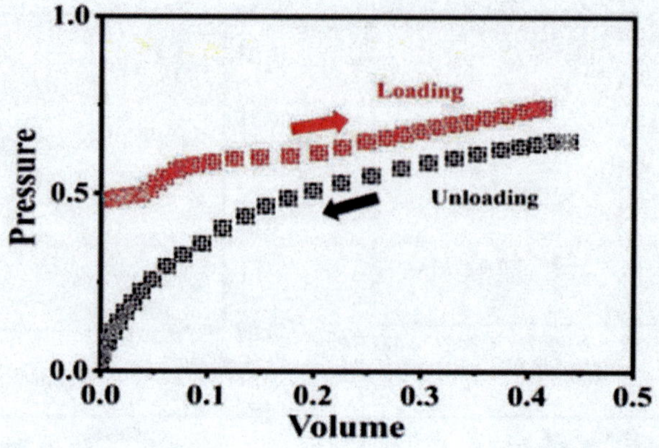

Fig. 8 P–V graph in phase space

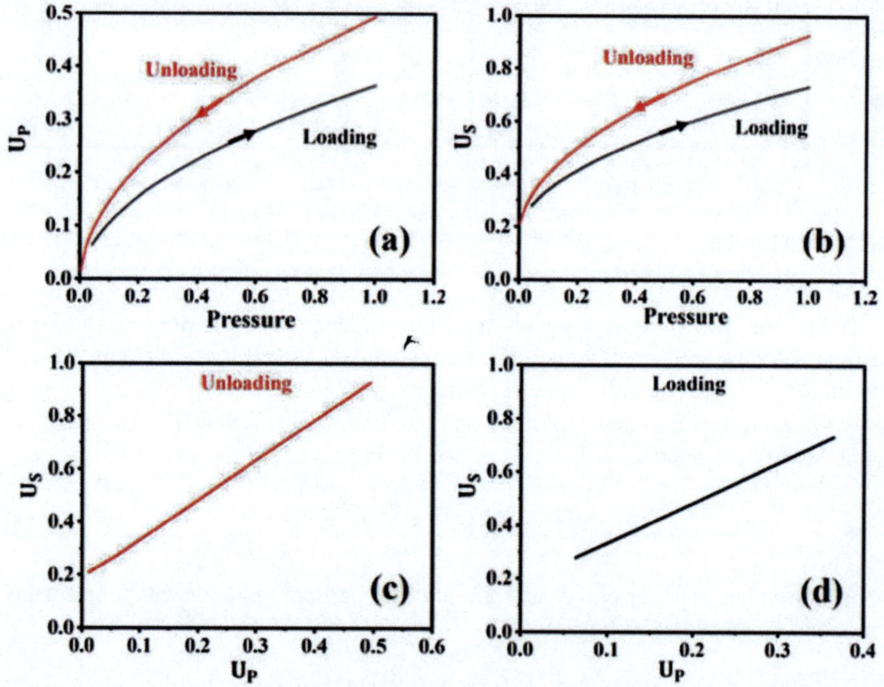

Fig. 9 Plots of particle velocity and shock velocity against pressure (U_s and U_p under normalized coordinate system)

which reduces the mass and subsequently increases the velocity while keeping the momentum constant. Figure 9c, d shows shock velocity versus particle velocity for both conditions which show linear characteristics.

4 Conclusion

The shock response of an epoxy resin has been measured by split-Hopkinson pressure Bar, using piezoresistive stress gauges at the strain rate of 1650 s^{-1}. The pressure–volume curve represents both the loading and unloading zones. The phase-space curves of shock velocity-particle velocity and pressure-particle velocity space have been plotted. Shock velocity-particle velocity curve is found to obey linear characteristics. Given that the particle velocity at the impact face would be half that of the impact, the Hugoniot of this material may be further studied using the appropriate equations. It has been demonstrated that as the longitudinal load on this material grows, so does its shear strength. At increasing impact loads, a reduction in the lateral stress behind the shock front suggests an improvement in shear strength. It is thought that these discoveries in this epoxy resin may reflect good correlation because similar behavior has been seen in PMMA, where it was stated that this may be the result of the material's viscoplastic nature. The graph is examined to determine HEL, which is 61.97 MPa. Similar results from PMMA imply that this result is reasonable even though no results from other researchers that corroborate it have been found. Additionally, the absence of spall has been noted. Further study requires experiments to be conducted at higher strain rates such as 2000–4000 s^{-1} to precisely measure the adiabatic temperatures during certain loading rates, as well as a failure threshold. Experimental data is being used to develop constitutive model for the Bisphenol A. Once the data is collected from samples embedded in the epoxy beads, its properties can be properly analyzed such that it can be determined whether the material can be used in shock absorption fields such as bulletproof vests, armor, vehicles and aircraft where fuel consumption is a major issue due to weight and evaluating the possibility of enhancement of other materials.

References

1. Ben-Dor G, Igra O, Elperin T (2000) Handbook of shock waves, three volume set. Elsevier
2. Chinke SL, Sandhu IS, Bhave TM, Alegaonkar PS (2020) Surface interactions of transonic shock waves with graphene-like nanoribbons. Surfaces 3(3):505–515
3. Chinke SL, Sandhu IS, Saroha DR, Alegaonkar PS (2018) Graphene-like nanoflakes for shock absorption applications. ACS Appl Nano Mater 1(11):6027–6037
4. Meyers MA (1994) Dynamic behavior of materials. Wiley
5. Centre JR, Health IF, Protection C (2010) Updated European Union risk assessment report: 4,4'-isopropylidenediphenol (bisphenol-A): environment addendum of February 2008. Publications Office
6. Vasiljevic T, Harner T (2021) Bisphenol A and its analogues in outdoor and indoor air: properties, sources and global levels. Sci Total Environ 789:148013. https://doi.org/10.1016/j.scitotenv.2021.148013
7. Uglea CV, Negulescu II (1991) Synthesis and characterization of oligomers. CRC Press
8. de Angelis A, Ingallina P, Perego C (2004) Solid acid catalysts for industrial condensations of ketones and aldehydes with aromatics. Ind Eng Chem Res 43(5):1169–1178. https://doi.org/10.1021/ie030429+

9. Serini V (2000) Polycarbonates. In: Ullmann's Encyclopedia of industrial chemistry
10. Ng F, Couture G, Philippe C, Boutevin B, Caillol S (2017) Bio-based aromatic epoxy monomers for thermoset materials 22(1):149
11. Gonçalves F, Kawano Y, Pfeifer C, Stansbury JW, Braga RR (2009) Influence of BisGMA, TEGDMA, and BisEMA contents on viscosity, conversion, and flexural strength of experimental resins and composites 117(4):442–446. https://doi.org/10.1111/j.1600-0722.2009.00636.x
12. Weinong W, Chen BS (2011) Split Hopkinson (Kolsky) bar, Springer
13. Kolsky H (1949) An investigation of the mechanical properties of materials at very high rates of loading. Proc Phys Soc Sect B 62(11):676–700. https://doi.org/10.1088/0370-1301/62/11/302
14. Millett J, Bourne N, Barnes NJJOAP (2002) The behavior of an epoxy resin under one-dimensional shock loading 92(11):6590–6594
15. Vignjevic R, Millett JCF, Bourne NK, Meziere Y, Lukyanov A (2006) The Behaviour of a carbon-fibre epoxy composite under shock loading. In: AIP Conf Proc 845(1):825–828. American Institute of Physics
16. Moy P, Gunnarsson CA, Weerasooriya T, Chen W (2011) Stress-strain response of PMMA as a function of strain-rate and temperature. In: Dynamic behavior of materials, vol 1. Springer, pp 125–133
17. Millett JCF, Bourne NK, GrayIii GT (2004) The response of polyether ether ketone to one-dimensional shock loading. J Phys D Appl Phys 37(6):942–947. https://doi.org/10.1088/0022-3727/37/6/021
18. Chinke SL, Sandhu IS, Bhave TM, Alegaonkar PS (2021) Shock wave hydrodynamics of nano-carbons. Mater Chem Phys 263:124337
19. Chinke SL, Sandhu IS, Alegaonkar PS (2021) Blast mitigation properties of porous nano-carbon. Diam Relat Mater 120:108691
20. Chinke S, Gawade R, Alegaonkar P (2019) Synthesis and characterization of graphene-like nano ribbons (GNR) using chemical vapor deposition for shock absorbent application. AIP Conf Proc 2142(1):060005. https://doi.org/10.1063/1.5122384

Paula Cerejeiras • Michael Reissig
Editors

Mathematical Analysis, its Applications and Computation

ISAAC 2019, Aveiro, Portugal, July 29–August 2

Springer

Editors
Paula Cerejeiras
Department of Mathematics
University of Aveiro
Aveiro, Portugal

Michael Reissig
Institute of Applied Analysis
TU Bergakademie Freiberg
Freiberg, Germany

ISSN 2194-1009 ISSN 2194-1017 (electronic)
Springer Proceedings in Mathematics & Statistics
ISBN 978-3-030-97126-7 ISBN 978-3-030-97127-4 (eBook)
https://doi.org/10.1007/978-3-030-97127-4

Mathematics Subject Classification: 43-06, 65-06

This Springer imprint is published by the registered company Springer Nature Switzerland AG
The registered company address is: Gewerbestrasse 11, 6330 Cham, Switzerland

Preface

The present volume is a collection of papers devoted to current research topics in mathematical analysis and its applications with emphasis on harmonic analysis, inverse problems, and its computational aspects. It originates from plenary lectures given at the 12th International ISAAC Congress, held during 29 July–2 August 2019 at the University of Aveiro, Portugal.

The chapters, authored by eminent specialists, aim at presenting to a large audience some of the attractive and challenging themes of modern analysis, its recent research, trends and future directions:

- The chapter by Afonso Bandeira and his co-authors provides a survey of a new and actual method, namely the so-called low degree method, with a view on understanding computational hardness of problems. This method is based on the idea that the second moment of the low-degree likelihood ratio provides information on the computational complexity of a given statistical task. The authors also discuss evidence for correctness of the low-degree method and present a formal connection between the low-degree lower bounds and the failure of spectral methods.
- Karlheinz Gröchenig presents an overview on recent progress in applications of totally positive functions in sampling theory and time-frequency analysis. The author provides results and gives relevant remarks on the ideas and methods of their proofs for their applications in different topics like Gabor frames, shift-invariant spaces, localization, and Beurling densities for interpolating and sampling sequences, thus providing a clear and concise overall picture.
- Roman Novikov presents a short review on the scattering problem for the stationary Schrödinger equation in a multidimensional setting, including its history and a survey on its actual state of art. He discusses efficient reconstructions of the potential from scattering data with a view on potential applications like phaseless inverse scattering, acoustic tomography, and tomographies using elementary particles.
- Durvudkhan Suragan provides a survey on the problem of Hardy type inequalities. The author also presents some recent results on this matter in the case of

homogeneous groups, with special emphasis to such inequalities on stratified groups and graded Lie groups. This chapter also contains several interesting examples, like the Heisenberg group.

The volume also includes a chapter outlining the mathematical contributions due to Bogdan Bojarski (1931–2018). He was an active member of the ISAAC society for many years. Special attention is paid to one of Bojarski's favorite topics, the theory of quasiconformal mappings. Other chapters, such as his work on boundary value problems, are also briefly presented.

Besides plenary talks, more than 430 scientific communications were delivered during the Aveiro ISAAC Congress. Their contributions are published in an independent volume. This congress, one of the largest congresses in ISAAC history, gave a clear demonstration of the major impact that ISAAC is having in many research areas of mathematical analysis and computation as well as in the integration of young and promising mathematicians from developing countries.

Aveiro, Portugal Paula Cerejeiras
Freiberg, Germany Michael Reissig

Contents

Notes on Computational Hardness of Hypothesis Testing: Predictions Using the Low-Degree Likelihood Ratio

Dmitriy Kunisky, Alexander S. Wein, and Afonso S. Bandeira

Abstract These notes survey and explore an emerging method, which we call the *low-degree method*, for understanding statistical-versus-computational tradeoffs in high-dimensional inference problems. In short, the method posits that a certain quantity—the second moment of the *low-degree likelihood ratio*—gives insight into how much computational time is required to solve a given hypothesis testing problem, which can in turn be used to predict the computational hardness of a variety of statistical inference tasks. While this method originated in the study of the sum-of-squares (SoS) hierarchy of convex programs, we present a self-contained introduction that does not require knowledge of SoS. In addition to showing how to carry out predictions using the method, we include a discussion investigating both rigorous and conjectural consequences of these predictions. These notes include some new results, simplified proofs, and refined conjectures. For instance, we point out a formal connection between spectral methods and the low-degree likelihood ratio, and we give a sharp low-degree lower bound against subexponential-time algorithms for tensor PCA.

D. Kunisky
Department of Computer Science, Yale University, New Haven, CT, USA

Department of Mathematics, Courant Institute of Mathematical Sciences, New York University, New York, NY, USA
e-mail: dmitriy.kunisky@yale.edu

A. S. Wein
Simons Institute for the Theory of Computing, UC Berkeley, Berkeley, CA, USA

Department of Mathematics, Courant Institute of Mathematical Sciences, New York University, New York, NY, USA
e-mail: awein@cims.nyu.edu

A. S. Bandeira (✉)
Department of Mathematics, ETH Zürich, Zürich, Switzerland

Department of Mathematics and Center for Data Science, Courant Institute of Mathematical Sciences, New York University, New York, NY, USA
e-mail: bandeira@math.ethz.ch

P. Cerejeiras, M. Reissig (eds.), *Mathematical Analysis, its Applications and Computation*, Springer Proceedings in Mathematics & Statistics 385,
https://doi.org/10.1007/978-3-030-97127-4_1

Keywords Statistical-to-computational gaps · Hypothesis testing · Low-degree likelihood ratio

Overview

Many problems in high-dimensional statistics are believed to exhibit gaps between what can be achieved *information-theoretically* (or *statistically*, i.e., with unbounded computational power) and what is possible with bounded computational power (e.g., in polynomial time). Examples include finding planted cliques [15, 34, 54, 83] or dense communities [28, 29, 49] in random graphs, extracting variously structured principal components of random matrices [22, 70, 71] or tensors [47, 50], and solving or refuting random constraint satisfaction problems [2, 60].

Although current techniques cannot prove that such average-case problems require super-polynomial time (even assuming $P \neq NP$), various forms of rigorous evidence for hardness have been proposed. These include:

- failure of Markov chain Monte Carlo methods [26, 54];
- failure of local algorithms [12, 24, 33, 44];
- methods from statistical physics which suggest failure of belief propagation or approximate message passing algorithms [28, 29, 70, 71] (see [106] for a survey or [21] for expository notes);
- structural properties of the solution space [2, 44–46, 61];
- geometric analysis of non-convex optimization landscapes [1, 78];
- reductions from planted clique (which has become a "canonical" problem believed to be hard in the average case) [7, 8, 22, 52, 102, 103];
- lower bounds in the statistical query model [30, 39, 42, 58, 63, 65];
- lower bounds against the sum-of-squares hierarchy [15, 34, 43, 47, 50, 83, 86, 98] (see [95] for a survey).

In these notes, we survey another emerging method, which we call the *low-degree method*, for understanding computational hardness in average-case problems. In short, we explore a conjecture that the behavior of a certain quantity—the second moment of the *low-degree likelihood ratio*—reveals the computational complexity of a given statistical task. We find the low-degree method particularly appealing because it is simple, widely applicable, and can be used to study a wide range of time complexities (e.g., polynomial, quasipolynomial, or nearly-exponential). Furthermore, rather than simply positing a certain "optimal algorithm," the underlying conjecture captures an interpretable structural feature that seems to dictate whether a problem is easy or hard. Finally, and perhaps most importantly, predictions using the low-degree method have been carried out for a variety of average-case problems, and so far have always reproduced widely-believed results.

Historically, the low-degree method arose from the study of the sum-of-squares (SoS) semidefinite programming hierarchy. In particular, the method is implicit in the *pseudo-calibration* approach to proving SoS lower bounds [15]. Two concurrent papers [47, 49] later articulated the idea more explicitly. In particular, Hopkins and Steurer [49] were the first to demonstrate that the method can capture sharp thresholds of computational feasibility such as the Kesten–Stigum threshold for community detection in the stochastic block model. The low-degree method was developed further in the PhD thesis of Hopkins [48], which includes a precise conjecture about the complexity-theoretic implications of low-degree predictions. In comparison to sum-of-squares lower bounds, the low-degree method is much simpler to carry out and appears to always yield the same results for natural average-case problems.

In these notes, we aim to provide a self-contained introduction to the low-degree method; we largely avoid reference to SoS and instead motivate the method in other ways. We will briefly discuss the connection to SoS in Sect. 4.2.2, but we refer the reader to [48] for an in-depth exposition of these connections.

These notes are organized as follows. In Sect. 1, we present the low-degree method and motivate it as a computationally-bounded analogue of classical statistical decision theory. In Sect. 2, we show how to carry out the low-degree method for a general class of additive Gaussian noise models. In Sect. 3, we specialize this analysis to two classical problems: the spiked Wigner matrix and spiked Gaussian tensor models. Finally, in Sect. 4, we discuss various forms of heuristic and formal evidence for correctness of the low-degree method; in particular, we highlight a formal connection between low-degree lower bounds and the failure of spectral methods (Theorem 8).

1 Towards a Computationally-Bounded Decision Theory

1.1 Statistical-to-Computational Gaps in Hypothesis Testing

The field of *statistical decision theory* (see, e.g., [68, 75] for general references) is concerned with the question of how to decide optimally (in some quantitative sense) between several statistical conclusions. The simplest example, and the one we will mainly be concerned with here, is that of *simple hypothesis testing*: we observe a dataset that we believe was drawn from one of two probability distributions, and want to make an inference (by performing a statistical *test*) about which distribution we think the dataset was drawn from.

However, one important practical aspect of statistical testing usually is not included in this framework, namely the *computational cost* of actually performing a statistical test. In these notes, we will explore ideas from a line of recent research about how one mathematical method of classical decision theory might be adapted to predict the capabilities and limitations of *computationally bounded statistical tests*.

The basic problem that will motivate us is the following. Suppose $\underline{\mathbb{P}} = (\mathbb{P}_n)_{n\in\mathbb{N}}$ and $\underline{\mathbb{Q}} = (\mathbb{Q}_n)_{n\in\mathbb{N}}$ are two sequences of probability distributions over a common sequence of measurable spaces $\mathcal{S} = ((\mathcal{S}_n, \mathcal{F}_n))_{n\in\mathbb{N}}$. (In statistical parlance, we will think throughout of $\underline{\mathbb{P}}$ as the model of the *alternative hypothesis* and $\underline{\mathbb{Q}}$ as the model of the *null hypothesis*. Later on, we will consider hypothesis testing problems where the distributions $\mathbb{P}$ include a "planted" structure, making the notation a helpful mnemonic.) Suppose we observe $\boldsymbol{Y} \in \mathcal{S}_n$ which is drawn from one of $\mathbb{P}_n$ or $\mathbb{Q}_n$. We hope to recover this choice of distribution in the following sense.

Definition 1 We say that a sequence of events $(A_n)_{n\in\mathbb{N}}$ with $A_n \in \mathcal{F}_n$ occurs with *high probability* (*in n*) if the probability of A_n tends to 1 as $n \to \infty$.

Definition 2 A sequence of (measurable) functions $f_n : \mathcal{S}_n \to \{\mathtt{p}, \mathtt{q}\}$ is said to *strongly distinguish*[1] $\mathbb{P}$ and $\mathbb{Q}$ if $f_n(\boldsymbol{Y}) = \mathtt{p}$ with high probability when $\boldsymbol{Y} \sim \mathbb{P}_n$, and $f_n(\boldsymbol{Y}) = \mathtt{q}$ with high probability when $\boldsymbol{Y} \sim \mathbb{Q}_n$. If such f_n exist, we say that $\mathbb{P}$ and $\mathbb{Q}$ are *statistically distinguishable*.

In our computationally bounded analogue of this definition, let us for now only consider polynomial time tests (we will later consider various other restrictions on the time complexity of f_n, such as subexponential time). Then, the analogue of Definition 2 is the following.

Definition 3 $\mathbb{P}$ and $\mathbb{Q}$ are said to be *computationally distinguishable* if there exists a sequence of measurable **and computable in time polynomial in *n*** functions $f_n : \mathcal{S}_n \to \{\mathtt{p}, \mathtt{q}\}$ such that f_n strongly distinguishes $\mathbb{P}$ and $\mathbb{Q}$.

Clearly, computational distinguishability implies statistical distinguishability. On the other hand, a multitude of theoretical evidence suggests that statistical distinguishability does not in general imply computational distinguishability. Occurrences of this phenomenon are called *statistical-to-computational (stat-comp) gaps*. Typically, such a gap arises in the following slightly more specific way. Suppose the sequence $\mathbb{P}$ has a further dependence on a *signal-to-noise* parameter $\lambda > 0$, so that $\mathbb{P}_\lambda = (\mathbb{P}_{\lambda,n})_{n\in\mathbb{N}}$. This parameter should describe, in some sense, the strength of the structure present under $\mathbb{P}$ (or, in some cases, the number of samples received). The following is one canonical example.

Example 1 (Planted Clique Problem [54, 64]) Under the null model $\mathbb{Q}_n$, we observe an n-vertex Erdős-Rényi graph $\mathcal{G}(n, 1/2)$, i.e., each pair $\{i, j\}$ of vertices is connected with an edge independently with probability $1/2$. The signal-to-noise parameter λ is an integer $1 \le \lambda \le n$. Under the planted model $\mathbb{P}_{\lambda,n}$, we first choose a random subset of vertices $S \subseteq [n]$ of size $|S| = \lambda$ uniformly at random. We then observe a graph where each pair $\{i, j\}$ of vertices is connected with probability 1 if $\{i, j\} \subseteq S$ and with probability $1/2$ otherwise. In other words, the planted model consists of the union of $\mathcal{G}(n, 1/2)$ with a planted *clique* (a fully-connected subgraph) on λ vertices.

[1] We will only consider this so-called *strong* version of distinguishability, where the probability of success must tend to 1 as $n \to \infty$, as opposed to the *weak* version where this probability need only be bounded above $\frac{1}{2}$. For high-dimensional problems, the strong version typically coincides with important notions of estimating the planted signal (see Sect. 4.2.6), whereas the weak version is often trivial.

As λ varies, the problem of testing between $\mathbb{P}_\lambda$ and $\mathbb{Q}$ can change from statistically impossible, to statistically possible but computationally hard, to computationally easy. That is, there exists a threshold λ_{stat} such that for any $\lambda > \lambda_{\mathsf{stat}}$, $\mathbb{P}_\lambda$ and $\mathbb{Q}$ are statistically distinguishable, but for $\lambda < \lambda_{\mathsf{stat}}$ are not. There also exists a threshold λ_{comp} such that for any $\lambda > \lambda_{\mathsf{comp}}$, $\mathbb{P}_\lambda$ and $\mathbb{Q}$ are computationally distinguishable, and (conjecturally) for $\lambda < \lambda_{\mathsf{comp}}$ are not. Clearly we must have $\lambda_{\mathsf{comp}} \geq \lambda_{\mathsf{stat}}$, and a stat-comp gap corresponds to strict inequality $\lambda_{\mathsf{comp}} > \lambda_{\mathsf{stat}}$. For instance, the two models in the planted clique problem are statistically distinguishable when $\lambda \geq (2+\varepsilon)\log_2 n$ (since $2\log_2 n$ is the typical size of the largest clique in $\mathscr{G}(n, 1/2)$), so $\lambda_{\mathsf{stat}} = 2\log_2 n$. However, the best known polynomial-time distinguishing algorithms only succeed when $\lambda = \Omega(\sqrt{n})$ [4, 64], and so (conjecturally) $\lambda_{\mathsf{comp}} \approx \sqrt{n}$, a large stat-comp gap.

The remarkable method we discuss in these notes allows us, through a relatively straightforward calculation, to predict the threshold λ_{comp} for many of the known instances of stat-comp gaps. We will present this method as a modification of a classical second moment method for studying λ_{stat}.

1.2 *Classical Asymptotic Decision Theory*

In this section, we review some basic tools available from statistics for understanding statistical distinguishability. We retain the same notations from the previous section in the later parts, but in the first part of the discussion will only be concerned with a single pair of distributions $\mathbb{P}$ and $\mathbb{Q}$ defined on a single measurable space $(\mathscr{S}, \mathscr{F})$. For the sake of simplicity, let us assume in either case that $\mathbb{P}_n$ (or $\mathbb{P}$) is absolutely continuous with respect to $\mathbb{Q}_n$ (or $\mathbb{Q}$, as appropriate).[2]

1.2.1 Basic Notions

We first define the basic objects used to make hypothesis testing decisions, and some ways of measuring their quality.

Definition 4 A *test* is a measurable function $f : \mathscr{S} \to \{\mathtt{p}, \mathtt{q}\}$.

Definition 5 The *type I error* of f is the event of falsely rejecting the null hypothesis, i.e., of having $f(\boldsymbol{Y}) = \mathtt{p}$ when $\boldsymbol{Y} \sim \mathbb{Q}$. The *type II error* of f is the event of falsely failing to reject the null hypothesis, i.e., of having $f(\boldsymbol{Y}) = \mathtt{q}$ when

[2] For instance, what will be relevant in the examples we consider later, any pair of non-degenerate multivariate Gaussian distributions satisfy this assumption.

$Y \sim \mathbb{P}$. The probabilities of these errors are denoted

$$\alpha(f) := \mathbb{Q}\left(f(Y) = \mathtt{p}\right),$$
$$\beta(f) := \mathbb{P}\left(f(Y) = \mathtt{q}\right).$$

The probability $1 - \beta(f)$ of correctly rejecting the null hypothesis is called the *power* of f.

There is a tradeoff between type I and type II errors. For instance, the trivial test that always outputs p will have maximal power, but will also have maximal probability of type I error, and vice-versa for the trivial test that always outputs q. Thus, typically one fixes a tolerance for one type of error, and then attempts to design a test that minimizes the probability of the other type.

1.2.2 Likelihood Ratio Testing

We next present the classical result showing that it is in fact possible to identify the test that is optimal in the sense of the above tradeoff.[3]

Definition 6 Let $\mathbb{P}$ be absolutely continuous with respect to $\mathbb{Q}$. The *likelihood ratio*[4] of $\mathbb{P}$ and $\mathbb{Q}$ is

$$L(Y) := \frac{d\mathbb{P}}{d\mathbb{Q}}(Y).$$

The *thresholded likelihood ratio test* with threshold η is the test

$$L_\eta(Y) := \begin{Bmatrix} \mathtt{p} : L(Y) > \eta \\ \mathtt{q} : L(Y) \le \eta \end{Bmatrix}.$$

Let us first present a heuristic argument for why thresholding the likelihood ratio might be a good idea. Specifically, we will show that the likelihood ratio is optimal in a particular "L^2 sense" (which will be of central importance later), i.e., when its quality is measured in terms of first and second moments of a testing quantity.

---

[3] It is important to note that, from the point of view of statistics, we are restricting our attention to the special case of deciding between two "simple" hypotheses, where each hypothesis consists of the dataset being drawn from a specific distribution. Optimal testing is more subtle for "composite" hypotheses in parametric families of probability distributions, a more typical setting in practice. The mathematical difficulties of this extended setting are discussed thoroughly in [75].

[4] For readers not familiar with the Radon–Nikodym derivative: if $\mathbb{P}$, $\mathbb{Q}$ are discrete distributions then $L(Y) = \mathbb{P}(Y)/\mathbb{Q}(Y)$; if $\mathbb{P}$, $\mathbb{Q}$ are continuous distributions with density functions p, q (respectively) then $L(Y) = p(Y)/q(Y)$.

Definition 7 For (measurable) functions $f, g : \mathscr{S} \to \mathbb{R}$, define the inner product and norm induced by $\mathbb{Q}$:

$$\langle f, g \rangle := \mathop{\mathbb{E}}_{Y \sim \mathbb{Q}} [f(Y)g(Y)],$$
$$\|f\| := \sqrt{\langle f, f \rangle}.$$

Let $L^2(\mathbb{Q})$ denote the Hilbert space consisting of functions f for which $\|f\| < \infty$, endowed with the above inner product and norm.[5]

Proposition 1 *If $\mathbb{P}$ is absolutely continuous with respect to $\mathbb{Q}$, then the unique solution f^* of the optimization problem*

$$\begin{aligned} &\textit{maximize} \quad \mathop{\mathbb{E}}_{Y \sim \mathbb{P}} [f(Y)] \\ &\textit{subject to} \quad \mathop{\mathbb{E}}_{Y \sim \mathbb{Q}} [f(Y)^2] = 1 \end{aligned}$$

is the (normalized) likelihood ratio

$$f^\star = L/\|L\|,$$

and the value of the optimization problem is $\|L\|$.

Proof We may rewrite the objective as

$$\mathop{\mathbb{E}}_{Y \sim \mathbb{P}} f(Y) = \mathop{\mathbb{E}}_{Y \sim \mathbb{Q}} [L(Y)f(Y)] = \langle L, f \rangle,$$

and rewrite the constraint as $\|f\| = 1$. The result now follows since $\langle L, f \rangle \leq \|L\| \cdot \|f\| = \|L\|$ by the Cauchy-Schwarz inequality, with equality if and only if f is a scalar multiple of L.

In words, this means that if we want a function to be as large as possible in expectation under $\mathbb{P}$ while remaining bounded (in the L^2 sense) under $\mathbb{Q}$, we can do no better than the likelihood ratio. We will soon return to this type of L^2 reasoning in order to devise computationally-bounded statistical tests.

The following classical result shows that the above heuristic is accurate, in that the thresholded likelihood ratio tests achieve the optimal tradeoff between type I and type II errors.

[5] For a more precise definition of $L^2(\mathbb{Q}_n)$ (in particular including issues around functions differing on sets of measure zero) see a standard reference on real analysis such as [100].

Lemma 1 (Neyman–Pearson Lemma [87]) *Fix an arbitrary threshold* $\eta \geq 0$. *Among all tests* f *with* $\alpha(f) \leq \alpha(L_\eta) = \mathbb{Q}(L(Y) > \eta)$, L_η *is the test that maximizes the power* $1 - \beta(f)$.

We provide the standard proof of this result in Appendix "Neyman-Pearson Lemma" for completeness. (The proof is straightforward but not important for understanding the rest of these notes, and it can be skipped on a first reading.)

1.2.3 Le Cam's Contiguity

Since the likelihood ratio is, in the sense of the Neyman–Pearson lemma, an optimal statistical test, it stands to reason that it should be possible to argue about statistical distinguishability solely by computing with the likelihood ratio. We present one simple method by which such arguments may be made, based on a theory introduced by Le Cam [69].

We will work again with sequences of probability measures $\underline{\mathbb{P}} = (\mathbb{P}_n)_{n\in\mathbb{N}}$ and $\underline{\mathbb{Q}} = (\mathbb{Q}_n)_{n\in\mathbb{N}}$, and will denote by L_n the likelihood ratio $d\mathbb{P}_n/d\mathbb{Q}_n$. Norms and inner products of functions are those of $L^2(\mathbb{Q}_n)$. The following is the crucial definition underlying the arguments to come.

Definition 8 A sequence $\underline{\mathbb{P}}$ of probability measures is *contiguous* to a sequence $\underline{\mathbb{Q}}$, written $\underline{\mathbb{P}} \triangleleft \underline{\mathbb{Q}}$, if whenever $A_n \in \mathscr{F}_n$ with $\mathbb{Q}_n(A_n) \to 0$ (as $n \to \infty$), then $\mathbb{P}_n(A_n) \to 0$ as well.

Proposition 2 *If* $\underline{\mathbb{P}} \triangleleft \underline{\mathbb{Q}}$ *or* $\underline{\mathbb{Q}} \triangleleft \underline{\mathbb{P}}$, *then* $\underline{\mathbb{Q}}$ *and* $\underline{\mathbb{P}}$ *are statistically indistinguishable (in the sense of Definition 2, i.e., no test can have both type I and type II error probabilities tending to 0).*

Proof We give the proof for the case $\underline{\mathbb{P}} \triangleleft \underline{\mathbb{Q}}$, but the other case may be shown by a symmetric argument. For the sake of contradiction, let $(f_n)_{n\in\mathbb{N}}$ be a sequence of tests distinguishing $\underline{\mathbb{P}}$ and $\underline{\mathbb{Q}}$, and let $A_n = \{Y : f_n(Y) = \mathtt{p}\}$. Then, $\mathbb{P}_n(A_n^c) \to 0$ and $\mathbb{Q}_n(A_n) \to 0$. But, by contiguity, $\mathbb{Q}_n(A_n) \to 0$ implies $\mathbb{P}_n(A_n) \to 0$ as well, so $\mathbb{P}_n(A_n^c) \to 1$, a contradiction.

It therefore suffices to establish contiguity in order to prove negative results about statistical distinguishability. The following classical second moment method gives a means of establishing contiguity through a computation with the likelihood ratio.

Lemma 2 (Second Moment Method for Contiguity) *If* $\|L_n\|^2 := \mathbb{E}_{Y\sim\mathbb{Q}_n}[L_n(Y)^2]$ *remains bounded as* $n \to \infty$ *(i.e.,* $\limsup_{n\to\infty} \|L_n\|^2 < \infty$*), then* $\underline{\mathbb{P}} \triangleleft \underline{\mathbb{Q}}$.

Proof Let $A_n \in \mathscr{F}_n$. Then, using the Cauchy–Schwarz inequality,

$$\mathbb{P}_n(A_n) = \mathop{\mathbb{E}}_{Y \sim \mathbb{P}_n} [\mathbb{1}_{A_n}(Y)] = \mathop{\mathbb{E}}_{Y \sim \mathbb{Q}_n} \left[L_n(Y)\mathbb{1}_{A_n}(Y)\right] \leq \left(\mathop{\mathbb{E}}_{Y \sim \mathbb{Q}_n} [L_n(Y)^2]\right)^{1/2} (\mathbb{Q}_n(A_n))^{1/2},$$

and so $\mathbb{Q}_n(A_n) \to 0$ implies $\mathbb{P}_n(A_n) \to 0$.

This second moment method has been used to establish contiguity for various high-dimensional statistical problems (see e.g., [20, 84, 90, 91]). Typically the null hypothesis $\mathbb{Q}_n$ is a "simpler" distribution than $\mathbb{P}_n$ and, as a result, $d\mathbb{P}_n/d\mathbb{Q}_n$ is easier to compute than $d\mathbb{Q}_n/d\mathbb{P}_n$. In general, and essentially for this reason, establishing $\underline{\mathbb{Q}} \triangleleft \underline{\mathbb{P}}$ is often more difficult than $\underline{\mathbb{P}} \triangleleft \underline{\mathbb{Q}}$, requiring tools such as the *small subgraph conditioning method* (introduced in [96, 97] and used in, e.g., [19, 80]). Fortunately, one-sided contiguity $\mathbb{P}_n \triangleleft \mathbb{Q}_n$ is sufficient for our purposes.

Note that $\|L_n\|$, the quantity that controls contiguity per the second moment method, is the same as the optimal value of the L^2 optimization problem in Proposition 1:

$$\left\{\begin{array}{l} \text{maximize } \mathbb{E}_{Y \sim \mathbb{P}_n}[f(Y)] \\ \text{subject to } \mathbb{E}_{Y \sim \mathbb{Q}_n}[f(Y)^2] = 1 \end{array}\right\} = \|L_n\|.$$

We might then be tempted to conjecture that $\underline{\mathbb{P}}$ and $\underline{\mathbb{Q}}$ are statistically distinguishable *if and only if* $\|L_n\| \to \infty$ as $n \to \infty$. However, this is incorrect: there are cases when $\underline{\mathbb{P}}$ and $\underline{\mathbb{Q}}$ are not distinguishable, yet a rare "bad" event under $\mathbb{P}_n$ causes $\|L_n\|$ to diverge. To overcome this failure of the ordinary second moment method, some previous works (e.g., [19, 20, 90, 91]) have used *conditional* second moment methods to show indistinguishability, where the second moment method is applied to a modified $\underline{\mathbb{P}}$ that conditions on these bad events not occurring.

1.3 Basics of the Low-Degree Method

We now describe the *low-degree* analogues of the notions described in the previous section, which together constitute a method for restricting the classical decision-theoretic second moment analysis to computationally-bounded tests. The premise of this *low-degree method* is to take low-degree multivariate polynomials in the entries of the observation Y as a proxy for efficiently-computable functions. The ideas in this section were first developed in a sequence of works in the sum-of-squares optimization literature [15, 47–49].

In the computationally-unbounded case, Proposition 1 showed that the likelihood ratio optimally distinguishes $\underline{\mathbb{P}}$ from $\underline{\mathbb{Q}}$ in the L^2 sense. Following the same heuristic, we will now find the low-degree polynomial that best distinguishes $\underline{\mathbb{P}}$ from $\underline{\mathbb{Q}}$ in the L^2 sense. In order for polynomials to be defined, we assume here that $\mathscr{S}_n \subseteq \mathbb{R}^N$ for some $N = N(n)$, i.e., our data (drawn from $\mathbb{P}_n$ or $\mathbb{Q}_n$) is a real-valued vector (which may be structured as a matrix, tensor, etc.).

Definition 9 Let $\mathcal{V}_n^{\leq D} \subset L^2(\mathbb{Q}_n)$ denote the linear subspace of polynomials $\mathcal{S}_n \to \mathbb{R}$ of degree at most D. Let $\mathcal{P}^{\leq D} : L^2(\mathbb{Q}_n) \to \mathcal{V}_n^{\leq D}$ denote the orthogonal projection[6] operator to this subspace. Finally, define the *D-low-degree likelihood ratio* (*D-LDLR*) as $L_n^{\leq D} := \mathcal{P}^{\leq D} L_n$.

We now have a low-degree analogue of Proposition 1, which first appeared in [47, 49].

Proposition 3 *The unique solution f^* of the optimization problem*

$$\begin{aligned} &\textit{maximize } \mathbb{E}_{Y\sim\mathbb{P}_n}[f(Y)] \\ &\textit{subject to } \mathbb{E}_{Y\sim\mathbb{Q}_n}[f(Y)^2] = 1, \\ &\qquad f \in \mathcal{V}_n^{\leq D}, \end{aligned} \tag{1}$$

is the (normalized) D-LDLR

$$f^\star = L_n^{\leq D} / \|L_n^{\leq D}\|,$$

and the value of the optimization problem is $\|L_n^{\leq D}\|$.

Proof As in the proof of Proposition 1, we can restate the optimization problem as maximizing $\langle L_n, f\rangle$ subject to $\|f\| = 1$ and $f \in \mathcal{V}_n^{\leq D}$. Since $\mathcal{V}_n^{\leq D}$ is a linear subspace of $L^2(\mathbb{Q}_n)$, the result is then simply a restatement of the variational description and uniqueness of the orthogonal projection in $L^2(\mathbb{Q}_n)$ (i.e., the fact that $L_n^{\leq D}$ is the unique closest element of $\mathcal{V}_n^{\leq D}$ to L_n).

The following informal conjecture is at the heart of the low-degree method. It states that a computational analogue of the second moment method for contiguity holds, with $L_n^{\leq D}$ playing the role of the likelihood ratio. Furthermore, it postulates that polynomials of degree roughly $\log(n)$ are a proxy for polynomial-time algorithms. This conjecture is based on [47–49], particularly Conjecture 2.2.4 of [48].

Conjecture 1 (Informal) For "sufficiently nice" sequences of probability measures $\underline{\mathbb{P}}$ and $\underline{\mathbb{Q}}$, if there exists $\varepsilon > 0$ and $D = D(n) \geq (\log n)^{1+\varepsilon}$ for which $\|L_n^{\leq D}\|$ remains bounded as $n \to \infty$, then there is no polynomial-time algorithm that strongly distinguishes (see Definition 2) $\underline{\mathbb{P}}$ and $\underline{\mathbb{Q}}$.

We will discuss this conjecture in more detail later (see Sect. 4), including the informal meaning of "sufficiently nice" and a variant of the LDLR based on

[6] To clarify, orthogonal projection is with respect to the inner product induced by $\mathbb{Q}_n$ (see Definition 7).

coordinate degree considered by [47, 48] (see Sect. 4.2.4). A more general form of the low-degree conjecture (Hypothesis 2.1.5 of [48]) states that degree-D polynomials are a proxy for time-$n^{\tilde{\Theta}(D)}$ algorithms, allowing one to probe a wide range of time complexities. We will see that the converse of these low-degree conjectures often holds in practice; i.e., if $\|L_n^{\leq D}\| \to \infty$, then there exists a distinguishing algorithm of runtime roughly n^D. As a result, the behavior of $\|L_n^{\leq D}\|$ precisely captures the (conjectured) power of computationally-bounded testing in many settings.

The remainder of these notes is organized as follows. In Sect. 2, we work through the calculations of L_n, $L_n^{\leq D}$, and their norms for a general family of additive Gaussian noise models. In Sect. 3, we apply this analysis to a few specific models of interest: the spiked Wigner matrix and spiked Gaussian tensor models. In Sect. 4, we give some further discussion of Conjecture 1, including evidence (both heuristic and formal) in its favor.

2 The Additive Gaussian Noise Model

We will now describe a concrete class of hypothesis testing problems and analyze them using the machinery introduced in the previous section. The examples we discuss later (spiked Wigner matrix and spiked tensor) will be specific instances of this general class.

2.1 The Model

Definition 10 (Additive Gaussian Noise Model) Let $N = N(n) \in \mathbb{N}$ and let $\boldsymbol{X}$ (the "signal") be drawn from some distribution $\mathcal{P}_n$ (the "prior") over $\mathbb{R}^N$. Let $\boldsymbol{Z} \in \mathbb{R}^N$ (the "noise") have i.i.d. entries distributed as $\mathcal{N}(0, 1)$. Then, we define $\mathbb{P}$ and $\mathbb{Q}$ as follows.

- Under $\mathbb{P}_n$, observe $\boldsymbol{Y} = \boldsymbol{X} + \boldsymbol{Z}$.
- Under $\mathbb{Q}_n$, observe $\boldsymbol{Y} = \boldsymbol{Z}$.

One typical situation takes $\boldsymbol{X}$ to be a low-rank matrix or tensor. The following is a particularly important and well-studied special case, which we will return to in Sect. 3.2.

Example 2 (Wigner Spiked Matrix Model) Consider the additive Gaussian noise model with $N = n^2$, $\mathbb{R}^N$ identified with $n \times n$ matrices with real entries, and $\mathcal{P}_n$ defined by $\boldsymbol{X} = \lambda \boldsymbol{x}\boldsymbol{x}^\top \in \mathbb{R}^{n \times n}$, where $\lambda = \lambda(n) > 0$ is a signal-to-noise parameter and $\boldsymbol{x}$ is drawn from some distribution $\mathcal{X}_n$ over $\mathbb{R}^n$. Then, the task of distinguishing $\mathbb{P}_n$ from $\mathbb{Q}_n$ amounts to distinguishing $\lambda \boldsymbol{x}\boldsymbol{x}^\top + \boldsymbol{Z}$ from $\boldsymbol{Z}$ where $\boldsymbol{Z} \in \mathbb{R}^{n \times n}$ has i.i.d. entries distributed as $\mathcal{N}(0, 1)$. (This variant is equivalent to the more standard model in which the noise matrix is symmetric; see Appendix "Equivalence of Symmetric and Asymmetric Noise Models".)

This problem is believed to exhibit stat-comp gaps for some choices of $\mathscr{X}_n$ but not others; see, e.g., [20, 66, 70, 71, 91]. At a heuristic level, the typical *sparsity* of vectors under $\mathscr{X}_n$ seems to govern the appearance of a stat-comp gap.

Remark 1 In the spiked Wigner problem, as in many others, one natural statistical task besides distinguishing the null and planted models is to non-trivially estimate the vector $\boldsymbol{x}$ given $\boldsymbol{Y} \sim \mathbb{P}_n$, i.e., to compute an estimate $\hat{\boldsymbol{x}} = \hat{\boldsymbol{x}}(\boldsymbol{Y})$ such that $|\langle \hat{\boldsymbol{x}}, \boldsymbol{x}\rangle|/(\|\hat{\boldsymbol{x}}\| \cdot \|\boldsymbol{x}\|) \geq \varepsilon$ with high probability, for some constant $\varepsilon > 0$. Typically, for natural high-dimensional problems, non-trivial estimation of $\boldsymbol{x}$ is statistically or computationally possible precisely when it is statistically or computationally possible (respectively) to strongly distinguish $\underline{\mathbb{P}}$ and $\underline{\mathbb{Q}}$; see Sect. 4.2.6 for further discussion.

2.2 Computing the Classical Quantities

We now show how to compute the likelihood ratio and its L^2-norm under the additive Gaussian noise model. (This is a standard calculation; see, e.g., [20, 84].)

Proposition 4 *Suppose* $\underline{\mathbb{P}}$ *and* $\underline{\mathbb{Q}}$ *are as defined in Definition 10, with a sequence of prior distributions* $(\mathscr{P}_n)_{n\in\mathbb{N}}$. *Then, the likelihood ratio of* $\mathbb{P}_n$ *and* $\mathbb{Q}_n$ *is*

$$L_n(\boldsymbol{Y}) = \frac{d\mathbb{P}_n}{d\mathbb{Q}_n}(\boldsymbol{Y}) = \mathop{\mathbb{E}}_{\boldsymbol{X}\sim\mathscr{P}_n}\left[\exp\left(-\frac{1}{2}\|\boldsymbol{X}\|^2 + \langle \boldsymbol{X}, \boldsymbol{Y}\rangle\right)\right].$$

Proof Write $\mathscr{L}$ for the Lebesgue measure on $\mathbb{R}^N$. Then, expanding the gaussian densities,

$$\frac{d\mathbb{Q}_n}{d\mathscr{L}}(\boldsymbol{Y}) = (2\pi)^{-N/2} \cdot \exp\left(-\frac{1}{2}\|\boldsymbol{Y}\|^2\right) \tag{2}$$

$$\begin{aligned}\frac{d\mathbb{P}_n}{d\mathscr{L}}(\boldsymbol{Y}) &= (2\pi)^{-N/2} \cdot \mathop{\mathbb{E}}_{\boldsymbol{X}\sim\mathscr{P}_n}\left[\exp\left(-\frac{1}{2}\|\boldsymbol{Y}-\boldsymbol{X}\|^2\right)\right] \\ &= (2\pi)^{-N/2} \cdot \exp\left(-\frac{1}{2}\|\boldsymbol{Y}\|^2\right) \cdot \mathbb{E}_{\boldsymbol{X}\sim\mathscr{P}_n}\left[\exp\left(-\frac{1}{2}\|\boldsymbol{X}\|^2 + \langle \boldsymbol{X}, \boldsymbol{Y}\rangle\right)\right],\end{aligned} \tag{3}$$

and L_n is given by the quotient of (3) and (2).

Proposition 5 *Suppose* $\mathbb{P}$ *and* $\mathbb{Q}$ *are as defined in Definition 10, with a sequence of prior distributions* $(\mathscr{P}_n)_{n\in\mathbb{N}}$. *Then,*

$$\|L_n\|^2 = \mathop{\mathbb{E}}_{X^1, X^2 \sim \mathscr{P}_n} \exp(\langle X^1, X^2 \rangle), \tag{4}$$

where X^1, X^2 *are drawn independently from* $\mathscr{P}_n$.

Proof We apply the important trick of rewriting a squared expectation as an expectation over the two independent "replicas" X^1, X^2 appearing in the result:

$$\begin{aligned} \|L_n\|^2 &= \mathop{\mathbb{E}}_{Y\sim\mathbb{Q}_n} \left[\left(\mathop{\mathbb{E}}_{X\sim\mathscr{P}_n} \exp\left(\langle Y, X\rangle - \frac{1}{2}\|X\|^2 \right) \right)^2 \right] \\ &= \mathop{\mathbb{E}}_{Y\sim\mathbb{Q}_n} \mathop{\mathbb{E}}_{X^1, X^2\sim\mathscr{P}_n} \exp\left(\langle Y, X^1 + X^2\rangle - \frac{1}{2}\|X^1\|^2 - \frac{1}{2}\|X^2\|^2 \right), \end{aligned}$$

where X^1 and X^2 are drawn independently from $\mathscr{P}_n$. We now swap the order of the expectations,

$$= \mathop{\mathbb{E}}_{X^1, X^2\sim\mathscr{P}_n} \left[\exp\left(-\frac{1}{2}\|X^1\|^2 - \frac{1}{2}\|X^2\|^2 \right) \mathop{\mathbb{E}}_{Y\sim\mathbb{Q}_n} \exp\left(\langle Y, X^1 + X^2 \rangle \right) \right],$$

and the inner expectation may be evaluated explicitly using the moment-generating function of a Gaussian distribution (if $y \sim \mathcal{N}(0,1)$, then for any fixed $t \in \mathbb{R}$, $\mathbb{E}[\exp(ty)] = \exp(t^2/2)$),

$$= \mathop{\mathbb{E}}_{X^1, X^2} \exp\left(-\frac{1}{2}\|X^1\|^2 - \frac{1}{2}\|X^2\|^2 + \frac{1}{2}\|X^1 + X^2\|^2 \right),$$

from which the result follows by expanding the term inside the exponential.[7]

To apply the second moment method for contiguity, it remains to show that (4) is $O(1)$ using problem-specific information about the distribution $\mathscr{P}_n$. For spiked matrix and tensor models, various general-purpose techniques for doing this are given in [90, 91].

[7] Two techniques from this calculation are elements of the "replica method" from statistical physics: (1) writing a power of an expectation as an expectation over independent "replicas" and (2) changing the order of expectations and evaluating the moment-generating function. The interested reader may see [82] for an early reference, or [21, 79] for two recent presentations.

2.3 Computing the Low-Degree Quantities

In this section, we will show that the norm of the LDLR (see Sect. 1.3) takes the following remarkably simple form under the additive Gaussian noise model.

Theorem 1 *Suppose* $\mathbb{P}$ *and* $\mathbb{Q}$ *are as defined in Definition 10, with a sequence of prior distributions* $(\mathscr{P}_n)_{n\in\mathbb{N}}$. *Let* $L_n^{\leq D}$ *be as in Definition 9. Then,*

$$\|L_n^{\leq D}\|^2 = \mathop{\mathbb{E}}_{X^1, X^2 \sim \mathscr{P}_n}\left[\sum_{d=0}^{D} \frac{1}{d!}\langle X^1, X^2\rangle^d\right], \tag{5}$$

where X^1, X^2 *are drawn independently from* $\mathscr{P}_n$.

Remark 2 Note that (5) can be written as $\mathbb{E}_{X^1,X^2}[\exp^{\leq D}(\langle X^1, X^2\rangle)]$, where $\exp^{\leq D}(t)$ denotes the degree-D truncation of the Taylor series of $\exp(t)$. This can be seen as a natural low-degree analogue of the full second moment (4). However, the low-degree Taylor series truncation in $\exp^{\leq D}$ is conceptually distinct from the low-degree projection in $L_n^{\leq D}$, because the latter corresponds to truncation in the Hermite orthogonal polynomial basis (see below), while the former corresponds to truncation in the monomial basis.

Our proof of Theorem 1 will follow the strategy of [47–49] of expanding L_n in a basis of orthogonal polynomials with respect to $\mathbb{Q}_n$, which in this case are the *Hermite polynomials*.

We first give a brief and informal review of the multivariate Hermite polynomials (see Appendix "Hermite Polynomials" or the reference text [101] for further information). The univariate Hermite polynomials[8] are a sequence $h_k(x) \in \mathbb{R}[x]$ for $k \geq 0$, with $\deg h_k = k$. They may be normalized as $\widehat{h}_k(x) = h_k(x)/\sqrt{k!}$, and with this normalization satisfy the orthonormality conditions

$$\mathop{\mathbb{E}}_{y\sim\mathcal{N}(0,1)} \widehat{h}_k(y)\widehat{h}_\ell(y) = \delta_{k\ell}. \tag{6}$$

The multivariate Hermite polynomials in N variables are indexed by $\boldsymbol{\alpha} \in \mathbb{N}^N$, and are merely products of the h_k: $H_{\boldsymbol{\alpha}}(\boldsymbol{x}) = \prod_{i=1}^N h_{\alpha_i}(x_i)$. They also admit a normalized variant $\widehat{H}_{\boldsymbol{\alpha}}(\boldsymbol{x}) = \prod_{i=1}^N \widehat{h}_{\alpha_i}(x_i)$, and with this normalization satisfy the orthonormality conditions

$$\mathop{\mathbb{E}}_{\boldsymbol{Y}\sim\mathcal{N}(0,\boldsymbol{I}_N)} \widehat{H}_{\boldsymbol{\alpha}}(\boldsymbol{Y})\widehat{H}_{\boldsymbol{\beta}}(\boldsymbol{Y}) = \delta_{\boldsymbol{\alpha\beta}},$$

which may be inferred directly from (6).

[8] We will not actually use the definition of the univariate Hermite polynomials (although we will use certain properties that they satisfy as needed), but the definition is included for completeness in Appendix "Hermite Polynomials".

The collection of those $\widehat{H}_{\boldsymbol{\alpha}}$ for which $|\boldsymbol{\alpha}| := \sum_{i=1}^{N} \alpha_i \leq D$ form an orthonormal basis for $\mathscr{V}_n^{\leq D}$ (which, recall, is the subspace of polynomials of degree $\leq D$). Thus we may expand

$$L_n^{\leq D}(\boldsymbol{Y}) = \sum_{\substack{\boldsymbol{\alpha} \in \mathbb{N}^N \\ |\boldsymbol{\alpha}| \leq D}} \langle L_n, \widehat{H}_{\boldsymbol{\alpha}} \rangle \widehat{H}_{\boldsymbol{\alpha}}(\boldsymbol{Y}) = \sum_{\substack{\boldsymbol{\alpha} \in \mathbb{N}^N \\ |\boldsymbol{\alpha}| \leq D}} \frac{1}{\prod_{i=1}^{N} \alpha_i!} \langle L_n, H_{\boldsymbol{\alpha}} \rangle H_{\boldsymbol{\alpha}}(\boldsymbol{Y}), \tag{7}$$

and in particular we have

$$\|L_n^{\leq D}\|^2 = \sum_{\substack{\boldsymbol{\alpha} \in \mathbb{N}^N \\ |\boldsymbol{\alpha}| \leq D}} \frac{1}{\prod_{i=1}^{N} \alpha_i!} \langle L_n, H_{\boldsymbol{\alpha}} \rangle^2. \tag{8}$$

Our main task is then to compute quantities of the form $\langle L_n, H_{\boldsymbol{\alpha}} \rangle$. Note that these can be expressed either as $\mathbb{E}_{\boldsymbol{Y} \sim \mathbb{Q}_n}[L_n(\boldsymbol{y}) H_{\boldsymbol{\alpha}}(\boldsymbol{Y})]$ or $\mathbb{E}_{\boldsymbol{Y} \sim \mathbb{P}_n}[H_{\boldsymbol{\alpha}}(\boldsymbol{Y})]$. We will give three techniques for carrying out this calculation, each depending on a different identity satisfied by the Hermite polynomials. Each will give a proof of the following remarkable formula, which shows that the quantities $\langle L_n, H_{\boldsymbol{\alpha}} \rangle$ are simply the moments of $\mathscr{P}_n$.

Proposition 6 *For any* $\boldsymbol{\alpha} \in \mathbb{N}^N$,

$$\langle L_n, H_{\boldsymbol{\alpha}} \rangle = \mathop{\mathbb{E}}_{\boldsymbol{X} \sim \mathscr{P}_n} \left[\prod_{i=1}^{N} X_i^{\alpha_i} \right].$$

Before continuing with the various proofs of Proposition 6, let us show how to use it to complete the proof of Theorem 1.

Proof of Theorem 1 By Proposition 6 substituted into (8), we have

$$\|L_n^{\leq D}\|^2 = \sum_{\substack{\boldsymbol{\alpha} \in \mathbb{N}^N \\ |\boldsymbol{\alpha}| \leq D}} \frac{1}{\prod_{i=1}^{N} \alpha_i!} \left(\mathop{\mathbb{E}}_{\boldsymbol{X} \sim \mathscr{P}_n} \left[\prod_{i=1}^{N} X_i^{\alpha_i} \right] \right)^2,$$

and performing the "replica" manipulation (from the proof of Proposition 5) again, this may be written

$$= \mathop{\mathbb{E}}_{\boldsymbol{X}^1, \boldsymbol{X}^2 \sim \mathscr{P}_n} \left[\sum_{\substack{\boldsymbol{\alpha} \in \mathbb{N}^N \\ |\boldsymbol{\alpha}| \leq D}} \frac{1}{\prod_{i=1}^{N} \alpha_i!} \prod_{i=1}^{N} (X_i^1 X_i^2)^{\alpha_i} \right]$$

$$= \mathop{\mathbb{E}}_{X^1, X^2 \sim \mathscr{P}_n} \left[\sum_{d=0}^{D} \frac{1}{d!} \sum_{\substack{\alpha \in \mathbb{N}^N \\ |\alpha| = d}} \binom{d}{\alpha_1 \cdots \alpha_N} \prod_{i=1}^{N} (X_i^1 X_i^2)^{\alpha_i} \right]$$

$$= \mathop{\mathbb{E}}_{X^1, X^2 \sim \mathscr{P}_n} \left[\sum_{d=0}^{D} \frac{1}{d!} \langle X^1, X^2 \rangle^d \right],$$

where the last step uses the multinomial theorem.

We now proceed to the three proofs of Proposition 6. For the sake of brevity, we omit here the (standard) proofs of the three Hermite polynomial identities these proofs are based on, but the interested reader may review those proofs in Appendix "Hermite Polynomials".

2.3.1 Proof 1: Hermite Translation Identity

The first (and perhaps simplest) approach to proving Proposition 6 uses the following formula for the expectation of a Hermite polynomial evaluated on a Gaussian random variable of non-zero mean.

Proposition 7 *For any $k \geq 0$ and $\mu \in \mathbb{R}$,*

$$\mathop{\mathbb{E}}_{y \sim \mathcal{N}(\mu, 1)} [h_k(y)] = \mu^k.$$

Proof of Proposition 6, Proof 1 We rewrite $\langle L_n, H_{\boldsymbol{\alpha}} \rangle$ as an expectation with respect to $\mathbb{P}_n$:

$$\begin{aligned}
\langle L_n, H_{\boldsymbol{\alpha}} \rangle &= \mathop{\mathbb{E}}_{\boldsymbol{Y} \sim \mathbb{Q}_n} [L_n(\boldsymbol{Y}) H_{\boldsymbol{\alpha}}(\boldsymbol{Y})] \\
&= \mathop{\mathbb{E}}_{\boldsymbol{Y} \sim \mathbb{P}_n} [H_{\boldsymbol{\alpha}}(\boldsymbol{Y})] \\
&= \mathop{\mathbb{E}}_{\boldsymbol{Y} \sim \mathbb{P}_n} \left[\prod_{i=1}^{N} h_{\alpha_i}(Y_i) \right]
\end{aligned}$$

and recall $\boldsymbol{Y} = \boldsymbol{X} + \boldsymbol{Z}$ for $\boldsymbol{X} \sim \mathscr{P}_n$ and $\boldsymbol{Z} \sim \mathcal{N}(\boldsymbol{0}, \boldsymbol{I}_N)$ under $\mathbb{P}_n$,

$$= \mathop{\mathbb{E}}_{\boldsymbol{X} \sim \mathscr{P}_n} \left[\mathop{\mathbb{E}}_{\boldsymbol{Z} \sim \mathcal{N}(\boldsymbol{0}, \boldsymbol{I}_N)} \prod_{i=1}^{N} h_{\alpha_i}(X_i + Z_i) \right]$$

$$= \mathop{\mathbb{E}}_{X \sim \mathscr{P}_n} \left[\prod_{i=1}^{N} \mathop{\mathbb{E}}_{z \sim \mathcal{N}(X_i,1)} h_{\alpha_i}(z) \right]$$

$$= \mathop{\mathbb{E}}_{X \sim \mathscr{P}_n} \left[\prod_{i=1}^{N} X_i^{\alpha_i} \right],$$

where we used Proposition 7 in the last step.

2.3.2 Proof 2: Gaussian Integration by Parts

The second approach to proving Proposition 6 uses the following generalization of a well-known integration by parts formula for Gaussian random variables.

Proposition 8 *If $f : \mathbb{R} \to \mathbb{R}$ is k times continuously differentiable and $f(y)$ and its first k derivatives are bounded by $O(\exp(|y|^\alpha))$ for some $\alpha \in (0, 2)$, then*

$$\mathop{\mathbb{E}}_{y \sim \mathcal{N}(0,1)} [h_k(y) f(y)] = \mathop{\mathbb{E}}_{y \sim \mathcal{N}(0,1)} \left[\frac{d^k f}{dy^k}(y) \right].$$

(The better-known case is $k = 1$, where one may substitute $h_1(x) = x$.)

Proof of Proposition 6, Proof 2 We simplify using Proposition 8:

$$\langle L_n, H_{\boldsymbol{\alpha}} \rangle = \mathop{\mathbb{E}}_{\boldsymbol{Y} \sim \mathbb{Q}_n} \left[L_n(\boldsymbol{Y}) \prod_{i=1}^{N} h_{\alpha_i}(Y_i) \right] = \mathop{\mathbb{E}}_{\boldsymbol{Y} \sim \mathbb{Q}_n} \left[\frac{\partial^{|\alpha|} L_n}{\partial Y_1^{\alpha_1} \cdots \partial Y_N^{\alpha_N}}(\boldsymbol{Y}) \right].$$

Differentiating L_n under the expectation, we have

$$\frac{\partial^{|\alpha|} L}{\partial Y_1^{\alpha_1} \cdots \partial Y_N^{\alpha_N}}(\boldsymbol{Y}) = \mathop{\mathbb{E}}_{X \sim \mathscr{P}_n} \left[\prod_{i=1}^{N} X_i^{\alpha_i} \exp\left(-\frac{1}{2} \|\boldsymbol{X}\|^2 + \langle \boldsymbol{X}, \boldsymbol{Y} \rangle \right) \right].$$

Taking the expectation over $\boldsymbol{Y}$, we have $\mathbb{E}_{\boldsymbol{Y} \sim \mathbb{Q}_n} \exp(\langle \boldsymbol{X}, \boldsymbol{Y} \rangle) = \exp(\frac{1}{2} \|\boldsymbol{X}\|^2)$, so the entire second term cancels and the result follows.

2.3.3 Proof 3: Hermite Generating Function

Finally, the third approach to proving Proposition 6 uses the following generating function for the Hermite polynomials.

Proposition 9 *For any* $x, y \in \mathbb{R}$,

$$\exp\left(xy - \frac{1}{2}x^2\right) = \sum_{k=0}^{\infty} \frac{1}{k!} x^k h_k(y).$$

Proof of Proposition 6, Proof 3 We may use Proposition 9 to expand L_n in the Hermite polynomials directly:

$$\begin{aligned} L_n(\boldsymbol{Y}) &= \mathop{\mathbb{E}}_{\boldsymbol{X}\sim\mathscr{P}_n} \left[\exp\left(\langle \boldsymbol{X}, \boldsymbol{Y}\rangle - \frac{1}{2}\|\boldsymbol{X}\|^2\right)\right] \\ &= \mathop{\mathbb{E}}_{\boldsymbol{X}\sim\mathscr{P}_n} \left[\prod_{i=1}^{N}\left(\sum_{k=0}^{\infty} \frac{1}{k!} X_i^k h_k(Y_i)\right)\right] \\ &= \sum_{\boldsymbol{\alpha}\in\mathbb{N}^N} \frac{1}{\prod_{i=1}^{N} \alpha_i!} \mathbb{E}_{\boldsymbol{X}\sim\mathscr{P}_n}\left[\prod_{i=1}^{N} X_i^{\alpha_i}\right] H_{\boldsymbol{\alpha}}(\boldsymbol{Y}). \end{aligned}$$

Comparing with the expansion (7) then gives the result.

Now that we have the simple form (5) for the norm of the LDLR, it remains to investigate its convergence or divergence (as $n \to \infty$) using problem-specific statistics of $\boldsymbol{X}$. In the next section we give some examples of how to carry out this analysis.

3 Examples: Spiked Matrix and Tensor Models

In this section, we perform the low-degree analysis for a particular important case of the additive Gaussian model: the *order-p spiked Gaussian tensor model*, also referred to as the *tensor PCA (principal component analysis)* problem. This model was introduced by [93] and has received much attention recently. The special case $p = 2$ of the spiked tensor model is the so-called *spiked Wigner matrix model* which has been widely studied in random matrix theory, statistics, information theory, and statistical physics; see [77] for a survey.

In concordance with prior work, our low-degree analysis of these models illustrates two representative phenomena: the spiked Wigner matrix model exhibits a sharp computational phase transition, whereas the spiked tensor model (with $p \geq 3$) has a "soft" tradeoff between statistical power and runtime which extends through the subexponential-time regime. A low-degree analysis of the spiked tensor model has been carried out previously in [47, 48]; here we give a sharper analysis that more precisely captures the power of subexponential-time algorithms.

In Sect. 3.1, we carry out our low-degree analysis of the spiked tensor model. In Sect. 3.2, we devote additional attention to the special case of the spiked Wigner model, giving a refined analysis that captures its sharp phase transition and applies to a variety of distributions of "spikes."

3.1 The Spiked Tensor Model

We begin by defining the model.

Definition 11 An n-dimensional order-p *tensor* $\boldsymbol{T} \in (\mathbb{R}^n)^{\otimes p}$ is a multi-dimensional array with p dimensions each of length n, with entries denoted $T_{i_1,\ldots,i_p}$ where $i_j \in [n]$. For a vector $\boldsymbol{x} \in \mathbb{R}^n$, the *rank-one tensor* $\boldsymbol{x}^{\otimes p} \in (\mathbb{R}^n)^{\otimes p}$ has entries $(\boldsymbol{x}^{\otimes p})_{i_1,\ldots,i_p} = x_{i_1} x_{i_2} \cdots x_{i_p}$.

Definition 12 (Spiked Tensor Model) Fix an integer $p \geq 2$. The order-p *spiked tensor model* is the additive Gaussian noise model (Definition 10) with $\boldsymbol{X} = \lambda \boldsymbol{x}^{\otimes p}$, where $\lambda = \lambda(n) > 0$ is a signal-to-noise parameter and $\boldsymbol{x} \in \mathbb{R}^n$ (the "spike") is drawn from some probability distribution $\mathscr{X}_n$ over $\mathbb{R}^n$ (the "prior"), normalized so that $\|\boldsymbol{x}\|^2 \to n$ in probability as $n \to \infty$. In other words:

- Under $\mathbb{P}_n$, observe $\boldsymbol{Y} = \lambda \boldsymbol{x}^{\otimes p} + \boldsymbol{Z}$.
- Under $\mathbb{Q}_n$, observe $\boldsymbol{Y} = \boldsymbol{Z}$.

Here, $\boldsymbol{Z}$ is a tensor with i.i.d. entries distributed as $\mathcal{N}(0, 1)$.[9]

Throughout this section we will focus for the sake of simplicity on the Rademacher spike prior, where $\boldsymbol{x}$ has i.i.d. entries $x_i \sim \mathsf{Unif}(\{\pm 1\})$. We focus on the problem of strongly distinguishing $\mathbb{P}_n$ and $\mathbb{Q}_n$ (see Definition 2), but, as is typical for high-dimensional problems, the problem of estimating $\boldsymbol{x}$ seems to behave in essentially the same way (see Sect. 4.2.6).

We first state our results on the behavior of the LDLR for this model.

Theorem 2 *Consider the order-p spiked tensor model with* $\boldsymbol{x}$ *drawn from the Rademacher prior,* $x_i \sim \mathsf{Unif}(\{\pm 1\})$ *i.i.d. for* $i \in [n]$. *Fix sequences* $D = D(n)$ *and* $\lambda = \lambda(n)$. *For constants* $0 < A_p < B_p$ *depending only on p, we have the following.*[10]

(i) If $\lambda \leq A_p n^{-p/4} D^{(2-p)/4}$ *for all sufficiently large n, then* $\|L_n^{\leq D}\| = O(1)$.
(ii) If $\lambda \geq B_p n^{-p/4} D^{(2-p)/4}$ *and* $D \leq \frac{2}{p} n$ *for all sufficiently large n, and* $D = \omega(1)$, *then* $\|L_n^{\leq D}\| = \omega(1)$.

(Here we are considering the limit $n \to \infty$ *with p held fixed, so* $O(1)$ *and* $\omega(1)$ *may hide constants depending on p.)*

---

[9] This model is equivalent to the more standard model in which the noise is symmetric with respect to permutations of the indices; see Appendix "Equivalence of Symmetric and Asymmetric Noise Models".

[10] Concretely, one may take $A_p = \frac{1}{\sqrt{2}} p^{-p/4-1/2}$ and $B_p = \sqrt{2} e^{p/2} p^{-p/4}$.

Before we prove this, let us interpret its meaning. If we take degree-D polynomials as a proxy for $n^{\tilde{\Theta}(D)}$-time algorithms (as discussed in Sect. 1.3), our calculations predict that an $n^{O(D)}$-time algorithm exists when $\lambda \gg n^{-p/4}D^{(2-p)/4}$ but not when $\lambda \ll n^{-p/4}D^{(2-p)/4}$. (Here we ignore log factors, so we use $A \ll B$ to mean $A \leq B/\text{polylog}(n)$.) These predictions agree precisely with the previously established statistical-versus-computational tradeoffs in the spiked tensor model! It is known that polynomial-time distinguishing algorithms exist when $\lambda \gg n^{-p/4}$ [3, 50, 51, 93], and sum-of-squares lower bounds suggest that there is no polynomial-time distinguishing algorithm when $\lambda \ll n^{-p/4}$ [47, 50].

Furthermore, one can study the power of subexponential-time algorithms, i.e., algorithms of runtime $n^{n^\delta} = \exp(\tilde{O}(n^\delta))$ for a constant $\delta \in (0, 1)$. Such algorithms are known to exist when $\lambda \gg n^{-p/4-\delta(p-2)/4}$ [11, 13, 94, 104], matching our prediction.[11] These algorithms interpolate smoothly between the polynomial-time algorithm which succeeds when $\lambda \gg n^{-p/4}$, and the exponential-time exhaustive search algorithm which succeeds when $\lambda \gg n^{(1-p)/2}$. (Distinguishing the null and planted distributions is information-theoretically impossible when $\lambda \ll n^{(1-p)/2}$ [57, 74, 90, 93], so this is indeed the correct terminal value of λ for computational questions.) The tradeoff between statistical power and runtime that these algorithms achieve is believed to be optimal, and our results corroborate this claim. Our results are sharper than the previous low-degree analysis for the spiked tensor model [47, 48], in that we pin down the precise constant δ in the subexponential runtime. (Similarly precise analyses of the tradeoff between subexponential runtime and statistical power have been obtained for CSP refutation [94] and sparse PCA [31].)

We now begin the proof of Theorem 2. Since the spiked tensor model is an instance of the additive Gaussian model, we can apply the formula from Theorem 1: letting $\boldsymbol{x}^1, \boldsymbol{x}^2$ be independent draws from $\mathscr{X}_n$,

$$\|L_n^{\leq D}\|^2 = \mathop{\mathbb{E}}_{\boldsymbol{x}^1,\boldsymbol{x}^2} \exp^{\leq D}(\lambda^2 \langle \boldsymbol{x}^1, \boldsymbol{x}^2\rangle^p) = \sum_{d=0}^{D} \frac{\lambda^{2d}}{d!} \mathop{\mathbb{E}}_{\boldsymbol{x}^1,\boldsymbol{x}^2}[\langle \boldsymbol{x}^1, \boldsymbol{x}^2\rangle^{pd}]. \tag{9}$$

We will give upper and lower bounds on this quantity in order to prove the two parts of Theorem 2.

3.1.1 Proof of Theorem 2: Upper Bound

Proof of Theorem 2(i) We use the moment bound

$$\mathop{\mathbb{E}}_{\boldsymbol{x}^1,\boldsymbol{x}^2}[|\langle \boldsymbol{x}^1, \boldsymbol{x}^2\rangle|^k] \leq (2n)^{k/2} k \Gamma(k/2) \tag{10}$$

[11] Some of these results only apply to minor variants of the spiked tensor problem, but we do not expect this difference to be important.

for any integer $k \geq 1$. This follows from $\langle \boldsymbol{x}^1, \boldsymbol{x}^2 \rangle$ being a *subgaussian* random variable with variance proxy n (see Appendix "Subgaussian Random Variables" for details on this notion, and see Proposition 12 for the bound (10)). Plugging this into (9),

$$\|L_n^{\leq D}\|^2 \leq 1 + \sum_{d=1}^{D} \frac{\lambda^{2d}}{d!}(2n)^{pd/2} pd\, \Gamma(pd/2) =: 1 + \sum_{d=1}^{D} T_d.$$

Note that $T_1 = O(1)$ provided $\lambda = O(n^{-p/4})$ (which will be implied by (11) below). Consider the ratio between successive terms:

$$r_d := \frac{T_{d+1}}{T_d} = \frac{\lambda^2}{d+1}(2n)^{p/2} p \frac{\Gamma(p(d+1)/2)}{\Gamma(pd/2)}.$$

Using the bound $\Gamma(x+a)/\Gamma(x) \leq (x+a)^a$ for all $a, x > 0$ (see Proposition 13), we find

$$r_d \leq \frac{\lambda^2}{d+1}(2n)^{p/2} p[p(d+1)/2]^{p/2} \leq \lambda^2 p^{p/2+1} n^{p/2} (d+1)^{p/2-1}.$$

Thus if λ is small enough, namely if

$$\lambda \leq \frac{1}{\sqrt{2}} p^{-p/4-1/2} n^{-p/4} D^{(2-p)/4}, \tag{11}$$

then $r_d \leq 1/2$ for all $1 \leq d < D$. In this case, by comparing with a geometric sum we may bound $\|L_n^{\leq D}\|^2 \leq 1 + 2T_1 = O(1)$.

3.1.2 Proof of Theorem 2: Lower Bound

Proof of Theorem 2(ii) Note that $\langle \boldsymbol{x}^1, \boldsymbol{x}^2 \rangle = \sum_{i=1}^n s_i$ where $s_1, \ldots, s_n$ are i.i.d. Rademacher random variables, so $\mathbb{E}_{\boldsymbol{x}^1,\boldsymbol{x}^2}[\langle \boldsymbol{x}^1, \boldsymbol{x}^2 \rangle^{2k+1}] = 0$, and

$$\mathop{\mathbb{E}}_{\boldsymbol{x}^1,\boldsymbol{x}^2}[\langle \boldsymbol{x}^1, \boldsymbol{x}^2 \rangle^{2k}] = \mathbb{E}\left[\left(\sum_{i=1}^n s_i\right)^{2k}\right] = \sum_{i_1,i_2,\ldots,i_{2k} \in [n]} \mathbb{E}[s_{i_1} s_{i_2} \cdots s_{i_{2k}}].$$

By counting only the terms $\mathbb{E}[s_{i_1} s_{i_2} \cdots s_{i_{2k}}]$ in which each s_i appears either 0 or 2 times, we have

$$\mathop{\mathbb{E}}_{\boldsymbol{x}^1,\boldsymbol{x}^2}[\langle \boldsymbol{x}^1, \boldsymbol{x}^2 \rangle^{2k}] \geq \binom{n}{k} \frac{(2k)!}{2^k}. \tag{12}$$

Let d be the largest integer such that $d \leq D$ and pd is even. By our assumption $D \leq \frac{2}{p}n$, we then have $pd/2 \leq n$. We now bound $\|L_n^{\leq D}\|^2$ by only the degree-pd term of (9), and using the bounds $\binom{n}{k} \geq (n/k)^k$ (for $1 \leq k \leq n$) and $(n/e)^n \leq n! \leq n^n$, we can lower bound that term as follows:

$$
\begin{aligned}
\|L_n^{\leq D}\|^2 &\geq \frac{\lambda^{2d}}{d!} \mathop{\mathbb{E}}_{x^1,x^2}[\langle x^1, x^2\rangle^{pd}] \\
&\geq \frac{\lambda^{2d}}{d!}\binom{n}{pd/2}\frac{(pd)!}{2^{pd/2}} \\
&\geq \frac{\lambda^{2d}}{d^d}\left(\frac{2n}{pd}\right)^{pd/2}\frac{(pd/e)^{pd}}{2^{pd/2}} \\
&= \left(\lambda^2 e^{-p} p^{p/2} n^{p/2} d^{p/2-1}\right)^d.
\end{aligned}
$$

Now, if λ is large enough, namely if

$$
\lambda \geq \sqrt{2} e^{p/2} p^{-p/4} n^{-p/4} D^{(2-p)/4}
$$

and $D = \omega(1)$, then $\|L_n^{\leq D}\|^2 \geq (2 - o(1))^d = \omega(1)$.

3.2 *The Spiked Wigner Matrix Model: Sharp Thresholds*

We now turn our attention to a more precise understanding of the case $p = 2$ of the spiked tensor model, which is more commonly known as the *spiked Wigner matrix model*. Our results from the previous section (specialized to $p = 2$) suggest that if $\lambda \gg n^{-1/2}$ then there should be a polynomial-time distinguishing algorithm, whereas if $\lambda \ll n^{-1/2}$ then there should not even be a subexponential-time distinguishing algorithm (that is, no algorithm of runtime $\exp(n^{1-\varepsilon})$ for any $\varepsilon > 0$). In this section, we will give a more detailed low-degree analysis that identifies the precise value of $\lambda\sqrt{n}$ at which this change occurs. This type of sharp threshold has been observed in various high-dimensional inference problems; another notable example is the Kesten-Stigum transition for community detection in the stochastic block model [28, 29, 76, 80, 81]. It was first demonstrated by [49] that the low-degree method can capture such sharp thresholds.

To begin, we recall the problem setup. Since the interesting regime is $\lambda = \Theta(n^{-1/2})$, we define $\hat{\lambda} = \lambda\sqrt{2n}$ and take $\hat{\lambda}$ to be constant (not depending on n). With this notation, the spiked Wigner model is as follows:

- Under $\mathbb{P}_n$, observe $\boldsymbol{Y} = \frac{\hat{\lambda}}{\sqrt{2n}}\boldsymbol{x}\boldsymbol{x}^\top + \boldsymbol{Z}$ where $\boldsymbol{x} \in \mathbb{R}^n$ is drawn from $\mathscr{X}_n$.
- Under $\mathbb{Q}_n$, observe $\boldsymbol{Y} = \boldsymbol{Z}$.

Here $\boldsymbol{Z}$ is an $n \times n$ random matrix with i.i.d. entries distributed as $\mathcal{N}(0, 1)$. (This asymmetric noise model is equivalent to the more standard symmetric one; see Appendix "Equivalence of Symmetric and Asymmetric Noise Models".) We will consider various spike priors $\mathscr{X}_n$, but require the following normalization.

Assumption 3 *The spike prior* $(\mathscr{X}_n)_{n \in \mathbb{N}}$ *is normalized so that* $\boldsymbol{x} \sim \mathscr{X}_n$ *satisfies* $\|\boldsymbol{x}\|^2 \to n$ *in probability as* $n \to \infty$.

3.2.1 The Canonical Distinguishing Algorithm: PCA

There is a simple reference algorithm for testing in the spiked Wigner model, namely *PCA (principal component analysis)*, by which we simply mean thresholding the largest eigenvalue of the (symmetrized) observation matrix.

Definition 13 The *PCA test* for distinguishing $\mathbb{P}$ and $\mathbb{Q}$ is the following statistical test, computable in polynomial time in n. Let $\overline{\boldsymbol{Y}} := (\boldsymbol{Y} + \boldsymbol{Y}^\top)/\sqrt{2n} = \frac{\hat{\lambda}}{n}\boldsymbol{x}\boldsymbol{x}^\top + \boldsymbol{W}$, where $\boldsymbol{W} = (\boldsymbol{Z} + \boldsymbol{Z}^\top)/\sqrt{2n}$ is a random matrix with the GOE distribution.[12] Then, let

$$f_{\hat{\lambda}}^{\mathsf{PCA}}(\boldsymbol{Y}) := \begin{Bmatrix} \mathtt{p} : \lambda_{\max}(\overline{\boldsymbol{Y}}) > t(\hat{\lambda}) \\ \mathtt{q} : \lambda_{\max}(\overline{\boldsymbol{Y}}) \le t(\hat{\lambda}) \end{Bmatrix}$$

where the threshold is set to $t(\hat{\lambda}) := 2 + (\hat{\lambda} + \hat{\lambda}^{-1} - 2)/2$.

The theoretical underpinning of this test is the following seminal result from random matrix theory, the analogue for Wigner matrices of the celebrated "BBP transition" [9].

Theorem 4 ([14, 41]) *Let* $\hat{\lambda}$ *be constant (not depending on* n*). Let* $\overline{\boldsymbol{Y}} = \frac{\hat{\lambda}}{n}\boldsymbol{x}\boldsymbol{x}^\top + \boldsymbol{W}$ *with* $\boldsymbol{W} \sim \mathsf{GOE}(n)$ *and arbitrary* $\boldsymbol{x} \in \mathbb{R}^n$ *with* $\|\boldsymbol{x}\|^2 = n$.

- *If* $\hat{\lambda} \le 1$, *then* $\lambda_{\max}(\overline{\boldsymbol{Y}}) \to 2$ *as* $n \to \infty$ *almost surely, and* $\langle \boldsymbol{v}_{\max}(\overline{\boldsymbol{Y}}), \boldsymbol{x}/\sqrt{n}\rangle^2 \to 0$ *almost surely (where* $\lambda_{\max}$ *denotes the largest eigenvalue and* $\boldsymbol{v}_{\max}$ *denotes the corresponding unit-norm eigenvector).*
- *If* $\hat{\lambda} > 1$, *then* $\lambda_{\max}(\overline{\boldsymbol{Y}}) \to \hat{\lambda} + \hat{\lambda}^{-1} > 2$ *as* $n \to \infty$ *almost surely, and* $\langle \boldsymbol{v}_{\max}(\overline{\boldsymbol{Y}}), \boldsymbol{x}/\sqrt{n}\rangle^2 \to 1 - \hat{\lambda}^{-2}$ *almost surely.*

Thus, the PCA test exhibits a sharp threshold: it succeeds when $\hat{\lambda} > 1$, and fails when $\hat{\lambda} \le 1$. (Furthermore, the leading eigenvector achieves non-trivial estimation of the spike $\boldsymbol{x}$ when $\hat{\lambda} > 1$ and fails to do so when $\hat{\lambda} \le 1$.)

---

12 Gaussian Orthogonal Ensemble (GOE): $\boldsymbol{W}$ is a symmetric $n \times n$ matrix with entries $W_{ii} \sim \mathcal{N}(0, 2/n)$ and $W_{ij} = W_{ji} \sim \mathcal{N}(0, 1/n)$, independently.

Corollary 1 *For any $\hat{\lambda} > 1$ and any spike prior family $(\mathscr{X}_n)_{n\in\mathbb{N}}$ valid per Assumption 3, $f_{\hat{\lambda}}^{\mathsf{PCA}}$ is a polynomial-time statistical test strongly distinguishing $\underline{\mathbb{P}}_{\lambda}$ and $\underline{\mathbb{Q}}$.*

For some spike priors $(\mathscr{X}_n)$, it is known that PCA is statistically optimal, in the sense that distinguishing (or estimating the spike) is information-theoretically impossible when $\hat{\lambda} < 1$. These priors include the prior with $\boldsymbol{x}$ drawn uniformly from the sphere of radius $\sqrt{n}$ and the priors with $\boldsymbol{x}$ having i.i.d. $\mathcal{N}(0, 1)$ or Rademacher (uniformly ± 1) entries [20, 25, 84, 91]. For these priors, we thus have $\hat{\lambda}_{\mathsf{stat}} = \hat{\lambda}_{\mathsf{comp}} = 1$, and there is no stat-comp gap.

A different picture emerges for other spike priors, such as the *sparse Rademacher prior*[13] with constant density $\rho = \Theta(1)$. If ρ is smaller than a particular small constant (roughly 0.09 [66]), it is known that $\hat{\lambda}_{\mathsf{stat}} < 1$. More precisely, an exponential-time exhaustive search algorithm succeeds in part of the regime where PCA fails [20]. For any given ρ, the precise threshold $\hat{\lambda}_{\mathsf{stat}}$ can be computed using the *replica-symmetric formula* from statistical physics (see, e.g., [10, 36–38, 57, 66, 70, 71, 73], or [77] for a survey). However, for any constant ρ, it is believed that $\hat{\lambda}_{\mathsf{comp}} = 1$, i.e., that no *polynomial-time* algorithm can "beat PCA." This has been conjectured in the same statistical physics literature based on failure of the *approximate message passing (AMP)* algorithm [66, 70, 71].

We will next give a low-degree analysis corroborating that conjecture: for a large class of spike priors (including the sparse Rademacher prior with constant ρ), our predictions suggest that any distinguishing algorithm requires nearly-exponential time whenever $\hat{\lambda} < 1$. One interesting case we will *not* cover is that of the sparse Rademacher prior with $\rho = o(1)$. In such models, there actually *are* subexponential-time algorithms that succeed for some $\hat{\lambda} < 1$; these are described in [31] along with matching lower bounds using the low-degree method. Furthermore, there are polynomial-time algorithms that can beat the PCA threshold once $\rho \lesssim 1/\sqrt{n}$; this is the much-studied "sparse PCA" regime (see, e.g., [5, 22, 31, 32, 55, 56, 62]).

3.2.2 Low-Degree Analysis: Informally, with the "Gaussian Heuristic"

We first give a heuristic low-degree analysis of the spiked Wigner model which suggests $\hat{\lambda}_{\mathsf{comp}} = 1$ (matching PCA) for sufficiently "reasonable" spike priors $(\mathscr{X}_n)$. In the next section, we will state and prove a rigorous statement to this effect.

Recall from (9) the expression for the norm of the LDLR:

$$\|L_n^{\leq D}\|^2 = \sum_{d=0}^{D} \frac{\lambda^{2d}}{d!} \mathop{\mathbb{E}}_{\boldsymbol{x}^1,\boldsymbol{x}^2}[\langle \boldsymbol{x}^1, \boldsymbol{x}^2\rangle^{2d}] = \sum_{d=0}^{D} \frac{1}{d!}\left(\frac{\hat{\lambda}^2}{2n}\right)^d \mathop{\mathbb{E}}_{\boldsymbol{x}^1,\boldsymbol{x}^2}[\langle \boldsymbol{x}^1, \boldsymbol{x}^2\rangle^{2d}].$$

[13] In the sparse Rademacher prior, each entry of $\boldsymbol{x}$ is nonzero with probability ρ (independently), and the nonzero entries are drawn uniformly from $\{\pm 1/\sqrt{\rho}\}$.

To predict $\hat{\lambda}_{\mathsf{comp}}$, it remains to determine whether $\|L_n^{\leq D}\|$ converges or diverges as $n \to \infty$, as a function of $\hat{\lambda}$. Recall that, per Assumption 3, $\mathscr{X}_n$ is normalized so that $\|\boldsymbol{x}\|^2 \approx n$. Thus when, e.g., $\boldsymbol{x}$ has i.i.d. entries, we may expect the following informal central limit theorem: "When $\boldsymbol{x}^1, \boldsymbol{x}^2 \sim \mathscr{X}_n$ independently, $\langle \boldsymbol{x}^1, \boldsymbol{x}^2 \rangle$ is distributed approximately as $\mathcal{N}(0, n)$." Assuming this heuristic applies to the first $2D$ moments of $\langle \boldsymbol{x}^1, \boldsymbol{x}^2 \rangle$, and recalling that the Gaussian moments are $\mathbb{E}_{g \sim \mathcal{N}(0,1)}[g^{2k}] = (2k-1)!! = \prod_{i=1}^{k}(2i-1)$, we may estimate

$$\|L_n^{\leq D}\|^2 \approx \sum_{d=0}^{D} \frac{1}{d!} \left(\frac{\hat{\lambda}^2}{2n}\right)^d \mathop{\mathbb{E}}_{g \sim \mathcal{N}(0,n)} [g^{2d}] = \sum_{d=0}^{D} \frac{1}{d!} \left(\frac{\hat{\lambda}^2}{2n}\right)^d n^d (2d-1)!! =: \sum_{d=0}^{D} T_d.$$

Imagine D grows slowly with n (e.g., $D \approx \log n$) in order to predict the power of polynomial-time algorithms. The ratio of consecutive terms above is

$$\frac{T_{d+1}}{T_d} = \hat{\lambda}^2 \cdot \frac{2d+1}{2(d+1)} \approx \hat{\lambda}^2,$$

suggesting that $\|L_n^{\leq D}\|$ should diverge if $\hat{\lambda} > 1$ and converge if $\hat{\lambda} < 1$.

While this style of heuristic analysis is often helpful for guessing the correct threshold, this type of reasoning can break down if D is too large or if $\boldsymbol{x}$ is too sparse. In the next section, we therefore give a rigorous analysis of $\|L_n^{\leq D}\|$.

3.2.3 Low-Degree Analysis: Formally, with Concentration Inequalities

We now give a rigorous proof that $\|L_n^{\leq D}\| = O(1)$ when $\hat{\lambda} < 1$ (and $\|L_n^{\leq D}\| = \omega(1)$ when $\hat{\lambda} > 1$), provided the spike prior is "nice enough." Specifically, we require the following condition on the prior.

Definition 14 A spike prior $(\mathscr{X}_n)_{n \in \mathbb{N}}$ admits a *local Chernoff bound* if for any $\eta > 0$ there exist $\delta > 0$ and $C > 0$ such that for all n,

$$\Pr\left\{|\langle \boldsymbol{x}^1, \boldsymbol{x}^2 \rangle| \geq t\right\} \leq C \exp\left(-\frac{1}{2n}(1-\eta)t^2\right) \quad \text{for all } t \in [0, \delta n]$$

where $\boldsymbol{x}^1, \boldsymbol{x}^2$ are drawn independently from $\mathscr{X}_n$.

For instance, any prior with i.i.d. *subgaussian* entries admits a local Chernoff bound; see Proposition 14 in Appendix "Subgaussian Random Variables". This includes, for instance, the sparse Rademacher prior with any constant density ρ. The following is the main result of this section, which predicts that for this class of spike priors, any algorithm that beats the PCA threshold requires nearly-exponential time.

Theorem 5 *Suppose $(\mathscr{X}_n)_{n \in \mathbb{N}}$ is a spike prior that (i) admits a local Chernoff bound, and (ii) has then $\|\boldsymbol{x}\|^2 \leq \sqrt{2}n$ almost surely if $\boldsymbol{x} \sim \mathscr{X}_n$. Then, for the* *spiked Wigner model with* $\hat{\lambda} < 1$ *and any* $D = D(n) = o(n/\log n)$*, we have* $\|L_n^{\leq D}\| = O(1)$ *as* $n \to \infty$.

Remark 3 The upper bound $\|\boldsymbol{x}\|^2 \leq \sqrt{2}n$ is without loss of generality (provided $\|\boldsymbol{x}\|^2 \to n$ in probability). This is because we can define a modified prior $\tilde{\mathscr{X}}_n$ that draws $\boldsymbol{x} \sim \mathscr{X}_n$ and outputs $\boldsymbol{x}$ if $\|\boldsymbol{x}\|^2 \leq \sqrt{2}n$ and $\boldsymbol{0}$ otherwise. If $(\mathscr{X}_n)_{n\in\mathbb{N}}$ admits a local Chernoff bound then so does $(\tilde{\mathscr{X}}_n)_{n\in\mathbb{N}}$. And, if the spiked Wigner model is computationally hard with the prior $(\tilde{\mathscr{X}}_n)_{n\in\mathbb{N}}$, it is also hard with the prior $(\mathscr{X}_n)_{n\in\mathbb{N}}$, since the two differ with probability $o(1)$.

Though we already know that a polynomial-time algorithm (namely PCA) exists when $\hat{\lambda} > 1$, we can check that indeed $\|L_n^{\leq D}\| = \omega(1)$ in this regime. For the sake of simplicity, we restrict this result to the Rademacher prior.

Theorem 6 *Consider the spiked Wigner model with the Rademacher prior:* $\boldsymbol{x}$ *has i.i.d. entries* $x_i \sim \mathsf{Unif}(\{\pm 1\})$*. If* $\hat{\lambda} > 1$*, then for any* $D = \omega(1)$ *we have* $\|L_n^{\leq D}\| = \omega(1)$.

The proof is a simple modification of the proof of Theorem 2(ii) in Sect. 3.1.2; we defer it to Appendix "Low-Degree Analysis of Spiked Wigner Above the PCA Threshold". The remainder of this section is devoted to proving Theorem 5.

Proof of Theorem 5 Starting from the expression for $\|L_n^{\leq D}\|^2$ (see Theorem 1 and Remark 2), we split $\|L_n^{\leq D}\|^2$ into two terms, as follows:

$$\|L_n^{\leq D}\|^2 = \mathop{\mathbb{E}}_{\boldsymbol{x}^1,\boldsymbol{x}^2}\left[\exp^{\leq D}\left(\lambda^2\langle \boldsymbol{x}^1, \boldsymbol{x}^2\rangle^2\right)\right] =: R_1 + R_2,$$

where

$$R_1 := \mathop{\mathbb{E}}_{\boldsymbol{x}^1,\boldsymbol{x}^2}\left[\mathbb{1}_{|\langle \boldsymbol{x}^1,\boldsymbol{x}^2\rangle| \leq \varepsilon n}\exp^{\leq D}\left(\lambda^2\langle \boldsymbol{x}^1, \boldsymbol{x}^2\rangle^2\right)\right],$$

$$R_2 := \mathop{\mathbb{E}}_{\boldsymbol{x}^1,\boldsymbol{x}^2}\left[\mathbb{1}_{|\langle \boldsymbol{x}^1,\boldsymbol{x}^2\rangle| > \varepsilon n}\exp^{\leq D}\left(\lambda^2\langle \boldsymbol{x}^1, \boldsymbol{x}^2\rangle^2\right)\right].$$

Here $\varepsilon > 0$ is a small constant to be chosen later. We call R_1 the *small deviations* and R_2 the *large deviations*, and we will bound these two terms separately.

We first consider bounding the large deviations. Using that $\|\boldsymbol{x}\|^2 \leq \sqrt{2}n$, that $\exp^{\leq D}(t)$ is increasing for $t \geq 0$, and the local Chernoff bound (taking ε to be a sufficiently small constant),

$$\begin{aligned}
R_2 &\leq \Pr\left\{|\langle \boldsymbol{x}^1, \boldsymbol{x}^2\rangle| > \varepsilon n\right\}\exp^{\leq D}\left(2\lambda^2 n^2\right) \\
&\leq C\exp\left(-\frac{1}{3}\varepsilon^2 n\right)\sum_{d=0}^{D}\frac{(\hat{\lambda}^2 n)^d}{d!}
\end{aligned}$$

and noting that the last term of the sum is the largest since $\hat{\lambda}^2 n > D$,

$$\begin{aligned}
&\leq C \exp\left(-\frac{1}{3}\varepsilon^2 n\right)(D+1)\frac{(\hat{\lambda}^2 n)^D}{D!} \\
&= \exp\left[\log C - \frac{1}{3}\varepsilon^2 n + \log(D+1) + 2D\log\hat{\lambda} + D\log n - \log(D!)\right] \\
&= o(1)
\end{aligned}$$

provided $D = o(n/\log n)$.

We now consider bounding the small deviations. We adapt an argument from [90]. Here we do not need to make use of the truncation to degree D at all, and instead simply use the bound $\exp^{\leq D}(t) \leq \exp(t)$ for $t \geq 0$. With this, we bound

$$\begin{aligned}
R_1 &= \mathop{\mathbb{E}}_{x^1,x^2}\left[\mathbb{1}_{|\langle x^1,x^2\rangle|\leq\varepsilon n}\exp^{\leq D}\left(\lambda^2\langle x^1,x^2\rangle^2\right)\right] \\
&\leq \mathop{\mathbb{E}}_{x^1,x^2}\left[\mathbb{1}_{|\langle x^1,x^2\rangle|\leq\varepsilon n}\exp\left(\lambda^2\langle x^1,x^2\rangle^2\right)\right] \\
&= \int_0^\infty \Pr\left\{\mathbb{1}_{|\langle x^1,x^2\rangle|\leq\varepsilon n}\exp\left(\lambda^2\langle x^1,x^2\rangle^2\right)\geq u\right\}du \\
&= 1 + \int_1^\infty \Pr\left\{\mathbb{1}_{|\langle x^1,x^2\rangle|\leq\varepsilon n}\exp\left(\lambda^2\langle x^1,x^2\rangle^2\right)\geq u\right\}du \\
&= 1 + \int_0^\infty \Pr\left\{\mathbb{1}_{|\langle x^1,x^2\rangle|\leq\varepsilon n}\langle x^1,x^2\rangle^2\geq t\right\}\lambda^2\exp\left(\lambda^2 t\right)dt \\
&\qquad\qquad\qquad\qquad\qquad\qquad\qquad\qquad (\text{where } \exp\left(\lambda^2 t\right) = u) \\
&\leq 1 + \int_0^\infty C\exp\left(-\frac{1}{2n}(1-\eta)t\right)\lambda^2\exp\left(\lambda^2 t\right)dt \\
&\qquad\qquad\qquad\qquad\qquad\qquad (\text{using the local Chernoff bound}) \\
&\leq 1 + \frac{C\hat{\lambda}^2}{2n}\int_0^\infty \exp\left(-\frac{1}{2n}(1-\eta-\hat{\lambda}^2)t\right)dt \\
&= 1 + C\hat{\lambda}^2(1-\eta-\hat{\lambda}^2)^{-1} \qquad\qquad (\text{provided } \hat{\lambda}^2 < 1-\eta) \\
&= O(1).
\end{aligned}$$

Since $\hat{\lambda} < 1$, we can choose $\eta > 0$ small enough so that $\hat{\lambda}^2 < 1-\eta$, and then choose ε small enough so that the local Chernoff bound holds. (Here, η and ε depend on $\hat{\lambda}$ and the spike prior, but not on n.)

4 More on the Low-Degree Method

In this section, we return to the general considerations introduced in Sect. 1.3 and describe some of the nuances in and evidence for the main conjecture underlying the low-degree method (Conjecture 1). Specifically, we investigate the question of what can be concluded (both rigorously and conjecturally) from the behavior of the low-degree likelihood ratio (LDLR) defined in Definition 9. We present conjectures and formal evidence connecting the LDLR to computational complexity, discussing various caveats and counterexamples along the way.

In Sect. 4.1, we explore to what extent the D-LDLR controls whether or not degree-D polynomials can distinguish $\underline{\mathbb{P}}$ from $\underline{\mathbb{Q}}$. Then, in Sect. 4.2, we explore to what extent the LDLR controls whether or not *any* efficient algorithm can distinguish $\underline{\mathbb{P}}$ and $\underline{\mathbb{Q}}$.

4.1 The LDLR and Thresholding Polynomials

Heuristically, since $\|L_n^{\leq D}\|$ is the value of the L^2 optimization problem (1), we might expect the behavior of $\|L_n^{\leq D}\|$ as $n \to \infty$ to dictate whether or not degree-D polynomials can distinguish $\underline{\mathbb{P}}$ from $\underline{\mathbb{Q}}$: it should be possible to strongly distinguish (in the sense of Definition 2, i.e., with error probabilities tending to 0) $\underline{\mathbb{P}}$ from $\underline{\mathbb{Q}}$ by *thresholding* a degree-D polynomial (namely $L_n^{\leq D}$) if and only if $\|L_n^{\leq D}\| = \omega(1)$. We now discuss to what extent this heuristic is correct.

Question 1 If $\|L_n^{\leq D}\| = \omega(1)$, does this imply that it is possible to strongly distinguish $\underline{\mathbb{P}}$ and $\underline{\mathbb{Q}}$ by thresholding a degree-D polynomial?

We have already mentioned (see Sect. 1.2.3) a counterexample when $D = \infty$: there are cases where $\underline{\mathbb{P}}$ and $\underline{\mathbb{Q}}$ are not statistically distinguishable, yet $\|L_n\| \to \infty$ due to a rare "bad" event under $\mathbb{P}_n$. Examples of this phenomenon are fairly common (e.g., [19, 20, 90, 91]). However, after truncation to only low-degree components, this issue seems to disappear. For instance, in sparse PCA, $\|L_n^{\leq D}\| \to \infty$ only occurs when either (i) there actually is an $n^{\tilde{O}(D)}$-time distinguishing algorithm, or (ii) D is "unreasonably large," in the sense that there is a trivial $n^{t(n)}$-time exhaustive search algorithm and $D \gg t(n)$ [31]. Indeed, we do not know any example of a natural problem where $\|L_n^{\leq D}\|$ diverges spuriously for a "reasonable" value of D (in the above sense), although one can construct unnatural examples by introducing a rare "bad" event in $\mathbb{P}_n$. Thus, it seems that for natural problems and reasonable growth of D, the smoothness of low-degree polynomials regularizes $L_n^{\leq D}$ in such a way that the answer to Question 1 is typically "yes." This convenient feature is perhaps related to the probabilistic phenomenon of *hypercontractivity*; see Appendix "Hypercontractivity" and especially Remark 5.

Another counterexample to Question 1 is the following. Take $\underline{\mathbb{P}}$ and $\underline{\mathbb{Q}}$ that are "easy" for degree-D polynomials to distinguish, i.e., $\|L_n^{\leq D}\| \to \infty$ and $\underline{\mathbb{P}}$, $\underline{\mathbb{Q}}$

can be strongly distinguished by thresholding a degree-D polynomial. Define a new sequence of "diluted" planted measures $\underline{\mathbb{P}}'$ where $\mathbb{P}_n'$ samples from $\mathbb{P}_n$ with probability $1/2$, and otherwise samples from $\mathbb{Q}_n$. Letting $L_n' = d\mathbb{P}_n'/d\mathbb{Q}_n$, we have $\|(L_n')^{\leq D}\| \to \infty$, yet $\underline{\mathbb{P}}'$ and $\underline{\mathbb{Q}}$ cannot be strongly distinguished (even statistically). While this example is perhaps somewhat unnatural, it illustrates that a rigorous positive answer to Question 1 would need to restrict to $\underline{\mathbb{P}}$ that are "homogeneous" in some sense.

Thus, while we have seen some artificial counterexamples, the answer to Question 1 seems to typically be "yes" for natural high-dimensional problems, so long as D is not unreasonably large. We now turn to the converse question.

Question 2 If $\|L_n^{\leq D}\| = O(1)$, does this imply that it is impossible to strongly distinguish $\underline{\mathbb{P}}$ and $\underline{\mathbb{Q}}$ by thresholding a degree-D polynomial?

Here, we are able to give a positive answer in a particular formal sense. The following result addresses the contrapositive of Question 2: it shows that distinguishability by thresholding low-degree polynomials implies exponential growth of the norm of the LDLR.

Theorem 7 *Suppose $\mathbb{Q}$ draws $\boldsymbol{Y} \in \mathbb{R}^N$ with entries either i.i.d. $\mathcal{N}(0,1)$ or i.i.d. $\mathsf{Unif}(\{\pm 1\})$. Let $\mathbb{P}$ be any measure on $\mathbb{R}^N$ that is absolutely continuous with respect to $\mathbb{Q}$. Let $k \in \mathbb{N}$ and let $f : \mathbb{R}^N \to \mathbb{R}$ be a polynomial of degree $\leq d$ satisfying*

$$\mathop{\mathbb{E}}_{\boldsymbol{Y}\sim\mathbb{P}}[f(\boldsymbol{Y})] \geq A \quad \textit{and} \quad \mathbb{Q}(|f(\boldsymbol{Y})| \geq B) \leq \delta \tag{13}$$

for some $A > B > 0$ and some $\delta \leq \frac{1}{2} \cdot 3^{-4kd}$. Then

$$\|L^{\leq 2kd}\| \geq \frac{1}{2}\left(\frac{A}{B}\right)^{2k}.$$

To understand what the result shows, imagine, for example, that $A > B$ are both constants, and k grows slowly with n (e.g., $k \approx \log n$). Then, if a degree-d polynomial (where d may depend on n) can distinguish $\underline{\mathbb{P}}$ from $\underline{\mathbb{Q}}$ in the sense of (13) with $\delta = \frac{1}{2}\cdot 3^{-4kd}$, then $\|L_n^{\leq 2kd}\| \to \infty$ as $n \to \infty$. Note though, that one weakness of this result is that we require the explicit quantitative bound $\delta \leq \frac{1}{2} \cdot 3^{-4kd}$, rather than merely $\delta = o(1)$.

The proof is an application of *hypercontractivity* (see e.g., [88]), a type of result stating that random variables obtained by evaluating low-degree polynomials on weakly-dependent distributions (such as i.i.d. ones) are well-concentrated and otherwise "reasonable." We have restricted to the case where $\mathbb{Q}$ is i.i.d. Gaussian or Rademacher because hypercontractivity results are most readily available in these cases, but we expect similar results to hold more generally. We review the necessary tools in detail in Appendix "Hypercontractivity".

Proof Consider the degree-$2kd$ polynomial f^{2k}. Using Jensen's inequality,

$$\mathbb{E}_{\mathbb{P}}[f(\boldsymbol{Y})^{2k}] \geq \left(\mathbb{E}_{\mathbb{P}}[f(\boldsymbol{Y})]\right)^{2k} \geq A^{2k}.$$

If $\mathbb{E}_{\mathbb{Q}}[f(\boldsymbol{Y})^{4k}] \leq B^{4k}$, then (14) holds. Otherwise, let $\theta := B^{4k}/\mathbb{E}_{\mathbb{Q}}[f(\boldsymbol{Y})^{4k}] \leq 1$. Then, we have

$$\begin{aligned}
\delta &\geq \mathbb{Q}(|f(\boldsymbol{Y})| \geq B) \\
&= \mathbb{Q}(f(\boldsymbol{Y})^{4k} \geq B^{4k}) \\
&= \mathbb{Q}(f(\boldsymbol{Y})^{4k} \geq \theta\, \mathbb{E}_{\mathbb{Q}}[f(\boldsymbol{Y})^{4k}]) \\
&\geq \frac{(1-\theta)^2}{3^{4kd}}. && \text{(by Proposition 2)}
\end{aligned}$$

Thus $\theta \geq 1 - 3^{2kd}\sqrt{\delta}$, implying

$$\mathbb{E}_{\mathbb{Q}}[f(\boldsymbol{Y})^{4k}] = \frac{B^{4k}}{\theta} \leq B^{4k}\left(1 - 3^{2kd}\sqrt{\delta}\right)^{-1}. \tag{14}$$

Using the key variational property of the LDLR (Proposition 1),

$$\|L^{\leq 2kd}\| \geq \frac{\mathbb{E}_{\mathbb{P}}[f(\boldsymbol{Y})^{2k}]}{\sqrt{\mathbb{E}_{\mathbb{Q}}[f(\boldsymbol{Y})^{4k}]}} \geq \left(\frac{A}{B}\right)^{2k}\sqrt{1 - 3^{2kd}\sqrt{\delta}} \geq \frac{1}{2}\left(\frac{A}{B}\right)^{2k}$$

since $\delta \leq \frac{1}{2} \cdot 3^{-4kd}$.

4.2 Algorithmic Implications of the LDLR

Having discussed the relationship between the LDLR and low-degree polynomials, we now discuss the relationship between low-degree polynomials and the power of *any* computationally-bounded algorithm.

Any degree-D polynomial has at most n^D monomial terms and so can be evaluated in time $n^{O(D)}$ (assuming that the individual coefficients are easy to compute). However, certain degree-D polynomials can of course be computed faster, e.g., if the polynomial has few nonzero monomials or has special structure allowing it to be computed via a spectral method (as in the *color coding* trick [6] used by [49]). Despite such special cases, it appears that for average-case high-dimensional hypothesis testing problems, degree-D polynomials are typically as powerful as general $n^{\tilde{\Theta}(D)}$-time algorithms; this informal conjecture appears as Hypothesis 2.1.5 in [48], building on the work of [15, 47, 49] (see also our previous discussion in Sect. 1.3). We will now explain the nuances and caveats of this conjecture, and give evidence (both formal and heuristic) in its favor.

4.2.1 Robustness

An important counterexample that we must be careful about is XOR-SAT. In the random 3-XOR-SAT problem, there are n $\{\pm 1\}$-valued variables $x_1, \ldots, x_n$ and we are given a formula consisting of m random constraints of the form $x_{i_\ell} x_{j_\ell} x_{k_\ell} = b_\ell$ for $\ell \in [m]$, with $b_\ell \in \{\pm 1\}$. The goal is to determine whether there is an assignment $x \in \{\pm 1\}^n$ that satisfies all the constraints. Regardless of m, this problem can be solved in polynomial time using Gaussian elimination over the finite field $\mathbb{F}_2$. However, when $n \ll m \ll n^{3/2}$, the low-degree method nevertheless predicts that the problem should be computationally hard, i.e., it is hard to distinguish between a random formula (which is unsatisfiable with high probability) and a formula with a planted assignment. This pitfall is not specific to the low-degree method: sum-of-squares lower bounds, statistical query lower bounds, and the cavity method from statistical physics also incorrectly suggest the same (this is discussed in [16, 23, 106]).

The above discrepancy can be addressed (see, e.g., Lecture 3.2 of [23]) by noting that Gaussian elimination is very brittle, in the sense that it no longer works to search for an assignment satisfying only a $1 - \delta$ fraction of the constraints (as in this case it does not seem possible to leverage the problem's algebraic structure over $\mathbb{F}_2$). Another example of a brittle algorithm is the algorithm of [105] for linear regression, which uses Lenstra-Lenstra-Lovász lattice basis reduction [72] and only tolerates an exponentially-small level of noise. Thus, while there sometimes exist efficient algorithms that are "high-degree", these tend not to be robust to even a tiny amount of noise. As with SoS lower bounds, we expect that the low-degree method correctly captures the limits of *robust* hypothesis testing [47] for high-dimensional problems. (Here, "robust" refers to the ability to handle a small amount of noise, and should not be confused with the specific notion of *robust inference* [47] or with other notions of robustness that allow adversarial corruptions [27, 40].)

4.2.2 Connection to Sum-of-Squares

The sum-of-squares (SoS) hierarchy [67, 89] is a hierarchy of increasingly powerful semidefinite programming (SDP) relaxations for general polynomial optimization problems. Higher levels of the hierarchy produce larger SDPs and thus require more time to solve: level d typically requires time $n^{O(d)}$. SoS lower bounds show that certain levels of the hierarchy fail to solve a given problem. As SoS seems to be at least as powerful as all known algorithms for many problems, SoS lower bounds are often thought of as the "gold standard" of formal evidence for computational hardness of average-case problems. For instance, if any constant level d of SoS fails to solve a problem, this is strong evidence that no polynomial-time algorithm exists to solve the same problem (modulo the robustness issue discussed above).

In order to prove SoS lower bounds, one needs to construct a valid primal certificate (also called a *pseudo-expectation*) for the SoS SDP. The *pseudo-calibration* approach [15] provides a strategy for systematically constructing a pseudo-expectation; however, showing that the resulting object is valid (in particular, showing that a certain associated matrix is positive semidefinite) often requires substantial work. As a result, proving lower bounds against constant-level SoS programs is often very technically challenging (as in [15, 47]). We refer the reader to [95] for a survey of SoS and pseudo-calibration in the context of high-dimensional inference.

On the other hand, it was observed by the authors of [15, 47] that the bottleneck for the success of the pseudo-calibration approach seems to typically be a simple condition, none other than the boundedness of the norm of the LDLR (see Conjecture 3.5 of [95] or Section 4.3 of [48]).[14] Through a series of works [15, 47–49], the low-degree method emerged from investigating this simpler condition in its own right. It was shown in [49] that SoS can be used to achieve sharp computational thresholds (such as the Kesten–Stigum threshold for community detection), and that the success of the associated method also hinges on the boundedness of the norm of the LDLR. So, historically speaking, the low-degree method can be thought of as a "lite" version of SoS lower bounds that is believed to capture the essence of what makes SoS succeed or fail (see [47–49, 95]).

A key advantage of the low-degree method over traditional SoS lower bounds is that it greatly simplifies the technical work required, allowing sharper results to be proved with greater ease. Moreover, the low-degree method is arguably more natural in the sense that it is not specific to any particular SDP formulation and instead seems to capture the essence of what makes problems computationally easy or hard. On the other hand, some would perhaps argue that SoS lower bounds constitute stronger evidence for hardness than low-degree lower bounds (although we do not know any average-case problems for which they give different predictions).

We refer the reader to [48] for more on the relation between SoS and the low-degree method, including evidence for why the two methods are believed to predict the same results.

4.2.3 Connection to Spectral Methods

For high-dimensional hypothesis testing problems, a popular class of algorithms are the *spectral methods*, algorithms that build a matrix $\boldsymbol{M}$ using the data and then

[14] More specifically, $(\|L_n^{\leq D}\|^2 - 1)$ is the variance of a certain pseudo-expectation value generated by pseudo-calibration, whose actual value in a valid pseudo-expectation must be exactly 1. It appears to be impossible to "correct" this part of the pseudo-expectation if the variance is diverging with n.

threshold its largest eigenvalue. (There are also spectral methods for estimation problems, usually extracting an estimate of the signal from the leading eigenvector of $\boldsymbol{M}$.) Often, spectral methods match the best[15] performance among all known polynomial-time algorithms. Some examples include the non-backtracking and Bethe Hessian spectral methods for the stochastic block model [18, 59, 76, 99], the covariance thresholding method for sparse PCA [32], and the tensor unfolding method for tensor PCA [50, 93]. As demonstrated in [50, 51], it is often possible to design spectral methods that achieve the same performance as SoS; in fact, some formal evidence indicates that low-degree spectral methods (where each matrix entry is a constant-degree polynomial of the data) are as powerful as any constant-degree SoS relaxation [47].[16]

As a result, it is interesting to try to prove lower bounds against the class of spectral methods. Roughly speaking, the largest eigenvalue in absolute value of a polynomial-size matrix $\boldsymbol{M}$ can be computed using $O(\log n)$ rounds of power iteration, and thus can be thought of as an $O(\log n)$-degree polynomial; more specifically, the associated polynomial is $\mathrm{Tr}(\boldsymbol{M}^{2k})$ where $k \sim \log(n)$. The following result makes this precise, giving a formal connection between the low-degree method and the power of spectral methods. The proof is similar to that of Theorem 7.

Theorem 8 *Suppose* $\mathbb{Q}$ *draws* $\boldsymbol{Y} \in \mathbb{R}^N$ *with entries either i.i.d.* $\mathcal{N}(0,1)$ *or i.i.d.* $\mathrm{Unif}(\{\pm 1\})$. *Let* $\mathbb{P}$ *be any measure on* $\mathbb{R}^N$ *that is absolutely continuous with respect to* $\mathbb{Q}$. *Let* $\boldsymbol{M} = \boldsymbol{M}(\boldsymbol{Y})$ *be a real symmetric* $m \times m$ *matrix, each of whose entries is a polynomial in* $\boldsymbol{Y}$ *of degree* $\le d$, *and let* $k \in \mathbb{N}$. *Suppose*

$$\mathop{\mathbb{E}}_{\boldsymbol{Y}\sim\mathbb{P}} \|\boldsymbol{M}\| \ge A \quad \text{and} \quad \mathbb{Q}(\|\boldsymbol{M}\| \ge B) \le \delta \tag{15}$$

(where $\|\cdot\|$ *denotes matrix operator norm) for some* $A > B > 0$ *and some* $\delta \le \frac{1}{2}\cdot 3^{-4kd}$. *Then*

$$\|L^{\le 2kd}\| \ge \frac{1}{2m}\left(\frac{A}{B}\right)^{2k}.$$

For example, suppose we are interested in polynomial-time spectral methods, in which case we should consider $m = \mathrm{poly}(n)$ and $d = O(1)$. If there exists a spectral method with these parameters that distinguishes $\underline{\mathbb{P}}$ from $\underline{\mathbb{Q}}$ in the sense of (15) for some constants $A > B$, and with $\delta \to 0$ faster than any inverse polynomial (in n),

[15] Here, "best" is in the sense of strongly distinguishing $\mathbb{P}_n$ and $\mathbb{Q}_n$ throughout the largest possible regime of model parameters.

[16] In [47], it is shown that for a fairly general class of average-case hypothesis testing problems, if SoS succeeds in some range of parameters then there is a low-degree spectral method whose maximum *positive* eigenvalue succeeds (in a somewhat weaker range of parameters). However, the resulting matrix could *a priori* have an arbitrarily large (in magnitude) negative eigenvalue, which would prevent the spectral method from running in polynomial time. For this same reason, it seems difficult to establish a formal connection between SoS and the LDLR via spectral methods.

then there exists a choice of $k = O(\log n)$ such that $\|L^{\leq O(\log n)}\| = \omega(1)$. And, by contrapositive, if we could show that $\|L^{\leq D}\| = O(1)$ for some $D = \omega(\log n)$, that would imply that there is no spectral method with the above properties. This justifies the choice of logarithmic degree in Conjecture 1. Similarly to Theorem 7, one weakness of Theorem 8 is that we can only rule out spectral methods whose failure probability is smaller than any inverse polynomial, instead of merely $o(1)$.

Remark 4 Above, we have argued that polynomial-time spectral methods correspond to polynomials of degree roughly $\log(n)$. What if we are instead interested in subexponential runtime $\exp(n^{\delta})$ for some constant $\delta \in (0, 1)$? One class of spectral method computable with this runtime is that where the dimension is $m \approx \exp(n^{\delta})$ and the degree of each entry is $d \approx n^{\delta}$ (such spectral methods often arise based on SoS [11, 13, 94]). To rule out such a spectral method using Theorem 8, we would need to take $k \approx \log(m) \approx n^{\delta}$ and would need to show $\|L^{\leq D}\| = O(1)$ for $D \approx n^{2\delta}$. However, Conjecture 1 postulates that time-$\exp(n^{\delta})$ algorithms should instead correspond to degree-n^{δ} polynomials, and this correspondence indeed appears to be the correct one based on the examples of tensor PCA (see Sect. 3.1) and sparse PCA (see [31]).

Although this seems at first to be a substantial discrepancy, there is evidence that there are actually spectral methods of dimension $m \approx \exp(n^{\delta})$ and *constant* degree $d = O(1)$ that achieve optimal performance among $\exp(n^{\delta})$-time algorithms. Such a spectral method corresponds to a degree-n^{δ} polynomial, as expected. These types of spectral methods have been shown to exist for tensor PCA [104].

Proof Let $\{\lambda_i\}$ be the eigenvalues of $\boldsymbol{M}$, with $|\lambda_1| \geq \cdots \geq |\lambda_m|$. Consider the degree-$2kd$ polynomial $f(\boldsymbol{Y}) := \mathrm{Tr}(\boldsymbol{M}^{2k}) = \sum_{i=1}^{m} \lambda_i^{2k}$. Using Jensen's inequality,

$$\mathbb{E}_{\mathbb{P}}[f(\boldsymbol{Y})] = \mathbb{E}_{\mathbb{P}} \sum_{i=1}^{m} \lambda_i^{2k} \geq \mathbb{E}_{\mathbb{P}}[\lambda_1^{2k}] \geq \left(\mathbb{E}_{\mathbb{P}}|\lambda_1|\right)^{2k} \geq A^{2k}.$$

If $\mathbb{E}_{\mathbb{Q}}[f(\boldsymbol{Y})^2] \leq m^2 B^{4k}$, then (16) holds. Otherwise, let $\theta := m^2 B^{4k} / \mathbb{E}_{\mathbb{Q}}[f(\boldsymbol{Y})^2] \leq 1$. Then, we have

$$\begin{aligned}
\delta &\geq \mathbb{Q}(\|\boldsymbol{M}\| \geq B) \\
&= \mathbb{Q}(\lambda_1^{2k} \geq B^{2k}) \\
&\geq \mathbb{Q}(f(\boldsymbol{Y}) \geq m B^{2k}) \\
&= \mathbb{Q}(f(\boldsymbol{Y})^2 \geq m^2 B^{4k}) \\
&= \mathbb{Q}(f(\boldsymbol{Y})^2 \geq \theta\, \mathbb{E}_{\mathbb{Q}}[f(\boldsymbol{Y})^2]) \\
&\geq \frac{(1-\theta)^2}{3^{4kd}}. \qquad \text{(by Proposition 2)}
\end{aligned}$$

Thus $\theta \geq 1 - 3^{2kd}\sqrt{\delta}$, implying

$$\mathbb{E}_{\mathbb{Q}}[f(Y)^2] = \frac{m^2 B^{4k}}{\theta} \leq m^2 B^{4k} \left(1 - 3^{2kd}\sqrt{\delta}\right)^{-1}. \tag{16}$$

Using the key variational property of the LDLR (Proposition 1),

$$\|L^{\leq 2kd}\| \geq \frac{\mathbb{E}_{\mathbb{P}}[f(Y)]}{\sqrt{\mathbb{E}_{\mathbb{Q}}[f(Y)^2]}} \geq \frac{1}{m}\left(\frac{A}{B}\right)^{2k} \sqrt{1 - 3^{2kd}\sqrt{\delta}} \geq \frac{1}{2m}\left(\frac{A}{B}\right)^{2k}$$

since $\delta \leq \frac{1}{2} \cdot 3^{-4kd}$.

4.2.4 Formal Conjecture

We next discuss the precise conjecture that Hopkins [48] offers on the algorithmic implications of the low-degree method. Informally, the conjecture is that for "sufficiently nice" $\mathbb{P}$ and $\mathbb{Q}$, if $\|L_n^{\leq D}\| = O(1)$ for some $D \geq (\log n)^{1+\varepsilon}$, then there is no polynomial-time algorithm that strongly distinguishes $\mathbb{P}$ and $\mathbb{Q}$. We will not state the full conjecture here (see Conjecture 2.2.4 in [48]) but we will briefly discuss some of the details that we have not mentioned yet.

Let us first comment on the meaning of "sufficiently nice" distributions. Roughly speaking, this means that:

1. $\mathbb{Q}_n$ is a product distribution,
2. $\mathbb{P}_n$ is sufficiently symmetric with respect to permutations of its coordinates, and
3. $\mathbb{P}_n$ is then perturbed by a small amount of additional noise.

Conditions (1) and (2) or minor variants thereof are fairly standard in high-dimensional inference problems. The reason for including (3) is to rule out non-robust algorithms such as Gaussian elimination (see Sect. 4.2.1).

One difference between the conjecture of [48] and the conjecture discussed in these notes is that [48] considers the notion of *coordinate degree* rather than *polynomial degree*. A polynomial has coordinate degree $\leq D$ if no monomial involves more than D variables; however, each individual variable can appear with arbitrarily-high degree in a monomial.[17] In [48], the low-degree likelihood ratio is defined as the projection of L_n onto the space of polynomials of coordinate degree $\leq D$. The reason for this is to capture, e.g., algorithms that preprocess the data by applying a complicated high-degree function entrywise. However, we are not aware of any natural problem in which it is important to work with coordinate degree

[17] Indeed, coordinate degree need not be phrased in terms of polynomials, and one may equivalently consider the linear subspace of $L^2(\mathbb{Q}_n)$ of functions that is spanned by functions of at most D variables at a time.

instead of polynomial degree. While working with coordinate degree gives lower bounds that are formally stronger, we work with polynomial degree throughout these notes because it simplifies many of the computations.

4.2.5 Empirical Evidence and Refined Conjecture

Perhaps the strongest form of evidence that we have in favor of the low-degree method is simply that it has been carried out on many high-dimensional inference problems and seems to always give the correct predictions, coinciding with widely-believed conjectures. These problems include planted clique [48] (implicit in [15]), community detection in the stochastic block model [48, 49], the spiked tensor model [47, 48], the spiked Wishart model [17], and sparse PCA [31]. In these notes we have also carried out low-degree calculations for the spiked Wigner model and spiked tensor model (see Sect. 3). Some of the early results [47, 49] showed only $\|L_n^{\leq D}\| = n^{o(1)}$ as evidence for hardness, which was later improved to $O(1)$ [48]. Some of the above results [47, 48] use coordinate degree instead of degree (as we discussed in Sect. 4.2.4). Throughout the above examples, the low-degree method has proven to be versatile in that it can predict both sharp threshold behavior as well as precise smooth tradeoffs between subexponential runtime and statistical power (as illustrated in the two parts of Sect. 3).

As discussed earlier, there are various reasons to believe that if $\|L_n^{\leq D}\| = O(1)$ for some $D = \omega(\log n)$ then there is no polynomial-time distinguishing algorithm; for instance, this allows us to rule out a general class of spectral methods (see Theorem 8). However, we have observed that in numerous examples, the LDLR actually has the following more precise behavior that does not involve the extra factor of $\log(n)$.

Conjecture 2 (Informal) Let $\mathbb{P}$ and $\mathbb{Q}$ be "sufficiently nice." Then, if there exists a polynomial-time algorithm to strongly distinguish $\mathbb{P}$ and $\mathbb{Q}$, then $\|L_n^{\leq D}\| = \omega(1)$ for any $D = \omega(1)$.

In other words, if $\|L_n^{\leq D}\| = O(1)$ for some $D = \omega(1)$, this already constitutes evidence that there is no polynomial-time algorithm. The above seems to be a cleaner version of the main low-degree conjecture that remains correct for many problems of practical interest.

4.2.6 Extensions

While we have focused on the setting of hypothesis testing throughout these notes, we remark that low-degree arguments have also shed light on other types of problems such as *estimation* (or *recovery*) and *certification*.

First, as we have mentioned before, non-trivial estimation[18] typically seems to be precisely as hard as strong distinguishing (see Definition 2), in the sense that the two problems share the same λ_{stat} and λ_{comp}. For example, the statistical thresholds for testing and recovery are known to coincide for problems such as the two-groups stochastic block model [76, 80, 81] and the spiked Wigner matrix model (for a large class of spike priors) [37, 38]. Also, for any additive Gaussian noise model, any lower bound against hypothesis testing using the second moment method (Lemma 2) or a conditional second moment method also implies a lower bound against recovery [20]. More broadly, we have discussed (see Sect. 4.2.3) how suitable spectral methods typically give optimal algorithms for high-dimensional problems; such methods typically succeed at testing and recovery in the same regime of parameters, because whenever the leading eigenvalue undergoes a phase transition, the leading eigenvector will usually do so as well (see Theorem 4 for a simple example). Thus, low-degree evidence that hypothesis testing is hard also constitutes evidence that non-trivial recovery is hard, at least heuristically. Note, however, that there is no formal connection (in either direction) between testing and recovery (see [20]), and there are some situations in which the testing and recovery thresholds differ (e.g., [85]).

In a different approach, Hopkins and Steurer [49] use a low-degree argument to study the recovery problem more directly. In the setting of community detection in the stochastic block model, they examine whether there is a low-degree polynomial that can non-trivially estimate whether two given network nodes are in the same community. They show that such a polynomial exists only when the parameters of the model lie above the problem's widely-conjectured computational threshold, the Kesten–Stigum threshold. This constitutes direct low-degree evidence that recovery is computationally hard below the Kesten–Stigum threshold.

A related (and more refined) question is that of determining the optimal estimation error (i.e., the best possible correlation between the estimator and the truth) for any given signal-to-noise parameter λ. Methods such as *approximate message passing* can often answer this question very precisely, both statistically and computationally (see, e.g., [10, 25, 35, 70], or [21, 77, 106] for a survey). One interesting question is whether one can recover these results using a variant of the low-degree method.

Another type of statistical task is *certification*. Suppose that $\boldsymbol{Y} \sim \mathbb{Q}_n$ has some property $\mathscr{P}$ with high probability. We say an algorithm *certifies* the property $\mathscr{P}$ if (i) the algorithm outputs "yes" with high probability on $\boldsymbol{Y} \sim \mathbb{Q}_n$, and (ii) if $\boldsymbol{Y}$ does not have property $\mathscr{P}$ then the algorithm *always* outputs "no." In other words, when the algorithm outputs "yes" (which is usually the case), this constitutes a proof that $\boldsymbol{Y}$ indeed has property $\mathscr{P}$. Convex relaxations (including SoS) are a common technique for certification. In [17], the low-degree method is used to argue that certification is computationally hard for certain structured PCA problems.

[18] Non-trivial estimation of a signal $\boldsymbol{x} \in \mathbb{R}^n$ means having an estimator $\hat{\boldsymbol{x}}$ achieving $|\langle \hat{\boldsymbol{x}}, \boldsymbol{x}\rangle|/(\|\hat{\boldsymbol{x}}\| \cdot \|\boldsymbol{x}\|) \geq \varepsilon$ with high probability, for some constant $\varepsilon > 0$.

The idea is to construct a *quiet planting* $\mathbb{P}_n$, which is a distribution for which (i) $\boldsymbol{Y} \sim \mathbb{P}_n$ never has property $\mathscr{P}$, and (ii) the low-degree method indicates that it is computationally hard to strongly distinguish $\mathbb{P}_n$ and $\mathbb{Q}_n$. In other words, this gives a reduction from a hypothesis testing problem to a certification problem, since any certification algorithm can be used to distinguish $\mathbb{P}_n$ and $\mathbb{Q}_n$. (Another example of this type of reduction, albeit relying on a different heuristic for computational hardness, is [102].)

Acknowledgments We thank the participants of a working group on the subject of these notes, organized by the authors at the Courant Institute of Mathematical Sciences during the spring of 2019. We also thank Samuel B. Hopkins, Philippe Rigollet, and David Steurer for helpful discussions.

DK was partially supported by NSF grants DMS-1712730 and DMS-1719545. ASW was partially supported by NSF grant DMS-1712730 and by the Simons Collaboration on Algorithms and Geometry. ASB was partially supported by NSF grants DMS-1712730 and DMS-1719545, and by a grant from the Sloan Foundation.

Appendix 1: Omitted Proofs

Neyman-Pearson Lemma

We include here, for completeness, a proof of the classical Neyman–Pearson lemma [87].

Proof of Lemma 1 Note first that a test f is completely determined by its *rejection region*, $R_f = \{\boldsymbol{Y} : f(\boldsymbol{Y}) = \mathbb{P}\}$. We may rewrite the power of f as

$$1 - \beta(f) = \mathbb{P}[f(\boldsymbol{Y}) = \mathbb{P}] = \int_{R_f} d\mathbb{P}(\boldsymbol{Y}) = \int_{R_f} L(\boldsymbol{Y}) d\mathbb{Q}(\boldsymbol{Y}).$$

On the other hand, our assumption on $\alpha(f)$ is equivalent to

$$\mathbb{Q}[R_f] \leq \mathbb{Q}[L(\boldsymbol{Y}) > \eta].$$

Thus, we are interested in solving the optimization

$$\begin{aligned} &\text{maximize } \textstyle\int_{R_f} L(\boldsymbol{Y}) d\mathbb{Q}(\boldsymbol{Y}) \\ &\text{subject to } R_f \in \mathscr{F}, \\ &\qquad\qquad \mathbb{Q}[R_f] \leq \mathbb{Q}[L(\boldsymbol{Y}) > \eta]. \end{aligned}$$

From this form, let us write $R_\star := \{\boldsymbol{Y} : L(\boldsymbol{Y}) > \eta\} = R_{L_\eta}$, then the difference of powers is

$$\begin{aligned}
(1 - \beta(L_\eta)) - (1 - \beta(f)) &= \int_{R_\star} L(\boldsymbol{Y}) d\mathbb{Q}(\boldsymbol{Y}) - \int_{R_f} L(\boldsymbol{Y}) d\mathbb{Q}(\boldsymbol{Y}) \\
&= \int_{R_\star \setminus R_f} L(\boldsymbol{Y}) d\mathbb{Q}(\boldsymbol{Y}) - \int_{R_f \setminus R_\star} L(\boldsymbol{Y}) d\mathbb{Q}(\boldsymbol{Y}) \\
&\geq \eta \left(\mathbb{Q}[R_\star \setminus R_f] - \mathbb{Q}[R_f \setminus R_\star] \right) \\
&= \eta \left(\mathbb{Q}[R_\star] - \mathbb{Q}[R_f] \right) \\
&\geq 0,
\end{aligned}$$

completing the proof.

Equivalence of Symmetric and Asymmetric Noise Models

For technical convenience, in the main text we worked with an asymmetric version of the spiked Wigner model (see Sect. 3.2), $\boldsymbol{Y} = \lambda \boldsymbol{x}\boldsymbol{x}^\top + \boldsymbol{Z}$ where $\boldsymbol{Z}$ has i.i.d. $\mathcal{N}(0, 1)$ entries. A more standard model is to instead observe $\widetilde{\boldsymbol{Y}} = \frac{1}{2}(\boldsymbol{Y} + \boldsymbol{Y}^\top) = \lambda \boldsymbol{x}\boldsymbol{x}^\top + \boldsymbol{W}$, where $\boldsymbol{W}$ is symmetric with $\mathcal{N}(0, 1)$ diagonal entries and $\mathcal{N}(0, 1/2)$ off-diagonal entries, all independent. These two models are equivalent, in the sense that if we are given a sample from one then we can produce a sample from the other. Clearly, if we are given $\boldsymbol{Y}$, we can symmetrize it to form $\widetilde{\boldsymbol{Y}}$. Conversely, if we are given $\widetilde{\boldsymbol{Y}}$, we can draw an independent matrix $\boldsymbol{G}$ with i.i.d. $\mathcal{N}(0, 1)$ entries, and compute $\widetilde{\boldsymbol{Y}} + \frac{1}{2}(\boldsymbol{G} - \boldsymbol{G}^\top)$; one can check that the resulting matrix has the same distribution as $\boldsymbol{Y}$ (we are adding back the "skew-symmetric part" that is present in $\boldsymbol{Y}$ but not $\widetilde{\boldsymbol{Y}}$).

In the spiked tensor model (see Sect. 3.1), our asymmetric noise model is similarly equivalent to the standard symmetric model defined in [93] (in which the noise tensor $\boldsymbol{Z}$ is averaged over all permutations of indices). Since we can treat each entry of the symmetric tensor separately, it is sufficient to show the following one-dimensional fact: for unknown $x \in \mathbb{R}$, k samples of the form $y_i = x + \mathcal{N}(0, 1)$ are equivalent to one sample of the form $\tilde{y} = x + \mathcal{N}(0, 1/k)$. Given $\{y_i\}$, we can sample $\tilde{y}$ by averaging: $\frac{1}{k}\sum_{i=1}^{k} y_i$. For the converse, fix unit vectors $\boldsymbol{a}_1, \ldots, \boldsymbol{a}_k$ at the corners of a simplex in $\mathbb{R}^{k-1}$; these satisfy $\langle \boldsymbol{a}_i, \boldsymbol{a}_j \rangle = -\frac{1}{k-1}$ for all $i \neq j$. Given $\tilde{y}$, draw $\boldsymbol{u} \sim \mathcal{N}(0, \boldsymbol{I}_{k-1})$ and let $y_i = \tilde{y} + \sqrt{1 - 1/k}\, \langle \boldsymbol{a}_i, \boldsymbol{u} \rangle$; one can check that these have the correct distribution.

Low-Degree Analysis of Spiked Wigner Above the PCA Threshold

Proof of Theorem 6 We follow the proof of Theorem 2(ii) in Sect. 3.1.2. For any choice of $d \le D$, using the standard bound $\binom{2d}{d} \ge 4^d/(2\sqrt{d})$,

$$\begin{aligned}
\|L_n^{\le D}\|^2 &\ge \frac{\lambda^{2d}}{d!} \mathop{\mathbb{E}}_{x^1,x^2}[\langle x^1, x^2\rangle^{2d}] \\
&\ge \frac{\lambda^{2d}}{d!}\binom{n}{d}\frac{(2d)!}{2^d} \qquad \text{(using the moment bound (12) from Section 3.1.2)} \\
&= \frac{\lambda^{2d}}{d!}\frac{n!}{d!(n-d)!}\frac{(2d)!}{2^d} \\
&= \lambda^{2d}\binom{2d}{d}\frac{n!}{(n-d)!2^d} \\
&\ge \lambda^{2d}\frac{4^d}{2\sqrt{d}}\frac{(n-d)^d}{2^d} \\
&= \frac{1}{2\sqrt{d}}\left(2\lambda^2(n-d)\right)^d \\
&= \frac{1}{2\sqrt{d}}\left(\hat{\lambda}^2\left(1-\frac{d}{n}\right)\right)^d.
\end{aligned}$$

Since $\hat{\lambda} > 1$, this diverges as $n \to \infty$ provided we choose $d \le D$ with $\omega(1) \le d \le o(n)$.

Appendix 2: Omitted Probability Theory Background

Hermite Polynomials

Here we give definitions and basic facts regarding the Hermite polynomials (see, e.g, [101] for further details), which are orthogonal polynomials with respect to the standard Gaussian measure.

Definition 15 The *univariate Hermite polynomials* are the sequence of polynomials $h_k(x) \in \mathbb{R}[x]$ for $k \ge 0$ defined by the recursion

$$\begin{aligned}
h_0(x) &= 1, \\
h_{k+1}(x) &= xh_k(x) - h_k'(x).
\end{aligned}$$

The *normalized univariate Hermite polynomials* are $\widehat{h}_k(x) = h_k(x)/\sqrt{k!}$.

The following is the key property of the Hermite polynomials, which allows functions in $L^2(\mathcal{N}(0, 1))$ to be expanded in terms of them.

Proposition 10 *The normalized univariate Hermite polynomials form a complete orthonormal system of polynomials for* $L^2(\mathcal{N}(0, 1))$.

The following are the multivariate generalizations of the above definition that we used throughout the main text.

Definition 16 The *N-variate Hermite polynomials* are the polynomials $H_{\boldsymbol{\alpha}}(\boldsymbol{X}) := \prod_{i=1}^{N} h_{\alpha_i}(X_i)$ for $\boldsymbol{\alpha} \in \mathbb{N}^N$. The *normalized N-variate Hermite polynomials in N variables* are the polynomials $\widehat{H}_{\boldsymbol{\alpha}}(\boldsymbol{X}) := \prod_{i=1}^{N} \widehat{h}_{\alpha_i}(X_i) = (\prod_{i=1}^{N} \alpha_i!)^{-1/2} \prod_{i=1}^{N} h_{\alpha_i}(X_i)$ for $\boldsymbol{\alpha} \in \mathbb{N}^N$.

Again, the following is the key property justifying expansions in terms of these polynomials.

Proposition 11 *The normalized N-variate Hermite polynomials form a complete orthonormal system of (multivariate) polynomials for* $L^2(\mathcal{N}(\boldsymbol{0}, \boldsymbol{I}_N))$.

For the sake of completeness, we also provide proofs below of the three identities concerning univariate Hermite polynomials that we used in Sect. 2.3 to derive the norm of the LDLR under the additive Gaussian noise model. It is more convenient to prove these in a different order than they were presented in Sect. 2.3, since one identity is especially useful for proving the others.

Proof of Proposition 8, Integration by Parts Recall that we are assuming a function $f : \mathbb{R} \to \mathbb{R}$ is k times continuously differentiable and f and its derivatives are $O(\exp(|x|^{\alpha}))$ for $\alpha \in (0, 2)$, and we want to show the identity

$$\mathbb{E}_{y\sim\mathcal{N}(0,1)}[h_k(y) f(y)] = \mathbb{E}_{y\sim\mathcal{N}(0,1)}\left[\frac{d^k f}{dy^k}(y)\right].$$

We proceed by induction. Since $h_0(y) = 1$, the case $k = 0$ follows immediately. We also verify by hand the case $k = 1$, with $h_1(y) = y$:

$$\begin{aligned}
\mathbb{E}_{y\sim\mathcal{N}(0,1)}[y f(y)] &= \frac{1}{\sqrt{2\pi}} \int_{-\infty}^{\infty} f(y) \cdot y e^{-y^2/2} dy \\
&= \frac{1}{\sqrt{2\pi}} \int_{-\infty}^{\infty} f'(y) e^{-y^2/2} dy \\
&= \mathbb{E}_{y\sim\mathcal{N}(0,1)}\left[f'(y)\right],
\end{aligned}$$

where we have used ordinary integration by parts.

Now, suppose the identity holds for all degrees smaller than some $k \geq 2$, and expand the degree k case according to the recursion:

$$\begin{aligned}\mathop{\mathbb{E}}_{y\sim\mathcal{N}(0,1)}[h_k(y)f(y)] &= \mathop{\mathbb{E}}_{y\sim\mathcal{N}(0,1)}[yh_{k-1}(y)f(y)] - \mathop{\mathbb{E}}_{y\sim\mathcal{N}(0,1)}[h'_{k-1}(y)f(y)] \\ &= \mathop{\mathbb{E}}_{y\sim\mathcal{N}(0,1)}[h'_{k-1}(y)f(y)] + \mathop{\mathbb{E}}_{y\sim\mathcal{N}(0,1)}[h_{k-1}(y)f'(y)] \\ &\qquad - \mathop{\mathbb{E}}_{y\sim\mathcal{N}(0,1)}[h'_{k-1}(y)f(y)] \\ &= \mathop{\mathbb{E}}_{y\sim\mathcal{N}(0,1)}[h_{k-1}(y)f'(y)] \\ &= \mathop{\mathbb{E}}_{y\sim\mathcal{N}(0,1)}\left[\frac{d^k f}{dy^k}(y)\right],\end{aligned}$$

where we have used the degree 1 and then the degree $k-1$ hypotheses.

Proof of Proposition 7, Translation Identity Recall that we want to show, for all $k \geq 0$ and $\mu \in \mathbb{R}$, that

$$\mathop{\mathbb{E}}_{y\sim\mathcal{N}(\mu,1)}[h_k(y)] = \mu^k.$$

We proceed by induction on k. Since $h_0(y) = 1$, the case $k = 0$ is immediate. Now, suppose the identity holds for degree $k-1$, and expand the degree k case according to the recursion:

$$\begin{aligned}\mathop{\mathbb{E}}_{y\sim\mathcal{N}(\mu,1)}[h_k(y)] &= \mathop{\mathbb{E}}_{y\sim\mathcal{N}(0,1)}[h_k(\mu+y)] \\ &= \mu \mathop{\mathbb{E}}_{y\sim\mathcal{N}(0,1)}[h_{k-1}(\mu+y)] + \mathop{\mathbb{E}}_{y\sim\mathcal{N}(0,1)}[yh_{k-1}(\mu+y)] \\ &\qquad - \mathop{\mathbb{E}}_{y\sim\mathcal{N}(0,1)}[h'_{k-1}(\mu+y)]\end{aligned}$$

which may be simplified by the Gaussian integration by parts to

$$\begin{aligned}&= \mu \mathop{\mathbb{E}}_{y\sim\mathcal{N}(0,1)}[h_{k-1}(\mu+y)] + \mathop{\mathbb{E}}_{y\sim\mathcal{N}(0,1)}[h'_{k-1}(\mu+y)] \\ &\qquad - \mathop{\mathbb{E}}_{y\sim\mathcal{N}(0,1)}[h'_{k-1}(\mu+y)] \\ &= \mu \mathop{\mathbb{E}}_{y\sim\mathcal{N}(0,1)}[h_{k-1}(\mu+y)],\end{aligned}$$

and the result follows by the inductive hypothesis.

Proof of Proposition 9, Generating Function Recall that we want to show the series identity for any $x, y \in \mathbb{R}$,

$$\exp\left(xy - \frac{1}{2}x^2\right) = \sum_{k=0}^{\infty} \frac{1}{k!} x^k h_k(y).$$

For any fixed x, the left-hand side belongs to $L^2(\mathcal{N}(0, 1))$ in the variable y. Thus this is merely a claim about the Hermite coefficients of this function, which may be computed by taking inner products. Namely, let us write

$$f_x(y) := \exp\left(xy - \frac{1}{2}x^2\right),$$

then using Gaussian integration by parts,

$$\begin{aligned}\langle f_x, \widehat{h}_k \rangle &= \frac{1}{\sqrt{k!}} \mathop{\mathbb{E}}_{y \sim \mathcal{N}(0,1)} [f_x(y) h_k(y)] \\ &= \frac{1}{\sqrt{k!}} \mathop{\mathbb{E}}_{y \sim \mathcal{N}(0,1)} \left[\frac{d^k f_x}{dy^k}(y)\right] \\ &= \frac{1}{\sqrt{k!}} x^k \mathop{\mathbb{E}}_{y \sim \mathcal{N}(0,1)} [f_x(y)] .\end{aligned}$$

A simple calculation shows that $\mathbb{E}_{y \sim \mathcal{N}(0,1)}[f_x(y)] = 1$ (this is an evaluation of the Gaussian moment-generating function that we have mentioned in the main text), and then by the Hermite expansion

$$f_x(y) = \sum_{k=0}^{\infty} \langle f_x, \widehat{h}_k \rangle \widehat{h}_k(y) = \sum_{k=0}^{\infty} \frac{1}{k!} x^k h_k(y),$$

giving the result.

Subgaussian Random Variables

Many of our rigorous arguments rely on the concept of *subgaussianity*, which we now define. See, e.g., [92] for more details.

Definition 17 For $\sigma^2 > 0$, we say that a real-valued random variable π is σ^2-*subgaussian* if $\mathbb{E}[\pi] = 0$ and for all $t \in \mathbb{R}$, the moment-generating function $M(t) = \mathbb{E}[\exp(t\pi)]$ of π exists and is bounded by $M(t) \leq \exp(\sigma^2 t^2/2)$.

Here σ^2 is called the *variance proxy*, which is not necessarily equal to the variance of π (although it can be shown that $\sigma^2 \geq \text{Var}[\pi]$). The name *subgaussian* refers to the fact that $\exp(\sigma^2 t^2/2)$ is the moment-generating function of $\mathcal{N}(0, \sigma^2)$.

The following are some examples of (laws of) subgaussian random variables. Clearly, $\mathcal{N}(0, \sigma^2)$ is σ^2-subgaussian. By Hoeffding's lemma, any distribution supported on an interval $[a, b]$ is $(b-a)^2/4$-subgaussian. In particular, the Rademacher distribution $\mathsf{Unif}(\{\pm 1\})$ is 1-subgaussian. Note also that the sum of n independent σ^2-subgaussian random variables is $\sigma^2 n$-subgaussian.

Subgaussian random variables admit the following bound on their absolute moments; see Lemmas 1.3 and 1.4 of [92].

Proposition 12 *If π is σ^2-subgaussian then*

$$\mathbb{E}[|\pi|^k] \leq (2\sigma^2)^{k/2} k \Gamma(k/2)$$

for every integer $k \geq 1$.

Here $\Gamma(\cdot)$ denotes the gamma function which, recall, is defined for all positive real numbers and satisfies $\Gamma(k) = (k-1)!$ when k is a positive integer. We will need the following property of the gamma function.

Proposition 13 *For all $x > 0$ and $a > 0$,*

$$\frac{\Gamma(x+a)}{\Gamma(x)} \leq (x+a)^a.$$

Proof This follows from two standard properties of the gamma function. The first is that (similarly to the factorial) $\Gamma(x+1)/\Gamma(x) = x$ for all $x > 0$. The second is *Gautschi's inequality*, which states that $\Gamma(x+s)/\Gamma(x) < (x+s)^s$ for all $x > 0$ and $s \in (0, 1)$.

In the context of the spiked Wigner model (Sect. 3.2), we now prove that subgaussian spike priors admit a local Chernoff bound (Definition 14).

Proposition 14 *Suppose π is σ^2-subgaussian (for some constant $\sigma^2 > 0$) with $\mathbb{E}[\pi] = 0$ and $\mathbb{E}[\pi^2] = 1$. Let $(\mathcal{X}_n)$ be the spike prior that draws each entry of $\boldsymbol{x}$ i.i.d. from π (where π does not depend on n). Then $(\mathcal{X}_n)$ admits a local Chernoff bound.*

Proof Since π is subgaussian, π^2 is *subexponential*, which implies $\mathbb{E}[\exp(t\pi^2)] < \infty$ for all $|t| \leq s$ for some $s > 0$ (see e.g., Lemma 1.12 of [92]).

Let π, π' be independent copies of π, and set $\Pi = \pi\pi'$. The moment-generating function of Π is

$$M(t) = \mathbb{E}[\exp(t\Pi)] = \mathbb{E}_\pi \mathbb{E}_{\pi'}[\exp(t\pi\pi')] \leq \mathbb{E}_\pi \left[\exp\left(\sigma^2 t^2 \pi^2/2\right)\right] < \infty$$

provided $\frac{1}{2}\sigma^2 t^2 < s$, i.e. $|t| < \sqrt{2s/\sigma^2}$. Thus $M(t)$ exists in an open interval containing $t = 0$, which implies $M'(0) = \mathbb{E}[\Pi] = 0$ and $M''(0) = \mathbb{E}[\Pi^2] = 1$ (this is the defining property of the moment-generating function: its derivatives at zero are the moments).

Let $\eta > 0$ and $f(t) := \exp\left(\frac{t^2}{2(1-\eta)}\right)$. Since $M(0) = 1$, $M'(0) = 0$, $M''(0) = 1$ and, as one may check, $f(0) = 1$, $f'(0) = 0$, $f''(0) = \frac{1}{1-\eta} > 1$, there exists $\delta > 0$ such that, for all $t \in [-\delta, \delta]$, $M(t)$ exists and $M(t) \le f(t)$.

We then apply the standard Chernoff bound argument to $\langle \boldsymbol{x}^1, \boldsymbol{x}^2 \rangle = \sum_{i=1}^n \Pi_i$ where $\Pi_1, \ldots, \Pi_n$ are i.i.d. copies of Π. For any $\alpha > 0$,

$$\begin{aligned}
\Pr\left\{\langle \boldsymbol{x}^1, \boldsymbol{x}^2 \rangle \ge t\right\} &= \Pr\left\{\exp(\alpha\langle \boldsymbol{x}^1, \boldsymbol{x}^2 \rangle) \ge \exp(\alpha t)\right\} \\
&\le \exp(-\alpha t)\mathbb{E}[\exp(\alpha\langle \boldsymbol{x}^1, \boldsymbol{x}^2 \rangle)] \qquad \text{(by Markov's inequality)} \\
&= \exp(-\alpha t)\mathbb{E}\left[\exp\left(\alpha \sum_{i=1}^n \Pi_i\right)\right] \\
&= \exp(-\alpha t)[M(\alpha)]^n \\
&\le \exp(-\alpha t)[f(\alpha)]^n \qquad \text{(provided } \alpha \le \delta\text{)} \\
&= \exp(-\alpha t)\exp\left(\frac{\alpha^2 n}{2(1-\eta)}\right).
\end{aligned}$$

Taking $\alpha = (1-\eta)t/n$,

$$\Pr\left\{\langle \boldsymbol{x}^1, \boldsymbol{x}^2 \rangle \ge t\right\} \le \exp\left(-\frac{1}{n}(1-\eta)t^2 + \frac{1}{2n}(1-\eta)t^2\right) = \exp\left(-\frac{1}{2n}(1-\eta)t^2\right)$$

as desired. This holds provided $\alpha \le \delta$, i.e. $t \le \delta n/(1-\eta)$. A symmetric argument with $-\Pi$ in place of Π holds for the other tail, $\Pr\left\{\langle \boldsymbol{x}^1, \boldsymbol{x}^2 \rangle \le -t\right\}$.

Hypercontractivity

The following hypercontractivity result states that the moments of low-degree polynomials of i.i.d. random variables must behave somewhat reasonably. The Rademacher version is the *Bonami lemma* from [88], and the Gaussian version appears in [53] (see Theorem 5.10 and Remark 5.11 of [53]). We refer the reader to [88] for a general discussion of hypercontractivity.

Proposition 15 (Bonami Lemma) *Let $\boldsymbol{x} = (x_1, \ldots, x_n)$ have either i.i.d. $\mathcal{N}(0, 1)$ or i.i.d. Rademacher (uniform ± 1) entries, and let $f : \mathbb{R}^n \to \mathbb{R}$ be a polynomial of* *degree k. Then*

$$\mathbb{E}[f(x)^4] \leq 3^{2k}\, \mathbb{E}[f(x)^2]^2.$$

We will combine this with the following standard second moment method.

Proposition 16 (Paley-Zygmund Inequality) *If $Z \geq 0$ is a random variable with finite variance, and $0 \leq \theta \leq 1$, then*

$$\Pr\{Z > \theta\, \mathbb{E}[Z]\} \geq (1-\theta)^2 \frac{\mathbb{E}[Z]^2}{\mathbb{E}[Z^2]}.$$

By combining Propositions 16 and 15, we immediately have the following.

Corollary 2 *Let $\boldsymbol{x} = (x_1, \ldots, x_n)$ have either i.i.d. $\mathcal{N}(0,1)$ or i.i.d. Rademacher (uniform ± 1) entries, and let $f : \mathbb{R}^n \to \mathbb{R}$ be a polynomial of degree k. Then, for $0 \leq \theta \leq 1$,*

$$\Pr\left\{f(x)^2 > \theta\, \mathbb{E}[f(x)^2]\right\} \geq (1-\theta)^2 \frac{\mathbb{E}[f(x)^2]^2}{\mathbb{E}[f(x)^4]} \geq \frac{(1-\theta)^2}{3^{2k}}.$$

Remark 5 One rough interpretation of Corollary 2 is that if f is degree k, then $\mathbb{E}[f(x)^2]$ cannot be dominated by an event of probability smaller than roughly 3^{-2k}.

References

1. A. Auffinger, G. Ben Arous, J. Černỳ, Random matrices and complexity of spin glasses. Commun. Pure Appl. Math. **66**(2), 165–201 (2013)
2. D. Achlioptas, A. Coja-Oghlan, Algorithmic barriers from phase transitions, in *2008 49th Annual IEEE Symposium on Foundations of Computer Science* (IEEE, IEEE, 2008), pp. 793–802
3. A. Anandkumar, Y. Deng, R. Ge, H. Mobahi, Homotopy analysis for tensor PCA (2016). arXiv preprint arXiv:1610.09322
4. N. Alon, M. Krivelevich, B. Sudakov, Finding a large hidden clique in a random graph. Random Struct. Algorithms **13**(3–4), 457–466 (1998)
5. A.A. Amini, M.J. Wainwright, High-dimensional analysis of semidefinite relaxations for sparse principal components, in *2008 IEEE International Symposium on Information Theory* (IEEE, Piscataway, 2008), pp. 2454–2458
6. N. Alon, R. Yuster, U. Zwick, Color-coding. J. ACM **42**(4), 844–856 (1995)
7. M. Brennan, G. Bresler, Optimal average-case reductions to sparse PCA: from weak assumptions to strong hardness (2019). arXiv preprint arXiv:1902.07380
8. M. Brennan, G. Bresler, W. Huleihel, Reducibility and computational lower bounds for problems with planted sparse structure (2018). arXiv preprint arXiv:1806.07508
9. J. Baik, G.Ben Arous, S. Péché, Phase transition of the largest eigenvalue for nonnull complex sample covariance matrices. Ann. Probab. **33**(5), 1643–1697 (2005)
10. J. Barbier, M. Dia, N. Macris, F. Krzakala, T. Lesieur, L. Zdeborová, Mutual information for symmetric rank-one matrix estimation: a proof of the replica formula, in *Proceedings of the*

30th International Conference on Neural Information Processing Systems (Curran Associates, 2016), pp. 424–432
11. V.V.S.P. Bhattiprolu, M. Ghosh, V. Guruswami, E. Lee, M. Tulsiani, Multiplicative approximations for polynomial optimization over the unit sphere. Electron. Colloq. Comput. Complexity **23**, 185 (2016)
12. G.Ben Arous, R. Gheissari, A. Jagannath, Algorithmic thresholds for tensor PCA (2018). arXiv preprint arXiv:1808.00921
13. V. Bhattiprolu, V. Guruswami, E. Lee, Sum-of-squares certificates for maxima of random tensors on the sphere (2016). arXiv preprint arXiv:1605.00903
14. F. Benaych-Georges, R. Rao Nadakuditi, The eigenvalues and eigenvectors of finite, low rank perturbations of large random matrices. Adva. Math. **227**(1), 494–521 (2011)
15. B. Barak, S. Hopkins, J. Kelner, P.K. Kothari, A. Moitra, A. Potechin, A nearly tight sum-of-squares lower bound for the planted clique problem. SIAM J. Comput. **48**(2), 687–735 (2019)
16. A. Blum, A. Kalai, H. Wasserman, Noise-tolerant learning, the parity problem, and the statistical query model. J. ACM **50**(4), 506–519 (2003)
17. A.S. Bandeira, D. Kunisky, A.S. Wein, Computational hardness of certifying bounds on constrained PCA problems (2019). arXiv preprint arXiv:1902.07324
18. C. Bordenave, M. Lelarge, L. Massoulié, Non-backtracking spectrum of random graphs: community detection and non-regular Ramanujan graphs, in *2015 IEEE 56th Annual Symposium on Foundations of Computer Science* (IEEE, Piscataway, 2015), pp. 1347–1357
19. J. Banks, C. Moore, J. Neeman, P. Netrapalli, Information-theoretic thresholds for community detection in sparse networks, in *Conference on Learning Theory* (2016), pp. 383–416
20. J. Banks, C. Moore, R. Vershynin, N. Verzelen, J. Xu, Information-theoretic bounds and phase transitions in clustering, sparse PCA, and submatrix localization. IEEE Trans. Inform. Theory **64**(7), 4872–4894 (2018)
21. A.S. Bandeira, A. Perry, A.S. Wein, Notes on computational-to-statistical gaps: predictions using statistical physics (2018). arXiv preprint arXiv:1803.11132
22. Q. Berthet, P. Rigollet, Computational lower bounds for sparse PCA (2013). arXiv preprint arXiv:1304.0828
23. B. Barak, D. Steurer, Proofs, beliefs, and algorithms through the lens of sum-of-squares. Course Notes (2016). http://www.sumofsquares.org/public/index.html
24. W.-K. Chen, D. Gamarnik, D. Panchenko, M. Rahman, Suboptimality of local algorithms for a class of max-cut problems. Ann. Probab. **47**(3), 1587–1618 (2019)
25. Y. Deshpande, E. Abbe, A. Montanari, Asymptotic mutual information for the two-groups stochastic block model (2015). arXiv preprint arXiv:1507.08685
26. M. Dyer, A. Frieze, M. Jerrum, On counting independent sets in sparse graphs. SIAM J. Comput. **31**(5), 1527–1541 (2002)
27. I. Diakonikolas, G. Kamath, D. Kane, J. Li, A. Moitra, A. Stewart, Robust estimators in high-dimensions without the computational intractability. SIAM J. Comput. **48**(2), 742–864 (2019)
28. A. Decelle, F. Krzakala, C. Moore, L. Zdeborová, Asymptotic analysis of the stochastic block model for modular networks and its algorithmic applications. Phys. Rev. E **84**(6), 066106 (2011)
29. A. Decelle, F. Krzakala, C. Moore, L. Zdeborová, Inference and phase transitions in the detection of modules in sparse networks. Phys. Rev. Lett. **107**(6), 065701 (2011)
30. I. Diakonikolas, D.M. Kane, A. Stewart, Statistical query lower bounds for robust estimation of high-dimensional Gaussians and gaussian mixtures, in *2017 IEEE 58th Annual Symposium on Foundations of Computer Science (FOCS)* (IEEE, Piscataway, 2017), pp. 73–84
31. Y. Ding, D. Kunisky, A.S. Wein, A.S. Bandeira, Subexponential-time algorithms for sparse PCA (2019). arXiv preprint
32. Y. Deshpande, A. Montanari, Sparse PCA via covariance thresholding, in *Advances in Neural Information Processing Systems* (2014), pp. 334–342
33. Y. Deshpande, A. Montanari, Finding hidden cliques of size $\sqrt{(N/e)}$ in nearly linear time. Found. Comput. Math. **15**(4), 1069–1128 (2015)

34. Y. Deshpande, A. Montanari, Improved sum-of-squares lower bounds for hidden clique and hidden submatrix problems, in *Conference on Learning Theory* (2015), pp. 523–562
35. D.L. Donoho, A. Maleki, A. Montanari, Message-passing algorithms for compressed sensing. Proc. Nat. Acad. Sci. **106**(45), 18914–18919 (2009)
36. A. El Alaoui, F. Krzakala, Estimation in the spiked Wigner model: a short proof of the replica formula, in *2018 IEEE International Symposium on Information Theory (ISIT)* (IEEE, 2018), pp. 1874–1878
37. A. El Alaoui, F. Krzakala, M.I. Jordan, Finite size corrections and likelihood ratio fluctuations in the spiked Wigner model (2017). arXiv preprint arXiv:1710.02903
38. A. El Alaoui, F. Krzakala, M.I. Jordan, Fundamental limits of detection in the spiked Wigner model (2018). arXiv preprint arXiv:1806.09588
39. V. Feldman, E. Grigorescu, L. Reyzin, S.S. Vempala, Y. Xiao, Statistical algorithms and a lower bound for detecting planted cliques. J. ACM **64**(2), 8 (2017)
40. U. Feige, J. Kilian, Heuristics for semirandom graph problems. J. Comput. Syst. Sci. **63**(4), 639–671 (2001)
41. D. Féral, S. Péché, The largest eigenvalue of rank one deformation of large Wigner matrices. Commun. Math. Phys. **272**(1), 185–228 (2007)
42. V. Feldman, W. Perkins, S. Vempala, On the complexity of random satisfiability problems with planted solutions. SIAM J. Comput. **47**(4), 1294–1338 (2018)
43. D. Grigoriev, Linear lower bound on degrees of Positivstellensatz calculus proofs for the parity. Theor. Comput. Sci. **259**(1–2), 613–622 (2001)
44. D. Gamarnik, M. Sudan, Limits of local algorithms over sparse random graphs, in *Proceedings of the 5th Conference on Innovations in Theoretical Computer Science*(ACM, New York, 2014), pp. 369–376
45. D. Gamarnik, I. Zadik, Sparse high-dimensional linear regression. algorithmic barriers and a local search algorithm (2017). arXiv preprint arXiv:1711.04952
46. D. Gamarnik I. Zadik, The landscape of the planted clique problem: Dense subgraphs and the overlap gap property (2019). arXiv preprint arXiv:1904.07174
47. S.B. Hopkins, P.K. Kothari, A. Potechin, P. Raghavendra, T. Schramm, D. Steurer, The power of sum-of-squares for detecting hidden structures, in *2017 IEEE 58th Annual Symposium on Foundations of Computer Science (FOCS)* (IEEE, Piscataway, 2017), pp. 720–731
48. S. Hopkins, *Statistical Inference and the Sum of Squares Method*. PhD thesis, Cornell University, August 2018
49. S.B. Hopkins, D. Steurer, Bayesian estimation from few samples: community detection and related problems (2017). arXiv preprint arXiv:1710.00264
50. S.B. Hopkins, J. Shi, D. Steurer, Tensor principal component analysis via sum-of-square proofs, in *Conference on Learning Theory* (2015), pp. 956–1006
51. S.B. Hopkins, T. Schramm, J. Shi, D. Steurer, Fast spectral algorithms from sum-of-squares proofs: tensor decomposition and planted sparse vectors, in *Proceedings of the Forty-Eighth Annual ACM Symposium on Theory of Computing* (ACM, New York, 2016), pp. 178–191
52. B. Hajek, Y. Wu, J. Xu, Computational lower bounds for community detection on random graphs, in *Conference on Learning Theory* (2015), pp. 899–928
53. S. Janson, *Gaussian Hilbert Spaces*, vol. 129 (Cambridge University Press, Cambridge, 1997)
54. M. Jerrum, Large cliques elude the Metropolis process. Random Struct. Algorithms **3**(4), 347–359 (1992)
55. I.M. Johnstone, A.Y. Lu, Sparse principal components analysis. Unpublished Manuscript (2004)
56. I.M. Johnstone, A.Y. Lu, On consistency and sparsity for principal components analysis in high dimensions. J. Am. Stat. Assoc. **104**(486), 682–693 (2009)
57. A. Jagannath, P. Lopatto, L. Miolane, Statistical thresholds for tensor PCA (2018). arXiv preprint arXiv:1812.03403
58. M. Kearns, Efficient noise-tolerant learning from statistical queries. J. ACM **45**(6), 983–1006 (1998)

59. F. Krzakala, C. Moore, E. Mossel, J. Neeman, A. Sly, L. Zdeborová, P. Zhang, Spectral redemption in clustering sparse networks. Proc. Nat. Acad. Sci. **110**(52), 20935–20940 (2013)
60. P.K. Kothari, R. Mori, R. O'Donnell, D. Witmer, Sum of squares lower bounds for refuting any CSP, in *Proceedings of the 49th Annual ACM SIGACT Symposium on Theory of Computing* (ACM, New York, 2017), pp. 132–145
61. F. Krzakała, A. Montanari, F. Ricci-Tersenghi, G. Semerjian, L. Zdeborová, Gibbs states and the set of solutions of random constraint satisfaction problems. Proc. Nat. Acad. Sci. **104**(25), 10318–10323 (2007)
62. R. Krauthgamer, B. Nadler, D. Vilenchik, Do semidefinite relaxations solve sparse PCA up to the information limit? Ann. Stat. **43**(3), 1300–1322 (2015)
63. A.R. Klivans, A.A. Sherstov, Unconditional lower bounds for learning intersections of halfspaces. Mach. Learn. **69**(2–3), 97–114 (2007)
64. L. Kučera, Expected complexity of graph partitioning problems. Discrete Appl. Math. **57**(2–3), 193–212 (1995)
65. R. Kannan, S. Vempala, Beyond spectral: Tight bounds for planted Gaussians (2016). arXiv preprint arXiv:1608.03643
66. F. Krzakala, J. Xu, L. Zdeborová, Mutual information in rank-one matrix estimation, in *2016 IEEE Information Theory Workshop (ITW)* (IEEE, Piscataway, 2016), pp. 71–75
67. J.B. Lasserre, Global optimization with polynomials and the problem of moments. SIAM J. Optim. **11**(3), 796–817 (2001)
68. L. Le Cam, *Asymptotic Methods in Statistical Decision Theory* (Springer, Berlin, 2012)
69. L. Le Cam, Locally asymptotically normal families of distributions. Univ. California Publ. Stat. **3**, 37–98 (1960)
70. T. Lesieur, F. Krzakala, L. Zdeborová, MMSE of probabilistic low-rank matrix estimation: Universality with respect to the output channel, in s*2015 53rd Annual Allerton Conference on Communication, Control, and Computing (Allerton)* (IEEE, 2015), pp. 680–687
71. T. Lesieur, F. Krzakala, L. Zdeborová, Phase transitions in sparse PCA, in *2015 IEEE International Symposium on Information Theory (ISIT)* (IEEE, Piscataway, 2015), pp. 1635–1639
72. A.K. Lenstra, H.W. Lenstra, L. Lovász, Factoring polynomials with rational coefficients. Math. Ann. **261**(4), 515–534 (1982)
73. M. Lelarge, L. Miolane, Fundamental limits of symmetric low-rank matrix estimation. Probab. Theory Related Fields **173**(3–4), 859–929 (2019)
74. T. Lesieur, L. Miolane, M. Lelarge, F. Krzakala, L. Zdeborová, Statistical and computational phase transitions in spiked tensor estimation, in *2017 IEEE International Symposium on Information Theory (ISIT)* (IEEE, Piscataway, 2017), pp. 511–515
75. E.L. Lehmann, J.P. Romano, *Testing Statistical Hypotheses* (Springer, Berlin, 2006)
76. L. Massoulié, Community detection thresholds and the weak Ramanujan property, in *Proceedings of the Forty-Sixth Annual ACM Symposium on Theory of Computing* (ACM, New York, 2014), pp. 694–703
77. L. Miolane, Phase transitions in spiked matrix estimation: information-theoretic analysis (2018). arXiv preprint arXiv:1806.04343
78. S.S. Mannelli, F. Krzakala, P. Urbani, L. Zdeborova, Passed & spurious: Descent algorithms and local minima in spiked matrix-tensor models, in *International Conference on Machine Learning* (2019), pp. 4333–4342
79. M. Mezard, A. Montanari, *Information, Physics, and Computation* (Oxford University Press, Oxford, 2009)
80. E. Mossel, J. Neeman, A. Sly, Reconstruction and estimation in the planted partition model. Probab. Theory Related Fields **162**(3–4), 431–461 (2015)
81. E. Mossel, J. Neeman, A. Sly, A proof of the block model threshold conjecture. Combinatorica **38**(3), 665–708 (2018)
82. M. Mézard, G. Parisi, M. Virasoro, *Spin Glass Theory and Beyond: An Introduction to the Replica Method and Its Applications*, vol. 9 (World Scientific Publishing Company, Singapore, 1987)

83. R. Meka, A. Potechin, A. Wigderson, Sum-of-squares lower bounds for planted clique, in *Proceedings of the Forty-Seventh Annual ACM Symposium on Theory of Computing* (ACM, New York, 2015), pp. 87–96
84. A. Montanari, D. Reichman, O. Zeitouni, On the limitation of spectral methods: from the Gaussian hidden clique problem to rank-one perturbations of gaussian tensors, in *Advances in Neural Information Processing Systems* (2015), pp. 217–225
85. L. Massoulié, L. Stephan, D. Towsley, Planting trees in graphs, and finding them back (2018). arXiv preprint arXiv:1811.01800
86. T. Ma, A. Wigderson, Sum-of-squares lower bounds for sparse PCA, in *Advances in Neural Information Processing Systems* (2015), pp. 1612–1620
87. J. Neyman, E.S. Pearson, IX. on the problem of the most efficient tests of statistical hypotheses. Philos. Trans. R. Soc. Lond. Ser. A Containing Papers Math. Phys. Charact. **231**(694–706), 289–337 (1933)
88. R. O'Donnell, *Analysis of Boolean Functions* (Cambridge University Press, Cambridge, 2014)
89. P.A. Parrilo, *Structured Semidefinite Programs and Semialgebraic Geometry Methods in Robustness and Optimization*. PhD thesis, California Institute of Technology, 2000
90. A. Perry, A.S. Wein, A.S. Bandeira, Statistical limits of spiked tensor models (2016). arXiv preprint arXiv:1612.07728
91. A. Perry, A.S. Wein, A.S. Bandeira, A. Moitra, Optimality and sub-optimality of PCA I: spiked random matrix models. Ann. Stat. **46**(5), 2416–2451 (2018)
92. P. Rigollet, J.-C. Hütter, High-dimensional statistics. Lecture Notes, 2018
93. E. Richard, A. Montanari, A statistical model for tensor PCA, in *Advances in Neural Information Processing Systems* (2014), pp. 2897–2905
94. P. Raghavendra, S. Rao, T. Schramm, Strongly refuting random CSPs below the spectral threshold, in *Proceedings of the 49th Annual ACM SIGACT Symposium on Theory of Computing* (ACM, New York, 2017), pp. 121–131
95. P. Raghavendra, T. Schramm, D. Steurer, High-dimensional estimation via sum-of-squares proofs (2018). arXiv preprint arXiv:1807.11419
96. R.W. Robinson, N.C. Wormald, Almost all cubic graphs are hamiltonian. Random Struct. Algorithms **3**(2), 117–125 (1992)
97. R.W. Robinson, N.C. Wormald, Almost all regular graphs are hamiltonian. Random Struct. Algorithms **5**(2), 363–374 (1994)
98. G. Schoenebeck, Linear level Lasserre lower bounds for certain k-CSPs, in *2008 49th Annual IEEE Symposium on Foundations of Computer Science* (IEEE, Piscataway, 2008), pp. 593–602
99. A. Saade, F. Krzakala, L. Zdeborová, Spectral clustering of graphs with the Bethe Hessian, in *Advances in Neural Information Processing Systems* (2014), pp. 406–414
100. E.M. Stein, R. Shakarchi, *Real Analysis: Measure Theory, Integration, and Hilbert Spaces* (Princeton University Press, Princeton, 2009)
101. G. Szegö, *Orthogonal Polynomials*, vol. 23 (American Mathematical Soc., 1939)
102. T. Wang, Q. Berthet, Y. Plan, Average-case hardness of RIP certification, in *Advances in Neural Information Processing Systems* (2016), pp. 3819–3827
103. T. Wang, Q. Berthet, R.J. Samworth, Statistical and computational trade-offs in estimation of sparse principal components. Ann. Stat. **44**(5), 1896–1930 (2016)
104. A.S. Wein, A. El Alaoui, C. Moore, The Kikuchi hierarchy and tensor PCA (2019). arXiv preprint arXiv:1904.03858
105. I. Zadik, D. Gamarnik, High dimensional linear regression using lattice basis reduction, in *Advances in Neural Information Processing Systems* (2018), pp. 1842–1852
106. L. Zdeborová, F. Krzakala, Statistical physics of inference: thresholds and algorithms. Adv. Phys. **65**(5), 453–552 (2016)

Totally Positive Functions in Sampling Theory and Time-Frequency Analysis

Karlheinz Gröchenig

Abstract Totally positive functions play an important role in approximation theory and statistics. In this chapter I will present recent new applications of totally positive functions in sampling theory and time-frequency analysis.

(i) We study the sampling problem for shift-invariant spaces generated by a totally positive function. These spaces are spanned by the integer shifts of a totally positive function and are often used as a substitute for the Paley-Wiener space of bandlimited functions. We give a characterization of sampling sets for a shift-invariant space with a totally positive generator of Gaussian type in the style of Beurling and Landau.

(ii) A related problem is the question of Gabor frames, i.e., the spanning properties of time-frequency shifts of a given function. For a subclass of totally positive functions their lattice shifts generate a frame, if and only if the density of the lattice exceeds 1. So far, totally positive functions seem to be the only window functions for which a complete characterization of the associated Gabor frames is tractable.

(iii) Yet another question in time-frequency analysis is the existence of zeros of the short-time Fourier transform. So far all examples of zero-free ambiguity functions are related to totally positive functions, e.g., the short-time Fourier transform of the Gaussian is zero free.

Keywords Totally positive function · Shift-invariant space · Sampling · Interpolation · Gabor frames · Riemann zeta function

Totally positive functions, Shift-invariant space, Sampling, Interpolation, Gabor frame, Riemann zeta function

K. Gröchenig (✉)
Faculty of Mathematics, University of Vienna, Vienna, Austria
e-mail: karlheinz.groechenig@univie.ac.at

P. Cerejeiras, M. Reissig (eds.), *Mathematical Analysis, its Applications and Computation*, Springer Proceedings in Mathematics & Statistics 385,
https://doi.org/10.1007/978-3-030-97127-4_2

1 Introduction

The motivation for the use of totally positive function is the eternal problem of sampling theory.

Given a set of data or measurements $(x_j, v_j)_{j\in J} \subseteq \mathbb{R}^d \times \mathbb{C}$,

find a functional relationship between x_j and v_j.

In other words, we try to find or learn or approximate a function $f : \mathbb{R}^d \to \mathbb{C}$ such that $f(x_j) \approx v_j$.

This question may be treated as a problem in statistics, or in a contemporary context it may be considered a problem in data science.

Clearly, as stated, this problem is not well defined. However, the sought-after function f is not completely arbitrary, and usually some a priori knowledge of f is available. This may be a parametric model for f, or some smoothness assumption on f, or a particular data model. In the mathematical analysis, one frequently assumes that f belongs to a linear subspace V of $L^2(\mathbb{R}^d)$ and possesses an expansion with respect to a suitable basis. In this case the expansion coefficients are the parameters to be determined. A frequent ansatz is to represent f as a linear combination of shifts of a single function ϕ as follows:

$$f(t) = \sum_{k=1}^{N} c_k \phi(t - y_k) . \tag{1}$$

This sum may be finite (realistic in applications) or infinite (interesting for mathematical analysis). For instance, in the theory of radial basis functions one studies approximation and interpolation by functions of the form $f(t) = \sum_{k=1}^N c_k e^{-\alpha(t-y_k)^2}$.

Assume now that $N \in \mathbb{N}$ and data $(x_j, v_j), j = 1, \ldots, N$, are given. To determine the coefficients $\{c_k : k = 1, ..., N\}$ and thus f, we need to solve the linear system $\sum_{k=1}^N c_k\phi(x_j - y_k) = v_j, j = 1, \ldots, N$. The intervening matrix is the sampling matrix

$$M = \big(\phi(x_j - y_k)\big)_{j,k=1,\ldots,N} \, . \tag{2}$$

For a unique recovery of $\{c_k\}$ and f, the sampling matrix M must be invertible, in other words, $\det M \neq 0$.

A small leap leads to the definition of totally positive functions: A function $\phi : \mathbb{R} \to \mathbb{R}$ is totally positive, if $\det M \geq 0$ for an arbitrary choice of points $x_1 < x_2 < \cdots < x_N$ and $y_1 < y_2 < \cdots < y_N$. Of course, for the sampling problem we need strict inequality $\det M > 0$.

This chapter is a survey of the role of totally positive functions in sampling theory and the related topic of time-frequency analysis. Both areas are vibrant areas of applied harmonic analysis with a impressive body of general results, structure theorems, numerical algorithms and qualitative assertions. However, the current state of the art offers few explicit and optimal results. This is where totally positive functions play a decisive role. Most, if not all, optimal results so far involve totally positive functions. The search for optimal results in sampling and in time-frequency analysis involves a fascinating interplay between different areas of analysis, notably complex analysis, approximation theory, sampling theory, time-frequency analysis. The goal of this chapter is to explain some of the recent results in sampling theory that use totally positive functions. As total positivity is essentially a one-dimensional concept, we will restrict ourselves exclusively to dimension $d = 1$.

This chapter is a version of my lecture at the 12th ISAAC conference in July 2019 in Aveiro. The chapter follows the flow of the lecture and is thus kept in an informal style with a minimum of prerequisites. The goal is to provide enough background for a more formal study of the technicalities of sampling and totally positive functions.

The chapter is organized as follows: In Sect. 2 we introduce rigorously the class of totally positive functions and formulate Schoenberg's characterization. In Sect. 3 we return to the sampling problem and define the shift-invariant spaces. In Sect. 4 we state the recent optimal sampling results, when the generating function is a totally positive function of Gaussian type. These results are in complete analogy with the known results about bandlimited functions, but they require a much bigger toolbox for their proofs. A few proof ideas and the decisive new observations are explained in detail. In Sect. 5 we discuss the basics of Gabor frames and explain the connection to sampling in shift-invariant spaces. Applying the optimal sampling results for totally positive generators, we obtain the known characterizations of Gabor frames with totally positive windows. In Sect. 6 we conclude with a curious problem about zero-free short-time Fourier transforms. Mysteriously, all known examples are related to totally positive functions. In Sect. 7 we use Schoenberg's factorization to connect Riemann's hypothesis with totally positive functions. Though this is not related to the main topic of this chapter, this facet of totally positive functions is amusing enough to be mentioned here.

2 Totally Positive Functions

The parametric model (1) and the required invertibility of the sampling matrix (2) motivate the following definition.

Definition 1 A function $\phi : \mathbb{R} \to \mathbb{R}$ is called totally positive, if for all finite sequences $x_1 < x_2 < \cdots < x_n$ and $y_1 < y_2 \cdots < y_n$ and $n \in \mathbb{N}$

$$\det\Big(\phi(x_j - y_k)\Big)_{j,k=1,\dots,n} \geq 0\,. \tag{3}$$

Strict total positivity, i.e., strict inequality in (3), is thus a guarantee that the matrix in (2) is invertible and that a function in the parametric model (1) can be recovered from its samples. In a sense we are cheating, because an appropriate definition does not solve the sampling problem. Definition 1 just guides us to ask the right questions, and we still need to obtain an understanding of total positivity.

Totally positive functions should not be mixed up with positive definite functions, where positive definite means that for every choice of $x_1, \dots, x_n$ the matrix $(\phi(x_j - x_k))_{j,k=1,\dots,N}$ is positive semi-definite. Although there is a common intersection of totally positive functions and positive definite functions, the definitions represent two rather different notions of positivity.

Total positivity per se is an important concept, and totally positive functions arise naturally in many areas.

- Spline theory: Totally positive functions were instrumental in the first development of spline theory between 1950 and 1980. One of their first applications was the general solution of the interpolation problem for polynomial splines [33].
- Statistics: totally positive functions play an important role in statistics, see for instance [9, 25].
- Sampling theory: Totally positive functions are natural generating functions in sampling models of the form (1). This aspect is the main topic of this chapter.
- Time-frequency analysis: Totally positive functions form a natural class of windows, and at this time, they are the *only* windows for which optimal results are known. See Sect. 5.
- Surprisingly, in a completely different context totally positive functions arise in the representation theory of infinite-dimensional motion groups [30].

We note that the class of totally positive functions is preserved under several natural transformations. If ϕ is totally positive, then the shift $x \mapsto \phi(x-u)$ and the dilation $x \mapsto \phi(ax)$ for $u \in \mathbb{R}$ and $a > 0$ are totally positive. Furthermore, after multiplication with an exponential $e^{ax}\phi(x)$ is again totally positive.

Examples Before we formulate a complete characterization of totally positive functions, we give some examples to show that totally positive functions are not terribly exotic. Indeed many common functions of analysis are totally positive.

A prototype of totally positive function is the one-sided exponential $\phi(x) = e^{-x/\nu}\chi_{[0,\infty)}(\nu x)$ for $\nu \neq 0$. The defining property (3) is easy to see. Assume $\nu > 0$, and let $D_1 = \mathrm{diag}\,(e^{-x_1}, \dots, e^{-x_n})$, $D_2 = \mathrm{diag}\,(e^{y_1}, \dots, e^{y_n})$, and

$$a_{jk} = \chi_{[0,\infty)}(\nu(x_j - y_k)) = \begin{cases} 1 & \text{if } y_k \le x_j \\ 0 & \text{if } y_k > x_j\,. \end{cases}$$

Then

$$M = \Big(e^{-(x_j-y_k)}\chi_{[0,\infty)}(\nu(x_j - y_k))\Big)_{j,k=1,\dots n} = D_1 A D_2\,,$$

and $\det M = \det D_1 \det A \det D_2 \ge 0$, if and only if $\det A \ge 0$. By inspection, we see that $\det A = 1$, if and only if A is a lower triangular matrix with ones on the diagonal, if and only if $y_j \le x_j < y_{j+1}$. Otherwise $\det A = 0$. It follows that $e^{-x/\nu}\chi_{[0,\infty)}(\nu x)$ is totally positive.

Further examples of totally positive functions are the symmetric exponential $\phi(x) = e^{-\nu|x|}$, $\nu > 0$ and the functions $\phi(x) = x^n e^{-\nu x}\chi_{[0,\infty)}(x)$, $\nu > 0$. The latter functions play an important role in establishing the fundamental properties of polynomial splines [33].

A quite general class of totally positive functions can be written as linear combinations of one-sided exponentials with rather special coefficients (that are obtained by a partial fraction decomposition of certain rational functions): let $\nu_j \in \mathbb{R}$, $\nu_j \neq 0$, $\nu_j \neq \nu_k$ for $j \neq k$, then

$$\phi(x) = \sum_{j=1}^{N}\left(\frac{1}{\nu_j}e^{-\frac{x}{\nu_j}}\chi_{[0,\infty)}(\nu_j x)\prod_{k=1,\,k\neq j}^{N}\left(1-\frac{\nu_k}{\nu_j}\right)^{-1}\right). \tag{4}$$

is totally positive.

Perhaps the most important totally positive function is the Gaussian $\phi(x) = e^{-\gamma x^2}$ for $\gamma > 0$. It will be our main test example to understand sampling and Gabor frames.

Finally we mention the hyperbolic secant $\phi(x) = \cosh(\beta x)^{-1} = (e^{-\beta x}+e^{\beta x})^{-1}$ and the exotic looking function $\phi(x) = \exp\left(ax - be^x\right)$ with $a, b > 0$.

The above examples may seem puzzling. Fortunately a complete characterization of all totally positive functions exists. This is a deep theorem of I. Schoenberg [32].

Theorem 1 *A function $\phi \in L^1(\mathbb{R})$ is totally positive, if and only if its Fourier transform possesses the factorization*

$$\hat{\phi}(\xi) = ce^{-\gamma\xi^2}e^{2\pi i\nu\xi}\prod_{j=1}^{N}(1+2\pi i\nu_j\xi)^{-1}e^{2\pi i\nu_j\xi} \tag{5}$$

with $c > 0$, $\nu, \nu_j \in \mathbb{R}$, $\gamma \ge 0$, $N \in \mathbb{N}\cup\{\infty\}$ and $0 < \gamma + \sum_j \nu_j^2 < \infty$.

The condition $\phi \in L^1(\mathbb{R})$ is not a serious restriction, as it can be shown that every totally function can be modified by multiplication with an exponential function to be integrable. We note that Schoenberg calls integrable totally positive functions Polya frequency function. Since we always assume integrability, we will not make this distinction.

Remark

1. The Fourier transform of $\phi(x) = \nu^{-1}e^{-x/\nu}\chi(\nu x)$ is $\hat{\phi}(\xi) = (1+2\pi i\nu\xi)^{-1}$. Thus the one-sided exponential is the simplest totally positive function corresponding to a single factor in the product (5). The general formula (4) is obtained from a finite product in (5) by a partial fraction expansion.

2. Since $\sum \nu_j^2$ is finite, the $\hat{\phi}$ in (5) can be extended to an meromorphic function with poles at $\frac{i}{2\pi\nu_j}$ on the imaginary axis. This implies that $\hat{\phi}$ is analytic in a horizontal strip containing the real axis. By a theorem of Paley-Wiener ϕ must therefore decay exponentially.

In our work it has turned out to be useful to distinguish certain subclasses of totally positive functions: a function ϕ is called

- totally positive of *finite type*, if $\gamma = 0$ and $N \in \mathbb{N}$ in the factorization (5),
- totally positive of *Gaussian type*, if $\gamma > 0$ and $N \in \mathbb{N}$,
- totally positive of *infinite type*, if $N = \infty$.

3 Back to Sampling: Shift-Invariant Spaces

Returning to sampling theory, we first fix the signal model. In signal processing, i.e., in the analysis and manipulation of one-dimensional signals, the standard models are the so-called shift-invariant spaces. These are function spaces that are generated by shifts of one or several generating functions.

Definition 2 Fix a non-zero function $\phi \in L^2(\mathbb{R})$. The shift-invariant space generated by ϕ is defined to be

$$V(\phi) = \{f \in L^2(\mathbb{R}) : f = \sum_{k\in\mathbb{Z}} c_k\phi(\cdot - k) \quad \text{with } c \in \ell^2(\mathbb{Z})\}$$

We always assume that the translates of ϕ are stable. This means that there exist constants $\alpha, \beta > 0$ such that

$$\alpha\|c\|_2 \le \|f\|_2 \le \beta\|c\|_2 \qquad \text{for all } c \in \ell^2(\mathbb{Z})\,. \tag{6}$$

Condition (6) implies that $V(\phi)$ is a closed subspace of $L^2(\mathbb{R})$. This stability assumption is satisfied for all reasonable generators. In fact, it is easy to see that the stability condition (6) is satisfied, if and only if for two constants $\alpha, \beta > 0$

$$\alpha \le \sum_{k\in\mathbb{Z}} |\hat{\phi}(\xi - k)|^2 \le \beta \quad \text{for almost all } \xi \in \mathbb{R}\,,$$

where $\hat{\phi}$ is the Fourier transform of ϕ. From the Fourier transform characterization of totally positive functions (5) is follows immediately that every totally positive function has stable translates, because such a $\hat{\phi}$ does not have any real zeros and is continuous.

In the literature shift-invariant spaces are also called spaces with finite rate of innovation (Vetterli [38]) or spline-type spaces (Feichtinger [10]). For an extended survey we refer to [3].

Note that the Fourier transform of $f \in V(\phi)$ is

$$\hat{f}(\xi) = \Big(\sum_{k\in\mathbb{Z}} c_k e^{-2\pi i k\xi}\Big)\hat{\phi}(\xi) = \mu(\xi)\hat{\phi}(\xi),$$

and thus factors into a 1-periodic function μ in $L^2([-1/2, 1/2])$ and $\hat{\phi}$.

The basic example of a shift-invariant space is the Paley-Wiener space with generator $\phi(x) = \frac{\sin \pi x}{\pi x}$. Since $\hat{\phi} = \chi_{[-1/2,1/2]}$, a function f is in $V(\phi)$, if and only if $\hat{f} = \mu\, \chi_{[-1/2,1/2]} \in L^2([-1/2, 1/2])$. In other words, $f \in V(\phi)$, if and only if supp $\hat{f} \subseteq [-1/2, 1/2]$. These are precisely the bandlimited functions of bandwidth 1. Signal processing in the past century was based almost exclusively the model of bandlimited functions.

Shift-invariant spaces may be seen as a generalization of the Paley-Wiener space of bandlimited functions. Assume that $\hat{\phi}$ has sufficient decay at infinity so that its essential support is concentrated near the origin. If $f \in V(\phi)$, then $\hat{f}(\xi) = \mu(\xi)\hat{\phi}(\xi)$ possesses the frequency envelope $\hat{\phi}$, and we may say that f is approximately bandlimited.

Our goal is to achieve a deeper understanding of sampling in shift-invariant spaces. First we need to clarify what is means that a function f is determined by and can be recovered from its sampled values on a set Λ. The standard notion of (stable) sampling is covered by the following definition.

Definition 3 A set $\Lambda \subseteq \mathbb{R}$ is a *set of stable sampling* for the closed subspace $V(\phi) \subseteq L^2(\mathbb{R})$, if there exist $A, B > 0$, such that

$$A\|f\|_2^2 \le \sum_{\lambda\in\Lambda} |f(\lambda)|^2 \le B\|f\|_2^2 \qquad \forall f \in V(\phi). \tag{7}$$

The (optimal) constants A and B are called the sampling constants, their ratio $\kappa = B/A$ is called the condition number of the sampling set.

In a sense, the sampling inequality (7) contains the solution of the sampling problem for $V(\phi)$. Clearly, if $f, g \in V(\phi)$ have the same samples $f(\lambda) = g(\lambda)$ for all $\lambda \in \Lambda$, then, since $f - g \in V(\phi)$, we have $\|f - g\|_2^2 \le A^{-1} \sum_{\lambda\in\Lambda} |f(\lambda) - g(\lambda)|^2 = 0$. This means that a function in the shift-invariant space $V(\phi)$ is uniquely determined by its samples on $\Lambda \subseteq \mathbb{R}$ and can therefore be reconstructed. Moreover, the left-side inequality in (7) asserts that the inverse mapping $f|_\Lambda \to f \in V(\phi)$ is continuous. Therefore a small measurement error of the sampled values will cause only a small error in the reconstruction.

While a sampling inequality (7) is the ultimate goal for the theoretical understanding of the sampling problem, in applications many additional aspects need to be studied. In particular, we must derive reconstruction procedures and understand how many samples are required to obtain a sampling inequality.

Numerical Reconstruction Algorithms At the end of the day we need a numerical reconstruction procedure for the recovery or the approximation of a function from given data. It is therefore important to observe that the sampling inequality (7) leads directly to several off-the-shelf reconstruction algorithms. More precisely, the right-hand side of (7) asserts that each point evaluation $f \to f(\lambda)$ is a bounded linear functional on $V(\phi)$ and are therefore of the form

$$f(\lambda) = \langle f, k_\lambda \rangle$$

for a unique kernel $k_\lambda \in V(\phi)$. This kernel can be expressed fairly explicitly in terms of the generator ϕ and is thus accessible for numerical computations. The sampling inequality (7) can then be rewritten as

$$A\|f\|_2^2 \le \sum_{\lambda\in\Lambda} |\langle f, k_\lambda\rangle|^2 \le B\|f\|_2^2 \qquad \forall f \in V(\phi)\,.$$

In the language of functional analysis, this means that the set of kernels $\{k_\lambda : \lambda \in \Lambda\}$ is a frame for $V(\phi)$, therefore any of the reconstruction algorithms from frame theory [8] can be used to recover $f \in V$ from its samples. A clean reconstruction formula can be obtained by means of the frame operator $Sf = \sum_{\lambda\in\Lambda} f(\lambda)k_\lambda$. Then $\langle Sf, f\rangle = \sum_{\lambda\in\Lambda} |f(\lambda)|^2 \ge A\|f\|_2^2$ for $f \in V(\phi)$ and thus S is invertible on $V(\phi)$. Setting $\widetilde{k_\lambda} = S^{-1}k_\lambda$ — this is the canonical dual frame — we can recover f from its samples by

$$f = S^{-1}Sf = \sum_{\lambda\in\Lambda} f(\lambda)S^{-1}k_\lambda = \sum_{\lambda\in\Lambda} f(\lambda)\widetilde{k_\lambda}\,.$$

Since the inversion of the frame operator may be difficult, one often resorts to iterative algorithms for the recovery.

Localization One of the drawbacks of bandlimited functions is the slow decay of the generator $\frac{\sin \pi x}{\pi x}$. This means that the size of a coefficient c_k has a significant influence on the value of $f(x)$ for x far apart from k. More explicitly, if $f(x) = \sum_{k\in\mathbb{Z}} c_k \frac{\sin \pi(x-k)}{\pi(x-k)}$ and $c_0 = f(0) = 1$, then the term $\frac{\sin \pi x}{\pi x}$ will still contribute to the value f at x with amplitude $1/|x|$. By contrast, if the generator ϕ of $V(\phi)$ possesses fast decay, then the correlation between the size of a coefficient c_k and the size of f is more localized. For simplicity, let us assume that ϕ decays exponentially, $|\phi(x)| \le Ce^{-\gamma|x|}$. Assume that $f(x) = \sum_{k\in\mathbb{Z}} c_k\phi(x-k) \in V(\phi)$ and $|x| \le M$. Then

$$\begin{aligned}\Big|\sum_{|k|>M+M_0} c_k\phi(x-k)\Big| &\le \|c\|_2\Big(\sum_{|k|>M+M_0} e^{-2\gamma|x-k|}\Big)^{1/2}\\ &\le \sqrt{2}\frac{e^{-\gamma M_0}}{(1-e^{-2\gamma})^{1/2}}\|c\|_2\\ &\le Ce^{-\gamma M_0}\|f\|_2\,.\end{aligned}$$

In other words, for $|x| \leq M$, we have

$$f(x) = \sum_{|k| \leq M+M_0} c_k \phi(x-k) + \mathscr{O}(e^{-\gamma M_0}),$$

and the coefficients c_k, $|k| \leq M + M_0$, determine the value of f on the interval $[-M, M]$ up to an exponentially small error. This is yet another reason why shift-invariant spaces have become an attractive signal model. Since every totally positive function decays exponentially, the associated shift-invariant spaces possess this useful form of localization.

Information Theory Clearly, a set of stable sampling must contain sufficiently many points. How much information is needed, is thus a question of information theory. It has turned out that the most useful quantity is the Beurling density.

Definition 4 The lower *Beurling density* of $\Lambda \subseteq \mathbb{R}$ is defined as

$$D^-(\Lambda) = \liminf_{r \to \infty} \inf_{x \in \mathbb{R}} \frac{\# \Lambda \cap [x, x+r]}{r}.$$

Analogously, the upper *Beurling density* is defined as

$$D^+(\Lambda) = \limsup_{r \to \infty} \sup_{x \in \mathbb{R}} \frac{\# \Lambda \cap [x, x+r]}{r}.$$

The Beurling density measures the average number of samples per time unit. We will usually assume that Λ is relatively separated, i.e., $\max_{x \in \mathbb{R}} \# \Lambda \cap [x, x+1] < \infty$. In terms of the Beurling densities this means that $D^+(\Lambda) < \infty$. This condition is clearly satisfied when Λ is separated, i.e., $\inf_{\lambda, \lambda' \in \Lambda} |\lambda - \lambda'| > 0$.

Let us recapitulate the prototype of all sampling theorems. These theorems are motivated by the Shannon-Whittaker-Kotelnikov sampling theorem and by engineering applications and are formulated for bandlimited functions. The mathematical theory developed by Landau, Beurling, and Kahane [7, 24, 27] offers a precise and general understanding of the connections between stability of sampling and the density of the sampling set.

Theorem 2 (Beurling-Kahane, Landau) *Let $\mathscr{B} = \{f \in L^2(\mathbb{R}) : \operatorname{supp} \hat{f} \subseteq [-1/2, 1/2]\} = V(\frac{\sin \pi x}{\pi x})$ be the Paley-Wiener space, and let $\Lambda \subseteq \mathbb{R}$ be relatively separated.*

(i) If $D^-(\Lambda) > 1$ and Λ is separated, then Λ is a set of stable sampling for $\mathscr{B}$.
(ii) If Λ is a set of stable sampling for $\mathscr{B}$, then $D^-(\Lambda) \geq 1$.

The necessary density condition in (ii) is proved with operator theory, and there are now several approaches to prove analogous density theorems for sampling in real and complex analysis. A general result for necessary density conditions that works in reproducing kernel Hilbert spaces is contained in [11].

By contrast, the sufficient condition (i) requires more tools and depends on the fine structure of the underlying function space. For the Paley-Wiener space one needs some complex analysis, precisely the analysis of zeros of entire functions of exponential type and Beurling's technique of weak limits of sets.

A similar result holds for interpolation in the Paley-Wiener space. We say that $\Lambda \subseteq \mathbb{R}$ is a set of interpolation for a subspace $V \subseteq L^2(\mathbb{R})$, if for every $a \in \ell^2(\Lambda)$ there exists a function $f \in V$, such that $f(\lambda) = a_\lambda$ for all $\lambda \in \Lambda$. The counter-part of Theorem 2 for interpolation asserts the following.

Theorem 3 *If $D^+(\Lambda) < 1$ and Λ is separated, then Λ is a set of interpolation for the Paley-Wiener space $\mathcal{B}$.*

Conversely, if Λ is a set of interpolation for the Paley-Wiener space $\mathcal{B}$, then $D^+(\Lambda) \leq 1$.

The cited results for bandlimited functions are the prototype of mathematical sampling results and have inspired several generations of mathematicians.

We now turn to the sampling problem in shift-invariant spaces. It is easy to derive necessary conditions for sampling in the style of Theorem 2 [2].

Theorem 4 *Assume that $\sum_{k\in\mathbb{Z}} \sup_{x\in[0,1]} |\phi(x+k)| < \infty$ (ϕ is in the amalgam space $W(\mathbb{R})$).*

If Λ is set of stable sampling for $V(\phi)$, then $D^-(\Lambda) \geq 1$.

The quest for sufficient conditions for stable sampling in a shift-invariant space is much harder. In general only qualitative results can be derived. In line with the notions of approximation theory, sufficient conditions are usually formulated with the maximum gap between samples, which is also known as the mesh size or the covering radius. Formally, the maximal gap of a set $\Lambda \subseteq \mathbb{R}$ is given by

$$\delta(\Lambda) = \sup_{\lambda\in\Lambda} \inf_{\mu\in\Lambda, \mu\neq\lambda} |\mu - \lambda| .$$

If the points in Λ are arranged by magnitude so that $\lambda_j < \lambda_{j+1}$ for all j, then $\delta(\Lambda) = \sup_{j\in\mathbb{Z}}(\lambda_{j+1} - \lambda_j)$.

The next theorem is an early qualitative sampling theorem for general shift-invariant spaces from [1].

Theorem 5 *For every ϕ satisfying $\sum_{k\in\mathbb{Z}} \sup_{x\in[0,1]} |\phi(x+k)| < \infty$, there exists maximum gap $\delta_0 = \delta_0(\phi) \leq 1$ with the following property:*

If $\delta(\Lambda) < \delta_0$ and Λ is relatively separated, then Λ is a set of stable sampling for $V(\phi)$.

In all available estimates the maximum gap depends on the smoothness of the generator ϕ and usually $\delta(\Lambda) << 1$. Since $D^-(\Lambda) \geq \delta(\Lambda)^{-1}$, most results in the spirit of Theorem 5 require high oversampling rates and thus are far from optimal and of limited use. Optimal sampling results with the covering density are known only for special generators. The following result summarizes our current knowledge.

Theorem 6 *Assume that either*

(i) ϕ *is a B-spline* $\phi = \chi_{[-1/2,1/2]} * \chi_{[-1/2,1/2]} * \cdots * \chi_{[-1/2,1/2]}$ *(n + 1-times), or*
(ii) ϕ *is an exponential B-spline:* $\phi = (e^{\lambda_1 \cdot}\chi_{[-1/2,1/2]}) * (e^{\lambda_2 \cdot}\chi_{[-1/2,1/2]}) * \dots * (e^{\lambda_n \cdot}\chi_{[-1/2,1/2]})$ *for* $\lambda_j \in \mathbb{R}$*, or*
(iii) $\hat{\phi}(\xi) = \prod_{j=1}^{n}(1 + 2\pi i \nu_j \xi)^{-1}$ *for* $\nu_j \in \mathbb{R}$.

Then: If $\delta(\Lambda) < 1$*, then* Λ *is a set of stable sampling for* $V(\phi)$.

The result about B-splines is from [2], the case of exponential B-splines is more recent [26]. According to the factorization (5), item (iii) covers precisely the totally positive functions of finite type, and Theorem 6(iii) was proved in [16] and [26]. All cases of Theorem 6 were proved with methods from approximation theory; specifically the Schoenberg-Whitney conditions for splines play a decisive role [33].

Remark

1. Theorem 6 is sharp because for every $\phi \in W(\mathbb{R})$ there exists an $x \in (0, 1)$ such that $x + \mathbb{Z}$ fails to be a set of sampling, see [21]. Thus the critical gap $\delta(\Lambda) = 1$ may not lead to sets of stable sampling even in the easy case of uniform sampling.
2. The cited results hold for more general sampling sets that allow a certain proportion of gaps to be greater than 1. See [2, 16] for the technical formulation of these results.

4 Totally Positive Generators of Gaussian Type

A completely different result holds for totally positive generators of Gaussian type. This result is in the spirit of the Beurling-Kahane-Landau results for bandlimited functions.

Theorem 7 *Assume that* $\hat{\phi}(\xi) = ce^{-\gamma\xi^2}\prod_{j=1}^{N}(1+2\pi i \nu_j \xi)^{-1}$ *with* $\gamma > 0$, $\nu_j \in \mathbb{R}$, $N \in \mathbb{N}$ *(in our terminology* ϕ *is totally positive of Gaussian type).*

If $\Lambda \subseteq \mathbb{R}$ *is separated and* $D^-(\Lambda) > 1$*, then* Λ *is a set of stable sampling for* $V(\phi)$.

Since $D^-(\Lambda) \geq 1$ is necessary for sets of stable sampling, Theorem 7 is almost a characterization of sampling sets. As in the case of the Paley-Wiener space [29], the case of critical sampling $D^-(\Lambda) = 1$ promises to be complicated. A set Λ at the critical density may or may not be sampling.

Theorem 7 is in complete analogy to the Paley-Wiener space, but it came as a big surprise to us, as characterizations of sampling sets by a Beurling density are extremely rare and so far were known only in complex analysis [35].

4.1 Sampling with Derivatives

With a bit more effort the proof of Theorem 7 can be extended to obtain an optimal theorem for sampling with derivatives. In this version of the sampling problem we assume that at every sampling point λ we measure $m(\lambda)$ derivatives of f. The appropriate density notion is a weighted Beurling density. Let $m : \Lambda \to \mathbb{N}$ denote the multiplicity function. Then

$$D^-(\Lambda, m) := \liminf_{r\to\infty} \inf_{y\in\mathbb{R}} \frac{1}{r} \sum_{x\in\Lambda\cap[y,y+r]} m(x).$$

is the corresponding density of Λ. Again, $D^-(\Lambda, m)$ is the average number of samples or data per unit time interval. In [15] we proved the following statement.

Theorem 8 *Let ϕ be a totally positive function of Gaussian type. Let $\Lambda \subseteq \mathbb{R}$ be a separated set and let $m : \Lambda \to \mathbb{N}$ be bounded.*

If $D^-(\Lambda, m) > 1$, then (Λ, m) is a sampling set for $V(\phi)$, i.e., there exist $A, B > 0$ such that

$$A\|f\|_2^2 \le \sum_{\lambda\in\Lambda} \sum_{k=0}^{m_\lambda - 1} |f^{(k)}(\lambda)|^2 \le B\|f\|_2^2 \qquad \textit{for all } f \in V(\phi)\,.$$

This result is optimal in view of the expected necessary sampling condition: *If (Λ, m) is a sampling set for $V^2(\phi)$, then $D^-(\Lambda, m) \ge 1$.*

The same result holds also for the Paley-Wiener space $\mathscr{B} = V(\mathrm{sinc})$. This fact is well known to complex analysts, but its first explicit formulation seems to be contained in [15]. Finally Theorem 8 also holds for the shift-invariant space whose generator is the hyperbolic secant $\psi(x) = \mathrm{sech}\,(\pi x) = \frac{2}{e^{\pi x}+e^{-\pi x}}$. Note that ψ is a totally positive function, but of infinite type. Currently, we do not know how to prove Theorems 7 and 8 for all totally positive functions.

In recent data science applications, sampling with derivatives has received renewed attention under the name event-based sampling or, in higher dimensions, sampling with gradient-augmented measurements.

4.2 Some Proof Ideas

Theorem 7 is far from trivial, and its proof requires methods from several branches of analysis. The proof uses ideas from

- complex analysis (counting density of zeros),
- spectral invariance in Banach algebra theory (the off-diagonal decay of infinite matrix is preserved by inversion),

- the connection to Gabor frames, and
- Beurling's technique of weak limits of sets.

In the following we will try to explain the main ideas that enter the proof.

We first consider the special case of the Gaussian. The sampling inequality (7) is a statement about both the uniqueness and the stability of the recovery. We therefore study first the properties of the zero set of functions in the shift-invariant space $V(\phi)$ with Gaussian generator. It is important to consider the shift-invariant space with generator ϕ and *bounded* coefficients rather than ℓ^2-coefficients. Formally, we define

$$V^\infty(\phi) = \{f(x) = \sum_{k\in\mathbb{Z}} c_k\phi(x-k) : (c_k) \in \ell^\infty(\mathbb{Z})\}.$$

Proposition 1 (Zero Sets for Gaussian Generator) *Let* $f = \sum_{k\in\mathbb{Z}} c_k e^{-\pi(x-k)^2} \in V^\infty(e^{-\pi x^2})$ *and* $N_f = \{x \in \mathbb{R} : f(x) = 0\}$ *be the set of* real *zeros of* f. *Then* $D^-(N_f) \le 1$.

Proof The decay of the Gaussian allows us to study the extension of f to a function on the complex plane $\mathbb{C}$. Let $z = x + iy$ and write the square in the exponent as $e^{-\pi(x+iy-k)^2} = e^{-\pi(x-k)^2} e^{\pi y^2} e^{-2\pi i x y} e^{2\pi i k y}$. Then the formula for the extension of f is

$$f(x+iy) = \sum_{k\in\mathbb{Z}} c_k e^{-\pi(x+iy-k)^2} = e^{\pi y^2} e^{-2\pi i x y} \sum_{k\in\mathbb{Z}} c_k e^{2\pi i k y} e^{-\pi(x-k)^2}.$$

We make two observations. First, the function $z \to f(z)$ possesses the growth

$$|f(x+iy)| \le C\|c\|_\infty e^{\pi y^2}. \tag{8}$$

Since the series converges uniformly on compact sets, f is an entire function.

Second,

$$\text{if } f(x) = 0, \text{ then } f(x+il) = 0 \qquad \text{for all } l \in \mathbb{Z}. \tag{9}$$

Thus the zero set of f as an entire function is periodic with period i in the complex plane. We now apply Jensen's formula to f. Let $n(r) = \#\{z \in \mathbb{C} : |z| \le r, f(z) = 0\}$ be the number of *complex* zeros of f in a disc of radius r, then

$$\int_0^R \frac{n(r)}{r}\, dr + \log|f(0)| = \frac{1}{2\pi}\int_0^{2\pi} \log|f(Re^{it})|\, dt.$$

Assuming without loss of generality that $f(0) \neq 0$, the growth estimate (8) implies that

$$\begin{aligned}\int_0^R \frac{n(r)}{r}\,dr &= \frac{1}{2\pi}\int_0^{2\pi} \log|f(Re^{it})|\,dt - \log|f(0)| \\ &\le \frac{1}{2\pi}\int_0^{2\pi} (\log C + \pi R^2 \sin^2 t)\,dt - \log|f(0)| \\ &= \log C' + \frac{\pi R^2}{2}\,. \end{aligned} \tag{10}$$

To treat the right-hand side of Jensen's formula, fix $\epsilon > 0$ and choose $R_0 > 0$ so that the number of real zeros in the interval $[-r, r]$ exceeds $2r(D^-(N_f) - \epsilon)$ for $r > R_0$. The distribution of the complex zeros in (9) implies that

$$n(r) \ge (D^-(N_f) - \epsilon)\pi r^2 \qquad \text{for all } r \ge R_0\,.$$

It follows that

$$\int_{R_0}^R \frac{n(r)}{r}\,dr \ge (D^-(N_f) - \epsilon)\pi\,\frac{R^2 - R_0^2}{2}\,. \tag{11}$$

The two inequalities (10) and (11) are compatible only when $D^-(N_f) \le 1$, which was to be proved. □

We may recast Proposition 1 as a uniqueness assertion in $V(\phi)$.

Corollary 1 *If $D^-(\Lambda) > 1$, then Λ is a uniqueness set for $V^\infty(e^{-\pi x^2})$, i.e., if $f \in V^\infty(e^{-\pi x^2})$ and $f(\lambda) = 0$ for all $\lambda \in \Lambda$, then $f \equiv 0$.*

Uniqueness Implies Sampling Morally, if $\Lambda \subseteq \mathbb{R}$ is uniqueness set of $V^\infty(\phi)$, then Λ is a sampling set of $V(\phi)$. The truth is more subtle and requires Beurling's technique of weak limits. Given a separated set $\Lambda \subseteq \mathbb{R}$, we consider all shifts $x + \Lambda$. A set $\Gamma \subseteq \mathbb{R}$ is a weak limit of Λ, if there exists a sequence x_n, such that $x_n + \Lambda$ converges locally to Γ in the Hausdorff topology. With this terminology the correct statement is the following.

Proposition 2 *Assume that $\sum_{k\in\mathbb{Z}} \sup_{x\in[0,1]} |\phi(x+k)| < \infty$. If every weak limit of $\Lambda \subseteq \mathbb{R}$ is a uniqueness set of $V^\infty(\phi)$, then Λ is a set of stable sampling for $V^\infty(\phi)$, i.e., $A\|f\|_\infty \le \sup_{\lambda\in\Lambda} |f(\lambda)|$ for all $f \in V^\infty(\phi)$ for some constant $A > 0$.*

The final ingredient is a non-commutative version of Wiener's lemma due to Sjöstrand [36] and others. It allows us to move from sets of uniqueness for $V^\infty(\phi)$ (bounded coefficients) to sets of stable sampling for $V(\phi)$ (ℓ^2-coefficients).

Proposition 3 *Assume that* $\sum_{k\in\mathbb{Z}} \sup_{x\in[0,1]} |\phi(x+k)| < \infty$. *Then a relatively separated set* $\Lambda \subseteq \mathbb{R}$ *is a set of stable sampling for* $V(\phi)$, *if and only if* Λ *is a set of stable sampling for* $V^\infty(\phi)$.

We note that the technical machinery holds for quite general class of generators. It follows from Propositions 2 and 3 that we do not need to establish a sampling inequality (7) for $V(\phi)$, but that it suffices to prove a uniqueness property, albeit on the larger space $V^\infty(\phi)$.

Propositions 2 and 3 now yield a proof of Theorem 7 for Gaussian generator. Assume that $\Lambda \subseteq \mathbb{R}$ is separated and $D^-(\Lambda) > 1$. It can be shown that every weak limit Γ is also separated and that $D^-(\Gamma) \geq D^-(\Lambda) > 1$. By Corollary 1 Γ is a uniqueness set for $V^\infty(e^{-\pi x^2})$. It follows from Propositions 2 and 3 that Λ is a set of stable sampling for $V(e^{-\pi x^2})$.

For the full details of this technique we refer to [14] and the corresponding paper on Gabor frames [13].

Comparison of Zero Sets for Different SI-Spaces For other generators, we need additional arguments. The following observation reduces the case of totally positive functions of Gaussian type to Gaussian generator ϕ.

Proposition 4 *Assume that* $\hat{\phi}(\xi) = ce^{-\gamma\xi^2}\prod_{j=1}^N(1+2\pi i\nu_j\xi)^{-1}$ *with* $\gamma > 0$, $\nu_j \in \mathbb{R}$, $N \in \mathbb{N}$.

Let $f = \sum_{k\in\mathbb{Z}} c_k\phi(x-k) \in V^\infty(\phi)$ *and* $N_f = \{x \in \mathbb{R} : f(x) = 0\}$ *be the set of real zeros of* f. *Then* $D^-(N_f) \leq 1$.

Proof Assume first that $\hat{\phi}(\xi) = e^{-\pi\xi^2}(1+2\pi i\delta\xi)^{-1}$. Then

$$(1+2\pi i\delta\xi)\,\hat{\phi}(\xi) = e^{-\pi\xi^2}$$

or, equivalently,

$$(1+\delta\frac{d}{dx})\phi(x) = e^{-\pi x^2}\,.$$

Let $h(x) = \sum_k c_k e^{-\pi(x-k)^2} \in V^\infty(e^{-\pi x^2})$. As noted above, we have

$$(1+\delta\frac{d}{dx})\sum_k c_k\phi(x-k) = \sum_k c_k e^{-\pi(x-k)^2} = h\,.$$

By writing $(1+\delta\frac{d}{dx})f(x) = \delta e^{-x/\delta}\frac{d}{dx}(e^{x/\delta}f(x))$ and applying Rolle's theorem, we see that between any two real zeros of f there is a real zero of h. Thus the zeros of f and h are interlaced, and we deduce that

$$D^-(N_f) \leq D^-(N_h)\,.$$

By Proposition 1 $D^-(N_h) \leq 1$ and consequently $D^-(N_f) \leq 1$.

If ϕ is a totally positive function of Gaussian type, we iterate the above argument N times and obtain the same conclusion. □

Applying Propositions 2 and 3 proves Theorem 7 for totally positive generators of Gaussian type.

5 Time-Frequency Analysis and Gabor Frames

Sampling theorems for shift-invariant spaces are closely connected with the theory of Gabor frames. Let us first set up the main dictionary of time-frequency analysis. Given a point $z = (x, \xi) \in \mathbb{R}^2$ in the time-frequency plane, the associated time-frequency shift (phase-space shift in the terminology of physics) is

$$M_\xi T_x g(t) = e^{2\pi i \xi t} g(t - x) \qquad x, \xi, t \in \mathbb{R} .$$

We fix a "window" function $g \in L^2(\mathbb{R})$, $g \neq 0$. The theory of Gabor frames studies the spanning properties of sets of time-frequency shifts $\{M_\xi T_x g : (x, \xi) \in \Lambda\}$. While this question is meaningful for arbitrary sets $\Lambda \subseteq \mathbb{R}^2$, we study only the case of time-frequency shifts over a rectangular lattice $\Lambda = \alpha\mathbb{Z} \times \beta\mathbb{Z}$ with lattice parameters $\alpha, \beta > 0$. A Gabor system is then the collection of all time-frequency shifts

$$\mathcal{G}(g, \alpha, \beta) = \{M_{\beta l} T_{\alpha k} g : k, l \in \mathbb{Z}\} .$$

A Gabor system $\mathcal{G}(g, \alpha, \beta)$ is a Gabor frame, if there exist $A, B > 0$, such that

$$A\|f\|_2^2 \leq \sum_{k,l \in \mathbb{Z}} |\langle f, M_{\beta l} T_{\alpha k} g\rangle|^2 \leq B\|f\|_2^2 \qquad \forall f \in L^2(\mathbb{R}) . \tag{12}$$

The field of Gabor analysis studies the question of

When is $\mathcal{G}(g, \alpha, \beta)$ a Gabor frame?

Although this question looks completely different from the sampling questions in the previous sections, the question of Gabor frames can be recast as a sampling problem. For this we consider the natural transform associated to time-frequency shifts. This is the short-time Fourier transform with respect to a non-zero window g defined as

$$V_g f(x, \xi) = \langle f, M_\xi T_x g\rangle = \int_{\mathbb{R}} f(t)\bar{g}(t - x)\, e^{-2\pi i \xi t}\, dt \qquad (x, \xi) \in \mathbb{R}^2 .$$

If g is a bump function, i.e., smooth with compact support in a neighborhood of the origin, then $V_g f$ is the Fourier transform of f smoothly truncated to a neighborhood of x. This explains why we refer to g as a "window" function.

Next, note that

$$\langle f, M_{\beta l} T_{\alpha k} g\rangle = V_g f(\alpha k, \beta l),$$

therefore the frame inequalities (12) express a sampling inequality on the function space $V = V_g(L^2)$ of all short-time Fourier transforms.

The analogy is even deeper than the formal similarity of the two definitions (7) and (12). The structure theory of Gabor frames over lattices connects their frame property (12) directly to sampling in shift-invariant spaces by the following theorem [22, 31].

Theorem 9 *The following are equivalent for* $g \in L^2(\mathbb{R})$ *and* $\Lambda \subseteq \mathbb{R}$*:*

(i) $\mathscr{G}(g, \alpha, 1)$ *is a frame for* $L^2(\mathbb{R})$*.*
(ii) The set $x + \alpha\mathbb{Z}$ *is a set of sampling for* $V(g)$ *for almost all* $x \in [0, 1)$ *(with uniform constants).*

Spelled out as inequalities, the equivalence of (i) and (ii) says that the frame inequalities

$$A\|f\|_2^2 \leq \sum_{k,l\in\mathbb{Z}} |\langle f, M_l T_{\alpha k} g\rangle|^2 \leq B\|f\|_2^2 \qquad \forall f \in L^2(\mathbb{R}^d)$$

hold, if and only if the sampling inequalities

$$A'\|c\|_2^2 \leq \sum_{j\in\mathbb{Z}} |\sum_{k\in\mathbb{Z}} c_k g(\alpha j + x - k)|^2 \leq B'\|c\|_2^2 \quad \forall c \in \ell^2(\mathbb{Z})$$

hold for almost all x with constants independent of $x \in [0, 1]$.

As with the sampling problem in shift-invariant spaces, one of the first questions is how much information is required for the frame inequality (12) to hold. The general theory of Gabor frames yields a necessary density condition in the style of Landau.

Proposition 5 *If* $\mathscr{G}(g, \alpha, \beta)$ *is frame, then* $(\alpha\beta)^{-1} \geq 1$*.*

Proof Fix $\beta = 1$ first. By Theorem 9 $x + \alpha\mathbb{Z}$ is a set of stable sampling for the shift-invariant space $V(\phi)$, consequently the necessary density condition of Theorem 4 implies that $D^-(x + \alpha\mathbb{Z}) = \alpha^{-1} \geq 1$.

To treat arbitrary rectangular lattices $\alpha\mathbb{Z} \times \beta\mathbb{Z}$, we use a scaling argument. Let $\gamma_\beta(t) = \beta^{-1/2}\gamma(t/\beta)$ so that $\|g_\beta\|_2 = \|g\|_2$. Then $\beta^{-1/2} M_{\beta l} T_{\alpha k} g(t/\beta) = M_l T_{\alpha\beta k} g_\beta(t)$. Therefore $\mathscr{G}(g, \alpha, \beta)$ is a Gabor frame, if and only if $\mathscr{G}(g_\beta, \alpha\beta, 1)$ is a Gabor frame. This implies that $(\alpha\beta)^{-1} \geq 1$. □

A slightly stronger version holds for "nice" windows: *Assume that g is continuous and* $\sum_{k\in\mathbb{Z}} \sup_{x\in[0,1]} |g(x+k)| < \infty$. *If* $\mathcal{G}(g,\alpha,\beta)$ *is frame, then* $(\alpha\beta)^{-1} > 1$. This result is known as the Balian-Low-Theorem [6].

Historically, the density theorem for Gabor frames preceeded the density theorem for shift-invariant spaces (Theorem 4) and was proved by different arguments. Necessary density conditions are an important aspect in sampling theory and have been the subject of numerous investigations. For further information we refer to the Heil's historical survey of the density theorem for Gabor frames [19] and to the general theory for sampling in reproducing kernel Hilbert space [11].

Gabor Frames and Totally Positive Functions The combination of the characterization of Theorem 9 and the results about sampling in shift-invariant spaces with totally positive generator now yields immediately the following characterization of Gabor frames with a totally positive window [14, 16].

Theorem 10 *Assume that* ϕ *is a totally positive function either of finite type or of Gaussian type, i.e.,* $\hat{\phi}(\xi) = e^{-\gamma\xi^2}\prod_{j=1}^{N}(1+2\pi i\nu_j\xi)^{-1}$ *with* $\nu_j \in \mathbb{R}$ *and either* $\gamma > 0$, $N \in \mathbb{N}$, *or* $\gamma = 0$, $N \geq 2$, *or let* $\phi = \operatorname{sech}(\pi x) = \frac{2}{e^{\pi x}+e^{-\pi x}}$.
Then $\mathcal{G}(\phi,\alpha,\beta)$ *is a frame for* $L^2(\mathbb{R})$ *if and only if* $\alpha\beta < 1$.

The proof follows by combining Theorem 7 with the characterization of Proposition 9.

For completeness we mention that much more general versions can be proved for nonuniform sets of time-frequency shifts of the form $\Lambda \times \beta\mathbb{Z}$.

For the Gaussian $\phi(x) = e^{-\gamma x^2}$, Theorem 10 is a special case of a groundbreaking result of Lyubarskii [28] and Seip [34]. The case of the hyperbolic secant $\operatorname{sech}(\pi x) = \frac{2}{e^{\pi x}+e^{-\pi x}}$ is due to Janssen and Strohmer [23]. Until 2013 there was only a handful examples, such as the one-sided or two-sided exponential functions and the hyperbolic cosine, for which the characterization of Theorem 10 was known to hold. The first hint towards Theorem 10 was the observation that all then known examples were totally positive functions. To this date, the characterization of Gabor frames via the density condition $\alpha\beta < 1$ is known only for a class of totally positive functions. Most likely, Theorem 10 holds for all totally positive functions, but so far all our "proofs" contained a gap, and the general statement is still open.

Note Added in Proof Belov, Kulikov, and Lyubarskii [5] have recently found a new class of window functions for which a complete characterization of Gabor frames is possible.

Let g be a Herglotz function, i.e., a function of the form

$$g(x) = \sum_{j=1}^{N} \frac{a_k}{x - i w_k}, \ a_k > 0, \ w_k > 0\,.$$

Then $\mathcal{G}(g,\alpha,\beta)$ *is a frame for* $L^2(\mathbb{R})$, *if and only if* $\alpha\beta \leq 1$.

The Fourier transform of g is a sum of one-sided exponentials similar to (4). However, in contrast to (4) all coefficients a_k are positive and the poles w_k all have the same sign. Herglotz functions (and their Fourier transforms) and totally positive functions of finite type are therefore different, complementary classes of functions.

6 Zero-Free Short-Time Fourier Transforms

The time-frequency analysis of so-called localization operators offers a curious and interesting question about the zeros of the short-time Fourier transform [4].

$$\boxed{\text{For which } g \text{ does } V_g g \text{ not have any zeros?}}$$

Recently this question also appeared in the problem of phase-retrieval from the spectrogram [18].

To get a feeling for this question, we compute the short-time Fourier transform of the Gaussian $g(t) = e^{-\pi t^2}$. After some manipulations with Gaussian integrals we obtain

$$V_g g(x, \xi) = e^{-2\pi i x \xi} \, e^{-\pi(x^2+\xi^2)/2} ,$$

and $V_g g$ clearly does not have any zeros.

For a long time we believed that the class of generalized Gaussian functions is the only class of functions whose short-time Fourier transform is zero-free. We were led to this belief based on the analogy to Hudson's theorem for positive Wigner distribution [20].

However, after trying a few other short-time Fourier transforms that can be calculated explicitly, the question acquired a different flavor. An elementary calculation implies that the short-time Fourier transform of the one-sided exponential $\eta_a(t) = e^{-at}\mathbf{1}_{(0,\infty)}(at)$ for $a \in \mathbb{R}$, $a \neq 0$, is given by

$$\begin{aligned} V_{\eta_a}\eta_b(x, \xi) &= e^{\pi i x \xi} \exp\left(-\frac{a-b}{2}x - \frac{a+b}{2}|x|\right) \frac{e^{-i\pi\xi|x|}}{a+b+2i\pi\xi} \\ &= \eta_{a,b}(x) \frac{e^{-i\pi\xi|x|}}{a+b+2i\pi\xi} , \end{aligned} \tag{13}$$

where $\eta_{a,b}(x) = (b-a)^{-1}\,\eta_a * \eta_{-b}(x) = \begin{cases} e^{-ax} & \text{when } x \geq 0 \\ e^{bx} & \text{when } x < 0 \end{cases}$. In particular $V_{\eta_a}\eta_a$ does not have any zero. A rather lengthy computation reveals that the short-time Fourier transform of $\eta_a * \eta_b$ for $b \neq -a$ is also zero-free [12].

A further example of a zero-free short-time Fourier transform is given by the function

$$g(x) = \exp(ax - be^x).$$

We now make an important observation. All these functions are totally positive. All other known examples of zero-free short-time Fourier transforms are also related to totally positive functions by applying a Fourier transform, shifts, or dilations.

These observations indicate that totally positive functions play a role in the investigation of zero-free short-time Fourier transforms. However, the situation might be rather complicated. The symmetric exponential function $e^{-a|x|} = (\eta_a * \eta_{-a})(x)$ is totally positive but its short-time Fourier transform possesses zeros. In general the short-time Fourier transform of every even (or odd) function must have a zero.

At this time it is a complete mystery which functions possess a zero-free short-time Fourier transform. A first investigation with many observations and partial results is contained in [12] with P. Jaming and E. Malinnikova.

7 Totally Positive Functions and the Riemann Hypothesis

Finally let us mention an amusing link between total positivity and number theory. Let $\zeta(s) = \sum_{n=1}^{\infty} n^{-s}$ for $s \in \mathbb{C}$, $\operatorname{Re} s > 1$, be the Riemann zeta function and let

$$\xi(s) = \tfrac{1}{2}s(s-1)\pi^{-s/2}\Gamma\Big(\frac{s}{2}\Big)\,\zeta(s)$$

be the Riemann xi-functions (where Γ is the usual gamma function). Using ξ, the functional equation for the Riemann zeta function can be conveniently formulated as the symmetry

$$\xi(s) = \xi(1-s) \qquad \text{for all } x \in \mathbb{C}. \tag{14}$$

In addition, ξ is an entire function with growth $|\xi(s)| \le Ce^{A|s|\log(1+|s|)}$ [37].

The Riemann hypothesis conjectures that all zeros of ξ lie on the critical strip $\{s \in \mathbb{C} : \operatorname{Re} s = 1/2\}$, so that the modified function $\xi(\frac{1}{2} + iz)$ will have only real zeros.

Let $\rho_j \in \mathbb{C}$ be an enumeration of the zeros of $\xi(\frac{1}{2} + iz)$. The Hadamard factorization theorem then yields the representation

$$\xi(\tfrac{1}{2} + iz) = e^{\delta z} \prod_{j=1}^{\infty} \big(1 - \frac{z}{\rho_j}\big)e^{z/\rho_j}$$

with a constant $\delta \in \mathbb{R}$. Now we change the variables and write $z = -2\pi i \tau$ and $\delta_j = 1/\rho_j$ and take reciprocals. We obtain that

$$\frac{1}{\xi(\frac{1}{2} - 2\pi\tau)} = e^{2\pi i \delta \tau} \prod_{j=1}^{\infty} \left(1 + 2\pi i \delta_j \tau\right)^{-1} e^{2\pi i \delta_j \tau} .$$

Here is the connection to Schoenberg's characterization of totally positive functions (5). If all zeros of $\xi(\frac{1}{2}+iz)$ are real, then $\xi(\frac{1}{2}+2\pi\tau)^{-1}$ is the Fourier transform of a totally positive function. Conversely, if $\xi(\frac{1}{2}+2\pi\tau)^{-1}$ is the Fourier transform of a totally positive function, then all zeros of $\xi(\frac{1}{2}+iz)$ are real. This connection therefore yields the following equivalence of Riemann's hypothesis.

Theorem 11 *The Riemann hypothesis holds, if and only if*

$$\Lambda(x) = \int_{-\infty}^{\infty} \frac{1}{\xi(\frac{1}{2} + 2\pi\tau)} e^{-2\pi i x \tau} \, d\tau$$

is a totally positive function.

Theorem 11 seems to be new. Note that (11) requires only *the values of* ζ *and* ξ *on the real line* to probe the secrets of ζ in the critical strip. For more background see the essay [17].

8 Summary

- Totally positive functions play an important role in sampling theory and time-frequency analysis.
- At this time totally positive functions are the only generators for which optimal results have been proved about sampling in shift-invariant spaces.
- Totally positive functions have led to progress in understanding the fine structure of Gabor frames.
- Despite some progress, many problems remain open. For example, we expect Theorems 7 and 10 to hold for all totally positive functions, but the case infinite type is not yet proved rigorously.

Acknowledgments The author was supported in part by the project P31887-N32 of the Austrian Science Fund (FWF). Most results were obtained in joint work with José Luis Romero, University of Vienna, and Joachim Stöckler, Technical University Dortmund.

References

1. A. Aldroubi, H.G. Feichtinger, Exact iterative reconstruction algorithm for multivariate irregularly sampled functions in spline-like spaces: the L^p-theory. Proc. Am. Math. Soc. **126**(9), 2677–2686 (1998)
2. A. Aldroubi, K. Gröchenig, Beurling-Landau-type theorems for non-uniform sampling in shift invariant spline spaces. J. Fourier Anal. Appl. **6**(1), 93–103 (2000)
3. A. Aldroubi, K. Gröchenig, Nonuniform sampling and reconstruction in shift-invariant spaces. SIAM Rev. **43**(4), 585–620 (2001)
4. D. Bayer, K. Gröchenig, Time-frequency localization operators and a Berezin transform. Integr. Equ. Oper. Theory **82**(1), 95–117 (2015)
5. Y. Belov, A. Kulikov, Y. Lyubarskii, Gabor Frames for rational functions. Preprint, arXiv:2103.08959
6. J.J. Benedetto, C. Heil, D.F. Walnut, Differentiation and the Balian–Low theorem. J. Fourier Anal. Appl. **1**(4), 355–402 (1995)
7. A. Beurling, The collected works of Arne Beurling, vol. 2, in *Harmonic Analysis*, ed. by L. Carleson, P. Malliavin, J. Neuberger, J. Wermer. Contemporary Mathematicians (Birkhäuser Boston, Boston, 1989)
8. R.J. Duffin, A.C. Schaeffer, A class of nonharmonic Fourier series. Trans. Am. Math. Soc. **72**, 341–366 (1952)
9. B. Efron, Increasing properties of Pólya frequency functions. Ann. Math. Stat. **36**, 272–279 (1965)
10. H.G. Feichtinger, Spline-type spaces in Gabor analysis, in *Wavelet Analysis (Hong Kong, 2001)*, vol. 1 of *Ser. Anal.* (World Sci. Publ., River Edge, 2002), pp. 100–122
11. H. Führ, K. Gröchenig, A. Haimi, A. Klotz, J.L. Romero, Density of sampling and interpolation in reproducing kernel Hilbert spaces. J. Lond. Math. Soc. (2) **96**(3), 663–686 (2017)
12. K. Gröchenig, P. Jaming, E. Malinnikova, Zeros of the Wigner distribution and the short-time Fourier transform. Rev. Mat. Complut. **33**(3), 723–744 (2020)
13. K. Gröchenig, J. Ortega-Cerdà, J.L. Romero, Deformation of Gabor systems. Adv. Math. **277**, 388–425 (2015)
14. K. Gröchenig, J.L. Romero, J. Stöckler, Sampling theorems for shift-invariant spaces, Gabor frames, and totally positive functions. Invent. Math. **211**(3), 1119–1148 (2018)
15. K. Gröchenig, J.L. Romero, J. Stöckler, Sharp results on sampling with derivatives in shift-invariant spaces and multi-window Gabor frames. Constr. Approx. **51**(1), 1–25 (2020)
16. K. Gröchenig, J. Stöckler, Gabor frames and totally positive functions. Duke Math. J. **162**(6), 1003–1031 (2013)
17. K. Gröchenig, Schoenberg's theory of totally positive functions and the Riemann zeta function. https://arxiv.org/pdf/2007.12889.pdf
18. P. Grohs, S. Koppensteiner, M. Rathmair, The mathematics of phase retrieval. SIAM Rev. **62**(2), 301–350 (2020)
19. C. Heil, History and evolution of the density theorem for Gabor frames. J. Fourier Anal. Appl. **13**(2), 113–166 (2007)
20. R.L. Hudson, When is the Wigner quasi-probability density non-negative? Rep. Math. Phys. **6**(2), 249–252 (1974)
21. A. Janssen, The Zak transform and sampling theorems for wavelet subspaces. IEEE Trans. Signal Process. **41**, 3360–3365 (1993)
22. A.J.E.M. Janssen, Duality and biorthogonality for Weyl-Heisenberg frames. J. Fourier Anal. Appl. **1**(4), 403–436 (1995)
23. A.J.E.M. Janssen, T. Strohmer, Hyperbolic secants yield Gabor frames. Appl. Comput. Harmon. Anal. **12**(2), 259–267 (2002)
24. J.-P. Kahane, Pseudo-périodicité et séries de Fourier lacunaires. Ann. Sci. École Norm. Sup. (3) **79**, 93–150 (1962)
25. S. Karlin, *Total Positivity*, vol. I (Stanford University Press, Stanford, 1968)

26. T. Kloos, J. Stöckler, Zak transforms and Gabor frames of totally positive functions and exponential B-splines. J. Approx. Theory **184**, 209–237 (2014)
27. H.J. Landau, Necessary density conditions for sampling and interpolation of certain entire functions. Acta Math. **117**, 37–52 (1967)
28. Y.I. Lyubarskiĭ, Frames in the Bargmann space of entire functions, in *Entire and Subharmonic Functions* (Amer. Math. Soc., Providence, 1992), pp. 167–180
29. J. Ortega-Cerdà, K. Seip, Fourier frames. Ann. of Math. (2) **155**(3), 789–806 (2002)
30. D. Pickrell, Mackey analysis of infinite classical motion groups. Pac. J. Math. **150**(1), 139–166 (1991)
31. A. Ron, Z. Shen, Weyl–Heisenberg frames and Riesz bases in $L_2(\mathbb{R}^d)$. Duke Math. J. **89**(2), 237–282 (1997)
32. I.J. Schoenberg, On Pólya frequency functions. I. The totally positive functions and their Laplace transforms. J. Anal. Math. **1**, 331–374 (1951)
33. I.J. Schoenberg, A. Whitney, On Pólya frequence functions. III. The positivity of translation determinants with an application to the interpolation problem by spline curves. Trans. Am. Math. Soc. **74**, 246–259 (1953)
34. K. Seip, Density theorems for sampling and interpolation in the Bargmann-Fock space. I. J. Reine Angew. Math. **429**, 91–106 (1992)
35. K. Seip, *Interpolation and Sampling in Spaces of Analytic Functions*, vol. 33 of *University Lecture Series* (American Mathematical Society, Providence, 2004)
36. J. Sjöstrand, Wiener type algebras of pseudodifferential operators, in *Séminaire sur les Équations aux Dérivées Partielles, 1994–1995* (École Polytech., Palaiseau, 1995), pp. Exp. No. IV, 21
37. E.C. Titchmarsh, *The Theory of the Riemann Zeta-Function*, 2nd edn. (The Clarendon Press, Oxford University Press, New York, 1986). Edited and with a preface by D. R. Heath-Brown
38. M. Vetterli, P. Marziliano, T. Blu, Sampling signals with finite rate of innovation. IEEE Trans. Signal Process. **50**(6), 1417–1428 (2002)

Multidimensional Inverse Scattering for the Schrödinger Equation

Roman G. Novikov

Abstract We give a short review of old and recent results on the multidimensional inverse scattering problem for the Schrödinger equation. A special attention is paid to efficient reconstructions of the potential from scattering data which can be measured in practice. In this connection our considerations include reconstructions from non-overdetermined monochromatic scattering data and formulas for phase recovering from phaseless scattering data. Potential applications include phaseless inverse X-ray scattering, acoustic tomography and tomographies using elementary particles. This paper is based, in particular, on results going back to M. Born (1926), L. Faddeev (1956, 1974), S. Manakov (1981), R. Beals, R. Coifman (1985), P. Grinevich, R. Novikov (1986), G. Henkin, R. Novikov (1987), and on more recent results of R. Novikov (1998–2019), A. Agaltsov, T. Hohage, R. Novikov (2019).

This paper is an extended version of the talk given at the 12th ISAAC Congress, Aveiro, Portugal, 29 July–2 August, 2019.

Keywords Schrödinger equation · Helmholtz equation · Phase retrieval · Tomography · Inverse scattering

1 Introduction

We consider the stationary Schrödinger equation:

$$-\Delta\psi + v(x)\psi = E\psi, \;\; x \in \mathbb{R}^d, \; d \geq 1, \; E > 0, \tag{1.1}$$

R. G. Novikov (✉)
CMAP, CNRS, Ecole Polytechnique, Institut Polytechnique de Paris, Palaiseau, France

IEPT RAS, Moscow, Russia
e-mail: novikov@cmap.polytechnique.fr

P. Cerejeiras, M. Reissig (eds.), *Mathematical Analysis, its Applications and Computation*, Springer Proceedings in Mathematics & Statistics 385,
https://doi.org/10.1007/978-3-030-97127-4_3

where v is a sufficiently regular function on $\mathbb{R}^d$ with sufficient decay at infinity, for example:

$$v \in L^\infty(\mathbb{R}^d), \quad supp\, v \subset D,$$

$$D \text{ is an open bounded domain in } \mathbb{R}^d, \tag{1.2a}$$

or

$$|v(x)| \leq q(1+|x|)^{-\sigma}, \quad x \in \mathbb{R}^d, \text{ for some } q \geq 0 \text{ and } \sigma > d. \tag{1.2b}$$

Equation (1.1), under assumptions (1.2*a*), can be used, in particular, for describing a quantum mechanical particle at fixed energy E interacting with a macroscopic object contained in D. In this case v is the potential of this interaction.

We recall also that the (time-dependent) Schrödinger equation is the quantum mechanical analogue of the Newton's second law.

Note that we assume that $\frac{\hbar^2}{2m} = 1$ (where $\hbar$ is the reduced Planck's constant, m is the mass of the particle) and Δ is the standard Laplacian in x in the Schrödinger equation (1.1).

For equation (1.1) we consider the scattering eigenfunctions $\psi^+(x,k)$, $k \in \mathbb{R}^d$, $k^2 = E$, specified by the following asymptotics as $|x| \to \infty$:

$$\psi^+(x,k) = e^{ikx} + c(d,|k|)\frac{e^{i|k||x|}}{|x|^{(d-1)/2}} f\big(k, |k|\frac{x}{|x|}\big) + o\big(\frac{1}{|x|^{(d-1)/2}}\big),$$

$$c(d,|k|) = -\pi i(-2\pi i)^{(d-1)/2}|k|^{(d-3)/2}, \tag{1.3}$$

for some a priori unknown f. The function f arising in (1.3) is the scattering amplitude for Eq. (1.1) for fixed E and is defined on

$$\mathcal{M}_E = \{k, l \in \mathbb{R}^d : \ k^2 = l^2 = E\} = \mathbb{S}^{d-1}_{\sqrt{E}} \times \mathbb{S}^{d-1}_{\sqrt{E}}. \tag{1.4}$$

We recall that function $\psi^+(x,k)$ at fixed k describes scattering of the incident plane wave described by e^{ikx} on the scatterer described by potential $v(x)$. In addition, the second term on the right-hand side of (1.3) describes the leading scattered spherical wave.

We also recall that in quantum mechanics the values of the functions ψ^+ and f with phase have no direct physical sense, whereas the phaseless values of $|\psi^+|^2$ and $|f|^2$ have probabilistic interpretations (the Born rule) and can be directly obtained in experiments; see [11, 21]. In particular, $|f(k,l)|^2$ is differential scattering cross section, describing probability density of scattering of particle with initial impulse k into direction $l/|l| \neq k/|k|$.

We consider, in particular, the following problems for Eq. (1.1):

Problem 1.1 Given v, find ψ^+ and f.

Problem 1.2a Reconstruct potential v from its scattering amplitude f.

Problem 1.2b Reconstruct potential v, under assumptions (1.2a), from ψ^+ appropriately given outside of D.

Problem 1.3a Reconstruct potential v from its phaseless scattering data $|f|^2$.

Problem 1.3b Reconstruct potential v, under assumptions (1.2a), from its phaseless scattering data $|\psi^+|^2$ appropriately given outside of D.

Problem 1.1 is the direct scattering problem for Eq. (1.1). This problem can be solved via the Lippmann-Schwinger integral equation (2.1) and formula (2.3); see Sect. 2.

Problem 1.2a is the inverse scattering problem (from far field) for Eq. (1.1).

Problem1.2b is the inverse scattering problem (from near field) for Eq. (1.1).

Problems 1.3a and 1.3b are the phaseless versions of the inverse scattering problems Problems 1.2a and 1.2b.

At this stage we do not specify which precise information about the functions f, ψ^+, $|f|^2$, $|\psi^+|^2$ is used in each of Problems 1.2a, 1.2b, 1.3a, 1.3b and, in particular, we do not specify whether E is fixed.

Note that earlier studies on inverse scattering for the Schrödinger equation (in fact, on Problem 1.2a) were essentially stimulated by the Heisenberg's publications [27]; see related discussion in [19]. In the mathematical literature the inverse scattering problem for the Schrödinger equation (in fact, Problem 1.2a) in dimension $d = 3$ without the spherically symmetric assumption on v was posed for the first time in [23]. At present, there are many important results on Problem 1.2a; see [3, 6, 7, 9, 12–20, 24, 26, 28, 29, 33, 34, 39, 40, 42, 45–53, 55, 57–60, 65, 68, 71, 73, 74, 76] and references therein. Formulas reducing Problem 1.2b to Problem 1.2a and vice versa are also known for a long time; see, for example, [9, 67].

On the other hand, in view of the Born rule, from applied point of view, Problems 1.3a, 1.3b and similar phaseless inverse scattering problems are much more important than Problems 1.2a, 1.2b and similar phased inverse scattering problems for the quantum mechanical Schrödinger equation. However, until recently, the mathematics of inverse wave propagation problems without phase information, in general, and of Problems 1.3a, 1.3b, in particular, were much less developed than for the phased case and an essential progress in this direction (phaseless case) was done during recent years; see [1, 3, 5, 14, 30–32, 35–38, 61–66, 69, 70, 72, 75] and references therein.

In the present paper we consider Problems 1.1–1.3 mainly for the multidimensional case (i.e. $d \geq 2$) for fixed E and, especially, for $d = 2$ or $d = 3$. More precisely, our further presentation can be described as follows.

In Sect. 2 we recall potential applications of results on Problems 1.1–1.3 at fixed E for $d = 2$ or $d = 3$.

In Sect. 3 we recall well-known results on Problem 1.1.

In Sect. 4 we formalise the main objective of Problem 1.2a at fixed and sufficiently large E. In Sect. 5 we recall an old classical result to this objective. And in Sects. 6, 8 we present results of [54–56, 60] which achieve this objective.

In Sect. 9 we recall examples of non-uniqueness for Problem 1.3a in its initial formulation and in Sects. 10, 11 we present results of [1, 64, 70] on modified Problem 1.3a with background scatterers.

In Sect. 12 we present formulas of [63, 66] reducing Problem 1.3b to Problem 1.2a.

2 Potential Applications

Results on Problems 1.1–1.3 at fixed E for $d = 2$ or $d = 3$ admit potential applications, in particular, in the following domains:

(i) Inverse problem of quantum scattering arising in nuclear physics and in tomographies using some elementary particles (see, for example, [14, 22]);
(ii) Acoustic tomography (see, for example, [7, 13]);
(iii) Coherent x-ray imaging (see, for example, [30, 32]).

As regards to quantum scattering we assume that this scattering is modeled using the Schrödinger equation (1.1).

As regards to acoustic tomography and coherent x-ray imaging, we assume that direct scattering is modeled using the Helmholtz equation

$$-\Delta\psi = \Big(\frac{\omega}{c(x)} + i\alpha(x,\omega)\Big)^2\psi, \quad x \in \mathbb{R}^d, \tag{2.1}$$

with velocity of wave propagation $c(x)$, absorption coefficient $\alpha(x,\omega)$, at fixed frequency ω, where

$$c(x) \equiv c_0, \quad \alpha(x,\omega) \equiv 0 \text{ for } |x| \geq r. \tag{2.2}$$

Equation (2.1), under conditions (2.2), can be written in the form of the Schrödinger equation (1.1), where

$$v = \frac{\omega^2}{c_0^2} - \Big(\frac{\omega}{c(x)} + i\alpha(x,\omega)\Big)^2, \quad E = \frac{\omega^2}{c_0^2}, \tag{2.3}$$

$$v = v(x,\omega) \equiv 0 \quad \text{for } |x| \geq r.$$

Therefore, reconstruction methods for Problems 1.2, 1.3 at fixed E can be also used for inverse scattering for the Helmholtz equation (2.2) at fixed ω, under conditions (2.2), for $d = 2$ or $d = 3$.

As it was already mentioned in the Introduction, from applied point of view, the inverse problem of quantum scattering is the most important in its phaseless versions. On the other hand, in acoustic or electrodynamic experiments phased scattering data like ψ^+ and f can be directly measured, at least, in principle. However, in many important cases of monochromatic electro-magnetic wave propagation described using Eq. (2.1) (e.g., X-rays and lasers) the wave frequency is so great that only phaseless scattering data like $|\psi^+|$ and $|f|$ can be measured in practice by modern technical devices; see, e.g., [30] and references therein.

3 Direct Scattering

The scattering eigenfunctions ψ^+ satisfy the Lippmann-Schwinger integral equation

$$\psi^+(x,k) = e^{ikx} + \int\limits_{\mathbb{R}^d} G^+(x-y,k)v(y)\psi^+(y,k)dy, \tag{3.1}$$

$$G^+(x,k) \overset{\text{def}}{=} -(2\pi)^{-d} \int\limits_{\mathbb{R}^d} \frac{e^{i\xi x}d\xi}{\xi^2 - k^2 - i0} = G_0^+(|x|,|k|), \tag{3.2}$$

where $x \in \mathbb{R}^d$, $k \in \mathbb{R}^d$, $k^2 = E$. Note that

$$G^+(x,k) = -\frac{i}{4}H_0^1(|x|\,|k|) \text{ for } d=2, \quad G^+(x,k) = -\frac{e^{i|k||x|}}{4\pi|x|} \text{ for } d=3, \tag{3.3}$$

where H_0^1 is the Hankel function of the first type.

For the scattering amplitude f the following formula holds:

$$f(k,l) = (2\pi)^{-d} \int\limits_{\mathbb{R}^d} e^{-ily}v(y)\psi^+(y,k)dy, \tag{3.4}$$

where $k \in \mathbb{R}^d$, $l \in \mathbb{R}^d$, $k^2 = l^2 = E$.

Equation (3.1) and formula (3.4) are a particular case of the equation and formulas produced in [43]. For basic mathematical results concerning (3.1), (3.4) we refer to [10, 20, 60, 77] and references therein.

Problem 1.1 can be solved via Eq. (3.1) and formula (3.4).

More precisely, we consider Problem 1.1 for those $E > 0$ that

$$\text{equation (3.1) is uniquely solvable for } \psi^+(\cdot, k) \in L^\infty(\mathbb{R}^d) \text{ for fixed } E > 0, \tag{3.5}$$

where $k \in \mathbb{R}^d$, $k^2 = E$. If, for example, v satisfies (1.2b) and is real-valued, then (3.5) is fulfilled automatically.

We also recall that for any $s > 1/2$ the following Agmon estimate holds:

$$\| < x >^{-s} G_0^+(E) < x >^{-s} \|_{L^2(\mathbb{R}^d) \to L^2(\mathbb{R}^d)} = O(E^{-1/2}), \ E \to +\infty, \tag{3.6}$$

where $< x >$ denotes the multiplication operator by the function $(1 + |x|^2)^{1/2}$, $G_0^+(E)$ denotes the operator such that

$$G_0^+(E)u(x) = \int\limits_{\mathbb{R}^d} G_0^+(|x - y|, \sqrt{E})u(y)dy, \tag{3.7}$$

where $G_0^+(|x|, \sqrt{E})$ is the function defined in (3.2), u is the test function. Estimate (3.6) was given implicitly in [4]. This estimate is very convenient for studies of Eq. (3.1) and formula (3.4) for large E; see, e.g., [60].

4 The Main Objective of Problem 1.2a at Fixed and Sufficiently Large E

In order to explain and justify the main objective of Problem 1.2a at fixed and sufficiently large E, for $d \geq 2$, we consider, first, Problems 1.1 and 1.2a in the Born approximation for $q \to 0$, where q is the number in (1.2b). In this approximation we have, in particular:

$$\psi^+(x, k) \approx e^{ikx}, \quad f(k, l) \approx \hat{v}(k - l), \tag{4.1}$$

where

$$\hat{v}(p) = (2\pi)^{-d} \int\limits_{\mathbb{R}^d} e^{ipx} v(x)dx, \quad p \in \mathbb{R}^d. \tag{4.2}$$

Note that

$$(k, l) \in \mathscr{M}_E \Rightarrow k - l \in \mathscr{B}_{2\sqrt{E}}, \tag{4.3}$$

$$p \in \mathscr{B}_{2\sqrt{E}} \Rightarrow \ \exists \ (k, l) \in \mathscr{M}_E \ \text{such that} \ p = k - l \ (\text{for } d \geq 2), \tag{4.4}$$

where $\mathcal{M}_E$ is defined by (1.4),

$$\mathcal{B}_r = \{p \in \mathbb{R}^d : \ |p| \le r\}. \tag{4.5}$$

Thus, in the Born approximation (for $q \to 0$), for $d \ge 2$, the scattering amplitude f on $\mathcal{M}_E$ is reduced to the Fourier transform $\hat{v}$ on $\mathcal{B}_{2\sqrt{E}}$.

Moreover, in this approximation, for $d \ge 2$, the scattering amplitude f on $\mathcal{M}_{[E_0,E]}$, $0 < E_0 \le E$, is also reduced to $\hat{v}$ on $\mathcal{B}_{2\sqrt{E}}$, where $\mathcal{M}_{[E_0,E]} = \cup_{\zeta\in[E_0,E]}\mathcal{M}_\zeta$.

Therefore, the most natural way for solving Problem 1.2a at fixed and sufficiently large E in the Born approximation (for $q \to 0$), for $d \ge 2$, consists in the following formulas:

$$v(x) = v^{lin}_{appr}(x, E) + v^{lin}_{err}(x, E), \tag{4.6}$$

$$v^{lin}_{appr}(x, E) = \int\limits_{|p|\le 2\sqrt{E}} e^{-ipx}\hat{v}(p)dp,$$

$$v^{lin}_{err}(x, E) = \int\limits_{|p|\ge 2\sqrt{E}} e^{-ipx}\hat{v}(p)dp.$$

Here, $v^{lin}_{appr}(x, E)$ is an approximate but stable reconstruction from f on $\mathcal{M}_E$ reduced to $\hat{v}$ on $\mathcal{B}_{2\sqrt{E}}$ and $v^{lin}_{err}(x, E)$ is the reconstruction error.

In addition, if $v \in W^{m,1}(\mathbb{R}^d)$ (m-times smooth functions in $L^1(\mathbb{R}^d)$), $m > d$, then

$$\|v^{lin}_{err}(\cdot, E)\|_{L^\infty(\mathbb{R}^d)} = O(E^{-(m-d)/2}), \quad E \to +\infty. \tag{4.7}$$

De facto, the main objective of Problem 1.2a at fixed and sufficiently large E, for $d \ge 2$, consisted in finding analogs for the general non-linearized case of the reconstruction result (4.6), (4.7) existing for the linearised case near zero potential.

This objective was achieved in [55, 56, 60]; see Sects. 6 and 8 below.

Of course, under condition (1.2*a*), $d \ge 2$, in the linearized case near zero potential, when f on $\mathcal{M}_E$ is reduced to $\hat{v}$ on $\mathcal{B}_{2\sqrt{E}}$, where $E \ge 0$, we have that $\hat{v}$ on $\mathcal{B}_{2\sqrt{E}}$ uniquely determines $\hat{v}$ on the entire $\mathbb{R}^d$ via analytic continuation. And, therefore, in this case f on $\mathcal{M}_E$ uniquely determines v. However, in contrast with (4.6) this reconstruction involves an analytic continuation and is rather unstable. For the general non-linearized case analogs of this uniqueness result were obtained in [12, 50–52]. However, despite their mathematical importance, we do not consider uniqueness theorems without sufficiently stable and accurate reconstruction as the main objective of inverse problems.

5 Old General Result on Problem 1.2a for $d \geq 2$

If v satisfies (1.2a), then

$$f(k, l) = \hat{v}(k - l) + O(E^{-1/2}),\ E \to +\infty,\ (k, l) \in \mathcal{M}_E, \tag{5.1}$$

where $\hat{v}$ is defined by (4.2). This result is known as the Born formula at high energies. As a mathematical theorem formula (5.1) goes back to [18]. At present, one can prove (5.1) using estimate (3.6); see, e.g., [60].

Using (5.1) for $d \geq 2$ with

$$k = k_E(p) = \frac{p}{2} + \eta_E(p),\ l = l_E(p) = -\frac{p}{2} + \eta_E(p), \tag{5.2}$$

where

$$\eta_E(p) = \big(E - \frac{p^2}{4}\big)^{1/2}\theta(p),\ |\theta(p)| = 1,\ \theta(p)p = 0,$$

one can reconstruct $\hat{v}(p)$ from f at high energies E for any $p \in \mathbb{R}^d$.

However, formula (5.1) gives no method to reconstruct v from f on $\mathcal{M}_E$ with the error smaller than $O(E^{-1/2})$ even if $v \in S(\mathbb{R}^d)$, where S stands for the Schwartz class.

Applying the inverse Fourier transform F^{-1} to both sides of (5.1), one can obtain an explicit linear formula for $u_1 = u_1(x, E)$ in terms of f on $\mathcal{M}_E$, where

$$u_1(x, E) = v(x) + O(E^{-\alpha_1}),\ E \to +\infty, \tag{5.3}$$

$$\alpha_1 = \frac{m - d}{2m} \text{ if } v \in W^{m,1}(\mathbb{R}^d).$$

One can see that

$$\alpha_1 \leq 1/2 \text{ even if } m \to +\infty. \tag{5.4}$$

Comparing (4.7) and (5.3), (5.4) one can see that the approximate reconstruction $u_1(x, E)$ is not optimal and does not achieve yet the objective formulated in Sect. 4.

6 Results of [55, 56]

Let

$$W_s^{m,1}(\mathbb{R}^d) = \{u : (1 + |x|)^s \partial^J v(x) \in L^1(\mathbb{R}^d),\ |J| \leq m\}, \tag{6.1}$$

where $m \in \mathbb{N} \cup 0,\ s > 0$.

In [55] for $v \in W_s^{m,1}(\mathbb{R}^2)$, $m > 2$, $s > 0$, for general nonlinearized case for $d = 2$, we succeeded, in particular, to give a stable reconstruction

$$f \text{ on } \mathcal{M}_E \overset{\text{stable reconstruction}}{\longrightarrow} v_{appr}(\cdot, E) \text{ on } \mathbb{R}^2 \tag{6.2}$$

such that

$$\|v - v_{appr}(\cdot, E)\|_{L^\infty(\mathbb{R}^2)} = O(E^{-(m-2)/2}) \text{ as } E \to +\infty. \tag{6.3}$$

For $d = 2$, this reconstruction result achieves the objective formulated in Sect. 4 in view of its stability and estimate (6.3) (which is similar to estimate (4.7) for $d = 2$).

Reconstruction (6.2) is based on Fredholm linear integral equations of the second type. Among these linear integral equations, the most important ones, historically, go back to the Gel'fand-Levitan integral equations of inverse scattering in dimension $d = 1$ and arise from a non-local Riemann-Hilbert problem for the Faddeev solutions ψ of the Schrödinger equation at fixed energy E. Riemann-Hilbert problems of such a type go back to [25, 44]. Definition of the Faddeev solutions ψ and some of their properties are recalled in Sect. 7.

For precise form of the equations and formulas involved into reconstruction (6.2) we refer to [55]. The main idea of [55] was published first in [54].

Reconstruction (6.2) together with its multifrequency generalization was implemented numerically in [13].

In [56] for $v \in W_s^{m,1}(\mathbb{R}^3)$, $m > 3$, $s > 0$, for general nonlinearized case for $d = 3$, we succeeded, in particular, to give a stable reconstruction

$$f \text{ on } \mathcal{M}_E \overset{\text{stable reconstruction}}{\longrightarrow} v_{appr}(\cdot, E) \text{ on } \mathbb{R}^3 \tag{6.4}$$

such that

$$\|v - v_{appr}(\cdot, E)\|_{L^\infty(\mathbb{R}^3)} = O(E^{-(m-3)/2} \ln E), \text{ as } E \to +\infty. \tag{6.5}$$

For $d = 3$, this reconstruction result achieves the objective formulated in Sect. 4 in view of its stability and estimate (6.5) (which is similar (in essence) to estimate (4.7) for $d = 3$).

Reconstruction (6.4) is based on linear and nonlinear integral equations. Among these integral equations, the most important are nonlinear ones arising from $\bar{\partial}$-approach to 3D inverse scattering at fixed energy. This $\bar{\partial}$-approach goes back to [8, 28] and involves different properties of the Faddeev generalized scattering amplitude h in complex domain at fixed energy E. Definition of the Faddeev generalized scattering amplitude h and some of its properties are recalled in Sect. 7.

For precise form of the equations and formulas involved into reconstruction (6.4) we refer to [56, 57].

Reconstruction (6.4) was implemented numerically in [7]. Some of these results of [7] are also presented in Sect. 4 of the survey paper [57].

However, the main disadvantage of reconstruction (6.4) is the overdetermination of $f\big|_{\mathcal{M}_E}$ for $d = 3$, required for this reconstruction. Indeed, $f\big|_{\mathcal{M}_E}$ is a function of 4 variables for $d = 3$ ($dim\,\mathcal{M}_E = 2d - 2 = 4$ for $d = 3$), whereas v is a function of 3 variables. This point was one of motivations for obtaining the results presented in Sect. 8.

7 Faddeev Functions

The results of [55, 56] presented in Sect. 6 are based on properties of the Faddeev's functions ψ, h, ψ_γ, h_γ (see [8, 20, 28, 55, 56]). Definitions and some properties of these functions are recalled below in this section.

The Faddeev solutions ψ of the Schrödinger equation are defined as the solutions of the integral equation (see [20, 28]):

$$\psi(x,k) = e^{ikx} + \int_{\mathbb{R}^d} G(x-y,k)v(y)\psi(y,k)dy, \tag{7.1}$$

$$G(x,k) = e^{ikx}g(x,k), \quad g(x,k) = -\left(\frac{1}{2\pi}\right)^d \int_{\mathbb{R}^d} \frac{e^{i\xi x}d\xi}{\xi^2 + 2k\xi}, \tag{7.2}$$

where $x \in \mathbb{R}^d$, $k \in \mathbb{C}^d\backslash\mathbb{R}^d$ (and at fixed k Eq. (7.1) is an equation for $\psi = e^{ikx}\mu(x,k)$, where μ is sought in $L^\infty(\mathbb{R}^d)$). The Faddeev function h is defined by the formula (see [20, 28]):

$$h(k,l) = \left(\frac{1}{2\pi}\right)^d \int_{\mathbb{R}^d} e^{-ilx}v(x)\psi(x,k)dx, \tag{7.3}$$

where $k, l \in \mathbb{C}^d\backslash\mathbb{R}^d$, $Im\,k = Im\,l$.

Here $\psi(x,k)$ satisfies (1.1) for $E = k^2$, and ψ, G and h are (nonanalytic) continuations of ψ^+, G^+ and f to the complex domain. In particular, $h(k,l)$ for $k^2 = l^2$ can be considered as the "scattering" amplitude in the complex domain for Eq. (1.1) for $E = k^2$.

Equation (7.1) and formulas (7.2), (7.3) are analogs in complex domain of Eq. (3.1) and formulas (3.2), (3.3).

The functions ψ_γ and h_γ are defined as follows (see [20, 28]):

$$\psi_\gamma(x,k) = \psi(x, k + i0\gamma), \quad h_\gamma(k,l) = h(k + i0\gamma, l + i0\gamma), \tag{7.4}$$

where $x, k, l, \gamma \in \mathbb{R}^d$, $|\gamma| = 1$. Note that

$$\psi^+(x,k) = \psi_{k/|k|}(x,k), \quad f(k,l) = h_{k/|k|}(k,l), \tag{7.5}$$

where $x, k, l \in \mathbb{R}^d$, $|k| > 0$.

The following relations are fulfilled (see [20, 28]):

$$\psi_\gamma(x,k) = \psi^+(x,k) + 2\pi i \int\limits_{\mathbb{R}^d} h_\gamma(k,m)\theta((m-k)\gamma)\delta(m^2-k^2)\psi^+(x,m)dm, \tag{7.6}$$

$$h_\gamma(k,l) = f(k,l) + 2\pi i \int\limits_{\mathbb{R}^d} h_\gamma(k,m)\theta((m-k)\gamma)\delta(m^2-k^2)f(m,l)dm, \tag{7.7}$$

where θ is the Heaviside step function, δ is the Dirac delta function, $x, k, l, \gamma \in \mathbb{R}^d$, $|\gamma| = 1$.

The following $\bar{\partial}$- equations and asymptotics hold (see [8, 28]):

$$\frac{\partial}{\partial \bar{k}_j}\mu(x,k) = -2\pi \int\limits_{\xi\in\mathbb{R}^d} \xi_j H(k,-\xi)e^{i\xi x}\delta(\xi^2+2k\xi)\mu(x,k+\xi)d\xi, \tag{7.8}$$

$$\mu(x,k) \to 1, \quad |k| \to \infty, \tag{7.9}$$

$$\frac{\partial}{\partial \bar{k}_j}H(k,p) = -2\pi \int\limits_{\xi\in\mathbb{R}^d} \xi_j H(k,-\xi)H(k+\xi,p+\xi)\delta(\xi^2+2k\xi)d\xi, \tag{7.10}$$

$$H(k,p) \to \hat{v}(p), \quad |k| \to \infty, \tag{7.11}$$

where

$$\mu(x,k) = e^{-ikx}\psi(x,k), \quad H(k,p) = h(k,k-p), \tag{7.12}$$

$\hat{v}$ is defined by (4.2), $x \in \mathbb{R}^d$, $k \in \mathbb{C}^d \backslash \mathbb{R}^d$, $|k| = \left((Rek)^2 + (Imk)^2\right)^{1/2}$, $p \in \mathbb{R}^d$, $j = 1, \ldots, d$.

The derivation of reconstruction (6.2) involves relations (7.6), (7.7), the $\bar{\partial}$—Eq. (7.8) and asymptotics (7.9), and some estimates on f and h, where $d = 2$.

The derivation of reconstruction (6.4) involves relations (7.7), the $\bar{\partial}$—Eq. (7.10) and asymptotics (7.11), and some estimates on f and h, where $d = 3$.

8 Results of [59, 60]

Let

$$\Gamma_E^\delta = \{k = k_E(p),\ l = l_E(p):\ \ p \in \mathscr{B}_{2\delta\sqrt{E}}\},\ 0 < \delta \le 1, \tag{8.1}$$

where $\mathscr{B}_r$ is defined by (4.5), $k_E(p)$ and $l_E(p)$ are defined as in (5.2), where θ is a piecewise continuous vector-function on $\mathbb{R}^d$, $d \ge 2$.

In this section we consider the following version of Problem 1.2a:

Reconstruct v on $\mathbb{R}^d$ from f on Γ_E^δ.

One can see that

$$\Gamma_E^\delta \subset \mathscr{M}_E, \tag{8.2a}$$

$$dim\, \mathscr{M}_E = 2d - 2,\ \ dim\, \Gamma_E^\delta = d \ \text{ for } \ d \ge 2, \tag{8.2b}$$

$$dim\, \mathscr{M}_E > d \ \text{ for } \ d \ge 3. \tag{8.2c}$$

Due to (8.2a), any reconstruction of v from f on Γ_E^δ is also a reconstruction of v from f on $\mathscr{M}_E$. In addition, due to (8.2b), (8.2c), the problem of finding v from f on $\mathscr{M}_E$ is overdetermined for $d \ge 3$, whereas the problem of finding v from f on Γ_E^δ is non-overdetermined.

In [60], for $d \ge 2$, we succeeded, in particular, to give by explicit formulas a stable iterative reconstruction

$$f \ \text{ on } \ \Gamma_E^\delta \ \xrightarrow{\text{stable reconstruction}} \ u_j(\cdot, E) \ \text{ on } \ \mathbb{R}^d,\ \ j = 1, 2, 3, \ldots \tag{8.3}$$

such that

$$\|u_j(\cdot, E) - v\|_{L^\infty(D)} = O(E^{-\alpha_j}) \ \text{ as } \ E \to +\infty, \tag{8.4}$$

$$\alpha_j = \big(1 - (\frac{m-d}{m})^j\big)\frac{m-d}{2d},\ \ j \ge 1,$$

under the assumptions that

$$v \ \text{ is a perturbation of some known background } \ v_0 \ \text{ satisfying } \ (1.2b), \tag{8.5}$$

$$\text{where } \ v - v_0 \in W^{m,1}(\mathbb{R}^d),\ m > d,\ \text{ and } \ supp\,(v - v_0) \subset D,$$

where D is an open bounded domain (which is fixed a priori).

One can see that:

$$\alpha_1 = \frac{m-d}{2m}, \tag{8.6a}$$

$$\alpha_j \to \frac{j}{2} \text{ if } m \to +\infty, \tag{8.6b}$$

$$\alpha_j \to \alpha_\infty = \frac{m-d}{2d} \text{ if } j \to +\infty, \tag{8.6c}$$

$$\alpha_\infty \to +\infty \text{ if } m \to +\infty. \tag{8.6d}$$

One can also see that α_1 is the number of (5.3) and that

$$\alpha_\infty < \frac{m-d}{2}, \tag{8.7}$$

where $(m-d)/2$ is the number of (4.7).

In view of (8.6), (8.7), the error estimate in (8.3), (8.4) is less perfect than in (6.2), (6.3) (for $d=2$) and in (6.4), (6.5) (for $d=3$). However, in (8.3) we reconstruct v from non-overdetermined data in contrast with (6.4) (for $d=3$). Besides, in fact, f on $\Gamma_E^{\delta(E)}$ only is used in (8.3), where

$$\delta(E) = \tau E^{-(d-1)/(2d)}, \quad \tau \in]0,1], \tag{8.8}$$

$$\delta(E) \to 0 \text{ as } E \to +\infty.$$

In any case, α_∞ in (8.6c) is comparable with $(m-d)/2$ in (4.7). Therefore, reconstruction (8.3), (8.4) can be also considered as an analog for the general non-linearized case of the reconstruction result (4.6), (4.7) existing for the linearised case near zero potential.

For precise form of formulas involved into reconstruction (8.3) we refer to [60].

Note that the iterative reconstruction (8.3), (8.4) of [60] was preceded by the following stability estimate obtained in [59]:

$$\|v_1 - v_2\|_{L^\infty(D)} \le C_1(q, D, \sigma) E^{1/2} \|f_1 - f_2\|_{C(\Gamma_E^{\delta(E)})} + \tag{8.9}$$

$$C_2(q, r, D, \sigma, m) E^{-(m-d)/(2d)}, \quad E \ge 1,$$

under the assumptions that v_1, v_2 satisfy (1.2b), $supp\,(v_1 - v_2) \subset D$, $\|v_1 - v_2\|_{W^{m,1}(\mathbb{R}^d)} \le r$, where q and σ are the numbers in (1.2b), $\delta(E)$ is defined by (8.8), $C_1 = C_1(q, D, \sigma)$, $C_2 = C_2(q, r, D, \sigma, m)$ are non-negative constants.

Note also that the proofs of (8.3), (8.4) and (8.9) are based on the Lippmann-Schwinger integral equation (3.1) for ψ^+, the integral formula (3.4) for the scattering amplitude f and the Agmon estimate (3.6) for the Green function G^+. These proofs are much simpler than the proofs of (6.2), (6.3) and (6.4), (6.5) and, in view of their simplicity, can be relatively easy adopted for other problems. In particular, the iterative approximate reconstruction (8.3), (8.4) was extended to the phaseless case in [1], where the extended reconstruction was also implemented numerically (see also Sect. 11).

9 Examples of Non-uniqueness for Problem 1.3a

It is known that $|f|^2$ even on $\mathcal{M}_{\mathbb{R}_+}$ does not determine v uniquely, in general, where $\mathcal{M}_{\mathbb{R}_+} = \cup_{E\in\mathbb{R}_+}\mathcal{M}_E$, $\mathbb{R}_+ =]0, +\infty[$.

In particular, $|f|^2$ on $\mathcal{M}_{\mathbb{R}_+}$ is invariant with respect to translations of v. More precisely, we have the formulas:

$$f_y(k,l) = e^{i(k-l)y} f(k,l), \tag{9.1}$$

$$|f_y(k,l)|^2 = |f(k,l)|^2, \tag{9.2}$$

$$y \in \mathbb{R}^d, \quad (k,l) \in \mathcal{M}_{\mathbb{R}_+},$$

where f is the scattering amplitude for v and f_y is the scattering amplitude for v_y, where

$$v_y(x) = v(x-y), \quad x \in \mathbb{R}^d, \quad y \in \mathbb{R}^d. \tag{9.3}$$

Formula (9.1) is known, in particular, from the theory of solitons; see, e.g., [58].

Besides, $|f|^2$ on $\mathcal{M}_{\mathbb{R}_+}$ does not determine uniquely even the point Bethe-Peierls-Fermi- Zel'dovich-Berezin-Faddeev potential $v_{0,\alpha}$ supported at $\{0\}$, for $d = 3$; see [65]. More precisely, we have the formulas:

$$f_\alpha(k,l) = \frac{1}{2\pi^2}\frac{1}{\alpha + i\sqrt{E}}, \tag{9.4}$$

$$|f_\alpha(k,l)|^2 = |f_{-\alpha}(k,l)|^2, \tag{9.5}$$

$$\alpha \in \mathbb{R}, \quad (k,l) \in \mathcal{M}_{\mathbb{R}_+},$$

where f_α is the scattering amplitude for the point potential $v_{0,\alpha}$, for $d = 3$. An analogous result is also valid in dimension $d = 1$.

10 Results of [64, 70] on Modified Problem 1.3a for $d \geq 2$

Let

$$S = \{|f|^2, |f_j|^2, \ j = 1, \ldots, n\}, \tag{10.1}$$

where f is the initial scattering amplitude for v satisfying (1.2a), f_j is the scattering amplitude for

$$v_j = v + w_j, \quad j = 1, \ldots, n, \tag{10.2}$$

where $w_1, \ldots, w_n$ are additional a priori known background scatterers such that

$$w_j \in L^\infty(\mathbb{R}^d), \quad supp\, w_j \subset \Omega_j, \tag{10.3}$$

$$\Omega_j \ \text{ is an open bounded domain in } \mathbb{R}^d, \quad \Omega_j \cap D = \emptyset,$$

$$w_j \neq 0, \quad w_{j_1} \neq w_{j_2} \ \text{ for } \ j_1 \neq j_2 \ \text{ in } \ L^\infty(\mathbb{R}^d),$$

$$j, j_1, j_2 \in \{1, \ldots, n\}.$$

In other words, S consists of the phaseless scattering data $|f|^2, |f_1|^2, \ldots, |f_n|^2$ measured sequentially for the unknown scatterer v and then for the unknown scatterer v in the presence of known scatterer w_j nonintersecting v for $j = 1, \ldots, n$.

In this section we consider the following modified version of Problem 1.3a:

Problem 10.1 Reconstruct potential v from its phaseless scattering data S for some appropriate a priori known background scatterers $w_1, \ldots, w_n$.

Problem 10.1 in dimension $d = 1$ for $n = 1$ was, actually, considered in [5]. However, to our knowledge, Problem 10.1 in dimension $d \geq 2$ was not considered in the literature before work [64].

Let the Fourier transforms $\hat{v}$ and $\hat{w}_j$ be represented as follows:

$$\hat{v}(p) = |\hat{v}(p)| e^{i\alpha(p)}, \quad \hat{w}_j(p) = |\hat{w}_j(p)| e^{i\beta_j(p)}, \tag{10.4}$$

where $p \in \mathbb{R}^d$, $j = 1, \ldots, n$.

In [64], for v, w_1, w_2 satisfying (1.2a), (10.3), we give, in particular, the following formulas for solving Problem 10.1, for $d \geq 2$, $n = 2$:

$$\begin{aligned} &|\hat{v}_j(p)|^2 = |f_j(k, l)|^2 + O(E^{-1/2}) \ \text{ as } \ E \to +\infty, \\ &p \in \mathbb{R}^d, \quad (k, l) \in \mathcal{M}_E, \quad k - l = p, \quad j = 0, 1, 2, \end{aligned} \tag{10.5}$$

where $v_0 = v$, $f_0 = f$, v_j is defined by (10.2), $j = 1, 2$;

$$|\hat{v}| \begin{pmatrix} \cos\alpha \\ \sin\alpha \end{pmatrix} = (2\sin(\beta_2 - \beta_1))^{-1} \times$$

$$\begin{pmatrix} \sin\beta_2 & -\sin\beta_1 \\ -\cos\beta_2 & \cos\beta_1 \end{pmatrix} \begin{pmatrix} |\hat{w}_1|^{-1}(|\hat{v}_1|^2 - |\hat{v}|^2 - |\hat{w}_1|^2) \\ |\hat{w}_2|^{-1}(|\hat{v}_2|^2 - |\hat{v}|^2 - |\hat{w}_3|^2) \end{pmatrix}, \tag{10.6}$$

where

$$|\hat{v}| = |\hat{v}(p)|, \;\; \alpha = \alpha(p), \;\; \beta_j = \beta_j(p), \;\; |\hat{w}_j| = |\hat{w}_j(p)|, \;\; |\hat{v}_j| = |\hat{v}_j(p)|, \;\; j = 1, 2,$$

$$\sin(\beta_2(p) - \beta_1(p)) \neq 0, \;\; |\hat{w}_1(p)| \neq 0, \;\; |\hat{w}_2(p)| \neq 0, \;\; p \in \mathbb{R}^d. \tag{10.7}$$

Formulas (10.5) follow from (5.1) and are the Born formulas for $|f|^2$, $|f_1|^2$, $|f_2|^2$ at high energies. Using (10.5) we find $|\hat{v}|$, $|\hat{v}_1|$, $|\hat{v}_2|$ from $|f|^2$, $|f_1|^2$, $|f_2|^2$ at high energies.

Using (10.4), (10.6) for appropriate w_1, w_2 we find $\hat{v}$ from $|\hat{v}|$, $|\hat{v}_1|$, $|\hat{v}_2|$ and from a priori known $\hat{w}_1$, $\hat{w}_2$.

In connection with conditions (10.7), for simplicity, we can also assume that, for example:

$$w_2(x) = w_1(x - y), \;\; y \in \mathbb{R}^d \backslash \{0\}. \tag{10.8}$$

In this case we have that

$$|\hat{w}_2(p)| = |\hat{w}_1(p)|, \;\; \beta_2(p) = \beta_1(p) + py, \;\; p \in \mathbb{R}^d, \tag{10.9}$$

and, therefore,

$$\begin{aligned} &\text{conditions (10.7) are fulfilled if and only if } p \in \mathbb{R}^d \backslash (A_y \cup Z), \\ &A_y = \{p \in \mathbb{R}^d : \; e^{2ipy} = 1\}, \;\; Z = \{p \in \mathbb{R}^d : \; |\hat{w}_1(p)| = 0\}. \end{aligned} \tag{10.10}$$

In addition:

$$A_y \text{ is closed and } Meas\, A_y = 0 \text{ in } \mathbb{R}^d, \;\; y \neq 0, \tag{10.11}$$

$$Z \text{ is closed and } Meas\, Z = 0 \text{ in } \mathbb{R}^d, \tag{10.12}$$

where properties (10.12) follow from (10.3).

Thus, using (10.5)–(10.7) and assuming, for example, (10.8) one can see that the phaseless scattering data $S = \{|f|^2, |f_1|^2, |f_2|^2\}$ at high energies in dimension $d \geq 2$ and the background scatterers w_1, w_2 uniquely determine v. More precisely, one can reconstruct $\hat{v}(p)$ from S at high energies and from $\hat{w}_1$, $\hat{w}_2$ for any $p \in$

$\mathbb{R}^d \backslash (A_y \cup Z)$, where A_y, Z are defined in (10.10), using formulas (10.5) with k, l as in (5.2) and then using formula (10.6).

Actually, the reconstruction result for Problem 10.1 consisted in formulas (10.5)–(10.7) is an analog of the old reconstruction result for Problem 1.2a consisted in formula (5.1).

While this paper was in press, further essential results on Problems 1.3a and 10.1 were obtained in [70]. In [70], in dimension $d \geq 2$, we show that $|f_1|^2$ at high energies uniquely determines v via explicit formulas, where f_1 is the scattering amplitude for $v + w_1$, w_1 is an a priori known nonzero background scatterer, under the condition that $supp\, v$ and $supp\, w_1$ are sufficiently disjoint. If this condition is relaxed, then we give similar formulas for finding v from $|f|^2, |f_1|^2$, where f is the scattering amplitude for v. In particular, by these results we continue the aforementioned studies of [64] and studies of [41] on reconstruction from phaseless Fourier transform.

11 Results of [1]

In [1], for $d \geq 2$, we succeeded, in particular, to give a stable iterative reconstruction

$$S \text{ on } \Gamma_E, \ w_1, w_2 \overset{\text{stable reconstruction}}{\longrightarrow} u_j(\cdot, E) \text{ on } \mathbb{R}^d, \ \ j = 1, 2, 3, \ldots \tag{11.1}$$

such that

$$\|u_j(\cdot, E) - v\|_{L^\infty(D)} = O(E^{-\alpha_j}) \text{ as } E \to +\infty, \tag{11.2}$$

$$\alpha_j = \frac{1}{2}\frac{m-d}{d + 2\rho + \frac{m-d}{2N+1}}\left(1 - \left(\frac{m-d}{m + 2\rho + \frac{m-d}{2N+1}}\right)^j\right), \ \ j \geq 1,$$

under the assumptions that v satisfies (1.2*a*), $v \in W^{m,1}(\mathbb{R}^d)$, $m > d$, $N \in \mathbb{N}$, w_1, w_2 satisfy (10.3) and

$$w_1(x) = w(x - T_1), \ \ w_2(x) = w(x - T_2), \ \ x \in \mathbb{R}^d, \tag{11.3}$$

$$\begin{gathered} w \in C(\mathbb{R}^d), \ \ w = \bar{w}, \ \ w(x) = 0 \ \text{ for } |x| > R, \\ \hat{w}(p) = \overline{\hat{w}(p)} \geq \kappa(1 + |p|)^{-\rho}, \ p \in \mathbb{R}^d, \end{gathered} \tag{11.4}$$

for some fixed $T_1, T_2 \in \mathbb{R}^d$, $T_1 \neq T_2$, $R > 0$, $\kappa > 0$, $\rho > d$. (A broad class of w satisfying (11.4) was constructed in Lemma 1 of [2].)

Here, S is defined as in (10.1) for $n = 2$ and Γ_E is defined as in (8.1) for $\delta = 1$. One can see that

$$\alpha_1 = \frac{1}{2}\frac{m-d}{m+2\rho+\frac{m-d}{2N+1}}, \tag{11.5a}$$

$$\alpha_j \to \frac{j}{2} \text{ if } m \to +\infty, \tag{11.5b}$$

$$\alpha_j \to \alpha_\infty = \frac{1}{2}\frac{m-d}{d+2\rho+\frac{m-d}{2N+1}} \text{ if } j \to +\infty, \tag{11.5c}$$

$$\alpha_\infty \to +\infty \text{ if } m \to +\infty,\ N \to +\infty. \tag{11.5d}$$

In [1], for $d \geq 2$, we also give, in particular, a stable iterative reconstruction

$$S \text{ on } \Gamma_E,\ w_1, w_2 \overset{\text{stable reconstruction}}{\longrightarrow} u_j(\cdot, E) \text{ on } \mathbb{R}^d,\quad j = 1, 2, 3, \ldots \tag{11.6}$$

such that

$$\|u_j(\cdot, E) - v\|_{L^\infty(D)} = O(E^{-\alpha_j}) \text{ as } E \to +\infty, \tag{11.7}$$

$$\alpha_j = \frac{1}{2}\frac{m-d}{d+2\rho}\left(1 - \left(\frac{m-d}{m+2\rho}\right)^j\right),\quad j \geq 1,$$

under the assumptions that v satisfies (1.2*a*), $v \in W^{m,1}(\mathbb{R}^d)$, $m > d$, w_1, w_2 satisfy (10.3) and

$$w_1(x) = w(x - T_1),\quad w_2(x) = iw(x - T_1),\quad x \in \mathbb{R}^d, \tag{11.8}$$

where w satisfies (11.4).

Here, S is defined as in (10.1) for $n = 2$ and Γ_E is defined as in (8.1) for $\delta = 1$. One can see that

$$\alpha_1 = \frac{1}{2}\frac{m-d}{m+2\rho}, \tag{11.9a}$$

$$\alpha_j \to \frac{j}{2} \text{ if } m \to +\infty, \tag{11.9b}$$

$$\alpha_j \to \alpha_\infty = \frac{1}{2}\frac{m-d}{d+2\rho} \text{ if } j \to +\infty, \tag{11.9c}$$

$$\alpha_\infty \to +\infty \text{ if } m \to +\infty. \tag{11.9d}$$

Note that an analog of u_1 of (11.1), (11.2) and an analog of u_1 of (11.6), (11.7) were constructed, first, in [2]. These u_1 are analogs for the phaseless case of u_1 of (5.3) for the phased case. In turn, reconstructions (11.1), (11.2) and (11.6), (11.7) are analogs for the phaseless case of reconstruction (8.3), (8.4) for the phased case.

In addition, in [1] we implemented numerically a version of reconstruction (11.1), (11.2) for the case of three background scatterers w_1, w_2, w_3 and we implemented numerically reconstruction (11.6), (11.7); see Section 4 of [1]. Numerical experiments show that it yields remarkably accurate approximations with small computational effort even for moderate energies.

12 Formulas of [63, 66] Reducing Problem 1.3b to Problem 1.2a

Let

$$f_1(k,l) = c(d,|k|)f(k,l), \quad (k,l) \in \mathcal{M}_E, \tag{12.1}$$

$$a(x,k) = |x|^{(d-1)/2}(|\psi^+(x,k)|^2 - 1), \quad x \in \mathbb{R}^d\backslash\{0\}, \quad k \in \mathbb{R}^d\backslash\{0\}, \tag{12.2}$$

where c is the constant of (1.3), ψ^+, f are the function of (1.3), (3.1), (3.4), $\mathcal{M}_E$ is defined by (1.4).

In [63], for v satisfying (1.2a), we succeeded, in particular, to give the following formulas reducing Problem 1.3b to Problem 1.2a, for $d \geq 2$:

$$\begin{pmatrix} Re\ f_1(k,l) \\ Im\ f_1(k,l) \end{pmatrix} = M\left(\begin{pmatrix} a(x_1,k) \\ a(x_2,k) \end{pmatrix} - \begin{pmatrix} \delta a(x_1,k) \\ \delta a(x_2,k) \end{pmatrix}\right) \tag{12.3}$$

$$M = \frac{1}{2\sin(\varphi_2 - \varphi_1)}\begin{pmatrix} \sin\varphi_2 & -\sin\varphi_1 \\ -\cos\varphi_1 & \cos\varphi_2 \end{pmatrix}, \tag{12.4}$$

$$x_1 = s\hat{l}, \quad x_2 = (s+\tau)\hat{l}, \quad \hat{l} = l/|l|, \tag{12.5}$$

$$\varphi_j = |k||x_j| - kx_j, \quad j = 1,2, \tag{12.6}$$

$$\varphi_2 - \varphi_1 = \tau(|k| - k\hat{l}), \tag{12.7}$$

$$\delta a(x_1,k) = O(s^{-\alpha}), \quad \delta a(x_2,k) = O(s^{-\alpha}) \text{ as } s \to +\infty \tag{12.8}$$

uniformly in $\hat{k} = k/|k|$, $\hat{l} = l/|l|$ and τ at fixed $E > 0$,

$$\alpha = 1/2 \text{ for } d = 2, \quad \alpha = 1 \text{ for } d \geq 3, \tag{12.9}$$

where

$$\sin(\varphi_2 - \varphi_1) \neq 0, \tag{12.10}$$

$(k, l) \in \mathscr{M}_E$, $s > 0$, $\tau > 0$.

Formulas (12.1), (12.3)–(12.10) are explicit two-point formulas for approximate finding phased $f(k, l)$ at fixed $(k, l) \in \mathscr{M}_E$, $k \neq l$, from phaseless $|\psi^+(x, k)|^2$ at two points $x = x_1, x_2$ defined in (12.5), where s is sufficiently large.

In turn, article [62] gives exact versions (without error terms) of formulas (12.1), (12.3)–(12.10) for the 3-point case for $d = 1$; see [61, 62] for details.

Detailed estimates for the error term $\delta a(x, k) = O(s^{-\alpha})$ in (12.3), (12.8), (12.9) are given in [69], for $d = 3$ and for $d = 2$. However, the main drawback of the two-point formulas (12.1), (12.3)–(12.10) for finding phased f from phaseless $|\psi^+|^2$ is a slow decay of the error as $s \to +\infty$; see (12.3), (12.8), (12.9). This drawback motivated our considerations given in [66].

In [66], for v satisfying (1.2a), for fixed $(k, l) \in \mathscr{M}_E$, $k \neq l$, for $d = 3$ and $d = 2$, we succeeded, in particular, to give

$$\text{formulas for finding } f(k, l) \text{ up to } O(s^{-n}) \text{ as } s \to +\infty \tag{12.11}$$

$$\text{from } |\psi^+(x, k)|^2 \text{ given at } 2n \text{ points } x = x_1(s), \ldots, x_{2n}(s),$$

where

$$x_i(s) = r_i(s)\hat{l}, \ \ i = 1, \ldots, 2n, \ \ \hat{l} = l/|l|, \tag{12.12}$$

$$r_{2j-1}(s) = \lambda_j s, \ \ r_{2j}(s) = \lambda_j s + \tau, \ \ j = 1, \ldots, n,$$

$$\lambda_1 = 1, \ \ \lambda_{j_1} < \lambda_{j_2} \text{ for } j_1 < j_2, \ \ \tau = \tau_{fixed} > 0.$$

The point is that in (12.11) we have a rapid decay of the error as $s \to +\infty$ if n is sufficiently large. For $d = 3$, $n = 1$, formulas (12.11), (12.12) reduce to (12.1), (12.3)–(12.10).

The general idea of obtaining the $2n$-point formulas (12.11), (12.12) can be described as follows.

We use that, under assumptions (1.2a), formula (1.3) admits the following much more precise version:

$$\psi^+(x, k) = e^{ikx} + \frac{e^{i|k||x|}}{|x|^{(d-1)/2}}\left(\sum_{j=1}^{n} \frac{f_j(k, |k|\frac{x}{|x|})}{|x|^{j-1}} + O\left(\frac{1}{|x|^n}\right)\right) \text{ as } |x| \to \infty, \tag{12.13}$$

where $x \in \mathbb{R}^d$, $k \in \mathbb{R}^d$, $k^2 = E > 0$, $n \in \mathbb{N}$.

Then, for fixed $(k, l) \in \mathcal{M}_E$, $k \neq l$, we look for formulas for finding $f_j(k, l)$ up to $O(s^{-(n-j+1)})$ as $s \to +\infty$, $j = 1, \ldots, n$, from $|\psi^+(x, k)|^2$ given at $2n$ points $x = x_1(s), \ldots, x_{2n}(s)$ of the form (12.12), where $f_j = f_j(k, l)$, $j = 1, \ldots, n$, are the functions arising in (12.13).

Using the later formulas for f_1 and using (12.1) we obtain formulas (12.11), (12.12).

For precise form of formulas (12.11), (12.12) we refer to [66].

The aforementioned formulas of [62, 63, 66] reducing Problem 1.3b to Problem 1.2a permit to apply to the phaseless inverse scattering Problem 1.3b well developed methods existing for the inverse scattering Problem 1.2a with phase information; see Sects. 4–8 for some of these methods.

References

1. A.D. Agaltsov, T. Hohage, R.G. Novikov, An iterative approach to monochromatic phaseless inverse scattering. Inverse Probl. **35**(2), 24001 (2019)
2. A.D. Agaltsov, R.G. Novikov, Error estimates for phaseless inverse scattering in the Born approximation at high energies. J. Geom. Anal. **30**(3), 2340–2360 (2020)
3. A.D. Agaltsov, R.G. Novikov, Examples of solution of the inverse scattering problem and the equations of the Novikov-Veselov hierarchy from the scattering data of point potentials. Uspekhi Mat. Nauk **74**(3), 3–16 (2019) (in Russian); English transl.: Russ. Math. Surv. **74**(3), 373–386 (2019)
4. S. Agmon, Spectral properties of Schrödinger operators and scattering theory. Ann. Scuola Norm. Sup. Pisa Cl. Sci. 4, **2**(2), 151–218 (1975)
5. T. Aktosun, P.E. Sacks, Inverse problem on the line without phase information. Inverse Probl. **14**, 211–224 (1998)
6. T. Aktosun, R. Weder, Inverse scattering with partial information on the potential. J. Math. Anal. Appl. **270**, 247–266 (2002)
7. N.V. Alexeenko, V.A. Burov, O.D. Rumyantseva, Solution of the three-dimensional acoustical inverse scattering problem. The modified Novikov algorithm. Acoust. J. **54**(3), 469–482 (2008) (in Russian); English transl.: Acoust. Phys. **54**(3), 407–419 (2008)
8. R. Beals, R.R. Coifman, Multidimensional inverse scattering and nonlinear partial differential equations. Proc. Symp. Pure Math. **43**, 45–70 (1985)
9. Yu.M. Berezanskii, On the uniqueness theorem in the inverse problem of spectral analysis for the Schrödinger equation. Tr. Mosk. Mat. Obshch. **7**, 3–62 (1958) (in Russian)
10. F.A. Berezin, M.A. Shubin, *The Schrödinger Equation*, vol. 66 of Mathematics and Its Applications (Kluwer Academic, Dordrecht, 1991)
11. M. Born, Quantenmechanik der Stossvorgange. Z. Phys. **38**(11–12), 803–827 (1926)
12. A.L. Buckhgeim, Recovering a potential from Cauchy data in the two-dimensional case. J. Inverse Ill-Posed Probl. **16**(1), 19–33 (2008)
13. V.A. Burov, N.V. Alekseenko, O.D. Rumyantseva, Multifrequency generalization of the Novikov algorithm for the two-dimensional inverse scattering problem. Acoust. J. **55**(6), 784–798 (2009) (in Russian); English transl.: Acoust. Phys. **55**(6), 843–856 (2009)
14. K. Chadan, P.C. Sabatier, *Inverse Problems in Quantum Scattering Theory*, 2nd edn (Springer, Berlin, 1989)
15. P. Deift, E. Trubowitz, Inverse scattering on the line. Commun. Pure Appl. Math. **32**, 121–251 (1979)

16. V. Enss, R. Weder, Inverse Potential Scattering: A Geometrical Approach. Mathematical Quantum Theory. II. Schrödinger Operators (Vancouver, BC, 1993), pp. 151–162. CRM Proc. Lecture Notes, vol. 8 (Amer.Math.Soc., Providence, 1995)
17. G. Eskin, *Lectures on Linear Partial Differential Equations, Graduate Studies in Mathematics*, vol. 123 (American Mathematical Society, 2011)
18. L.D. Faddeev, Uniqueness of the solution of the inverse scattering problem. Vest. Leningrad Univ. **7**, 126–130 (1956) (in Russian)
19. L.D. Faddeev, The inverse problem in the quantum theory of scattering. Uspehi Mat. Nauk **14**(4), 57–119 (1959) (in Russian); English transl.: J. Math. Phys. **4**, 72–104 (1963)
20. L.D. Faddeev, Inverse problem of quantum scattering theory. II, Itogy Nauki i Tekh. Ser. Sovrem. Probl. Mat. **3**, 93–180 (1974) (in Russian); English transl.: J. Sov. Math. **5**, 334–396 (1976)
21. L.D. Faddeev, S.P. Merkuriev, *Quantum Scattering Theory for Multi-Particle Systems*. Mathematical Physics and Applied Mathematics, vol. 11 (Kluwer Academic Publishers Group, Dordrecht, 1993)
22. D. Fanelli, O. Öktem, Electron tomography: a short overview with an emphasis on the absorption potential model for the forward problem. Inverse Probl. **24**(1), 013001 (2008)
23. I.M. Gel'fand, Some problems of functional analysis and algebra, in *Proceedings of the International Congress of Mathematicians*, vol. I (Amsterdam, 1954), pp. 253–276
24. P.G. Grinevich, The scattering transform for the two-dimensional Schrödinger operator with a potential that decreases at infinity at fixed nonzero energy. Uspekhi Mat. Nauk **55**(6), 3–70 (2000); English transl.: Russ. Math. Surv. **55**(6), 1015–1083 (2000)
25. P.G. Grinevich, R.G. Novikov, Analogues of multisoliton potentials for the two-dimensional Schrödinger equations and a nonlocal Riemann problem. Sov. Math. Dokl. **33**(1), 9–12 (1986)
26. P. Hähner, T. Hohage, New stability estimates for the inverse acoustic inhomogeneous medium problem and applications. SIAM J. Math. Anal. **33**(3), 670–685 (2001)
27. W. Heisenberg, Die "beobachtbaren Grossen" in der Theorie der Elementarteilchen. Z. Phys. **120**, 513–538, 673–702 (1943)
28. G.M. Henkin, R.G. Novikov, The $\bar{\partial}$-equation in the multidimensional inverse scattering problem. Uspekhi Mat. Nauk **42**(3), 93–152 (1987) (in Russian); English transl.: Russ. Math. Surv. **42**(3), 109–180 (1987)
29. T. Hohage. On the numerical solution of a three-dimensional inverse medium scattering problem. Inverse Probl. **17**(6), 1743–1763 (2001)
30. T. Hohage, R.G. Novikov, Inverse wave propagation problems without phase information. Inverse Probl. **35**(7), 070301 (2019)
31. O. Ivanyshyn, R. Kress, Identification of sound-soft 3D obstacles from phaseless data. Inverse Probl. Imaging **4**, 131–149 (2010)
32. P. Jonas, A.K. Louis, Phase contrast tomography using holographic measurements. Inverse Probl. **20**(1), 75–102 (2004)
33. M.I. Isaev, Exponential instability in the inverse scattering problem on the energy interval. Funkt. Anal. Prilozhen. **47**(3), 28–36 (2013) (in Russian); English transl.: Funct. Anal Appl. **47**, 187–194 (2013)
34. M.I. Isaev, R.G. Novikov, New global stability estimates for monochromatic inverse acoustic scattering. SIAM J. Math. Anal. **45**(3), 1495–1504 (2013)
35. M.V. Klibanov, P.E. Sacks, Phaseless inverse scattering and the phase problem in optics. J. Math. Phys. **33**, 3813–3821 (1992)
36. M.V. Klibanov, Phaseless inverse scattering problems in three dimensions. SIAM J. Appl. Math. **74**(2), 392–410 (2014)
37. M.V. Klibanov, N.A. Koshev, D.-L. Nguyen, L.H. Nguyen, A. Brettin, V.N. Astratov, A numerical method to solve a phaseless coefficient inverse problem from a single measurement of experimental data. SIAM J. Imaging Sci. **11**(4), 2339–2367 (2018)
38. M.V. Klibanov, V.G. Romanov, Reconstruction procedures for two inverse scattering problems without the phase information. SIAM J. Appl. Math. **76**(1), 178–196 (2016)

39. E. Lakshtanov, R.G. Novikov, B.R. Vainberg, A global Riemann-Hilbert problem for two-dimensional inverse scattering at fixed energy. Rend. Istit. Mat. Univ. Trieste **48**, 21–47 (2016)
40. E. Lakshtanov, B.R. Vainberg, Recovery of L^p-potential in the plane. J. Inverse Ill-Posed Probl. **25**(5), 633–651 (2017)
41. B. Leshem, R. Xu, Y. Dallal, J. Miao, B. Nadler, D. Oron, N. Dudovich, O. Raz, Direct single-shot phase retrieval from the diffraction pattern of separated objects. Nature Commun. **7**(1), 1–6 (2016)
42. B.M. Levitan, Inverse Sturm-Liuville Problems, VSP, Zeist, 1987
43. B.A. Lippmann, J. Schwinger, Variational principles for scattering processes. I. Phys. Rev. Lett. **79**(3), 469–480 (1950)
44. S.V. Manakov, The inverse scattering transform for the time dependent Schrödinger equation and Kadomtsev-Petviashvili equation. Phys. D, **3**(1–2), 420–427 (1981)
45. V.A. Marchenko, *Sturm-Liuville Operators and Applications* (Birkhäuser, Basel, 1986)
46. R.B. Melrose, *Geometric scattering theory*. Stanford Lectures (Cambridge University Press, Cambridge, 1995)
47. C.J. Merono, Recovery of the singularities of a potential from backscattering data in general dimension. J. Differ. Equ. **266**(10), 6307–6345 (2019)
48. H.E. Moses, Calculation of the scattering potential from reflection coefficients. Phys. Rev. **102**, 559–567 (1956)
49. R.G. Newton, *Inverse Schrödinger Scattering in Three Dimensions* (Springer, Berlin, 1989)
50. R.G. Novikov, Multidimensional inverse spectral problem for the equation $-\Delta\psi + (v(x) - Eu(x))\psi = 0$. Funkt. Anal. Prilozhen. **22**(4), 11–22 (1988) (in Russian); English transl.: Funct. Anal. Appl. **22**, 263–272 (1988)
51. R.G. Novikov, The inverse scattering problem at fixed energy level for the two-dimensional Schrödinger operator. J. Funct. Anal. **103**, 409–463 (1992)
52. R.G. Novikov, The inverse scattering problem at fixed energy for Schrödinger equation with an exponentially decreasing potential. Commun. Math. Phys. **161**, 569–595 (1994)
53. R.G. Novikov, Inverse scattering up to smooth functions for the Schrödinger equation in dimension 1. Bull. Sci. Math. **120**, 473–491 (1996)
54. R.G. Novikov, Rapidly converging approximation in inverse quantum scattering in dimension 2. Phys. Lett. A **238**, 73–78 (1998)
55. R.G. Novikov, Approximate inverse quantum scattering at fixed energy in dimension 2. Proc. Steklov Inst. Math. **225**, 285–302 (1999)
56. R.G. Novikov, The $\bar{\partial}$-approach to approximate inverse scattering at fixed energy in three dimensions. Int. Math. Res. Pap. **2005**(6), 287–349 (2005)
57. R.G. Novikov, The $\bar{\partial}$-approach to monochromatic inverse scattering in three dimensions. J. Geom. Anal. **18**, 612–631 (2008)
58. R.G. Novikov, Absence of exponentially localized solitons for the Novikov-Veselov equation at positive energy. Phys. Lett. A **375**, 1233–1235 (2011)
59. R.G. Novikov, Approximate Lipschitz stability for non-overdetermined inverse scattering at fixed energy. J. Inverse Ill-Posed Probl. **21** (6), 813–823 (2013)
60. R.G. Novikov, An iterative approach to non-overdetermined inverse scattering at fixed energy. Sbornik Math. **206**(1), 120–134 (2015)
61. R.G. Novikov, Inverse scattering without phase information. Séminaire Laurent Schwartz - EDP et Applications (2014–2015), Exp. No16, 13p.
62. R.G. Novikov, Phaseless inverse scattering in the one-dimensional case. Eur. J. Math. Comput. Appl. **3**(1), 63–69 (2015)
63. R.G. Novikov, Formulas for phase recovering from phaseless scattering data at fixed frequency. Bull. Sci. Math. **139**(8), 923–936 (2015)
64. R.G. Novikov, Explicit formulas and global uniqueness for phaseless inverse scattering in multidimensions. J. Geom. Anal. **26**(1), 346–359 (2016)
65. R.G. Novikov, Inverse scattering for the Bethe-Peierls model. Eur. J. Math. Comput. Appl. **6**(1), 52–55 (2018)

66. R.G. Novikov, Multipoint formulas for phase recovering from phaseless scattering data. J. Geom. Anal. **31**(2), 1965–1991 (2021)
67. R.G. Novikov, Multipoint formulas for scattered far field in multidimensions. Inverse Probl. **36**(9), 095001 (2020)
68. R.G. Novikov, Multipoint formulas for inverse scattering at high energies. Uspekhi Mat. Nauk **76**(4), 177–178 (2021) (in Russian); English transl.: Russ. Math. Surv. **76**(4), 723–725 (2021)
69. R.G. Novikov, V.N. Sivkin, Error estimates for phase recovering from phaseless scattering data. Eur. J. Math. Comput. Appl. **8**(1), 44–61 (2020)
70. R.G. Novikov, V.N. Sivkin, Phaseless inverse scattering with background information. Inverse Probl. **37**(5), 055011 (2021)
71. N.N. Novikova, V.M. Markushevich, On the uniqueness of the solution of the inverse scattering problem on the real axis for the potentials placed on the positive half-axis. Comput. Seismol. **18**, 176–184 (1985) (in Russian)
72. V. Palamodov, A fast method of reconstruction for X-ray phase contrast imaging with arbitrary Fresnel number (2018). arXiv:1803.08938v1
73. Rakesh, M. Salo, Fixed angle inverse scattering for almost symmetric or controlled perturbations. SIAM J. Math. Anal. **52**(6), 5467–5499 (2020)
74. T. Regge, Introduction to complex orbital moments. Nuovo Cimento **14**, 951–976 (1959)
75. V.G. Romanov, Inverse problems without phase information that use wave interference. Sib. Math. J. **59**(3), 494–504 (2018)
76. P. Stefanov, Stability of the inverse problem in potential scattering at fixed energy. Ann. Inst. Fourier **40** (4), 867–884 (1990)
77. G. Vainikko, Fast solvers of the Lippmann–Schwinger equation, in *Direct and inverse problems of mathematical physics* (Newark, DE, 1997). Int. Soc. Anal. Appl. Comput. vol. 5, (Kluwer, Dordrecht, 2000), pp. 423–440

A Survey of Hardy Type Inequalities on Homogeneous Groups

Durvudkhan Suragan

Abstract In this review paper, we survey Hardy type inequalities from the point of view of Folland and Stein's homogeneous groups. Particular attention is paid to Hardy type inequalities on stratified groups which give a special class of homogeneous groups. In this environment, the theory of Hardy type inequalities becomes intricately intertwined with the properties of sub-Laplacians and more general subelliptic partial differential equations. Particularly, we discuss the Badiale-Tarantello conjecture and a conjecture on the geometric Hardy inequality in a half-space of the Heisenberg group with a sharp constant.

Keywords Hardy inequality · Sub-Laplacian · Heisenberg group · Stratified group · Homogeneous group · Nilpotent Lie group

1 Introduction

In 1918, G. H. Hardy proved an inequality (discrete and in one variable) [25] now bearing his name, which in $\mathbb{R}^n$ can be formulated as

$$\left\| \frac{f}{\|x\|} \right\|_{L^2(\mathbb{R}^n)} \leq \frac{2}{n-2} \|\nabla f\|_{L^2(\mathbb{R}^n)}, \quad n \geq 3,$$

where ∇ is the standard gradient in $\mathbb{R}^n$, $f \in C_0^\infty(\mathbb{R}^n)$, $\|x\|$ is the Euclidean norm, and the constant $\frac{2}{n-2}$ is known to be sharp. Note that the multidimensional version of the Hardy inequality was proved by J. Leray [34].

D. Suragan (✉)
Department of Mathematics, Nazarbayev University, Nur-Sultan City, Republic of Kazakhstan
e-mail: durvudkhan.suragan@nu.edu.kz

P. Cerejeiras, M. Reissig (eds.), *Mathematical Analysis, its Applications and Computation*, Springer Proceedings in Mathematics & Statistics 385,
https://doi.org/10.1007/978-3-030-97127-4_4

The Hardy inequality has many applications in the analysis of linear and nonlinear PDEs, for example, in existence and nonexistence results for second order partial differential equations of the form:

$$\left.\begin{matrix} 0 \\ u_t \\ u_{tt} \end{matrix}\right\} - \Delta u = \lambda \frac{|u|^s}{\|x\|^2}.$$

The criteria on existence (or nonexistence) of a solution depends on a relation between the constants λ and $\frac{2}{n-2}$.

In the equation, instead of $\frac{1}{\|x\|^2}$ (Hardy potential) one may have more general function which motivates to study weighted versions of the Hardy inequality. On the other hand, one may consider some different operators instead of the Laplacian, say, the p-Laplacian.

The L^p-version of the Hardy inequality (used e.g. for p-Laplacian) takes the form

$$\left\| \frac{f}{\|x\|} \right\|_{L^p(\mathbb{R}^n)} \leq \frac{p}{n-p} \|\nabla f\|_{L^p(\mathbb{R}^n)}, \quad 1 < p < n,$$

where the constant $\frac{p}{n-p}$ is sharp. There are already many excellent presentations on the (classical) Hardy inequalities and their extensions, see e.g. [5, 11, 36], and [40] as well as references therein.

The purpose of the present review paper is to offer a brief survey of Hardy type inequalities on homogeneous groups. These groups give one of most general classes of noncommutative nilpotent Lie groups. The intersection of analysis of Hardy type inequalities and theory of homogeneous groups is a beautiful area of mathematics with links to many other subjects.

Since L. Hörmander's fundamental work [26] the operators of the type sum of squares of vector fields have been studied intensively, and today's literature on the subject is quite large. Much of the development in the field has been connected to the development of analysis on the homogeneous groups, following the ideas of E. Stein's talk at ICM 1970 [52]. Since then continuing through the rest of his life, a substantial part of E. Stein's research is related to analysis on homogeneous groups (see [16]).

Among many, one of the motivations behind doing analysis on the homogeneous groups is the "distillation of ideas and results of harmonic analysis depending only on the group and dilation structures" [18].

In the 1990s, a lot of work concerning Hardy inequalities was already developed in the context of the elliptic operators, but not very much had been done in the framework of (nonelliptic) subellipticity, in particular, for the Heisenberg sub-Laplacians.

Note that the sub-Laplacian on the nilpotent Lie groups are (left invariant homogeneous) subelliptic differential operators and it is known that it is elliptic if and only if the Lie group is Abelian (Euclidean).

N. Garofalo and E. Lanconelli [23] have an important contribution to the development of Hardy inequalities on the Heisenberg group with their original approach, which is based on a far-reaching use of the fundamental solutions. Later this idea was extended to general stratified groups by several authors. There is also another approach in the theory of Hardy inequalities on stratified groups, the so-called horizontal estimates, which was suggested by L. D'Ambrosio on the Heisenberg group [7] and on stratified groups [8]. We give further discussions in this direction in Sect. 2. The general theory of Hardy type inequalities in the setting of homogeneous groups is reviewed in Sect. 3.

These are notes mainly from my lecture at the 12th ISAAC Congress in Aveiro (2019). The lecture was partially based on our recent open access book with the title "Hardy inequalities on homogeneous groups" with Michael Ruzhansky [50].

2 Hardy Type Inequalities on Stratified Groups

A (connected and simply connected) Lie group $\mathbb{G}$ is graded if its Lie algebra:

$$\mathfrak{g} = \bigoplus_{i=1}^{\infty} \mathfrak{g}_i,$$

where $\mathfrak{g}_1, \mathfrak{g}_2, \ldots,$ are vector subspaces of $\mathfrak{g}$, only finitely many not $\{0\}$, and

$$[\mathfrak{g}_i, \mathfrak{g}_j] \subset \mathfrak{g}_{i+j} \quad \forall i, j \in \mathbb{N}.$$

If $\mathfrak{g}_1$ generates the Lie algebra $\mathfrak{g}$ through commutators, the group is said to be stratified (see, e.g. [14]).

Example 1 (Abelian Case) The Euclidean group $(\mathbb{R}^n, +)$ is graded: its Lie algebra $\mathbb{R}^n$ is trivially graded. Obviously, it is also stratified.

Example 2 (Heisenberg Group) The Heisenberg group $\mathbb{H}^n$ is stratified: its Lie algebra $\mathfrak{h}_n$ can be decomposed as $\mathfrak{h}_n = \mathfrak{g}_1 \oplus \mathfrak{g}_2$ where $\mathfrak{g}_1 = \oplus_{j=1}^{n} \mathbb{R}X_j \oplus \mathbb{R}Y_j$ and $\mathfrak{g}_2 = \mathbb{R}T$, where

$$X_j = \partial_{x_j} - \frac{y_j}{2}\partial_t, \quad Y_j = \partial_{y_j} + \frac{x_j}{2}\partial_t, \quad j = 1, \ldots, n, \quad T = \partial_t.$$

Note that the concept of stratified groups was introduced for the first time by Gerald Folland in 1975 [15]. However, in the literature of sub-Riemannian geometry, stratified groups are commonly called (homogeneous) Carnot groups.

Let $\mathbb{G}$ be a stratified group, i.e. there is $\mathfrak{g}_1 \subset \mathfrak{g}$ (the first stratum), with its basis $X_1, \ldots, X_N$ generating its Lie algebra $\mathfrak{g}$ through their commutators. Then the sub-Laplacian

$$\mathcal{L} := X_1^2 + \cdots + X_N^2$$

is subelliptic, and $\nabla_H = (X_1, \ldots, X_n)$ is the so-called horizontal gradient.

Folland proved that $\mathcal{L}$ has a fundamental solution $= Cd(x)^{2-Q}$ for some homogeneous quasi-norm $d(x)$ called the $\mathcal{L}$-gauge. Here Q is homogeneous dimension of $\mathbb{G}$.

As we mentioned in the introduction, by using the fundamental solution (i.e. the $\mathcal{L}$-gauge) of the sub-Laplacian $\mathcal{L}$, in 1990 Garofalo and Lanconelli [23] proved the Hardy inequality on the Heisenberg group,[1] then in 2005 L. D'Ambrosio [8] and independently in 2008 Goldstein and Kombe [24] established the Hardy inequality on the general stratified groups:

$$\left\| \frac{f}{d(x)} \right\|_{L^2(\mathbb{G})} \leq \frac{2}{Q-2} \|\nabla_H f\|_{L^2(\mathbb{G})}, \quad Q \geq 3. \tag{1}$$

In [43], we obtained the Hardy inequality on the general stratified groups with boundary terms. In general, this approach can be called a *fundamental solution approach*. On the other hand, there is another approach to obtain the Hardy inequality on the stratified groups which is a *horizontal estimate approach*. In 2004, L. D'Ambrosio proved a "horizontal" version of the Hardy inequality on the Heisenberg group [7] (see also [8]). In 2017, we extended this result for the general stratified groups [44] in the form

$$\left\| \frac{f}{|x'|} \right\|_{L^2(\mathbb{G})} \leq \frac{2}{N-2} \|\nabla_H f\|_{L^2(\mathbb{G})}, \quad N \geq 3, \tag{2}$$

with the Euclidean norm $|x'|$ on the first stratum and $x = (x', x'')$. Here N is (topological) dimension of the first stratum of $\mathbb{G}$. The L^p-version of (1) on the Heisenberg group and the stratified groups were proved by using different methods by Niu et al. in 2001 [39], L. D'Ambrosio in 2004 [7] and in 2005 [8], Adimurthi and Sekar in 2006 [1], Danielli et al. in 2011 [10], as well as Jin and Shen in 2011 [27]:

$$\left\| \frac{f}{d(x)} \right\|_{L^p(\mathbb{G})} \leq \frac{p}{Q-p} \|\nabla_H f\|_{L^p(\mathbb{G})}, \quad Q \geq 3, \ \ 1 < p < Q. \tag{3}$$

[1] In the case of the Heisenberg group, $\mathcal{L}$-gauge is called a Kaplan distance.

Moreover, the L^p-version of (2) was proved by L. D'Ambrosio on the Heisenberg group [7] and then it extended on the general stratified groups in [8]. We obtained its extension on the general stratified groups [44] by a different method in the form

$$\left\| \frac{f}{|x'|} \right\|_{L^p(\mathbb{G})} \leq \frac{p}{N-p} \|\nabla_H f\|_{L^p(\mathbb{G})}, \quad Q \geq 3, \quad 1 < p < N. \tag{4}$$

Both constants in (3) and (4) are sharp for functions from, say, $C_0^\infty(\mathbb{G})$. In particular, a special case of the horizontal estimate implies the Badiale-Tarantello conjecture.

Badiale-Tarantello Conjecture Let $x = (x', x'') \in \mathbb{R}^N \times \mathbb{R}^{n-N}$. Badiale and Tarantello [2] proved that for $1 < p < N \leq n$ there exists a constant $C_{n,N,p}$ such that

$$\left\| \frac{1}{|x'|} f \right\|_{L^p(\mathbb{R}^n)} \leq C_{n,N,p} \|\nabla f\|_{L^p(\mathbb{R}^n)}. \tag{5}$$

Clearly, for $N = n$ this gives the classical Hardy inequality with the best constant

$$C_{n,p} = \frac{p}{n-p}.$$

It was conjectured by Badiale and Tarantello that the best constant in (5) is given by

$$C_{N,p} = \frac{p}{N-p}. \tag{6}$$

This conjecture was proved by Secchi, Smets and Willen in 2003 [53].

As a consequence of the horizontal estimate techniques, we also gave a new proof of the Badiale-Tarantello conjecture in [44].

Hardy Inequality in Half-Space Let us recall the Hardy inequality in a half-space of $\mathbb{R}^n$:

$$\left(\frac{p-1}{p}\right)^p \int_{\mathbb{R}^n_+} \frac{|u|^p}{x_n^p} dx \leq \int_{\mathbb{R}^n_+} |\nabla u|^p dx, \quad p > 1, \tag{7}$$

for every function $u \in C_0^\infty(\mathbb{R}^n_+)$, where ∇ is the usual Euclidean gradient and

$$\mathbb{R}^n_+ := \{(x', x_n) | x' := (x_1, \ldots, x_{n-1}) \in \mathbb{R}^{n-1}, x_n > 0\}, \ n \in \mathbb{N}.$$

There is a number of studies related to inequality (7) by Maz'ya, Davies, Opic-Kufner (see, e.g. [11, 36], and [40]) and others.

Filippas et al. in 2007 [13] established the Hardy-Sobolev inequality in the following form

$$C\left(\int_{\mathbb{R}^n_+}|u|^{p^*}dx\right)^{\frac{1}{p^*}} \le \left(\int_{\mathbb{R}^n_+}|\nabla u|^p dx - \left(\frac{p-1}{p}\right)^p\int_{\mathbb{R}^n_+}\frac{|u|^p}{x_n^p}dx\right)^{\frac{1}{p}}, \tag{8}$$

for all function $u \in C_0^\infty(\mathbb{R}^n_+)$, where $p^* = \frac{np}{n-p}$ and $2 \le p < n$. For a different proof of this inequality, see Frank and Loss [20]. Obviously, (7) implies from (8).

Version on the Heisenberg Group Let $\mathbb{H}^n$ be the Heisenberg group, that is, the set $\mathbb{R}^{2n+1}$ equipped with the group law

$$(z,t)\circ(\widetilde{z},\widetilde{t}) := (z+\widetilde{z}, t+\widetilde{t}+2\,\mathrm{Im}\langle z,\widetilde{z}\rangle),$$

where $(z,t) = (z_1,\dots,z_n,t) = (x_1,y_1,\dots,x_n,y_n,t) \in \mathbb{H}^n$, $x \in \mathbb{R}^n$, $y \in \mathbb{R}^n$, $t \in \mathbb{R}$ and $\mathbb{R}^{2n}$ is identified by $\mathbb{C}^n$.

A Hardy inequality in a half-space of the Heisenberg group was shown by Luan and Yang in 2008 [33] in the form

$$\int_{\mathbb{H}^n_{t>0}}\frac{|x|^2+|y|^2}{t^2}|u|^2 d\xi \le \int_{\mathbb{H}^n_{t>0}}|\nabla_H u|^2 d\xi, \tag{9}$$

for every function $u \in C_0^\infty(\mathbb{H}^n_{t>0})$.

In 2016, Larson generalised the above inequality to any half-space of the Heisenberg group [31]:

$$\frac{1}{4}\int_{\mathbb{H}^+}\frac{\sum_{i=1}^n\langle X_i(\xi),\nu\rangle^2+\langle Y_i(\xi),\nu\rangle^2}{dist(\xi,\partial\mathbb{H}^+)^2}|u|^2 d\xi \le \int_{\mathbb{H}^+}|\nabla_H u|^2 d\xi, \tag{10}$$

where X_i and Y_i (for $i = 1,\dots,n$) are left-invariant vector fields on the Heisenberg group, ν is the unit vector. However, the following conjecture remained open (see, e.g. [31]).

Conjecture The L^p-version of the above Hardy inequality (in a half-space) should be valid, that is,

$$\left(\frac{p-1}{p}\right)^p\int_{\mathbb{H}^+}\frac{\mathscr{W}(\xi)^p}{dist(\xi,\partial\mathbb{H}^+)^p}|u|^p d\xi \le \int_{\mathbb{H}^+}|\nabla_H u|^p d\xi.$$

Also, the constant should be sharp.

Here we have the so-called angle function (see [22])

$$\mathscr{W}(\xi) := \sqrt{\sum_{i=1}^n\langle X_i(\xi),\nu\rangle^2+\langle Y_i(\xi),\nu\rangle^2}.$$

The following theorem approves the above conjecture:

Theorem 1 *[47] Let* $\mathbb{H}^+$ *be a half-space of the Heisenberg group* $\mathbb{H}^n$*. For* $2 \leq p < Q$ *with* $Q = 2n + 2$*, there exists some* $C > 0$ *such that for every function* $u \in C_0^\infty(\mathbb{H}^+)$ *we have*

$$C\left(\int_{\mathbb{H}^+} |u|^{p^*} d\xi\right)^{\frac{1}{p^*}}$$

$$\leq \left(\int_{\mathbb{H}^+} |\nabla_H u|^p d\xi - \left(\frac{p-1}{p}\right)^p \int_{\mathbb{H}^+} \frac{\mathscr{W}(\xi)^p}{dist(\xi, \partial\mathbb{H}^+)^p} |u|^p d\xi\right)^{\frac{1}{p}}, \tag{11}$$

where $p^* := Qp/(Q-p)$ *and* $dist(\xi, \partial\mathbb{H}^+) := \langle \xi, \nu \rangle - d$*.*

Horizontal Poincaré Inequality The "horizontal" approach also implies the following Poincaré type inequality [44] on stratified groups:

$$\frac{|N-p|}{Rp} \|f\|_{L^p(\Omega)} \leq \|\nabla_H f\|_{L^p(\Omega)}, \quad 1 < p < \infty, \tag{12}$$

for $f \in C_0^\infty(\Omega \backslash \{x' = 0\})$ and $R = \sup\limits_{x \in \Omega} |x'|$.

For example, let us consider the blow-up solutions to the p-sub-Laplacian heat equation on the stratified group, that is,

$$\begin{cases} u_t(x,t) - \mathscr{L}_p u(x,t) = f(u(x,t)), & (x,t) \in \Omega \times (0, +\infty), \\ u(x,t) = 0, & (x,t) \in \partial\Omega \times [0, +\infty), \\ u(x,0) = u_0(x) \geq 0, & x \in \overline{\Omega}, \end{cases} \tag{13}$$

where f is locally Lipschitz continuous on $\mathbb{R}$, $f(0) = 0$, and such that $f(u) > 0$ for $u > 0$. Here $\mathscr{L}_p$ is the p-sub-Laplacian. By using (12) it can be proved that nonnegative solution to (13) blows up at a finite time T^*. Thus, inequality (12) is a powerful tool proving the existence or/and nonexistence (blow-up) of the solution of subelliptic partial differential equations.

However, in general, the constant (12) is not optimal. When $p = 2$ the optimal constant can be expressed in terms of the positive ground state (if it exists on a stratified group). For general stratified groups the question about positivity of the ground state is open.

It is important to note that for the first time the horizontal Poincaré inequality (12) appears in [8, Theorem 2.12] which is valid for general vector fields (including general stratified groups setting).

Let $\Omega \subset \mathbb{G}$ be an open set and we denote its boundary by $\partial\Omega$. The notation $u \in C^1(\Omega)$ means $\nabla_H u \in C(\Omega)$. Let $\Omega \subset \mathbb{G}$ be a set supporting the divergence formula on $\mathbb{G}$. Let $u \in C_0^1(\Omega)$ and $0 < \phi \in C^2(\Omega)$. We have

$$\left|\nabla_H u - \frac{\nabla_H \phi}{\phi} u\right|^2 = |\nabla_H u|^2 - \frac{\nabla_H \phi}{\phi} \nabla_H u^2 + \frac{|\nabla_H \phi|^2}{\phi^2} u^2 \tag{14}$$

and

$$-\frac{\nabla_H \phi}{\phi} \nabla_H u^2 = -\nabla_H \cdot \left(\frac{\nabla_H \phi}{\phi} u^2\right) + \frac{\mathcal{L}\phi}{\phi} u^2 - \frac{|\nabla_H \phi|^2}{\phi^2} u^2. \tag{15}$$

These imply

$$\left|\nabla_H u - \frac{\nabla_H \phi}{\phi} u\right|^2 = |\nabla_H u|^2 - \nabla_H \cdot \left(\frac{\nabla_H \phi}{\phi} u^2\right) + \frac{\mathcal{L}\phi}{\phi} u^2, \tag{16}$$

that is,

$$\int_\Omega \left|\nabla_H u - \frac{\nabla_H \phi}{\phi} u\right|^2 dx = \int_\Omega \left(|\nabla_H u|^2 - \nabla_H \cdot \left(\frac{\nabla_H \phi}{\phi} u^2\right) + \frac{\mathcal{L}\phi}{\phi} u^2\right) dx. \tag{17}$$

Now by using the divergence formula (see, e.g. [50]) to the second term in the right hand side, we arrive at

$$0 \leq \int_\Omega \left|\nabla_H u - \frac{\nabla_H \phi}{\phi} u\right|^2 dx = \int_\Omega \left(|\nabla_H u|^2 + \frac{\mathcal{L}\phi}{\phi} |u|^2\right) dx \tag{18}$$

for any $u \in C_0^1(\Omega)$ and $0 < \phi \in C^2(\Omega)$. Here the equality case holds if and only if u is proportional to ϕ. Indeed, we have the equality case if and only if

$$0 = \left|\nabla_H u - \frac{\nabla_H \phi}{\phi} u\right|^2 = \left|\nabla_H \left(\frac{u}{\phi}\right)\right|^2 \phi^2,$$

that is, $X_k \left(\frac{u}{\phi}\right) = 0,\ k = 1, \ldots, N$. Since any left invariant vector field of $\mathbb{G}$ can be represented by Lie brackets of $\{X_1, \ldots, X_N\}$, we conclude that $\frac{u}{\phi}$ is a constant if and only if $\left|\nabla_H \left(\frac{u}{\phi}\right)\right| = 0$.

Consider the following (spectral) problem for the minus Dirichlet sub-Laplacian:

$$-\mathcal{L}\phi(\xi) = \mu\phi(\xi), \xi \in \Omega, \quad \Omega \subset \mathbb{G}, \tag{19}$$
$$\phi(\xi) = 0, \xi \in \partial\Omega.$$

Let both $\mu > 0$ and $\phi > 0$ satisfy (19), that is, $\frac{\mathscr{L}\phi}{\phi} = -\mu$. Then (18) implies the sharp Poincaré (or Steklov) inequality on a stratified group $\mathbb{G}$

$$\int_{\Omega} |u|^2 dx \leq \frac{1}{\mu} \int_{\Omega} |\nabla_H u|^2 dx.$$

Note that if Ω is an open smooth bounded set of the Heisenberg group $\mathbb{H}^n$, then there exist $\mu > 0$ and $\phi > 0$, which are the first eigenvalue and the corresponding eigenfunction of the minus Dirichlet sub-Laplacian, respectively.

One can iterate the above process to obtain higher order versions of equality (18). It is discussed for general real smooth vector fields in [41]. In turn, this equality follows the proof of the Poincaré inequality, characterization of the best constant and its existence as well as characterization of nontrivial extremizers and their existence. Now we restate some results from [41] in terms of stratified groups. We also briefly recall their proofs.

Theorem 2 *[41] Assume that $\varphi > 0$ is a positive eigenfunction of $-\mathscr{L}$ with an eigenvalue λ, that is, $-\mathscr{L}\varphi = \lambda\varphi$ in $\Omega \subset \mathbb{G}$. For every $u \in C_0^\infty(\Omega)$ the following identities are valid:*

$$\begin{aligned} &\left|\nabla_H^{2m} u\right|^2 - \lambda^{2m}|u|^2 \\ &= \sum_{j=0}^{m-1} \lambda^{2(m-1-j)} \left(\left|\mathscr{L}^{j+1}u + \lambda\mathscr{L}^j u\right|^2 + 2\lambda \left|\nabla_H \mathscr{L}^j u - \frac{\nabla_H \varphi}{\varphi}\mathscr{L}^j u\right|^2 \right) \\ &+ \sum_{j=0}^{m-1} 2\lambda^{2(m-1-j)+1} \nabla_H \cdot \left(\frac{\nabla_H \varphi}{\varphi} (\mathscr{L}^j u)^2 - \mathscr{L}^j u \nabla_H \mathscr{L}^j u \right), \quad (20) \end{aligned}$$

where $m = 1, 2, \ldots$, and

$$\begin{aligned} &\left|\nabla_H^{2m+1} u\right|^2 - \lambda^{2m+1}|u|^2 = \left|\nabla_H \mathscr{L}^m u - \frac{\nabla_H \varphi}{\varphi}\mathscr{L}^m u\right|^2 \\ &+ \sum_{j=0}^{m-1} \lambda^{2(m-j)-1} \left(\left|\mathscr{L}^{j+1}u + \lambda\mathscr{L}^j u\right|^2 + 2\lambda \left|\nabla_H \mathscr{L}^j u - \frac{\nabla_H \varphi}{\varphi}\mathscr{L}^j u\right|^2 \right) \\ &+2 \sum_{j=0}^{m-1} \lambda^{2(m-j)} \nabla_H \cdot \left(\frac{\nabla_H \varphi}{\varphi} \left(\mathscr{L}^j u\right)^2 - \mathscr{L}^j u \nabla_H \mathscr{L}^j u \right) \\ &+\nabla_H \cdot \left(\frac{\nabla_H \varphi}{\varphi} \left(\mathscr{L}^m u\right)^2 \right), \quad (21) \end{aligned}$$

where $m = 0, 1, 2, \ldots$.

Theorem 2 has the following interesting consequence in the Euclidean setting.

Theorem 3 *[41] Let $\Omega \subset \mathbb{R}^n$ be a connected domain, for which the divergence theorem is true. Then we have the remainder of the higher order Poincaré inequality*

$$\int_\Omega \left|\nabla^{2m} u\right|^2 dx - \lambda_1^{2m} \int_\Omega |u|^2 dx$$

$$= \sum_{j=0}^{m-1} \lambda_1^{2(m-1-j)} \left(\int_\Omega \left|\Delta^{j+1} u + \lambda_1 \Delta^j u\right|^2 dx + 2\lambda_1 \int_\Omega \left|\nabla \Delta^j u - \frac{\nabla u_1}{u_1} \Delta^j u\right|^2 dx \right) \geq 0, \tag{22}$$

where $m = 1, 2, \ldots,$ and

$$\int_\Omega \left|\nabla^{2m+1} u\right|^2 dx - \lambda_1^{2m+1} \int_\Omega |u|^2 dx = \int_\Omega \left|\nabla \Delta^m u - \frac{\nabla u_1}{u_1} \Delta^m u\right|^2 dx$$

$$+ \sum_{j=0}^{m-1} \lambda_1^{2(m-j)-1} \left(\int_\Omega \left|\Delta^{j+1} u + \lambda_1 \Delta^j u\right|^2 dx + 2\lambda_1 \int_\Omega \left|\nabla \Delta^j u - \frac{\nabla u_1}{u_1} \Delta^j u\right|^2 dx \right) \geq 0, \tag{23}$$

where $m = 0, 1, \ldots,$ for all $u \in C_0^\infty(\Omega)$. Here u_1 is the ground state of the Dirichlet Laplacian $-\Delta$ in Ω and λ_1 is the corresponding eigenvalue. The equality cases hold if and only if u is proportional to u_1.

Proof of Theorem 2 For $m = 1$, a direct computation yields

$$\left|\mathscr{L} u - \frac{\mathscr{L}\varphi}{\varphi} u\right|^2 = |\mathscr{L} u|^2 - 2\frac{\mathscr{L}\varphi}{\varphi} u \mathscr{L} u + \left(\frac{\mathscr{L}\varphi}{\varphi}\right)^2 |u|^2$$

$$= |\mathscr{L} u|^2 - \frac{\mathscr{L}\varphi}{\varphi}\left(\mathscr{L}\left(|u|^2\right) - 2|\nabla_H u|^2\right) + \left(\frac{\mathscr{L}\varphi}{\varphi}\right)^2 |u|^2,$$

$$-\frac{\mathscr{L}\varphi}{\varphi} \mathscr{L}\left(|u|^2\right) = 2\lambda \nabla_H \cdot (u \nabla_H u) \tag{24}$$

and

$$2\frac{\mathscr{L}\varphi}{\varphi}|\nabla_H u|^2 = 2\frac{\mathscr{L}\varphi}{\varphi}\left(-\frac{\mathscr{L}\varphi}{\varphi}|u|^2 + \left|\nabla_H u - \frac{\nabla_H \varphi}{\varphi} u\right|^2 + \nabla_H \cdot \left(\frac{\nabla_H \varphi}{\varphi}|u|^2\right)\right).$$

With $-\mathscr{L}\varphi = \lambda\varphi$ these follow that

$$\left|\mathscr{L}u - \frac{\mathscr{L}\varphi}{\varphi}u\right|^2 = |\mathscr{L}u|^2 - \left(\frac{\mathscr{L}\varphi}{\varphi}\right)^2 |u|^2$$

$$+2\frac{\mathscr{L}\varphi}{\varphi}\left(\left|\nabla_H u - \frac{\nabla_H\varphi}{\varphi}u\right|^2 + \nabla_H\cdot\left(\frac{\nabla_H\varphi}{\varphi}|u|^2\right)\right) + 2\lambda\nabla_H\cdot(u\nabla_H u)$$

$$= |\mathscr{L}u|^2 - \lambda^2|u|^2 - 2\lambda\left(\left|\nabla_H u - \frac{\nabla_H\varphi}{\varphi}u\right|^2 + \nabla_H\cdot\left(\frac{\nabla_H\varphi}{\varphi}|u|^2\right)\right) + 2\lambda\nabla_H\cdot(u\nabla_H u),$$

that is,

$$|\mathscr{L}u|^2 - \lambda^2|u|^2 = |\mathscr{L}u + \lambda u|^2 + 2\lambda\left(\left|\nabla_H u - \frac{\nabla_H\varphi}{\varphi}u\right|^2 + \nabla_H\cdot\left(\frac{\nabla_H\varphi}{\varphi}|u|^2 - u\nabla_H u\right)\right).$$

By the induction $m \Rightarrow m+1$, we establish

$$\begin{aligned}
&\left|\mathscr{L}^{m+1}u\right|^2 - \lambda^{2(m+1)}|u|^2 \\
&= \left|\mathscr{L}\mathscr{L}^m u\right|^2 - \lambda^2\left|\mathscr{L}^m u\right|^2 + \lambda^2\left(\left|\mathscr{L}^m u\right|^2 - \lambda^{2m}|u|^2\right) \\
&= \left|\mathscr{L}^{m+1}u + \lambda\mathscr{L}^m u\right|^2 + 2\lambda\left|\nabla_H\mathscr{L}^m u - \frac{\nabla_H\varphi}{\varphi}\mathscr{L}^m u\right|^2 \\
&\quad + 2\lambda\nabla_H\cdot\left(\frac{\nabla_H\varphi}{\varphi}\left(\mathscr{L}^m u\right)^2 - \mathscr{L}^m u\nabla_H\mathscr{L}^m u\right) \\
&\quad + \sum_{j=0}^{m-1}\lambda^{2(m-j)}\left(\left|\mathscr{L}^{j+1}u + \lambda\mathscr{L}^j u\right|^2 + 2\lambda\left|\nabla_H\mathscr{L}^j u - \frac{\nabla_H\varphi}{\varphi}\mathscr{L}^j u\right|^2\right) \\
&\quad + 2\sum_{j=0}^{m-1}\lambda^{2(m-j)}\left(\lambda\nabla_H\cdot\left(\frac{\nabla_H\varphi}{\varphi}\left(\mathscr{L}^j u\right)^2 - \mathscr{L}^j u\nabla_H\mathscr{L}^j u\right)\right) \\
&= \sum_{j=0}^{m}\lambda^{2(m-j)}\left(\left|\mathscr{L}^{j+1}u + \lambda\mathscr{L}^j u\right|^2 + 2\lambda\left|\nabla_H\mathscr{L}^j u - \frac{\nabla_H\varphi}{\varphi}\mathscr{L}^j u\right|^2\right) \\
&\quad + 2\sum_{j=0}^{m}\lambda^{2(m-j)+1}\nabla_H\cdot\left(\frac{\nabla_H\varphi}{\varphi}\left(\mathscr{L}^j u\right)^2 - \mathscr{L}^j u\nabla_H\mathscr{L}^j u\right).
\end{aligned}$$

It proves formula (20).

Now it remains to show relation (21). We have

$$\left|\nabla_H u - \frac{\nabla_H \varphi}{\varphi} u\right|^2 = |\nabla_H u|^2 - 2\frac{\nabla_H \varphi}{\varphi} u \nabla_H u + \frac{|\nabla_H \varphi|^2}{\varphi^2}|u|^2$$
$$= |\nabla_H u|^2 - \frac{\nabla_H \varphi}{\varphi}\nabla_H |u|^2 + \frac{|\nabla_H \varphi|^2}{\varphi^2}|u|^2, \tag{25}$$

and

$$-\frac{\nabla_H \varphi}{\varphi}\nabla_H |u|^2 = -\nabla_H \cdot \left(\frac{\nabla_H \varphi}{\varphi}|u|^2\right) + \frac{\mathcal{L}\varphi}{\varphi}|u|^2 - \frac{|\nabla_H \varphi|^2}{\varphi^2}|u|^2. \tag{26}$$

Equalities (25) and (26) imply

$$|\nabla_H u|^2 - \lambda |u|^2 = \left|\nabla_H u - \frac{\nabla_H \varphi}{\varphi} u\right|^2 + \nabla_H \cdot \left(\frac{\nabla_H \varphi}{\varphi}|u|^2\right). \tag{27}$$

It gives (21) when $m = 0$.

Now by using the scaling

$$u \mapsto \mathcal{L}^m u$$

to (27), we obtain

$$\left|\nabla_H \mathcal{L}^m u\right|^2 = \left|\nabla_H \mathcal{L}^m u - \frac{\nabla_H \varphi}{\varphi}\mathcal{L}^m u\right|^2 + \nabla_H \cdot \left(\frac{\nabla_H \varphi}{\varphi}\left(\mathcal{L}^m u\right)^2\right) + \lambda \left(\mathcal{L}^m u\right)^2. \tag{28}$$

Finally, by using formula (20), we establish

$$\begin{aligned}
&\left|\nabla_H \mathcal{L}^m u\right|^2 - \lambda^{2m+1}|u|^2 \\
&= \left|\nabla_H \mathcal{L}^m u - \frac{\nabla_H \varphi}{\varphi}\mathcal{L}^m u\right|^2 + \nabla_H \cdot \left(\frac{\nabla_H \varphi}{\varphi}\left(\mathcal{L}^m u\right)^2\right) \\
&\quad + \lambda\left(\left(\mathcal{L}^m u\right)^2 - \lambda^{2m}|u|^2\right) \\
&= \left|\nabla_H \mathcal{L}^m u - \frac{\nabla_H \varphi}{\varphi}\mathcal{L}^m u\right|^2 + \nabla_H \cdot \left(\frac{\nabla_H \varphi}{\varphi}\left(\mathcal{L}^m u\right)^2\right)
\end{aligned}$$

$$+\sum_{j=0}^{m-1}\lambda^{2(m-j)-1}\left(\left|\mathscr{L}^{j+1}u+\lambda\mathscr{L}^{j}u\right|^{2}+2\lambda\left|\nabla_{H}\mathscr{L}^{j}u-\frac{\nabla_{H}\varphi}{\varphi}\mathscr{L}^{j}u\right|^{2}\right)$$

$$+2\sum_{j=0}^{m-1}\lambda^{2(m-j)}\nabla_{H}\cdot\left(\frac{\nabla_{H}\varphi}{\varphi}\left(\mathscr{L}^{j}u\right)^{2}\right).$$

□

Picone type representation formula:

Theorem 4 *[41] For all $u\in C^{1}(\Omega)$ and $\varphi\in C^{2}(\Omega)$ with $\varphi>0$, we have*

$$|\nabla_{H}u|^{p_m}+\left(\frac{\mathscr{L}\varphi}{\varphi}+\sigma_m\right)u^{p_m}=\sum_{j=1}^{m-1}\left||\nabla_{H}\left(u^{p_{m-j-1}}\right)|^{p_j}-2^{p_j-1}u^{p_{m-1}}\right|^{2}$$

$$+\left|\nabla_{H}\left(u^{p_{m-1}}\right)-\frac{\nabla_{H}\varphi}{\varphi}u^{p_{m-1}}\right|^{2}+\nabla_{H}\cdot\left(\frac{\nabla_{H}\varphi}{\varphi}u^{p_m}\right)\ \text{in}\ \Omega\subset\mathbb{G},\qquad(29)$$

where m is a nonnegative integer. Here $p_m=2^{m}$, $m\geq 0$, and

$$\sigma_m=\frac{1}{4}\sum_{j=1}^{m-1}4^{p_j},\ m\geq 1.$$

Theorem 4 has the following interesting consequence in the Euclidean setting.

Theorem 5 ([41]) *Let $\Omega\subset\mathbb{R}^{n}$ be a connected domain, for which the divergence theorem is true. For all $u\in C_0^1(\Omega)$, we have*

$$\int_{\Omega}|\nabla u|^{p_m}dx-(\lambda_1-\sigma_m)\int_{\Omega}|u|^{p_m}dx$$

$$=\sum_{j=1}^{m-1}\int_{\Omega}\left||\nabla\left(u^{p_{m-j-1}}\right)|^{p_j}-2^{p_j-1}u^{p_{m-1}}\right|^{2}dx+\int_{\Omega}\left|\nabla\left(u^{p_{m-1}}\right)-\frac{\nabla u_1}{u_1}u^{p_{m-1}}\right|^{2}dx\geq 0,\qquad(30)$$

where $\sigma_m=\frac{1}{4}\sum_{j=1}^{m-1}4^{p_j}$, $m\in\mathbb{N}$, $p_j=2^{j}$, u_1 is the ground state of the minus Dirichlet Laplacian $-\Delta$ in Ω and λ_1 is the corresponding eigenvalue.

Note that, for $m=1$ the sigma notation term in (30) disappears as usual (since the lower index is greater than the upper one). It is important to observe that (30) can be considered as a remainder term for some L^p-Poincaré inequalities (which are also commonly called as L^p-Friedrichs inequalities). In general, determining the sharp constant in the L^p-Friedrichs inequality is an open problem.

Proof of Theorem 4 When $m = 1$, we have $p_1 = 2$, $\sigma_1 = 0$, and

$$|\nabla_H u|^2 + \frac{\mathcal{L}\varphi}{\varphi}u^2 = \left|\nabla_H u - \frac{\nabla_H \varphi}{\varphi}u\right|^2 + \nabla_H \cdot \left(\frac{\nabla_H \varphi}{\varphi}u\right).$$

When $m = 2$, we have $p_2 = 4$, $\sigma_2 = \frac{1}{4}4^{p_1} = 4$, and

$$|\nabla_H u|^4 + \left(\frac{\mathcal{L}\varphi}{\varphi} + 4\right)u^4$$
$$= ||\nabla_H u|^2 - 2|u|^2|^2 + \left|\nabla_H\left(|u|^2\right) - \frac{\nabla_H \varphi}{\varphi}|u|^2\right|^2 + \nabla_H \cdot \left(\frac{\nabla_H \varphi}{\varphi}u^4\right).$$

In order to use the induction process, we observe

$$\left||\nabla_H u|^{p_m} - 2^{p_m - 1}u^{p_m}\right|^2 = |\nabla_H u|^{p_{m+1}} - 2^{p_m}u^{p_m}|\nabla_H u|^{p_m} + \frac{1}{4}4^{p_m}u^{p_{m+1}}.$$

Plugging in $|u|^2$ instead of u in (29) we get

$$\left|\nabla_H\left(|u|^2\right)\right|^{p_m} + \left(\frac{\mathcal{L}\varphi}{\varphi} + \sigma_m\right)\left(|u|^2\right)^{p_m}$$
$$= \sum_{j=1}^{m-1}\left|\left|\nabla_H\left(|u|^2\right)^{p_{m-j-1}}\right|^{p_j} - 2^{p_j - 1}\left(|u|^2\right)^{p_{m-1}}\right|^2$$
$$+ \left|\nabla_H\left(|u|^2\right)^{p_{m-1}} - \frac{\nabla_H \varphi}{\varphi}\left(|u|^2\right)^{p_{m-1}}\right|^2 + \nabla_H \cdot \left(\frac{\nabla_H \varphi}{\varphi}\left(|u|^2\right)^{p_m}\right),$$

where

$$\left|\nabla_H\left(|u|^2\right)\right|^{p_m} + \left(\frac{\mathcal{L}\varphi}{\varphi} + \sigma_m\right)\left(|u|^2\right)^{p_m} = 2^{p_m}u^{p_m}|\nabla_H u|^{p_m} + \left(\frac{\mathcal{L}\varphi}{\varphi} + \sigma_m\right)u^{p_{m+1}}$$

and

$$\sum_{j=1}^{m-1}\left|\left|\nabla_H\left(|u|^2\right)^{p_{m-j-1}}\right|^{p_j} - 2^{p_j - 1}\left(|u|^2\right)^{p_{m-1}}\right|^2$$
$$+ \left|\nabla_H\left(|u|^2\right)^{p_{m-1}} - \frac{\nabla_H \varphi}{\varphi}\left(|u|^2\right)^{p_{m-1}}\right|^2 + \nabla_H \cdot \left(\frac{\nabla_H \varphi}{\varphi}\left(|u|^2\right)^{p_m}\right)$$

$$= \sum_{j=1}^{m-1} \left| |\nabla_H \left(u^{p_{m-j}}\right)|^{p_j} - 2^{p_j-1} u^{p_m} \right|^2$$

$$+ \left| \nabla_H \left(u^{p_m}\right) - \frac{\nabla_H \varphi}{\varphi} u^{p_m} \right|^2 + \nabla_H \cdot \left(\frac{\nabla_H \varphi}{\varphi} u^{p_{m+1}} \right).$$

It yields

$$|\nabla_H u|^{p_{m+1}} = 2^{p_m} u^{p_m} |\nabla_H u|^{p_m} + \left| |\nabla_H u|^{p_m} - 2^{p_m-1} u^{p_m} \right|^2 - \frac{1}{4} 4^{p_m} u^{p_{m+1}}$$

$$= - \left(\frac{\mathcal{L}\varphi}{\varphi} + \sigma_m + \frac{1}{4} 4^{p_m} \right) u^{p_{m+1}} + \left| |\nabla_H u|^{p_m} - 2^{p_m-1} u^{p_m} \right|^2$$

$$+ \sum_{j=1}^{m-1} \left| |\nabla_H \left(u^{p_{m-j}}\right)|^{p_j} - 2^{p_j-1} u^{p_m} \right|^2$$

$$+ \left| \nabla_H \left(u^{p_m}\right) - \frac{\nabla_H \varphi}{\varphi} u^{p_m} \right|^2 + \nabla_H \cdot \left(\frac{\nabla_H \varphi}{\varphi} u^{p_{m+1}} \right)$$

$$= - \left(\frac{\mathcal{L}\varphi}{\varphi} + \sigma_{m+1} \right) u^{p_{m+1}} + \left| |\nabla_H u|^{p_m} - 2^{p_m-1} u^{p_m} \right|^2$$

$$+ \sum_{j=1}^{m-1} \left| |\nabla_H \left(u^{p_{m-j}}\right)|^{p_j} - 2^{p_j-1} u^{p_m} \right|^2$$

$$+ \left| \nabla_H \left(u^{p_m}\right) - \frac{\nabla_H \varphi}{\varphi} u^{p_m} \right|^2 + \nabla_H \cdot \left(\frac{\nabla_H \varphi}{\varphi} u^{p_{m+1}} \right).$$

□

Finally, note that in addition to the above discussed two approaches (the fundamental solution and horizontal estimate approach) there is another interesting approach for Hardy-type inequalities on Heisenberg group so called the *CC-distance approach*. For discussions in this direction we suggest [3, 4, 6, 19, 32] and [55] as well as references therein.

3 Hardy Type Inequalities on Homogeneous Groups

By the definition, there is no homogeneous (horizontal) gradient on non-stratified graded groups, so there is no horizontal estimates. A non-stratified graded group $\mathbb{G}$ may not have a homogeneous sub-Laplacian, but it always has so-called Rockland operators, which are left-invariant homogeneous subelliptic differential operators on $\mathbb{G}$. Therefore, the fundamental solution approach can be applied to general graded groups.

Beyond Graded Groups There is no invariant homogeneous subelliptic differential operator on non-graded homogeneous groups; in particular, no fundamental solution.

Question is even:

- How to formulate Hardy's inequality on non-graded homogeneous groups?

A systematic analysis towards an answer to this question was presented recently in the book form [50]. One of the key ideas was consistently working with the quasi-radial derivative operator $\mathscr{R}_{|x|} := \frac{d}{d|x|}$ and with the Euler operator $\mathbb{E} := |x|\frac{d}{d|x|}$ to obtain homogeneous group analogues of the Hardy type inequalities.

Actually, it can be shown that any (connected, simply connected) nilpotent Lie group is some $\mathbb{R}^n$ with a polynomial group law: $\mathbb{R}^n$ with linear group law, $\mathbb{H}^n$ with quadratic group law, etc. So we can identify $\mathbb{G}$ with $\mathbb{R}^n$ (topologically).

Definition 1 If a Lie group (on $\mathbb{R}^n$) $\mathbb{G}$ has a property that there exist n-real numbers $\nu_1, \ldots, \nu_n$ such that the dilation

$$D_\lambda(x) := (\lambda^{\nu_1} x_1, \ldots, \lambda^{\nu_n} x_n), \quad D_\lambda : \mathbb{R}^n \to \mathbb{R}^n,$$

is an automorphism of the group $\mathbb{G}$ for each $\lambda > 0$, then it is called a homogeneous group.

Let us fix a basis $\{X_1, \ldots, X_n\}$ of the Lie algebra $\mathfrak{g}$ of the homogeneous group $\mathbb{G}$ such that X_k is homogeneous of degree ν_k.

Then the homogeneous dimension of $\mathbb{G}$ is

$$Q = \nu_1 + \cdots + \nu_n.$$

A class of homogeneous groups is one of most general subclasses of nilpotent Lie groups, that is, the class of homogeneous groups gives almost the class of all nilpotent Lie groups but is not equal to it. In 1970, Dyer gave an example of a (nine-dimensional) nilpotent Lie group that does not allow for any family of dilations [12].

Special cases of the homogeneous groups:

- the Euclidean group $(\mathbb{R}^n; +)$,
- H-type groups,
- stratified groups,
- graded (Lie) groups.

We also recall the standard Lebesgue measure dx on $\mathbb{R}^n$ is the Haar measure for $\mathbb{G}$. It makes the class of homogeneous groups convenient for analysis. One also can assume that the origin 0 of $\mathbb{R}^n$ is the identity of $\mathbb{G}$, If it is not, then by using a smooth diffeomorphism one can obtain a new (isomorphic) homogeneous group which has the identity 0. For further discussions in this direction we refer to a recent open access book [50].

Let $\mathbb{G}$ be a homogeneous group of homogeneous dimension Q. Then for all $f \in C_0^\infty(\mathbb{G}\backslash\{0\})$ and for any homogeneous quasi-norm $|\cdot|$, we have the following

Hardy inequality on homogeneous groups [43]:

$$\left\| \frac{f}{|x|} \right\|_{L^p(\mathbb{G})} \leq \frac{p}{Q-p} \|\mathcal{R}_{|x|} f\|_{L^p(\mathbb{G})}, \quad 1 < p < Q,$$

for all $f \in C_0^\infty(\mathbb{G}\backslash\{0\})$, where $\mathcal{R}_{|x|} := \frac{d}{d|x|}$. Moreover, the constant above is sharp and is attained if and only if $f = 0$.

The main idea to prove these type of inequalities is consistently to work with radial derivative $\mathcal{R}_{|x|} := \frac{d}{d|x|}$ and with the Euler operator $\mathbb{E} := |x|\frac{d}{d|x|}$, so to obtain relations on homogeneous groups in terms of $\mathcal{R}_{|x|}$ or/and $\mathbb{E}$. These yield many inequalities: Hardy, Rellich, Caffarelli-Kohn-Nirenberg, Sobolev type, ..., with best constants for any homogeneous quasi-norm.

Theorem 6 ([48]) *Let $\mathbb{G}$ be a homogeneous group of homogeneous dimension $Q \geq 3$. Then for every complex-valued function $f \in C_0^\infty(\mathbb{G}\backslash\{0\})$ and any homogeneous quasi-norm $|\cdot|$ on $\mathbb{G}$ we have*

$$\left\| \frac{1}{|x|^\alpha} \mathcal{R} f \right\|^2_{L^2(\mathbb{G})} - \left(\frac{Q-2}{2} - \alpha \right)^2 \left\| \frac{f}{|x|^{\alpha+1}} \right\|^2_{L^2(\mathbb{G})} = \left\| \frac{1}{|x|^\alpha} \mathcal{R} f + \frac{Q-2-2\alpha}{2|x|^{\alpha+1}} f \right\|^2_{L^2(\mathbb{G})} \tag{31}$$

for all $\alpha \in \mathbb{R}$.

As a consequence of (31), we obtain the weighted Hardy inequality on the homogeneous group $\mathbb{G}$, with the sharp constant:

For all complex-valued functions $f \in C_0^\infty(\mathbb{G}\backslash\{0\})$ we have

$$\frac{|Q-2-2\alpha|}{2} \left\| \frac{f}{|x|^{\alpha+1}} \right\|_{L^2(\mathbb{G})} \leq \left\| \frac{1}{|x|^\alpha} \mathcal{R} f \right\|_{L^2(\mathbb{G})}, \quad \forall \alpha \in \mathbb{R}. \tag{32}$$

If $\alpha \neq \frac{Q-2}{2}$, then constant in (32) is sharp for any homogeneous quasi-norm $|\cdot|$ on $\mathbb{G}$, and inequality (32) is attained if and only if $f = 0$.

In the Euclidean case $\mathbb{G} = (\mathbb{R}^n, +)$, $n \geq 3$, we have $Q = n$, so for any homogeneous quasi-norm $|\cdot|$ on $\mathbb{R}^n$, (32) implies a new inequality with the optimal constant:

$$\frac{|n-2-2\alpha|}{2} \left\| \frac{f}{|x|^{\alpha+1}} \right\|_{L^2(\mathbb{R}^n)} \leq \left\| \frac{1}{|x|^\alpha} \frac{df}{d|x|} \right\|_{L^2(\mathbb{R}^n)}, \tag{33}$$

for all $\alpha \in \mathbb{R}$. We observe that this inequality holds for any homogeneous quasi-norm on $\mathbb{R}^n$.

Note that the constant in (33) is optimal for any homogeneous quasi-norm.

For the standard Euclidean distance $\|x\| = \sqrt{x_1^2 + \ldots + x_n^2}$, by using Schwarz's inequality, this implies (with the optimal constant):

$$\frac{|n-2-2\alpha|}{2}\left\|\frac{f}{\|x\|^{\alpha+1}}\right\|_{L^2(\mathbb{R}^n)} \le \left\|\frac{1}{\|x\|^{\alpha}}\nabla f\right\|_{L^2(\mathbb{R}^n)}, \tag{34}$$

for all $f \in C_0^\infty(\mathbb{R}^n\backslash\{0\})$.

With $\alpha = 0$ we have

$$\left\|\frac{f}{\|x\|}\right\|_{L^2(\mathbb{R}^n)} \le \frac{2}{n-2}\|\nabla f\|_{L^2(\mathbb{R}^n)}, \quad n \ge 3.$$

Moreover, these also can be extended to the weighted L^p-Hardy inequalities [46].

Let $\mathbb{G}$ be a homogeneous group of homogeneous dimension Q and let $\alpha \in \mathbb{R}$. Then for all complex-valued functions $f \in C_0^\infty(\mathbb{G}\backslash\{0\})$, and any homogeneous quasi-norm $|\cdot|$ on $\mathbb{G}$ for $\alpha p \neq Q$ we have

$$\left\|\frac{f}{|x|^{\alpha}}\right\|_{L^p(\mathbb{G})} \le \left|\frac{p}{Q-\alpha p}\right| \left\|\frac{1}{|x|^{\alpha}}\mathbb{E}f\right\|_{L^p(\mathbb{G})}, \quad 1 < p < \infty.$$

If $\alpha p \neq Q$ then the constant $\left|\frac{p}{Q-\alpha p}\right|$ is sharp.

For $\alpha p = Q$ we have the critical case (believe to be already new in $\mathbb{R}^n$)

$$\left\|\frac{f}{|x|^{\frac{Q}{p}}}\right\|_{L^p(\mathbb{G})} \le p\left\|\frac{\log|x|}{|x|^{\frac{Q}{p}}}\mathbb{E}f\right\|_{L^p(\mathbb{G})},$$

where the constant p is sharp.

Rellich Inequalities on Homogeneous Groups Let $\mathbb{G}$ be a homogeneous group of homogeneous dimension $Q \ge 5$. Let $|\cdot|$ be any homogeneous quasi-norm on $\mathbb{G}$.

Then for every $f \in C_0^\infty(\mathbb{G}\backslash\{0\})$ [42]:

$$\begin{aligned}\left\|\mathscr{R}^2 f + \frac{Q-1}{|x|}\mathscr{R}f + \frac{Q(Q-4)}{4|x|^2}f\right\|^2_{L^2(\mathbb{G})} &+ \frac{Q(Q-4)}{2}\left\|\frac{1}{|x|}\mathscr{R}f + \frac{Q-4}{2|x|^2}f\right\|^2_{L^2(\mathbb{G})} \\ &= \left\|\mathscr{R}^2 f + \frac{Q-1}{|x|}\mathscr{R}f\right\|^2_{L^2(\mathbb{G})} - \left(\frac{Q(Q-4)}{4}\right)^2\left\|\frac{f}{|x|^2}\right\|^2_{L^2(\mathbb{G})},\end{aligned}$$

which implies the (quasi-radial) Rellich inequality

$$\left\| \frac{f}{|x|^2} \right\|_{L^2(\mathbb{G})} \leq \frac{4}{Q(Q-4)} \left\| \mathcal{R}^2 f + \frac{Q-1}{|x|} \mathcal{R} f \right\|_{L^2(\mathbb{G})}, \quad Q \geq 5.$$

The constant $\frac{4}{Q(Q-4)}$ is sharp and it is attained if and only if $f = 0$.

After our paper [42] the Rellich inequality was extended (see, [37] and [38]) to the range $1 < p < Q/2$:

$$\left\| \frac{f}{|x|^2} \right\|_{L^p(\mathbb{G})} \leq \frac{p^2}{Q(p-1)(Q-2p)} \left\| \mathcal{R}_{|x|}^2 f + \frac{Q-1}{|x|} \mathcal{R}_{|x|} f \right\|_{L^p(\mathbb{G})},$$

for all $f \in C_0^\infty(\mathbb{G}\backslash\{0\})$. The constant is sharp and it is attained if and only if $f = 0$.

One can also obtain the Sobolev type inequalities through identities:

Theorem 7 ([45]) *Let $\mathbb{G}$ be a homogeneous group of homogeneous dimension Q. Then for all $f \in C_0^\infty(\mathbb{G}\backslash\{0\})$ and $1 < p < Q$*

$$\left\| \frac{p}{Q} \mathbb{E} f \right\|_{L^p(\mathbb{G})}^p - \|f\|_{L^p(\mathbb{G})}^p = p \int_{\mathbb{G}} I_p \left(f, -\frac{p}{Q} \mathbb{E} f \right) \left| f + \frac{p}{Q} \mathbb{E} f \right|^2 dx, \tag{35}$$

where $\mathbb{E} := |x| \frac{d}{d|x|}$ is the Euler operator, and I_p is given by

$$I_p(h, g) = (p-1) \int_0^1 |\xi h + (1-\xi) g|^{p-2} \xi d\xi.$$

The identity (35) implies, for all $f \in C_0^\infty(\mathbb{G}\backslash\{0\})$, the L^p-Sobolev type inequality on $\mathbb{G}$:

$$\|f\|_{L^p(\mathbb{G})} \leq \frac{p}{Q} \|\mathbb{E} f\|_{L^p(\mathbb{G})}, \quad 1 < p < \infty,$$

where the constant $\frac{p}{Q}$ is sharp.

In the Euclidean case $\mathbb{G} = (\mathbb{R}^n; +)$, we have $Q = n$, so for $n \geq 1$ it implies the Sobolev type inequality:

$$\|f\|_{L^p(\mathbb{R}^n)} \leq \frac{p}{n} \|x \cdot \nabla f\|_{L^p(\mathbb{R}^n)}, \quad 1 < p < \infty.$$

Above ideas can be extended to other weighted identities on $L^2(\mathbb{G})$: e.g. for all $k \in \mathbb{N}$ and $\alpha \in \mathbb{R}$, for every $f \in C_0^\infty(\mathbb{G}\backslash\{0\}), \alpha \in \mathbb{R}$, and any homogeneous quasi-norm $|\cdot|$ on $\mathbb{G}$ we have:

$$\left\|\frac{1}{|x|^{\alpha}}\mathcal{R}_{|x|}^{k}f\right\|_{L^2(\mathbb{G})}^2=\left[\prod_{j=0}^{k-1}\left(\frac{Q-2}{2}-(\alpha+j)\right)^2\right]\left\|\frac{f}{|x|^{k+\alpha}}\right\|_{L^2(\mathbb{G})}^2$$

$$+\sum_{l=1}^{k-1}\left[\prod_{j=0}^{l-1}\left(\frac{Q-2}{2}-(\alpha+j)\right)^2\right]\left\|\frac{1}{|x|^{l+\alpha}}\mathcal{R}_{|x|}^{k-l}f+\frac{Q-2(l+1+\alpha)}{2|x|^{l+1+\alpha}}\mathcal{R}_{|x|}^{k-l-1}f\right\|_{L^2(\mathbb{G})}^2$$

$$+\left\|\frac{1}{|x|^{\alpha}}\mathcal{R}_{|x|}^{k}f+\frac{Q-2-2\alpha}{2|x|^{1+\alpha}}\mathcal{R}_{|x|}^{k-1}f\right\|_{L^2(\mathbb{G})}^2.$$

$$\left\|\frac{1}{|x|^{\alpha}}\mathbb{E}f\right\|_{L^2(\mathbb{G})}^2=\left(\frac{Q}{2}-\alpha\right)^2\left\|\frac{f}{|x|^{\alpha}}\right\|_{L^2(\mathbb{G})}^2+\left\|\frac{1}{|x|^{\alpha}}\mathbb{E}f+\frac{Q-2\alpha}{2|x|^{\alpha}}f\right\|_{L^2(\mathbb{G})}^2.$$

The following lemma allows to obtain fractional orders of previous inequalities.

Lemma 1 *The operator $\mathbb{A}=\mathbb{E}\mathbb{E}^*$ is Komatsu-non-negative in $L^2(\mathbb{G})$:*

$$\|(\lambda+\mathbb{A})^{-1}\|_{L^2(\mathbb{G})\to L^2(\mathbb{G})}\le\lambda^{-1},\ \forall\lambda>0. \tag{36}$$

Since $\mathbb{A}$ is Komatsu-non-negative, we can define fractional powers of the operator $\mathbb{A}$ as in [35] and we denote

$$|\mathbb{E}|^{\beta}=\mathbb{A}^{\frac{\beta}{2}},\qquad \beta\in\mathbb{C}.$$

For example, we have the following Hardy inequality with the fractional Euler operator.

Theorem 8 ([49]) *Let $\mathbb{G}$ be a homogeneous group of homogeneous dimension Q, $\beta\in\mathbb{C}_+$ and let $k>\frac{\mathrm{Re}\beta}{2}$ be a positive integer. Then for all complex-valued functions $f\in C_0^{\infty}(\mathbb{G}\backslash\{0\})$ we have*

$$\|f\|_{L^2(\mathbb{G})}\le C\left(k-\frac{\beta}{2},k\right)\left(\frac{2}{Q}\right)^{\mathrm{Re}\beta}\left\||\mathbb{E}|^{\beta}f\right\|_{L^2(\mathbb{G})},\ Q\ge 1, \tag{37}$$

where

$$C(\beta,k)=\frac{\Gamma(k+1)}{|\Gamma(\beta)\Gamma(k-\beta)|}\frac{2^{k-\mathrm{Re}\beta}}{\mathrm{Re}\beta(k-\mathrm{Re}\beta)}. \tag{38}$$

Stein-Weiss Type Inequalities on Homogeneous Groups Let $0<\lambda<Q$ and $\frac{1}{q}=\frac{1}{p}+\frac{\lambda}{Q}-1$ with $1<p<q<\infty$. Then the following inequality is valid on $\mathbb{G}$

of homogeneous dimension Q:

$$\left|\int_{\mathbb{G}}\int_{\mathbb{G}}\frac{f(y)h(x)}{|y^{-1}x|^{\lambda}}dxdy\right| \leq C\|f\|_{L^{p}(\mathbb{G})}\|h\|_{L^{q'}(\mathbb{G})}, \tag{39}$$

for all $f \in L^{p}(\mathbb{G})$ and $h \in L^{q'}(\mathbb{G})$. The Euclidean version of this inequality is called the Hardy-Littlewood-Sobolev (HLS) inequality. In 1958, Stein and Weiss established a two-weight extension of the (Euclidean) HLS inequality [54] (see also [21]). Nowadays, the two-weight extension of the HLS inequality is called the Stein-Weiss inequality. Note that Folland and Stein obtained the HLS inequality on the Heisenberg groups [17]. On stratified groups a version of the Stein-Weiss inequality was obtained in [28]. The Stein-Weiss inequality was extended to graded groups in [51]:

Theorem 9 ([51]) *Let $\mathbb{G}$ be a graded group of homogeneous dimension Q and let $|\cdot|$ be an arbitrary homogeneous quasi-norm. Let $1 < p, q < \infty$, $0 \leq a < Q/p$ and $0 \leq b < Q/q$. Let $0 < \lambda < Q$, $0 \leq \alpha < a + Q/p'$ and $0 \leq \beta \leq b$ be such that*

$$(Q - ap)/(pQ) + (Q - q(b - \beta))/(qQ) + (\alpha + \lambda)/Q = 2$$

and $\alpha + \lambda \leq Q$, where $1/p + 1/p' = 1$. Then for all $f \in \dot{L}^{p}_{a}(\mathbb{G})$ and $h \in \dot{L}^{q}_{b}(\mathbb{G})$ we have

$$\left|\int_{\mathbb{G}}\int_{\mathbb{G}}\frac{\overline{f(x)}h(y)}{|x|^{\alpha}|y^{-1}x|^{\lambda}|y|^{\beta}}dxdy\right| \leq C\|f\|_{\dot{L}^{p}_{a}(\mathbb{G})}\|h\|_{\dot{L}^{q}_{b}(\mathbb{G})} \tag{40}$$

where C is a positive constant independent of f and h. Here $\dot{L}^{p}_{a}(\mathbb{G})$ stands for a homogeneous Sobolev space of order a over L^{p} on the graded Lie group $\mathbb{G}$.

In [29], the Stein-Weiss inequality was extended to general homogeneous groups.

Theorem 10 ([29]) *Let $|\cdot|$ be an arbitrary homogeneous quasi-norm on $\mathbb{G}$ of homogeneous dimension Q. Let*

$$0 < \lambda < Q, \quad 1 < p \leq q < \infty,$$

$$\alpha < \frac{Q}{p'}, \quad \beta < \frac{Q}{q}, \quad \alpha + \beta \geq 0,$$

$$\frac{1}{q} = \frac{1}{p} + \frac{\alpha + \beta + \lambda}{Q} - 1, \quad \frac{1}{p} + \frac{1}{p'} = 1, \quad \frac{1}{q} + \frac{1}{q'} = 1.$$

Then for all $f \in L^{p}(\mathbb{G})$ and $h \in L^{q'}(\mathbb{G})$ we have

$$\left|\int_{\mathbb{G}}\frac{f(y)h(x)}{|x|^{\beta}|y^{-1}x|^{\lambda}|y|^{\alpha}}dxdy\right| \leq C\|f\|_{L^{p}(\mathbb{G})}\|h\|_{L^{q'}(\mathbb{G})}, \tag{41}$$

where C is a positive constant independent of f and h.

The reverse Stein-Weiss inequality also holds on $\mathbb{G}$.

Theorem 11 ([30]) *Let $\mathbb{G}$ be a homogeneous group of homogeneous dimension Q and let $|\cdot|$ be an arbitrary homogeneous quasi-norm on $\mathbb{G}$. Let*

$$0 \leq \alpha < -\frac{Q}{q} \quad 0 \leq \beta < -\frac{Q}{p'}, \quad q, p' \in (0,1),$$

$$\frac{1}{q'} + \frac{1}{p} = \frac{\alpha+\beta+\lambda}{Q} + 2, \; 0 < \lambda < \infty, \; \frac{1}{p} + \frac{1}{p'} = 1, \; \frac{1}{q} + \frac{1}{q'} = 1.$$

Then for all non-negative functions $f \in L^{q'}(\mathbb{G})$ and $h \in L^p(\mathbb{G})$ we have

$$\int_{\mathbb{G}} \int_{\mathbb{G}} |x|^{\alpha} |y^{-1}x|^{\lambda} f(x) h(y) |y|^{\beta} dx dy \geq C \|f\|_{L^{q'}(\mathbb{G})} \|h\|_{L^p(\mathbb{G})}, \tag{42}$$

where C is a positive constant independent of f and h.

In the Euclidean (Abelian) case $\mathbb{G} = (\mathbb{R}^N, +)$, hence $Q = N$ and $|\cdot|$ can be any homogeneous quasi-norm on $\mathbb{R}^N$, particularly with the standard Euclidean distance, that is, with $|\cdot| = \|\cdot\|_E$ it was studied in [9].

An improved version of the reverse Stein-Weiss inequality can be also proved [30]: Let $|\cdot|$ be an arbitrary homogeneous quasi-norm on $\mathbb{G}$ of homogeneous dimension Q. Let $\frac{1}{q'} + \frac{1}{p} = \frac{\alpha+\beta+\lambda}{Q} + 2$ with $p, q' \in (0,1)$, where $0 < \lambda < \infty$, $\frac{1}{p} + \frac{1}{p'} = 1$ and $\frac{1}{q} + \frac{1}{q'} = 1$. Then for all non-negative functions $f \in L^{q'}(\mathbb{G})$ and $h \in L^p(\mathbb{G})$, inequality (42) holds, that is,

$$\int_{\mathbb{G}} \int_{\mathbb{G}} |x|^{\alpha} |y^{-1}x|^{\lambda} f(x) h(y) |y|^{\beta} dx dy \geq C \|f\|_{L^{q'}(\mathbb{G})} \|h\|_{L^p(\mathbb{G})}$$

if $0 \leq \alpha < -\frac{Q}{q}$ or $0 \leq \beta < -\frac{Q}{p'}$.

References

1. Adimurthi, A. Sekar, Role of the fundamental solution in Hardy-Sobolev-type inequalities. Proc. R. Soc. Edinb. Sect. A **136**(6), 1111–1130 (2006)
2. N. Badiale, G. Tarantello, A Sobolev-Hardy inequality with applications to a nonlinear elliptic equation arising in astrophysics. Arch. Ration. Mech. Anal. **163**, 259–293 (2002)
3. H. Bahouri, J.-Y. Chemin, I. Gallagher, Refined Hardy inequalities. Sup. Pisa Cl. Sci **5**, 375–391 (2006)
4. H. Bahouri, J.Y. Chemin, C.J. Xu, Trace and trace lifting theorems in weighted Sobolev spaces. J. Inst. Math. Jussieu **4**(4), 509–552 (2007)
5. A.A. Balinsky, W.D. Evans, R.T. Lewis, *The Analysis and Geometry of Hardy's Inequality*. Universitext (Springer, Cham, 2015)

6. L. Capogna, D. Danielli, S.D. Pauls, J.T. Tyson, *An Introduction to the Heisenberg Group and the Sub-Riemannian Isoperimetric Problem*. Progress in Mathematics, vol. 314 (Birkhäuser, Basel, 2007)
7. L. D'Ambrosio, Some Hardy inequalities on the Heisenberg group. Differ. Equ. **40**(4), 552–564 (2004)
8. L. D'Ambrosio, Hardy-type inequalities related to degenerate elliptic differential operators. Ann. Sc. Norm. Super. Pisa Cl. Sci. (5) **4**(3), 451–486 (2005)
9. L. Chen, G. Lu, C. Tao. Reverse Stein-Weiss inequalities on the upper half space and the existence of their extremals. Adv. Nonlinear Stud. **19**(3), 475–494 (2019)
10. D. Danielli, N. Garofalo, N. C. Phuc, Hardy-Sobolev type inequalities with sharp constants in Carnot-Carathéodory spaces. Potential Anal. **34**, 223–242 (2011)
11. E.B. Davies, The Maz'ya anniversary Collection. OT Adv. Appl. 55–67 (1999)
12. J.L. Dyer, A nilpotent Lie algebra with nilpotent automorphism group. Bull. Am. Math. Soc. **76**, 52–56 (1970)
13. S. Filippas, V. Maz'ya, A. Tertikas, Critical Hardy-Sobolev inequalities. J. Math. Pures Appl. **87**, 37–56 (2007)
14. V. Fischer, M. Ruzhansky, *Quantization on Nilpotent Lie Groups*. Progress in Mathematics, vol. 314 (Birkhäuser, Basel, 2016)
15. G. Folland, Subelliptic estimates and function spaces on nilpotent Lie groups. Arkiv Mat. **13**(2), 161–207 (1975)
16. G. Folland, The Heisenberg group and its relatives in the work of Elias M. Stein. J. Geom. Anal. **13**(2), 161–207 (2019)
17. G.B. Folland, E.M. Stein, Estimates for the $\overline{\partial_b}$ complex and analysis on the Heisenberg group. Commun. Pure Appl. Math. **27**, 429–522 (1974)
18. G.B. Folland, E.M. Stein, *Hardy Spaces on Homogeneous Groups*. Mathematical Notes, vol. 28 (Princeton University Press, Princeton; University of Tokyo Press, Tokyo, 1982)
19. V. Franceschi, D. Prandi, Hardy-type inequalities for the Carnot-Carathéodory distance in the Heisenberg group. J. Geom. Anal. (2020). https://doi.org/10.1007/s12220-020-00360-y. arXiv:1903.08486 (2019)
20. R.L. Frank, M. Loss, The Hardy-Sobolev-Maz'ya inequalities for arbitrary domains. J. Math. Pures Appl. **97**, 39–54 (2012)
21. R.L. Frank, E.H Lieb, Sharp constants in several inequalities on the Heisenberg group. Ann. Math. **176**, 349–381 (2012)
22. N. Garofalo, Geometric second derivative estimates in Carnot groups and convexity. Manuscr. Math. **126**, 353–373 (2008)
23. N. Garofalo, E. Lanconelli, Frequency functions on the Heisenberg group, the uncertainty principle and unique continuation. Ann. Inst. Fourier (Grenoble) **40**, 313–356 (1990)
24. J.A. Goldstein, I. Kombe, The Hardy inequality and non-linear parabolic equations on Carnot groups. Nonlinear Anal. **69**(12), 4643–4653 (2008)
25. G.H. Hardy, Notes on some points in the integral calculus (51). Messenger Math. **48**, 107–112 (1918)
26. L. Hörmander, Hypoelliptic second order differential equations. Acta Math. **119**, 147–171 (1967)
27. Y. Jin, S. Shen, Weighted Hardy and Rellich inequality on Carnot groups. Arch. Math. (Basel) **96**(3), 263–271 (2011)
28. V. Guliyev, R. Mustafayev, A. Serbetci, Stein-Weiss inequalities for the fractional integral operators in Carnot groups and applications. Complex Var. Elliptic Equ. **55**(8–10), 847–863 (2010)
29. A. Kassymov, M. Ruzhansky, D. Suragan, Hardy-Littlewood-Sobolev and Stein-Weiss inequalities on homogeneous Lie groups. Integr. Transforms Special Funct. **30**, 643–655 (2019)
30. A. Kassymov, M. Ruzhansky, D. Suragan, Reverse Stein-Weiss, Hardy-Littlewood-Sobolev, Hardy, Sobolev and Caffarelli-Kohn-Nirenberg inequalities on homogeneous groups (2019). arXiv:1912.01460

31. S. Larson, Geometric Hardy inequalities for the sub-elliptic Laplacian on convex domain in the Heisenberg group. Bull. Math. Sci. **6**, 335–352 (2016)
32. J. Lehrbäck, Hardy inequalities and Assouad dimensions. J. Anal. Math. **131**(1), 367–398 (2017)
33. J. Luan, Q. Yang, A Hardy type inequality in the half-space on R^n and Heisenberg group. J. Math. Anal. Appl. **347**, 645–651 (2008)
34. J. Leray, Étude de diverses équations intégrales non linéaires et de quelques problemes que pose l'hydrodynamique. J. Math. Pures Appl. **12**, 1–82 (1933)
35. C.C. Martinez, M.A. Sanz, *The Theory of Fractional Powers of Operators*. North-Holland Mathematics Studies (North-Holland Publishing, Amsterdam, 2001)
36. V.G. Maz'ya, *Sobolev Spaces* (Springer, Berlin, 1985)
37. D.T. Nguyen, N. Lam-Hoang, T.A. Nguyen, Hardy and Rellich inequalities with exact missing terms on homogeneous groups. J. Math. Soc. Jpn. **71**(4), 1243–1256 (2019)
38. D.T. Nguyen, N. Lam-Hoang, T.A. Nguyen, Hardy-Rellich identities with Bessel pairs. Arch. Math. **113**, 95–112 (2019)
39. P. Niu, H. Zhang, Y. Wang, Hardy type and Rellich type inequalities on the Heisenberg group. Proc. Am. Math. Soc. **129**(12), 3623–3630 (2001)
40. B. Opic, A. Kufner, *Hardy-Type Inequalities* Pitman Research Notes, vol. 219 (1990)
41. T. Ozawa, D. Suragan, Representation formulae for the higher order Steklov and L^{2^m}-Friedrichs inequalities. preprint, 2019
42. M. Ruzhansky, D. Suragan, Hardy and Rellich inequalities, identities, and sharp remainders on homogeneous groups. Adv. Math. **317**, 799–822 (2017)
43. M. Ruzhansky, D. Suragan, Layer potentials, Kac's problem, and refined Hardy inequality on homogeneous Carnot groups. Adv. Math. **308**, 483–528 (2017)
44. M. Ruzhansky, D. Suragan, On horizontal Hardy, Rellich, Caffarelli-Kohn-Nirenberg and p-sub-Laplacian inequalities on stratified groups. J. Differ. Equ. **262**, 1799–1821 (2017)
45. M. Ruzhansky, D. Suragan, N. Yessirkegenov, Sobolev type inequalities, Euler-Hilbert-Sobolev and Sobolev-Lorentz-Zygmund spaces on homogeneous groups. Integr. Equ. Oper. Theory **90**, 10 (2018)
46. M. Ruzhansky, D. Suragan, N. Yessirkegenov, Extended Caffarelli-Kohn-Nirenberg inequalities, and remainders, stability, and superweights for L_p-weighted Hardy inequalities Trans. Am. Math. Soc. Ser. B. **5**, 32–62 (2018)
47. M. Ruzhansky, B. Sabitbek, D. Suragan, Geometric Hardy and Hardy-Sobolev inequalities on Heisenberg groups. Bull. Math. Sci. **10**(3), 2050016 (2020)
48. M. Ruzhansky, D. Suragan, Anisotropic L2-weighted Hardy and L2-Caffarelli-Kohn-Nirenberg inequalities. Commun. Contemp. Math. **19**(6), 1750014 (2017)
49. M. Ruzhansky, D. Suragan, N. Yessirkegenov, Hardy-Littlewood, Bessel-Riesz, and fractional integral operators in anisotropic Morrey and Campanato spaces. Fract. Calc. Appl. Anal. **21**(3), 577–612 (2018)
50. M. Ruzhansky, D. Suragan, *Hardy Inequalities on Homogeneous Groups*. Progress in Math., vol. 327 (Birkhäuser, Basel, 2019), 588 pp
51. M. Ruzhansky, N. Yessirkegenov, Hypoelliptic functional inequalities (2018). arXiv:1805.01064v1
52. E.M. Stein, Some problems in harmonic analysis suggested by symmetric spaces and semi-simple groups. Actes Congres Intern. Math. Nice **1**, 179–189 (1970)
53. S. Secchi, D. Smets, M. Willen, Remarks on a Hardy-Sobolev inequality. C. R. Acad. Sci. Paris Ser. I **336**, 811–815 (2003)
54. E.M. Stein, G. Weiss, Fractional integrals on n-dimensional Euclidean Space. J. Math. Mech. **7**(4), 503–514 (1958)
55. Q.-H. Yang, Hardy type inequalities related to Carnot-Carathéodory distance on the Heisenberg group. Proc. Am. Math. Soc. **141**, 351–362 (2013)

Bogdan Bojarski in Complex and Real Worlds

Gia Giorgadze and Vladimir Mityushev

Abstract This paper outlines the famous mathematical contributions due to Bogdan Bojarski (1931–2018), an active member of the ISAAC society. His results constitute a part of the modern complex analysis. We outline the Gokhberg-Krein-Bojarski criteria of the stability of partial indices. This result makes clear the algebraic and topological structure of boundary value problems. The special attention is payed to one of the favorite topics discussed by Bojarski during his life, quasiconformal mappings. His results on boundary value problems including the Riemann-Hilbert problem for a multiply connected domain, and their applications to composites are briefly presented. The paper contains unique, perhaps first published, pictures.

Keywords Complex analysis · Partial indices · Quasiconformal mapping · Boundary value problem · Riemann-Hilbert problem

1 Scientific Career

The name of an outstanding Polish mathematician Bogdan Bojarski (1931–2018) is associated with revival of the Polish science after World War II. Bogdan Bojarski graduated in mathematical studies at the University of Lodz in 1951. He presented his PhD thesis at the Moscow State University in 1954. The extraordinary scientific career of Bojarski took off under the guidance of Ilia N. Vekua at the Moscow State University. It was the time when the theory of generalized analytic functions and quasiconformal mappings was being created. In 1955–1965 Bodgan Bojarski

G. Giorgadze
Vekua Inst. Appl. Math., Tbilisi, Georgia
e-mail: gia.giorgadze@tsu.ge

V. Mityushev (✉)
Cracow University of Technology, Krakow, Poland
e-mail: wladimir.mitiuszew@pk.edu.pl

P. Cerejeiras, M. Reissig (eds.), *Mathematical Analysis, its Applications and Computation*, Springer Proceedings in Mathematics & Statistics 385,
https://doi.org/10.1007/978-3-030-97127-4_5

contributed his seminal papers [4–16] and at the age of 28 he was granted the Doctor of Sciences degree (habilitation) by the Steklov Mathematical institute in Moscow, one of the prominent world mathematical centers.

From the end of the 1950s, Georgia (then Soviet republic) attracted Bogdan Bojarski not only by the great achievements of Georgian mathematicians. Up to early 1960s, Bojarski visited his supervisor Vekua in Tbilisi and presented an impressive talk on the stability of partial indices of the boundary value problem at the famous Nikoloz Muskhelishvili's seminar. He kept friendly contacts with Georgian mathematicians G. Manjavidze, N. Vakhania, N. Berikashvili, G. Janashia, G. Khimshiashvili, GG (the author of this paper) and others. His favorite toast in the circle of friends was the dialog between two guys: "Do you know Lenin?" The answer was "No". "Do you know Stalin?"—-"No. But I do know Givi, Saba and have many other good friends". Next, Bogdan suggested to drink for the friends who really know and support each other.

It was quite amazing to one of us (VM) that during our visit in Tbilisi in 2011 (International Conference dedicated to the 120th birthday anniversary of Muskhelishvili) Bogdan, perfectly speaking in Russian, had his accent vary from Moscow dialect to Georgian (Figs. 1 and 2).

From 1970, he worked at the University of Warsaw. In 1970–1981 he was the director of the Mathematical Institute of the University of Warsaw, in 1986–2002 the director of the Mathematical Institute of the Polish Academy of Sciences, and in 1993–2002 the director of the International Mathematical Center named Banach. From 1973, he was a correspondent member, from 1986 an ordinary member of the Polish Academy of Sciences, from 2000 a correspondent member of the Polish

Fig. 1 Bogdan Bojarski at conference devoted to the 70th anniversary of the Georgian National Academy of Sciences and the 120th birthday anniversary of its first President, Academician N. Muskhelishvili, Tbilisi, Georgia, 2011

Fig. 2 Bogdan Bojarski with his Georgian friends Gia Giorgadze, and his wife Grazyna (in right), Georgia, 2011

Fig. 3 Bogdan Bojarski in the opening of the 9th ISAAC Congress, August 5, 2013, Krakow, Poland

Academy of Arts. Bogdan Bojarski was a Board Member of the ISAAC society (Fig. 3).

Below, we outline his famous works having impacted onto mathematics and its applications. We concentrate our attention onto his results initiated new trends including our investigations and to problems discussed with him.

2 The Partial Indices of Matrix Function

First, we outline the report by Bojarski at Muskhelishvili's seminar mentioned above. It is known now as the Gokhberg-Krein-Bojarski criteria of the stability of partial indices. This result makes clear the algebraic and topological structure of boundary value problems. In particular, he demonstrated equivalence of the following three problems: closed form solution to the matrix boundary value problem, an analytical representation of the solution of system of singular integral equations and a formula for the partial indices expressed by means of invariants of the problem. Since in general there is no analytic solution to these problems, the construction of an approximate solution arises naturally. The key point in Bojarski's investigations was a relation between the behavior of partial indices and the approximation process. After unsuccessful attempts to solve a matrix boundary value problem in closed form, it was thought that one could approximate a matrix-function coefficient by a rational one and apply the simple factorization method for rational matrix-functions to find partial indices and solution. However, instability of partial indices established by Bojarski and Manjavidze questioned and revised approximation methods [36].

The present constructive applications of this theory to diffraction problems based on the Helmholtz equation should use the stability criterion for partial indices. It seems to us that the modern authors tacitly use the stability without proper mathematical justifications. The considered problems of diffraction perhaps are stable and confirmed by physical intuition, but we do not see a clear mathematical explanation in modern applied works.

We now proceed to discuss the **Riemann-Hilbert boundary value problem** called also by **$\mathbb{C}$-linear problem** (see the further discussion concerning the terminology in Sect. 4). Let Γ be a smooth closed positively oriented loop in $\mathbb{CP}^1$ which separates $\mathbb{CP}^1$ into two connected domains U^+ and U^-. Suppose $0 \in U^+$ and $\infty \in U^-$. Let us denote by Ω the space of all Hölder-continuous matrix functions $f : \Gamma \to GL_n(\mathbb{C})$ with the natural topology.

Find a piecewise holomorphic vector function $\varphi(t)$ in $U^+ \cup U^-$, which admits continuous boundary values on Γ and satisfies on Γ the boundary condition

$$\varphi^+(t) = f(t)\varphi^-(t), \quad t \in \Gamma, \tag{1}$$

and has finite order at ∞.

Let $\Omega^+ = \{f \in \Omega : f$ be the boundary value of the matrix function holomorphic in $U^+\}$, and $\Omega^- = \{f \in \Omega : f$ be the boundary value of the matrix function holomorphic in U$^-$ and regular at infinity where $f(\infty) = \mathbf{1}\}$.

Theorem 1 (Factorization Theorem) *Any matrix function $f \in \Omega$ can be represented as*

$$f(t) = f^-(t)d_K f^+(t), \quad t \in \Gamma, \tag{2}$$

where $f^{\pm} \in \Omega^{\pm}$ *and* d_K *is a diagonal matrix* $d_K = diag(t^{k_1}, \dots, t^{k_n})$ *with integer powers satisfying the condition* $k_1 \geq \dots \geq k_n$.

The diagonal matrix d_K will be called the *characteristic loop* of the corresponding matrix function, $K = (k_1, k_2, \dots, k_n)$ called the *characteristic multi-index* or *partial indices* of f. Two matrix functions $f, g \in \Omega$ will be called equivalent, if f and g have identical characteristic multi-indices.

For $K = (k_1, k_2, \dots, k_n)$, denote by Ω_K the set of equivalence classes of loops Ω. The representation (2) is not unique, but if one fixes f^+ (or f^-) then f^- (respectively f^+) will be uniquely defined.

The factorization theorem for matrix functions plays a crucial role in the matrix boundary value problems. It was applied to systems of differential and integral equations, as well as to the classification problem of holomorphic bundles over Riemann surfaces. Theorem 1 is formulated for the Hölder continuous matrix functions but its modifications hold for wider classes of matrix functions.

In general, there exist many different factorizations of the matrix function similar to (2). Seemingly the first such representation has been given by L. Sauvage in the analytic theory of differential equations. In different areas of mathematics analogical factorizations were used by Schlesinger, Hilbert, Birkhoff, Wiener and Hopf, Grothendieck, Simonenko (see modern overview of the matrix factorization theory [31]).

The partial indices of a matrix function are called *stable*, if in its sufficiently small neighborhood all matrix functions have the same partial indices.

The topological space Ω can be decomposed onto a countable number of open components

$$\Omega^k = \{f \in \Omega, \Delta_\Gamma \arg(\det f(t)) = 2\pi k\}, \quad \Omega = \cup_k \Omega^k \quad (k \in \mathbb{Z}). \tag{3}$$

The set $\Omega^k = \cup_K \Omega_K$ is connected.

Theorem 2 ([8]) *The set of partial indices is stable iff* $|k_i - k_j| \leq 1,\ i, j = 1, 2, \dots, n$.

The partial indices completely describe the solvability properties of the considered boundary value problem. They also determine solvability conditions for the Riemann-Hilbert monodromic problem (Hilbert 21st problem (see [26])). The global index is the only topological invariant of the Riemann-Hilbert problem in the sense that each Ω^k is connected. Thus in view of Theorem 3 one can say that generally the topological invariant completely describes the qualitative character of the solutions of the problem [18, 19].

Theorem 3 ([8]) *The matrix functions* $f_1(t)$ *and* $f_2(t)$ *belong to the same* Ω^k *iff* $f_1(t)$ *and* $f_2(t)$ *are homotopic.*

It follows from the above theorems that for every k, there exists a diagonal matrix with stable partial indices $(p+1, p+1, \dots, p+1, p, p, \dots, p)$, where $k = np+r$, $0 \leq r < n$ and every matrix function can be transformed into this stable diagonal matrix by elementary operations. Besides, the multi-index $K = (k_1, \ldots, k_n)$ as a function of $f \in \Omega^k$ has discontinuities only on the strata Ω_K.

Theorem 4 ([8, 19]) *Let $0 < k < n$,. Then in Ω^k, among the strata Ω_K the only ones which are open and dense subspaces are the ones with $K = (1, 1, \ldots, 1, 0, 0, \ldots 0)$, i. e. for such K, $\Omega^k \setminus \Omega_K$ does not contain interior points.*

In the particular case, when $k = np$ and $k = 0$ we get

Corollary 1

(1) If $k = np$, the stability of the partial indices implies that $K = (p, p, \ldots, p)$.
(2) If $k = 0$ and $K = (k_1, \ldots, k_n)$ is stable, then $K = (0, 0, \ldots, 0)$.

Consider the holomorphic vector bundle on $\mathbb{CP}^1$ which is obtained by the covering of the Riemann sphere $\mathbb{CP}^1$ by three open sets $\{U^+, U^-, U_3 = \mathbb{CP}^1 \setminus \{0, \infty\}\}$, with transition functions

$$g_{13} = f^+ : U^+ \cap U_3 \to GL_n(\mathbb{C}), \quad g_{23} = f^- d_K : U^- \cap U_3 \to GL_n(\mathbb{C}).$$

It is denoted by $E \to \mathbb{CP}^1$.

Theorem 5 ([26]) *Every holomorphic vector bundle splits into direct sum of the line bundles*

$$E \cong E(k_1) \oplus \ldots \oplus E(k_n). \tag{4}$$

The numbers $k_1, \ldots, k_n$ are the Chern numbers of the line bundles $E(k_1), \ldots, E(k_n)$ and satisfy the conditions $k_1 \geq \ldots \geq k_n$. The integer-valued vector $K = (k_1, \ldots, k_n) \in \mathbb{Z}^n$ is called the *splitting type* of the holomorphic vector bundle E. It defines uniquely the holomorphic type of the bundle E.

Connection between partial indices of the boundary value problem, *characteristic multi-index* of the matrix-function $f \in \Omega$ and *splitting type* of the holomorphic vector bundle E are presented in the following summarizing theorem:

Theorem 6 ([26]) *There is a one-to-one correspondence between the strata Ω_K and holomorphic vector bundles on $\mathbb{CP}^1$.*

Denote by $\mathcal{O}(E)$ the sheaf of germs of holomorphic sections of the bundle E. The solutions of the RHbvp are elements of the zeroth cohomology group $H^0(\mathbb{CP}^1, \mathcal{O}(E))$, therefore the number l of the linearly independent solutions of problem I is $\dim H^0(\mathbb{CP}^1, \mathcal{O}(E))$. Since the Chern number $c_1(E)$ of the bundle E is equal to the index of $\det G(t)$, we have obtained the known criterion of solvability of the Riemann-Hilbert monodromic problem. In particular, the following theorem is true:

Fig. 4 Atiyah and Bojarski

Theorem 7 *The Riemann-Hilbert boundary problem has solutions if and only if* $c_1(E) \geq 0$, *and the number* l *of linearly independent solutions is*

$$l = \dim H^0(\mathbb{CP}^1, \mathscr{O}(E)) = \sum_{k_i > 0} k_i + 1.$$

One of the preconditions of I. Gelfand's conjecture that the index of elliptic boundary value problem depends only on the homotopy class of the boundary curve (known today as Atiyah-Singer index theorem) was Bojarski's work [4]. Even after publication of Atiyah-Singer theorem, Bojarski's papers [15, 16] dedicated to the topological nature of the index of boundary value problem, are important not only from the historical viewpoint. They may be considered as the maximally possible extension of the classic methods applied to partial differential equations and to boundary value problems as well (Fig. 4).

To summarize this section we conclude that Bojarski described and developed boundary value problems by means of the modern language of mathematics connecting geometric and algebraic features of the subject in the papers [19] and [26].

3 Quasiconformal Mapping

Consider general system of partial differential equations of two variables in the complex form [7]

$$w_{\overline{z}} - \mu w_z - \nu \overline{w}_z = Aw + B\overline{w} + C. \tag{5}$$

This is an $\mathbb{R}$-linear differential equation.

Consider the following important particular cases of Eq. (5):

- *Beltrami* equation

$$w_{\overline{z}} - \mu(z) w_z = 0, \tag{6}$$

- *general Beltrami* equation

$$w_{\overline{z}} - \mu w_z - \nu(z)\overline{w}_z = 0 \tag{7}$$

- *Carleman-Bers-Vekua* equation

$$w_{\overline{z}} = Aw + B\overline{w}. \tag{8}$$

Equations (6), (7) at a point z are elliptic if

$$|\mu(z)| < 1 \quad \text{or} \quad |\mu(z)| + |\nu(z)| < 1.$$

The ellipticity is uniform in a subdomain U of the complex plane $\mathbb{C}$ if

$$|\mu(z)| + |\nu(z)| \leq k < 1, \quad z \in U$$

for some constant k.

The functions μ and ν are assumed to be measurable complex valued functions in z, and $A, B, C \in L_p(U)$, where $p > 2$ is a given number.

Generalized regular solution $w(z)$ of (5) is a function which is continuous in U and possesses generalized L_p derivatives w_z and $w_{\overline{z}}$ in U, i.e., $w(z) \in W_p(U)$. The functions $w_z, w_{\overline{z}}$ and w satisfy (5) almost everywhere in U. For each entire function $f(z)$ there exists a unique generalized regular solution $w(z)$ such that $w(z)$ is continuous in the entire plane $\mathbb{C}$ analytic in the complement of $\overline{U}$, $w(z) \in W_p(\mathbb{C})$, and $w(z) \sim f(z)$ as $z \to \infty$.

The following integral representation was applied to solve this problem

$$w(z) = f(z) - \frac{1}{\pi}\int\int \frac{\omega(t)}{t-z} dU \equiv f(z) + T(\omega). \tag{9}$$

It is readily shown that ω must satisfy the following equation

$$\omega - q_1 S(\omega) - q_2\overline{S(\omega)} = AT(\omega) + B\overline{T(\omega)} + C_0,$$

where C_0 is known and vanishes outside U. The integral

$$S(\omega) = -\frac{1}{\pi}\int_{\mathbb{C}}\int \frac{\omega(t)}{(t-z)^2} d\sigma_t.$$

can be estimated by the Calderron-Zygmund inequality. Therefore, the problem is ultimately reduced to the fixed point theorem in a Banach space.

The Beltrami equation, which expresses the conformality of a mapping with respect to the Riemann metric, may be written in the complex form (6) when μ is a Hölder continuous function. This has been done by Vekua. Bojarski extended Vekua's note to a bounded measurable μ [7]. After this the integral

$$w(z) = z - \frac{1}{\pi} \int\int \frac{\omega(t)}{t - z} dU$$

is called principal homeomorphism by terminology of Vekua and Bojarski.

One of the possible statements of the main results on quasiconformal mapping is known now as *Ahlfors-Bers-Bojarski* theorem given below.

Theorem 8 ([7]) *Let D be the unit disc and a function μ satisfies the Hölder condition at the origin. Let $f(\chi)$ be an analytic function defined in the unit disc. Then there exists such a homeomorphism $\chi(z)$ of the unit disc onto itself which keeps the origin fixed and the function (6) is a solution of (9).*

The proof is based on solution to the nonlinear differential equation on the function χ

$$\chi_{\bar{z}} = q(z) \frac{\overline{f'(\chi)}}{f'(\chi)} \overline{\chi_{\bar{z}}}. \tag{10}$$

A solution is sought in the form $\chi(z) = e^{\psi}$, where ψ has to be imaginary on the unit circle. Equation (10) can be reduced to an integral equation which can be solved by means of the fixed-point theorem. An priori estimation be a consequence of the Calderron-Zygmund inequality.

Furthermore, straightforward application of the chain rule for derivatives and a formula for the derivative of inverse function imply that every solution of the Beltrami system is an analytic function and that the Jacobian of the constructed mapping does not vanish almost everywhere. This construction also implies that the derivatives of the solution w are absolutely integrable in some power $p > 2$ [7].

For the behavior of the complex dilatation $\mu_f = \frac{w_{\bar{z}}}{w_z}$ under composition of the quasiconformal mappings $f = w \circ v^{-1}$ expressed by simple but important formula

$$\mu_f = \left[\frac{\mu_w - q_v}{1 - \bar{\mu}_v \mu_w} \frac{v_z}{\overline{v_z}} \right] \circ v^{-1}.$$

Bojarski also showed, that the solution w maps measurable sets onto measurable sets, the sets of zero measure onto the sets of zero measure, and continuous functions having generalized L_2 derivatives onto functions with the same property. The constructed solution gives a homeomorphic mapping of the whole plane onto itself.

Theorem 9 ([7]) *Every solution of (6) may be written in the form*

$$w(z) = f[\chi(z)] \tag{11}$$

where f is analytic and χ is a homeomorphism of the whole plane onto itself satisfying together with its inverse a uniform Hölder condition.

Bojarski stressed in [24] the advantages of the Vekua school in the development of new approach for the Beltrami equation and its solutions. Bojarski's research gave great impetus to the understanding of the quasi-conformal mappings and the geometric structure of the sets of functions satisfying partial differential equations [1].

Concerning the contribution of Vekua and Bojarski to the theory of quasiconformal mappings (Theorems 1 and 2) in 1978 Lars V. Ahlfors [3] writed: "It must be clear that I am condensing years of research into minutes. The fact is that the post-Teichmüller era of quasiconformal mappings did not start seriously until 1954. In 1957 I.N. Vekua in the Soviet Union proved the existence and uniqueness theorem for the Beltrami equation, and in the same year L. Bers discovered that the theorem had been proved already in 1938 by C. Morrey. The great difference in language and emphasis had obscured the relevance of Morrey's paper for the theory of q.c. mappings. The simplest version of the proof is due to B. V. Bojarski who made it a fairly straightforward application of the Calderron-Zygmund theory of singular integral transforms."

By analogy with the one dimensional case, consider regular solutions of systems of $2n$ elliptic partial differential equations presented in complex form.

A vector $w(z) = (w_1, \ldots, w_n)$ is called a *generalized analytic vector* [9, 12, 13] in the domain U if it is a solution of an elliptic system

$$\partial_{\bar{z}} w - Q(z)\partial_z w + A(z)w + B(z)\bar{w} = 0, \tag{12}$$

where $A(z)$, $B(z)$ are given quadratic matrices of order n of the class $L_{p_0}(U)$, $p_0 > 2$, and $Q(z)$ is a matrix of the following special form: it is quasidiagonal and every block $Q^r = (q^r_{ik})$ is a lower (upper) triangular matrix satisfying the conditions

$$q^r_{11} = \ldots = q^r_{m_r,m_s} = q^r, \quad |q^r| \le q_0 < 1,$$
$$q^r_{ik} = q^r_{i+s,k+s} \quad (i + s \le n, k + s \le n).$$

Moreover, we suppose that $Q(z) \in W^1_p(\mathbb{C})$, $p > 2$, and $Q(z) = 0$ outside of a circle.

If $A(z) \equiv B(z) \equiv 0$, Eq. (12) become

$$\partial_{\bar{z}} w - Q(z)w_z = 0. \tag{13}$$

The solutions of Eq. (13) are called *$Q-$ holomorphic vectors.*

Theorem 10 ([9]) *Equation (13) has a solution of the form*

$$\zeta(z) = zI + T\omega, \tag{14}$$

where I is the unit matrix and $\omega(z)$ is a solution of equation

$$\omega(z) + Q(z)S\omega = Q(z)$$

belonging to $L_p(\mathbb{C})$ for $p > 2$.

The solution (14) of Eq. (13) is analogous to the *fundamental (principal) homeomorphism* of the Beltrami equation.

Consider a boundary value problem for the nonhomogeneous elliptic system corresponding to (12)

$$w_{\bar{z}} - Qw_z = Aw + Bw + F. \tag{15}$$

The unknown w is a complex vector-function of dimension n. Q is a triangular matrix vanishing outside of the sufficiently large circle and admitting generalized derivatives Q_z, $Q_{\bar{z}}$ in L_p for $p > 2$. The elements of the matrices A and B are functions in L_p. Let U be a multiply connected domain in the complex plane with the boundary consisting of $m+1$ Lyapunov curves. It is assumed that the system (15) is strongly elliptic in U.

In a natural way, Eqs. (5) and (12) are considered in a Sobolev space. This yields profound investigations of Sobolev spaces. From this point of view, the area of Bojarski's interest during his whole scientific activity was the investigation of the interior structure of the Sobolev spaces, and of their analytic and geometric properties. The obtained results [20–23, 25, 27, 29, 30] are very deep and elegant. They naturally fill up the theory of elliptic differential equations.

At the end of this section we outline Bojarski's contribution to the deformation of surfaces and corresponding problems of mechanics. It was one of the most important research directed by I. Vekua and his school based on the application of the Beltrami equation to isothermal coordinate systems on surfaces. We note the first work with I. Vekua [10] and the second one with the famous geometer V. Efimov [11]. As is well known, for a regular surface $z = z(x, y)$, no part of which is planar, the component ζ (along the z-axis) of an infinitesimal bending field $\tau(\xi, \eta, \zeta)$ attains its maximum and its minimum on the boundary of the surface. The rigidity of piecewise regular closed convex surfaces of non negative curvature is proved, and the rigidity of closed convex piecewise twice differentiable surfaces is established in [10, 11, 17].

4 Boundary Value Problems

The term "Riemann-Hilbert problem" for the problem (1) was used by Muskhelishvili, Vekua and Bojarski due to the general note by Riemann [46] devoted to boundary value problems and the results by Hilbert [34] devoted to singular integral equations. Later F.D. Gakhov [32] just for convenience introduced the terms "Riemann problem" and "Hilbert problem" in order to divide two different boundary value problems (16) and (17). With the further development the terms like "generalized boundary value problem" arisen. G.S. Litvinchuk and E.I. Zverovich noted that any generalized boundary value problem can be generalized again. They asked to think about new sensible terminology. The appropriate terms "$\mathbb{C}$-linear and $\mathbb{R}$-linear problems" were proposed by VM for the problems (1) and (19), respectively. This terminology was introduced by association with the $\mathbb{C}$-linear condition $W = aZ$ and the $\mathbb{R}$-linear condition $W = aZ + b\overline{Z}$ between two complex values W and Z.

Consider the *Riemann-Hilbert problem* on the N-dimensional vector-functions w satisfying the Beltrami equation in U and Hölder continuous in $U \cup \Gamma$

$$\mathrm{Re}\,[G(t)\; w(t)] = h(t), \quad t \in \Gamma. \tag{16}$$

This problem can be reduced to the $\mathbb{C}-$linear problem

$$w^+(t) = G(t)w^-(t) + h(t), \quad t \in \Gamma. \tag{17}$$

The vector-functions $w^+(t)$ and $w^-(t)$ satisfy (15) in $U \cup \Gamma$ and $\mathbb{C}\backslash(U \cup \Gamma)$, respectively. These problems are reduced to singular integral equations.

Let l and l' be the numbers of the $\mathbb{R}-$linearly independent solutions of the homogeneous problems (16) when $h(t) = 0$ and its adjoint, respectively. Bojarski [13] established a relation between the total index (winding number) $\varkappa = \frac{1}{2\pi}\Delta_L \arg(\det\ G(t))$, the connectivity of the domain n and the dimension N. Let $(f, g) = \sum_{j=1}^{N} f_j \bar{g}_j$ denote the scalar product of two Hölder continues vector-functions f and g; Q^T a matrix transposed with Q.

Theorem 11 ([13]) *Necessary and sufficient solvability conditions for inhomogeneous Riemann-Hilbert problem have the form*

$$Im \int_L (h(t), (Edt + Q^T dt) f) = 0 \tag{18}$$

for all solutions f of the adjoint homogeneous Riemann-Hilbert problem. Moreover,

$$l - l' = 2\varkappa - N(n-1).$$

The problems (16) and (17) sometimes are called two-element. The presented above results are the benchmark of the Nöther theory of the two-element boundary value problems developed by a group of mathematicians with Bojarski. We now proceed to discuss a three-element scalar problem for analytic functions.

Let given functions $a(t)$, $b(t)$ and $c(t)$ are Hölder continuous on Γ. It is required to find a function $\varphi(z)$ analytic in U^+ and U^-, continuous in the closures of the considered domains with the $\mathbb{R}$-linear conjugation condition [43]

$$\varphi^+(t) = a(t)\varphi^-(t) + b(t)\overline{\varphi^-(t)} + c(t), \quad t \in \Gamma. \tag{19}$$

Here, $\varphi^+(t)$ and $\varphi^-(t)$ are the limit values of $\varphi(z)$ when $z \in U^+$ tends to $t \in \Gamma$ and $z \in U^-$ tends to $t \in \Gamma$, respectively. In particular, the domain U^+ can consist in mutually disjoint domains D_k $(k = 1, 2, \ldots, n)$ and $\Gamma = -\cup_{k=1}^n \partial D_k$. The domain U^- is multi-connected.

If the functions $a(t)$ and $b(t)$ are constant at each component ∂D_k, and $c(t) \equiv 0$ the $\mathbb{R}$-linear problem is equivalent to the transmission problem from the theory of composites [33]

$$u^+(t) = u^-(t),\ \lambda_k \frac{\partial u^+}{\partial n}(t) = \lambda \frac{\partial u^-}{\partial n}(t), \quad t \in \partial D_k \quad (k = 1, 2, \ldots, n). \tag{20}$$

Here, the real function $u(z)$ is harmonic in U^+ and continuously differentiable in $D_k \cup \partial D_k$ $(k = 1, 2, \ldots, n)$ and in $U^+ \cup \Gamma$, $\frac{\partial}{\partial n}$ is the normal derivative to Γ. The conjugation conditions express the perfect contact between materials with different conductivities λ_k and λ. The functions $\varphi(z)$ and $u(z)$ are related by equations

$$u(z) = \mathrm{Re}\ \varphi(z),\ z \in U^-,\ u(z) = \frac{2\lambda}{\lambda_k + \lambda} \mathrm{Re}\ \varphi(z),\ z \in D_k\ (k = 1, 2, \ldots, n). \tag{21}$$

The coefficients are related by formulae (for details see [42, Sec. 2.12] and [33, Chapter 2, Sec.2.1]).

$$a(t) = 1,\ b(t) = \rho_k := \frac{\lambda_k + \lambda}{\lambda_k - \lambda}, \quad t \in \partial D_k. \tag{22}$$

The constants ρ_k are called the contrast parameters. In this case, the problem (19) becomes

$$\varphi^-(t) = \varphi_k(t) - \rho_k \overline{\varphi_k(t)} - f(t), \quad t \in \partial D_k\ (k = 1, 2, \ldots, n), \tag{23}$$

where the external field is modeled by a given Hölder continuous function $f(t)$.

In 1932, having used the theory of potentials N.I. Muskhelishvili [44] (see also [45], p.522) reduced the problem (20) to a Fredholm integral equation and proved that it has a unique solution in the case $\lambda^{\pm} > 0$.

In 1946 Markushevich [37] had stated the $\mathbb{R}$–linear problem in the form (19) and studied it in the case $a(t) = 0$, $b(t) = 1$, $c(t) = 0$. Later N.I. Muskhelishvili [45]

(p. 455) did not determined whether (19) was his problem (20) discussed in 1932 in terms of harmonic functions.

Scientific seminar "Boundary value problems" at Belorussian State University (Minsk, Belarus) was headed by F.D. Gakhov and after his dead in 1980 by E.I. Zverovich who stressed an elegant short note [14] by B. Bojarski on the $\mathbb{R}$-linear problem published in 1960. It was proved in [14] that in the case $|b(t)| < |a(t)|$ with $a(t)$, $b(t)$ belonging to the Hölder class $\mathscr{H}^{1-\varepsilon}$ with sufficiently small $\varepsilon > 0$, the $\mathbb{R}$-linear problem (19) qualitatively was similar to the $\mathbb{C}$-linear problem (see Theorem 12 below)

$$\varphi^{+}(t) = a(t)\varphi^{-}(t) + c(t), \quad t \in \varGamma. \tag{24}$$

Let $\varkappa = wind_L a(t)$ denote the winding number (index) of $a(t)$ along L. Mikhailov [39] reduced the problem (19) to an integral equation and justified the absolute convergence of the method of successive approximation for the later equation in the space $\mathscr{L}^p(L)$ for $wind_L a(t) = 0$ and under the restriction

$$(1 + \|S_p\|)|b(t)| < 2|a(t)|, \tag{25}$$

where $\|S_p\|$ is the norm of the singular integral in $\mathscr{L}^p(\varGamma)$. Later Mikhailov [39] (first published in [38]) developed this result to continuous coefficients $a(t)$ and $b(t)$; $c(t) \in \mathscr{L}^p(\varGamma)$. The case $|b(t)| < |a(t)|$ was called the elliptic case. It corresponds to the particular case of the real constant coefficients a and b considered by Muskhelishvili [44].

Theorem 12 ([14, 28]) *Let the coefficients of the problem (19) satisfy the inequality*

$$|b(t)| < |a(t)|. \tag{26}$$

If $\varkappa \geq 0$, the problem (19) is solvable and the homogeneous problem (19) ($c(t) = 0$) has $2\varkappa$ $\mathbb{R}$-linearly independent solutions vanishing at infinity. If $\varkappa < 0$, the problem (19) has a unique solution if and only if $|2\varkappa|$ $\mathbb{R}$-linearly independent conditions on $c(t)$ are fulfilled.

The condition (25) is stronger than (26) since always $\|S_p\| \geq 1$ [42].

Theorem 13 ([28]) *Let the coefficients of the problem (19) satisfy the inequality (26) and $\varkappa = 0$. Then, the problem (19) has a unique solution vanishing at infinity that can be found by uniformly convergent successive approximations.*

The successive approximations mentioned in Theorem 13 can be interpreted as the generalized alternating method of Schwarz called shortly Schwarz's method below. This method is effective in study of composites, when non–overlapping inclusions are embedded in a host material which occupies a domain $\varOmega$. Then, the inclusions occupying the disjoint domains $\varOmega_1$ and $\varOmega_2$ interact with each other through the host material. Schwarz's method can be considered as a decomposition method [47] used in numerical solution to PDE.

There are two different methods of integral equations associated to boundary value problems. The first method is known as the method of potentials. In complex analysis, it is equivalent to the method of singular integral equations [45, 48]. Schwarz's method can be presented as a method of integral equations of another type [40]. Let ∂D_k be Lyapunov's simple closed curves. It is assumed that each ∂D_k leaves the inclusion D_k on the left.

Introduce a space $\mathscr{H}(U^+)$ consisting of functions analytic in $U^+ = \cup_{k=1}^n D_k$ and Hölder continuous in the closure of U^+ endowed the norm

$$||\omega|| = \sup_{t\in L} |\omega(t)| + \sup_{t_{1,2}\in L} \frac{|\omega(t_1)| - \omega(t_2)|}{|t_1 - t_2|^\alpha}, \tag{27}$$

where $0 < \alpha \leq 1$. The space $\mathscr{H}(U^+)$ is Banach, since the norm in $\mathscr{H}(U^+)$ coincides to the norm of functions Hölder continuous on Γ (sup on $D^+ \cup \Gamma$ in (27) is equal to sup on Γ). It follows from Harnack's principle that convergence in the space $\mathscr{H}(D^+)$ implies the uniform convergence in the closure of D^+.

The conjugation condition (23) can be written in the form

$$\varphi_k(t) - \varphi^-(t) = \rho_k \overline{\varphi_k(t)} + f(t), \quad t \in \partial D_k \ (k = 1, 2, \ldots, n), \tag{28}$$

A difference of functions analytic in D^+ and in D is in the left hand part of the later relation. Then application of Sochocki's formulae yield

$$\varphi_k(z) = \sum_{m=1}^n \frac{\rho_m}{2\pi i} \int_{L_m} \frac{\overline{\varphi_m(t)}}{t - z} dt + f_k(z), \quad z \in D_k \ (k = 1, 2, \ldots, n), \tag{29}$$

where the function

$$f_k(z) = \frac{1}{2\pi i} \sum_{m=1}^n \int_{\partial D_m} \frac{f(t)}{t - z}\, dt$$

is analytic in D_k and Hölder continuous in its closure. The integral equations (29) can be continued to ∂D_k as follows

$$\varphi_k(z) = \sum_{m=1}^n \rho_m \left[\frac{\overline{\varphi_k(z)}}{2} + \frac{1}{2\pi i} \int_{\partial D_m} \frac{\overline{\varphi_m(t)}}{t - z} dt \right] + f_k(z), \tag{30}$$
$$z \in \partial D_k \ (k = 1, 2, \ldots, n).$$

One can consider Eqs. (29), (30) as an equation with linear bounded operator in the space $\mathscr{H}(U^+)$.

Equations (29), (30) correspond to Schwarz's method. Write, for instance, Eq. (29) in the form

$$\varphi_k(z) - \frac{\rho_k}{2\pi i}\int_{\partial D_k} \frac{\overline{\varphi_k(t)}}{t-z}dt = \sum_{m\neq k} \frac{\rho_m}{2\pi i}\int_{\partial D_m} \frac{\overline{\varphi_m(t)}}{t-z}dt + f_k(z), \tag{31}$$

$$z \in D_k \ (k = 1, 2, \ldots, n),$$

At the zeroth approximation we arrive at the problem for the single inclusion D_k $(k = 1, 2, \ldots, n)$

$$\varphi_k(z) - \frac{\rho_k}{2\pi i}\int_{\partial D_k} \frac{\overline{\varphi_k(t)}}{t-z}dt = f_k(z),\ z \in D_k. \tag{32}$$

Let the problem (32) is solved. Further, its solution is substituted into the right hand part of (31). Then we arrive at the first order problem etc. Therefore, Schwarz's method can be considered as a method of implicit iterations applied to integral equations (29), (30).

Theorem 13 says that Schwarz's method in the presented form always uniformly converges, i.e., these geometrical restrictions are redundant. This is an interesting example of the difference between absolute and uniform convergence which shows that estimations on the absolute values or on the norm are too strong in comparison to the study of the uniform convergence. This means that Schwarz's method for boundary value problems with zero winding number always uniformly converges.

5 Riemann-Hilbert Problem for a Multiply Connected Domain

I.N. Vekua included Bojarski's results in his famous book [48]. In 1958 he asked Bojarski and other colleagues to read a draft of the book and made use of their remarks. The famous supplement "On the special case of the Riemann-Hilbert problem" to chapter "A Survey of Hardy Type Inequalities on Homogeneous Groups" of the Vekua book devoted the classic scalar Riemann-Hilbert problem for a multiply connected domain was written by Bojarski. The problem (16) considered in this supplement has the form

$$\mathrm{Re}\,[\overline{G(t)}\varphi(t)] = 0, \quad t \in \Gamma. \tag{33}$$

Here, we modify the designation of the previous section following Bojarski's work assuming that closed simple smooth curves Γ_k form the boundary of the multiply connected bounded domain U^+ and $\Gamma = \cup_{k=0}^{n}\Gamma_k$. The curve Γ_0 is an exterior curve to U^+ oriented counterclockwise, Γ_k $(k = 1, 2, \ldots, n)$ lie interior of Γ_0 and oriented clockwise.

The adjoint problem to (33) has the form

$$\operatorname{Re}\,[it'(s)G(t)\varphi(t)] = 0, \quad t \in \Gamma, \tag{34}$$

where $t(s)$ denotes a complex parametric equation of the components of the smooth contour Γ and $t'(s) = \exp[i\theta(s)]$. Here, $\theta(s)$ is the angle formed by the tangent to Γ and the real axis.

Introduce the index along each component of Γ

$$\varkappa_k = \operatorname{ind}_{\Gamma_k} G(t) = \frac{1}{2\pi} \arg\; G(t)|_{\Gamma_k} \tag{35}$$

and the index of the coefficient $G(t)$

$$\varkappa = \sum_{k=1}^{n} \varkappa_k. \tag{36}$$

For instance, the index of the function $t'(s)$ is equal to the number of rotations of the tangent line to Γ, i.e. $\operatorname{ind}_\Gamma t'(s) = -n + 1$.

Let ℓ and ℓ' denote the number of the $\mathbb{R}$-linear independent solutions of the problems (33) and (34), respectively. Vekua [48] derived the formula $\ell - \ell' = 2\varkappa - n + 1$ and found that

$$\ell = \max(0, 2\varkappa - n + 1), \tag{37}$$

when $\varkappa < 0$ or $\varkappa > n - 1$. Examples [32] demonstrated that the numbers ℓ and ℓ' do not depend on the index only when

$$0 \leq \varkappa \leq n - 1. \tag{38}$$

That is why the case (38) is called special.

The complete investigation of the special case was given by Bojarski [48]. First, Bojarski noted that the problem (33) by means of the conformal mapping can be reduced to a circular domain $\mathbb{U}^+$ when $\Gamma_k = \{t \in \mathbb{C} : \; |t - a_k| = r_k\}$. Without loss of generality one can assume that Γ_0 is the unit circle and the point $z = 0$ belongs to $\mathbb{U}^+$. Bojarski called such a domain canonical. Next, Bojarski reduced the problem (33) to the problem

$$\operatorname{Re}\,[\exp(-i\pi\alpha_k)\varphi(t)] = 0, \quad |t - a_k| = r_k \; (k = 0, 1, \ldots, n), \tag{39}$$

where the function $\varphi(z)$ is analytic in U^+ except at zero where $\varphi(z)$ may have a pole of order $\varkappa$. The given constants α_k are real and $\alpha_0 = 0$.

One can consider the set $\boldsymbol{\alpha} = (\alpha_1, \alpha_2, \ldots, \alpha_n)$ as a point of the n-dimensional torus T_n with unit periods. Let $R_\varkappa$ denote a subset of T_n for which (37) holds and $CR_\varkappa$ be the complement of $R_\varkappa$ to T_n. It follows from Vekua's result that the set $CR_\varkappa$ is empty when $\varkappa < 0$ or $\varkappa > n-1$. Bojarski demonstrated that the set $CR_\varkappa$ is not empty in the special case (38) and describe its structure. In particular, CR_0 consists of one point $(0, 0, \ldots, 0)$; CR_1 is a 2-dimensional manifold on the torus T_n. The sets $CR_\varkappa$ and $CR_{n-1-\varkappa}$ have the same structures. Bojarski proved that $R_\varkappa$ is the set of zeros of an analytic function in the real domain with respect to the point $\boldsymbol{\alpha} \in T_n$. Moreover, Bojarski described a system of linear algebraic equations associated to solvability of the problem (39).

Two problems (33) with different $G(t)$ and $\tilde{G}(t)$ are called sufficiently closed if $|G(t) - \tilde{G}(t)| < \varepsilon$ for sufficiently small $\varepsilon > 0$. Bojarski proved that sufficiently closed problems to a fixed problem (33) have the same number of $\mathbb{R}$-linearly independent solutions if and only if the formula (37) holds.

Bojarski's investigation was completed 40 years after. Solution to the Riemann-Hilbert problem (33) was described in [42] and papers cited therein. First, the general problem was reduced to the problem similar to (33) as in Bojarski's work. Next, the function $\varphi(z)$ was exactly written in the form of the uniformly convergent Poincaré type series. The famous Bojarski system of linear algebraic equations associated to solvability of the problem was explicitly written. The convergence of the Poincaré type series was based on other Bojarski's paper [14] devoted to the $\mathbb{R}$-linear problems. Thus, two independent Bojarski's works played the crucial role in solution to the scalar Riemann-Hilbert problem for a multiply connected domain (Fig. 5).

Fig. 5 Bojarski, Nirenberg and Lax

6 Conclusion

The bright mathematical contributions due to Bogdan Bojarski constitute a part of the modern complex analysis, in particular, the theory of quasiconformal mappings and boundary value problems. His key ideas are surfaced again and applied in various topics.

Bogdan Bojarski addressed to applications of Chen's iterated integral to the matrix Beltrami equation. He wanted to represent the Carleman-Bers-Vekua equation as the differential-geometric connectivity of the corresponding vector bundle on the Riemann sphere. In general, he suggested to rewrite the theory of generalized analytic functions in an invariant form. Bogdan Bojarski thought that this approach could be interesting not only in pure mathematics but could help us to understand the problems of modern theoretical physics. He called this conception by *Complex Geometry of Real World.* When Bojarski was young his mind was occupied to the topological invariance of index. This approached him to the famous Atiyah-Singer index theorem [2]. Bojarski's thoughts on this question were published in the paper [15].

The scalar Riemann-Hilbert problem (33) for a multiply connected domain was solved 40 years after his seminal results [14] and [48, Supplement] which were the basis of this complete solution.

Though Bojarski did not address directly to engineering problems, the considered by him boundary value problems have applications in continuum mechanics. The $\mathbb{R}$-linear problem (19) expresses the perfect contact conditions between two different materials. The Riemann-Hilbert problem describe stationary two-dimensional physical fields (heat conduction, filtration, diffusion, electric conductivity etc.) in media with cavities. His fundamental Theorem 8 on quasiconformal mapping yielded new bounds for the effective properties of anisotropic composites [35, 41]. During our recent meetings in 2008–2013 Bogdan Bojarski posed a number of fascinating ideas concerning polyanalytic functions.

Frequently, a mathematician apply significant efforts uses advanced complicated manipulations to get a new result. Bojarski in his contributions tried rather to present a fresh idea leading to an elegant result. Some his papers were short and pithy. They demonstrated deep relations between analysis and geometry, and initiated modern lines ingrained in the modern complex analysis.

References

1. K. Astala, T. Iwaniec, G. Martin, *Elliptic Partial Differential Equations and Quasiconformal Mappings in the Plane* (Princeton University Press, Princeton, 2009)
2. M.F. Atiyah, I.M. Singer, The index of elliptic operators on compact manifolds. Bull. Am. Math. Soc. **69**(3), 422–433 (1963)
3. M. Atiyah, D. Iagolnitzer (eds.), *Fields Medalists Lectures* (World Scientific, Singapore, 1997)

4. B. Bojarski, On a boundary problem for a system of elliptic first-order partial differential equations. Dokl. Akad. Nauk SSSR **102**, 201–204 (1955) [in Russian]
5. B. Bojarski, Homeomorphic solutions of Beltrami systems. Dokl. Akad. Nauk SSSR **102**, 661–664 (1955) [in Russian]
6. B. Bojarski, On solutions of a linear elliptic system of differential equations in the plane. Dokl. Akad. Nauk SSSR **102**, 871–874 (1955) [in Russian]
7. B. Bojarski, Generalized solutions of a system of differential equations of first order and elliptic type with discontinuous coefficients. Mat. Sb. **43**(85), 451–563 (1957) [in Russian]
8. B. Bojarski, Stability of the Hilbert problem for a holomorphic vector. Bull.Georgian Acad. Sci. **21**, 391–398 (1958) [in Russian]
9. B. Bojarski, On a boundary value problem of the theory of analytic functions. Dokl. AN SSSR **119**, 199–202 (1958) [in Russian]
10. B. Bojarski, I.N. Vekua, Proof of the rigidity of piecewise regular closed convex surfaces of non-negative curvature. Izv. Akad. Nauk SSSR Ser. Mat. **22**, 165–176 (1958) [in Russian]
11. B. Bojarski, N.V. Efimov, The maximum principle for ifinitesimal bending of piecewise-regular convex surfaces. Uspehi Mat. Nauk **14**(6), 147–153 (1959)
12. B. Bojarski, Some boundary value problems for $2n$ system of elliptic differential equations on plane. Dokl. Akad. Nauk SSSR **124**, 543–546 (1959) [in Russian]
13. B. Bojarskii, The Riemann-Hilbert problem for a holomorphic vector. Dokl. Akad. Nauk SSSR **126**, 695–698 (1959) [in Russian]
14. B. Bojarski, On generalized Hilbert boundary value problem. Soobsch. AN Gruz SSR **25**, 385–390 (1960) [in Russian]
15. B. Bojarski, On the index problem for systems of singular integral equations, Bull. Pol. Acad. Sc. **10**, 653–655 (1965)
16. B. Bojarski, On the index problem for systems of singular integral equations. Bull. Ac. Sci. Polon. **9**, 627–631 (1965)
17. B. Bojarski, Subsonic flow of compressible fluid. Arch. Mech. Stos. **18**, 497–520 (1966)
18. B. Bojarski, Direct approach to the theory of systems of singular integral equations, in *Chapter in the Monograph of Muskelishvili N.I., Singular Integral Equations* (Nauka, Moscow, 1968), pp. 478–488 [in Russian]
19. B. Bojarski, Connection between complex and global analysis: analytical and geometrical aspects of the Riemann-Hilbert transition problem, in *Complex Analysis, Methods, Application* (Berlin, A.V. 1983)
20. B. Bojarski, Pointwise differentiability of weak solutions of elliptic divergence type equations. Bull. Pol. Acad. Sci. Math. **1**, 16 (1985)
21. B. Bojarski, C. Sbordone, I. Wik, The Muckenhoupt class $A_1(R)$. Stud. Math. **101**(2), 155–163 (1992)
22. B. Bojarski, P. Hajlasz, Pointwise inequalities for Sobolev functions and some applications. Stud. Math. **106**, 77–92 (1993)
23. B. Bojarski, P. Hajlasz, P. Strzelecki, Sard's theorem for mappings in Hölder and Sobolev spaces. Manuscr. Math. **118**, 383–397 (2005)
24. B. Bojarski, On The Beltrami Equation, once again: 54 years later. Ann. Acad. Sci. Fennicæ Math. **35**, 59–73 (2010)
25. B. Bojarski, Taylor expansion and Sobolev spaces. Bull. Georgian Natl. Acad. Sci. (N.S.) **5**, 510 (2011)
26. B. Bojarski, G. Giorgadze, Some analytical and geometric aspects of stable partial indices. Proc. Vekua Inst. Appl. Math. **61–62**, 14–32 (2011/2012)
27. B. Bojarski, L. Ihnatsyeva, J. Kinnunen, How to recognize polynomials in higher order Sobolev spaces. Math. Scand. **112**, 161–181 (2013)
28. B. Bojarski, V. Mityushev, $\mathbb{R}$-linear problem for multiply connected domains and alternating method of Schwarz. J. Math. Sci. **189**, 68–77 (2013)
29. B. Bojarski, J. Kinnunen, Th. Zürcher, Higher order Sobolev-type spaces on the real line. J. Funct. Spaces **3**, 1–13 (2014)
30. B. Bojarski, Sobolev spaces and averaging I. Proc. A. Razmadze Math. Inst. **164**, 19–44 (2014)

31. I. Gohberg, M. Kasshoek, I. Spitkovski, An overview matrix factorization theory and operator application. Oper. Theory Adv. Appl. **141**, 1–102 (2003)
32. F.D. Gakhov, *Boundary Value Problems*, 3rd rev. edn. (Elsevier, Amsterdam, 1966; Nauka, Moscow, 1977)
33. S. Gluzman, V. Mityushev, W. Nawalaniec, *Computational Analysis of Structured Media* (Elsevier, Amsterdam, 2018)
34. D. Hilbert, Grundzüge einer allgemeinen Theorie der linearen Integralgleichungen. Chelsea, reprint (1953)
35. L. Leonetti, V. Nesi, Quasiconformal solutions to certain first order systems and the proof of a conjecture of GW Milton. J. Math. Pures et Appl. **76**, 109–124 (1997)
36. G.F. Manjavidze, Approximate solution of boundary problems of the theory of analytic functions. Issledovaniya po sovremennym problemam teorii funkcii kompleksnogo peremennogo (Gosudarstv. Izdat. Fiz.-Mat. Lit., Moscow 1960), pp. 365–370
37. A.I. Markushevich, On a boundary value problem of analytic function theory. Uch. zapiski MGU. **1**, 20–30 (1946) [in Russian]
38. L.G. Mikhailov, On a boundary value problem. DAN SSSR **139**, 294–297 (1961) [in Russian]
39. L.G. Mikhailov, *New Class of Singular Integral Equations and its Applications to Differential Equations with Singular Coefficients*, 2nd edn. (Akademie Verlag, Berlin, 1970)
40. S.G. Mikhlin, *Integral Equations and Their Applications to Certain Problems in Mechanics, Mathematical Physics and Technology*, 2nd rev. edn. (Macmillan, New York, 1964)
41. G.W. Milton, *The Theory of Composites* (Cambridge University Press, Cambridge 2002)
42. V.V. Mityushev, S.V. Rogosin, *Constructive Methods for Linear and Non-linear Boundary Value Problems of the Analytic Function*. Theory and applications (Chapman & Hall/CRC, Boca Raton, 2000)
43. V.V. Mityushev, $\mathbb{R}$-linear and Riemann-Hilbert problems for multiply connected domains, in *Developments in Generalized Analytic Functions and Their Applications*, ed. by G. Giorgadze (Tbilisi State University Publ., Tbilisi, 2011)
44. N.I. Muskhelishvili, To the problem of torsion and bending of beams constituted from different materials. Izv. AN SSSR **7**, 907–945 (1932)
45. N.I. Muskhelishvili, *Singular Integral Equations*, 3rd edn. (Moscow, Nauka, 1968)
46. B. Riemann *Collected Works* (Dover, 1953). Reprint
47. B. Smith, P. Bjorstad, W. Gropp, *Domain Decomposition: Parallel Multilevel Methods for Elliptic Partial Differential Equations* (Cambridge University Press, Cambridge, 1996)
48. I.N. Vekua, *Generalized Analytic Functions* (Elsevier, Pergamon Press, 1962)